Semiconductor Technologies

Edited by
Jan Grym

Semiconductor Technologies
Edited by Jan Grym

Republished by InTech
Janeza Trdine 9, 51000 Rijeka, Croatia

Technical Editor Sonja Mujacic

Cover Designer Dino Smrekar

Additional hard copies can be obtained from orders@intechopen.com

Semiconductor Technologies,
Edited by Jan Grym
p. cm.
ISBN 978-953-307-080-3

For EU product safety concerns:
IN TECH d.o.o., Prolaz Marije Krucifikse Kozulić 3, 51000 Rijeka, Croatia,
info@intechopen.com or visit our website at intechopen.com.

Preface

Semiconductor technologies continue to evolve and amaze us. New materials, new structures, new manufacturing tools, and new advancements in modelling and simulation form a breeding ground for novel high performance electronic and photonic devices. This book covers all aspects of semiconductor technology concerning materials, technological processes, and devices, including their modelling, design, integration, and manufacturing.

High costs, long manufacturing cycles, and enormous increase in computing power are behind a recent rapid progress of the modelling, simulation, optimisation, and design of semiconductor devices. The first two chapters present the state-of-the-art in modelling of semiconductor processes and devices. Several examples are given: simulation of the switching characteristics of SiC GTO, MOSFET DC modelling for distortion analysis, or high-k dielectric-semiconductor modelling.

Continuous advancement of semiconductor technology and growth of semiconductor industry impose new requirements on semiconductor manufacturing. Semiconductor manufacturing belongs to the most challenging and complicated production systems involving huge capital investment and cutting-edge technologies. Chapter 3 discusses automation and integration in semiconductor manufacturing; chapter 4 is devoted to the contamination monitoring and analysis. Chapter 5 covers advanced plasma processing techniques and their emerging applications of etching, deposition, and surface modification of semiconductor materials. Chapters 6 and 7 concern themselves with oxidation techniques of III-V compounds and their application in MOS-based structures and gas sensors.

Tremendous interest in gallium nitride for high-frequency and high-power applications stems mainly from its wide and direct energy bandgap, thermal and chemical stability, and highelectron drift velocity. Chapter 8 is a comprehensive presentation of the GaN-based MOS devices with the emphasis on the description of various deposition methods of the dielectric film.

In chapter 9, the authors address two novel concepts for a mid-to-high voltage power semiconductor switch directly addressing the limitations of current IGBT and SJ MOSFET technologies. Chapter 10 is devoted to the study of the external optical feedback in nanostructure-based semiconductor lasers. In chapter 11, the authors investigate the influence of the electron transport on the optical properties of quantum-cascade structures.

Chapter 12 is dedicated to the preparation of transparent conductive oxide based on aluminumdoped ZnO for solar cells. Chapter 13 summarizes the preparation of high purity III-V layers grown by liquid phase epitaxy from rare-earth treated melts.

Optical technologies are the future of communication systems. Chapter 14 is a review of a device engineering method to provide high functionality of passive nonlinear vertical-cavity devices exploiting saturable absorption in semiconductor MQWs. Chapter 15 is a summary of the stateof-the-art of all-optical flip-flops based on semiconductor technologies. Chapter 16 reviews the current development of optical detection technologies on silicon photonics platform. In chapter 17, the authors describe the design, fabrication technology, and device performance of InP Mach-Zehnder modulator monolithically integrated with semiconductor optical amplifier. In chapter 18, the authors propose a new approach to ultra-fast all-optical signal processing based on quantum dot devices. Chapter 19 discusses present status and future direction of all-optical digital processing through semiconductor optical amplifiers.

Finally, chapter 20 presents a new approach to biomedical monitoring and analysis of selected human cognitive processes.

Jan Grym

Contents

1

SEMICONDUCTOR PROCESSES AND DEVICES MODELLING

Florin Babarada
Faculty of Electronics Telecommunications and Information Technology
University Politehnica of Bucharest, Romania

1. Introduction

The advancement of knowledge in the electronic design is strongly influenced by Technology Computer Aided Design–TCAD. Here is an interesting positive feedback, because the computing power helps the designers to perform modelling, simulation, optimisation and design of the new devices with improved performance, which have the capability to increase the computing power. The chapter present the basic aspects and the state-of-the-art of processes and devices modelling completed with the new aspects presented by the author in their last few year's papers like the results of his researches.

2. Models in micro and nanoelectronics

2.1 Modelling simulation and analysis

Due to high costs and long manufacturing cycle the modelling, simulation and optimisation or simply TCAD is the foundation of micro/nanoelectronics rapidly progress.
The analysis involves the separation of the entire in component parts, characterization and judgment them and also the examination of the elements of system and the relations between them to understand.
Simulation is the imitative representation of the operation of a system or process through the operation of other's or the examination of a problem without experimentation. If the analysis can be precise about the simulation we accept the idea of approximation.
Modelling is the production of a representation or simulation of a problem, process or device, making a description or analogies to help the visualization of the aspects that can't be directly observed. Modelling is a need for analysis, simulation and design optimisation.
A model for a pure simulation, such as that produced by fitting of the curves, is usually much simpler than a model for analysis, which should reflect the physical aspects in a qualitative manner. An example is the application of Monte Carlo method, which is equivalent to producing an imitative representation of the system functioning. We must always know the limitations of a model in order to don't interpret to naive the obtained results only through the improper application of the model. Also must have experience from previous experiments and simulations.

Development of electronic devices involves many tests and scraps for manufacturing until the scope respectively the designed operation parameters are performed. Implementation of device models, simulation and analysis, can now and in the future, to decrease substantially the number of iterations during development. A rough estimation of the development effort saved by analysis and simulation is of the order of 40%. This percentage depends heavily on the conditions of each individual project. The complete elimination of the tests and scraps of development is not possible today due to the uncertainty of many parameters of the models available that are already very sophisticated and too large. It is expected that modelling of devices, especially those with high scaling factor, at which the quantum mechanical aspects become predominate, to become more important in the future. This prediction is supported by the decreased cost of computing resources and special increases the cost of experimental investigation. Numerical modelling of the devices becomes more important for miniaturized models for which only large models are the existing and imaginable tools for precise prediction and analysis of device performances.

2.2 Technological models

The technological model is a schematic or analog description of the phenomenon or system that matter for his knowledge or associated properties and could be used to the further study of its properties. When possible the models used in microelectronics are physical models that mean the modelled phenomena are represented by physical effects well understood. When the studied phenomena are unknown is calling to the empirical models. In this case the relationships between phenomenon variables are experimental determined. Quantitative empirical model in this case is a mathematical expression that fits to the experimental data. Models can contribute to technological progress as follows:

- Physical models can produce explanations or views of the phenomenon or device studied even the studied effect is not directly observable
- Physical and empirical models serve as vehicles for processes or devices studied; certain aspects of real devices or processes can be examined by studying the characteristics or the respectively model operation by simulation

Although the simulation does not replace the manufacturing, it reduces the test time and errors to a stable and optimised process. Often, a simulation allows studying windows of process to help optimise the device structure by setting the process conditions. The biggest challenge is to develop models that can quickly, cheaply and accurately to simulate processes and phenomena of advanced semiconductor devices.

3. Physical models development

The development of physical models is generally in the following stages:

- Making the model qualitatively
- Making the model quantitatively
- Solving equations of quantitative model and centering process

Achieving quality model is the design, mechanisms deduction and relationships that comprise the essence of the phenomenon observed. In this form the model can realize the visualisation of the phenomena or devices studied. This is particularly important in microelectronics, where the phenomena being investigated are not usually directly observable. An example is the description of the Bohr atom model, the nuclei surrounded by electrons occupying orbits well defined. Another example is the description of the drift current in semiconductors as the movement of discrete particles of electric charge, respectively the electrons and holes, moving under the influence of electric field. In this case qualitative model can be translated into a set of equations or computing operations, resulting in a quantitative model. Finally the solution equations representing the quantitative model must be found and compared with experimental data to ensure that the simulation correctly emulates the phenomenon observed. In microelectronics quantitative physical models often take the form of equations with partial derivatives, Partial Differential Equations-PDE.

Simulation of micro/nanoelectronics devices and processes equations requires the evaluation model, computing operations, numerical analysis and advanced computer graphics (Rusu, 1990). Simulation helps manufacture of devices by increasing the success probability of the first experiment, for new products or processes. Existing manufacturing processes can be improved by centering process, ie finding the process combination of conditions for that we have the smallest results at the changing of the process conditions. Thus, circuit simulating behaviour modification based on conditions changing of manufacturing can identify the process tolerances.

In general, semiconductor devices can be simulated more precisely as the manufacturing processes of devices and integrated circuits because the physics of semiconductor devices it is better known. In comparison, several physical processes manufacturing of integrated circuits is still not well understood and must rely on the empirical models.

4. Empirical models

4.1 Introduction

The empirical models are only representations of experimental data and have a little or nothing physical background. Experimental data are used to create an empirical model as follows:

- Experimental results are stored in a database in the computer but are not provided information's about the interpolation results for approximation of unknown values between two points
- It used a mathematical function, which is adjusted by experimental data

To adjust the experimental data can use the next methods:

- Is used a polynomial function to pass through all points, leading to very complex functions and some experimental points may be wrong
- At one set of data graphically represented is choose a close mathematical function, which don't passes through all points and which can be adjusted

The most popular method of adjustment is the method of least squares, respectively the minimal sum of squares differences between points and the curve. If a data set match on a straight line $y=a+bx$, the process of finding the coefficients a and b, known as regression coefficients, is called linear regression. If it adopts a non-linear functional approximation, is used non-linear regression to find the regression coefficients.

4.2 Empirical models in semiconductor simulation

The empirical models are used to simulate semiconductor because:

- We don't have other option when the physical background is not yet known

- If incorporated as part of a program for simulation of a process or device, empirical models can serve as a tool for storing experimental data
- The simulation results for these models is fast and direct
- The empirical models can provide accurate simulations for some particular experimental conditions
- If the simulated conditions are between the experimental data, the interpolation results can be found with reasonable accuracy

Sometimes, if the individuals do not produce quantitative expressions, which constitute these models, this major limitation makes it impossible to extrapolate to conditions out of the experimented field.

Semiempirical models are the models in which phenomena are modelled by equations based on physical parameters corresponding to these phenomena. Most models used in simulation of semiconductor devices and processes are semiempirical models. Thus, the silicon oxidation in dry oxygen, at thickness of less than 350Å does not correspond Deal-Grove model. Nicollian and Reisman created a model for this area $t_{ox}=a(t+\tau)^b$ with a, b=constant, t=time of growth and τ=time required to raise an initial layer thickness x_i. Other example is the boron implantation effects arising from sewage, ie penetration of boron ions deeper than monocrystalline silicon. For modelling this effect, an exponential portion was added to Pearson IV model. The length of the falling exponential part is determined empirically to the value of 450Å. This empirical model is available in TSUPREM III and IV.

5. Design of experiment

Simulation of the manufacturing process using process simulators and extracting electrical characteristics using device simulators allow prediction of the behaviour and characteristics of the circuit from the design phase. The problem arises is that every attempt to obtain a performance model that is capable to incorporate the change effects in a broader set of parameters, is hit by hinder or even impossible to generate an analytical model that can be used effectively in design. Such as particularly important are the following aspects:

- Choosing a set of factors as more comprehensive
- Choosing the set of responses that characterize the best performance expected from the product design

The problem is usually solved iterative following the overlapping findings resulting from a series of individual experiments. Optimising a design involves finding a complete set of factors chosen so that the founding responses to have a high degree of confidence. Also, the sensitivity of the responses, given by the statistical nature of the technological implementation steps, is of particular importance in assessing the limits of tolerance of the desired response. Thus, by using based models simulation, can identify the input variables that allow the attainment of targets. More precisely, starting from the process variables such as temperature, time, energy and dose of implantation, may control the threshold voltage of the MOS transistors, parameter which influences the shape of the IV characteristics and so on the device parameters used in the circuit simulation and finally influences the circuit performances. Design of Experiment-DOE goal is to minimize the number of experiments in parallel with extraction of maximum information useful to designers. In this respect distinguish the following stages of analysis:

- Defining a set of factors that are considered to be sufficient for analysing the performance parameters of the requisite responses; the choice is based on previous knowledge
- Choosing a field of operation and a nominal value for each factor, considered acceptable in terms of tolerance of the technological process
- Defining a matrix corresponding with the DOE strategy selected
- Experimentation in selected points and collecting the results for each response
- Building a response surface model and analysis the conclusions from that study
- The revaluation of the set factors and the strategy of experimentation
- Obtaining the final response surface model

6. Response surface modelling

Design of experiments is the key point of the optimisation process design. Results of experiments are used to generate the Response Surface Model-RSM. Are taken into account three model categories:

- Linear models, where the responses are linear functions of factors
- Models of order two for higher order design, the answers are obtained as functions of parabolic factors; these models constitute the standard in RSM techniques
- Transcendental models for higher order design, which provides improved techniques for analysing data and are used in analysis of amended RSM

Linear models assume that response R_i is a linear combination of factors $f_1, f_2, \ldots f_n$. When using these models only one factor is change in each run. The experiments are chosen in star. These are easily designed and expanded to higher dimensions.

The square models guess that the response R_i is a square combination of input factors with two power grade of the factors and products of factors. In this way is take into account the interactions between factors. Strategies to design experiments in this case are different: full-factorial type and fractional-factorial type. Full-factorial strategies take into consideration all possible combinations of factors. This approach provides more information but also presents the inconvenience of requiring a long running time. Fractional-factorial strategies select a subset of the experimental points from the set full-factorial. Presents the advantage of data reuse at increasing of the problem size and easy change to full-factorial analyse.

The transcendental models assume the existence of a mechanism for transforming an initial set of factors in a modified set used for RSM.

The data obtained by experiments and those obtained by simulation are used to build a RSM, from which analysis may conclude a set of information about:

- Main effects, linear or nonlinear
- The interaction of factors
- Various factors importance in the evolution of a response
- Sensitivities of response to some factor
- Comparing the effect of a factor with the others, etc.

These results are iterative used for adjustment of coefficients, which are the input data of the RSM.

Like example, for the case of MOSFET technology flow the threshold voltage VT is the output data and the input data are the following factors: oxide thickness TOX, Nsub concentration of substrate, the peak concentration of channel implant for threshold voltage adjustment VTPEAK, the peak concentrations of LDD source and drain NLDDpeak, distance between the windows of the source and drain Lgate.

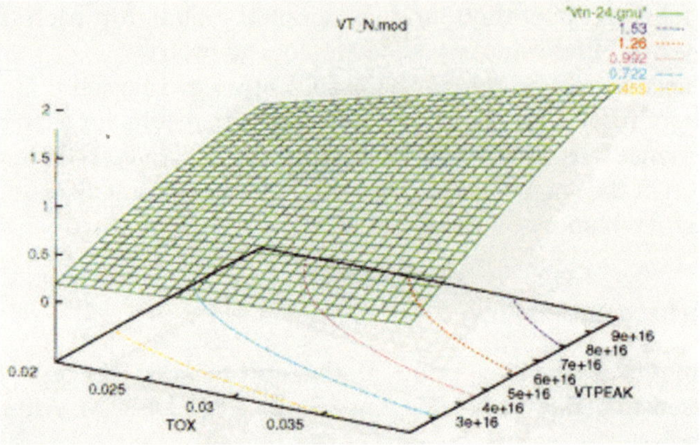

Fig. 1. RSM results for MOSFET threshold voltage VT

Were identified the main parameters that determines the threshold voltage of MOSFET like the MOS oxide thickness and the implantation dose for threshold adjustment, fig. 1.

7. Process optimisation

7.1 Introduction

TCAD software packages first need equipment models using configuration and settings as input parameters in order to obtain the process environment and process model to create the wafer data characteristics. Next using TCAD process simulator obtains the wafer state and using TCAD device simulator obtain device performance and the input data for circuit simulation. A final challenge and grand opportunity for future process modelling is to implement the accurate atomic scale reaction models respectively reaction energies, rates, products and process equipment models respectively gas flows, reactant concentrations and temperatures versus equipment settings.

In the state of the art devices small geometry effects including hot electron transport, punch-trough, avalanche multiplication, drain induced barrier lowering, oxide and junction breakdown, leakage currents, grain size effects and discrete doping elements effects are of great importance (Veendrick, 2008). Devices are also starting to exhibit significant quantum effects including gate oxide and bandgap tunnelling, inversion layer quantization, quantum transport and carrier density smoothing.

7.2 Optimisation strategies

Optimising a process technology or a device parameter involves an optimum set of factor setting such that a number of relevant results meet predefined targets. This problem is solved using the concepts of statistical Design of Experiments-DOE, for planning a number of experiments for different settings of input factors.

The simulation of process, device and circuit are performed in specific points respectively specific values for input factors for which the simulation are running. The results of experiments are analysed for each of the responses as a function of the input factors and we obtain a response surface model-RSM. The DOE/RSM concept guarantees that with a minimum number of experiments we obtain a maximum information respectively detection of the important main effects, factor interaction effects or which factor are the most important. The RSM models are used to find factor settings that produce devices with desired specifications (Govoreanu, 2002).

7.3 Example

Process optimisation example refers to n-type MOSFET realized in 0,5µm technology using Taurus-workbench software package from Synopsys. We start with substrate <100> boron doped at $5x10^{18}$. Then epitaxial growth of 6µm silicon layer, 0,2µm oxide layer and 0,15µm nitride and in the last two layers is successively configured ISLAL and NWELL and phosphorus is implanted with $2x10^{12}$ dose and 300KeV energy. After nitride removing and oxide configuration the threshold voltage adjusting doping is performed in two steps VTN implant with boron and PUNCH implant with boron at $5x10^{11}$ dose and 50KeV energy.
The gate oxide is grown, the polysilicon gate is deposited configured and implanted with phosphorus at $5x10^{15}$ dose and 45KeV energy. After NLDD implant in the gate and source/drain area the deep implant for source/drain configuration with phosphorus at $4x10^{15}$ dose and 80KeV energy is performed.
The next process steps perform the contact and interconnection between devices and the circuit protection layer.

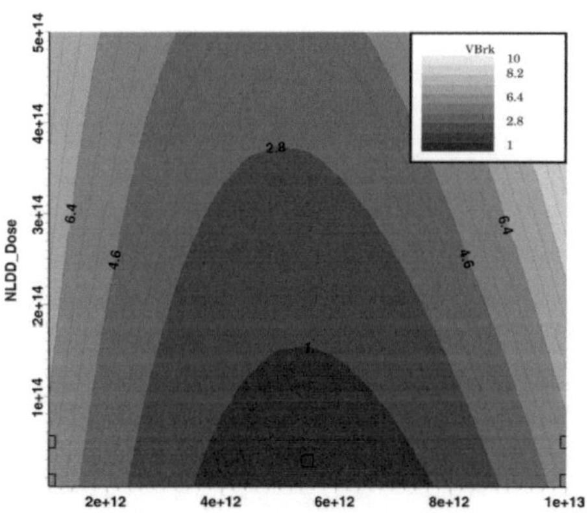

Fig. 2. RSM-V_{BRK} versus NLDD, VTN

Wafer	Units	Samp.1	Samp.2	Samp.3	Samp.4
VTN_Dose	Dose	1E+13	1.0E+13	1.0E+12	1.0E+12
PNCH_Dose	Dose	5E+11	5.0E+11	5.0E+11	5.0E+11
PNCH_Energy	Energy	50	50	50	50
tox	nm	11.9786	11.9786	11.996	11.996
NLDD_Dose	Dose	7E+12	5.0E+13	7.0E+12	5.0E+13
Xj	nm	192.945	231.634	214.388	347.719
Lchan	nm	538.034	393.185	376.42	254.542
Vt	V	0.9348	0.9338	0.3487	0.3222
IDSat	A/um	1.8E-04	3.3E-04	5.1E-04	6.7E-04
VBrk	V	5.57	5.57	5.91	5.73

Table 1. The samples parameters

Using DOE and RSM techniques the most sensitive process steps were identified respectively V_t adjusting implant and NLDD implant. These two parameters were modified successively.
The RSM results indicate a high dependence of breakdown voltage function of NLDD implant dose and a strong decreasing around $5x10^{12}$ VTN implant dose (fig. 2.), a big dependence of threshold voltage versus PUNCH implant dose and a small dependence versus VTN implant dose and a high dependence of saturation current (I_{DSS}) function of NLDD implant dose and a decreasing around $8x10^{12}$ VTN implant dose.

The increase of NLDD implant dose at Sample 2 and 4 reduces the polysilicon depletion effect, by reducing the voltage drop across the polysilicon gate and improving the device transconductance (the higher slope of transfer characteristics for Sample 2 and 4), fig. 7.

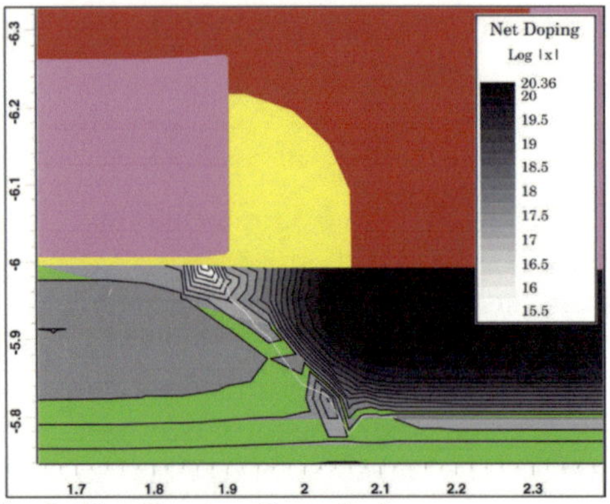

Fig. 3. Net Doping Sample 1

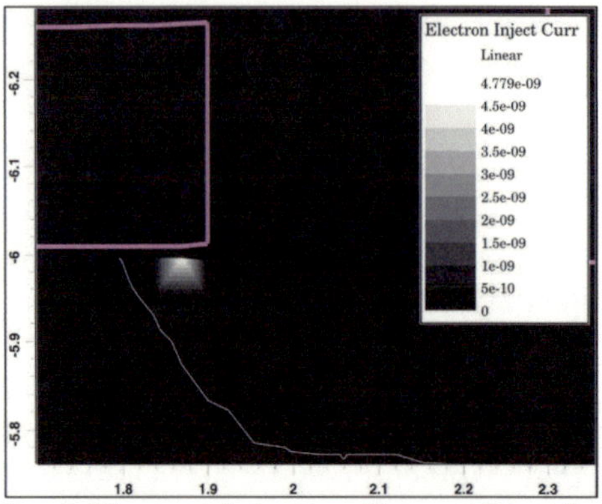

Fig. 4. Electron Injection Current Sample 2

According to Table 1 and fig. 3 to 6 the breakdown voltage is proportionally with the radius of source/drain junction (Kwong, 2002). Output resistance is reduced by decreasing the VTN boron adjusting implant dose (Sample 3 and 4) and can be explained by higher electron concentration in the channel, which allows a shorter pinchoff region. A shorter pinchoff region gives rise to a much larger magnitude of the Early voltage.

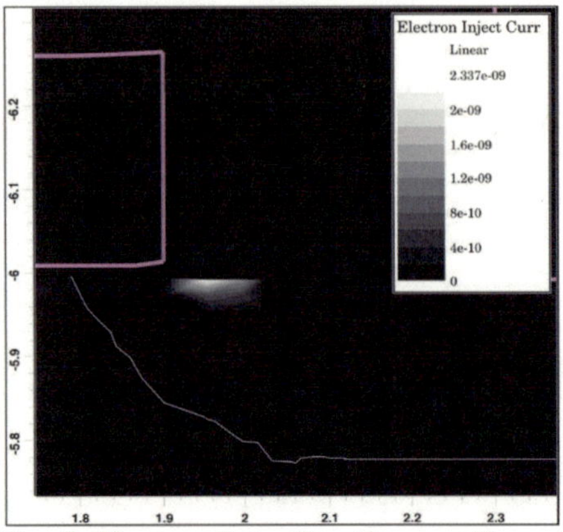

Fig. 5. Electron Injection Current Sample 3

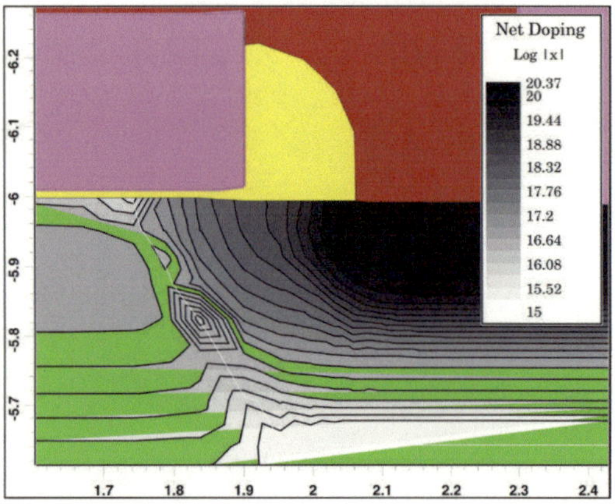

Fig. 6. Net Doping Sample 4

The decrease of NLDD dose in Sample 3, fig. 5, comparing with Sample 2 fig. 4 move the electron injection current from gate oxide to spacer decreasing the gate oxide breakdown possibility and reduces electron injection concentration which improve reliability. For the all four samples the transfer characteristics are presented in fig. 7 and the external characteristics in fig. 8 (Campian, 2003).

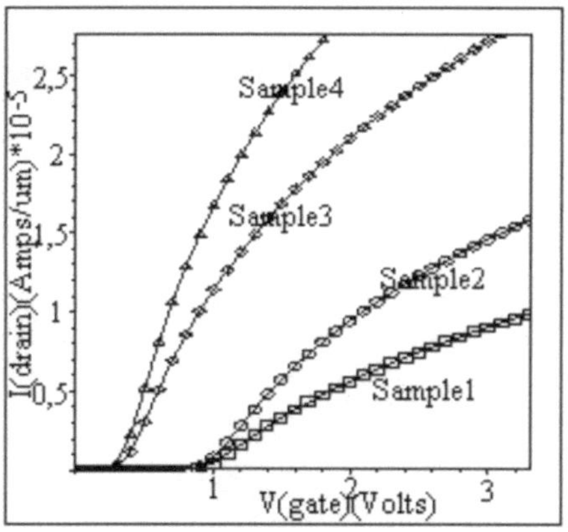

Fig. 7. I_D-V_{GS} Characteristics

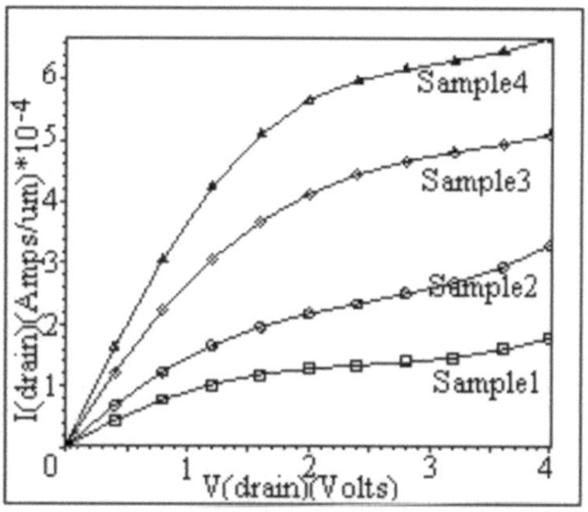

Fig. 8. I_D-V_{DS} Characteristics

The higher electron concentration in the channel gives a large Early voltage very useful in analog circuits. Increasing of lateral source/drain slope lowers also the series resistance, which improves the drive current, but for very abrupt profile junction the improvement is paid by degradation in leakage current due to more severe short channel effects.

8. MOSFET DC modelling for distortions analysis

8.1 Introduction

The scaling-down evolution of semiconductor devices will ultimately attend fundamental limits as transistor reach the nanoscale aria. In this context the MOSFET models must give the process variations and the relevant characteristics like current, conductance, transconductance, capacitances, flicker thermal or high frequency noise and distortion (Ytterdal, 2003). The new challenge of nanotechnology needs very accurate models for active devices (Scholten, 2009). The design of linear analog circuits lacks models for state-of-the-art MOS transistors to accurately describe distortion effects. This is produced by the inaccurate modelling of the second order effects induced by high vertical gate field such as mobility degradation and series resistance and second order effects induced by parallel drain field like velocity saturation in the ohmic region, channel length modulation, static feedback, weak avalanche and selfheating in the saturation region. After a rigorous description of transistor transconductance and channel conductance in ohmic and saturation region we included these effects in the MOSFET model, using a compact drain current expression for time computation reasons.

8.2 Gate induced distortions modelling

Carriers mobility degradation modelling

The channel mobility must be treated quantum-mechanically because the thickness of the inversion layer is in the order of a few Å, smaller than the De Broglie wavelength of the carriers. Quantum-mechanical calculations show that energy subbands of electrons and holes are formed in different energy valleys. The spacing of these subbands increases with the normal electric field Ex. In the weak inversion region where many subbands are occupied, quantum effects can be neglected, but in the strong inversion region where only few subbands are occupied, quantum effects become important. In most cases more than one subband is filled and the modelling must give the right approximation of very complex scattering processes in the inversion layer. The mobility can be described by considering three mechanisms, which dominate the scattering of charge carriers in the inversion layer at the Si-Si0$_2$ interface.

***Coulomb Scattering* - μ_C**

Charged centres near the Si-Si0$_2$ interface may be of the same charge type as the mobile inversion charge leading to Coulomb repulsion. This results in scattering, which is important for lightly inverted surfaces, high surface-charge densities or substrate doping concentrations, and less important for heavily inverted surfaces due to carrier screening. Coulomb scattering limited mobility μ_C is given by:

$$\mu_C \cong \frac{Q_{inv}}{N_A} \tag{1}$$

The above type of scattering has influence only in the weak and moderate inversion region when the drain current is dominated by the exponential dependence of inversion layer charge Q_{inv} on the gate voltage.

***Phonon Scattering* - μ_{ph}**

Surface phonons or surfons from the quantum vibrations of the crystal lattice scatter the mobile charge carriers. Under the assumption that carriers in the inversion layer only occupy the lowest subband, the mobility determined by acoustic phonon scattering is described by:

$$\mu_{ph} \cong \left(\frac{11}{32}Q_{inv}+Q_{dep}\right)^{-1/3} \tag{2}$$

Experimentally it was found for both holes and electrons that: $\mu_{ph} \cong E_{eff}^{-1/3}$ (3)

Expression (3) deviates slightly from (2), which is ascribed to the fact that electrons occupy several subbands at intermediate values of effective field (Babarada, 2003).

***Surface Roughness Scattering* - μ_{sr}**

The interface between the silicon crystal and the gate oxide is not atomically smooth. The above interface roughness scatters the mobile charge carriers. This type of scattering is especially important under strong inversion conditions because the strength of the interaction is governed by the distance of the carriers to the surface. The carriers, which are to the surface, will have the stronger scattering due to surface roughness. Under the assumption of single subband occupation and a Gaussian type autocorrelation function of interface roughness, μ_{sr} can be described by:

$$\frac{1}{\mu_{sr}} \cong E_{av}^2 \cdot \int_0^{\pi}(1-\cos\theta)\cdot \exp\left[-\frac{1}{2}\cdot k_w^2 \cdot L_c^2 \cdot (1-\cos\theta)\right]\cdot d\theta \tag{4}$$

where L_C is the correlation length of interface roughness, k_w is the carrier wave vector and E_{av} is the normal field averaged over the inversion layer. For a uniform doping profile the average field E_{av} can be calculated to be equal to the effective field E_{eff} with $\eta = ½$. Non-uniform doping profiles will lead to different values of η, which is an empirical parameter.

In the limit that the correlation length is much smaller than the carrier wavelength ($L_C <<1/ k_w$), the mobility limited by surface roughness scattering μ_{sr} reduces to:

$$\mu_{sr} \cong E_{eff}^{-2} \tag{5}$$

The above dependence of surface roughness scattering corresponds to the experimentally found dependence of electron mobility on effective field. For larger values of correlation length L_C, mobility deviates from the inversely quadratic dependence on E_{eff} owing to the fact that the integral term in (4) depends on Q_{inv}. For holes, was found experimentally:

$$\mu_{sr} \cong E_{eff}^{-1} \tag{6}$$

The difference between equations (5) and (6) for electrons and holes, respectively, is often ascribed to the fact that at high transverse fields holes tend to congregate further away from the interface than electrons do. The larger average distance leads to a reduced influence of the interface roughness and thus to less surface roughness scattering for holes.
The above-described mechanisms can be incorporated into one channel mobility μ, as follow:

$$\frac{1}{\mu} = \frac{1}{\mu_0} + \frac{1}{\mu_{sr}} + \frac{1}{\mu_{ph}} \tag{7}$$

where μ_0 is the carrier mobility limited by ionised impurity scattering and acoustic phonon scattering in the bulk material. Equation (7) leads to:

$$\mu = \frac{\mu_0}{\theta_c + \theta_{ps}\sqrt{\theta_{ph} E_{eff}^{2/3} + \theta_{sr} E_{eff}^{2n}}} \tag{8}$$

where θ_c, θ_{ps}, θ_{ph} and θ_{sr} are empirical parameters, and n = 2 for electrons and n = 1 for holes. Expression (8) assumes that the surface roughness scattering limited mobility is more important than phonon scattering limited mobility and leads to a more accurate description of high-order derivatives.

Series resistance modelling

For DC model only the resistive element is important out of MOSFET source and drain junction parasitic components. For short-channel devices the drain current may reach very high values. At large current values the series resistance is no longer negligible and has to be taken into account using the source resistance R_S and the drain resistance R_D like in the equivalent circuit given in fig. 9.

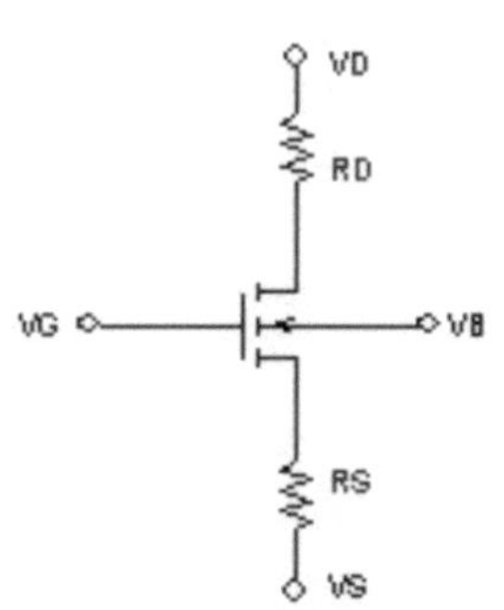

Fig. 9. The equivalent MOSFET

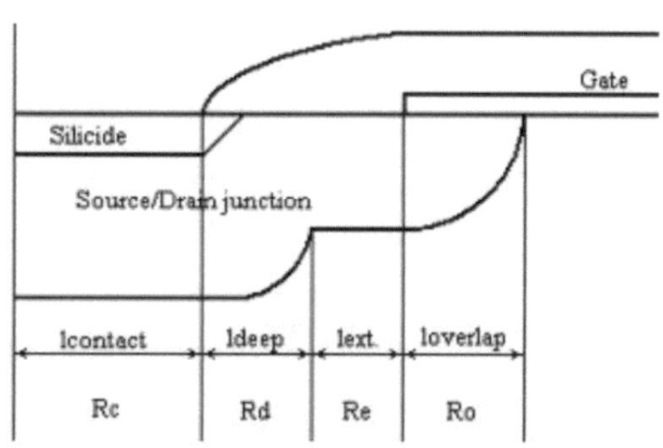

Fig. 10. Parasitic series resistance components

Including the source series resistance R_S and drain series resistance R_D in the MOSFET model we don't include additional nodes and the computation efficiency is better.

According to fig. 10, the series resistance can be divided in four components: overlap resistance R_o, extension resistance R_e, deep resistance R_d and silicide-diffusion contact resistance R_c .

Overlap resistance

With scaling-down the transistor dimensions the source-drain extension to gate overlap in the gate length is increasing around 35% and this resistance become more important and bias dependent. The overlap resistance is closely dependent on doping concentration in overlap region because the current spreading and the accumulation carrier charge density is dependent on the overlap doping concentration. The lateral slope of overlap doping profile is one of the most important parameter to controlling the short channel effect and finally the nanotransistor characteristics and for these reasons the modelling of the overlap resistance is very important for simulation and design.

The approximation of exponentially sloped doping profile and the constant accumulation charge density in the overlap region are suited for the computation reasons but usually needs fitting parameters and has good precision only near the metallurgical junction. Better results of the accumulation and spreading current are given by Gaussian doping gradient in the lateral direction of the overlap region. Because the most important component of current is distributed in the lateral junction, the vertical doping gradient of the junction is not so important. The transistor has an accumulation layer under the gate oxide in the overlap region and because the nanotransistors have very shallow junctions with high doping level and large depletion width the current is spread in the bulk region along the depletion boundary of source-drain extension junction. As a result the overlap resistance can be modelled by combination of resistances R_{OS1} series with R_{OS2} in parallel with R_{OP} as given in fig. 11. R_{OS1} and R_{OS2} are the resistances corresponding to accumulation layer in the entire surface under the gate oxide in the overlap region and R_{OP} is the spreading resistance in the neutral bulk region. The current spreading in the overlap region is modelled based on the depletion approximation and spreading resistance is calculated versus the depletion width in overlap region and the spreading angle.

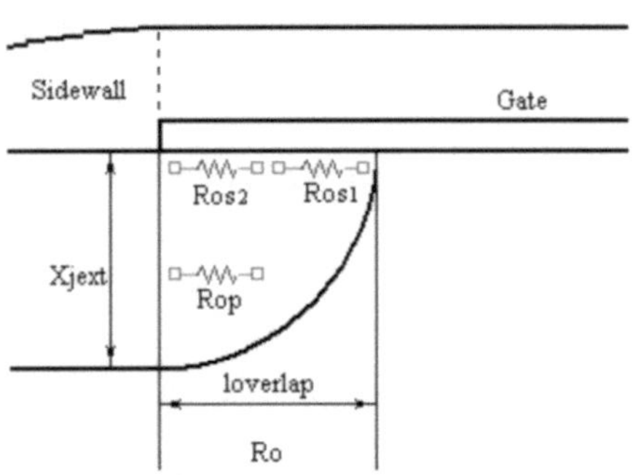

Fig. 11. Overlap resistance components components

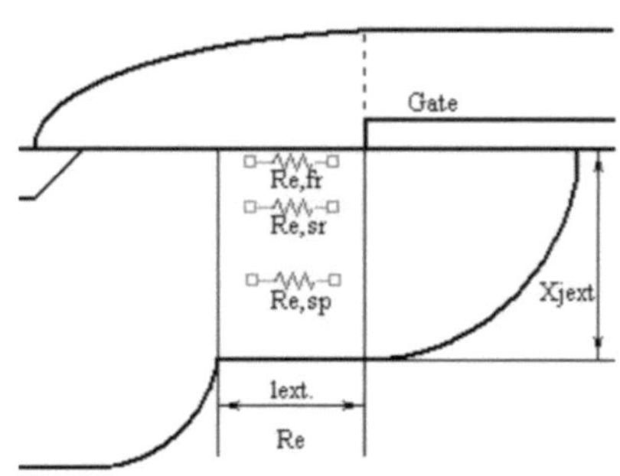

Fig. 12. Extension resistance

Extension resistance

The extension resistance corresponds to the region from the gate end to beginning of deep source/drain junction like in fig. 12. Three component resistances connected in parallel compose the extension resistance: fringing resistance, surface resistance and spreading resistance. The fringing resistance taken into account to characterize the fringing field effect on the surface extension region and can be calculated by iterative solution of the potential relationship of gate-sidewall-silicon system. The surface resistance characterize the uniform doping region in the extension region and is calculated in function of rezistivity, doping concentration and dimensions of these region. The spreading resistance in the extension region can be expressed function of vertically graded Gaussian doping profile and the spreading angle.

Deep resistance

The deep resistance corresponds to lateral deep diffusion of source/drain-gate structure and its contribution to series resistance is small because of heavily doped deep junction, fig. 13. In the deep region the modulation effect of the gate fringing field is negligible because the distance to gate and because the deep region is heavily doped to be modulated. The surface resistance in deep region include the lateral extension of silicide layer. The spreading resistance in the deep region can be calculated considering one similar angle like in the extension region.

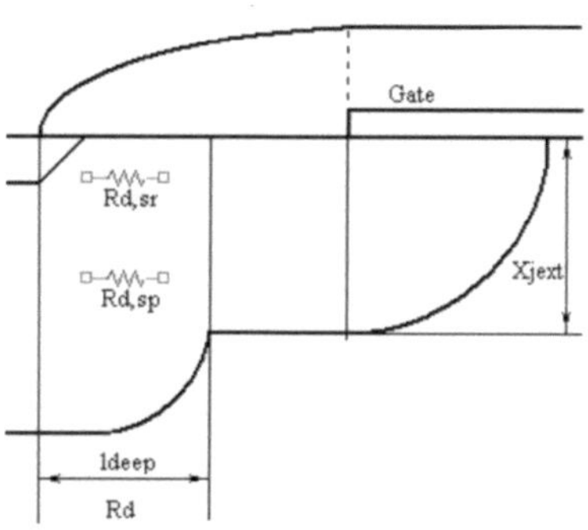

Fig. 13. Deep resistance components

Contact resistance

The contact resistance refers at resistance between the silicide contact and the diffusion layer and is strong dependent on the silicide layer thickness, the junction doping concentration and the silicide material. Under the silicide-diffusion contact the current flow increases with the increase of the silicide thickness and a significant current is pushed in the silicon region under the silicide. Because the total contact silicide-diffusion resistance has a minor effect in total series resistance we can neglect this effect. Now the source or drain series resistance is given by:

$$R_{S,D}=Ro+Re+Rd+Rc=(R_{O,S1}+R_{O,S2}\parallel R_{O,Sp})+(Re,fr\parallel Re,sr\parallel Re,sp)+(Rd,sr+Rd,sp)+Rc \quad (9)$$

In the ohmic region the drain series resistance value is equal with the source series resistance. At high level of lateral electric field if the carriers attend the saturated velocity at the drain end, the drain series resistance and source series resistance become different. In the saturation region the drain series resistance may reach several times higher values than the source series resistance but because the drain current is strongly dependent on drain voltage, the effect of drain series resistance is low.

The extension resistance, deep resistance and contact resistance are independent of terminal voltages and are inversely proportional to channel width. The overlap resistance due to overlap of polysilicon gate on the source/drain region, an accumulation layer is formed.

Because the accumulation layer charge is directly dependent on the gate voltage the overlap resistance is dependent on the gate voltage too.
From computing reasons we used a very compact form of series resistance versus gate voltage dependent, given by:

$$R_S = r_1 \cdot \left(1 + \frac{r_2}{r_3 + V_{GS} - V_T}\right), \tag{10}$$

where, r_1, r_2 and r_3 are empirical parameters.

The current expression

In order to include the series resistance R_S and R_D in MOS transistor model we must replace the new expressions including the resistances R_S and R_D in the drain current in the ohmic region.

$$V_{GS,1} = V_{GS} - I_D \cdot R_S \tag{11}$$

$$V_{DS,1} = V_{DS} - I_D \cdot (R_D + R_S) \tag{12}$$

$$V_{SB,1} = V_{SB} + I_D \cdot R_S \tag{13}$$

$$V_{T,1} = V_{FB} + \Phi_B + \gamma\sqrt{\Phi_B + V_{SB} + I_D \cdot R_S} \cong V_{FB} + \Phi_B + \gamma\sqrt{\Phi_B + V_{SB}} + \delta \cdot I_D \cdot R_S \tag{14}$$

V_{FB} is the flat-band voltage, Φ_B is the surface potential at onset of strong inversion, γ is the body effect coefficient $\gamma = \frac{\sqrt{2 \cdot q \cdot \varepsilon_{Si} \cdot N_A}}{C_{ox}}$, δ describe the influence of non-uniform doping concentration $\delta = \frac{\gamma}{2\sqrt{\Phi_B + V_{SB}}}$, R_S is the series resistance corresponding to drain or source resistance. The drain current in the ohmic region is given by:

$$I_D = \frac{\mu \cdot C_{ox} \cdot W}{L} \cdot \left[V_{GS,1} - V_{T,1} - \frac{1}{2}(1+\delta) \cdot V_{DS,1}\right] \cdot V_{DS,1} \tag{15}$$

The finally drain current expression is given by:

$$I_D = \frac{K\left[V_{GS} - V_T - \frac{1}{2}(1+\delta)V_{DS}\right] \cdot V_{DS}}{1 + K \cdot 2R_S \cdot \left[V_{GS} - V_T - \frac{1}{2}(1+\delta)V_{DS}\right]} \tag{16}$$

$$\text{where} \quad K = \frac{\mu \cdot C_{ox} \cdot W}{L}, \tag{17}$$

μ is the mobility given by expression (8), C_{ox} is the gate oxide capacitance per unit area $C_{ox} = \frac{\varepsilon_{ox}}{t_{ox}}$, W is the effective channel width and L is the effective channel length.

8.3 Drain induced distortions modelling

In the analog circuit design the MOS transistor can be driven as well by the drain terminal. By example a drain voltage driven MOSFET operating in the ohmic region like a gate voltage controlled resistor with the resistor linearity limited by the transistor drain voltage induced distortions. Other example can be the load device in amplifier circuits where the drain terminal drives the MOSFET in saturation.

Ohmic region

Drain Current

The drain current expression for strong inversion long channel devices in compact models is obtained by drift current expression function of surface potential. Using a first-order Taylor expansion around V_{SB}, with the source as reference, the drain current is:

$$I_D = \mu C_{ox}\frac{W}{L}\left[V_{GS} - V_T - \frac{1}{2}(1+\delta)\cdot V_{DS}\right]\cdot V_{DS} \tag{18}$$

where mobility μ is bias dependent due to mobility degradation. The equation (18) is quadratic expression in drain voltage, implying that the third-order derivative $\frac{\partial^3 I_D}{\partial V_{DS}^3}$ is equal to zero for all bias conditions, which is non-physical and also is non-symmetrical, which leads to a discontinuity in the higher-order derivatives at V_{DS}=0. For more accurate and symmetrical description of the higher-order derivatives drift current expression function of surface potential can be expanded around $V_{SB}+\frac{1}{2}V_{DS}$ and results (Babarada, 2005):

$$I_D = \mu C_{ox}\frac{W}{L}\cdot V_{GS\#}\cdot V_{DS} = K\cdot V_{GS\#}\cdot V_{DS} \tag{19}$$

$$V_{GS\#} = V_{GB} - V_{T0} - \frac{1}{2}(V_{DB}+V_{SB}) - \gamma\left(\sqrt{\frac{1}{2}(V_{DB}+V_{SB})+\Phi_B} - \sqrt{\Phi_B}\right) \tag{20}$$

The equation (20) is valid for both positive and negative values of V_{DS} and consequently the second-order and higher-order derivatives are continuous at V_{DS}=0. The mobility is given by expression (8). The simulation results of second and third-order harmonic amplitude, using the drain current expression (19) gives better fit to the measurements.

Velocity Saturation

For short-channel transistors equation (19) becomes inaccurate due to the effect of velocity saturation. The usual description of the carrier velocity is expressed by an empirical relation. For a more precise description of hole velocity saturation and prevent discontinuities at V_{DS}=0, used the next adjusted expression:

$$v = \frac{\mu\frac{\partial\psi_S}{\partial y}}{\sqrt{1+\left[\left(\frac{\mu}{v_C}\frac{\partial\psi_S}{\partial y}\right)^2 / \sqrt{P^2+\left(\frac{\mu}{v_C}\frac{\partial\psi_S}{\partial y}\right)^2}\right]+\left(\frac{\mu}{v_{sat}}\frac{\partial\psi_S}{\partial y}\right)^2}} \tag{21}$$

where v_c is saturation velocity limited by acoustic phonon scattering, P is a fitting parameter and v_{sat} is saturation velocity limited by optical phonon scattering. Assuming first-order approximation of $\frac{\partial\psi_S}{\partial y}$ equals with $\frac{V_{DS}}{L}$, this equation can be implemented like the empirical relation of the carrier velocity, in the drain current expression, including also the series resistance effect:

$$I_D = \frac{\mu_0\cdot C_{ox}\cdot W\cdot V_{GS\#}V_{DS}}{\sqrt{\left(\frac{\mu_0 L}{\mu}\right)^2+\left[\left(\frac{\mu_0}{v_c}V_{DS}\right)^2 / \sqrt{P^2+\left(\frac{\mu_0}{v_c}\frac{V_{DS}}{L}\right)^2}\right]+\left(\frac{\mu_0}{v_{sat}}V_{DS}\right)^2}+\beta_R V_{GS\#}} \tag{22}$$

where $\beta_R = 2\mu_0 C_{ox} W R_S$ and R_S is given by equation (10). The harmonic distortion results of equation (22) are accurate also for p-type transistors and have no discontinuities at V_{DS}=0.

Saturation region

The drain current expression (22) gives an accurate description of the DC behaviour in the ohmic region, which can be extended to the saturation region by replacing the drain voltage V_{DS} by the saturation voltage V_{DSsat}. For long-channel devices the saturation voltage V_{DSsat} can be calculated in good approximation from the zero value of the first-order derivative of the drain current expression (19) to drain voltage V_{DS}. For short-channel devices the calculation of V_{DSsat} is less evident, since the saturated drain conductance may be much larger than zero. The saturated drain current can by written:

$$I_{Dsat} = -W \cdot Q_{invL} \cdot v_{sat} \tag{23}$$

where Q_{invL} is the inversion layer charge at the drain side. Using the current expression (22) and (23) and neglecting the influence of mobility degradation, the saturation voltage is:

$$V_{DSsat} = V_{DSsat\_i}\left[1-\frac{1}{2}\frac{\left(\sqrt{\beta_n^2+\beta_p^2}-\frac{1}{2}\beta_R\right)}{\sqrt{\left(L/V_{DSsat\_i}\right)^2+\beta_n^2+\beta_p^2}+\frac{1}{2}\beta_R}\right] \tag{24}$$

where $\beta_n=\mu_0/v_{sat}$, $\beta_p = \mu_0/\left(v_c \cdot \sqrt{P}\right)$ and V_{DSat_i} is the saturation voltage for an ideal long-channel transistor.

Channel length modulation

When V_{DS} is increased beyond V_{DSsat}, the velocity saturation point moves towards the source, causing effectively that the channel length L is shortened by a length ΔL. That made the conductance to be non-zero in the saturation region and the drain current function of I_{Dsat}, the drain current at $V_{DS}=V_{DSsat}$, and L_{eff}, the effective channel length defined by:

$$L_{eff} = \mathrm{L} + \frac{\mu}{v_{sat}} \cdot V_{DSsat} \tag{25}$$

Velocity saturation results in an effective increase of the channel length, which may become important for short-channel devices. Using the velocity saturation expression (21) the effective channel length is:

$$L_{eff} = \sqrt{L^2 + \frac{\left(\frac{\mu}{v_c}\cdot V_{DSsat}\right)^2}{\sqrt{P^2+\left(\frac{\mu}{v_c}\cdot\frac{V_{DSsat}}{L}\right)^2}} + \left(\frac{\mu}{v_{sat}}\cdot V_{DSsat}\right)^2} \tag{26}$$

The channel length modulation ΔL expression is:

$$\Delta L = l_c \cdot ln\left(\frac{V_{DS}-V_{DSsat}+\sqrt{\left(V_{DS}-V_{DSsat}\right)^2+E_{sat}^2\cdot l_c^2}}{E_{sat}\cdot l_c}\right) \tag{27}$$

where E_{sat} is the lateral electric field for which the carrier velocity saturates and is assumed to be constant and lc is length parameter.

Static feedback

Because the electrostatic coupling between drain and channel region an extra mobile charge ΔQ_{inv} is injected in the channel when the drain bias is increased beyond saturation. The effect of ΔQ_{inv} on drain current can be modelled as a decrease of the threshold voltage, where σ_{sf} is a parameter:

$$\Delta V_{Tsf} = \frac{\sigma_{sf}\cdot\sqrt{V_{DSsat}}\cdot V_{DS}}{L} \tag{28}$$

Self-Heating

The effective working temperature in the inversion layer is: $T = T_0 + R_{Th} \cdot P_{dis}$ (29)

where T_0 is the ambient temperature, the dissipated power is: $P_{dis} = I_D \cdot V_{DS}$ (30)

and thermal device resistance is:

$$R_{Th} = \frac{X_{sub}}{\lambda_{Th} \cdot W \cdot L} \tag{31}$$

where X_{sub} is the thickness of the substrate and λ_{Th} is the thermal conductivity of silicon. The dependence of drain current function of temperature is:

$$I_D = I_{Dsat} / \left(1 + R_{Th} \cdot I_{Dsat} \cdot V_{DS} / T_0\right) \tag{32}$$

8.4 Smoothing function

Usual is assumed that drift current can be neglected in weak inversion and diffusion current can be neglected in strong inversion. Around threshold voltage is considered the moderate inversion region, where both the drift current and diffusion current are important.

The transition from the ohmic region to the saturation region is continuous for the drain current. For analog circuit modelling, this transition should also be continuous for the higher-order derivatives of drain current to drain voltage, or the model should be continuous. To arise these requires we replace drain-source voltage V_{DS} with an empirical function V_{DSsf} that changes smoothly from V_{DS} in the ohmic region to V_{DSsat} in the saturation region. This empirical function V_{DSsf} is usual named smoothing function.

In order to preserve the model symmetry respectively the discontinuities of higher-order derivatives at V_{DS}=0V, we choose a smoothing function for which the derivate reported to V_{DS} is equal to unity, at V_{DS}=0V:

$$V_{DSsf} = \frac{V_{DS} \cdot V_{DSsat}}{\left(V_{DS}^{2m} + V_{DSsat}^{2m}\right)^{1/2m}} \tag{33}$$

where: m is an empirical parameter, which can be integer only; V_{DS} is the drain-source voltage and V_{Dssat} is the drain-source saturation voltage.

8.5 The current expression

Distinction between the drift and the diffusion component of the drain current should be maintained in all inversion regions, for an accurate description of the moderate inversion region (Gildenblat, 2009):

$$I_D = I_{drift} + I_{dif} \tag{34}$$

For model symmetry the drift current expression is given by usual formula, and making a Taylor expansion around $V_{SB} + \frac{1}{2} V_{DS}$, like in equation (19):

$$I_{drift} = \beta \left(V_{GBeff} + \Delta V_G - \frac{\psi_{sL} - \psi_{s0}}{2} - \gamma \sqrt{\frac{\psi_{sL} + \psi_{s0}}{2}} \right) (\psi_{sL} - \psi_{s0}) \tag{35}$$

where: β is the gain factor; V_{GBeff}=V_{GS}+ V_{SB}- V_{FB}, ΔV_G is determined by transition from drain induced barrier lowering-DIBL in weak inversion to static feedback in strong inversion; ψ_{sL}

is the surface potential at the drain side; ψ_{S0} is the surface potential at the source side; γ is the body effect coefficient. Similar the diffusion current is:

$$I_{dif}=\beta\cdot\gamma\cdot u_T\left[\sqrt{\psi_{s0}+u_T\exp\left(\frac{\psi_{s0}-V_{DS}-2\Phi_F}{u_T}\right)}-\sqrt{\psi_{s0}}-\left(\sqrt{\psi_{sL}+u_T\exp\left(\frac{\psi_{sL}-V_{DS}-V_{SB}-2\Phi_F}{u_T}\right)}-\sqrt{\psi_{sL}}\right)\right] \quad (36)$$

and the total drain current expression, taking into account: drain saturation voltage, drain induced barrier lowering, mobility degradation, series resistance, velocity saturation, channel length modulation, static feedback and weak avalanche, is given by:

$$I_D=\frac{L(I_{drift}+I_{dif})}{(\mu_0/\mu)(L_{eff}-\Delta L)+(\beta_R+\beta_{Th}\cdot V_{DS}\cdot V_{DSsf})V_{GS-2}}(1+W_{av}) \quad (37)$$

where: L is the channel length; μ_0 is low field bulk mobility; μ is carriers mobility (Babarada, 2003); L_{eff} is the effective channel length; ΔL is the channel length modulation; β_R includes the series resistance; β_{Th} includes self heating; W_{av} is the weak avalanche, V_{GS-2} has the expressions:

$$V_{GS-2}=\frac{1}{2}V_{GS-1}+\frac{1}{2}\sqrt{V_{GS-1}^2+4\cdot\varepsilon^2} \quad (38)$$

where ε is a smoothing factor and V_{GS-1} has the next expressions:

$$V_{GS-1}=V_{GS}-V_{FB}-2\Phi_F-V_{DSsf}/2-\gamma\sqrt{2\Phi_F+V_{SB}+V_{DSsf}/2} \quad (39)$$

8.6 Results

The higher-order derivatives have been performed by applying a sinusoidal signal to the terminal under investigation and by measuring the higher-order harmonics in the drain current, like in the configuration from fig. 14. The signal frequency is a few KHz in order to neglect the influence of capacitances. In this situation the distortions can be completely determined by the MOS transistor steady state.

The drain current I_D can be expanded in a Taylor series:

$$I_D=d_0+d_1\cdot v_i+d_2\cdot v_i^2+d_3\cdot v_i^3+\ldots=\sum_{i=0}^{\infty}d_i\cdot v_i^i \quad (40)$$

v_i is a sinusoidal signal $v_i=V\cdot\sin(\omega t)$ and the coefficients $d_i=\frac{1}{i!}\cdot\left.\frac{\partial^i I_D}{\partial V_{G(D)S}^i}\right|_{V_{Do},V_{Go},V_{Bo}}$.

The drain current expression can be rewritten in terms of sin($n\omega t$):

$$I_D=a_o+a_1\cdot\sin(\omega t)+a_2\cdot\cos(2\omega t)+a_3\cdot\sin(3\omega t)+\ldots \quad (41)$$

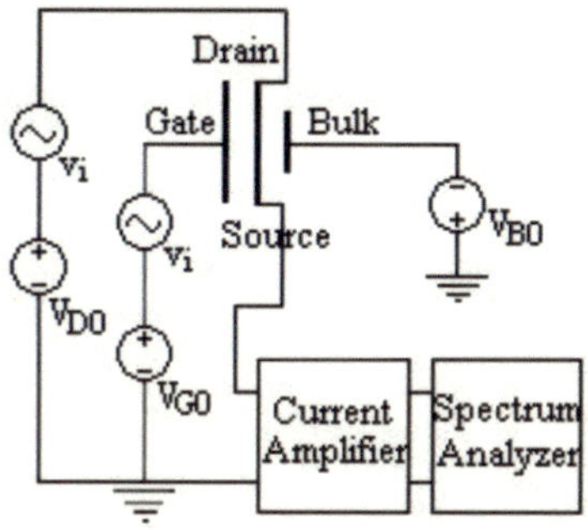

Fig. 14. Measurement circuit

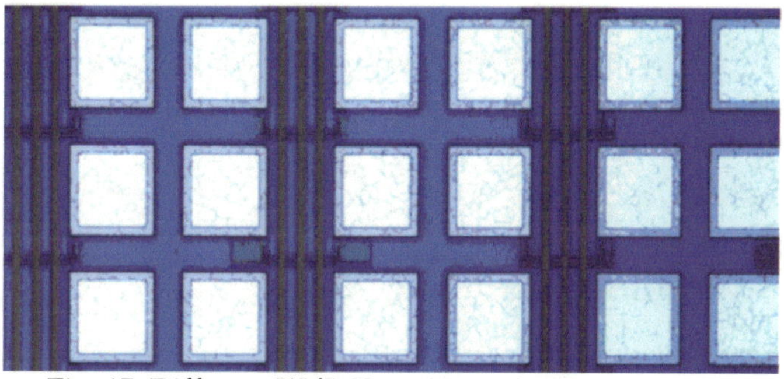
Fig. 15. Different W/L Transistors Array

The coefficients $|a_1|$, $|a_2|$ and $|a_3|$ are the signal harmonics amplitudes and were measured with the spectrum analyser, fig. 14. The coefficients can be written as:

$$a_1 = d_1 \cdot V + \frac{3}{4} \cdot d_3 \cdot V^3 + \frac{5}{8} \cdot d_5 \cdot V^5 + \ldots$$
$$a_2 = -\frac{1}{2} \cdot d_2 \cdot V^2 - \frac{1}{2} \cdot d_4 \cdot V^4 - \frac{15}{32} \cdot d_6 \cdot V^6 - \ldots \tag{42}$$
$$a_3 = -\frac{1}{4} \cdot d_3 \cdot V^3 - \frac{5}{16} \cdot d_5 \cdot V^5 - \frac{21}{64} \cdot d_7 \cdot V^7 - \ldots$$

For small enough values of the signal amplitude V, the coefficients (42) reduce to:

$$a_1 \approx d_1 \cdot V \;;\; a_2 \approx -\frac{1}{2} \cdot d_2 \cdot V^2 \;;\; a_3 \approx -\frac{1}{4} \cdot d_3 \cdot V^3 \tag{43}$$

In this way the high order derivatives reported to V_{GS} notated g_{mi} or reported to V_{DS} notated g_{di} can be calculated or extracted from the measured harmonics amplitude.
Using the harmonic amplitude notation HD_i the second-order and third-order harmonic amplitude are:

$$HD_2 = a_2 \approx \frac{1}{2} \cdot d_2 \cdot V^2 = \frac{1}{4} \cdot \frac{\partial^2 I_D}{\partial V^2_{G(D)S}} \cdot V^2 \tag{44}$$

$$HD_3 = a_3 \approx \frac{1}{4} \cdot d_3 \cdot V^2 = \frac{1}{24} \cdot \frac{\partial^3 I_D}{\partial V^3_{G(D)S}} \cdot V^3 \tag{45}$$

Fig. 15 show device geometries array with W=10μm and different L like were used in the next figures. Fig. 16 present for n-MOSFET the simulated (lines) and measured (symbols) values of gate induced distortion $g_{m3} = (\partial^3 I_D)/(\partial V_{GS3})$ at low drain bias (V_{DS}=50mV) and fig. 17 present the drain induced distortion $g_{d3} = (\partial^3 I_D)/(\partial V_{DS3})$ at V_{GS}=1V.

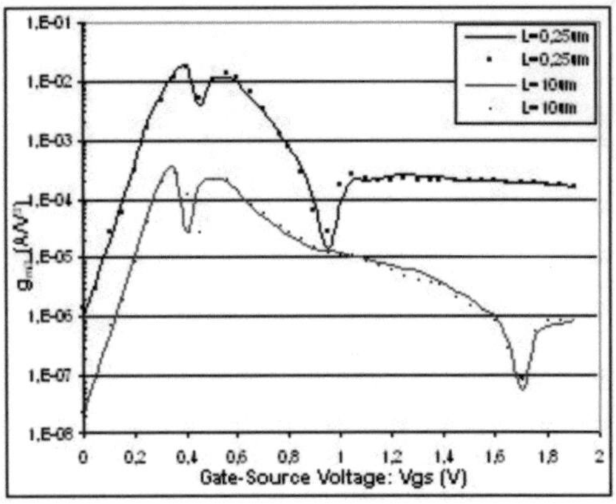

Fig. 16. g_{m3} for n-type MOS Transistor

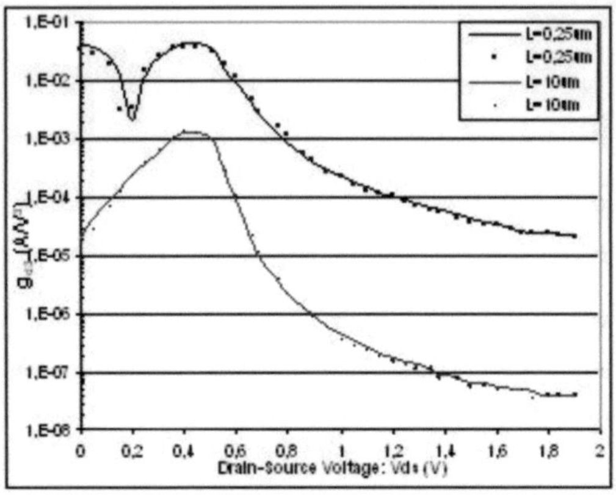

Fig. 17. g_{d3} for n-type MOS Transistor

Fig. 18 present for p-MOSFET the simulated (lines) and measured (symbols) values of gate induced distortion $g_{m3} = (\partial^3 I_D)/(\partial V_{GS3})$ at low drain bias (V_{DS}=-50mV) and fig. 19 present the drain induced distortion $g_{d3} = (\partial^3 I_D)/(\partial V_{DS3})$ at V_{GS}=-1V.

A good fit between simulated and experimental results can be observed.

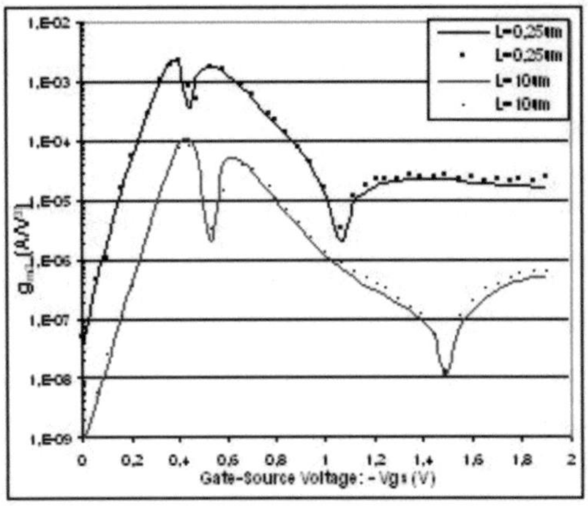

Fig. 18. g_{m3} for p-type MOS Transistor

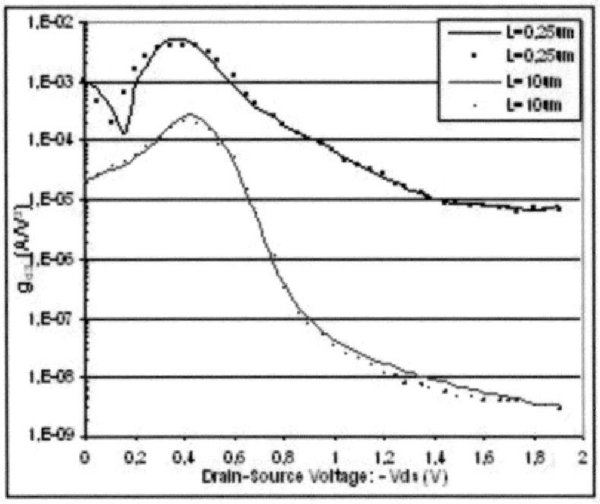

Fig. 19. g_{d3} for p-type MOS Transistor

The transistors array is performed in 0,25μm technology with transistors n and p enhanced type and n^+ polysilicon gate for transistors type n and p^+ polysilicon gate for transistors type p. The gate oxide thickness is about 4nm and the maximum allowed supply voltage V_{DD} is 2,5V. The LDD structure is no longer used and description of gate voltage dependent series resistance become redundant and the model will be a constant source and drain resistance.

8.7 Conclusions

Precise physical description of mobility degradation in circuit-level MOSFET models is essential for distortion analysis. In strong-inversion both hole and electron mobility is mainly limited by two scattering mechanisms: phonon scattering and surface roughness scattering, but they still exhibit a different dependence on effective normal field E_{eff}. For long channel MOS transistors at low drain voltage V_{DS}, the I_D current expression and its higher-order derivatives is accurate for a large region of gate voltages and bulk voltages. The results can be extended to short channel length transistors, when the gate voltage dependency of series resistance was also incorporated in the models.

Gate induced distortion increases with decreasing channel length as series resistance becomes more important. To minimize distortion in circuits, it is often important to make the third-order derivative zero. This is possible for long channel n-type MOS transistors, where mobility degradation is more dominant than the effect of series resistance.

The DC-biasing point for which g_{m3} becomes zero is essentially determined by the presence of the E^2_{eff} term and can be accurately predicted using equation (37). So the designers of analog integrated circuits, where the sensitivity to the modelling details and the feedback between components is more important than for digital electronics can simulate, optimise and design more efficiently.

PMOS transistors are the better choice for drain terminal driven low-distortion applications, because is less affected by the harmonic distortion in the ohmic region and from weak-avalanche effects, compared with n-type transistors.

The unified model gives an accurate description of harmonic distortion in all inversion. The model has a low number of parameters for both n or p type of transistors. The parameters extraction requires the measured current-voltage characteristics, which can be performed with usual testing tools. The unified model is well scalable and the drain current and its higher order derivatives are precisely described over a large geometries range and terminal applied voltage values. The model can be used for modern CMOS technologies in which both the n-channel and p-channel transistors are of the enhancement-type and LDD-structures are no used, it was found that an adjusted expression of hole mobility and a constant series resistance have to be used.

9. High k dielectric-semiconductor interface modelling and analysing

9.1 Introduction

High-k dielectric oxides are presently investigated as alternative gate dielectric films for complementary metal oxide silicon-CMOS transistors. The continuous decrease of ultra-large-scale integration dimensions determines that SiO_2 gate dielectric attends critical dimensions of tens of Å and reaches fundamental limitations in preventing current leakage from the gate into the channel (Lo, 1997). One possible solution is to replace SiO_2 with higher permittivity insulator materials. Among the promising candidate materials investigated are HfO_2, $HfSiO_4$, $ZrSiO_4$, La_2O_3 and Y_2O_3. Presently are important issues related to the integration of these materials and that makes highly desirable to previously understand how the new materials properties affects the functionality of CMOS devices. A correct quantum-mechanical model must properly evaluate the channel charge distribution and the leakage current flowing between the gate and the channel through tunnelling. Consequently the gate, the insulator and the substrate semiconductor channel must be considered like a single space region accessible to all free charge carriers. Looking from the gate this high quality thin oxide is responsible for the continued increase of the gate leakage, which increase the power consumption of integrated circuits. The understanding of the

MOS system begins very important for research the tunnelling current in EEPROM devices and also in high performance MOS devices with ultra thin oxides (Cassan, 2000).

9.2 The gate leakage currents

The charge distribution and quantum-mechanical leakage currents in ultra thin metal-insulator-semiconductor gate stacks composed of several layers materials are very important (Yeo, 2002). Considering all the capacitor like a single quantum mechanical quantity the effective mass approximation for the electrons in the different valley and the Hartree approximation for the electron-electron interaction in inversion layer, the Schrödinger-Poisson equation can be solved. Because the insulating layer is relatively thin but the energy barriers separating the inversion layer from the gate electrode is high enough to prevent the flow of electrons to the gate, the potential well host the majority of inversion layer electrons and the channel is coupled only weakly with the gate (Magnus, 2000).

9.3 The iterative approximation method

The first fully numerical self-consistent results of the inverted MOS structure were mainly attributed to Stern. Then the self–consistent solution has been extended to holes in inverted pMOS structure by Moglestue. The quantum mechanical treatment of the MOS structure in the accumulation regime was described by Sune (Sune, 1992). The self-consistent Schrödinger-Poisson equations were applicable to an inverted structure in the next approximations: the effective mass approximation, the ideal interface semiconductor-oxide and interruption of wave function at interface semiconductor-oxide. The time-independent Schrödinger equation in 3D space, using the position vector $\mathbf{R}=(\mathbf{r}, z)$ can be formally written:

$$H\psi(\mathbf{r}, z) = E\psi(\mathbf{r}, z), \tag{46}$$

where $\psi(\mathbf{r},z)$ is the wave function, E is the eigenvalue energy, H is the system Hamiltonian, composed from kinetic energy T and potential energy W. For long channel device the potential profile is mainly one dimensional and the drain and source regions can be considered like electrons reservoirs for the inversion layer. The 1D simplification allows using the wave operator like a function of the z coordinate only:

$$\psi(\mathbf{r},z) = \phi(z)e^{i\mathbf{k}\cdot\mathbf{r}}, \tag{47}$$

where $\mathbf{k}=(k_x,k_y)$ is the wave vector in the (x,y) plane. So the carrier are quantized in the z direction and are free to move in the $\mathbf{r}=(x,y)$ plane, with a continuous energy component. After phase transformation and imposing the constraint of vanishing for the first derivative of the wave function, the envelope 1D time-independent reduced equation (46) is:

$$-\frac{\hbar^2}{2m_{zz}}\psi''(z) + W\psi = E_z\psi(z) \tag{48}$$

where $\hbar$ is reduced Planck constant, m_{zz} is the effective masses in m_o units, W is potential energy, $\Psi(z)$ is the 1D envelope wave functions and E_z is the eingenvalue energy.

Considering the MOS structure a quantum mechanical system, an externally applied gate bias induces a potential well that confines carriers in the region of the semiconductor-oxide interface. The electrostatic potential and charge respect the Poisson equation in any z direction from silicon region:

$$\frac{d^2V(z)}{dz^2} = -\frac{1}{k_{Si}\varepsilon_0}\rho(z) \tag{49}$$

where $V(z)$ is the electrostatic potential, $\rho(z)$ is the charge density, k_{Si} is the Si relative dielectric constant. Assuming the p-type substrate with completely ionized impurities and neglecting the hole concentration in inversion can approximate the charge density:

$$\rho(z) = \rho_{depl}(z) - qn(z), \tag{50}$$

where ρ_{depl} is the depletion layer charge and $n(z)$ is the carrier's distribution.
Close to the interface the electrons have a position dependent concentration proportional with the probability density and a sum of each energy valley and subband.

$$n(z) = \sum_{i,j} n_{i,j}(z) = \sum_{i,j} N_{ij}^{(2D)}(E_{z,ij}, E_F)|\psi(z)|^2 \tag{51}$$

where $N_{ij}^{(2D)}$ is the subband population which integrates the all possible energies of a subband of the 2D density of states, $|\psi(z)|^2$ is the probability density, $E_{z,ij}$ is the solution of 1D Schrödinger equation (48) and represents the discrete bottom level of a particular energy subband j, for each valley i and E_F is Fermi energy level. The carrier's distribution can be more detailed using the valley and spin degeneracy and Fermi-Dirac statistics. The assumption that the silicon-oxide interface is ideally, was technologically realized by election the [001] surface orientation that minimizes the dangling bonds at the interface, resulting a high quality interface after passivation (Babarada, 2008).
Considering the quantization effects of silicon-insulator interface an approximate geometrical solution to calculate the charge densities and subband energy levels reduces consistently the computational complexity for leakage current evaluation. Using the same effective mass approximation the areal density of charge in the inversion layer is:

$$N_{inv} = \sum_{i,j} \oint_z N_{ij}^{(2D)}(E_{z,ij}, E_F)|\psi(z)|^2 dz = \sum_{i,j} N_{ij}^{(2D)}(E_{z,ij}, E_F) \tag{52}$$

Using the geometrical approximation of Si band bending in inversion (Muller, 1997) the energy level is:

$$E_{z,ij} = \left(\frac{\hbar^2}{2m_{z,i}}\right)^{1/3} \left(\pi q F_{ef} \frac{3}{2}\left(j + \frac{3}{4}\right)\right)^{2/3} \tag{53}$$

and the subband charge is:

$$q_{ij} = \frac{2E_{z,ij}}{3qF_{ef}}, \tag{54}$$

where F_{ef} is the $E_{z,ij}$ corresponding effective electric field. Then the inversion charge is:

$$q_{inv} = \sum_{i,j} q_{i,j} \frac{N_{i,j}^{(2D)}}{N_{inv}}, \tag{55}$$

and the total silicon surface bending:

$$\Psi_S = \Psi_D + q\frac{N_{inv} q_{inv}}{k_{Si}\varepsilon_0} + \frac{k_B T}{q} \tag{56}$$

where Ψ_S is the surface potential, Ψ_D is the drop voltage at surface due to space charge region. The last term is the influence of doping concentration to charge region (Muller, 1997). Using the charge boundary conditions the equations can be iteratively solved to attain the convergence in the next sequence:

- Guess the initial N_{inv}, Ψ_S and Ψ_D
- Consider charge boundary condition $N_{inv\text{-}bc}$
- Iterate Ψ_S with condition $N_{inv}(\Psi_S)/N_{inv\text{-}bc} \to 1$
- Iterate Ψ_D with condition $\Delta\Psi_D \to 0$

- Compute the potential distribution

We have possible loops from out to input, of step 3 and 4 and from out of step 4 to input of step 3. The method can be used also for tunnelling based leakage currents in high-k dielectric stach.

9.4 Results

For numerical simulations we used the ATLAS devices simulator software package from Silvaco. The main module program used is presented in fig. 20, in order to generate the MOS structure presented in fig. 21.

Then was calculated the gate current, fig. 22 and the capacity from gate to substrate, fig. 23, function of polysilicon doping concentrations $10^{19}cm^{-3}$, $10^{20}cm^{-3}$ and $10^{21}cm^{-3}$.

```
mesh
x.mesh loc=-0.01 spac=0.01
x.mesh loc=0.01  spac=0.01
y.mesh loc=-0.04 spac=0.001
y.mesh loc=0.02  spac=0.001
region number=1 x.min=-0.01 x.max=0.01 y.min=-0.04 y.max=-0.03  \
       material=aluminum
region number=2 x.min=-0.01 x.max=0.01 y.min=-0.03 y.max=-0.005 \
       material=poly
region number=3 x.min=-0.01 x.max=0.01 y.min=-0.005 y.max=0.0   \
       material=oxide
region number=4 x.min=-0.01 x.max=0.01 y.min=0.0 y.max=0.02     \
       material=silicon
electrode x.min=-0.01 x.max=0.01 y.min=-0.04 y.max=-0.03 name=gate
electrode bottom name=substrate
doping region=2 p.type concentration=1e19 uniform
doping region=4 p.type concentration=1e17 uniform
solve init
solve vgate=-1.5
solve vgate=-3
save outfile=mos2ex15-3V19.str
# tonyplot mos2ex15-3V19.str -set mos2ex15_BD.set
log outfile=mos2ex15_CV19.log
solve vgate=-2.8 vstep=0.2 vfinal=3.0 name=gate ac freq=1e6 previous
tonyplot mos2ex15_CV19.log -set mos2ex15_CV.set
```

Fig. 20. The main ATLAS program module

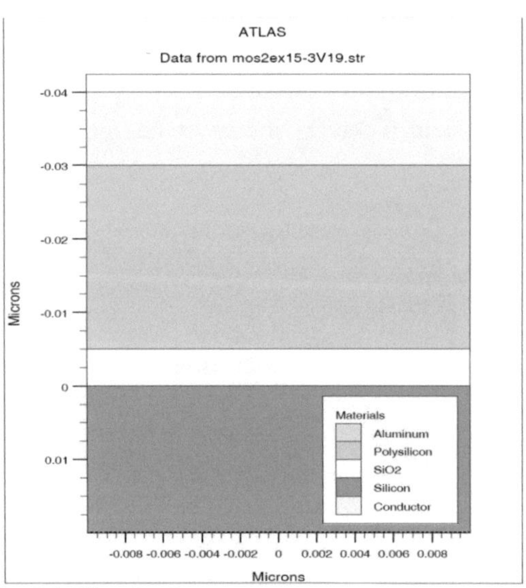

Fig. 21. The device structure

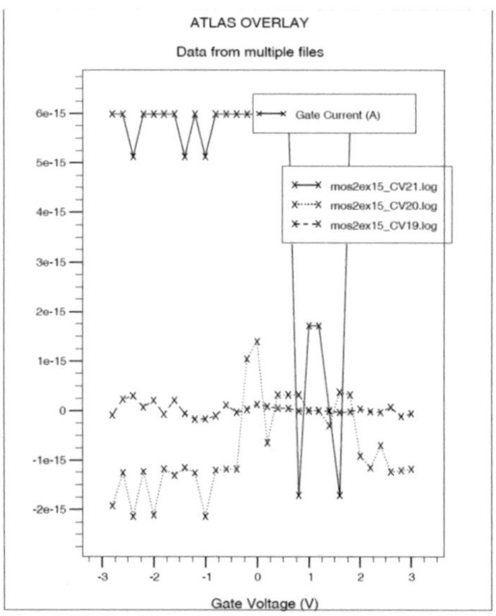

Fig. 22. The Gate Current

The first numerical simulations proves the dependence of leakage current, fig. 22 and depletion effect fig. 23, function of doping concentration like considered in chapter 9.3. Using the barrier height of 3.1eV, substrate doping $5x10^{17}cm^{-3}$, effective silicon oxide mass of $0.5m_o$ and donor poly doping $6x10^{19}$ the results of short computation iterative approximation, fig. 24, of silicon oxide current gate density calculated (1.5nm and 2nm) was in good agreement with experimental gate current density curves presented in (Yang, 2000) and noted [9] (1.41nm[9] and 1.95nm[9]).

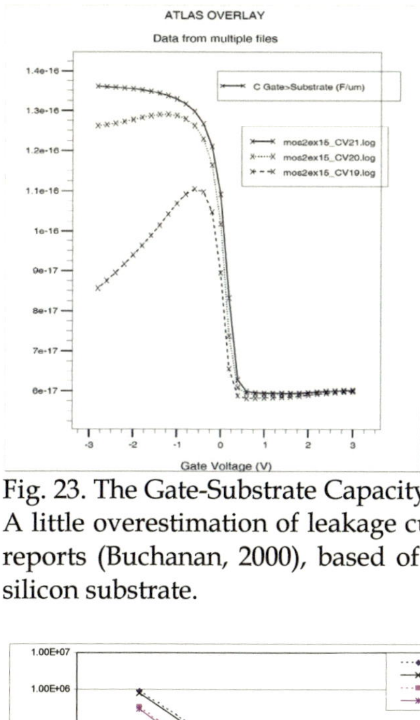

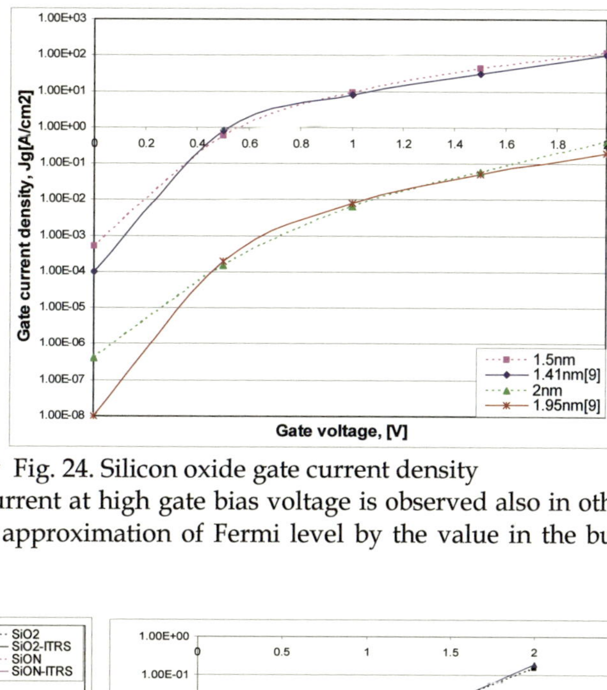

Fig. 23. The Gate-Substrate Capacity Fig. 24. Silicon oxide gate current density

A little overestimation of leakage current at high gate bias voltage is observed also in other reports (Buchanan, 2000), based of approximation of Fermi level by the value in the bulk silicon substrate.

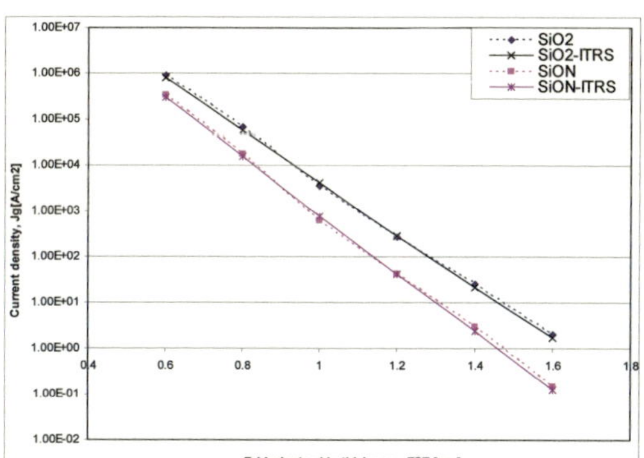

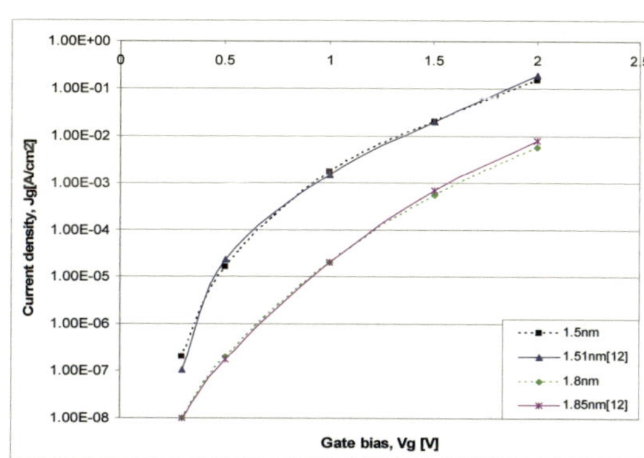

Fig. 25. Oxide and oxynitride leakage current Fig. 26. Al_2O_3 high-k stacks leakage currents

The polysilicon doping level suppresses the gate leakage current for gate bias in inversion because the additional voltage drops over the depleted layer (Yang, 2000). This solution decreases the drive capacitance and the device performances. The substrate doping level affects the leakage current through the surface potential of the channel. Because increasing the physical thickness of gate dielectric affects the device parameters like drive current, a compromise solution is to increase the dielectric constant using the SiON layer with dielectric constant up to 7.6 for Si_3N_4. The performances of SiON like gate dielectric are better than SiO_2 as in fig. 25, according with simulations at Vg=1V and ITRS. Comparing the calculated data with gate leakage current through Al_2O_3 high-k dielectric stacks presented in (Buchanan, 2000) a good fit was obtained, fig. 26.

9.5 Conclusions

High-k atomic layer deposition stacks like insulating in the metal-insulating-semiconductor structure was studied. An iterative approximate method to calculate the 1D MOS structures main electric parameters without using the Schrödinger-Poisson equations was used. This method is based on approximation of effective field function of doping parameters. The tunnelling currents can be calculated more rapidly and the study for different gate dielectric stacks can be made. The precision can be increased by 2D or 3D analysis of Schrödinger-Poisson equations. The main application is to calculate the direct tunnelling current due to the thin oxide layers. The method is extensible to high-k dielectric stacks in order to study the influence of several material parameters like the impact of layer thickness on gate leakage and the approach of gate stack scalability. The results obtained using numerical calculation show that the increase of the gate dielectric constant has a very important effect in reducing the leakage currents. Comparing the results from fig. 24 and fig. 26 for 1V gate bias and 1.5nm thickness the increase of dielectric constant to 7 reduce the leakage current with 4 order of magnitude. Other simulations show that the leakage current decrease significant when the interfacing oxide is completely eliminated. Future works will be focus of other high-k dielectric stacks like HfO_2, $HfSiO_4$, $ZrSiO_4$, La_2O_3, and Y_2O_3.

10. References

Babarada, F.; et all. (2003). Carrier Mobility and Series Resistance MOSFET Modelling, *BioMEMS and Nanotechnology*, vol. 5275-49, pp. 354-363, SPIE's, Perth, Australia

Babarada, F.; et all. (2005). MOSFET Conductance Modelling Including Distortion Analysis Aspects, *Proceedings of the International Conference*, Sinaia, România

Babarada, F.; Plugaru, R.; Rusu, A. (2008). Electrical characterization of atomic layer high-k dielectic gate for advanced CMOS devices, *Proceedings of the International Conference*, pp. 363-366, ISBN 978-1-4244-2004-9, Sinaia, România

Buchanan, D.A.; et all. (2000). 80 nm poly-silicon gated n-FET with ultra thin Al_2O_3 gate dielectric for ULSI applications, *IEDM Tech. Dig.*, pp. 223-226, San Francisco, USA

Bucher, M.; et all. (2007). A Scalable Advanced RF IC Design-Oriented MOSFET Model, *Int. Journal of RF and Microwave Computer-Aided Engineering*, DOI 10.1002/mmce

Campian, I.; Profirescu, O.; Babarada, F.; Lakatos, E. (2003). MOSFET Simulation-TCAD Tools/Packages, *Proceedings of the Int. Conf.*, pp. 235-238, Sinaia, România

Cassan, E. (2000). On the reduction of direct tunnelling leakage through ultra-thin gate oxides by one-dimensional Schr.-Poisson solver, *J. Appl. Phys.* 87(11), pp. 7931-7939

Gildenblat, G.; Zhu, Z.; McAndrew, C. (2009). Surface potential equation for bulk MOSFET, *Solid-State Electronics*, 53: pp. 11-13

Govoreanu, B.; et all. (2002). On the use of Bayesian Neural Networks for TCAD Empirical Modeling, *Romanian Journal Science and Technology*, pp.329-338, vol 5, no 4, România

Kwong, M.; Kasnavi, R.; Grifin, P.; Duton, R. (2002). Impact of Lateral Source/Drain Abruptness on Device Performance, *IEEE Trans. on El. Dev.*, vol. 49, no. 11.

Lo, S.; Buchanan, D.; Taur, Y.; Wang, W (1997). Quantum mechanical modelling of electron tunnelling current from the inversion layer of ultra-thin-oxide nMOSFET, *IEEE El. Dev. Lett.*, 18(5), pp. 209-211

Magnus, W.; Schoenmaker, W.; (2000). Full quantum mechanical model for the charge distribution and the leakage currents in ultrathin metal-insulator-semiconductor capacitors, *J. Appl. Phys.* 88(10), pp. 5833-5842

Muller, H.; Schultz, M. (1997). Simplified method to calculate the band bending and the subband energies in MOS capacitors, *IEEE Trans. El. Dev.* 44(9), pp. 1539-1543

Rusu, A. (1990). *Microelectronics Active Components Modelling*, Editura Academiei Române

Scholten, A.J.; et all. (2009). The new CMC standard compact MOS model PSP: advantages for RF applications, *IEEE Journal of Solid-State Circuits*, vol. 44, no. 5, pp. 1415-1424

Sune, J.; Olivio, P.; Ricco, B. (1992). Quantum mechanical modelling of accumulation layers in MOS structures, *IEEE Trans. El. Dev.*, 39(7), pp. 1732-1739

Veendrick, H. (2008). *Nanometer CMOS ICs: From Basics to ASICs*, Springer, ISBN 978-1-4020-8332-7, Netherland

Yang, N.; Henson, W.; Wortman, J. (2000). A comparative study of gate dielectric tunnelling and drain leakage currents in n-MOSFET with sub-2-nm gate oxides, *IEEE Trans. El. Dev.* 47(8), pp. 1636-1644

Yeo, Y.; King, T.; Hu, C.; (2002). Direct tunnelling leakage current and scalability of alternative gate dielectrics, *Appl. Phys. Lett.*, 81(11), pp. 2091-2093

Ytterdal, T.; Cheng, Y.; Fjeldly, T. (2003). *Device Modeling for Analog and RF CMOS Circuit Design*, J. Wiley, ISBN 0-471-49869-6, England

2

Iterative Solution Method in Semiconductor Equations

Norainon Mohamed, Muhamad Zahim Sujod
and Mohamad Shawal Jadin
Universiti Malaysia Pahang, Lebuhraya Tun Razak, 26300 Kuantan, Pahang
Malaysia

1. Introduction

The FEM (sometimes referred to as finite element analysis (FEA)) is a numerical technique for finding approximate solutions of partial differential equation as well as of integral equations. The solution approach is based either an approximating system of ordinary differential equations, which are then solved using standard techniques such as Newton Method. It is the objective of this paper to describe the application of the method to device simulation. The device which described in this paper is Silicon Carbide Gate Turn-Off Thyristor (SiC-GTO Thyristor). The doping profile with the material properties of the device can be modelled. This paper specifically focuses on the numerical simulation of the device compare with the common Silicon GTO Thyristor.

The main advantages of the FEM are that conservation laws (e.g., current conservation) are exactly satisfied even by coarse approximations, it is easy to treat irregular geometries, the computational mesh can be graded to be fine in regions to rapid change, local mesh refinement is easier to implement than finite difference method (FDM).

In the following sections, the finite element equations which are arise from the semiconductor equations are derived and it is shown the equations are the base of semiconductor device simulations. The implementation of finite element equations will be discussed in the next section. For detailed discussion of the numerical simulation, it is in the results and discussion section.

2. Numerical Method

2.1 Semiconductor Equations

The semiconductor equations are a set of five equations that govern the behavior of semiconductor materials and devices. The set of equations composed of:

Poisson's equation

$$\nabla^2\psi = \frac{q}{\varepsilon}\left(n - p - N_d\right) \tag{1}$$

Current Continuity equations

$$q\frac{\partial p}{\partial t} = -\nabla \cdot \mathbf{J}_{\mathbf{p}} - qR \tag{2}$$

$$q\frac{\partial n}{\partial t} = \nabla \cdot \mathbf{J}_{\mathbf{n}} - qR \tag{3}$$

Drift-Diffusion equations

$$\mathbf{J}_{\mathbf{p}} = q\left(p\widetilde{v}_p - D_p \nabla p\right) \tag{4}$$

$$\mathbf{J}_{\mathbf{n}} = q\left(p\widetilde{v}_n - D_n \nabla n\right) \tag{5}$$

In these equations, the three unknown quantities are the space-charge potential (ψ), the electron (n) and hole (p) densities, N_d is the doping densities, the constant q is the magnitude of electronic charge and ε is the dielectric permittivity. $\mathbf{J}_{\mathbf{p}}$ and $\mathbf{J}_n$ are the hole and electron current densities. R is the recombination rate. $\widetilde{v}_p$ and $\widetilde{v}_n$ are the hole and electron drift velocities. D_p and D_n are the hole and electron diffusion coefficients.

The diffusion coefficients and drift velocities are electric field dependent and so the equations are nonlinear. The recombination term which is also nonlinear may be approximated by its thermal equilibrium value (Shockley Read Hall Theory).

2.2 Finite Element Equations

To solve (1) to (5), boundary conditions for the space-charge potential and electron and hole charge carrier densities are required. The finite element equations are derived from (1) to (3) by multiplying them by $\Phi_i(x,y)$ and integrating over the region Ω occupied by the device[4].

$$\begin{aligned} &\int_{\Omega} \Phi_i(x,y) \nabla^2 \psi ds = \\ &-\frac{q}{\varepsilon} \int_{\Omega} (x,y)(p-n-N_d) ds \end{aligned} \tag{6}$$

$$\frac{1}{q}\int_{\Omega}\Phi_i(x,y)\nabla\ \mathbf{J}_p ds = \int_{\Omega}\Phi_i(x,y)(\frac{\partial p}{\partial t}+R)ds \tag{7}$$

$$-\frac{1}{q}\int_{\Omega}\Phi_i(x,y)\nabla\ \mathbf{J}_n ds = \int_{\Omega}\Phi_i(x,y)(\frac{\partial n}{\partial t}+R)ds \tag{8}$$

2.3 Final Form of Equations

In computer solution by the finite element method there are four stages:

1. Read in (or generate internally) material properties- Si and SiC) and element connectivity (mesh).
2. Assemble the equations (6), (7) and (8) which the finite element equations and inserting boundary conditions.
3. Solve the resulting linear equations
4. Repeat 2 and 3 iteratively for nonlinear and/or time dependent problems.

3. Simulation Flow

The simulation systems have been implemented by using MATLAB/Simulink surrounding. The simulation process is used Poisson's equation together with current continuity and drift-diffusion equations to simulate the performances of SiC GTO thyristor. Figure 1 shows the schematic structure of the simulator. Each phase describes complex process which involves the physical models along with the basic semiconductor equations as the basis to simulate the GTO performances.

The simulation process is controlled by the Material Input Database in each phase. The red line indicates the connection with the material database. Material Input Database is initialized at the initialization process. The basic structure of SiC GTO thyristor is initialized. The device structure and circuit definitions and additional information like material properties are loaded from the Material Input Database.

In the next step, the device or the circuit and its embedded devices are loaded and analyzed. For each segment of each device the material is determined. In the calculations steps, the basic semiconductor equations along with the physical models are solved by using numerical method, finite element method. The method is a powerful method for solving partial differential equations which involves lots of integral and differential. The method is used because of its approximation to the solution of the equation. In the postprocessing, the output quantities are calculated from the computed solution.

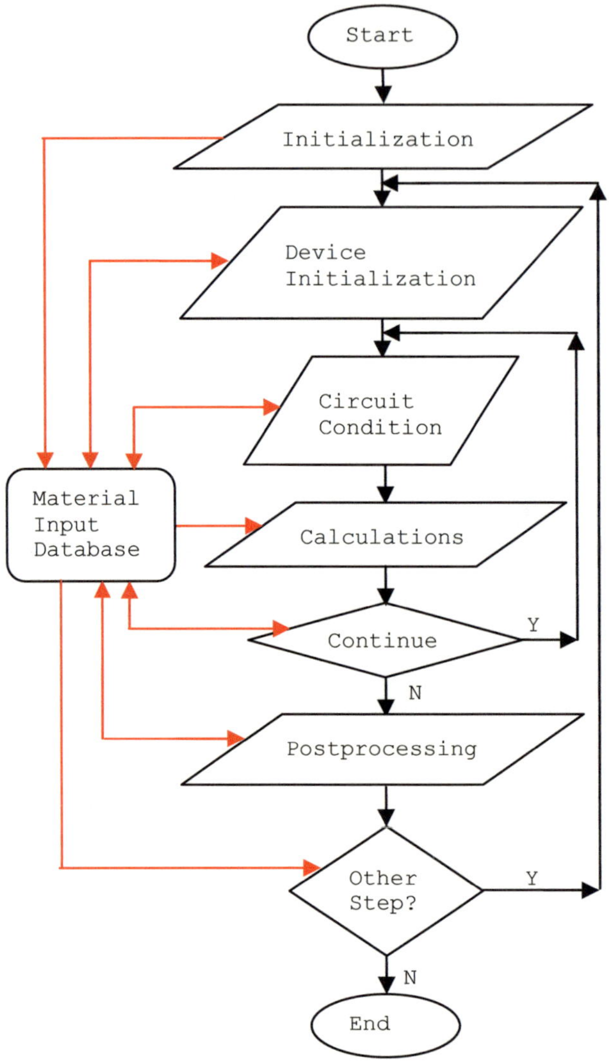

Fig. 1. Simulation flow of the device.

4. Calculation Method

The full set of semiconductor equations are solved numerically. As for discretization of space, the Scharfetter-Gummel scheme and the standard three-point formula are used formula are used for the Poisson's Equation and the continuity equation, respectively. These difference equations are solved based on Newton method.

5. Results and Discussions

5.1 Turn-on Characteristics

Figure 2 shows the single-shot GTO thyristor turn-on voltage and current waveforms. These waveforms show the GTO thyristor's switching characteristics such as turn-on delay and turn-on rise time. The turn-on delay, is defined when the gate current, Ig, rises to 10% of its peak and when the GTO thyristor anode voltage, Va, falls to 90% of its initial value. From the figure 2, GTOs are turned on when the anode current is increased, the anode voltage is decreased. Then they are turned off by the negative gate pulse.

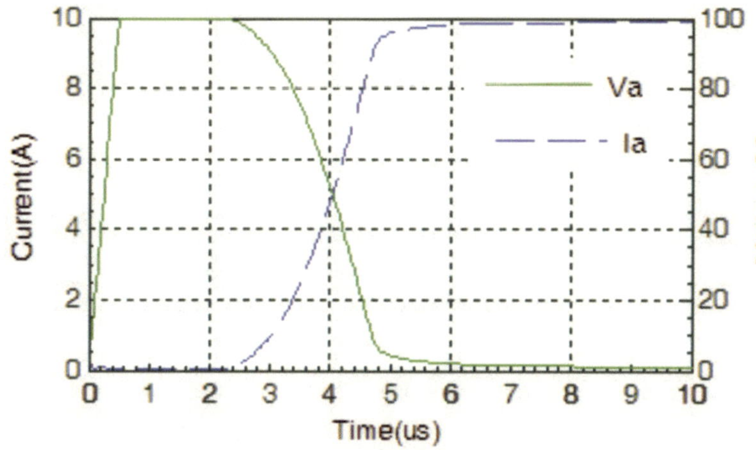

(a)

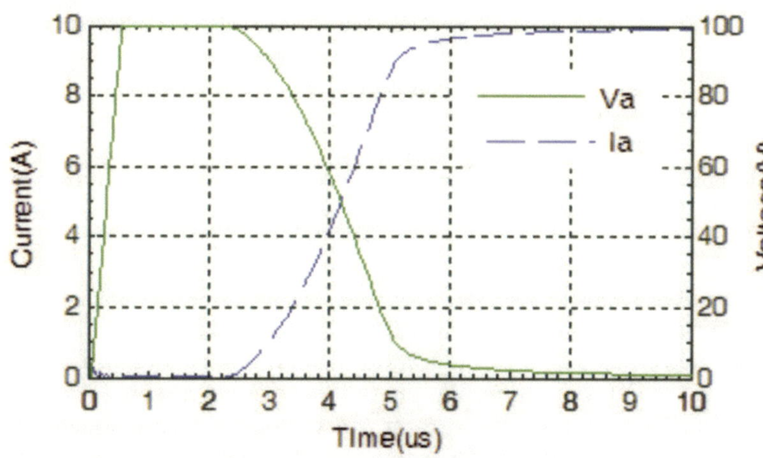

(b)

Fig. 2. Single-shot GTO thyristor turn-on characteristics (a) Si GTO thyristor anode voltage and current (b) SiC GTO thyristor anode voltage and current.

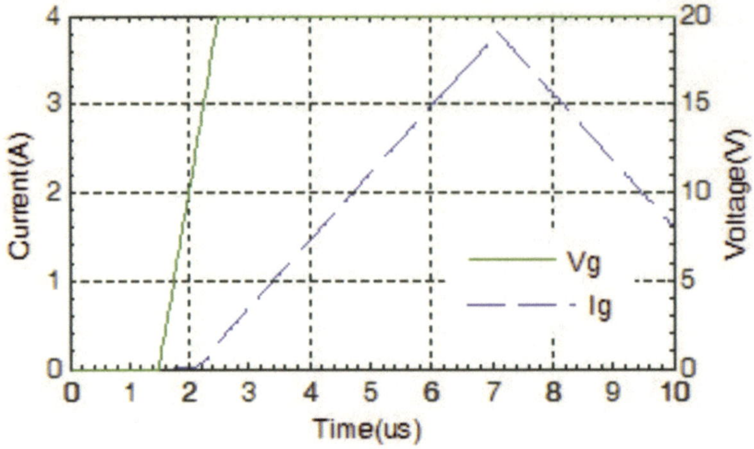

(c)

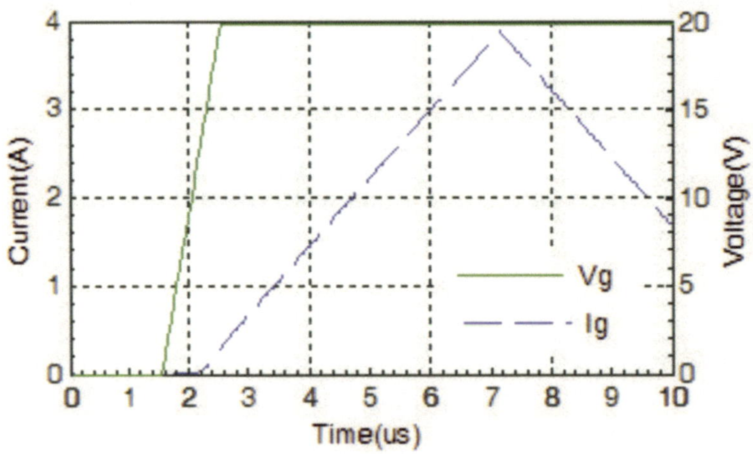

(d)

Fig. 2. Single-shot GTO thyristor turn-on characteristics (c) Si GTO thyristor gate voltage and current (d) SiC GTO thyristor gate voltage and current.

5.2 Turn-off Characteristics

The GTO thyristor turn-off as a function of time is given in figure 5. The GTO thyristor turn-off time was investigated as a function time. We can see the large difference at turn-off time of SiC GTO thyristor waveforms. We know that turn off time of SiC GTO thyristor is better than that Si GTO thyristor. The turn-on and turn-off time are shown in Table II (all units in us). Result show that switching time of SiC-GTO is decreased extremely and the performance of SiC in GTO is in the storage time, fall time and tail time.

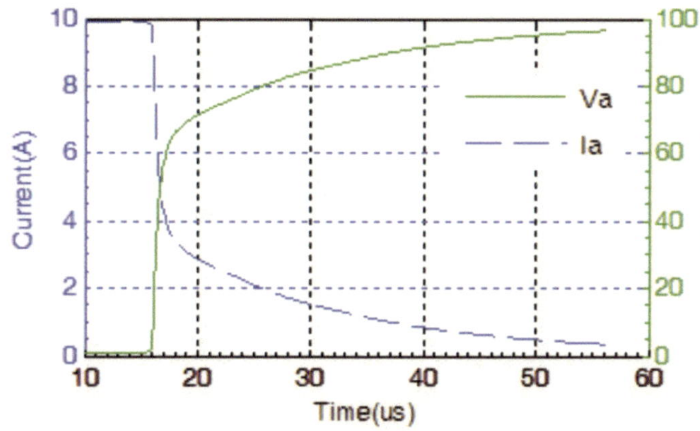

(a)

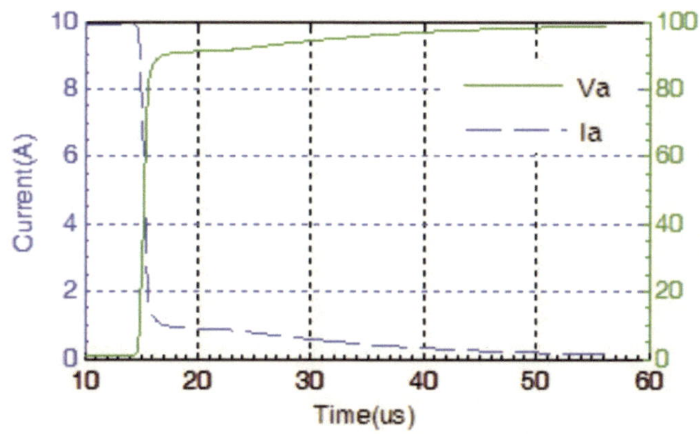

(b)

Fig. 3 Single-shot GTO thyristor turn-off characteristics (a) Si GTO thyristor anode voltage and current (b) SiC GTO thyristor anode voltage and current

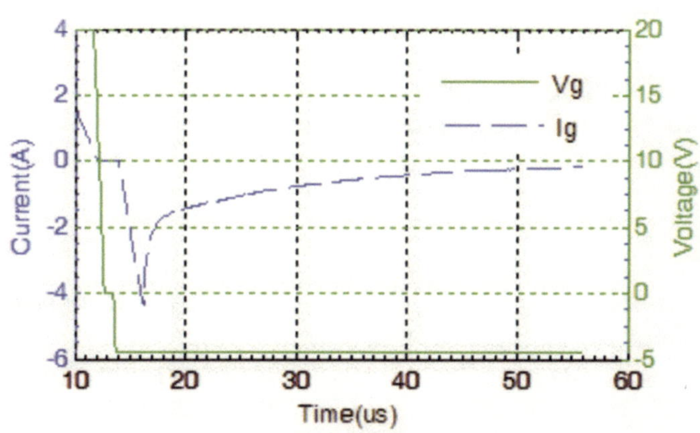

(c)

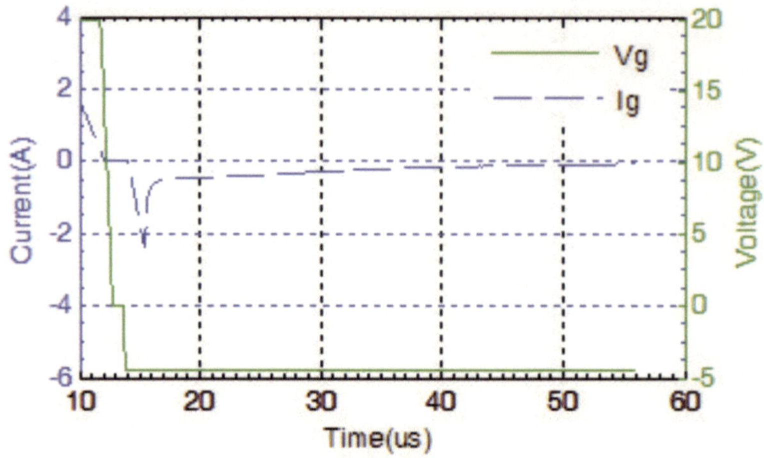

(d)

Fig. 3 Single-shot GTO thyristor turn-off characteristics (c) Si GTO thyristor gate voltage and current (d) SiC GTO thyristor gate voltage and current.

	Si GTO	SiC GTO
Turn-on time (us)	3.00	3.00
Delay Time	1.45	1.45
Rise Time	1.55	1.55
Turn off time (us)	82.5	62.2
Storage time	15.9	14.7
Fall time	17.3	15.4
Tail time	49.4	32.1
Switching Time (us)	**85.5**	**32.1**

Table 1. Switching time of Si and SiC GTO Thyristors.

6. Conclusion

We compared the switching waveforms of usual Si GTO thyristor and new SiC GTO thyristor under inductive load. Turn off time is smaller in the case of SiC GTO thyristor than in that Si GTO thyristor.

7. References

A. R. Powell & L. B. Rowland, "SiC Materials progress, Status & Potential Roadblocks," Proc. IEEE, vol. 90, no. 6, pp. 942-955, 2002.

J. A. Cooper, JR., and A. Agrawal, "SiC Power Switching Devices The Second Electronic Revolution?" Proc. of the IEEE, vol. 90, pp. 956-968, 2002.

A. K. Agarwal, P. **A.** Ivanov, M. E. Levinshtein, J. **W.** Palmour, **S.** L. Rumyantsev, *S.* H. Ryu and M. *S*.Shur, "Turn-off Performance of a 2.6 kV 4H-SiC Asymmetrical GTO Thyristor," Material ScienceForum, vol. 353-356, pp 743-746,2001.

R. **R.** Siergiej, J. B. Casady, **A. K.** Agarwal, L. B.Rowland. *S.* Seshadri, *S.* Mani, P. A. Sanger and C.D. Brandt, "1OOOV 4H-Sic Gate Turn Off (GTO)Thyristor," Compound Semiconductors, IEEEInternational Symposium, pp. 363-3156. 1997.

A. K. Agarwal. P. A. Ivanov, **M.** E. Levinshteitl, J.W. Palmour, *S.* L. Rumyantsev, *S.* H. Ryu, and M. *S* .Shur, "Turn-off performance of a 2.6 kV 4H-SiC Asymmetrical GTO Thyristor," Material Science Forum, pp. 353-356.

D.L. Scharfetter and H.K. Gummel, *Large-Signal Analysis of A Silicon,* IEEE Trans. Electron Devices, vol. ED-16,pp 64-77, Jan 1969.

J. B. Fedison, "High Voltage Silicon Carbide Junction Rectifiers and GTO Thyristors", PhD Thesis, Rennsealer Polytecnic Institute, New York, May 2001.

H. Sakata, M. Zahim, "Device Simulation of SiC-GTO", IEEE Power Conversion Conference", vol. 1, April 2002, pp. 220-225.

3

Automation and Integration in Semiconductor Manufacturing

Da-Yin Liao
Applied Wireless Identifications (AWID)
U.S.A.

1. Introduction

Semiconductor manufacturing spans across many manufacturing areas, including *wafer manufacturing* where electronic circuitry is built layered on a wafer, *chip manufacturing* that involves circuit probing and testing, and *product manufacturing* from which the final IC (integrated circuits) products are assembled, and finally tested. Semiconductor manufacturing is well known as the most challenging and complicated production systems that involve huge capital investment and advanced technologies. Fabrication of semiconductor products demands sophisticated control on quality, variability, yield, and reliability. It is crucial to automate all the semiconductor manufacturing processes to ensure the correctness and effectiveness of process sequences and the corresponding parameter settings, and to integrate all the fab (semiconductor factory) activities to provide the efficiency, reliability, and availability of semiconductor manufacturing. Automation and integration are the keys to success in modern semiconductor manufacturing. This chapter deals with the automation and integration problems in semiconductor manufacturing.

Automation plays an increasingly important role in daily operations of semiconductor manufacturing. Like in the other industry, automation in semiconductor manufacturing originated from replacing human operators in tasks that are routine but tedious, or that should be done in dangerous, hazard environments. The ultimate goal of automation in semiconductor manufacturing is to eliminate the need of humans in fab operations. Depending on different degrees of operator attention and automatic control, fab operations are usually classified into three modes: *Manual, Semi-Automated,* and *Fully Automated.* Traditional manual mode of operations where fab tools (semiconductor equipment) are operated without computer assistance is very scarce to find in existing commercial fabs. Semi-automated operations are still quite popular in 6- and 8-in fabs where processing tools are automated and controlled by computers, but fab operators are responsible for the movement of materials from and to the tools. Fully automated mode is now well established in 12-in (300-mm) fab operations where there are complete computer-controlled processing and handling. Automation in semiconductor fabs has saved billions of dollars by eliminating and reducing misprocessed products, and improved operational efficiency by reducing human times and costs spent in data entry and product movement.

Automation in semiconductor manufacturing has to provide the intelligence and control to drive the operations of semiconductor fabrication processes, in which layers of materials are deposited on substrates, doped with impurities, and patterned using photolithography to generate integrated circuits. Automation in semiconductor industry adopts the hierarchical machine control architecture that allows for quick insertion into current fabrication facilities. In the architecture, the lower-level of the hierarchy includes embedded controllers to provide real-time control and analysis of fabrication equipment where sensors are installed for in situ monitoring and characterization. At the higher-level, more complex, context-dependent combination of process or metrology operations or materials movements is handled, sequenced, and executed.

Contemporary semiconductor manufacturing increasingly uses cluster tools, each of which consists of several single-wafer processing chambers, for diverse semiconductor fabrication processes, shorter cycle time, faster process development, and better yield for less contamination. To illustrate the automation in semiconductor fabrication equipment, we adopt a PDV (Physical Vapour Deposition) cluster tool as an example to convey the idea of hierarchical architecture and the associated communication protocols, intelligent job scheduler/dispatcher, as well as process modelling, monitoring, diagnosis and control.

Semiconductor manufacturing integration encompasses the allocation, coordination and mediation among system dynamics and flows of information, command, control, communication, and materials, in a timely and effective way. Because of the ever-increasing complexity of semiconductor devices and their manufacturing processes, computer or CIM (Computer-Integrated Manufacturing) systems are essential for the smooth integration of semiconductor manufacturing. However, CIM systems generally are loosely coupled, monolithic, and difficult to extend to support the new needs. Researchers and practitioners have been devoted to build an integration framework with a common, modular, flexible, and integrated object model to tackle the critical problems in semiconductor manufacturing integration: islands of automation, emergence of new applications, distributed systems, as well as data integrity.

Automatic Materials Handling System (AMHS) is considered as a must in modern semiconductor manufacturing environment. In a large-scaled AMHS, there are usually hundreds of OHT (Overhead Hoist Transport) vehicles running in dozens of loops. The management and control of even a single AMHS loop has proved to be crucial but difficult (Liao, 2005). The transport requirements of AMHS vehicles among different loops are usually changing from time to time, according to the dynamic WIP (Wafers in Process) distribution, process conditions, and equipment capacity. It is therefore needed an effective methodology to integrate AMHS with other CIM systems to cope with the dynamic changes on the material handling services. We propose an intelligent AMHS management framework to optimize and manage the integration of fab operations with AMHS.

Development of automation and integration usually requires the help of system definition, validation or verification techniques. To the large dynamic systems like semiconductor manufacturing, it is always difficult and challenging to define, validate, and verify their system dynamics, not to say, to consider their various and changing control and managerial policies. In this chapter, we adopt Petri-net techniques (Zhou & Jeng, 1998; Liao *et al.*, 2007) to build models for a PVD cluster tool. Mathematical analysis and computer simulation are conducted to verify and validate the correctness of the automation and integration in the developed models.

This chapter is organized as follows: Section 1 describes the need of automation and integration in semiconductor manufacturing. In Section 2, automation in semiconductor manufacturing is detailed. Section 3 gives an illustrating example of automation of a representative cluster tool in semiconductor manufacturing. Section 4 discusses the integration problems and issues in semiconductor manufacturing. An intelligent, integrated framework is presented in Section 5. Section 6 deals with the modelling, validation and verification of processing and material handling systems in semiconductor manufacturing. Finally, Section 7 concludes this chapter with some visions and challenges to the automation and integration in future semiconductor manufacturing.

2. Automation in Semiconductor Manufacturing

2.1 Considerations of Semiconductor Manufacturing Automation

Reasons for fab automation are from many aspects, including lower costs, increasing fab performance, reliability and product quality. Very basically, fab automation should execute fab operations which are sequences or collection of the following activities:

- *Lot selection* (or *dispatching*) to determine which lot to process next
- *Transport* to locate and move the lot
- *Setting of process condition and recipe* to setup processing conditions
- *Process start* to initiate processing
- *Process data collection* to record and report measurement data during processing
- *Go/No-Go quality gating* to determine the acceptance of the processing results
- *Exception handling* to handle and solve production exceptions
- *Alarm handling* to handle and react predefined alarms

In addition to automate the above fab activities, automation in semiconductor fabs should also avoid or prevent frauds or problems in daily fab operations. Common problems in fab operations are listed as below:

- Wrong lot goes to the tool,
- Unable to get the lot when required,
- Unable to get the reticle (photolithography mask) when required,
- Wrong recipe is used,
- Inefficient recipe setting or tool setup,
- Errors or incomplete data are collected,
- Tools are not well monitored,
- Tool capacity is not fully utilized, and so on.

Semiconductor manufacturing automation usually involves business, technical, and economic issues. In addition, the following considerations must be addressed:

- Message sequencing standards between a tool and the host computer
- Load/unload port design
- Materials handling
- Wafer cassette/pod identification
- Recipe ID and recipe body check
- Process control
- Engineering review and control
- Manual override

For decades, semiconductor manufacturing operations have evolved from manual, semi-automation to fully automation. Considerations of automation are no longer on the issues in adoption of automation or not or full support from the management, because automation is considered as mandatory and must-have in contemporary fab operations. Semiconductor manufacturing arose from the interface and control of lot track in/out operations between processing tool and the host computer, MES (Manufacturing Execution System). Such centralized systems are proprietary, not flexible and very expensive to sustain the operations and reliability due to the weakness of single point of failures. Thanks to the advance of computer and network technology, modern fab automation moves toward a hierarchical and distributed architecture.

2.2 Hierarchical, Distributed Automation Architecture

Semiconductor manufacturing operations are inherently distributed. Most applications take place at physically separated locations where local decisions are made and executed. Modern distributed computing techniques enable semiconductor manufacturing to automate its processes in an open, transparent, and scalable way. The distributed automation architecture is drastically more fault tolerant and more powerful than stand-alone mainframe systems.

Due to the complexity of shop floor operations in semiconductor manufacturing, semiconductor manufacturing automation is hierarchically decomposed into three levels of control modules, each of which is linked by means of a hierarchical integrative automation system. In the automation hierarchy, flow of control is strictly vertical and between adjacent levels; however, data are shared across one or more levels. Each control module decomposes an input command from its supervisor into: (1) procedures to be executed at that level; (2) subcommands to be issued to one or more subordinate modules; and (3) status feedback sent back to the supervisor. This decomposition process is repeated until a sequence of primitive actions is generated. Status data are provided by each subordinate to its supervisor to close the control loop and to support adaptive actions.

In view of equipment functionality or process consistency, a fab can be considered as being composed of a series of manufacturing cells. Within each cell, there is a computer system for planning, controlling, and executing the production activities in the cell. Such manufacturing cells are autonomous, i.e., having the power to self-government. Each cell is capable of managing the fabrication of wafers within it, involving automatically distributing jobs to all workstations and equipment in the cell, monitoring the states of each workstation and equipment, and feeding back these states to its upper-level supervisor systems. Fig. 1 depicts the three-levelled hierarchical, distributed architecture of semiconductor manufacturing automation.

Automation in semiconductor manufacturing comprises three categories: *Tool Automation*, *Cell Automation*, and *Fab Automation*. Tool Automation includes automation of dry and wet atmospheric and vacuum wafer handling systems, integrated front-end modules, load ports, FOUP (Front Opening Unified Pod) tracking, alignment, calibration and e-diagnostics.

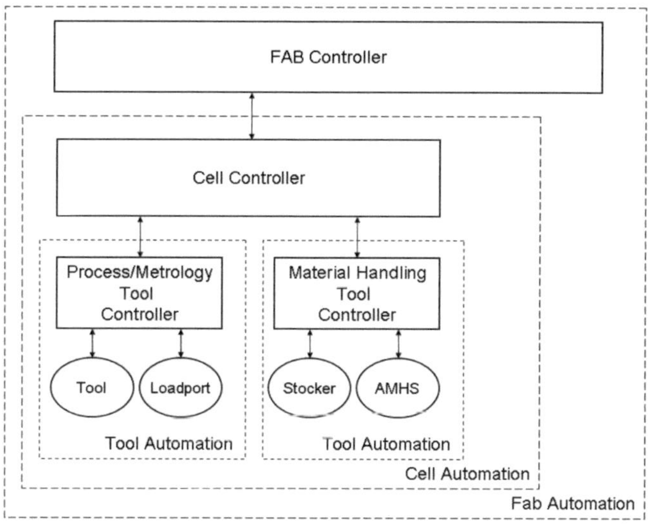

Fig. 1. The Three-levelled Hierarchical, Distributed Automation Architecture

Tool Automation also consists of wafer sorters, reticle inspection tools, reticle stockers, wafer stockers, and Automated Materials Handling Systems (AMHS). Cell Automation manages materials movement and control, tool connectivity, station control, and advanced process control (APC). Fab Automation covers system integration, manufacturing execution, scheduling and dispatching, activity management, and preventive maintenance.

3. Tool Automation

3.1 Interfacing to Semiconductor Tools

Escalating device complexity and cost have driven the demand for increased levels of automation and isolation in modern fabs. The goal of tool automation is to enable seamless integration among process control, auto identification (ID), load ports, environment control, data collection, and advanced robotics for wafer movement. However, the very challenge arose from interfacing the many and various semiconductor tools.

In 1978, Hewlett-Packard (HP) proposed to Semiconductor Equipment and Materials International (SEMI) to establish standards for communications among various semiconductor manufacturing tools (equipment). SEMI later published the SECS-1 standards in 1980 and the SECS-II standards in 1982. SECS is a point-to-point protocol via RS-232 communication. SECS is also a layered protocol consisting of three levels: Message Protocol, Block Transfer Protocol, and Physical Link (RS-232). The Message Protocol is used to send SECS-II messages between the host computer and the tool. Each SECS-II message, also referred to as a transaction, contains a primary message and an optional secondary reply message. SECS-II messages are referred to as *Streams* and *Functions*. Each message has a *Stream* value (*Sx*) and a *Function* value (*Fy*), where *Streams* are categories of messages and *Functions* are specific messages within the category. The *Function* value is always an odd number in a primary message, and even, or one greater, in the associated secondary reply. Fig. 2 illustrates the sequence diagram of an example of the message of *Stream* 1 *Function*, *S1F1* ("Are You There"). Note that in Fig. 2, the host computer sends the message

S1F1 to the tool to query the equipment status. The tool then replies to the host computer with a message of *S1F2* after receiving the *S1F1* message.

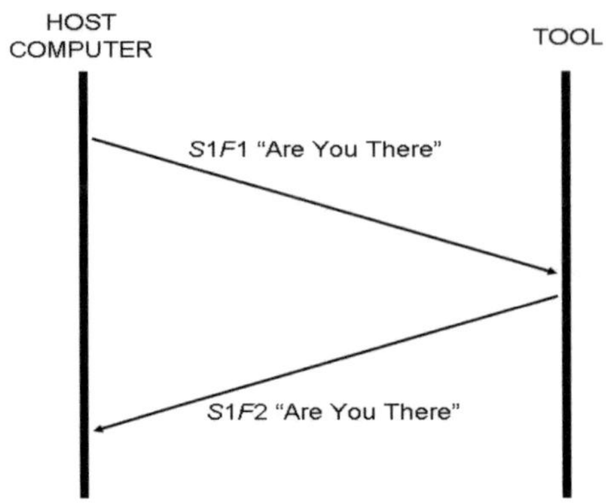

Fig. 2. Sequence Diagram of A *S1F1* Transaction

The structure (or layout) of a SECS-II message defines all the data items for the message. The layout of a SECS-II message is what follows the *S*tream and *F*unction notation. An example of the message layout of *S2F11* is given as below:

```
S2F11
<L
        <A "START">
        <L>
>.
```

Note that the above *S2F11* message is represented in SML (SECS Message Language) format. Similar to the notation used in SEMI Standards, SML is a more precise and regular notation language for describing SECS-II messages and is often used in semiconductor tool manuals.
The Block Transfer Protocol (SECS-I) is used to establish the direction of communication and provide an environment for passing message blocks. Due to the data size limitation in the SECS-I protocol, a SECS-II message may not fit into one SECS-I transaction, i.e., over-sized. The SECS-II message is then divided into smaller blocks, and sent in one block at a time, which is referred as multi-block messaging. As general communication protocols, SECS-I defines four different timeouts during the handshaking process: T1 (inter-character timeout), T2 (protocol timeout), T3 (reply timeout), and T4 (inter-block timeout). No interleaved blocks are allowed from the tool to the host. That is, the tool always sends all blocks of one message before sending the first block of the next message. This simplifies the job of the host. However, the tool allows the host to send interleaved blocks, if it so chooses.
The tool may initiate several simultaneous outstanding SECS transactions by sending a secondary message before the host has sent the reply to a previous message. This occurs when the tool reports alarms and events.
Before SECS-II messages can be sent between the host computer and the tool, communications must be first established by a *S1F13* (Establish Communications Request)

message, which is sent following an initial setup or after a long period of not communicating.

Contemporary semiconductor manufacturing adopts the Generic Model for Communications and Control of Manufacturing Equipment (GEM) standards so that fab host software can communicate with the manufacturing tool for monitoring and controlling purposes. The GEM standard, frequently referred to as the GEM or SECS/GEM standard, is formally designated and referred to as SEMI Standard E30. GEM defines messages, state machines and scenarios to enable fab software to control and monitor manufacturing tools.

SEMI Standard High Speed Message Service—Single Session (HSMS-SS) defines TCP/IP networking communication protocols for host software and a GEM tool. All GEM compliant manufacturing tools use a consistent interface to communicate with a GEM capable host either via TCP/IP (the HSMS-SS standard, SEMI E37.1) or RS-232 (the SECS-I standard, SEMI E4) protocols.

In order to facilitate the integration of automation of all the tools, contemporary semiconductor fabs demand single communication line between every tool to the host. The Equipment Front End Module (EFEM) must be integrated through the tool rather than connected directly to the host. Tool supplier must provide hardware on the tool to connect to the fab local area network (LAN). This communication connection must comply with HSMS protocol and be able to transmit and receive all SECS-II messages. Fig. 3 shows the idea of single communication link.

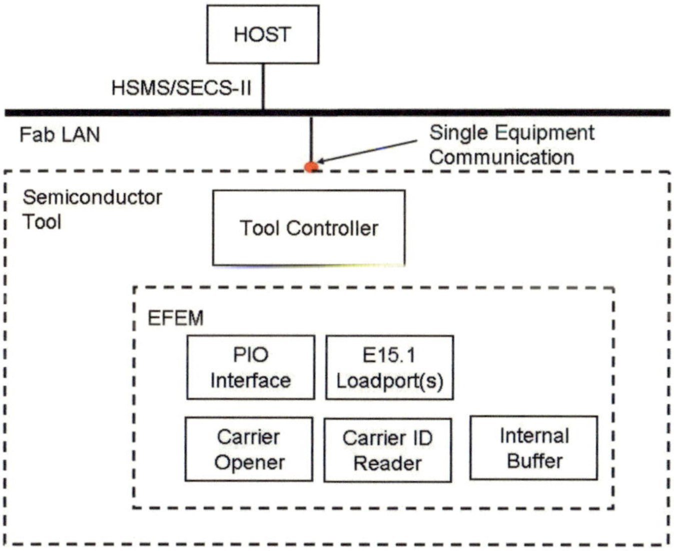

Fig. 3. Single Communication Link

3.2 Automation in Cluster Tools

Semiconductor manufacturing operations are inherently distributed. Most applications take place at physically separated locations where local decisions are made and executed.

Modern distributed computing techniques enable semiconductor manufacturing to automate its processes in an open, transparent, and scalable way. The distributed automation architecture is drastically more fault tolerant and more powerful than stand-alone mainframe systems.

Contemporary semiconductor manufacturing increasing uses cluster tools, each of which consists of several single-wafer processing chambers, for diverse semiconductor fabrication processes, shorter cycle time, faster process development, and better yield for less contamination. To illustrate the automation in semiconductor fabrication equipment, we adopt a PDV (Physical Vapour Deposition) cluster tool as an example to convey the idea of hierarchical architecture and the associated communication protocols, intelligent job scheduler/dispatcher, as well as process modelling, monitoring, diagnosis and control.

PVD cluster tools are used for vacuum film deposition on semiconductor wafers and are widely used in the fabrication of modern VLSI (Very Large Scaled Integration) circuits. The films provide conducting regions within the device, electrical insulation between metals, and protection from the environment. As PDV techniques provide more precise controls such as uniform film thickness, better crystal structure especially for compound semiconductor, PVD clusters are widely applied in contemporary fabs.

A PVD cluster tool is a fully automated system using a single wafer processing, multi-chambered design. Each single-wafer processing chamber performs a unique process without chamber redundancy. After being process, a wafer will be held by the process chamber for further pickup by a transporter. Wafer input and output are through cassette loadlocks. Integration among different process modules and allowing simultaneous processing of different routes significantly increase the operational complexity and cost. Wafer operations of different process flows compete for the use of functional modules of a PVD cluster tool such as robot transporters, buffer space and processing chambers.

The multi-chambered design of the PVD cluster tool allows for precise control over all process parameters to enhance consistency and uniformity among wafers. Major components of a PVD cluster tool include mainframe, process, transport, and cassette modules. Each module has its specific function and is mechanically linked together to form an integrated environment to execute a defined sequence of flows. Fig. 4 demonstrates an example of 300-mm PVD cluster tool configuration.

The mainframe module consists of two major chambers: transfer chamber and buffer chamber, each of which is with a robot of transfer modules. Each process module performs a unique process. Each process chamber has a wafer lid to facilitate wafer exchange with the wafer handling robot. A chamber must be at the atmospheric pressure level before the lid can be opened. Recipe change within a process chamber is allowed and it takes time to setup. A chamber can be switched to various processes, but sometimes the required setup time is significant and usually need to do some testing after switching. Therefore, a process chamber is usually fixed to some specific process only. Cassette modules include cassette loadlocks that provide access to the cluster tool system while isolating wafer process routes from atmosphere. A PVD cluster tool is usually equipped with two cassette loadlocks. The cassette loadlock provides a storage and indexing capability for programmable wafer processing sequences. Two loadlocks can operate independently to increase system throughputs and flexibility. Integration among various process modules has advantages such as cycle time reduction, footprint reduction, and so on. However, along with the flexibility, the operational complexity increases significantly.

The PVD cluster tool is a single-wafer processing tool where each chamber can accommodate at most only one wafer. Wafer movement is done mechanically by one robot in the transfer chamber and one robot in the buffer chamber. After a FOUP arrives at a loadport of the cluster tool, the cassette is loaded into a cassette loadlock which is then pumped down to vacuum. The buffer chamber robot picks a wafer from the cassette and places it in the degas chamber where the wafer is re-oriented and degassed. After being degassed, the buffer chamber robot then takes the wafer from the degas chamber and places it in a preclean chamber for preclean with plasma etching. After completing the preclean process, the transfer chamber robot picks the wafer from the preclean chamber and places it on one process chamber for deposition of aluminium (Al), titanium (Ti), or titanium nitride (TiW), as specified by the processing recipe of the wafer. After completing the deposition process, the wafer is carried by the transfer chamber robot again from the process chamber and places it in a cooldown chamber in which the wafer is cooled down.

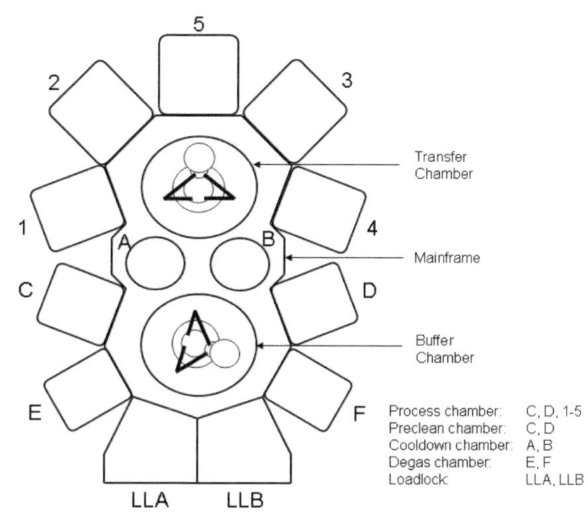

Fig. 4. An Example of 300-mm PVD Cluster Tool Configuration

Once the temperature of the wafer reduces to the specific degree, the buffer chamber robot brings the wafer from the cooldown chamber and places it back to the same cassette from which the wafer is removed. After all wafers in the cassette complete the processing and return to the cassette, the loadlock chamber raises its pressure to atmospheric pressure and returns the cassette to the FOUP in the loadport. This then completes the entire process. Arrows in Fig. 5 indicates an example of the process flows executed in the PVD cluster tool, where the process starts at s1 (arrival at the loadlock), then goes to s2 (degassed), s3 (cooling), s4 (deposition), s5 (cooling), and then return to the loadlock to complete the process.

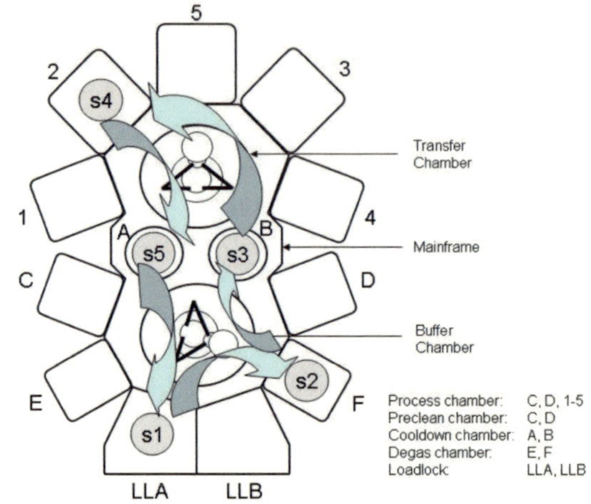

Fig. 5. An Example of Process Flows in the PVD Cluster Tool

The configuration of a PVD cluster tool allows itself to execute multiple process flows simultaneously. This capability of parallel processing of multiple process flows facilitates better utilization of tool capacity. However, operational complexity and difficulty in its tool automation become increased and challenging. Due to the potentially circular wait of shared resources (for example, a wafer is a process chamber is waiting to be transferred to a cooldown chamber where there is another wafer waiting to be the process chamber), the cluster tool system is inherently subject to deadlocks, which may cause the entire system to halt and thus ill-functioned. Tool automation in such a system is no longer simply the implementation of a predefined sequence of logics, code, or programs. It involves the knowledge and techniques from operation research, optimization, system engineering, and so on.

4. Computer-Integrated Semiconductor Manufacturing

Semiconductor manufacturing integration encompasses the allocation, coordination and mediation among system dynamics and flows of information, command, control, communication, and materials, in a timely and effective way. According to the definition of ITRS 2007 Roadmap (ITRS 2007), semiconductor fab integration is divided into five thrusts: *Fab Operations, Production Equipment, Materials Handling, Fab Information & Control Systems*, and *Facilities*. Among these five thrusts, Fab Operations is the key driver of requirements and actions for the other four thrusts, while Fab Information & Control Systems is the facilitator to the integration in semiconductor manufacturing.

Because of the ever-increasing complexity of semiconductor devices and their manufacturing processes, computer or CIM (Computer-Integrated Manufacturing) systems are essential for the smooth integration of semiconductor manufacturing. However, CIM systems generally are loosely coupled, monolithic, and difficult to extend to support the new needs. Researchers and practitioners have been devoted to build an integration framework with a common, modular, flexible, and integrated object model to tackle the critical problems in semiconductor manufacturing integration: islands of automation, emergence of new applications, distributed systems, as well as data integrity.

SEMATECH (1995, 1998) created a CIM Framework based on the Microelectronics Manufacturing Science and Technology (MMST) Project in Texas Instruments (TI), a member company of SEMATECH. The CIM Framework intends to promote integration on the shop floor, reduce costs, and increase reuse through object-oriented technology. Based on the definition of SEMATECH, a framework is a *software infrastructure* that creates a common environment for integrating applications and sharing information in a given domain. The CIM Framework is a framework of components that provide the functionality common across applications (programs consisting of a collection of interoperating objects). The CIM Framework also enables integration of those applications.

A set of functional components are defined and designed in the SEMATECH CIM Framework Specification to work together to form an integrated manufacturing system. The functional components of the CIM Framework Specification are grouped by application areas. The CIM Framework adopts CORBA (Object Management Group, 1999) as the common interface that defines the object-oriented architecture of an object request broker, which enables and manages interoperability between objects and applications across heterogeneous computer boundaries. The functional components in each application area are tabulated in Table 1. Fig. 6 demonstrates the integration of the SEMATECH CIM Framework.

Application Area	Functional Component
Factory Services	▪ Document Management ▪ Version Management ▪ History Management ▪ Event Broker
Factory Management	▪ Factory ▪ Product Release ▪ Factory Operations ▪ Product Request
Factory Labour	▪ Person Management ▪ Skill Management
Process Specification Management	▪ Process Specification ▪ Process Capability
Schedule Management	▪ Dispatching
Machine Control	▪ Machine Management ▪ Recipe Management ▪ Resource Tracking
Material Management	▪ Product Management ▪ Durable Management ▪ Consumable Management ▪ Inventory Region ▪ Product Specification ▪ Bill of Material
Material Movement	▪ Material Movement
Advanced Process Control	▪ PlugIn Management ▪ PlugIn Execution ▪ Control Management ▪ Control Execution ▪ Control Database ▪ Data Collection Plan

Table 1. Functional Components by Application Area in SEMATECH CIM Framework

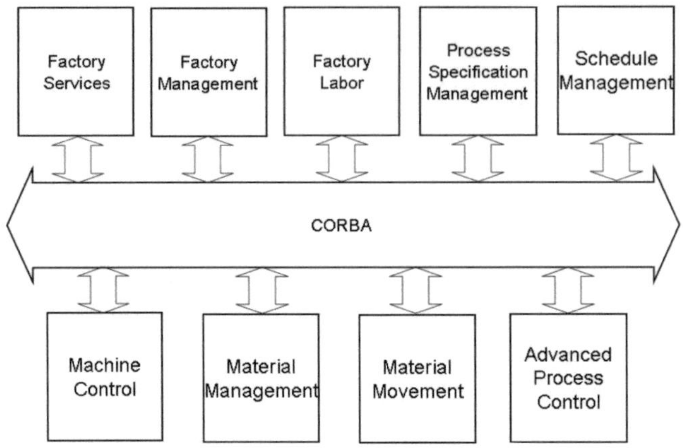

Fig. 6. SEMATECH CIM Framework

5. Intelligent and Computer-Integrated Framework for Prioritized Manufacturing Services

Automatic Materials Handling System (AMHS) is considered as a must in modern semiconductor manufacturing environment (International SEMATECH, 1999). There are usually hundreds of vehicles running in dozens of loops in an AMHS. The transport requirements of AMHS vehicles among different loops are usually changing from time to time, according to the dynamic WIP distribution, process conditions, and equipment capacity. It is therefore needed an effective methodology to integrate AMHS with other CIM systems to cope with the dynamic changes on the materials handling services. We propose an intelligent AMHS management framework to optimize and manage the integration of fab operations with AMHS.

Daily fab operations are usually differentiated to several levels of priorities to cope with frequent process changes, engineering experiments, or pilot production. In a fab, a lot is granted as high priority (named as Hot Lot or Super Hot Lot) if either it is going to execute several operations for experiments or inspections on process conditions; or it was borne as a pilot or risk lot for process characterization or design validation before releasing a new product for production (Liao & Tsai, 2006). Hot lots are given to higher priority to reduce their cycle time. Operations of high priority lots can be either pre-emptive against normal operations, or capacity-reserved for no-wait services (Liao & Wang, 2004; Liao & Wang, 2006). The introduction of high priority lots has the potential to shuffle the regular production and should be well managed (Ehteshami *et al.*, 1992). The AMHS management framework should be effective to minimize the transport time of hot lots while optimizing the use of underlying resources for regular production.

In order to support prioritized automated materials handling services in semiconductor manufacturing, Liao proposed a management and control framework (Liao, 2002) as depicted in Fig. 7, where there are four main modules in the framework: Service Level Agreement Management (SLAM), AMHS Traffic Forecast (AMHSTF), Dynamic OHT Allocation (DOHTA), and Dynamic OHT Dispatching (DOHTD). Several QoS (Quality of Service) levels are first defined to provide differentiated automated materials handling services to transport jobs. Service-level specifications (SLS) are used to describe the appropriate QoS parameters which the AMHS should take into account when transporting prioritized jobs. The SLS also specifies or guarantees an upper bound of a performance measure, such as average delivery time. In order to achieve QoS guarantees, the framework plans and manages the requirements for service subscription according to available resources. The dynamic OHT allocation and dispatching modules of the framework is responsible for managing the allocation of OHT vehicles and for controlling the transport sequence to meet the SLS demands provided by the SLA management. With the integration of dynamic OHT allocation and dispatching functions, the proposed framework ensures that agreed-upon SLS are adequately provisioned.

In the prioritized AMHS service framework, SLAM receives transport service requests from lot dispatching/scheduling function of fab CIM systems. Assessments are then made with traffic forecast to evaluate the impact on transport performance due to these requests for high priority materials handling services. A SLA is negotiated and contracted to the lot dispatching/scheduling functions (Liao *et al.*, 1996) to provide a guaranteed service. With this precise prediction on wafer transport times, lot dispatching/scheduling functions are more effective to cope with the real-time fab dynamics. SLS are then interpreted according to the SLA and are translated into commands to dynamic OHT allocation and dispatching to meet the fluctuating requirements specified in the SLS. A simulation model (Liao & Fu, 2004) based on the dispatching policies and the allocation method is used for fab traffic forecast.

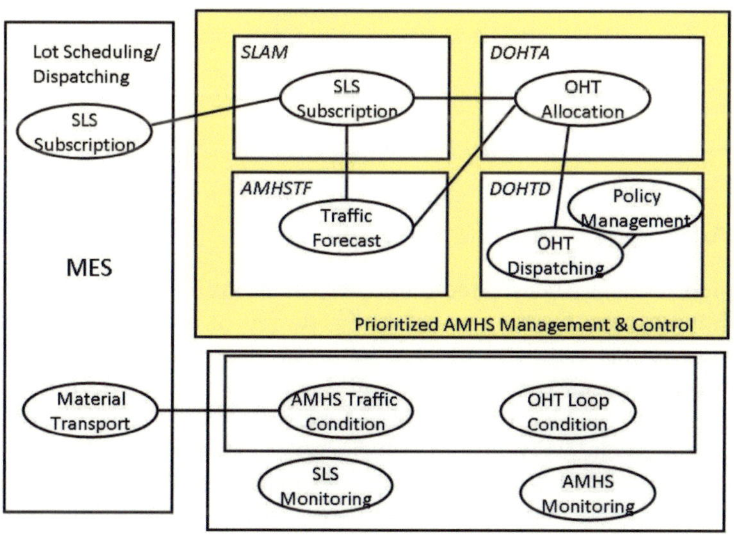

Fig. 7. The Management and Control Framework for Prioritized AMHS Services

Although the average (or static) loading of the AMHS has been optimized in the stage of fab layout design, the transport loadings of different interbay/intrabay loops are usually various and changing from time to time. For an interbay/intrabay loop, its transport requirements are dynamic according to the varying WIP distribution and the fluctuating processing capacity of tools within the loop. Such requirements are usually local and urgent. They demand timely and flexible autonomous responses and actions immediately, which now can be achieved with the help of intelligent agents (Jennings & Wooldridge, 1998). Exploiting the agent-based technology, we implement the Prioritized Management and Control Framework with agents for OHT dispatching, resource management, traffic monitoring, and policy management applications, that completes the intelligent, computer-integration framework for prioritized semiconductor manufacturing services. Fig. 8 depicts the integration of the agent-based components with the other fab CIM systems.

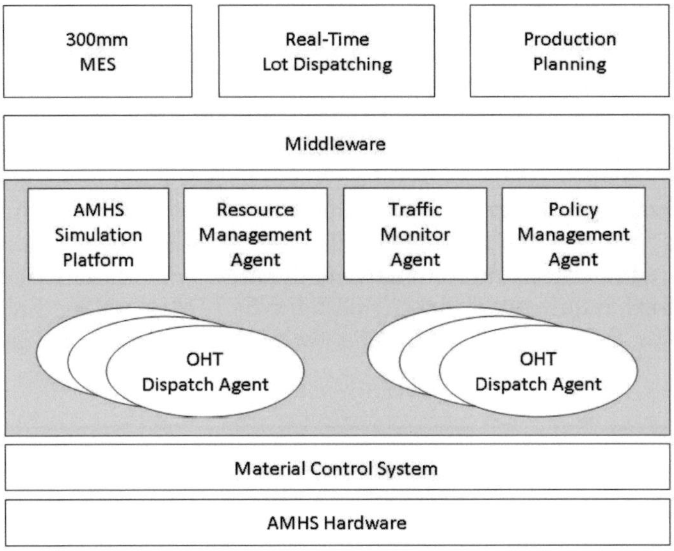

Fig. 8. Intelligent, Computed-integrated Framework for Prioritized Services

6. Design of Automation and CIM Systems

Systematic design and analysis methodologies, like system definition, validation or verification techniques, are always needed in the design of automation and CIM systems. During the design phase, the most tedious job is to implement the dynamic behaviours between system components and objects for all manufacturing applications involved. To the large dynamic systems like semiconductor manufacturing, it is always difficult and challenging to define, validate, and verify their system dynamics, not to say, to consider their various and changing control and managerial policies. In this Chapter, we adopt Petri-net techniques to build models for both PVD cluster tool and AMHS. Mathematical analysis and computer simulation are conducted to verify and validate the correctness of the automation and integration in the developed model.

A Petri net (PN) (Peterson, 1981; Murata, 1989) is a special kind of directed bipartite graph that consists of nodes as places and transitions. Directed arcs in a PN are either from a place to a transition or from a transition to a place. Each place may hold either none or a positive number of tokens. In a place, tokens are used to represent the number of available resources or to check whether a condition is satisfied or not. When all the input places of a transition hold enough number of tokens, the transition is enabled. A transition is firing at an enabled transition if firing conditions are satisfied. Such a firing changes the token distribution in places of the PN, which are usually to model the change in system states (markings). Pictorially, places in a PN are depicted by circles and transitions by bars. A TPN is a PN where either zero or positive time delays are associated with places, transitions, and/or arcs. Mathematically, a TPN C is defined as follows:

$$C = (\mathbf{P}, \mathbf{T}, I, O, m)$$

Where

$\mathbf{P} = \{p_1, p_2, \ldots, p_i\}$ is the finite set of places, where $i > 0$;

$\mathbf{T} = \{t_1, t_2, \ldots, t_j\}$ is the finite set of places, where $j > 0$; with $\mathbf{P} \cup \mathbf{T} \neq \emptyset$ and $\mathbf{P} \cap \mathbf{T} = \emptyset$;

$I : \mathbf{P} \times \mathbf{T} \rightarrow N$ is the input function defining the set of directed arcs from **P** to **T** with $N = \{0, 1, 2, \ldots\}$;

$O : \mathbf{T} \times \mathbf{P} \rightarrow N$ is the output function defining the set of directed arcs from **T** to **P** with $N = \{0, 1, 2, \ldots\}$;

$m : \mathbf{P} \rightarrow N$ is the marking representing the number of tokens in the places.

Consider the PVD cluster tool described in Section 3.2. Each process chamber (C, D, E, F, 1-5) can be in one of the following states:

- *Idle*: chamber is free to be accessed
- *Move In*: chamber is reserved to move in wafer
- *Processing*: chamber is reserved to process wafer
- *Wait/Move Out*: chamber is reserved to move out wafer

Based on the definition of chamber states, each process chamber can be modelled as a Petri net. The chamber state could become *Move In* only when it is in the state of *Idle*. After the specific duration of move-in time, the chamber changes its state from *Move In* to *Processing*. After the processing time elapses, the chamber again changes its state from Processing to *Wait/Move Out*, where physically wafer is wait for Transport Chamber to move out the wafer from the process chamber. Finally, the process chamber becomes *Idle* after the wafer is moved out. In a process chamber Petri net, the places represent states of the chamber and the transitions represents the change of states with associated time delays. The token in the Petri net represents the availability of state and there is only one token in the Petri net, i.e., m=1.

Different to the Petri net model for process chamber, the Petri net model for wafers is determined by the process flow of the wafers, which is specified in the recipe and received from fab MES systems via tool automation. Each process flow involves a sequence of operations as well as processing requirements. A Petri net model for the process flow in Fig. 5 is shown in Fig 9, where the Petri net models of process chambers F and 2, preclean chamber B, cooldown chambers A, are all combined as an integrated model. Mathematically, the process flow Petri net is the union of these Petri nets of chambers F, 2, B and A.

The graphical presentation of Petri nets helps not only modelling a system, but also validating the system. It is easy to trace all the possible states of the PVD cluster tool when wafers are processed in the tool. The Petri model can be verified with the analysis of its reachability, liveness, safeness, and so forth (Srinivasan, 1998). Mathematically, it can be proved that the process flow Petri net in Fig. 9 is live and safe. That is, for any wafer lot to go with the process flow will complete the entire process.

Automation and integration in semiconductor manufacturing usually involve discrete event systems that exhibit sequential, concurrent, and conflicting relations among the events and operations. The evolution is dynamic over time. A formal approach such as Petri nets enables one to describe complex discrete event systems precisely and thus allows one to perform both qualitative and quantitative analysis, scheduling and control of the automation and integration systems.

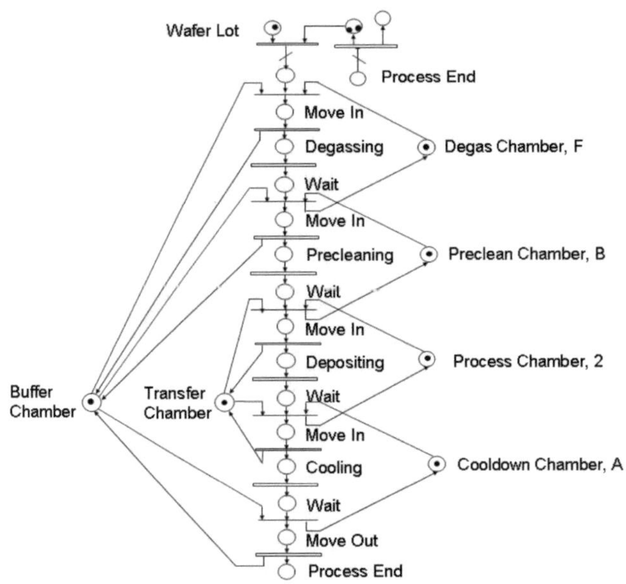

Fig. 9. Petri Net Model of Process Flow in Fig. 5

7. Conclusion

Semiconductor automation originates from the prevention and avoidance of frauds in daily fab operations. As semiconductor technology and business continuously advance and grow, manufacturing systems must aggressively evolve to meet the changing technical and business requirements in this industry. Semiconductor manufacturing has been suffering pains from islands of automation. The problems associated with these systems are limited flexibility and functionality, low level of integration, and high cost of ownership. Thanks to the recent technological advances that can provide significant approaches in dealing with these problems, we are able to realize the promise of semiconductor manufacturing with sound automation and integration.

In this Chapter, we have reviewed the need of automation and integration in semiconductor manufacturing. Some considerations in fab automation are addressed. The three-levelled hierarchical, distributed automation architecture is discussed, where automation in semiconductor manufacturing is classified into tool, cell, and fab automation. Three popular protocols, SECS, GEM, and HSMS, for interfacing to semiconductor tools, are reviewed. In addition, the concept of single communication link is highlighted due to its importance in the design of modern tool automation. Specially, we take the PVD cluster tool as the study vehicle for tool automation. We have reviewed the SEMATECH CIM Framework. We have proposed an intelligent and integrated CIM framework for prioritized manufacturing services, where the management and control to AMHS services are discussed and intelligent and autonomous agents are used to facilitate the prioritized services in modern semiconductor manufacturing. Finally, we adopt the Petri net technology to go through the modelling, validation, and verification in the design of automation and integration systems in semiconductor manufacturing.

This Chapter adopts Petri nets to demonstrate the techniques for system modelling, validation and verification of automation and integration in semiconductor manufacturing. Some useful approaches like Unified Modelling Language (UML), computer simulation, queueing network analysis, mathematical programming, and so on, are also frequently used in system analysis of fab automation and integration applications.

The automation and integration in semiconductor manufacturing must continue to evolve to meet the needs of the competitive and vital industry.

8. References

Ehteshami, B. ; Petrakian, R. G. & Shabe, P. M. (1992). Trade-Offs in Cycle Time Management : Hot Lots. *IEEE Transactions on Semiconductor Manufacturing*, Vol. 5, No. 2, May 1992 101-106

International SEMATECH (1999). *Automated Material Handling System (AMHS) Framework User Requirements Document : Version 1.0,* International SEMATECH, 1999

ITRS (2007). *International Technology Roadmap for Semiconductors : 2007 Edition,* http://www.itrs.net/reports.html

Jennings, N. R. & Wooldridge, M. J. (1998). *Agent Technology – Foundations, Applications, and Markets.* Springer, 1998

Liao, D. (2002). A Management and Control Framework for Prioritized Automated Materials Hnadling Services in 300mm Waer Fourndry. *Proceedings of 2002 IEEE International Conference on Systems, Man, and Cybernetics, Hammamet, Tunisia,* October 2002 18-23

Liao, D. (2005). Vehicle Clustering Phenomenon in Automatic Materials Handling Systems in 300mm Semiconductor Manufacturing. *Journal of Material Science Forum, Progress on Advanced Manufacture for Micro/Nano Technology 2005, Part 2,* December 2005 1129-1134

Liao, D.; Chang, S. ; Pei, K. & Chang, C. (1996). Daily Scheduling for R&D Semiconductor Fabrication. *IEEE Transactions on Semiconductor Manufacturing,* Vol. 9, No. 4, November 1996 550-561

Liao, D. & Fu, H. (2004). A Simulation-Based, Two-Phased Approach for Dynamic OHT Allocation and Dispatching in Large-Scaled 300mm AMHS Management. *IEEE Robotics & Automation Magazine,* Vol. 11, Issue 3, September 2004 22-32

Liao, D.; Jeng, M. & Zhou, M. (2007). Application of Petri Nets and Lagrangian Relaxation to Scheduling Automatic Materials Handling Vehicles in 300-mm Semiconductor Manufacturing. *IEEE Transactions on Systems, Man, and Cybernets – Part C,* July 2007 1-13

Liao, D. & Tsai, M. (2006). A Quota-Constrained, Speed Control Model for Production Scheduling in Semiconductor Manufacturing. *International Journal of Manufacturing Technology and Management,* Vol. 9, No. 3/4, 2006 294-308

Liao, D. & Wang, C. (2004). Neural-Network-Based Delivery Time Estimates for Prioritized 300mm Automatic Material Handling Operations. *IEEE Transactions on Semiconductor Manufacturing,* Vol. 17, No. 3, August 2004 324-332

Liao, D. & Wang, C. (2006). Differentiaed Preemptive Dispatching for Automatic Materials Handling Services in 300mm Semiconductor Foundry. *International Journal of Advanced Manufacturing Technology,* Vol. 29, No. 9-10, February 2006 890-896

Murata, T. (1989). Petri Nets: Properties, Analysis and Applications, *Proceedings of the IEEE,* Vol. 77, No. 4, April 1989, 541-580

Object Management Group. (1999). *The Common Object Request Broker: Architecture and Specification*, Rev. 2.3, Needham, MA, USA, http://www.omg.org/

Peterson, J. L. (1981). *Petri Net Theory and the Modelling of System*, Addison-Wesley, 1981

SEMATECH. (1995). *Computer Integrated Manufacturing (CIM) Application Framework Specification 1.2,* SEMATECH Technology Transfer#93061697E-ENG, 1995, Austin, TX 78741, http://www.sematech.org/

SEMATECH. (1998). *Computer Integrated Manufacturing (CIM) Framework Specification Version 2.0,* SEMATECH Technology Transfer#93061697J-ENG, January 31, 1998, Austin, TX 78741, http://www.sematech.org/

SEMI. http://www.semi.org/

Srinivasan, R. S. (1998). Modeling and Performance Analysis for Cluster Tools Using Petri Nets. *IEEE Transactions on Semiconductor Manufacturing,* Vol. 11, No. 3, August 1998, 394-403

Zhou, M.-C. & Jeng, M.-D. (1998). Modeling, Analysis, Simulation, Scheduling, and Control of Semiconductor Manufacturing Systems : A Petri Net Approach. *IEE Transactions on Semiconductor Manufacturing*, Vol. 11, No. 3, August 1998, 333-357

4

Contamination monitoring and analysis in semiconductor manufacturing

Baltzinger Jean-Luc and Delahaye Bruno
Altis Semiconductor
France

1. Introduction: Contamination on wafers

1.1 Definition of the different type of contamination

Contamination is defined as a foreign material at the surface of the silicon wafer or within the bulk of the silicon wafer. The contamination can be particles or ionic contamination, liquid droplets… The mechanism of contamination of silicon wafer is summarized on figure 1 (Leroy, 1999):

- The source of contamination
- The transportation of the contamination
- The location of the contamination: surface, bulk
- The evolution of the contamination: how to remove it? Does the cleaning remove the contamination? Does the cleaning bring the contamination?

The chemistries of the cleaning solutions which are described within the figure 1 are able to remove particles or metallic contamination. They can also bring both of these contaminants. In this discussion, we just want to underline the source of contamination, and the way to measure it. Another way to consider wafer contamination source is the environment of the wafer (Pic, 2006):

- Contact with the wafer: chemicals, Gases, Ultra pure Deionised Water, resist, ionic implantation, deposition layers, etching process
- Environment for the process: tool, network for gases and chemicals distribution, boxes for wafer handling and transportation.
- General environment: facilities, human, external pollution (traffic, industrial)

Semiconductor devices are sensitive to the contamination, due to different possible root causes: device size reduction, device sensitivities on some process steps, cross contamination induced by chemicals, ultra pure water and gases. The environment is also contributing to the contamination effect on the wafer as tools, transportation boxes, and clean-room. Contamination can be divided in three categories: ionic contamination, airborne molecular contamination (AMC) and particles (defect density).

In this chapter, after a short description of the different contamination impact on wafers, we focus on metallic and anions contamination measurements with some examples. Then the second part of this chapter will consider the particle monitoring on bare wafers and patterned wafers.

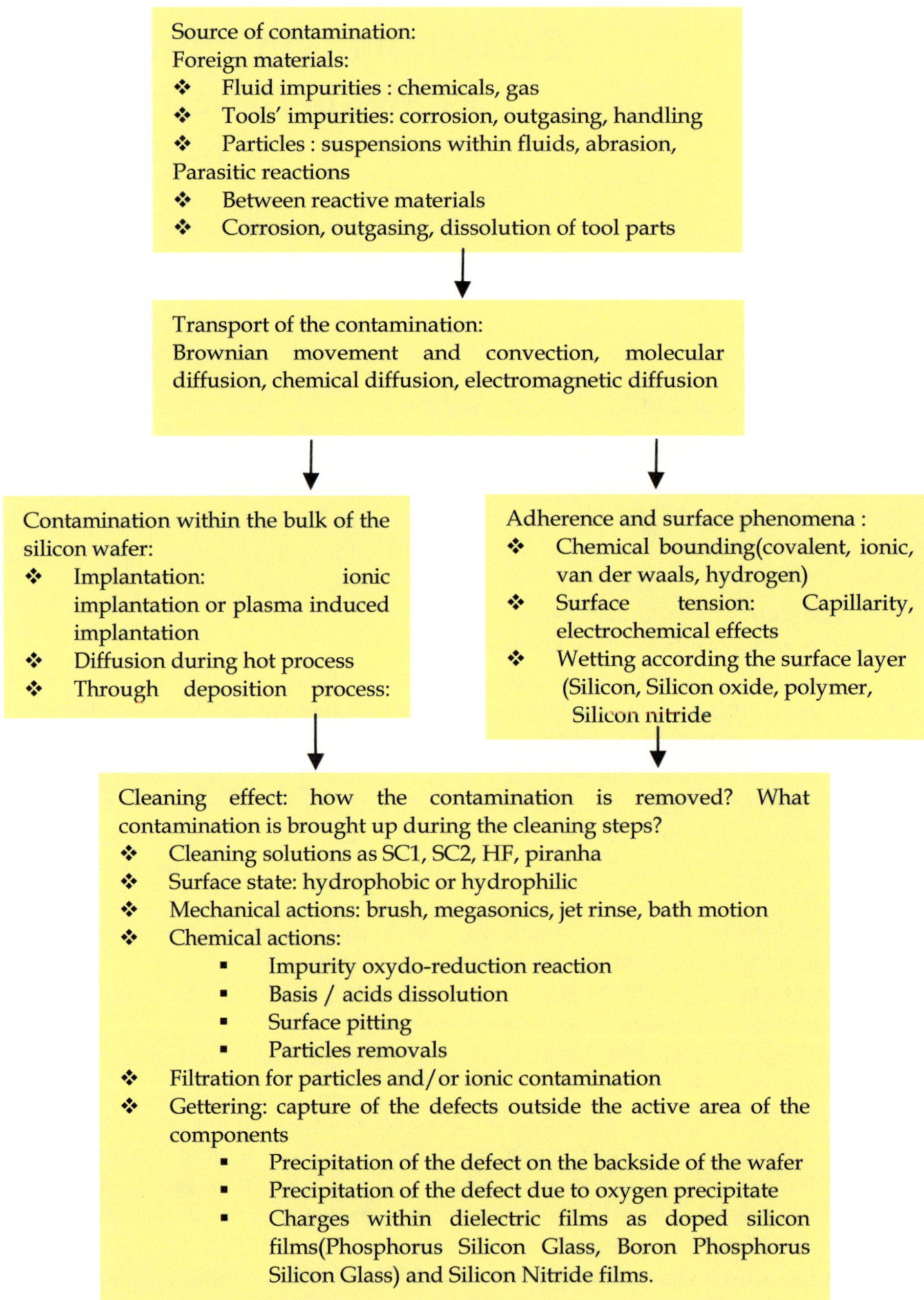

Fig. 1. Contamination workflow: mechanism and questions.

1.2 Contamination impact on wafers

The contamination impacts of the three different contaminants are summarized in table 1

Contamination Classification	Elements	Sources	Wafer effects
Ionic contaminant	Alkaline Na,K	Human pollution Works Chemical and gases	Electrical instability ▪ gate oxide leakage ▪ retention
Ionic contaminant	Transition Metals Ni,Co,Fe, …	Human pollution Works Chemical and gases Networks- tools-process	Gate oxide integrity (GOI) degradation
Ionic contaminant	Dopants Al, P, In, Ga, As, B,…	Process: wet processes, implantation / Works Material out gassing Chemicals and gases	Shift of voltage threshold of the transistor device.
Ionic contaminant & Air molecular contamination	Acids F-, Cl- ,CH3COO-,Br- , PO4--,SO4--	Process pollution: etch, wet process, Chemical Vapor Deposition (CVD) Works Material out-gassing Traffic pollution Industrial pollution	Pad corrosion Aluminum corrosion Defectivity on Deep UV (DUV) and Mid UV (MUV) resist Salt deposition on lens, masks, wafers
Ionic contaminant & Air molecular contamination	Bases NH3 Amines	Process pollution: etch, wet process, CVD deposition. Works Material out-gassing Traffic pollution Industrial pollution	Footing on DUV resist Salt deposition on lens, masks, wafers Photolithography activation especially with 193 nm process
Organics	Organics	Process pollution: Wet process and lithography process	Photolithography activation especially with 193 nm process. Eg: contamination with solvent on resist
Particles	Organics	Process Pollution: dry etch polymers, resist strip, wet process, Material out gassing Chemicals and gases	Gate oxide integrity High resistivity contact Deposition on surface, lens degradation Defectivity with opens or shorts on pattern wafers
Particles	inorganic	Process Pollution: dry etch polymers, resist strip, wet process, Material out gassing Chemicals and gases	Gate oxide integrity High resistivity contact Deposition on surface, lens degradation Defectivity with opens or shorts on pattern wafers

Table 1. Description of Contamination source and wafer effects

2. Contamination analysis and monitoring

2.1 Measurement techniques

The analytical techniques for measurements of the different contaminants defined in the table 1 are break down within four categories (Galvez 2006)

- metallic contamination analysis
- Anions impurities analysis with ion chromatography
- Chemical composition analysis as gas chromatography, (GC), Total Organic Compound (TOC) Analyser for Deionized water (DI water)…
- Liquid particle measurement with liquid particle counters for particle size above or equal 0.1 μm diameter for chemicals. Tools for the characterization of the particles size distribution are also interesting, but not in the scope of this presentation.

In this chapter, we focus on metallic contamination in silicon which represents one of the major causes for low yields and poor performance of semiconductor devices. Transition metals in silicon have deleterious effects on device characteristics. Airborne molecular contamination affects key process steps, as gate oxide quality.

Measurement techniques of metallic contamination are divided in two categories:

- Inline measurement technique: direct measurement on the wafer without any sample preparation
- Off line measurement technique: Either the technique, or the sample preparation pre-treatment before measurement, involves the analysis within a laboratory environment.

All these measurement technique have performance defined by parameters as :

- Detection Limit (DL) is the capability to distinguish a signal from the noise of the measurement system. Typically, Signal to Noise Ratio (SNR) is needed to be greater than 3.
- Quantification Limit (QL): It is defined as QL = A x DL, where A is integer number. Its value depends on analytical conditions.
- Surface analysis: the spot size of the analytical technique. Sample preparation as Vapor phase Decomposition (VPD) is able to increase the surface analysis, by etching the contaminants at the surface of the silicon wafer or within the bulk of the oxide film deposited at the surface of a wafer. Then the droplet is either used for analysis on ICP-MS measurement, either dried for TXRF measurement
- Probing depth of the analytical method: the volume of material probed during the analysis
- Time response: delay between the sampling and the analytical response. It depends on the sensitivity requested, as Quantification Limit can be improved by accumulation or concentration steps, the measurement time is increasing.
- Analytical coverage: metallic elements which are detected.

Sample Preparation as VPD is pushing detection limit by one to two order of magnitude according elements, but it has a clear impact on the time response. A compromise has to be found between the different parameters.

The in line measurement techniques are surface analysis as EDX or TXRF or SPV described in table 2. The off-line measurement techniques are installed within laboratory. Surface, film or bulk characterizations can be run on different surface analysis tool as Atomic absorption Spectroscopy (AAS), VPD-TXRF (a tool available for manufacturing environment is already available) , VPD ICPMS, SIMS, Auger, XPS. It is described in table 3 and 4.

In Line Measurement technique	EDX	SPV	TRXF
Physical Principle	Energy Dispersive X-ray Spectroscopy: X Ray of elements contained within samples	Measurement of minority carrier diffusion length linked to lifetime	X-Ray fluorescence of elements at the surface of the sample after excitation with X ray at a grazing angle
Impact on sample of the measurement	None, not destructive	None, not destructive	None, not destructive
Surface analysis	Few nm	1 mm	1 cm2
Probing depth	10E2 to 10E4 nm	10 – 150 µm	1 nm
Analytical coverage	Elements after Na within periodic table	All metals electrically active in bulk All charge in the silicon oxide	Elements after Na within periodic table
Detection limit	Qualitative results as main compounds of particles until composition of one percent, are identified	5 E9 At/cm3	Fe : 5E9 At/cm2
Sample characteristics	Bare wafer/ patterned wafers Need localization of particles for composition characteristics	Bare wafer But need activation. Fe can be identified if measurement pre and post anneal is done	Bare wafer
Results	X ray spectrum of elements contains within the material	Diffusion length, not qualitative except on Fe with P substrate Points/Mapping	Surface concentration Points/ Mapping

Table 2. parameters description of metallic measurement with in line techniques

IC	: Ion Chromatography
TXRF	: Total X-ray Reflection Fluorescence
SPV	: Surface Photo Voltage analysis
AAS	: Atomic Absorption Spectroscopy
ICP MS	: Inductively Coupled Plasma Mass Spectroscopy
VPD TXRF	: Vapour Phase Decomposition TXRF
VPD ICP MS	: Vapour Phase Decomposition ICP MS
ppb	: part per billion typically ng/g for metallic impurities in chemicals
ppt	: part per billion typically pg/g for metallic impurities in chemicals

Off Line Measurement technique	IC	AAS	ICP MS	VPD TXRF	VPD-ICPMS
Physical Principle	Variable Retention Time of anions on column	Wavelength absorption specific according elements	Mass Spectrometer coupled to an Inductively Coupled Plasma source	Same as TXRF with VPD preparation for integration of the surface of the wafer	Same as ICPMS with VPD preparation for integration of the surface of the wafer
Impact on sample of the measurement	Destructive as the liquid containing the liquid is analyzed	Destructive as the liquid containing the metallic elements is analyzed	Destructive as the liquid containing the metallic elements is analyzed	Destructive as the liquid containing the metallic elements is analyzed	Destructive as the liquid containing the metallic elements is analyzed
Surface analysis	sample preparation	sample preparation	sample preparation	Bare wafer	Bare wafer
Probing depth	None	None	None	1 nm to 1 µm	1 nm to 1 µm
Analytical coverage	Anions: F-,Cl-, NO3-,PO4--, and acetate	All elements, mainly Alkaline as Na,K	All elements within periodic elements	Elements after Na within periodic table	All elements within periodic elements
Detection limit	Few ppt depending on sample preparation	Few ppt depending on sample preparation	Few ppt depending on sample preparation	Fe: 10E7 At/cm2	Fe: 10E7 At/cm2
Sample characteristics	Chemicals, extraction from materials Air Molecular Contamination	Chemicals, sample preparation needed with matrix removal for better sensitivity	Chemicals, sample preparation needed with matrix removal for better sensitivity	Bare wafer with native oxide or thicker oxide with sample preparation by HF Vapors dissolution of Silicon dioxide	Bare wafer with native oxide or thicker oxide with sample preparation by HF vapors dissolution of Silicon dioxide
Results	Concentration of contaminants within solution in ppt or ppb	Concentration of contaminants within solution in ppt or ppb	Concentration of contaminants within solution in ppt or ppb	Average value of metallic contamination on wafer	Average value of metallic contamination on wafer

Table 3. parameters description of metallic measurement with off line techniques part 1

Off Line Measurement technique	SIMS	XPS	Auger
Physical Principle	Ar Sputtering and Ionization of Species within Sample, Mass analyzer	X Ray photoelectron spectroscopy of chemical compounds, bounding of species impacts response	Auger electron emission characteristic of the species within the sample.
Impact on sample of the measurement	Destructive as sputtering of Sample	Not always destructive	Not always destructive
Surface analysis	> 10 µm2	15 µm	8 nm spot size
Probing depth	20 nm to 10 µm	0.4 to 10 nm. Sputtering of the sample is also possible for profiling	0.4 to 10 nm. Sputtering of the sample is also possible for profiling
Analytical coverage	All	All	All
Detection limit	sensitivity changes according to elements : ppb range to ppm	>0.5 % atomic weight	>0.5 % atomic weight
Sample characteristics	Bare wafers with implants, films or patterned wafers if specific macros are forecast, Small samples	Bare/patterned wafers/small sample (KLA file recognition)	Bare/patterned wafers/small sample
Results	Elemental, quantification with standard.	Point or Surface or Elemental composition, chemical maps Chemical state for bounding between elements	Point or Surface or Elemental composition, chemical maps,

Table 4. parameters description of metallic measurement with off line techniques part 2

SIMS : Secondary Ion Mass Spectroscopy
XPS : X-ray Photoelectron Spectroscopy
ppm : part per million, typically µg/g

2.2 monitoring of Main topics: AMC, Chemicals

2.2.1 AMC

Air Molecular Contamination monitoring scheme is based on collection of contamination on beakers, bubblers or directly on wafers. The measurements are then done by IC, or TXRF for measurement on the wafer. The time of collection will be able to enhance the sensitivity. A deposition rate is then calculated.

ITEMS	AMC Monitoring (Molecular acids, bases)	Parameter value
AMC monitoring	Frequency	1/4 weeks
	Method	beakers
	Sampling time	22h
	Analytical Method	IC for beakers TXRF for wafers
	Method	beakers
		Impinger
		deposition rate on bare wafers
control limit unit - pptM (Part Per trillion molar)	F-	1200
	Cl-	400
	NO3-	1900
	NO4-	1400
	PO4(3-)	900
	SO4 (2-)	900
	NH4+ (ppbM)	0.16

Table 5. Description of AMC monitoring

NH4+ has a specific monitoring for litho tools. For example in table 6, results for different location and Litho Tool set show that the value is greater than the action limit. Then the tool is stopped and root cause analyses are done. The measurements have been done with an Ion Chromatography (IC) by Balazs laboratory from Air Liquide Electronics Europe.

Location	[NH4+] in ppbM measured by IC
MUV Tool A	12,80
DUV Tool A	0,55
DUV Tool B	0,85
DUV Tool C	1,60
Clean Room 1	3,05
DUV Tool D	0,28
MUV	< QL
QL in ppb M	0.16

Table 6. Measurement of [NH4+] in different locations

2.2.2 Chemicals

Quality of chemicals and Ultra pure water monitoring depends on the flow of the chemicals through the chemical supply, from the tank to the wafer. For Chemicals, the sampling can be done at the delivery of the products before the central chemical supply (in incoming inspection): the Point of Entry (POE). It can also be done on the process tool, at the point of Use (POU). Chemicals at the POE can be measured by ICPMS. At POU, bare wafers which are processed with a complete recipeare then measured by TXRF. At POU another approach is the sampling of chemicals at POU ICPMS analysis. Results at POE and POU measured by ICPMS are presented in Table 7.

Elements	Element	QL in ppt	Ammonia POE A Tank 1	Ammonia POE A Tank 1	H2O2 POE B	POU SC1 in tool bath	Spécification POE and POU
Sodium	Na	5	17	40	121	63	1000 ppt
Magnesium	Mg	5	12	6	24	NA	1000 ppt
Aluminium	Al	5	62	7	35	66	1000 ppt
Potassium	K	5	12	47	16	NA	1000 ppt
Calcium	Ca	5	41	56	113	93	1000 ppt
Chrome	Cr	5	< QL	< QL	6	< QL	1000 ppt
Manganése	Mn	5	< QL	< QL	< QL	NA	1000 ppt
Fer	Fe	5	7	9	59	53	1000 ppt
Nickel	Ni	5	13	10	< QL	< QL	1000 ppt
Cobalt	Co	5	< QL	< QL	< QL	NA	1000 ppt
Copper	Cu	5	< QL	8	< QL	< QL	1000 ppt
Zinc	Zn	5	< QL	16	17	< QL	1000 ppt
Argent	Ag	5	< QL	< QL	< QL	NA	1000 ppt
Plomb	Pb	5	< QL	< QL	8	NA	1000 ppt

NA: Not analysed / **ppt** : part per trillion, typically pg/g for metallic contamination.
Table 7. Metallic measurements on chemicals at POE and POU

The measurements have been done with an ICPMS by Balazs laboratory from Air Liquide Electronics Europe.

2.3 Sampling and confidence level on monitoring scheme

Monitoring of the semiconductor manufacturing line is done on the product wafers, or on the facilities as ultra pure water, chemicals or gases. Measurements on a product wafer can address impact of metallic contamination on gate oxide from hot, implant processes.

The question related to sampling is "why do we need to monitor defect?" In the case of metallic contamination, it is not such easy. Metallic effects are known, but the analytical tools have time response much slower than for the defect density tools. Then, the monitoring scheme of metallic contamination needs to be think according pragmatic approach. First the line is divided in two parts:

- Front End Of Line : Device construction
- Back End Of Line : Connection with metal line

TXRF, VPD TXRF and SPV measurement technique are used for standard monitoring, but also after maintenance procedure, or any troubleshooting. Decision tree and clear instruction are also needed in order to help manufacturing running the tool properly.
In addition the monitoring of the chemicals, Gas and DI Water before the POU is indicating the quality level of the facilities. This monitoring scheme is summarized in table 8.

Items	Monitoring	Analytical Tool	Frequency	Process Tool Facilities
FEOL	Standard	SPV TXRF VPD TXRF	Periodic according risk	All, Wet tool, Hot process…
BEOL	Troubleshooting	SPV TXRF VPD TXRF VPD ICP MS	Define within action plan	All, Wet tool, Hot process, Etch…
BEOL	Standard	SPV TXRF VPD TXRF	Periodic according risk	Wet process Cleaning tool
BEOL	Troubleshooting	SPV TXRF VPD TXRF VPD ICP MS	Define within action plan	Wet process Cleaning tool
Chemicals	Standard	ICP MS TXRF	Audit mode	Chemical supply
Chemicals	Troubleshooting	ICP MS VPD ICPMS TXRF VPD TXRF	Define within action plan	Chemical supply
AMC	Standard	IC	Periodic according risk	Clean Room
AMC	Troubleshooting	IC	Define within action plan	Clean Room

Table 8. Monitoring scheme of metallic contamination

3. Impact of metallic contamination through examples

3.1 Metallic in wet chemistry

On a cleaning tool working with continuous flow chemistry process (Sanogo 2008), vibrations have loosened a screw which was maintaining the Vessel as shown in Fig 2. This has been dissolved by the different chemistry of the cleaning process SC1, SC2, HF, before Gate oxide growth. Monitoring measurement with dark field inspection tool on product wafers has identified particles. EDX analysis on these particles has identified Fe and Ni compounds

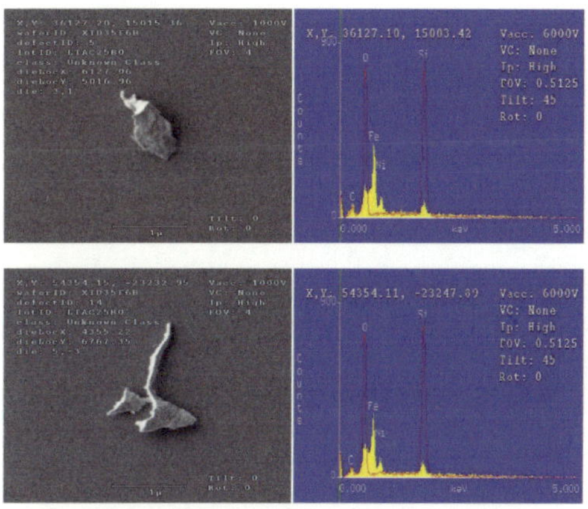

Particles localized with Dark Field inspection tool and EDX spectrum : Fe, Ni elements identified

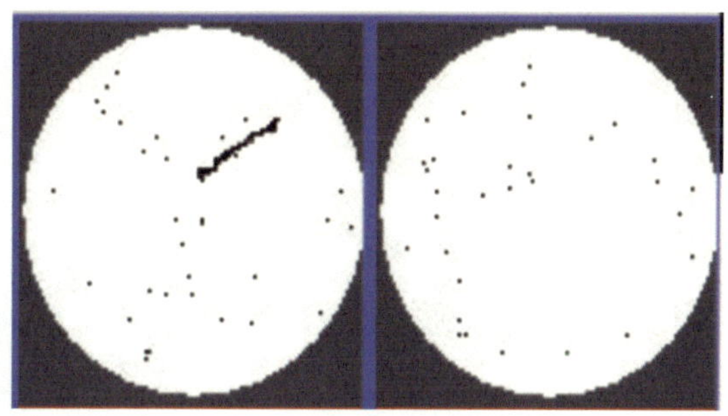

Particles Map measured with dark field inspection tool

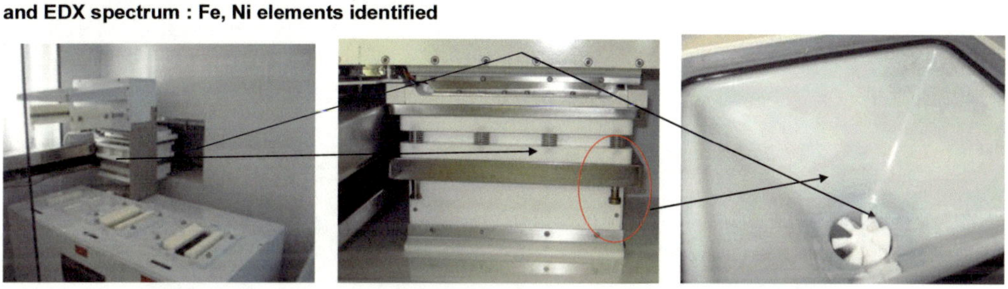

Fig. 2. Metallic contamination on Wet process tool, EDX identification

3.2 Metallic in Implant Process

For an Ionic Implant Tool, the plasma is generated within an Arc chamber in order to do the ionisation of the different species before going trough the mass spectrometer filter for implantation on the wafer. The wall of this Arc chamber can be made within two metals, either Molybdenum, either Tungsten. During the implantation of the BF2 species for the device channel implant, Mo++ has been implanted with BF2 implant (Demarest 2009). For information, AMU of BF2 is 49, and the isotopic value of Mo ++ around AMU 49 is AMU = 48,5 ==> $^{97}Mo^{++}$ =9,5% and AMU = 49 ==>^{98} Mo^{++}= 24,4%.

W wall material is double cost compared to Mo. The concentration of molybdenum within the bulk has been measured with SIMS technique. The quantity of molybdenum is increasing with higher current as it is needed for increasing implantation doses.

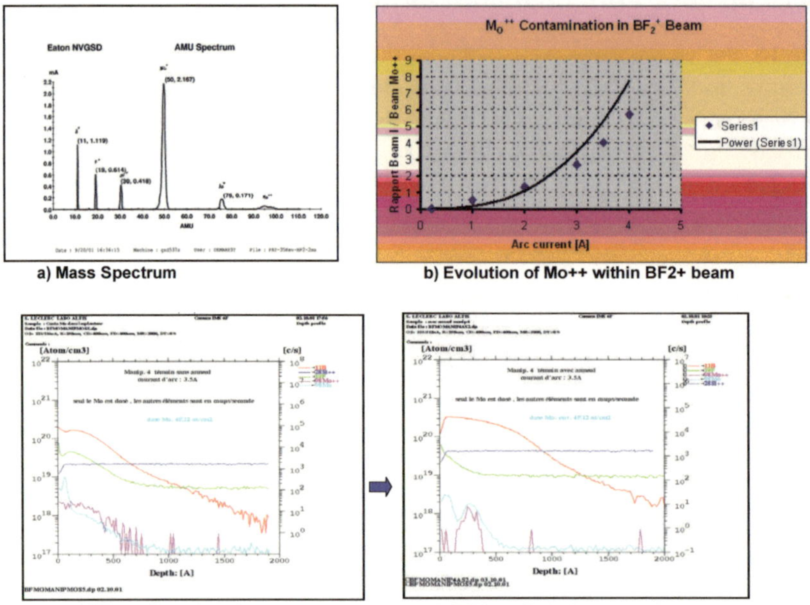

a) Mass Spectrum

b) Evolution of Mo++ within BF2+ beam

c) SIMS profile before Anneal

d) SIMS profile before Anneal

Fig. 3. Mo Contamination Within Wafer during BF2 implant

3.3 Furnace Contamination

The monitoring of Furnace oxidation process with SPV has been evaluated to catch Na contamination in case the handling procedure would not be followed. In Fig 4, the trace of the finger touching the wafer through gloves is detected (Garroux 2005)

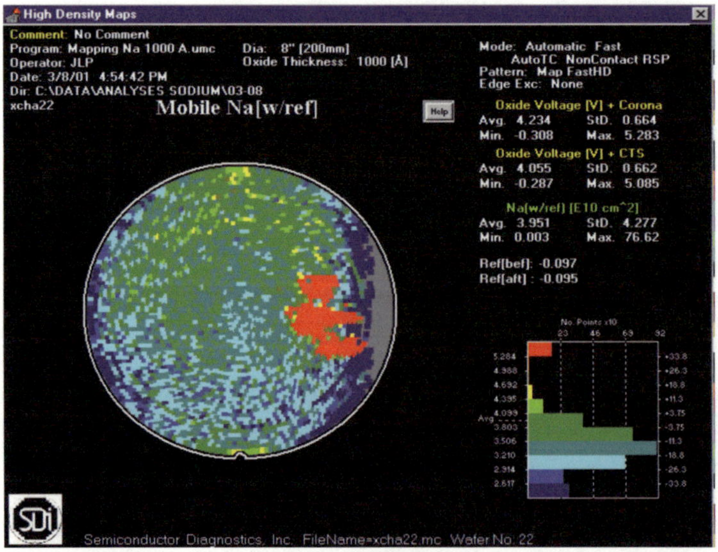

Fig. 4. SPV measurement on bare wafer post oxidation.

4. Defect density on product wafers

Defect density is one of the main detractors of the final test yield in semi-conductor manufacturing, and the impact of the particles on the device functionality is even more critical for sub-micron designs. It is the reason why the investment for defect density measurement increased for the last years: yield prediction through in-line defect inspection is requested to improve yield learning on new product and each node generation.
In this chapter, we will describe the latest tool set available in the manufacturing lines and the detection capabilities of the bright field, dark field and SEM (Scanning Electron Microscope) inspection tools and optical / SEM review tools. The sampling strategy for the defect review and the automatic defect binning are optimised to improve the classification of the defects of interest.
Defect classification accuracy and the defect size influence on chip functionality will be presented through the critical area definition and die to die yield calculation. The methodology for yield prediction through defect density inspection and classification will be described. The confidence level of yield prediction depends on the inspection tool capabilities and sampling strategy, the defect size and killer ratio calculation for each defect type.

4.1 Defect inspection

3 types of inspection tools on product wafers are proposed for defect density analysis:
- Bright field inspection tools: using standard light or UV light for sub micron design inspection. Sensitive to image differences, detect deformed designs as micro masking, embedded or surface foreign materials, scratches, mainly defects providing a good image contrast.
- Dark field inspection tool: using a laser, will detect easily surface defects. Tools covering both dark and bright field inspections mode are now available.
- Scanning Electron Microscope (SEM) inspection tool. This tool compares SEM images to detect small defects (0.1 µm), charge contrast defects (as contact open, line shorts, device leakages), or defects in high aspect ratio structures (Baltzinger et al., 2004; Hong Xiao et al, 2009).

These tools will compare images from one die to an adjacent one. If any difference is detected, the tool will check the image with another die. The die different compared to the other will be considered as defective. For memory products, the sensitivity and throughput of the tool can be improved by comparison of memory blocks inside of the dies. The inspection tools provide defect coordinates on a wafer map. Some tools are able to classify the defects to facilitate the defect density analysis. The sampling for defect analysis review can be more efficient by removing non killer defects, nuisances, detected and classified by the inspection tool.
The choice of one of these tools is driven by the in-line inspection strategy. This strategy is built with the following information:

- Pareto of the defects to be detected
- Information of the final test analysis and failure analysis.
- Manufacturability of the in-line controls (scan time, resources for classification)

Today's recipes are generally 100 % surface scan of the chip to inspect exhaustively all of the active structures of the product. It allows the detection of all type of defects on the different structures of the chip, but sensitivity of the inspection tools in the array is reduced with random mode inspection. Defect size distribution depends on image filtering, detection threshold, pixel (smallest image size for die comparison) chosen in the recipe. These parameters are adjusted to keep a count of defects affordable for manufacturing inspection. So, the recipe will be built to avoid encroaching and saturation concerns. Focus parameter will be adjusted to catch surface or embedded defects. An example of a defect size distribution is given in the Fig. 5.

General law for the defect distribution is: $D=A/X^n$ (1)

With:

- X is the defect size
- D is the particle count
- A and n are constants (n used to be closed to the value of 3)

A log/log graph will give a straight line where the slope is n.

After wafer inspection, defect map and chip yield is provided. The chip yield or defect count depends on the sensitivity of the recipe. In the case of the previous graph most of defects under 0.3 µm are not detected by the KLA 2135 using the pixel 0.39 µm (random mode). Using the pixel of 0.25 µm allows the detection of defect size of 0.18 µm, but will increase the total of the detected defects on the wafer.

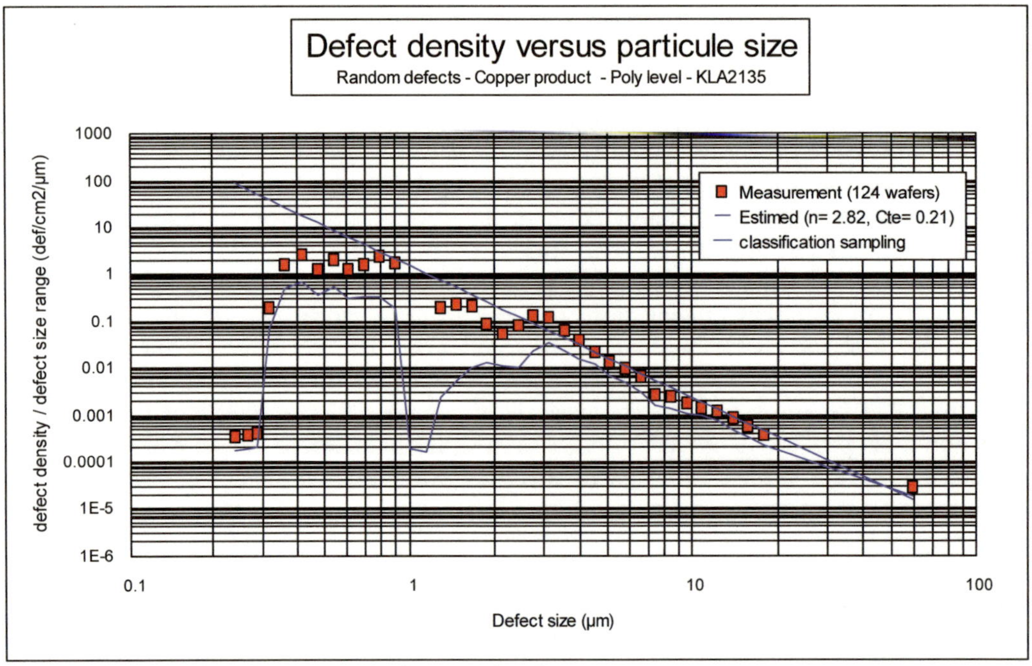

Fig. 5. defect size distribution on 0.18 µm technology

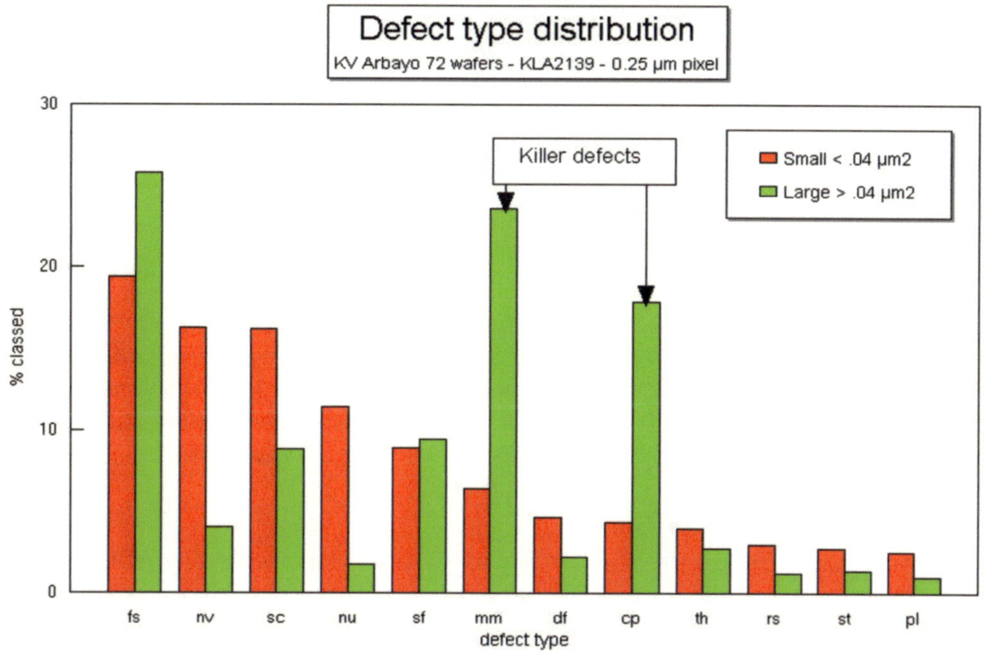

Fig. 6. Ratio of classified defects, comparison with 2 inspection recipes: the killer defects (MM, micro masking and CP, silicon pitting) are under sampled using 0.25 µm inspection recipe (red bars), which detect small defects (under 0.04 μm^2), compared to the 0.39 µm recipe (green bars).

The aim of the defect inspection is to detect most of the killer defects. The recipes using smaller pixel size will grow the total defect count mainly with small defects (Fig. 6), which are not the main detractors at final test yield. This will enlarge the width of the defect size distribution and could cause the lost of the defects of interest review. It is the reason why inspection tools are providing today a previous rough binning to improve the efficiency of the defect classification sampling and the review.

4.2 Defect review

Defect review is required to identify defects of interest and to address the root cause of each defect type. The defect review is processed using 2 types of tools:

- Optical review allows the classification of large defects (more than 1 µm). The benefit of optical review is to get pictures of embedded defects. Confocal microscopes provide topological information.
- SEM review allows the classification of smaller defects, but embedded defects will not be systematically redetected, because SEM is sensitive to the surface only. Last generation of review SEM is able to redetect, focus and take automatically a picture of the defects, to improve the throughput of the review. EDS (Energy Dispersive Spectroscopy) can be added to have elemental analysis of particles.

4.3 Defect sampling strategy for defect classification

All the detected defects are not reviewed on optical or SEM tools because the amount of the defect is generally too high for the review tool capacity. So a sampling is applied on the total inspected defects, with the following possible methodologies:

- remove previous inspected layers defects to classify only current level defects
- take only 2 or 3 images of large defects (clusters)
- classify randomly failing dies
- classify largest defects to improve sampling of killer defects
- classify a sampling of proposed the defects binned by the inspection tool

Sampled defects will be automatically classified (ADC: Automatic Defect Classification proposed for SEM or Optical review tools) or manually classified with an operator. The SEM review is more accurate due to its better resolution, but is not able to detect some embedded defects. The measurement of the efficiency of the defect classification (ADC for this example) is given by the following 2 parameters (Chen-Ting Lin et al., 2001):

¶

$$\text{Accuracy} = \text{Total correctly classified by ADC/ Total classified by the expert} \quad (2)$$

¶

$$\text{Purity} = \text{Total correctly classified by ADC / Total classified by ADC} \quad (3)$$

Accuracy gives the capability of the classifier to detect a given defect type. Purity gives a measurement of the "noise" of the classification. ADC classification goal is to obtain in general more than 80 % for accuracy and purity. A trained operator achieves more than 90 %. Defined defects classes provided to ADC or an operator has to be consistent with:

- Process root causes of the defect
- Size and possible impact of the defect at final test
- Defect should be easily recognizable by ADC or an operator to get good level of accuracy and purity

4.4 Final test yield prediction from in-line defect inspection data

PLY (Photo Limited Yield) calculation from in-line wafer inspection and defect classification will provide an estimated final test yield of a wafer. The PLY calculation for the defect j for one inspected level is the following (semi-deterministic model):

$$PLY_j = 100*(1 - P_j * C_j * DC / NTC) \quad (4)$$

Where :

P_j : probability of fail of the defect j
C_j : chips classified with the defect j
DC : total defective dies
NTC : total dies on the wafer

PLY of one inspection level is the product of PLY_j of the j defects classified on the wafer. The overall estimated yield is the product of PLY of the inspected levels. The aim of the following part is to discuss about the reliability, the accuracy and the precision of the PLY data. When PLY trend degradation is observed, we need to know the accuracy of the measure and the assumptions taking in account in the calculation to be sure that what is measured is a real process concern.

Probability of fail calculation

When a defect is classified, a probability of fail is associated, depending on the impact of this defect on the chip functionality. Different methods are used for the calculation of this killer ratio. The most frequently used is STPLY (Statistical Test PLY). It is a chip to chip correlation, between failing chips seen with the in line inspection tools and the final test yield (Grolier, 2000). For a given defect type i, the calculated killer factor is:

$$\text{Killer factor} = \text{final test failing chip with the defect i} / \text{total defect i found} \quad (5)$$

Some error on the calculation can be done, because some killer defects not detected with in line inspection tools can match with detected defects without any electrical impact. Some "noise" subtraction is proposed. STPLY allows the probability of fail calculation of all types of semiconductors, memories and logics.

Another method called ETPLY (Electronic Test PLY) is to overlay the PLY defects map and the bit fail map given by the final test of the memory products. This method of killer factor calculation allows a better accuracy than the previous method because there is very few of "random hits", even with an overlay specification of 100 µm. Nevertheless, this method is only applicable for memories (Fig. 7).

The manual classification does not report accurately the size and the impact of the defect on the design. Some classification like small embedded, embedded and large embedded are dependant on the operator; large embedded with a killer factor of 1 is a given size defect or a defect connecting 2 structures. Some defect codes have a killer factor of 0, as Nuisance, Non visible, Discoloration, Fill Shape (Defect in non electrically active area). to give the most accurate predicted yield. The inspection recipes have to be optimised to reduce the amount of such defects. Nevertheless, the size of the defect has a strong impact on the killer ratio (Fig. 8), and the interaction between defect size and product design has been studied to estimate the impact of the defect density on final test yield.

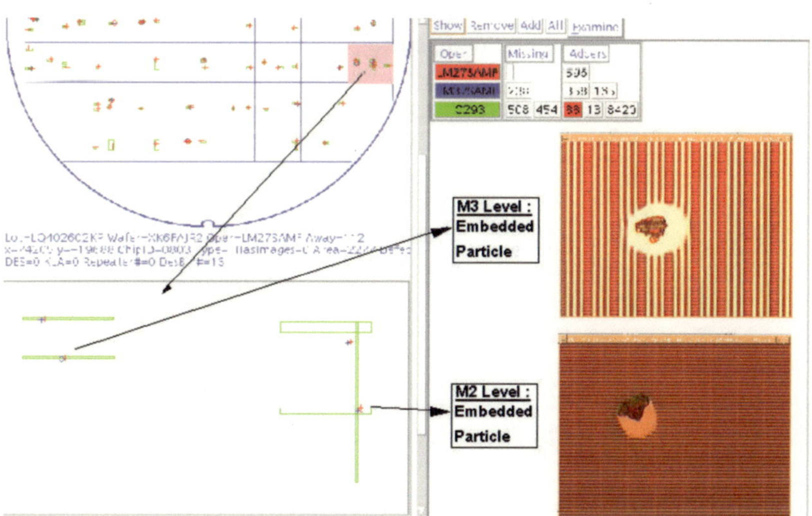

Fig. 7. Bit fail map correlation (electrical fails are green rectangles) with physical defect detected on KLA 2135 post copper CMP Metal 2 and Metal 3

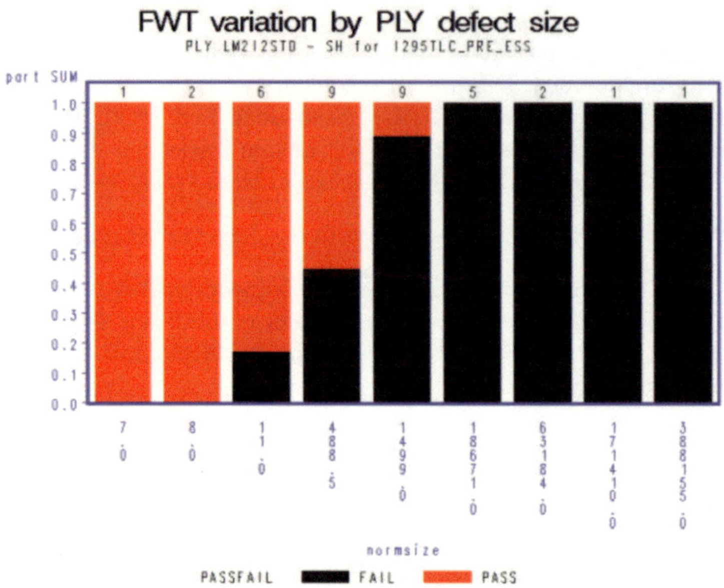

Fig. 8. killer ratio calculation by size (normsize , in µm²) given by STPLY method.

Defect impact on the product: critical area definition

For a given defect density yield models are able to propose a corresponding yield calculation as binomial, Poisson laws (Fig. 9). Nevertheless, these laws are not taking in account the product complexity and device redundancies (Donovan, R. P., 1988). Some corrective factor can be added to improve the predicted yield, but the more precise estimation can be given by software including the design descriptions for all the layers of the product and the modelization of the defect density.

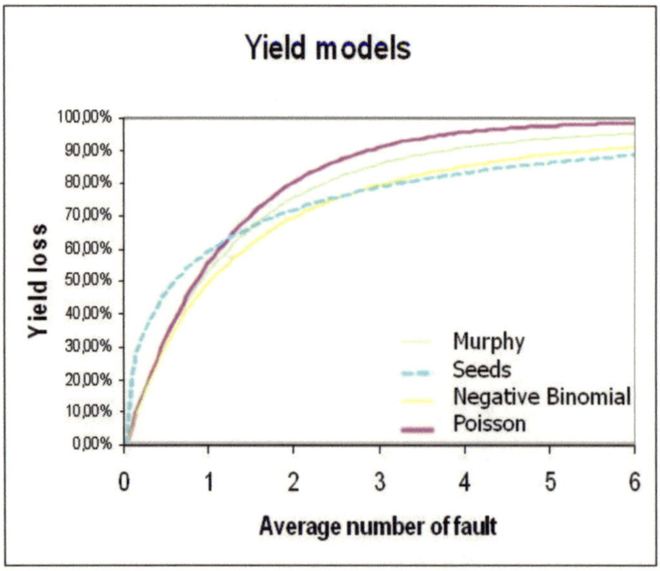

Fig. 9. Yield calculation from different models

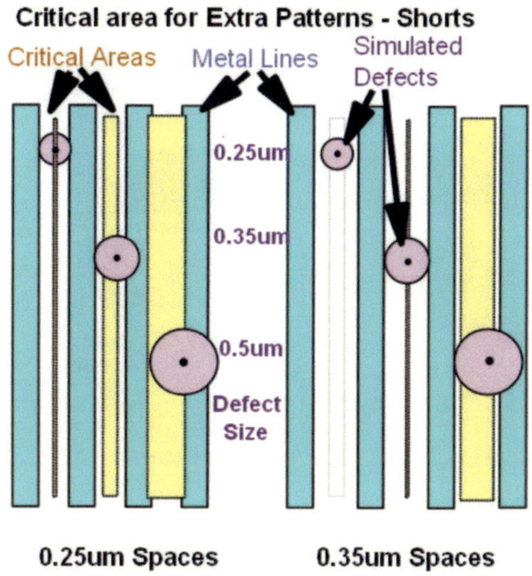

Fig. 10. Blue bars are metal lines. Critical area is the yellow surface. If the centre of a circle particle is inside the yellow surface, the particle will cause a fail (metal short).

This final test yield estimation is based on the critical area calculation (Fig.10). The critical area is the surface where the centre of a particle will cause a failure (Barberan & Duvivier , 1996). The critical area depends on the particle size, the product design and the impact of the particle on the design. As all these information are available, yield estimation can be calculated (Allan & Walton , 1996) with the following law (Fig. 11):

$$Y_{event} = \exp\left(-\int_{x_0}^{\infty} CA_{event,i}(x) DSD_i(x) dx \right)$$

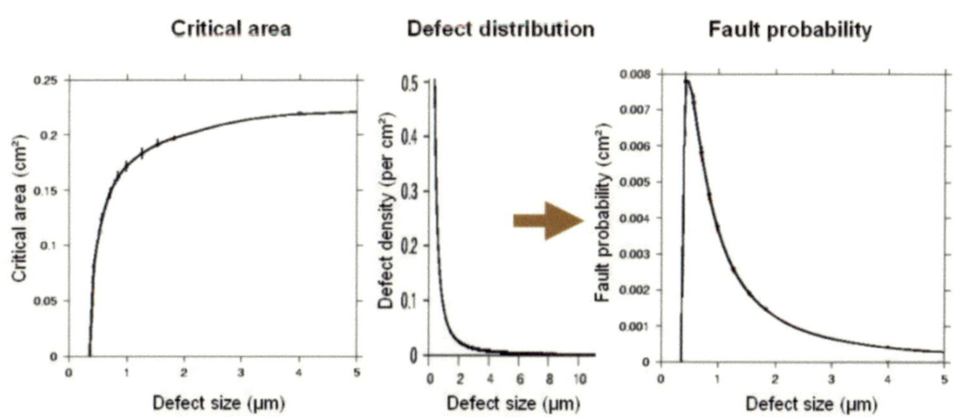

Fig. 11. Defect density $DSD_i(x)$ and Critical area $CA_{event,i}(x)$ of a given size x defect will provide a yield loss $Y_{event,i}$ corresponding to the surface under the fault probability curve.

Most critical part of the product design or layers can be highlighted and corrected to improve final test yield (Fig. 12). Redundancies as contacts can also be added.

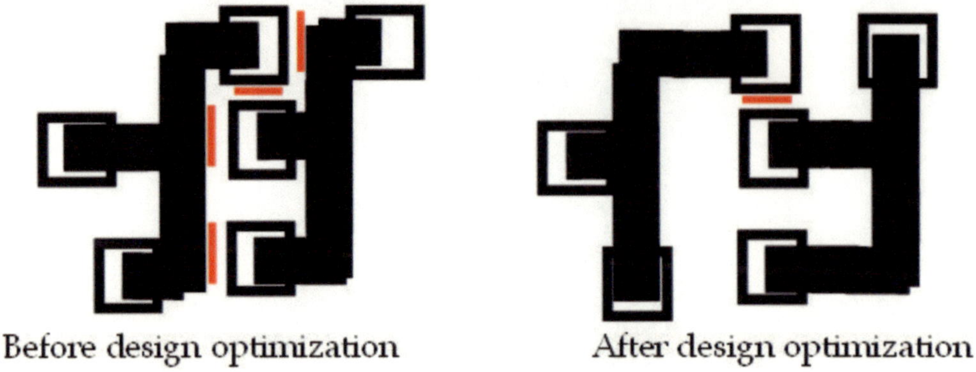

Fig. 12. Critical area reduction with design optimisation

Wafer / lot / defect sampling

The PLY result depends on the sampling strategy. The more defects are classified; the better will be the confidence level on PLY data. This can be modelized with a binomial law (see Fig. 13) as far as we suppose that a defect frequency follows a Gaussian distribution (6):

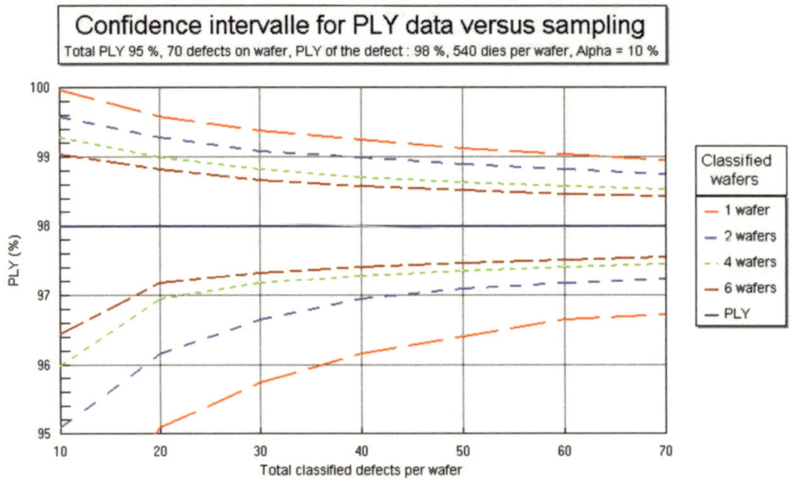

Fig. 13. confidence level of PLY data depending on the defect sampling

In this case, the cumulated wafers data are supposed to have the same defect distribution. This is consistent with the simulation of different defect sampling proposed in 1997 by J-L. Grolier and J. Combronde.

Actual sampling is 25 % of the production lot, 2 wafers per lot, and 50 classified defects maximum per wafers, according to the previous study. At this time, some tool are proposed to define the best sampling depending on amount of defect type, the stability of the process and the required confidence level.

To improve the sampling efficiency and PLY results accuracy, the recipes have to be optimised to reduce the amount of prior level defects, nuisances and non visible defects (false defects ...). Inspection level detection is chosen to detect killer defects. Generally post

STI (silicon trench Isolation) module, PC (poly gate etch), Contact, metal layers are the most common level used for inspection.

Predicted Final Test Yield

At the end, overall PLY calculated with the final test date for each lot will give a prediction of the final test yield induced by the defect density (PLY = multiplication of all the defects yields for all the levels of inspection). The following graph shows the overall PLY calculated at final test and the final test results week by week (Fig. 10). Some errors induced by the overall PLY calculation can occur when lots are not crossing the process flow at the same time. Process issues (CD variations, resistive vias ...) will be estimated with another calculation to give a better final test yield prediction.

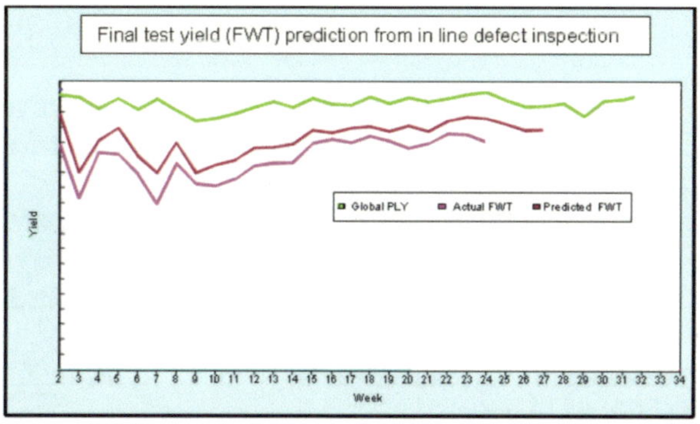

Fig. 14. PLY to Final test yield correlation week by week. PLY was able to detect the down trend weeks 3 and 7, and the yield improvement starting week 10.

Conclusion: Overall PLY accuracy

To get an overall PLY accuracy estimation, killer factors calculated each month with SPLY method can be reported on a graph. A sigma can be estimated for each defect type, as a critical process parameter of the line. In general, the higher killer factors have a lower standard deviation for a given probability of fail calculation (Baltzinger, 2009). For the low killer factor defects, the "noise" impact on the calculation is higher. In this case, the defect density engineer has to understand the root cause of this high variability to improve the level of confidence:

- Inspection recipes
- Sampling and classification accuracy
- Process changes

5. References

Leroy B., Contamination, Internal communication, 1999

Pic N. and Martin C, Review of in-line and off-line analytical techniques used to monitor airborne and wafer surface contamination, ARCSIS Conference, Nov 2006

C Galvez, MV Deydier, Contaminants analyses in semiconductor: the expertise of a chemist, ARCSIS Conference, Nov 2006

Sanogo M., Metallic Contamination on continous flow chemistry process, Internal communication, Altissemiconductor, 2008

Demarest P., Molybdenum contamination during BF2 implantation process, Internal communication, Altissemiconductor, 2009

Garroux D., Monitoring of Alkaline contamination on furnace oxidation process, Internal communication, Altissemiconductor, 2005

Allan G.A. and Walton A.J., Yield prediction by sampling with the EYES tool, *IEEE International Symposium on defect and fault tolerance in VLSI systems*, Boston, 1996

Allan G.A. and Walton A.J., Yield prediction for ULSI, *VLSI multilevel metal interconnection conference,* pages 207-212, Santa Clara, California, June 1996

A Normograph of the cumulative binomial distribution. *Industrial, Quality Control 23* (1966/67), p. 270

Baltzinger J-L., Desmercière S., Lasserre S, Champonnois P., Mercier M., E-beam inspection of dislocations: product monitoring and process change validation, *Proceedings of IEEE/SEMI advanced semiconductor manufacturing conference,* page 361, Boston, may 2004.

Barberan S. And Duvivier F., Management of critical area and defective data for yield trend modeling. *IEEE International Symposium on defect and fault tolerance in VLSI systems,* pages 17-25, Austin, Texas, (November 1998).

Chen-Ting Lin and al. Defect reduction in a high-volume fab. *Semiconductor International* (July 2001)

Donovan R. P., Particle control for semiconductor manufacturing, *The Semiconductor Research Corporation*, SCR technical report T88105, November 1988.

Ferris-Prabhu A.V., Modeling the critical area in yield forecasts, *IEEE journal of solid-state circuits,* vol. SC-20, #4, August 1985.

Grolier JL, Combronde J. Sampling strategy / cost of manual classification. *Internal publication* (1997).

Grolier JL. SPLY and probability of fail calculation *Altis Technology day* (June 2000).

Michel Tenenhaus. La régression PLS. *Edition Technip*, Paris 1998

Simca-P for windows graphical soft of multivariate process modelling. *Umetri AB* 1996, Box 7960, S90719 Umea, Sweden.

Xiao, H; Ma L; Wang F, Zhao Y, Jau J. (2009) Study of devices leakage of 45 nm node with different SRAM layout using an advanced e-beam inspection systems, *Proceedings of SPIE,* San José, February 2009, 7272-55.

J.-L. Baltzinger, F. Rodier, J-L. Grolier, J. Lafarge, F. Poli "Final test yield estimation from critical area calculation", 12th Technical and Scientific Meeting of ARCSIS, September 2009, ST university Fuveau, France.

5

Advanced Plasma Processing: Etching, Deposition, and Wafer Bonding Techniques for Semiconductor Applications

Michael Shearn, Xiankai Sun, M. David Henry, Amnon Yariv, and Axel Scherer
Thomas J. Watson, Sr., Laboratory of Applied Physics
California Institute of Technology
Pasadena, California 91125
USA

1. Introduction

Plasma processing techniques are one of the cornerstones of modern semiconductor fabrication. Low pressure plasmas in particular can achieve high radical density, high selectivity, and anisotropic etch profiles at low temperatures and mild voltages. This gentle processing environment prevents unwanted diffusion and degradation of materials due to heat and lattice damage from ion bombardment. Plasma treatments have a minimal effect on existing wafer structure, which is a key requirement for large scale integration schemes such as CMOS. In addition, recent progress in plasma-assisted wafer bonding has demonstrated low temperature, low pressure recipes utilizing O_2 plasma surface treatment for joining dissimilar semiconductor materials, such as silicon (Si) and indium phosphide (InP) (Fang et al., 2006).

In this chapter, we will cover the applications of plasmas to etching and depositing materials, as well as novel processing modalities such as surface treatments in preparation for wafer bonding. All these processes rely on the inductively coupled plasma reactive ion etcher (ICP-RIE) used in the integrated electronics industry, which we will explain in detail. The unique chemical environment of ICP-RIE generated plasmas gives process engineers new capabilities that are not found in other techniques that are compatible with existing architecture requirements.

After an overview of the principles of the ICP-RIE, we will describe our work in novel mask materials and processing conditions in plasma etching and deposition. High aspect ratio nanopillars have recently been fabricated using this technique, with features as small as 22 nm etched over 1.25 μm deep. In particular, our use of Al_2O_3 as a mask material along with cryogenic wafer temperatures has demonstrated to increase the etch selectivity of silicon over mask to more than 5000:1, enabling ultrathin masks for nanoscale pattern transfer (Henry et al., 2009a). After patterning, in situ deposition can encapsulate these structures in preparation for further processing.

We will also discuss our work in hybrid integration of Si/III–V materials for on-chip optics applications using plasma etching and bonding techniques. Plasma-assisted wafer bonding has realized hybrid Si/III–V structures, using the III–V material as an efficient gain medium while maintaining the economic and electronic integration benefit of a silicon platform (Sun et al., 2009). In addition, smooth etches have realized low loss waveguides in silicon. Finally, unique planar microcoils have been made using the deep etching and in situ deposition capabilities in the ICP-RIE.

2. Plasmas in Electronics Processing

Plasmas are found in a wide range of industrial applications, including ashing, sputtering, etching, and chemical vapor deposition. However, the types of plasmas used in each process vary greatly. In this section, we will describe the important parameters of plasmas for understanding ICP-RIE etching and deposition, and establish how they relate to our desired processing results.

2.1 Figures of Merit

Central to semiconductor processing is the high fidelity transfer of a pattern onto a substrate through addition, modification, or removal of material. In order to quantify the ability to accomplish this selective processing, it is useful to have a few figures of merit to describe the process, namely:

- *Etch rate* – controllable and robust to small deviation in processing conditions. Depending on the application, one may want a higher etch rate for increased throughput or a lower etch rate for precision
- *Uniformity* – both at each feature and across the wafer
- *Selectivity* – the ability to etch only the desired material, relative to the etching of mask and other substrate materials
- *Anisotropy* – the verticality of the etch profile. Also, the nanostructure of this vertical surface is important in many applications, particularly waveguides
- *Damage* – any surface or substrate damage acquired from the processing technique.

In addition, the processing environment is important, as some conditions will have a deleterious effect on existing wafer structure. Chief among these is the temperature of processing. Heating (and cooling) can cause many problems, including thin film delamination due to thermal expansion coefficient mismatch, unwanted dopant diffusion, and other negative effects (Schmidt, 1998). In the CMOS industry, the wafer's tolerance to temperature fluctuations is often captured as a thermal budget (Takeuchi et al., 2005), which means that minimizing use during one process step can give more latitude in other steps.

2.2 Plasma Characteristics

Plasma is a partially ionized gas with a combination of free electrons, ions, radicals, and neutral species. To create and sustain a plasma in the laboratory, energy input is required. Generally, this energy is transferred via coupling of an external electromagnetic field to the plasma constituents. Different coupling methods generate plasmas with different characteristics. Most useful to reactive ion etching (RIE) are those generated by glow discharge plasmas (GDP), capacitive coupled plasmas (CCP), inductively coupled plasmas (ICP), or some combination thereof.

In a GDP process, electromagnetic energy is delivered as a voltage applied between two conducting plates, known as the cathode and the anode. The applied voltage is usually constant or in the low frequency regime, such that the characteristic time of field variation is longer than the response time of the system. The potential difference generates an electric field across the gases in the chamber. Plasma initiation occurs when a small initial population of charged species is accelerated through the electric field and collides with other molecules, causing them to ionize. A relatively high voltage is required to initiate and sustain the plasma, which is a severe processing drawback. High voltage will cause the resultant energy of incident ions on the cathode to be high, favoring rough, physical processes (sputtering) over smooth, chemical processes (surface reactions). This will lower the selectivity to masking materials and cause sidewall roughening due to mask erosion. For these reasons, GDP sources are often used to sputter materials rather than etch anisotropically (Chapman, 1980).

In a CCP process, energy is again supplied as a voltage between an anode and a cathode plate, but in a time-varying fashion. Most commonly, a radio frequency (RF) voltage is applied to the plates. The frequency of operation is often at 13.56 MHz, which is a band reserved for industrial use by the Federal Communications Commission in the United States. In this time-varying field, electrons in the plasma oscillate between the anode and the cathode plates. Collisions of rapidly moving electrons with the slowly moving ions cause further ionizations. However, massive ions are less mobile and cannot track the rapidly oscillating electric field changes. By placing a capacitor between the anode plate and the RF supply, negative charge accumulates on the plate (typically referred to as the table). The resulting potential difference between the plasma and the negatively charged plate is called the self-bias V_b. The electric field due to V_b drives the positive ions in plasma towards the negatively charged table. This is the basis for traditional reactive ion etching.

In an ICP process, the excitation is again a time-varying RF source, but is delivered inductively, instead of capacitively, resulting in a changing magnetic field. This changing magnetic field, through the Maxwell-Faraday equation, induces an electric field that tends to circulate the plasma in the plane parallel to the CCP plates. Similarly to a CCP, collisions of the rapidly moving electrons with the slowly moving ions cause further ionizations. Loss of electrons from the plasma through the grounded chamber walls tends to create a static voltage, deemed the plasma voltage V_{plasma}. This is distinct from the self bias V_b, as will be examined later. Inductive coupling is generally realized through a large 4 to 5 turn coil encircling the plasma chamber. In the typical geometry, this means that one is able to change ion density and other plasma parameters without significantly perturbing the incident energy of the ions.

The experimental results discussed in this chapter are realized on Oxford Systems Plasma Lab 100 ICP-RIE 380 systems, which utilize a CCP and an ICP power source, as seen in Fig. 1. Throughout the text, we will frequently refer to the CCP power as the "forward power" in order to distinguish it from the ICP power and to emphasize its role in driving ions toward the sample surface. This dual plasma powering affords the greatest flexibility in altering plasma characteristics such as ion density and bias voltage independently of each other. These systems have been extensively studied, particularly for silicon etching (Jansen et al., 2009).

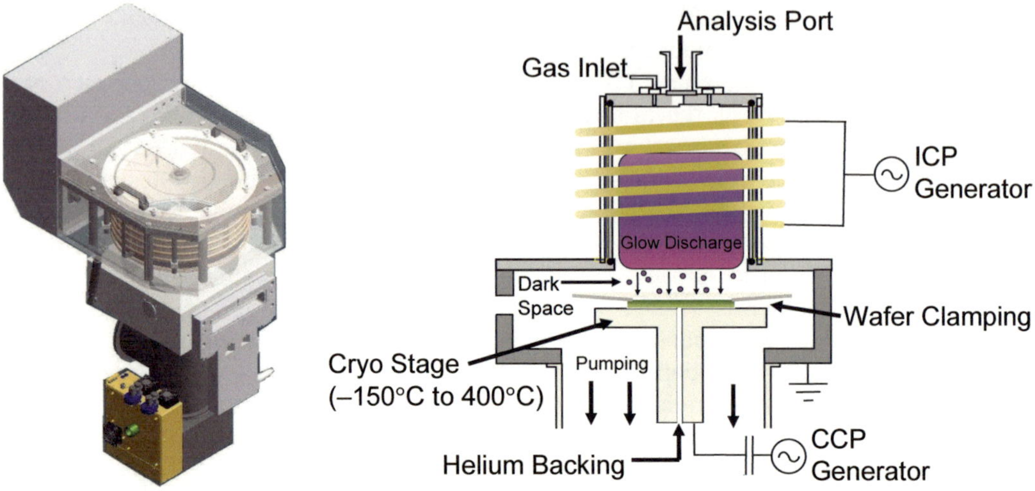

Fig. 1. Isometric (left) and cross-sectional view (right) of an Oxford Instruments ICP-RIE

2.3 Processing Parameters

There are a few important features of an ICP-RIE plasma that have an effect on etching. Most noticeable during operation is the region of glow discharge, where visible light emission occurs from a cloud of energetic ions and electrons. As the gas particles move in the plasma, some collisions occur which transfer energy to bound electrons. When these electrons return to their ground state, a photon may be emitted. The color of the plasma is characteristic of the excited gas species, because the photon energy is a function of the electronic structure of the gas molecules and their interactions with surrounding molecules (Hodoroaba et al., 2000). This can be a good diagnostic for incorrect plasma striking conditions or other adverse changes in your plasma. For example, in a multiple gas recipe, sometimes the emission looks like only one of your gas species, instead of the average of the colors. This happens when the other species are not being ionized, and thus will cause the process to take on a completely different character from a calibrated recipe.

Beneath the glow discharge region is a dark space, where atoms are no longer excited into emitting photons due to the depletion of electrons. This dark space is also the part of the plasma that most directly affects the paths of incoming ions that will accomplish the etching. Neutral atoms and other ions will tend to scatter the otherwise straight path of the ions from the edge of the glow discharge to the cathode.

We can characterize this spreading in both energy and trajectory into probability distributions, called the ion angular distribution function (IADF) and the ion energy distribution function (IEDF) (Jansen et al., 2009). These distributions, depicted in Fig. 2, describe the likelihood that an incident ion has a given energy and trajectory. IADF strongly affects the sidewall profile, as a wider IADF corresponds to a higher flux of ions reaching the sidewall. Similarly, the IEDF controls the types of processes the ions can be engaged in when they reach the surface, including removing passivating species, overcoming activation energies for chemical reactions, and enhancing sputtering yield. These processes determine the performance characteristics of an etch, so understanding these effects and recognizing associated faults are paramount to optimizing a recipe. Parameters controlling the IADF and IEDF include the bias voltage V_b, the ion density, the gas composition, and the mean free path (which also depends on the aforementioned parameters).

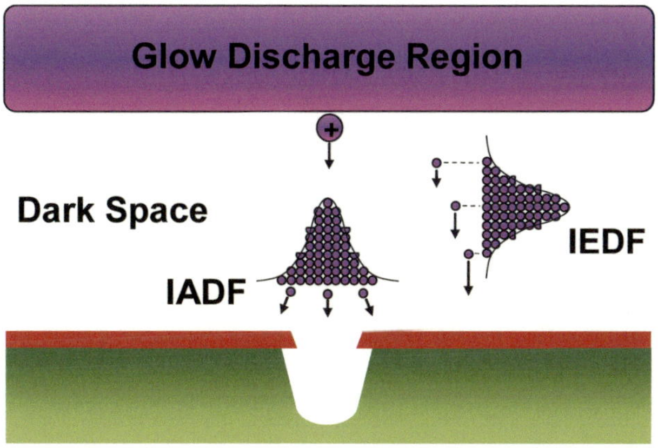

Fig. 2. Illustration of the ion angular and ion energy distribution functions, with hypothetical resultant etched profile distortion. Points in IEDF correspond to different ion kinetic energies, while points in the IADF correspond to different angles of incidence

2.4 Etch Reaction Dynamics

In wet chemical processes, etching is accomplished through physical dissolution or reaction-specific dissolution (Reinhardt & Kern, 2008). This takes place at any exposed surface and thus results in isotropic etching, although the etch rate can vary along different crystalline orientations due to the bonding state variation of the surfaces. A good example of crystalline anisotropy in Si wet processing is potassium hydroxide (KOH) etching, which is widely used for making MEMS structures that capitalize on the direction-dependent etch rate of KOH (Wolf & Tauber, 2000). However, in a myriad of planar processes that are utilized in the semiconductor industry, an anisotropic etching profile with sidewalls perpendicular to the wafer surface is frequently required for effective pattern transfer.

In order to prevent the isotropic or crystalline anisotropic behavior of our processing gases, the sidewalls must be protected from further etching. This is accomplished by forming a passivating or inhibiting layer on the sidewall, in one of the following ways:

- Surface passivation
  - inserting gases in the plasma which react with wafer materials and forming involatile compounds (Legtenberg et al., 1995)
  - freezing volatile reaction products at the structure's walls using, e.g., cryogenic wafer cooling (Aachboun et al., 2000)
- Inhibitor deposition
  - using polymer precursor gases to form physical barrier layers (e.g., C_4F_8) (Kenoyer et al., 2003)
  - eroding and redepositing inert mask materials

All of these processes are important to consider when evaluating an etch, as there may be problems with the etch profile related to the deleterious effect of one of these regimes. We use both surface passivation and inhibitor deposition techniques in the following etch descriptions.

2.5 Time-Dependent Processes

In addition to the previously discussed processing parameters, we have one additional variable at our disposal: time. A notable example of using time as an etching parameter is the Bosch silicon etch process, which occurs in a time-multiplexed manner, or "pulsed mode," using an etching plasma followed immediately by a deposition plasma. Alternatively, we can try to accomplish the etching and deposition simultaneously by using a plasma that contains both etching and deposition gases. This is called a "mixed mode" process. Finally, we can also tune our processes to change continuously over time in response to the changing surface condition of our wafer, or to compensate for a negative effect due to the initial conditions of the wafer.

2.6 Conclusion

As we have seen in this section, plasma processes depend on a large number of variables, which accounts for both their sensitivity and their flexibility. By having basic knowledge of the underlying physical processes, diagnosing your processes becomes more intuitive and makes recipe invention and refinement much easier. In the following sections, we will refer to many of the concepts covered here to explain results and understand how we arrived at a given recipe. However, there is still no replacement for hands-on experimentation for building an even greater understanding of ICP-RIE processing.

3. Deep Silicon Etching

Silicon is the workhorse of the semiconductor industry, and thus etching of Si is one of the most frequent processes used in a fab. In order to achieve deep etches in silicon using an ICP-RIE, three basic etch requirements must be met. First, the etch must have a relatively high etch rate. A slow etch rate is cost prohibitive in a high throughput, industrial process and has the potential for the introduction of process variations, leading to etch failures. Second requirement is that the etch must have a high selectivity, or preference, to etch the silicon as compared to the etch mask. Insufficient selectivity limits the maximum etch depth or requires complicated thick masks to compensate for erosion, limiting the minimum feature size. Finally, the etch must remain anisotropic throughout the etching process. If lateral etching occurs, pattern transfer begins to fail as the etching continues vertically.

To date, only two etching modalities have the potential to stand up to these rigorous requirements: pulsed mode and mixed mode silicon etches. Both etch schemes employ forms of etching combined with passivation that actively protect sidewalls during etching and improve anisotropy. Each has their own advantages and disadvantages which will become clear during the discussion. To illustrate the differences between the two modes of etches, two widely used etches will be discussed here. For the pulsed mode etch we describe the chopping Bosch silicon etch, which uses gas "chopping" to alternately etch and deposit inhibitor on your surface, and for the mixed mode etch, we demonstrate the cryogenic silicon etch, which uses a different gas chemistry to form passivating compounds at the sidewalls at the same time as etching. Note that both gas chemistries reviewed here can be used in either pulsed or mixed mode.

As mentioned, the chopping Bosch etch requires two alternating plasma steps. The first step etches the silicon for a short period then rapidly shuts off the gas and plasma. The second step then initiates a plasma that deposits an inhibitor film on exposed surfaces. This alternating sequence continues as the etch progresses. Inherent in the discreteness of the etching is notching on the sidewalls that occurs every step. The duty cycle between steps controls the etch angle and the total length of the combined steps controls the depth of the notching.

In contrast, cryogenic silicon etching combines the discrete etch and passivation steps into a single continuous etch. By using cryogenic temperatures from –80 °C to –140 °C, improvements in etch mask selectivity and passivation effects are enabled. Both of these etching chemistries, mask selections, and characteristics will be reviewed here along with their applications or demonstrations.

3.1 Gas Chemistries

Chopping Bosch etching utilizes sulfur hexafluoride, SF_6, as the etching gas and octafluorocyclobutane, C_4F_8, as the passivation gas. As described earlier, when the SF_6 is injected into the chamber, the plasma ionizes and radicalizes the gas molecules to create a mixture of SF_x and F_y ions and neutrals, where x and y range from 0 to 6 and 1 to 2, respectively (Cliteur et al., 1999). The potential established between the plasma and the substrate, due in part to the ICP and the CCP power, causes the electric field that drives the ions down to the substrate. The unmasked silicon then bonds to the fluorine atoms to create the volatile tetrafluorosilane (SiF_4) etch product which is then pumped away from the chamber. The etch becomes a combination of chemical bonding and mechanical milling; the milling is established from the momentum imparted to the ions from the electric field. While the chemical etching is essentially isotropic in nature, the mechanical milling is anisotropic. After a few seconds of etch time, the SF_6 flow is rapidly terminated and the C_4F_8 gas is then injected into the chamber for the passivation step. During this step, the C_4F_8 fragments into smaller CF_x ions which act as film precursors (Takahashi et al., 2000). A Teflon-like film forms on the substrate, on both the vertical and horizontal surfaces. The thickness of the protective layer is dependent on the passivation step time. Once the deposition is complete and the subsequent etch step begins, the ions first mill away the horizontal passivation layers and then begin again with the silicon etching. This cyclic process of etching followed by passivating continues on until the etching is terminated, leaving the etched silicon structures coated with the passivation polymer.

The cryogenic silicon etch also utilizes the SF_6 chemistry similar to that of the chopping Bosch. However, by lowering the substrate's temperature, and by simultaneously injecting SF_6 and oxygen gas, O_2, a passivation layer is created simultaneously as the silicon is etched. The current understanding of the chemical process is that oxygen ions combine with the fluorine bonded to the silicon surface prior to the silicon's removal and forms a SiO_xF_y layer. The exact composition of this layer is a topic of current research (Mellhaoui et al., 2005). In a manner similar to the chopping Bosch passivation, the SiO_xF_y passivation layer protects the exposed vertical silicon while the unmasked horizontal silicon is etched way. To make this passivation process as energetically favorable as the chemical reaction of making SiF_4, the substrate temperature is required to be cooler than approximately –80 °C. When the silicon is warmed back up to room temperature, the SiO_xF_y becomes volatile and leaves the sample (Pereira et al., 2009).

3.2 Mask Selection

The ultimate test of a mask is the fidelity of pattern transfer into the silicon over the entire etching period. Since the mask interacts with the etching process parameters, it is vital to understand which masks to use for different etches. As stated earlier, if the selectivity is too low a thicker mask is required to achieve the desired etch depths. Furthermore, as the edge of the mask erodes it will impart undesired slope or features to the sidewalls of the etched structure, often referred to as mask-induced roughness. For these reasons, deep silicon etching requires higher selectivity masks. Conventional silicon etch masks are metal, oxides, and resist.

Metal masks, such as chrome, offer the advantage of high selectivity as high as thousands to one. This is primarily due to their lack of chemical reactivity with the etch gas molecules and their mechanical strength. However, metal masks typically induce detrimental effects such as notching at the top of the etched structures, due to image forces, and unwanted masking due to redeposited metal introduced by ion sputtering. A particular problem with chrome during the cryogenic etch is that oxygen radicals appear to be locally deactivated around the mask reducing the silicon passivation layer near the top of the mask (Jansen et al., 2009). Silicon dioxide masks typically offer high selectivity (150:1 for Bosch and 200:1 for cryogenic etching) with the added cost of more complicated patterning. The oxide layer must be grown or deposited, followed by pattern transfer from another material or resist into the oxide mask. Increasing the number of processing steps increases the effort needed for accurate pattern transfer as well as the potential for reduction in mask fidelity. Resist masks offer the simplicity of a single processing step along with good selectivity (approximately 75:1 for Bosch and 100:1 for cryogenic etching). These selectivity values highly depend on process conditions and are seen to widely vary. Jansen et al. have commented that sidewall protection using resist is better than that using oxide masks due to the erosion of the resist mask providing additional material to protect the etched walls (Jansen et al., 2009).

Several new masks have recently demonstrated improvements both in selectivity and in ease of integration. Sputtered aluminum oxide, or alumina, provides mask selectivity greater than 5000:1 for cryogenic etching. Because of the extremely high selectivity, only a thin layer is required for masking. This makes the film easily patterned via resist liftoff, instead of traditional ion milling for hardmask pattern transfer. Patterning difficulty is only slightly increased as compared to traditional resist processing. Starting with a photoresist mask, a thin layer of alumina is sputtered onto the sample. This is followed by liftoff of the undesired alumina and the resist using acetone. Due the brittle nature of the material, the alumina cleanly fractures and easily lifts off. Furthermore, since the alumina is electrically insulating, image force effects and undercutting seen in metal masks are not seen with this mask. Removal of the alumina mask is easily achieved using buffered hydrofluoric acid or ammonium hydroxide combined with hydrogen peroxide, both of which do not significantly etch silicon.

A second new mask innovation is using gallium (Ga) to mask silicon (Chekurov et al., 2009). The Ga mask is implanted by a focused ion beam (FIB), where a gallium beam is focused on a silicon sample and writes out the pattern in a similar way to other direct-write lithography techniques. The dose can be accurately controlled by manipulating the time the beam spends focused on the silicon as well as the accelerating voltage of the beam. This offers the advantage of high mask resolution on small feature sizes (~40 nm) without needing a polymer to be patterned or a developer to be used. Typical dosing creates masks around 30 nm in thickness and offers greater than 1000:1 selectivity in a cryogenic etch. Unfortunately, mask removal poses a problem since the gallium atoms are implanted in the silicon and damage to the silicon surface has not yet been characterized. Since using the Ga mask is, in a sense, a maskless and resistless technique, pattern definition can take place on any surface upon which the beam can be focused. This presents the opportunity of multidimensional patterning, such as patterning on a pre-etched sidewall to create a lateral mask.

3.3 Etching Conditions and Optimization

Control over the etch rates, selectivity, sidewall profile, and etch roughness is achieved through tuning process parameters. The major controllable parameters include ICP power, forward power, temperature, chamber pressure, and gas flow rates. While this list is not all inclusive, these parameters directly control the state of the chamber and therefore the plasma. Many subtleties also play an important role in the etch process. This list would include silicon loading, chamber conditioning, and chemical interactions in the gas chemistry and with the mask. Each etch process will have optimization parameters that will be reactor specific, but this section will assist in building intuition for both the Bosch and the cryogenic etch.

The ICP power controls the amount of ionization occurring for a given gas flow rate and chamber pressure. Typically, as the ICP power is increased, the amount of ions created will also increase. This will increase the chemical etch rate, both vertically and laterally, increase the milling etch rate, reduce the selectivity by milling the mask away faster, and reduce the effect of passivation by bombarding the sidewalls more due to the ion angular dispersion effect. If the vacuum pumping rate does not change, e.g., when controlling the throttle valve position instead of the chamber pressure, then when increasing the ICP power one can measure the fact that more gas is ionized by measuring the chamber pressure. It should be noted that increasing the ICP power does not increase the etch rate infinitely. In fact there is an optimum ICP power for a given etch gas flow rate. These trends apply for both Bosch and cryogenic etching for the SF_6 chemistry. Increasing the ICP power for the passivation step of chopping Bosch, similarly to the etching, will increase the thickness of the passivation for a given passivation time. A subtle effect of increasing the ICP power is that it also slightly increases the bias between the plasma and the electrode. For the Bosch etch, the bias from the forward power is typically much greater in magnitude than plasma potential increase from the ICP change and the effect is generally unnoticed. Since the cryogenic etch uses very little forward power, applying more ICP power can significantly increase the amount of milling occurring. Another subtle effect is that a higher etch rate also increases the substrate's temperature. For the cryogenic etch, it is estimated that the exothermic formation of SiF_4 releases 2 W/cm^2 for an 8 $\mu m/min$ etch rate. For an unmasked 6" Si wafer, this results in approximately 360 W of exothermic heating.

Increasing the forward power establishes a larger electric field between the plasma and the table electrode. By imparting more momentum to the ions, the silicon milling rate increases. This usually increases just the vertical etch rate, but due to the IAD (ion angular distribution) effect the lateral etching does slightly increase. Since the milling action increases, the erosion rate of the mask also increases, thereby reducing the selectivity. Similar to the temperature effect from increasing ICP power, increasing the forward power increases the rate and energy of ion bombardment to the substrate. This effect is easily calculated from the potential difference and the ion flux for the cryogenic etch and is estimated around 0.5 W/cm^2.

The Bosch etch is typically insensitive to temperature effects, while the cryogenic etch is extremely responsive to any temperature changes. Since the Bosch etch is performed at 20 °C, the polymer passivation layer is far from both the melting and freezing regimes. However, the high temperature dependence of the passivation reaction during the cryogenic etch means even small temperature fluctuations change the etching profile. Heating by as little as 5 °C during the cryogenic etch reduces the passivation rate and thereby induces undercutting due to image force effects. Passivation during the cryogenic etch roughly begins to occur around -85 °C. However, if the wafer is too cold, SF_x etch gases and SiF_x product gases can freeze on the sample sidewalls, adding to the SiO_xF_y passivation layer. Variations in table temperature by 5 °C due to oscillations in the table temperature controller have been seen to change the profile of deep etches adding a sinusoidal curvature to the sidewalls. Temperature is typically controlled by cooling the stage with liquid nitrogen or water and thermally connecting the wafer to the table by flowing helium between them. When silicon samples smaller than a full wafer are etched, they require thermal conductivity to the carrier wafer. This is accomplished by using thermal grease or Fomblin pump oil on the backside of the wafer to the substrate. Removal of the thermal grease is done with trichloroethylene and the Fomblin is easily removed by isopropanol.

Chamber pressure controls the amount of gas in the chamber for ionization. As noted during the ICP power discussion, changing the amount of incident ions controls both etch rate and selectivity. For a given ICP power, there is an optimum gas flow rate for SF_6. Increasing the pressure can be accomplished by shutting the throttle valve or by injecting more gas. A subtle effect of increasing chamber pressure is that it also increases the scattering collisions of ions traversing the Faraday dark space. This creates a larger angular spread in incident ions to the substrate, or increases the IAD. This increases the amount of undercut or lateral etch.

Other parameters which can alter both the Bosch and the cryogenic etch are not necessarily due to changing a mechanical feature on the reactor. Changing the amount of exposed silicon can also change etch results. Increasing the ratio of exposed silicon to masked silicon changes the amount of ions needed for etching and will significantly reduce the etch rate. As explained earlier, the exothermic nature of etching more silicon also induces an increase in substrate heating. A positive effect, however, is that for large silicon loading, slight changes in mask patterning have relatively minor effects in etch results. This is a convenient feature for establishing an etch for a wide range of users. It also reduces the effect of changing the etch as the etch goes deeper into the silicon and effectively exposes more silicon surface. Cleanliness of the chamber can also change the effects of etches. Since the plasma interacts with the sidewalls as well as the substrate, residual molecules on the sidewalls can be redeposited on the etched surface, causing micromasking, or can chemically react with the etch gas. For this reason, it is highly recommended that good chamber cleans followed by chamber conditioning be performed prior to etching.

3.4 Application: High Aspect Ratio Pillars and Metallization Liftoff

Using the high selectivity of photoresist for the cryogenic silicon etch, fabrication of high aspect ratio micropillars was demonstrated (Henry et al., 2009a) and serves as an example of achievable profiles using the mixed mode etching process. These pillars were utilized for validating theories concerning radial *p–n* junctions for applications of solar cells (Kayes et al., 2008). The patterns transferred to a 1.6 μm thick photoresist on a silicon substrate were groups of disks 5, 10, 20, and 50 μm in diameter in a hexagonal array. The spacing between each disk grouping was equivalent to the diameter of the disks, i.e., each 5 μm disk was separated from its nearest neighbor by 5 μm, the 10 μm disks by 10 μm, etc.

Fig. 3. High aspect ratio silicon micropillars: This cross-sectional SEM of 5 μm wide and 83 μm tall silicon micropillars demonstrates the cryogenic silicon etch using a 110 nm thick alumina etch mask. The very tops of the pillars indicate that mask erosion is beginning

Concluding etch profile optimization, multiple samples of the patterns were etched for varying times. Since each etch had an array of the four different diameters, a direct study of aspect ratio, i.e., ratio of the etched depth to width, dependence upon etch depth was made. Assuming that the etch rate was comprised of the etch rate of silicon with no structures (zero aspect ratio) minus a linear dependence on aspect ratio, a simple differential equation may be solved to yield the following:

$$d(t) = \frac{E_0 \cdot w}{b}\left[1 - \exp\left(-\frac{bt}{w}\right)\right].$$

Here E_0 and b are the zero aspect ratio etch rate and the aspect ratio dependent etch rate, respectively. The equation solves for the etch depth d given the etched trench width w and the etching time t. Using this equation, etches were performed achieving an aspect ratio of 17.5:1. The angle of the micropillars' sidewalls was controlled by varying the oxygen flow rate, which allowed for passivation rates to be controlled and consequently changing the angle of the profile up to 6°. This number appears small at first but when deep etches are being performed, controlling the angle can prove critical to not etching the base of the pillars to a point.

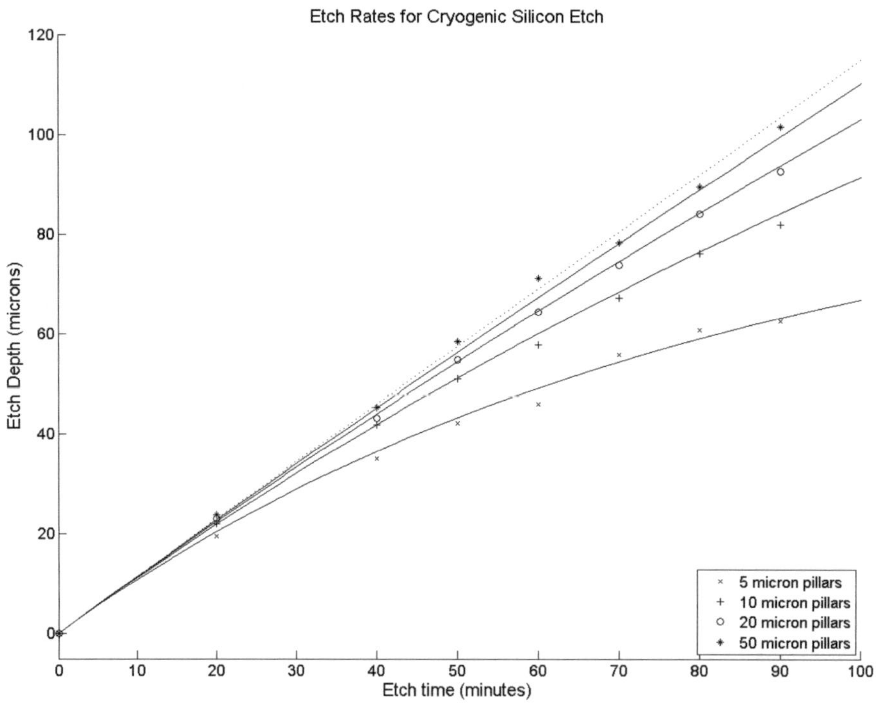

Fig. 4. Etch rates and aspect ratio dependence: This graph contains data points taken from etches creating the silicon micropillar arrays of 5, 10, 20, and 50 μm diameter pillars as well as the solutions to the solved differential equation for the various widths. It becomes evident that as the aspect ratio of the etched trench increases, the etch rate slows down. This is the so-called "Aspect Ratio Dependent Etching" or ARDE effect

A second use of the cryogenic etch is based on the high selectivity of the etch mask. Since very little resist is eroded away during etching, the remaining etch mask becomes useful as a layer for metallization liftoff. This fabrication sequence was employed for creating silicon microcoils (Henry et al., 2009b). Using a 1.6 μm thick patterned photoresist, a cryogenic etch was used to etch highly doped silicon. The structures then had varying thicknesses of chemically vapor deposited (CVD) amorphous silicon dioxide. Following the deposition, copper was thermally evaporated into the trenches with thickness up to 15 μm. Liftoff of the silicon dioxide and metal using acetone was then performed. Typically for conventional metallization, resist heights are required to be 3-4 times thicker than the metal being deposited with necessary rigorous sidewall profile control. Here, since the metal is approximately 10 times thicker than the resist, the depth of the cryogenic etch can replace the thick resist requirements as well as reliably accomplishing the profile requirements needed for the thick metal deposit. This fabrication sequence created planar copper microcoils embedded in silicon and insulated from the substrate using silicon dioxide.

Fig. 5. Planar copper microcoils in cross section. Coils are copper 10 μm thick embedded in silicon and insulated using a 1 μm thick CVD oxide. Using the high selectivity of the cryogenic silicon etch, thick copper metallization is possible with liftoff achieved using the etch mask

4. Nanoscale Silicon Etching

Unlike deep silicon etching, nanoscale etching requires neither extraordinary selectivity nor large etch rates. On the contrary, moderate selectivity of 5:1 is acceptable and slower etch rates, 100–200 nm/min, are more useful for accuracy of etch depths. Further, Bosch etching and cryogenic etching prove to be unsuitable for very small structures due to the notching and lateral etching of the two chemistries respectively. In general, nanoscale etch properties should include smooth and highly controllable sidewalls, slow etch rates, and low undercutting effects. To meet the first two requirements, mixed mode gas chemistries become useful due to the simultaneous etching and passivating. Proper choices in masks can reduce undercutting effects. This section will discuss several emerging mask technologies and demonstrate nanoscale etching using SF_6/C_4F_8, termed as the Pseudo Bosch etch here.

4.1 Gas Chemistries

Although the cryogenic etch creates very smooth sidewalls, its inherent undercut is typically too much for the nanoscale regime. Furthermore, the etch rates are too high for accurate control on the nanoscale. A combination of the Bosch gases introduced in a mixed mode process creates an ideal etch recipe which has allowed silicon nanopillars with an aspect ratio of 60:1 and diameters down to 20 nm. To etch the silicon, SF_6 is again used while C_4F_8 is used to passivate simultaneously. Since ions are constantly needing to mill the continuously deposited fluorocarbon polymer layer, the etch rate significantly reduces to 200–300 nm/min. Etch recipe parameters are similar to the cryogenic etch and are around 1200 W for the ICP power and 20 W for the forward power. The advantage of using the C_4F_8 as the passivation gas also extends to not requiring cryogenic temperatures.

4.2 Mask Selection

Typical masks for nanoscale etches are based on the difficult patterning requirements. To define structures down to 20 nm, e-beam resists such as polymethylmethacrylate (PMMA) are employed with thicknesses ranging from 500 nm down to 30 nm. The advantage of using this as the etch mask is the simplicity in pattern transfer: once the e-beam patterning is complete, the resist can be developed leaving the patterned etch mask. The disadvantage is that typical selectivity values range from 4:1 to 0.5:1. This implies that only very shallow etches can be performed on the very small structures since thicker e-beam resists are difficult to expose for small structures. However, a great advantage is achieved by using alumina etch masks with this etch. A thin layer of alumina, approximately 30 nm thick, can serve as an etch mask yielding selectivity of better than 60:1. This allows for e-beam resists, with thickness to be patterned and developed, followed by having the alumina sputter deposited. After liftoff in acetone, the alumina pattern remains on the silicon. Another common etch mask is nickel, which is patterned similarly to that of sputtered alumina. Sputtered nickel offers good selectivity with the disadvantage of increased mask undercutting due to image forces.

We recently have also demonstrated using implanted gallium as an etch mask for silicon nanostructures. With this method, Ga ions are implanted in the silicon substrate using a focused ion beam. The dwell time of the beam combined with the current determines the dosage while the beam accelerating voltage determines the depth and spread of the mask. Typical threshold dosages are about 10^{16} ions/cm^2 or 2000 μC/cm^2. For comparison, typical resist sensitivities range from 200–1200 μC/cm^2 when exposed on a 100 keV electron beam lithography system. Using a 30 kV beam, we estimate the Ga layer to be approximately 20 nm thick. Using the Pseudo Bosch etch, selectivities greater than 50:1 have been demonstrated using a Ga mask with resolution of better than 60 nm. At this point, we suspect that the resolution has not reached the intrinsic limit imposed by the implantation process, and is instead limited by our beam optics.

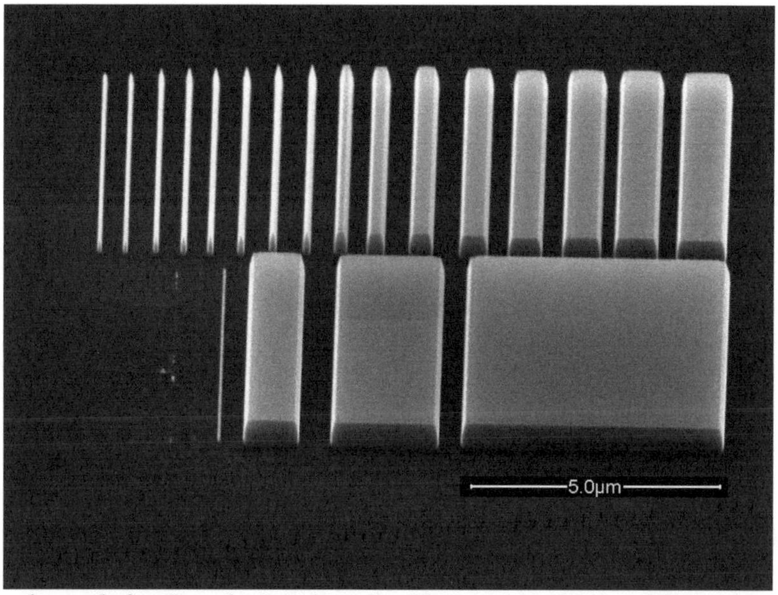

Fig. 6. Ga etch mask for Pseudo Bosch etch: This cross-sectional SEM, taken at 45°, of a series of blocks etched to 700 nm demonstrates focused ion beam implanted Ga acting as an etch mask for the Pseudo Bosch etch. The smallest resolvable feature here is 80 nm; however mask erosion did occur for the 2×10^{16} ions/cm^2 Ga dose. The simulated Ga implantation depth is 27 nm with a longitudinal spread of 9 nm

4.3 Etching Conditions and Optimization

By changing the ratio of the etch gas to passivation gas, SF_6:C_4F_8, the sidewall profile can be controlled. A typical ratio is 1:3 with the absolute gas flow rates dependent upon chamber volume, as sufficient flow is required to establish a chamber pressure of 10 mTorr; a good starting point is roughly 30 and 90 sccm respectively. Increasing the ratio improves the etch rate, reduces the selectivity, and drives the sidewall to be reentrant. Typical ICP power is around 1200 W combined with a slightly higher forward power than that of the cryogenic etch of around 20 W. Increasing the forward power again reduces the selectivity with a slight improvement in etching rates. Unlike cryogenic mixed mode, this etch is typically performed at room temperature or 15–20 °C.

4.4 Application: Waveguides and Nanopillars

Since passivation occurs during etching, very straight and smooth sidewalls can be fabricated on nanoscale structures. In particular, combining this feature of the Pseudo Bosch etch with the high selectivity of the alumina etch mask, impressive 60:1 aspect ratio nanopillars have been demonstrated. Pillars were created by first patterning PMMA using a 100 kV electron beam and developing the pattern using methyl isobutyl ketone (MIBK) and isopropanol solution. A 30 nm thick alumina layer was then sputtered and lifted off leaving the alumina mask on silicon. The Pseudo Bosch etch was then performed with an etch rate of 250 nm/min leaving well defined arrays of silicon nanopillars. The smallest diameter created was 22 nm for a pillar standing 1.26 μm tall.

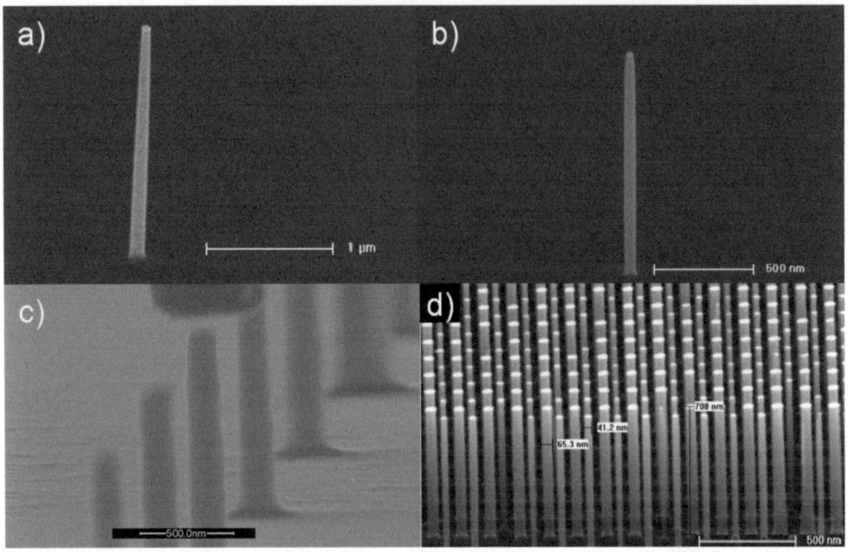

Fig. 7. High aspect ratio silicon nanopillars: These cross-sectional SEMs, taken at 45°, of a) 73 nm diameter and 2.8 μm tall silicon nanopillar, b) 40 nm diameter and 1.75 μm tall silicon nanopillar, c) tungsten probe contacting a 200 nm diameter and 1.25 μm tall silicon nanopillar, and d) array of alternating 40 nm and 65 nm diameter pillars 1 μm tall demonstrate the Pseudo Bosch silicon etch using a 30 nm thick alumina etch mask

5. Nanoscale Indium Phosphide Etching

In contrast to the previously discussed fluorine-based etch recipes, many III–V materials require the use of chlorine-based chemistries. This is due to the difference in chemical properties of the etch products. As seen in the previous section, the proposed mechanism for Ga masking of Si is the formation of involatile GaF_x compounds that prevent further etching. Thus, for etching of Ga and other similar compounds, we expect that a Cl_2-based etch will result in faster etching rate and smoother sidewalls from the readily removed etch products. In this section, we will discuss an InP etch that uses a hybrid gas mixture of Cl_2, CH_4, and H_2.

5.1 Gas Chemistries

The gas composition of this etching recipe is a hybrid between two established InP recipes. Specifically, high etch rate recipes with Cl_2- and Cl_2/Ar-based plasmas are well known but suffer from sidewall roughness and require high processing temperatures to volatilize $InCl_x$ species (Yu & Lee, 2002), as illustrated in Fig. 8. Smooth etch recipes with CH_4/H_2 plasmas have also been studied but have prohibitively slow etch rates. In this case, the smoothness is a result of two factors. Firstly, the likely etching mechanism of InP is the evolution of volatile products PH_3 and $In(CH_3)_3$, which can be controlled by adjusting the gas flow rates (Feurprier et al., 1998). Secondly, the deposition of CH films from the source gases serves to protect the sidewalls (von Keudell & Möller, 1994). In our etch, we utilize a precise ratio of source gases that balances all these properties and takes interactions into account, such as removal of H and Cl ions by formation of HCl.

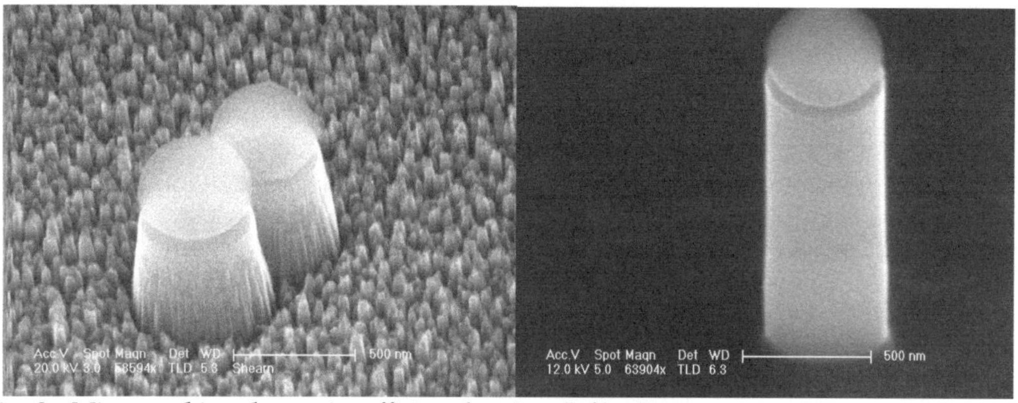

Fig. 8. Micromasking due to insufficient heating (left). By increasing ICP power and thus raising sample temperature, micromasking is removed (right)

5.2 Mask Selection

Appropriate masks for the InP etch are metals and dielectrics. This is due to the high rate of mask erosion inherent in the etching conditions. The forward bias and thus bias voltage that drive ions toward the wafer surface are much higher than those found in the SF_6-based silicon etches in previous sections. This will make the etch more milling, and will help to maintain the same etch characteristics in other stoichiometries of interest, such as InGaAsP compounds. We utilized masks of silicon dioxide spheres and evaporated Au layers in the etching experiments. The selectivity of oxide was approximately 10:1; however, faceting occurred before the mask was completely eroded, limiting the useful selectivity to a more modest 4:1. In deeper nanoscale etching applications, a silicon nitride or metal mask is preferred as it has high selectivity and does not suffer from faceting as readily as oxide. As seen in Fig. 9, the metal hardmask has eliminated most pattern-induced roughness.

Fig. 9. Anisotropic InP etch using a metal hardmask. Smoothness is only limited by mask irregularities

5.3 Etching Conditions and Optimization

Our etch had Cl_2:H_2:CH_4 ratio of 8:7:4 with actual gas flows of 32 sccm Cl_2, 28 sccm H_2, and 16 sccm CH_4 and a chamber pressure of 4 mTorr. The table was heated to 60 °C to reduce polymer deposition, and no helium backing was applied in order to have the plasma heat the sample. This heating is key to proper etch characteristics, as too little heat will cause micromasking due to involatile gas products such as $InCl_x$. The forward power was 180 W, found experimentally by varying until an anisotropic profile was achieved without excessive mask erosion. This resulted in a cathode bias of approximately 200 V. ICP power was 2200 W, also found experimentally by monitoring the transition of "black" InP to smooth InP due to the cessation of micromasking during etching. The etch rate of pure InP was measured to be 1.2 μm/min.

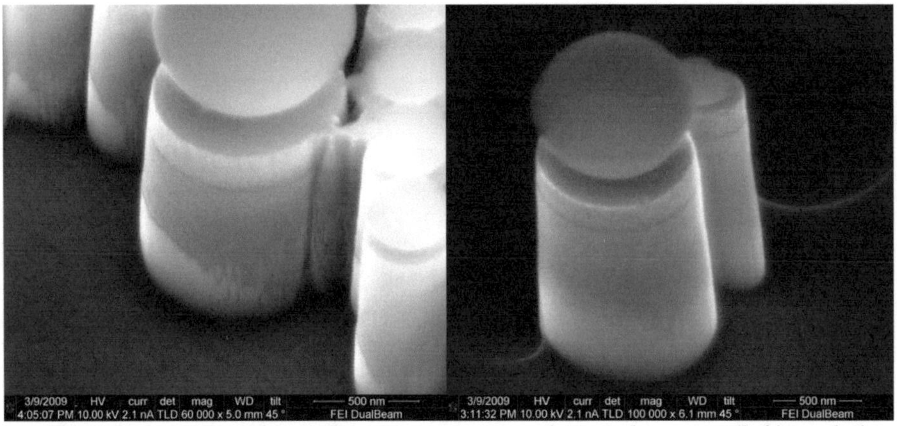

Fig. 10. InGaAsP on InP etching showing excessive forward power (left) and the correct amount of forward power (right). The features at the bottom of the pillar are due to faceting and redeposition of mask materials

During some etches with identical conditions, a roughening of the bottom surface was noticed due to chamber cleanliness. The sensitivity of this etch to chamber condition is not as high as the cryogenic Si etch described earlier, but reproducible results require a regular cleaning schedule to return the chamber to a known "clean" state. This is best implemented by running a short, minute-long SF_6 cleaning plasma just before etching to remove any contaminants that are readily incorporated into the plasma. For long term cleanliness, periodic hour-long SF_6/O_2 plasma is run. The frequency depends on what other etches have been performed previously, but is typically one hour of cleaning per three to four hours of etching. In an industrial setting, this could be done in shorter periods between each wafer to maintain a constant chamber state.

6. Inductively Coupled Plasma Chemical Vapor Deposition

ICP-RIE systems have been demonstrated in this chapter to be a gentle environment for etching both silicon and III–V materials. Over the last decade, research has extended this useful environment to the deposition of thin dielectric films and conductive silicon layers.

A common film deposition method is low pressure chemical vapor deposition, which is typically performed at temperatures around 600 °C. The low pressure limits unwanted gas phase reactions, while the high temperature ensures that there are adequate diffusion and energy to overcome any activation barrier in the desired reactions. By using plasma enhanced chemical vapor deposition (PECVD), typical deposition temperature can be lowered to the range of 300 °C to 400 °C. By adding a gas ring to improve gas uniformity in an ICP-RIE, the new operation of ICP chemical vapor deposition, or ICP-CVD, is added to the machine. In ICP-CVD, depositions may proceed with temperatures in the range of 50 °C to 150 °C. Densities of layers deposited at 70 °C using ICP-CVD have now become comparable to PECVD at 350 °C. This illustrates the key advantage of ICP-CVD over traditional CVD processes: high density films deposited at lower temperatures.

A typical PECVD reactor has a single power source establishing both the plasma density and the ion flux in a method similar to that of an RIE (although new PECVD reactors can have a 13.56 MHz source combined with a kHz source, where both of the sources create RIE plasmas albeit at different frequencies). A unique advantage of ICP-CVD reactors over PECVD is that both the ion density and the ion flux can be independently controlled. In this case, the ICP power changes the ion density while the forward power controls the ion flux (Lee et al., 2000). This provides another method for tuning film parameters such as optical index, deposition rate, and density.

Since the ICP-RIE has improved efficiency ionizing the gas over an RIE-only reactor, and due to the lower operating pressures of ICP-RIEs over that of PECVD, significantly less gas is required for depositions. Typical flow rates for gases are 20 to 100 sccm. However, typical deposition rates for ICP-CVD are significantly lower than PECVD, ranging from 6–30 nm/min whereas PECVD rates range from 60–250 nm/min. The slower rates in the ICP-CVD allow for precise control over thin films with the intended use as dielectrics. Using the same method for creation of thicker films for etch masks becomes impractical.

6.1 Gas Chemistries

Types of films available using ICP-CVD include SiO_2, SiN_x, $SiON_x$, a-Si, and SiC. All of these silicon containing compounds require the source gas silane (SiH_4). Because silane is pyrophoric and will combust spontaneously in air, it is typically diluted to 2–10% levels using inert carrier gases such as nitrogen, helium, or argon. The silane flows into the reactor through a gas ring surrounding the table. A second gas is injected from the top of the chamber where the ICP ionizes the gas, creating various radicals and ions. In the same manner as in etching processes, the electrostatic potential difference between the plasma and the table drives the ions to the wafer surface, where they chemically combine with silane to create the film. Typical additive gases are N_2O for silicon dioxide, NH_3 and N_2 for silicon nitride, both N_2O and NH_3 for silicon oxy-nitride, and CH_4 for silicon carbide. The ratio of additive gas flow rates to the silane flow rate into the chamber can be used to change the index of the deposited material by making nonstoichiometric films. For a silicon dioxide film with refractive index of 1.44, the required ratio of pure silane to N_2O ranges from 0.3 to 0.4. For a silicon nitride film with refractive index of 2.0, a ratio of pure silane to NH_3 ranges from 1.18 to 1.28.

6.2 Process Tuning

A deposited film can be evaluated by several basic material properties: refractive index (indicative of stoichiometry), density, and stress levels. Evaluation of refractive index is easily performed using an ellipsometer which usually can also determine thickness simultaneously. This type of measurement also allows for deposition rate to be directly measured. Measurement of the density of the film is performed by wet etching in buffered hydrofluoric acid with the figure of merit being the etch rate; slower rates imply denser films. Stress is usually measured using an interferometer to determine the radius of curvature of wafer bowing.

Control over the refractive index is coarsely adjusted by setting the ratio of flow rates of gases into the reactor. Although increasing the chamber pressure can raise the index of the film, a more linear increase in index can be achieved by increasing the forward power of the deposition. Film density is most directly controlled through the plate's temperature, where higher temperatures correspond with higher densities. However, increasing the plate temperature above 150 °C for silicon dioxide deposition and above 70 °C for silicon nitride has little additional effect on film density. Another method for increasing density is by increasing the forward power and thus ion flux. Although control over stress in silicon dioxide is difficult, silicon nitride stress can be tuned using chamber pressure. Increasing chamber pressure can move stress values from negative to positive, and thus zero stress silicon nitride films are possible. For further quantization of process trends of silicon nitride and silicon dioxide, see (Lee et al., 2000).

6.3 Applications of ICP-CVD films

As seen earlier, in situ deposition of these films immediately after etching is possible. Deposition of a cladding layer or even a dielectric stack after etching an optical structure is possible, due to the refractive index tunability of the films. Particularly as the thermal budgets of many wafers shrink due to the use of exotic materials such as HfO_2 (He et al., 2005), low temperature processes such as ICP-CVD will become more valuable.

7. Plasma-Assisted Wafer Bonding

In previous sections we have focused individually on the processing of Si-based materials and InP-based III–V materials. Silicon systems make modern computing and data processing possible, while III–V gain materials generate and amplify light for transmission over optical fibers. Silicon is an indirect semiconductor and thus a very poor convertor of electricity to light while III–V's are direct semiconductors and thus efficient convertors of injected electrons to photons. Now it is widely recognized that major future progress in both computers and optical communication will require chip-scale integration of these two types of materials (Mathine, 1997).

Heteroepitaxy by molecular beam epitaxy (MBE) or metalorganic chemical vapor deposition (MOCVD) has proved to be very successful for lattice-matched material systems such as GaAs- and InP-based ternary or quaternary compounds. However, due to the large lattice mismatch (4.1% between Si and GaAs and 8.1% between Si and InP), heteroepitaxial growth of III–V's on silicon has not produced high-quality material for practical applications. The high density of threading dislocations in the epitaxial layers greatly reduces the lifetime of fabricated devices (Pearton et al., 1996).

7.1 Direct Wafer Bonding

As an alternative to epitaxial growth, direct wafer bonding provides a way to join together two flat and clean semiconductor surfaces at room temperature without the restriction of matching lattice constants. The intermolecular and interatomic forces bring the two wafers together and the bonds form at the interface. By introducing a superlattice defect-blocking layer, dopants and defects are prevented from migrating from the bonding interface to the active region so that the luminescence from the multiple quantum well structure can be preserved (Black et al., 1999). To increase the bond strength, a high-temperature post-bonding annealing step is usually required. However, this high-temperature annealing step induces material degradation and is incompatible with backend Si CMOS processing. For this purpose, many efforts have been put into reducing the annealing temperature while keeping a strong bonding (Takagi et al., 1996; Krauter et al., 1997; Berthold et al., 1998).
For Si-to-InP wafer bonding, a pre-bonding oxygen plasma treatment for both wafer surfaces has been demonstrated to yield a very spontaneous bonding at room temperature (Pasquariello & Hjort, 2002). Similar to previously discussed etching plasmas, the pre-bonding plasma aims to have a high density of chemically active species arrive at the surface with a low incident power to minimize surface damages, such as dislocations, that work against bond formation. The post-bonding annealing temperature can be below 200 °C while the interface strength can be as high as the bulk fracture energy of InP. As mentioned in previous sections, the plasma affects the bonding surfaces both physically and chemically. The oxygen plasma is used to remove hydrocarbon and water molecules so as to reduce the probability of the formation of interfacial bubbles and voids during post-bonding annealing. Additionally, the plasma treatment generates a very smooth and reactive thin oxide layer which helps in bonding process.

7.2 Bonding Procedure

We have succeeded in transferring InGaAsP epifilms to Si substrates using oxygen plasma assisted wafer bonding technique. We start with a Si wafer and an InP wafer with InGaAsP epitaxial film. The InGaAsP epifilm consists of an InGaAs contact layer at the top, a *p*-InP upper cladding layer at middle followed by an InGaAsP active layer and then an *n*-InP lower cladding layer at the bottom. The total thickness of the epifilm is ~2 μm. The bonding procedure begins with solvent cleaning of both surfaces. A 10-nm-thick oxide layer is grown on top of the Si wafer to enhance the bonding strength. The surfaces of the two wafers are then activated through exposure to oxygen plasma, and bonded together under a pressure of 0.1 MPa at 150 °C for 2 h. Following the bonding process, the InP substrate is removed by HCl wet etching. Fig. 11 clearly shows the cross-sectional structure consisting of the remaining InGaAsP epifilm bonded onto the Si substrate. The bonding interface between the epifilm and Si is thin and smooth. Fig. 11(a) focuses on one end of the epifilm: it is evident that the top InGaAs layer is protrusive at the end due to its different composition from the *p*-InP layer below. Fig. 11(b) focuses on a middle part of the epifilm. The pyramids at the top of *p*-InP layer are results of HCl wet etching during InP substrate removal due to the damaged outer InGaAs layer.

Fig. 11. SEM cross-sectional view of InGaAsP epifilm on Si by wafer bonding after the step of InP substrate removal. The InGaAsP epifilm consists of an InGaAs contact layer at the top, a *p*-InP upper cladding layer at middle followed by an InGaAsP active layer and then an *n*-InP lower cladding layer at the bottom. (a) Zoom at one end of the epifilm. The top InGaAs layer is protrusive at the end due to its different composition from the *p*-InP layer below. (b) Zoom at middle of the epifilm. The pyramids at the top of *p*-InP layer are results of HCl wet etching during InP substrate removal due to the damaged outer InGaAs layer.

7.3 Application: Hybrid Si/III–V Optoelectronic Devices

With this epilayer transferring technology, the two disparate materials Si and III–V's now can be brought together to realize a variety of active devices on Si, such as lasers (Fang et al., 2006; Fang et al., 2008; Sun et al., 2009), amplifiers (Park et al., 2007a), modulators (Chen et al., 2008; Kuo et al., 2008), and detectors (Park et al., 2007b). To take the hybrid Si lasers for an example: as seen in Fig. 12, the hybrid Si/III-V structure consists of a prepatterned SOI wafer and a III-V epilayer bonded together. The III-V epilayer has been reported to be AlGaInAs (Fang et al., 2006) or InGaAsP (Sun et al., 2009) quaternary semiconductor compounds, either of which can be epigrown onto an InP substrate with very high crystal quality with current state of the art. In our work (Sun et al., 2009) the thicknesses of the buried SiO_2 layer and the undoped Si device layer are respectively 2.0 μm and 0.9 μm. The Si waveguide is defined using e-beam lithography and SF_6/C_4F_8 plasma reactive ion etching as described in detail in previous sections. The Si to the two sides of the waveguide is entirely etched down to the SiO_2 layer, and the waveguide width ranges between 0.9 μm and 1.3 μm. A 5-μm-wide center current channel by means of proton implantation on its two sides is created to enable efficient current injection. In a working device, the injected current starts from the top *p*-side contact, passes through the center current channel in the *p*-InP cladding and the InGaAsP active region, then bifurcates in the *n*-InP layer until it reaches the *n*-side contacts on both sides (not shown in Fig. 12). This hybrid structure is designed to support a joint optical mode, whose profile overlaps both materials. The modal gain is obtained by the evanescent penetration of the joint mode into the III–V active region. The devices are referred to as "hybrid Si evanescent lasers" (Fang et al., 2006). Single facet output power from the Si waveguide can exceed 10 mW at room temperature, making these devices ready for practical use (Sun et al., 2009).

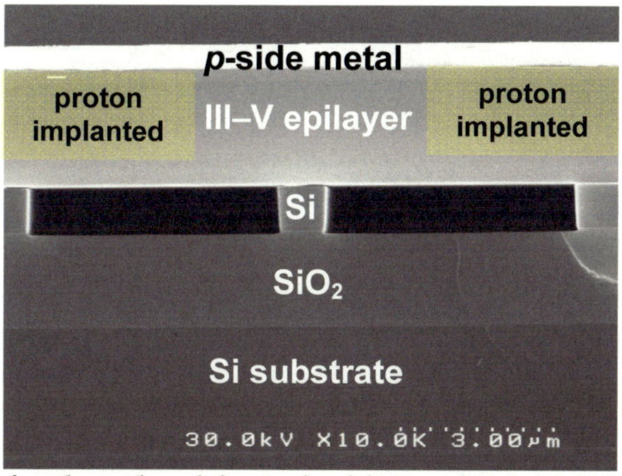

Fig. 12. SEM view of a cleaved end facet of a fabricated hybrid Si/III-V laser. This is a close-up at the center Si waveguide region. Approximate proton implanted regions are superimposed on the image for illustration. The III-V epilayer can be either AlGaInAs or InGaAsP quaternary semiconductor compounds.

8. Conclusion

In this chapter, we have described and explored some of the emerging plasma applications of etching, deposition, and surface modification to semiconductor materials. The process latitude available in modern ICP-RIE systems has enabled these novel processes. This is a direct consequence of the multitude of changes one can effect on a plasma by changing the pressure, driving fields, gases, temperature, and other parameters as discussed in the introduction. In particular, temperature-controlled stages capable of cryogenic cooling have given process engineers the ability to tune the types and rates of chemical reactions that occur on their samples. This was strongly illustrated in our discussion of cryogenic silicon etching, but a similar scheme could be imagined for other materials, given appropriate gas selection. We demonstrated both deep Si etching appropriate for MEMS as well as nanoscale Si etching, and discussed the difference in processing details between these two regimes.

New applications of these etches were discussed. By combining a chemically inert mask (Al_2O_3) with the nanoscale Pseudo Bosch silicon etch, high aspect ratio nanopillars were demonstrated, along with the Ga implantation masking concept that also relies on the chemical inertness of the mask material to etch gases. Similarly, by taking an established mask technology (photoresist) and combining it with the highly selective, fluoropolymer-free deep cryogenic silicon etch and in situ deposition, novel inductive elements were fabricated.

Next, applications of III-V etching and wafer bonding were discussed. By using a plasma treatment of both surfaces, we were able to join together two dissimilar semiconductors because this method is not affected by lattice constant mismatch. This bonding technique showed room temperature spontaneous bonding which can be annealed at 200 °C or less while keeping a high interfacial strength.

Finally, the culmination of all of the previously discussed processing techniques in Si and III-V's was the demonstration of a hybrid evanescent Si/III-V laser, made using Si etching and wafer bonding. The III–V active material was used as an efficient gain medium while the Si provides the guidance and feedback for the lasing modes and also serves to couple the light out to the rest of the on-chip photonic circuit. Lasers had room temperature output of more than 10 mW emitted from a single facet of the Si waveguide. The authors expect further innovations in the plasma processing field and hope these examples have demonstrated the great versatility of this research.

9. Acknowledgments

Michael Shearn thanks the National Science Foundation for support under their Graduate Research Fellowship Program. M. David Henry gratefully acknowledges the John and Fannie Hertz Foundation for support. The authors acknowledge Andrew Homyk for his substantial work on silicon etching and Sameer Walavalkar who made substantial contributions on the alumina silicon etch mask work. Thanks also to the Kavli Nanoscience Institute at Caltech where much of the experimental work was done.

10. References

Aachboun, S.; Ranson, P.; Hilbert, C. & Boufnichel, M. (2000). Cryogenic etching of deep narrow trenches in silicon. *J. Vac. Sci. Technol. A*, Vol. 18, No. 4, pp. 1848-1852

Berthold, A.; Jakoby, B. & Vellekoop, M. J. (1998). Wafer-to-wafer fusion bonding of oxidized silicon to silicon at low temperatures. *Sens. Actuator A*, Vol. 68, No. 1-3, pp. 410-413

Black, K. A.; Abraham, P.; Karim, A.; Bowers, J. E. & Hu, E. L. (1999). Improved luminescence from InGaAsP/InP MQW active regions using a wafer fused superlattice barrier, *Proceedings of the 11th International Conference on Indium Phosphide and Related Materials (IPRM 1999)*, pp. 357-360, Davos, Switzerland, May 1999, IEEE

Chapman, B. (1980). *Glow Discharge Processes: Sputtering and Plasma Etching*. Wiley, New York

Chekurov, N.; Grigoras, K.; Peltonen, A.; Franssila, S. & Tittonen, I. (2009). The fabrication of silicon nanostructures by local gallium implantation and cryogenic deep reactive ion etching. *Nanotechnology*, Vol. 20, No. 6, pp. 065307

Chen, H. W.; Kuo, Y. H. & Bowers, J. E. (2008). A hybrid silicon-AlGaInAs phase modulator. *IEEE Photon. Technol. Lett.*, Vol. 20, No. 21-24, pp. 1920-1922

Cliteur, G. J.; Suzuki, K.; Tanaka, Y.; Sakuta, T.; Matsubara, T.; Yokomizu, Y. & Matsumura, T. (1999). On the determination of the multi-temperature SF_6 plasma composition. *J. Phys. D*, Vol. 32, No. 15, pp. 1851-1856

Fang, A. W.; Park, H.; Cohen, O.; Jones, R.; Paniccia, M. J. & Bowers, J. E. (2006). Electrically pumped hybrid AlGaInAs-silicon evanescent laser. *Opt. Express*, Vol. 14, No. 20, pp. 9203-9210

Fang, A. W.; Lively, E.; Kuo, H.; Liang, D. & Bowers, J. E. (2008). A distributed feedback silicon evanescent laser. *Opt. Express*, Vol. 16, No. 7, pp. 4413-4419

Feurprier, Y.; Cardinaud, C.; Grolleau, B. & Turban, G. (1998). Proposal for an etching mechanism of InP in CH_4-H_2 mixtures based on plasma diagnostics and surface analysis. *J. Vac. Sci. Technol. A*, Vol. 16, No. 3, pp. 1552-1559

He, G.; Liu, M.; Zhu, L. Q.; Chang, M.; Fang, Q. & Zhang, L. D. (2005). Effect of postdeposition annealing on the thermal stability and structural characteristics of sputtered HfO_2 films on Si (100). *Surf. Sci.*, Vol. 576, No. 1-3, pp. 67-75

Henry, M. D.; Walavalkar, S.; Homyk, A. & Scherer, A. (2009a). Alumina etch masks for fabrication of high-aspect-ratio silicon micropillars and nanopillars. *Nanotechnology*, Vol. 20, No. 25, pp. 255305

Henry, M. D.; Welch, C. & Scherer, A. (2009b). Techniques of cryogenic reactive ion etching in silicon for fabrication of sensors. *J. Vac. Sci. Technol. A*, Vol. 27, No. 5, pp. 1211-1216

Hodoroaba, V. D.; Hoffmann, V.; Steers, E. B. M. & Wetzig, K. (2000). Emission spectra of copper and argon in an argon glow discharge containing small quantities of hydrogen. *J. Anal. At. Spectrom.*, Vol. 15, No. 8, pp. 951-958

Jansen, H. V.; de Boer, M. J.; Unnikrishnan, S.; Louwerse, M. C. & Elwenspoek, M. C. (2009). Black silicon method X: a review on high speed and selective plasma etching of silicon with profile control: an in-depth comparison between Bosch and cryostat DRIE processes as a roadmap to next generation equipment. *J. Micromech. Microeng.*, Vol. 19, No. 3, pp. 033001

Kayes, B. M.; Filler, M. A.; Henry, M. D.; Maiolo, J. R.; Kelzenberg, M. D.; Putnam, M. C.; Spurgeon, J. M.; Plass, K. E.; Scherer, A.; Lewis, N. S. & Atwater, H. A. (2008). Radial pn junction, wire array solar cells, *Proceedings of the 33rd IEEE Photovoltaic Specialists Conference*, San Diego, CA, USA, May 2008, IEEE

Kenoyer, L.; Oxford, R. & Moll, A. (2003). Optimization of Bosch etch process for through wafer interconnects, *Proceedings of the 15th Biennial University/Government/Industry Microelectronics Symposium (2003)*, pp. 338-339, Boise, ID, USA, Jun. 2003, IEEE

Krauter, G.; Schumacher, A.; Gosele, U.; Jaworek, T. & Wegner, G. (1997). Room temperature silicon wafer bonding with ultra-thin polymer films. *Adv. Mater.*, Vol. 9, No. 5, pp. 417-420

Kuo, Y. H.; Chen, H. W. & Bowers, J. E. (2008). High speed hybrid silicon evanescent electroabsorption modulator. *Opt. Express*, Vol. 16, No. 13, pp. 9936-9941

Lee, J. W.; Mackenzie, K. D.; Johnson, D.; Sasserath, J. N.; Pearton, S. J. & Ren, F. (2000). Low temperature silicon nitride and silicon dioxide film processing by inductively coupled plasma chemical vapor deposition. *J. Electrochem. Soc.*, Vol. 147, No. 4, pp. 1481-1486

Legtenberg, R.; Jansen, H.; de Boer, M. & Elwenspoek, M. (1995). Anisotropic reactive ion etching of silicon using $SF_6/O_2/CHF_3$ gas mixtures. *J. Electrochem. Soc.*, Vol. 142, No. 6, pp. 2020-2028

Mathine, D. L. (1997). The integration of III-V optoelectronics with silicon circuitry. *IEEE J. Sel. Top. Quantum Electron.*, Vol. 3, No. 3, pp. 952-959

Mellhaoui, X.; Dussart, R.; Tillocher, T.; Lefaucheux, P.; Ranson, P.; Boufnichel, M. & Overzet, L. J. (2005). SiO_xF_y passivation layer in silicon cryoetching. *J. Appl. Phys.*, Vol. 98, No. 10, pp. 104901

Park, H.; Fang, A. W.; Cohen, O.; Jones, R.; Paniccia, M. J. & Bowers, J. E. (2007a). A hybrid AlGaInAs-silicon evanescent amplifier. *IEEE Photon. Technol. Lett.*, Vol. 19, No. 2-4, pp. 230-232

Park, H.; Fang, A. W.; Jones, R.; Cohen, O.; Raday, O.; Sysak, M. N.; Paniccia, M. J. & Bowers, J. E. (2007b). A hybrid AlGaInAs-silicon evanescent waveguide photodetector. *Opt. Express*, Vol. 15, No. 10, pp. 6044-6052

Pasquariello, D. & Hjort, K. (2002). Plasma-assisted InP-to-Si low temperature wafer bonding. *IEEE J. Sel. Top. Quantum Electron.*, Vol. 8, No. 1, pp. 118-131

Pearton, S. J.; Abernathy, C. R. & Ren, F. (1996). *Topics in Growth and Device Processing of III-V Semiconductors*. World Scientific

Pereira, J.; Pichon, L. E.; Dussart, R.; Cardinaud, C.; Duluard, C. Y.; Oubensaid, E. H.; Lefaucheux, P.; Boufnichel, M. & Ranson, P. (2009). In situ x-ray photoelectron spectroscopy analysis of SiO_xF_y passivation layer obtained in a SF_6/O_2 cryoetching process. *Appl. Phys. Lett.*, Vol. 94, No. 7, pp. 071501

Reinhardt, K. A. & Kern, W. (2008). *Handbook of Silicon Wafer Cleaning Technology*. William Andrew Publishing, Norwich, NY

Schmidt, M. A. (1998). Wafer-to-wafer bonding for microstructure formation. *Proc. IEEE*, Vol. 86, No. 8, pp. 1575-1585

Sun, X. K.; Zadok, A.; Shearn, M. J.; Diest, K. A.; Ghaffari, A.; Atwater, H. A.; Scherer, A. & Yariv, A. (2009). Electrically pumped hybrid evanescent Si/InGaAsP lasers. *Opt. Lett.*, Vol. 34, No. 9, pp. 1345-1347

Takagi, H.; Kikuchi, K.; Maeda, R.; Chung, T. R. & Suga, T. (1996). Surface activated bonding of silicon wafers at room temperature. *Appl. Phys. Lett.*, Vol. 68, No. 16, pp. 2222-2224

Takahashi, K.; Itoh, A.; Nakamura, T. & Tachibana, K. (2000). Radical kinetics for polymer film deposition in fluorocarbon (C_4F_8, C_3F_6 and C_5F_8) plasmas. *Thin Solid Films*, Vol. 374, No. 2, pp. 303-310

Takeuchi, H.; Wung, A.; Sun, X.; Howe, R. T. & King, T. J. (2005). Thermal budget limits of quarter-micrometer foundry CMOS for post-processing MEMS devices. *IEEE Trans. Electron Devices*, Vol. 52, No. 9, pp. 2081-2086

von Keudell, A. & Möller, W. (1994). A combined plasma-surface model for the deposition of C:H films from a methane plasma. *J. Appl. Phys.*, Vol. 75, No. 12, pp. 7718-7727

Wolf, S. & Tauber, R. N. (2000). *Silicon Processing for the VLSI Era, Vol. 1: Process Technology*, 2nd edn., Lattice Press, Sunset Beach, CA

Yu, J. S. & Lee, Y. T. (2002). Parametric reactive ion etching of InP using Cl_2 and CH_4 gases: effects of H_2 and Ar addition. *Semicond. Sci. Technol.*, Vol. 17, No. 3, pp. 230-236

6

Wet thermal oxidation of GaAs and GaN

Ryszard Korbutowicz and Joanna Prażmowska
Wroclaw University of Technology
Poland

1. Introduction

The chapter is devoted to the thermal wet oxidation of AIIIBV semiconductor compounds, mainly to gallium arsenide and gallium nitride. It has been divided into several topics, containing of monoclinic gallium oxide[1] β-Ga_2O_3 properties data, techniques of oxide fabrication and application description. In the first part, properties of mentioned semiconductor's oxides are characterized. Then methods of manufacturing with a special attention for wet thermal oxidation are described. After that, applications of gallium oxide structures in electronics are given. It focuses also on the semiconductor structures dedicated for gas sensors application while gallium oxide layers improve significantly the most critical parameters of the detector compared to those containing of e.g. SnO_2.

AIIIBV and AIIIN semiconductors compounds are wide known as materials for optoelectronics devices. They are used often also to the construction of high temperature and microwave devices or chemical gas sensors. In these applications dielectric layers are necessary. There is a possibility of using their own oxides - Ga_2O_3 gives a chance to manufacture many different devices - MOS structures (Metal-Oxide-Semiconductor). It can be MOS capacitors, power Metal Oxide Semiconductor Field Effect Transistors (MOSFETs), high mobility GaAs MOSFETs or gate turn-off thyristors and, probably, CMOS applications (Pearton et al., 1999; Wu et al., 2003). The MOS-gate version of the HEMT has significantly better thermal stability than a metal-gate structure and is well suited to gas sensing (Schweben et al., 1998; Baban et al., 2005; Hong et al., 2007).

2. Properties of β-Ga_2O_3

Gallium oxide β-Ga_2O_3 is a wide band gap material that ensures deep-UV transparency. Appropriately doped could reach conductive properties thus is included to the TCO's (transparent conductive oxides) materials like ITO or ZnO which are the state-of-the-art materials in optoelectronics.

Gallium oxide occurs in various structures like α, β, γ, σ, ε types (Kim & Kim, 2000). Among many polymorphs, monoclinic β-Ga_2O_3 is considered to be the equilibrium phase (Battiston et al., 1996; Chen et al., 2000; Vı´llora et al., 2004). It is stable thermally and chemically

---

[1] The typical name for Ga_2O_3 are: digallium trioxide, gallium(III) oxide, gallium trioxide, gallium oxide. We use in the text term: gallium oxide.

(Battiston et al., 1996; Vı´llora et al., 2004). The thermal stability of β-Ga_2O_3 reaches nearly melting point reported as 1740 °C (Orita et al., 2004) and 1807 °C (Tomm et al., 2000) or 2000 K (Vı´llora et al., 2004) what determines also possibility of working at high temperature. β-Ga_2O_3 in monoclinic structure has a elemental unit dimensions as follows: a=12.214 Å, b=3.0371 Å, c=5.7981 Å and β=103.83° (Tomm et al., 2000) or a=12.23 Å, b=3.04 Å, c=5.8 Å and β=103.7° (Vı´llora et al., 2004). Cleavage along (100) plane (Tomm et al., 2000; Ueda a et al., 1997; Vı´llora et al., 2004) and (001) (Vı´llora et al., 2004) are highly preferred.
The space group of β-Ga_2O_3 is C2/m (C^3_{2h}) where GaO_6 share octahedral sites along b and are connected by GaO_4 tertrahedra thus anisotropy of optical as well as electrical properties is expected depending on the direction to the chains – perpendicular or parallel (Ueda b et al., 1997). The β-Ga_2O_3 unit cell along b, c and a-axis could be found in (Vı´llora et al., 2004).

2.1 Electrical properties

At room temperature β-Ga_2O_3 is an insulating material, above 500 °C has a semiconductor properties (Fleischer & Meixner, 1993; Battiston et al., 1996; Frank et al., 1996; Orita et al., 2004). Although electrically conductive crystals of β-Ga_2O_3 have been also reported, see Table 1 (Vı´llora et al., 2004).

direction	resistivity	mobility	carrier concentration
	Ω·cm	$cm^2V^{-1}s^{-1}$	cm^{-3}
<100>	0.11	83	$7x10^{17}$
<010>	0.19	78	$4x10^{17}$
<001>	0.08	93	$9x10^{17}$

Table 1. Electrical properties measured for β-Ga_2O_3 single crystal along certain direction

2.1.1 Electrical conductance

The tetravalent tin ion Sn^{4+} is most often chosen as a donor dopant (Orita et al., 2000; Orita et al., 2004) because its ionic radius is close to that of Ga^{3+} and simultaneously Sn^{4+} ions prefer sixfold coordination. This causes substituting of Ga^{3+} octahedral sites and results in formation of shallow donors (Orita et al., 2004). Additionally formation of oxygen vacancies in the layer provides an occurrence of shallow levels as reported for β-Ga_2O_3 crystals (Ueda a et al., 1997). Thus much emphasize has been placed on the optimization of the deposition process conditions. Alteration of ambient atmosphere and substrate temperature in e.g. PLD (Pulsed Lased Deposition) technology had significant impact on the properties of the layer. In order to assure formation of oxygen vacancies and doping by Sn^{4+} low partial pressure of oxygen and elevation of substrate temperature to 880 °C were applied. It could increase chemical potential of oxygen in the lattice what introduces oxygen vacancies and solution of tin ions to the lattice. Reported mobility of carriers was 0.44 $cm^2V^{-1}s^{-1}$ and maximum electrical conductivity 1.0 $S{\cdot}cm^{-1}$. These parameters were achieved for layers deposited on substrates maintained at 880 °C under pressure equal to $6x10^{-5}$ Pa. Increase of oxygen pressure to $1.3x10^{-2}$ Pa led to lowering of conductivity to $3.6x10^{-3}$ $S{\cdot}cm^{-1}$ (Orita et al., 2000). This effect was confirmed by Ueda et al. (Ueda b et al., 1997) for crystals obtained in floating zone technique; increase of oxygen flow rate significantly affected electrical conductivity of investigated material (see Fig. 1(a)). Under O_2 atmosphere undoped crystals were insulating

$\sigma<10^{-9}\ \Omega^{-1}cm^{-1}$. With addition of N_2 to the atmosphere the conductivity increased and reached 0.63 $\Omega^{-1}cm^{-1}$. However the N_2 content in the growth ambient is limited by the stability of crystals (Ueda a et al., 1997). The maximum obtained electrical conductivity was 38 $\Omega^{-1}cm^{-1}$ for sample grown in gas mixture of N_2/O_2 with partial pressure ratio of 0.4/0.6. To achieve enhancement in conductivity of gallium oxide also Ti^{4+} and Zr^{4+} donor dopants in polycrystalline films for application in gas sensors were used (Frank et al., 1996). Unexpected only slight increase in conductivity and decrease in sensitivity were obtained. Thus SnO_2 doping was applied by Frank et al. The highest conductivity was reached for 0.5% SnO_2 (see Fig. 1 (b)). Doping possibility seems to be restricted due to solution in lattice limit (Frank b et al. 1998).

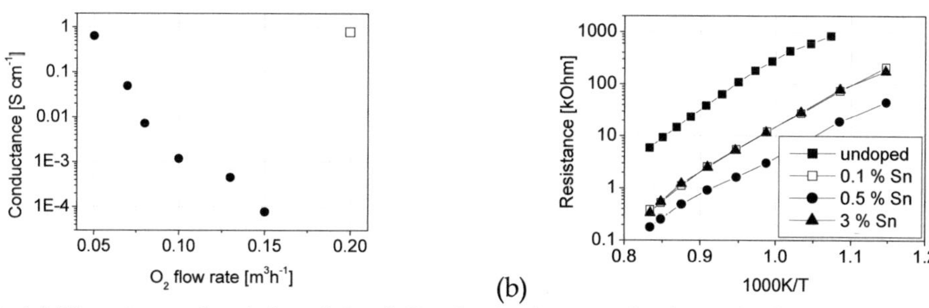

(a) (b)

Fig. 1. (a) Electric conductivity of the β-Ga_2O_3 single crystals along the b-axis as a function of the O_2 flow rate. The closed circles - samples grown from undoped Ga_2O_3 rods; open square - sample grown from Sn-doped Ga_2O_3 rods (Ueda a et al., 1997); (b) Resistance in wet synthetic air of undoped and SnO_2 doped thin films (Frank b et al., 1998)

Depending on orientation, crystals grown in floating zone technique, had resistivities as follows: 0.11 <100>, 0.19 <010> and 0.08 <001> Ω·cm (Vı́llora et al., 2004) and those obtained by Ueda et al. were 0.026 Ω·cm (b-axis) and 0.45 Ω·cm (c-axis) (Ueda b et al., 1997). Conductivity did not depend on temperature in the range of 0 – 300 K as shown in Fig 2.

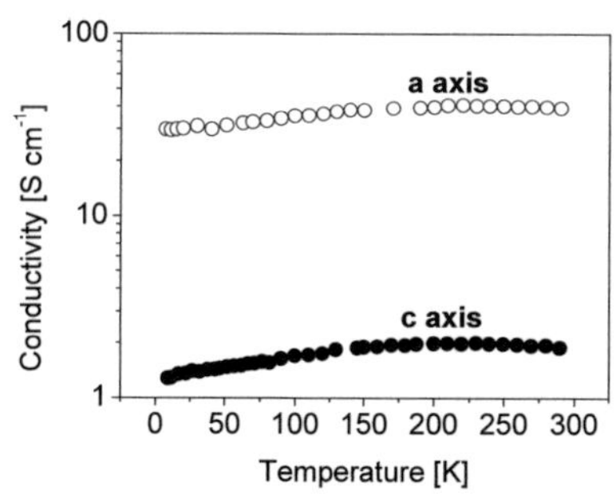

Fig. 2. Temperature dependence of the electrical conductivity along b- and a -axis of β-Ga_2O_3 single crystals (Ueda b et al., 1997)

The Mg^{2+} was used as an acceptor dopant by Frank et al. (Frank et al., 1996) to achieve conversion to p-type semiconductive material from intrinsic n-type. Layers sequence consisting of dopant and Ga_2O_3 was deposited on quartz glass substrates by reactive sputtering and subsequently annealed at temperatures up to 1200°C. Strong decrease of

conductivity was observed, but values of this parameters were comparable for 0.3 and 3% of MgO what indicated on achievement of solubility limit already by 0.3% of MgO.
Bulk crystals of β-Ga_2O_3 are electrically conductive but achievement of conductive layers has required many efforts. Application of proper substrate is crucial. The octahedral Ga^{3+} chains are responsible for generating paths for electrons thus growth of (-201) β-Ga_2O_3 in which b axis is parallel to the substrates are considered as promising (Orita et al., 2004). This requirement could be fulfilled by application of (0006) Al_2O_3 substrates.

2.1.2 Carriers mobility

Villora et al. have measured carriers mobilities of three different oriented samples. Results were 83, 78 and 98 $cm^2V^{-1}s^{-1}$ for <100>, <010> and <001> directions, respectively (Vı´llora et al., 2004). Proper optimization of process in floating zone technology led to the achievement of carrier mobility of 46 $cm^2V^{-1}s^{-1}$ for single crystals (Ueda b et al., 1997). Electron mobility in single crystals and polycrystalline, inhomogeneous, porous ceramics of Ga_2O_3 was also investigated by M. Fleischer et al. at temperature from 800 °C to 1000 °C as in this range gas sensors operate (Fleischer & Meixner, 1993). Electron mobility determined from Hall measurement was nearly the same for both samples - 10 $cm^2V^{-1}s^{-1}$. Also increase of mobility was observed with increasing the temperature of measurement. Ueda et al. (Ueda b et al., 1997) have measured mobilities along b and c directions for floating zone obtained (100) platelets. Mobility along b direction was of one order of magnitude larger than that for c direction. Obtained values were as follows: 46 $cm^2V^{-1}s^{-1}$ (b-axis) and 2.6 $cm^2V^{-1}s^{-1}$ (c-axis).

2.1.3 Dielectric constant and breakdown field

Dielectric constant of gallium oxide is rather rarely investigated. Value reported in (Zhou et al., 2008) is of about 10.2. Shan et al. (Shan et al., 2005) have measured dielectric constant for thin layers deposited on p-Si (100) and sapphire (0001) substrates by PEALD (Plasma Enhanced Atomic Layer Deposition). Substrates temperature during the process was maintained at 200 °C. As-deposited layers were amorphous and had large leakage current. Annealing at 700 or 900 °C has improved that parameter but simultaneously layers dielectric constants decreased. Figure 3 presents dielectric constants of as-deposited and annealed at various temperatures Ga_2O_3 thin films.
The breakdown field/voltage depends significantly on technology of fabrication of the material. Values reported for thermally oxidized layers were 0.05-0.1 MV/cm (Readinger et al., 1999), 0.65 MV/cm (Zhou et al., 2008), 1 MV/cm (Lin et al., 2006) and 3.85 MV/cm (Kim et al., 2001) when those for PEALD were 1-1.5 MV/cm (Shan et al., 2005) and e-beam evaporation 3.6 MV/cm (Passlack et al., 1995).

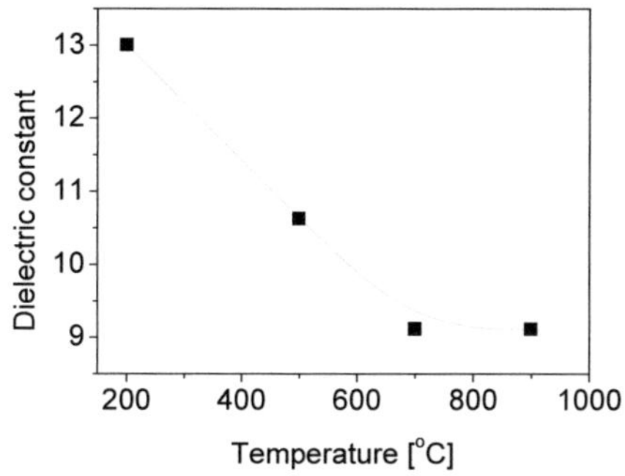

Fig. 3. Dielectric constants of as-deposited Ga_2O_3 thin film and of those annealed at various RTA temperatures (Shan et al., 2005)

2.2 Optical properties

Gallium oxide is highly transparent in visible and near UV range of wavelengths (Vı́llora et al., 2004). Transmittances in the range of visible light and UV range exceeding 80% were reported by (Matsuzuki et al., 2006; Orita et al., 2004; Ueda b et al., 1997) and that reaching nearly 100% by (Shan et al., 2005). Enhancement of transmittance could be obtained by increase of oxide layer deposition temperature (Orita et al., 2004; Orita et al., 2000) or appropriate annealing of crystals (Ueda a et al., 1997).

Because of the anisotropy of optical properties of β-Ga_2O_3 the absorption edge depends on the angle of incident beam of light to the c or b direction of the crystal. Assuming that φ is an angle between c-axis and electric field vector E Ueda et al. have investigated the transmission of floating zone obtained (001) β-Ga_2O_3 platelets with thickness of 0.159 mm (Ueda 2 et al., 1997). Figure 4 (a) presents the optical transmission spectra of an insulating β-Ga_2O_3 for various φ. The dichroism of the transmission spectra is remarkable. It was also indentified by Tippins (Tippins, 1965) and Ueda (Ueda a et al., 1997) which observed absorption edge at 255 nm and additional shoulder at 275 nm. The explanation of the dichroism could be found at (Ueda b et al., 1997).

Band gap energy E_g of gallium oxide varies in a wide range from 4.23 to 5.24 eV depending on the parameters of applied technology, see Table 2 and Fig. 4 (b). One of the parameters influencing the band gap is temperature of annealing of deposited layer Fig. 4 (b).

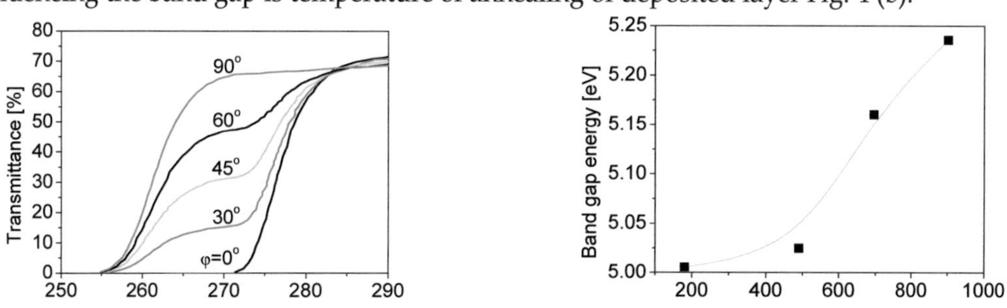

(a) (b)

Fig. 4. (a) Optical transmission spectra of an insulating β-Ga_2O_3 for various φ (Ueda b et al., 1997) and (b) band gap energy of the as-deposited Ga_2O_3 and annealed at various temperatures thin films (Shan et al., 2005)

Refractive indexes of gallium oxide are generally in the range of 1.8-1.9. Ueda et al. have reported in (Ueda a et al., 1997) for single crystals values of 1.84 and 1.88 at 980 nm, those for e-beam evaporated layers were 1.841-1.885 at 980 nm (Passlack et al., 1995), for PEALD 1.89 (Shan et al., 2005), and for bulk 1.91 (Passlack et al., 1995).

E_g (eV)	substrate	technology	reference
5.04	fused silica	e-beam evaporation under oxygen partial pressure of $5x10^{-4}$ mbar	(Al-Kuhaili et al. 2003)
5.00	Al_2O_3 (0001)	plasma enhanced ALD	(Shan et al. 2005)
4.9	α-Al_2O_3, silica	PLD	(Matsuzuki et al. 2006) (Orita et al. 2000)
4.84	fused silica	e-beam evaporation with no excess of oxygen	(Al-Kuhaili et al. 2003)
4.79	-	floating zone (along b-axis)	(Ueda b et al. 1997)
4.75	quartz	spray pyrolysis	(Hao & Cocivera 2002)
4.74	Si (001)	rf magnetron sputtering	(Rebien et al. 2002)
4.72	GaAs	e-beam evaporation	(Rebien et al. 2002)
4.6	-	Czochralski	(Tomm et al. 2000)
4.52	-	floating zone (along c-axis)	(Ueda b et al. 1997)
4.48	quartz	spray pyrolysis (annealed at 600 °C)	(Hao & Cocivera 2002)
4.44	quartz	spray pyrolysis (annealed at 900 °C)	(Hao & Cocivera 2002)
4.23	borosilicate glass	spray pyrolysis	(Kim & Kim 2000)

Table 2. Bang gap energy values of gallium oxide for various techniques

2.3 Crystalline quality and morphology

Depending on deposition technology gallium oxide layers are generally amorphous as reported for samples obtained by PEALD (Shan et al., 2005), MOVPE (Metalorganic Vapor Phase Epitaxy) (Battiston et al., 1996; Kim & Kim, 2004) or PLD (Orita et al., 2004). Crystallization of films could be assured by its annealing at high temperature (Battiston et al., 1996; Kim & Kim, 2004; Shan et al., 2005), increase of substrates temperature (Orita et al., 2004) or oxidation at high temperatures. For example amorphous layers of β-Ga_2O_3 deposited on (0001) Al_2O_3 substrates in PLD at 325 °C were obtained by Orita et al.. Diffraction peaks appeared in layers deposited at temperatures above 380 °C (Orita et al., 2004). In case of CVD (Chemical Vapor Deposition) β-Ga_2O_3 layers annealing process at temperatures above 700 °C assured crystallization.

Crystalline properties of gallium oxide are usually determined based on XRD (X-Ray Diffraction) with CuK_α radiation measurement of 2Θ patterns. The most commonly appearing peaks are (-202), (400) and (-111) at 2Θ angle of about 100°, 54° and 17° (Battiston et al. 1996) or (11-3) and (30-6) at 39° and 59° (Chen et al., 2000) as well as (-206) and (-306) at 49.6° and 59.1° (Kim et al., 2001). Except of crystallites directions determination, XRD method is applied to the estimation of crystallites sizes. For this purpose broadening of peak widths is investigated and Sherrer formula applied. Crystallites sizes are usually in the range from 20 to 50 nm (Battiston et al., 1996; Frank et al., 1996) and could be enlarged by temperature of annealing increase. The root mean square of layers varies in the range from 5 to 25 Å. Reported values were as follows: 5 Å (Shan et al., 2005), 8-13 Å (Wolter et al., 2000), 10-20 Å (Matsuzaki et al., 2006) and 12-23 Å (Kim & Kim, 2004).

There is a variety of other parameters of oxides apart from those described and analyzed above. Authors refer to photoluminescence studies to (Hao & Cocivera, 2002), density of interface states (Lin et al., 2005; Lin et al., 2006; Nakano b et al., 2003; Zhou et al., 2008) or oxidation states as well as etching behavior (Passlack et al., 1995).

3. Fabrication of the gallium oxide layers

There are several methods of manufacturing of gallium oxide for semiconductor devices applications. One can, after C.W. Wilmsen (Wilmsen, 1985), say that the main ways are: (a) chemical oxidation, (b) thermal oxidation, (c) anodic oxidation, (d) plasma oxidation, and (e) other methods. Dry or wet thermal oxidation are also applied. Since several years bulk crystals of Ga_2O_3 are available (Tomm et al., 2000; Vı´llora et al., 2004) as well.
Thermal oxidation of AIIIBV and AIIIN compounds – dry (Wilmsen, 1985; Lin et al., 2000) or wet (Readinger et al., 1999) – is unfortunately not similar to silicon oxidation. It is caused by other structure – these compounds consist of two or more elements which variously react with oxygen and water. In addition, AIIIBVs and AIIINs are very thermodynamically unstable – especially As and N have high partial pressures. In spite of these difficulties in many laboratories one carry on thermal oxidation studies because this technique is relatively cheap and can give good results. The most known is AlAs oxidation for diode lasers (especially VCSEL lasers), electroluminescent diodes and detectors with Bragg reflectors (Geib et al., 1997; Pucicki et al., 2004).

3.1 Chemical oxidation

Typical chemical oxidizers for GaAs are HNO_3 with H_2O_2 and H_2O. Chemical oxidation method is used as a part of surface GaAs substrates cleaning and rather is not used for making gallium oxide layers for devices (Ghidaoui, 2002).

3.2 Electrochemical (anodic) oxidation

Oxidation of semiconductors electrochemical anodic process is similar to typical process of oxide layer manufacturing in electrolyte. Anodic process is more efficient if in semiconductor materials holes occur – type p of electrical conduction. For type n it is necessary to generate holes in order to facilitate oxidation. The easier way is illumination by the appropriate light – with photon energy in the range from 1.4 eV to 5 eV (Wilmsen, 1985). The level of the photon energies depends on energy bandgap of the semiconductor.
The process is simple in the theory. Problems appear with increasing of the oxide thickness – both, electrical resistivity of oxide layer and diffusion rate of the oxidizing factor in the layer are variable. Problem that could appear is a choice of the appropriate electrolyte – it depends on the material, anodic equipment etc. One should to make electrical contacts to the semiconductor: permanent or temporary, which have to be stable in the applied electrolyte. Oxides after anodic process of GaAs are a mixture of gallium and arsenic oxides. These layers are not tight and ought to be annealed in high temperature (more than 450 °C). At this temperature arsenic oxides dissociate and gallium oxide will remain crystalline.
The anodic oxidation of n-type GaN (1.7 μm thick layer on sapphire substrate with carrier concentration of 4.6×10^{18} cm^{-3}) under laboratory illumination at a constant current density of 5 $mA \cdot cm^{-2}$ in sodium tungstate electrolyte (0.1 $MNa_2WO_4 \cdot 2H_2O$) at 298 K was made by

Pakes et al. (Pakes et al., 2003). They have observed local oxidation and that the oxidation has occurred at troughs in the faceted GaN layers. Near the peaks in the faceted surface oxidation was negligible. The localized nature of the oxidation of the GaN is presumed, after authors, to be related to the strength of the Ga-N bond and non-uniform distributions of impurity, non-stoichiometry or defects in the substrate (Pakes et al., 2003). The oxide was non-uniform and textured with pore-like features. The absence of a compact anodic film is probably due to extensive generation of nitrogen during anodic oxidation which disrupts development of a uniform anodic film.
Peng et al. (Peng et al., 2001) have patented the method of nitride material oxidation enhanced by illumination with UV light at room temperature. Authors used 254-nm UV light to illuminate the GaN crystals to generate electron-hole pairs. The pH value of the electrolyte was in the range of approximately 3 to 10, preferably about 3.5. The authors (Peng et al., 2001) claim that: "*This invention allows the rapid formation of gallium oxide at room temperature, and it is possible to monitor the thickness of the oxide in-situ by means of measuring the loop current.*".

3.3 Plasma oxidation

By plasma oxidation of GaAs gaseous plasma containing oxygen are used. The sources of oxygen are O_2, N_2O or CO_2, and it is excited by a RF coil (Wilmsen, 1985; Hartnagel & Riemenschnieder, 1999). A DC bias oxidation takes place in a similar way to the wet anodization process. In the oxide layers without thermal treatment Ga_2O_3 and As_2O_3 almost in equal proportions were found. Ions which attacked substrate can sputter the surface, and thus lead to a reduced growth rate and to a modification of surface stoichiometry due to a preferential sputtering of the arsenic component (Hartnagel & Riemenschnieder, 1999). The plasma parameters (RF frequency, RF power and gas pressure) may not affect the oxide growth, but they do affect the degree of GaAs surface degradation during the initial stage of oxide formation. In contrast, wet anodic oxidations give almost damage-free oxides.

3.4 Dry thermal oxidation

Dry thermal oxidation processes of GaAs and GaN are carried out in ambient of oxygen or mixture of nitrogen and oxygen. Dry oxidation of GaAs is made rather seldom. Processes are very complicated because of problems with arsenic and its low thermal stability. Typical top oxide layers on GaAs surface consist of mixture: Ga_2O_3 + $GaAsO_4$ + As_2O_3 and are rough. Near the interface of oxide-gallium arsenide occur Ga_2O_3 and elemental As (after: Wilmsen, 1985). These layers are amorphous. By higher oxidation temperature (above 500 °C) oxides are polycrystalline and also rather rough. They contain mainly Ga_2O_3 but $GaAsO_4$ was also observed. The elemental As, small crystallites of As_2O_5 and As_2O_3 appeared in layers as well (after: Pessegi et al., 1998). Arsenic oxides have low thermal stability and during annealing processes oxides undergo decomposition releasing arsenic which escapes from the samples.
Thermal oxidation of GaAs technique has more than thirty years. Thermal oxidation of GaN epilayers is a considerably younger – it is a matter of last ten years.
Gallium nitride needs higher temperature as GaAs or AlAs: typical range of dry oxidation is between 800 and 1100 °C (Chen et al., 2000). Processes are carried out usually in atmosphere of oxygen (Chen et al., 2000; Lin et al., 2006). Chen at al. (Chen et al., 2000) described several

experiments with GaN layers on sapphire substrates. Authors made oxidation of GaN samples in dry oxygen. Time of oxidation was changed from 20 min to 8 h by the flow of O_2 of about 1 slm. Temperature was changed from 800 to 1100 °C. They have observed two different courses for temperatures of over 1000 °C: very rapid oxidation process in the initial stage of oxidation and then, after about 1 h, followed by a relatively slow process. Authors have deliberated after Wolter et al. (Wolter et al., 1998) the reaction rate constant and have concluded that in the first step of oxidation (rapid process) the oxide creation reaction is limited by the rate of reaction on GaN-oxide interface. In second step (slow process by thicker oxide layers) the oxide creation reaction is determined by the diffusion-controlled mechanism (transition from reaction-controlled mechanism to the diffusion-controlled mechanism). They have supposed GaN decomposition at high temperature (over 1000 °C) which can speed up the gallium oxidation (Chen et al., 2000). The authors also have observed volume increase of about 40% after oxidation.

Similar experiments were made by Zhou et al. (Zhou et al., 2008) by oxidation of GaN powder and GaN free-standing substrates with Ga-terminated surface (front side) from HVPE epitaxial processes. They have used dry oxygen as a reactor chamber atmosphere only and have changed time (from 4 to 12 hours) and temperature (850, 900, 950 and 950 °C) of oxidation. According to authors, oxidation rate in temperature below 750 °C is negligible. They have made similar analysis as Chen et al. (Chen et al., 2000) after Wolter et al. (Wolter et al., 1998) and observed similar dependence of the oxide thickness versus time process. In GaN dry oxidation processes one could observe two zones: interfacial reaction-controlled and diffusion-controlled mechanism for low and high temperature, respectively (Zhou et al., 2008). Authors of this paper have wrote about "*thermally grown gallium oxide on (...) GaN substrate*". It is typical for many authors although all of them described oxidation process.

3.5 Wet thermal oxidation

Problems in wet thermal oxidation of GaAs processes are very similar to those which occur during dry oxidation. Arsenic in GaAs has low thermal stability in high temperature and it is rather difficult to carry out oxidation process at the temperature higher than 600 °C. The applied temperatures from the range below 600 °C gave not rewarding results. The obtained by Korbutowicz et al. (Korbutowicz et al., 2008) gallium oxide layers have been very thin and had have weak adhesion.

Processes of wet thermal GaN oxidation are carried out more often. Gallium nitride has better thermal stability than gallium arsenide and one can apply higher temperature to obtained Ga_2O_3 is thicker and has better parameters.

Typical apparatus for wet thermal oxidation of GaAs or GaN is very similar to that which is applied to wet thermal oxidation of AlAs or $Al_xGa_{1-x}As$. It can be: Closed Chamber System CCS (a) or Open Chamber System OCS (b). The open systems are more often used as the systems with closed tube one.

3.5.1 Close chamber systems

Choe et al. have described in their paper (Choe et al., 2000) CCS equipment for AlAs oxidation which was schematically depicted in Figure 5 a. It also can be applied to GaAs oxidation. The quartz reaction (oxidation) chamber had two temperature zones – the upper and lower zone, one for the sample and second for the water source. It was small chamber –

3 cm in diameter by 30 cm in length. Typical amount of water was about 2 cm^3. Chamber with sample and water was closed and the air was evacuated using a pump. After this hermetically closed chamber was inserted into a furnace. During the heating, water was expanded as a vapour and filled whole volume of the quartz ampoule. Typical temperature in the upper zone was 410 °C and in the lower zone was varied from 80 °C to 220 °C. In this apparatus the oxidation process is controlled by two parameters: temperature of oxidation and temperature of water source.

These systems have some advantages: reaction kinetics in controlled by two temperatures: oxidation and water vapour creation, there is a small demand of oxidizing agent - water and no carrier gas. A considerable inconvenience is the necessity of vacuum pumps application.

3.5.2 Open chamber systems

Open chamber system for GaAs and GaN oxidation looks like silicon oxidation system. It consists of horizontal (very often) quartz tube, water bubbler and source of the gases: carrier - nitrogen N_2 or argon Ar and (sometimes) oxygen O_2 (Choquette et al., 1997; Readinger et al., 1999; Pucicki et al., 2004; Geib et al., 2007; Korbutowicz et al., 2008). The three-zone resistant furnace works as a system heating (Fig. 5 b). Korbutowicz et al. (Korbutowicz et al., 2008) have used the bubbler (in the heating jacket with a temperature control) with deionized water H_2O as a source of oxidizing agent and nitrogen N_2 as a main gas and the initial water level was the same in all experiments to keep the same conditions of the carrier gas saturation.

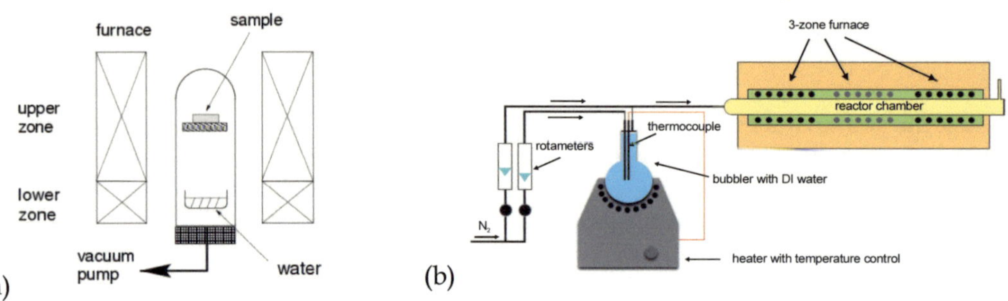

Fig. 5. (a) A schematic diagram of the CCS for wet thermal oxidation (Choe et al., 2000); (b) typical apparatus for GaAs and GaN wet thermal oxidation

The open systems are cheaper as the closed ones. The work with the OCS's are more complicated - one need to take into consideration numerous parameters: source water temperature, reaction temperature, main gas flow and flow of the carrier gas through the bubbler, kind of gases and using or not of oxygen. The significant water consumption during oxidation and the requirement of the water source temperature stabilization also constitute problems. But the valuable advantage of open systems is their simple construction.

Thermal wet oxidation method as a more frequently applied way to get gallium oxide layers will be wider described now.

Reaction kinetics of thermal wet oxidation and reaction results depend on several parameters: a zone reaction temperature (a), a water source temperature (water bubbler) (b), a flow of a main currier gas (c), a flow of a carrier gas through the water bubbler (d), time of the reaction (e) and type of currier gas (f).

Korbutowicz et al. (Korbutowicz et al., 2008) have described processes of the GaAs and GaN thermal wet oxidation - GaAs wafers and GaN layers manufactured by MOVPE and HVPE (Hydride Vapor Phase Epitaxy) on sapphire substrates were used in these studies. GaAs in form of bare wafers (one side polished, Te doped) or wafers with epilayers (Si doped) were employed in investigations. A range of oxidation temperature was between 483 and 526 °C. Time was varied from 60 to 300 minutes. Typical main flow of nitrogen was 2 800 sccm/min and typical flows through the water bubbler were 260 and 370 sccm/min.

Thicknesses of the gallium oxides layers grown on gallium arsenide substrates surface were uneven - it was visible to the naked eye: one can observed variable colors on the surface (see Fig. 6 (a)). Defects are preferable points to create oxide - from these spots started the oxidation process (Fig. 6 (b)). Authors were able to obtain thin layers only, since by longer process duration oxide layers were cracked and exfoliated. In Fig. 6 (c) one can see that oxide layers were thin and transparent. Occurring cracks show that in interface region of GaAs-oxide exists a considerable strain.

(a) 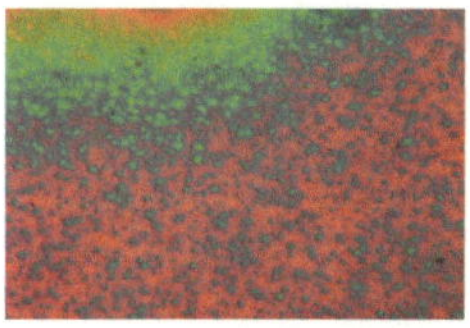 (b) 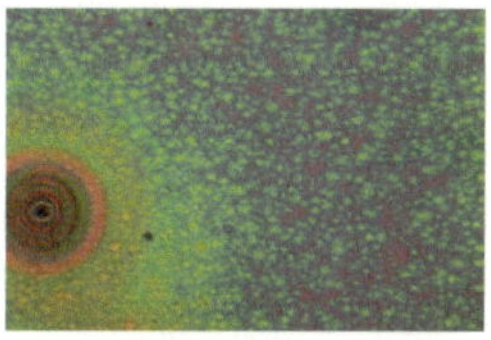 (c) 

Fig. 6. Views of oxide surface's layers from optical microscope: variable colors of gallium oxide (a); substrate's defect and oxide (b); cracked and exfoliated oxide layer (c)

Two kinds of GaN samples have been used - GaN epilayers deposited on sapphire substrates - thin layers from MOVPE and thick layers from HVPE with surface as grown. Temperature of oxidation was higher as for GaAs samples and was as follows: 755, 795 and 827 °C. Typical water temperature was 95 or 96 °C. The main flows of nitrogen were varied from 1 450 to 2 800 sccm/min and the flows through the water bubbler were altered from 260 to 430 sccm/min. The total gas flow in the reactor chamber was about 3 000 sccm/min.

In order to determine suitable parameters, temperature of water source and temperature of reaction (oxidation) zone were changed. Gas flows and time of the process were varied also. The obtained thicknesses of gallium oxide were from several nanometers up to hundreds of nanometers. The MOVPE GaN layers has much more smoother surface as from HVPE ones. The influence of this difference one can remark after oxidation.

Optical observations by using naked eyes and optical microscope gave a lot of information about morphology of surface with oxide. One can observe (Fig. 7.) e.g. smoothing of GaN hexagonal islands. Wet oxidation of gallium arsenide appeared to be more difficult than that of GaN. The Ga_2O_3 layers which were obtained by Korbutowicz et al. were heterogeneous (see below results from X-ray diffraction - Fig. 8).

Fig. 7. HVPE GaN layer surface after wet thermal oxidation

Figure 8. shows x-ray spectrum of gallium compounds on sapphire substrate (G32 sample). One can remark that oxidized surface layer contained GaN, Ga_2O_3 and Ga_xNO_y.

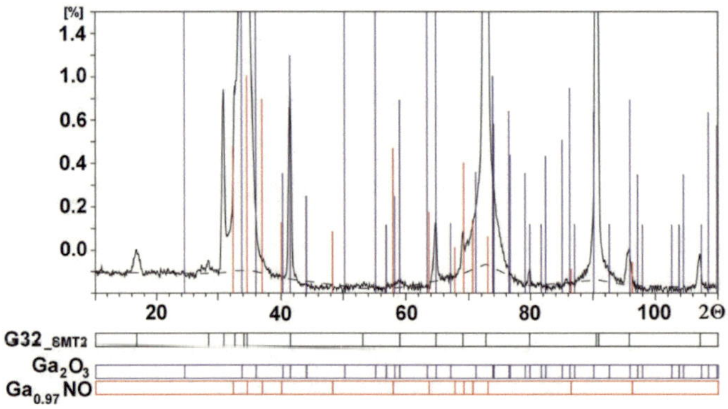

Fig. 8. X-ray diffraction spectrum of oxidized GaN on sapphire from HVPE; G32_SMT2 – spectrum from thick GaN layer

The MOVPE GaN crystals had smoother surface as HVPE crystals and were more resistant for oxidation. In Figure 9 results of AFM (Atomic Force Microscope) observations of the surface and profile of MOVPE sample, thickness of 880 (nm) (a) and HVPE sample, thickness of 12 (μm) (b) are shown. Both samples were oxidized in the same conditions: reaction temperature of 827 °C, water source temperature of 95 °C, process time of 120 min and the same water vapour concentration. The initial surface of MOVPE sample was smooth, while the surface of HVPE thick layers was rather rough. The oxidation process was faster by HVPE crystals because at these crystals surfaces was more developed. The surface of oxidized GaN from MOVPE remained smooth, whereas on the surface of the sample from HVPE one could observe typical little bumps.

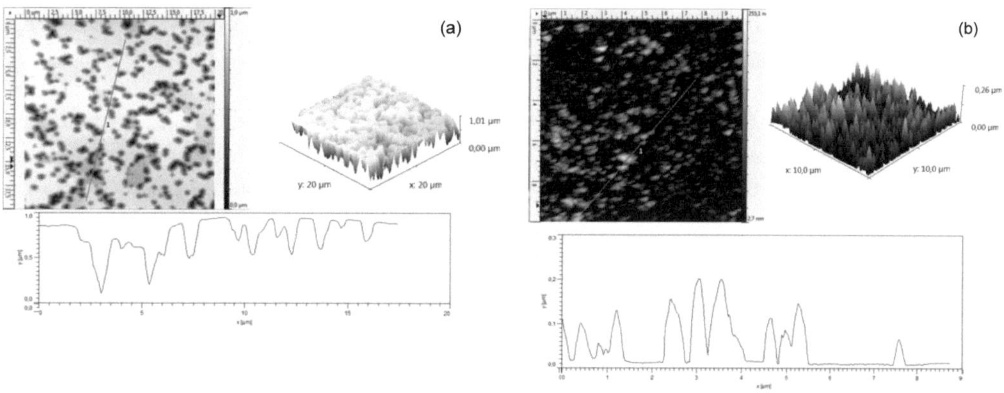

Fig. 9. AFM images of the surface of $GaN_{(MOVPE)}$ sample (a) and $GaN_{(HVPE)}$ sample (b)

Readinger et al. (Readinger et al., 1999) have carried out processes applying GaN powder and GaN thick layers on sapphire from vertical HVPE. Atomic percentage of water vapor in carrier gas (O_2, N_2, and Ar) was maintained on the same level (77%±8%) for all furnace temperatures (700, 750, 800, 850 and 900 °C) and carrier gas combinations. For comparison purposes authors have prepared a dry oxidation processes (in dry oxygen) for the same samples. Sample's surfaces after wet oxidation were much smoother as from dry process. The authors have observed that below 700 °C in which GaN has a good stability in oxidizing environments. They also have found that in ambient of oxygen (dry or wet) the oxidation had faster rate as in wet nitrogen or argon atmosphere. Thicknesses of gallium oxide layers in wet O_2 process revealed linear dependence on duration of oxidation. Wet oxidation have given even poorer electrical results than dry oxidation. The authors have judged that electrical parameters deterioration aroused from very irregular morphology at the wet oxide/GaN interface.

3.6 Other oxidation methods

These above mentioned oxidation methods are not the only ways to get gallium oxide. There are several others ones:

- ion-beam induced oxidation (after: Hartnagel & Riemenschnieder, 1999),
- laser assisted oxidation (Bermudez, 1983),
- low-temperature oxidation (after: Hartnagel & Riemenschnieder, 1999),
- photowash oxidation (Offsay et al., 1986),
- oxidation by an atomic oxygen beam (after: Hartnagel & Riemenschnieder, 1999),
- UV/ozone oxidation (after: Hartnagel & Riemenschnieder, 1999),
- vacuum ultraviolet photochemical oxidation (Yu et al., 1988).

3.7 Summary

Apart from above mentioned methods are several other ways to obtain or manufacture gallium oxide layers. One can deposited by Chemical Vapour Deposition CVD, Physical Vapour Deposition PVD or Physical Vapour Transport PVT methods. One can use Local Anodic Oxidation LAO by applying AFM equipment (Matsuzaki et al., 2000; Lazzarino et al., 2005; Lazzarino et al., 2006) to GaAs or GaN surface oxidizing and creating small regions covered by gallium oxide. As was told earlier in chapter 2, the best parameters for semiconductor devices has monoclinic β-Ga_2O_3. This type of oxide is easy to obtain by thermal oxidation: dry or wet. These methods also give possibility to selective oxidation using dielectric mask (e.g. SiO_2). Despite the difficulties and problems on account of numerous process parameters which ought to be taken into consideration, wet thermal oxidation of GaAs and GaN processes seem to be the best way for making oxide layers for devices applications.

4. Applications of gallium oxide structures in electronics

Due to existent of native silicon oxide domination of silicon in electronics lasts many years. Semiconductor compounds as AIIIBV or AIIIN have very good parameters which just predestine to work in a region of high frequencies and a high temperature with a high power: insulating substrates, high carrier mobility and wide bandgap. These all give a big advantage over Si and their alloys. But silicon still dominates. Why?

SiO_2 is an amorphous material which does not bring strain in underlying silicon. Gallium arsenide GaAs applied in semiconductor devices technology has cubic crystal structure (as other AIIIBV compounds) and typical surface orientation (100). Gallium oxide with monoclinic structure, which is the only variety of Ga_2O_3 stable in high temperature that stays stable after cooling, is strongly mismatched to GaAs. It causes bad relationships between GaAs epitaxial layers and oxide. In addition, gallium oxide growth on a surface of gallium arsenide is in a reality a mixture of Ga_2O_3, As_2O_3, As_2O_5 and elemental As, as was mentioned above. This mixture is unstable at elevated temperature and has poor dielectric parameters. In order to avoid problems with the growth of Ga_2O_3 on GaAs surface some of researches have applied thin dielectric layer of Al_2O_3 in GaAs MOSFET structure (e.g. Jun, 2000) but it is not a matter of our consideration.

By GaN oxidation is other situation than by GaAs treatment. Gallium nitride applied in electronics has hexagonal structure and is better matched. GaN, in comparison to GaAs, is more chemical, thermal and environmental resistant. Therefore nitrides are more often used to construction of numerous devices with a oxide-semiconductor structure: MOS diodes and transistors, gas and chemical sensors.

Silicon electronics supremacy was a result of, among others, applying of silicon oxide SiO_2 possibility. Properties of interface silicon oxide and silicon are just excellent. This fact allows manufacturing of very-large scale integration circuits with Complementary Metal Oxide Semiconductor (CMOS) transistors (Hong, 2008). But silicon devices encounter difficulties going to nanoscale - very thin dielectric gate layers is too thin and there is no effect: charge carriers can flow through the gate dielectric by the quantum mechanical tunnelling mechanism. Leakage current is too high - Si devices need dielectrics with higher electrical permittivity k. Also power devices made from silicon and their alloys operate in smaller range of power and frequency. One can draw a conclusion: MOS devices need high k gate dielectric and carriers with higher mobility in channels of transistors as in silicon's ones. Whole microelectronics requires something else, for example indium phosphide, diamond, silicon carbide, gallium arsenide or gallium nitride and their alloys (see Fig. 10 (Kasu, 2004)).

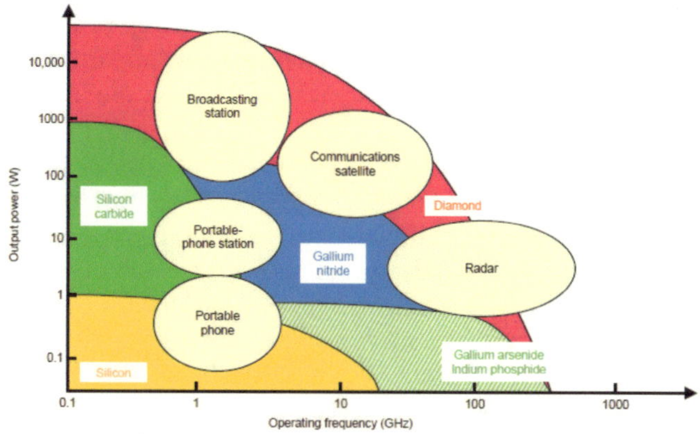

Fig. 10. Demand for high-frequency high-power semiconductors to support the rise in communication capacity (Kasu, 2004)

Despite very good properties, AIIIBV and AIIIN have problems to become commonly used, especially in power applications. A big obstacle is a lack of high quality stable gate dielectrics with high value of dielectric constant. In opinion Ye (Ye, 2008): "*The physics and chemistry of III–V compound semiconductor surfaces or interfaces are problems so complex that our understanding is still limited even after enormous research efforts.*" and that can be the purpose although first GaAs MOSFETs was reported by Becke and White in 1965 (after: Ye, 2008) still there are problems with wide scale production.

One can deposit silicon dioxide, silicon nitride and similar dielectrics but these materials have relatively small dielectric constant. SiO_2 has dielectric constant equal to 3.9, Si_3N_4 has constant = 7.5, but silicon nitride is not easy in a treatment. Typical value of dielectric constant given in literature for Ga_2O_3 is in a range from 9.9 to 14.2 (Passlack et al., 1995; Pearton et al., 1999).

4.1 Metal Oxide Semiconductor devices

The first thermal-oxide gate GaAs MOSFET was reported in the work of Takagi et al. in 1978 (Takagi et al., 1978). The gate oxide, which has been grown by the new GaAs oxidation technique in the As_2O_3 vapor, was chemically stable. Oxidation process was carried out in a closed quartz ampoule. Temperature of liquid arsenic trioxide was equal to 470 °C and temperature of GaAs (gallium oxide growth) was 500 °C. Authors supposed that this method can be used in large scale as a fabrication process. But up to now it is not the typical manufacture technique.

Typical GaAs MOSFET has the gate dielectric in the form of oxides mixture: $Ga_2O_3(Gd_2O_3)$. This mixture comes not from oxidation but from UHV deposition (e.g. Passlack, et al. 1997; Hong et al., 2007; Passlack et al., 2007). Practically almost all papers of Passlack's team from the last twenty years have described oxide structures this type: $Ga_2O_3(Gd_2O_3)$ which were made in UHV apparatus.

Difficulties with obtaining good Ga_2O_3 layers on GaAs from thermal oxidation inclined researches to make GaAs MOS structures with oxidized thin layer of AlGaAs or InAlP but then aluminium is oxidized, not gallium (e.g. Jing et al., 2008).

Matter of the GaN MOS structures looks similar and different too. In many cases gate dielectric is Gadolinium Gallium Garnet (GGG) $Gd_3Ga_5O_{12}$ called also Gadolinium Gallium

Oxide (GGO), a synthetic crystalline material of the garnet group or $Ga_2O_3(Gd_2O_3)$ (e.g. Gila et al., 2000) as by GaAs MOSFETs. Some researches tried to make Ga_2O_3 layer on GaN as dielectric film for MOS applications: MOS capacitors (Kim et al., 2001; Nakano & Jimbo, 2003) or MOS diodes (Nakano a et al., 2003).

Kim et al. (Kim et al., 2001) were studied dry thermal oxidation of GaN in ambient of oxygen. It was a furnace oxidation at 850 °C for 12 h which resulted in the formation of monoclinic β-Ga_2O_3 layer, 88 nm in thickness. Authors have analyzed the structural properties of the oxidized sample by SEM (scanning electron microscopy), XRD and AES (Auger Electron Spectroscopy) measurements. In order to develop the electrical characteristics of the thermally oxidized GaN film, a MOS capacitor was fabricated. Based on observations and measurements, authors have found that: (i) the formation of monoclinic β-Ga_2O_3 occurred, (ii) the breakdown field strength of the thermal oxide was 3.85 $MVcm^{-1}$ and, (iii) the C–V curves showed a low oxide charge density (N_f) of 6.77×10^{11} cm^{-2}. After Kim et al. it suggests that the thermally grown Ga_2O_3 is promising for GaN-based power MOSFET applications (Kim et al.; 2001).

Nakano & Jimbo (Nakano & Jimbo, 2003) have described their study on the interface properties of thermally oxidized n type GaN metal–oxide–semiconductor capacitors fabricated on sapphire substrates. A 100 nm thick β-Ga_2O_3 was grown by dry oxidation at 880 °C for 5 h. After epitaxial growth, authors have made typical lateral dot-and-ring β-Ga_2O_3/GaN MOS capacitors by a thermal oxidation method. In order to reach this aim a 500 nm thick Si layer was deposited on the top surface of the GaN sample as a mask material for thermal oxidation. Formation of monoclinic β-Ga_2O_3 was confirmed by XRD. They have also observed from SIMS (secondary ion mass spectrometry) measurements, an intermediate Ga oxynitride layer with graded compositions at the β-Ga_2O_3/GaN interface (see Fig. 11). The presence of GaNO was remarked by Korbutowicz et al. (Korbutowicz et al., 2008) in samples from the wet thermal oxidation after XRD measurements as well. Nakano & Jimbo (Nakano & Jimbo, 2003) have not observed in the C– t and DLTS (Deep Level Transient Spectroscopy) measurements discrete interface traps. They have judged that it is in reasonable agreement with the deep depletion feature and low interface state density of 5.53×10^{10} $eV^{-1}cm^{-2}$ revealed by the C–V measurements. They have supposed that the surface Fermi level can probably be unpinned at the β-Ga_2O_3/GaN MOS structures fabricated by a thermal oxidation technique. The authors have compared as well the sputtered SiO_2/GaN MOS and β-Ga_2O_3/GaN MOS samples in DLTS measurements. In Fig. 12 results of this study were shown. In contrast to the β-Ga_2O_3/GaN MOS structure, SiO_2/GaN MOS sample has a large number of interface traps may induce the surface Fermi-level pinning at the MOS interface, resulting in the capacitance saturation observed in the deep depletion region of the C–V curve (Nakano & Jimbo, 2003).

In slightly later publication of Nakano et. al. (Nakano a et al., 2003) have described electrical properties of thermally oxidized p-GaN MOS diodes with n^+ source regions fabricated on Al_2O_3 substrates. Oxide was grown in the same way as in paper (Nakano & Jimbo, 2003). Results obtained by authors in this study have suggested that the thermally grown β-Ga_2O_3/p-GaN MOS structure is a promising candidate for inversion-mode MOSFET.

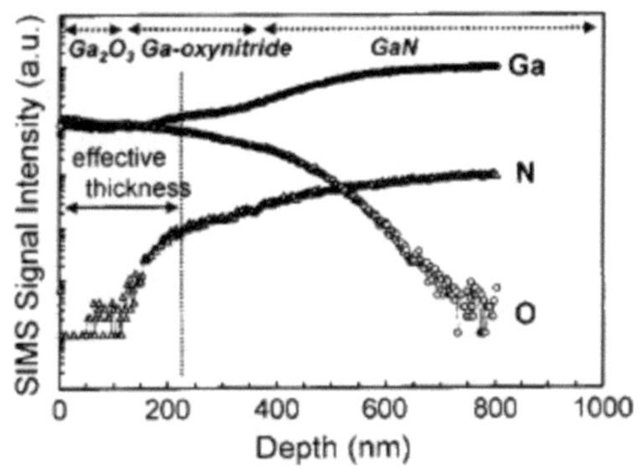

Fig. 11. SIMS profiles of Ga, N, and O atoms in the thermally oxidized β-Ga_2O_3/GaN MOS structure (Nakano & Jimbo, 2003).

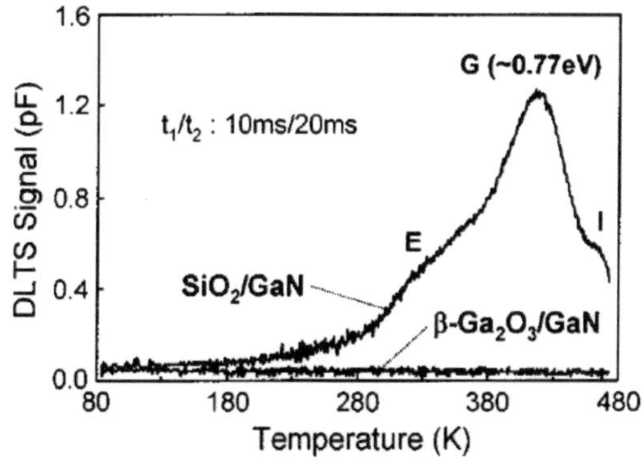

Fig. 12. Typical DLTS spectra at a rate window t_1/t_2 of 10 ms/20 ms for the thermally oxidized β-Ga_2O_3/GaN MOS and sputtered SiO_2/n-GaN MOS structures after applying the bias voltage of 225 V (Nakano & Jimbo, 2003).

Lin et al. (Lin et al., 2006) have studied the influence of oxidation and annealing temperature on quality of Ga_2O_3 grown on GaN. GaN wafers were oxidized at 750 °C, 800 °C and 850 °C. Authors have measured the electrical characteristics and interface quality of the resulting MOS capacitors have compared. The process steps for making GaN MOS capacitor is shown in Fig. 13. The 300-nm SiO_2 layer was deposited on the GaN surface by radio-frequency sputtering to play as a mask for oxidation.

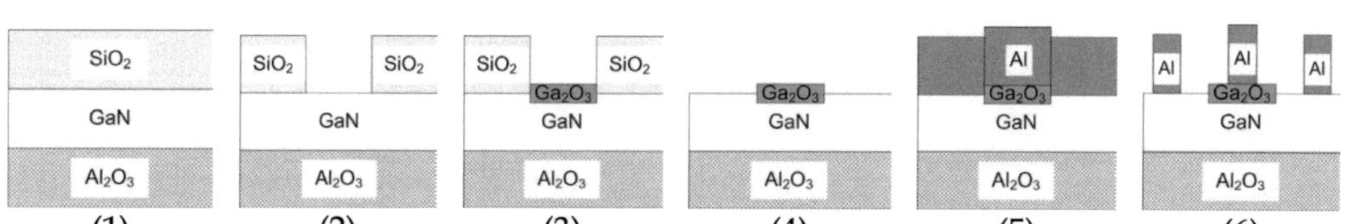

Fig. 13. Process flow for GaN MOS capacitor (Lin et al., 2006)

Oxidation was carried out in dry oxygen ambient and followed by a 0.5 h annealing in argon at the same temperature as oxidation. GaN oxidized at a higher temperature of 850 °C presented better interface quality because less traps were formed at the interface between GaN and the oxide due to more complete oxidation of GaN at higher temperature. But the best current-voltage characteristics and C-V characteristics in accumulation region and surface morphology had the sample from 800 °C oxidation process (Lin et al., 2006).

4.2 Gas sensors

Metal oxides Ga_2O_3 gas sensors operating at high temperatures are an alternative for widely used SnO_2 based sensors. Both types of sensors are not selective but react for a certain group of gasses depending on the temperature of operation. Responses on oxygen, NO, CO, CH_4, H_2, ethanol and acetone are most often investigated. Ga_2O_3 sensors exhibit faster response and recovery time, and lower cross-sensitivity to humidity than SnO_2 based sensors, see Fig. 14 (Fleischer & Meixner, 1999). Additional advantages are long-term stability and no necessity of pre-ageing. Ga_2O_3 sensors show stability in atmospheres with low oxygen content what make them suitable for exhaust gas sensing. There is also no necessity of degassing cycles in contrary to SnO_2 sensors. Disadvantages are lower sensitivity and higher power consumption due to high temperature operation (Hoefer et al., 2001).

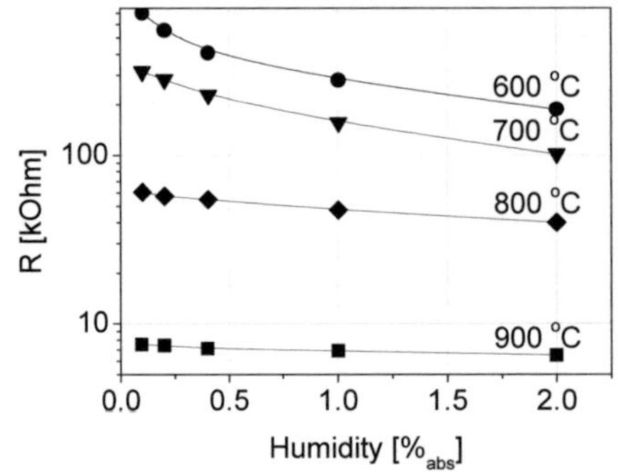

Fig. 14. Temperature dependence of the effect of humidity on the conductivity of Ga_2O_3 thin films, measured in synthetic air (Fleischer & Meixner, 1999)

Typical structure of a gas sensor consists of interdigital electrode (Fig. 16. Type A) (usually platinum) deposited on the sensing layer composed of polycrystalline Ga_2O_3 with grain sizes of 10 and 50 nm (Fleischer a et al., 1996) or 50–100 nm (Schwebel et al., 2000; Fleischer & Meixner et al., 1995).

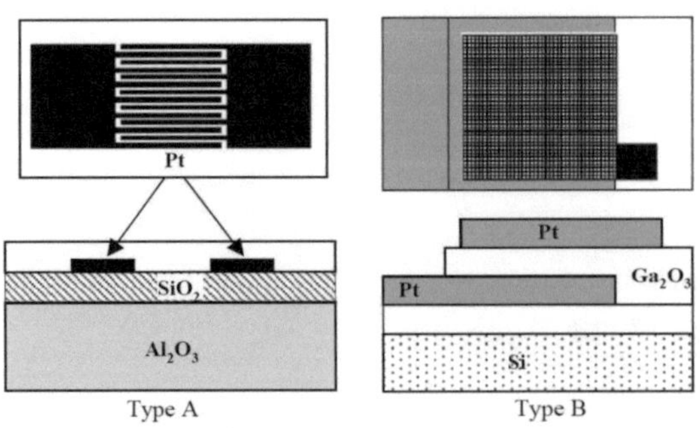

Fig. 15. Typical interdigital oxide sensor (Type A) and modified mesh structure (Type B) (Baban et al., 2005)

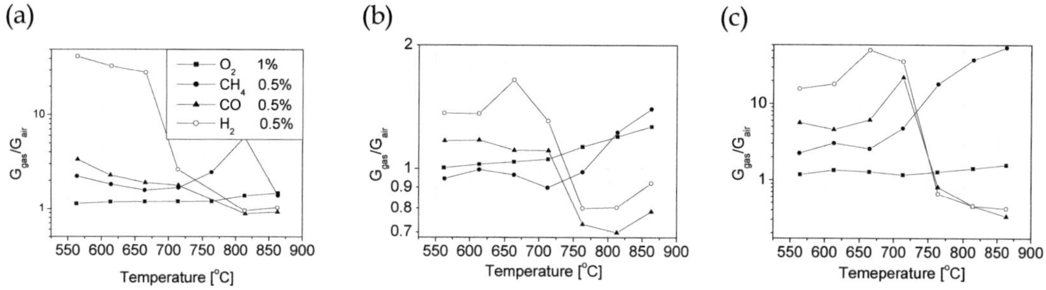

Fig. 16. Comparison of the gas sensitivity of three different morphologies of β-Ga_2O_3: (a) single crystals, (b) bulk ceramics with closed pore structure, and (c) polycrystalline thin film (Fleischer & Meixner, 1999)

However, sensitivities of three different morphologies of β-Ga_2O_3 as single crystals, bulk ceramics with closed pore structure and polycrystalline thin film were also investigated (see Fig. 16) (Fleischer & Meixner, 1999).

Baban et al. proposed sandwich structure with double Ga_2O_3 layer and mesh double Pt electrode layer (Fig. 15. Type B), nevertheless, that device did not achieve neither higher sensitivity nor fast response time, but it helped to conclude about the mechanism of detection (Baban et al., 2005). The most commonly applied fabrication technique is sputtering of thin Ga_2O_3 and its subsequent annealing in order to achieve crystallization of the layer. Although low-cost, screen printed, thick Ga_2O_3 layers with sensing properties similar to that based on thin layers could be also used (Frank a et al., 1998).

Sensing mechanism is assumed to be based on charge carrier exchange of adsorbed gas with the surface of the sensing layer. Resistance modulation is a consequence of the change of free charge carrier concentration resulted from the alteration of acceptor concentration on the surface raising from the reaction of molecules with adsorbed oxygen ions when exposed to oxygen containing ambient (Hoefer et al., 2001).

Generally adsorbed reducing or oxidizing gas species inject electrons into or extract electrons from semiconducting material (Li et al., 2003) thus changing material conductivity. Gallium oxide exhibits gas sensitivity at temperature range from 500 °C to 1000 °C. At lower temperatures reducing gases sensitivity occurred. In the range from 900 °C to 1000 °C the detection mechanism is bound to O_2 defects equilibrium in the lattice (Fleischer b et al., 1996).

Modification of sensor parameters, such as sensitivity, selectivity (cross-sensitivity) and response as well as recovery times for certain gas, could be assured by three ways: temperature modulation, deposition of appropriate filter layer/clusters on the active layer or by its doping. As described in (Fleischer a et al., 1995) gallium oxide layers of 2 μm deposited by sputtering technique (grain sizes typically 50-100 nm) exhibited response to reducing gases in the range of 500 - 650 °C of operating temperatures. Increase of temperature caused decrease of the sensitivity to these gases and simultaneous enhancement of response to NH_4. Temperatures of 740 - 780 °C assured suppression of reducing gases sensitivity leading to the selectivity to NH_4.

Cross-sensitivity of ethanol and other organic solvents to methane were restricted by application of filter layer of porous β-Ga_2O_3 deposited on thin sensing Ga_2O_3 layer Fig. 17 (Flingelli et al., 1998).

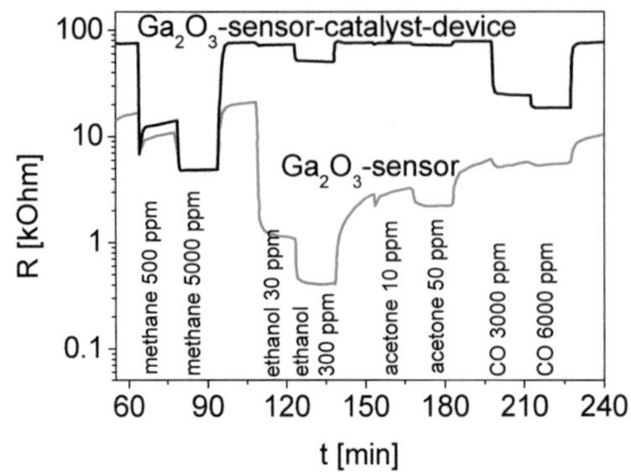

Fig. 17. Response of a pure Ga_2O_3 sensor and a sensor catalyst device (hybrid research type) to methane, ethanol, acetone and CO in wet synthetic air at 800 °C (Flingelli et al., 1998)

Fleischer et al. (Fleischer b et al., 1996) have investigated application of amorphous SiO_2 layer covering Ga_2O_3 on the sensitivity, selectivity and stability of hydrogen sensor. Polycrystalline, 2 µm thick gallium oxide layers were deposited by sputtering technique and subsequently heated at 850 °C for 15 hours or 1100 °C for 1 hour. Crystallites sizes were 10 and 50 nm, respectively. Sensors sensitivity was investigated for: NO (300 ppm by vol.), CO (100 ppm by vol.), CH_4 (1% by vol.), H_2 (1000 ppm by vol.), ethanol (15 ppm by vol.) and acetone (15 ppm by vol.) In order to avoid cross-sensitivity the measurements were prepared in 0.5% of humidity; also influence of humidity reduction to 0.025% by vol. as well as O_2 content from 20 to 1% was evaluated. Uncoated Ga_2O_3 sensor responded by decrease of the conduction of the layer for reducing gases. At lower temperatures stronger response was to more chemically reactive gases in contrary to higher temperatures where significant response to chemically stable gasses was observed. Detection time of H_2 strongly depended on the operating temperature of the sensor. Response time at 600 °C was 10 min and 30 s at above 700 °C. Temperatures of 900 °C and above assured rapid decrease in conductivity of layer. All responses were reversible. To prevent the formation of oxygen on the Ga_2O_3 surface during the oxidation process, what would exclude this kind of layers from the application for H_2 sensing, additional SiO_2 layers were used. Use of 30 nm SiO_2 layer caused lowering of response to reducing gases at temperatures of 900 °C and below, except of H_2. The optimal operation temperature for H_2 detection was 800 °C. Silicon dioxide layers of 300 nm thick have suppressed responses to all gasses at all temperatures except to H_2. In this case optimal temperature of operation was 700 °C. Gallium oxide sensor with SiO_2 cap layer could be used as a selective, high temperature hydrogen sensor (Fleischer b et al., 1996). To assure of oxygen selectivity in oxygen-rich atmospheres Schwebel et al. (Schwebel et al., 2000) have applied catalytically active oxides. Modification materials like CeO_2, Mn_2O_3 and La_2O_3 were deposited on the surface of 2 µm thick Ga_2O_3 sputtered on ceramic substrates and annealed at 1050 °C for 10 hours (crystallite sizes 50–100 nm). Sensors with surface modified by La_2O_3 or CeO_2 responded only to oxygen changes in the ambient, in contrary to uncoated Ga_2O_3 sensor, which reacts with variety of gases. Modification of the surface with Mn_2O_3 caused insensitivity to any gases and thus could be used as reference sensor for

compensation of temperature influence in double sensor construction because of similar values of thermal activation energy for conduction (Schwebel et al., 2000).
Gallium oxide sensors are sensitive for strongly reducing gases. Thus detection of NO_3, NH_3 or CO_2 is considerably restricted. To investigate their influence on the selectivity various layers like Ta_2O_5, WO_3, NiO, $AlVO_4$, $SrTiO_2$, TiO_2 and Ta_2O_3 were deposited on properly prepared sensors consisting of 2 μm thick gallium oxide obtained by sputtering technique and subsequently annealed. Application of TiO_2 and $SrTiO_2$ did not improve the selectivity to O_2 or eliminate the cross-sensitivity to reducing gases. Modification of the surface with WO_3 gave a strong reaction to NH_3 at 600 °C and NO at 350 °C compared to bare Ga_2O_3. In case of NiO coating suppression of reaction with methane was revealed at 600-700 °C. That effect could be used as a reference in double sensor construction. Using of $AlVO_4$ assured selectivity for O_2 when operating at 700 °C and insensitivity to gases at temperature above 900 °C (Fleischer a et al., 1996).
Lang et al. have applied modification of Ga_2O_3:SnO_2 sensing layer surface by iridium, rhodium and ruthenium clusters. Ruthenium modified layers exhibited significant increase of response on ethanol, when iridium modified sensor demonstrated enhanced sensitivity to hydrogen at lower operating temperature. Sensitivity was 80 at 550 °C (3000 ppm H_2) compared to unmodified sensor which sensitivity was 20 at 700 °C (3000 ppm H_2). Measurements of as low concentration as 30 ppm were possible. Rhodium modified sensor could be used only as a detector of presence of ethanol (Lang et al., 2000).
Dopants such as ZrO_2, TiO_2 and MgO were applied in sandwich structure of sensor containing as follows: substrate/Pt interdigital structure/Ga_2O_3/dopant/Ga_2O_3/dopant/ Ga_2O_3 in order to investigate their influence on the sensitivity. However, no influence on the sensitivity to O_2 was reported. Additionally, response decrease to CH_4 for ZrO_2 doping and slight increase for MgO doping was observed (Frank et al., 1996).
Sensitivity to CO and CH_4 was achieved by application of SnO_2 doping in the sandwich structure. The highest response was for 0.1% at. for both gases. However no influence of doping on oxygen sensitivity was observed (Frank b et al., 1998).
Responses on oxygen of Ga_2O_3 semiconducting thin films doped with Ce, Sb, W and Zn were investigated by Li et al. 2003 (Li et al., 2003). Films doped with Zn exhibited the largest responses for gas concentrations as follows: 100 ppm, 1000 ppm and 10000 ppm. The optimum operation temperature was 420 °C. On the other hand Ce doped gallium oxide samples responded promptly to the gas induced. The reaction time was less than 40 s, when that for Zn doped layer was 100 s. Baban et al. have obtained response times on oxygen of 14 and 27 s for ordinary interdigital platinum structure and newly proposed sandwich structure, respectively (Baban et al., 2005). Li et al. have also investigated stability and repeatability of the sensors. Responses of all sensors were relatively reproducible, see Fig. 18 (Li et al., 2003).

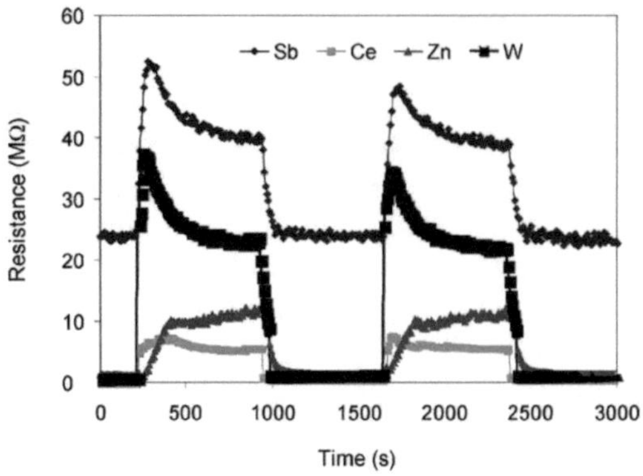

Fig. 18. Electrical response of doped Ga_2O_3 films at temperature of 500 °C (1000 ppm O_2) (Li et al., 2003)

Sensors doped with Sb and W after exposure to the analyzed gas exhibited initial growth of resistance followed by its exponential decrease.

5. Conclusion

Gallium oxide appeared to be a good candidate for optoelectronic and electronic applications. Intrinsic Ga_2O_3 layers have insulating nature, but after appropriate modification could reach conductive parameters. Very interesting effect is n-type semiconducting behavior at elevated temperatures originating from oxygen deficiencies in Ga_2O_3. Gallium oxide is a material included to the group of transparent conductive oxides (TCOs) that are of great interest. Among all TCOs, e.g. ITO or ZnO, β-Ga_2O_3 has the largest value of band-gap what assures high transparency in the range from visible to deep-UV wavelengths. Additionally β-Ga_2O_3 is chemically and thermally stable. That all advantages make β-Ga_2O_3 to be intensively investigated although there is a lot of issues that should researched.

In the chapter main focus was placed on the monoclinic gallium oxide and its most widely applied fabrication methods. There is also a large part devoted to the application of that material. Metal Oxide Semiconductor transistors and gas sensors, based on pure and doped gallium oxide, principles of operation and parameters were described.

Parameters of Ga_2O_3 chosen to the analysis and discussion were selected concerning possible application of that material. Influence of parameters of process of layers deposition or crystal growth on the electrical as well as optical parameters of gallium oxide was included. Possible ways of modification of layers properties are also embraced.

6. Refereces

Al-Kuhaili, M.F.; Durrani, S.M.A. & Khawaja, E.E. (2003). Optical properties of gallium oxide films deposited by electron beam evaporation. *Applied Physics Letters*, Vol. 83, No. 22, (December 2003) 4533-4535, ISSN: 0003-6951

Baban, C.; Toyoda, Y. & Ogita, M. (2005). Oxygen sensing at high temperatures using Ga_2O_3 films. *Thin Solid Films*, Vol. 484, No. 1-2, (July 2005) 369-373, ISSN: 0040-6090

Battiston, G.A.; Gerbasi, R.; Porchia, M.; Bertoncello, R. & Caccavale, F. (1996). Chemical vapour deposition and characterization of gallium oxide thin films. *Thin Solid Films*, Vol. 279, No. 1-2, (June 1996) 115-118, ISSN: 0040-6090

Bermudez, V. M. (1983). Photoenhanced oxidation of gallium arsenide. *Journal of Applied Physics*, Vol. 54, No. 11, (November 1983) 6795-6798, ISSN: 0021-8979

Chen, P.; Zhang R.; Xu X.F.; Chen Z.Z.; Zhou Y.G.; Xie S.Y.; Shi Y.; Shen B.; Gu S.L.; Huang Z.C.; Hu J. & Zheng Y.D. (2000). Oxidation of gallium nitride epilayers in dry oxygen. *Journal of Applied Physics A: Materials Science & Processing*, Vol. 71, No. 2, (August 2000) 191-194, ISSN: 09478396

Choe, J.-S.; Park S.-H.; Choe B.-D. & Jeon H. (2000). Lateral oxidation of AlAs layers at elevated water vapour pressure using a closed-chamber system. *Semiconductor Science and Technology*, Vol. 15, No. 10, (October 2000) L35-L38, ISSN: 0268-1242

Fleischer, M. & Meixner, H. (1993). Electron mobility in single- and polycrystalline Ga_2O_3. *Journal of Applied Physics*, Vol. 74, No. 1, (July 1993) 300-305, ISSN: 0021-8979

Fleischer, M. & Meixner, H. (1995). Sensitive, selective and stable CH_4 detection using semiconducting Ga_2O_3 thin films. *Sensors and Actuators B*, Vol. 26, No. 1, (May 1995) 81-84, ISSN: 0925-4005

Fleischer a, M.; Seth, M.; Kohl, C.-D. & Meixner, H. (1996). A study of surface modification at semiconducting Ga_2O_3 thin film sensors for enhancement of the sensitivity and selectivity. *Sensors and Actuators B*, Vol. 35-36, No. 1-3, (October 1996) 290-296, ISSN: 0925-4005

Fleischer b, M.; Seth, M.; Kohl, C.-D. & Meixner, H. (1996). A selective H_2 sensor implemented using Ga_2O_3 thin-films which were covered with a gas filtering SiO_2 layer. *Sensors and Actuators B*, Vol. 35-36, No. 1-3, (October 1996) 297-302, ISSN: 0925-4005

Fleischer, M. & Meixner, H. (1999). Thin-film gas sensors based on high-temperature-operated metal oxides. *Journal of Vacuum Science and Technology A*, Vol. 14, No. 4, (July/August 1999) 1866-1872, ISSN: 0734-2101

Flingelli, G.K.; Fleischer, M.M. & Meixner, H. (1998). Selective detection of methane in domestic environments using a catalyst sensor system based on Ga_2O_3. *Sensors and Actuators B*, Vol. 48, No. 1, (May 1998) 258-262, ISSN: 0925-4005

Frank, J.; Fleischer, M. & Meixner, H. (1996). Electrical doping of gas-sensitive, semiconducting Ga_2O_3 thin films. *Sensors and Actuators B*, Vol. 34, No. 1, (August 1996) 373-377, ISSN: 0925-4005

Frank a, J.; Fleischer, M. & Meixner, H. (1998). Gas-sensitive electrical properties of pure and doped semiconducting Ga_2O_3 thick films. *Sensors and Actuators B*, Vol. 48, No. 1, (May 1998) 318-321, ISSN: 0925-4005

Frank b, J.; Fleischer, M.; Meixner, H.; & Feltz, A. (1998). Enhancement of sensitivity and conductivity of semiconducting Ga_2O_3 gas sensors by doping with SnO_2. *Sensors and Actuators B*, Vol. 49, No. 1, (June 1998) 110-114, ISSN: 0925-4005

Ghidaoui, D.; Lyon, S. B.; Thomson, G. E. & Walton, J. (2002). Oxide formation during etching of gallium arsenide. *Corrosion Science*, Vol. 44, No. 3, (March 2002) 501-509, ISSN: 0010-938X

Gila, B. P.; Lee, K. N.; Johnson, W.; Ren, F.; Abernathy, C. R.; Pearton, S. J.; Hong, M.; Kwo, J.; Mannaerts, J. P.; Anselm, K. A. (2000). A comparison of gallium gadolinium oxide and gadolinium oxide for use as dielectrics in GaN MOSFETs; *High Performance Devices, Proceedings of IEEE/Cornell Conference 2000*, pp. 182-191, August 2000, Ithaca, NY, USA, ISBN: 0-7803-6381-7, IEEE CNF

Hao, J. & Cocivera, M. (2002). Optical and luminescent properties of undoped and rare-earth-doped Ga_2O_3 thin films deposited by spray pyrolysis. *Journal of Physics D: Applied Physics*, Vol. 35, No. 5, (March 2002) 433-438, ISSN: 0022-3727

Hoefer, U.; Franf, J. & Fleischer, M. (2001). High temperature Ga_2O_3-gas sensors and SnO_2-gas sensors: a comparison. *Sensors and Actuators B*, Vol. 78, No. 1, (August 2001) 6-11, ISSN: 0925-4005

Hong, M.; Lee, W. C.; Huang, M. L.; Chang, Y. C.; Lin, T. D.; Lee, Y. J.; Kwo, J.; Hsu, C. H. & Lee, H. Y. (2007). Defining new frontiers in electronic devices with high kappa dielectrics and interfacial engineering. *Thin Solid Films,* Vol. 515, No. 14, (May 2007) 5581-5586, ISSN: 0040-6090

Jing, Z.; Kosel, T. H.; Hall, D. C. & Fay, P. (2008). Fabrication and Performance of 0.25-μm Gate Length Depletion-Mode GaAs-Channel MOSFETs With Self-Aligned InAlP Native Oxide Gate Dielectric. *Electron Device Letters, IEEE*, Vol. 29, No. 2, (February 2008) 143-145, ISSN: 0741-3106

Jun, B. K.; Kim D.;H.; Leem, J. Y.; Lee, J. H.; Lee, Y. H. (2000), Fabrication of a depletion mode GaAs MOSFET using Al_2O_3 as a gate insulator through the selective wet oxidation of AlAs, *Thin Solid Films*, Vol. 360, No. 1-2, (February 2000), 229-232, ISSN: 0040-6090

Kasu, M. (2004), Selected Papers: Advanced Materials and Their Applications to Future Functional Devices - Microwave Operation of Diamond Field-Effect Transistor, *NTT Technical Review*, Vol. 2, No. 6, (June 2004), 19-24, ISSN: 13483447

Kim, H-G. & Kim, W-T. (2000). Optical properties of β-Ga_2O_3 and α-Ga_2O_3:Co thin films grown by spray pyrolysis, *Journal of Applied Physics*, Vol. 62, No. 5, (September 1987) 2000-2002, ISSN: 0021-8979

Kim, H.; Park, S.-J. & Hwang, H. (2001). Thermally oxidized GaN film for use as gate insulators. *Journal of Vacuum Science & Technology B*, Vol. 19, No. 2., (March/April 2001) 579-581, ISSN: 1071-1023

Kim, H.W. & Kim, N.H. (2004). Annealing effects on the properties of Ga_aO_3 thin films grown on sapphire substrates by metal organic chemical vapor deposition. *Applied Surface Science,* Vol. 230, No. 1-4, (May 2004) 301-306, ISSN: 0169-4332

Korbutowicz, R.; Prażmowska, J.; Wągrowski, Z.; Szyszka, A. & Tłaczała, M. (2008), Wet thermal oxidation for GaAs, GaN and Metal/GaN device applications, *The Seventh International Conference on Advanced Semiconductor Devices and Microsystems.* ASDAM 2008, pp. 163-166, ISBN: 978-1-4244-2325-5, Smolenice, Slovakia, October 2008, Ed. by S. Hašcik, J. Osvald, NJ: IEEE CNF, cop. 2008, Piscataway

Lang, A.C.; Fleischer, M. & Meixner, H. (2000). Surface modifications of Ga_2O_3 thin film sensors with Rh, Ru and Ir clusters. *Sensors and Actuators B*, Vol. 66, No. 1, (July 2000) 80-84, ISSN: 0925-4005

Lazzarino, M.; Padovani, M.; Mori, G.; Sorba, L.; Fanetti, M. & Sancrotti, M. (2005). Chemical composition of GaAs oxides grown by local anodic oxidation: a spatially resolved Auger study. *Chemical Physics Letters*, Vol. 402, No. 1-3, (January 2005) 155-159, ISSN: 0009-2614

Lazzarino, M.; Mori, G.; Sorba, L.; Ercolani, D.; Biasiol, G.; Heun, S. & Locatelli, A. (2006). Chemistry and formation process of Ga(Al)As oxide during local anodic oxidation nanolithography. *Surface Science*, Vol. 600, No. 18, (September 2006), 3739-3743, ISSN: 0039-6028

Li, Y.; Trinchi, A.; Włodarski, W.; Galatsis, K. & Kalantar-zadeh, K. (2003). Investigation of the oxygen gas sensing performance of Ga_2O_3 thin films with different dopants, *Sensors and Actuators B*, Vol. 93, No. 1 , (August 2003) 431 - 434, ISSN: 0925-4005

Lin, L.; Luo, Y.; Lai, P.T. & Lau, K. M. (2005). Effects of oxidation temperature on Ga_2O_3 film thermally grown on GaN, *Proceedings of* 2005 IEEE Conference on Electron Devices and Solid-State Circuits, pp. 605-608, ISBN: 0-7803-9339-2, Hong Kong, December 2005

Lin, L. M.; Luo, Y.; Lai, P. T. & Lau, K. M. (2006). Influence of oxidation and annealing temperatures on quality of Ga_2O_3 film grown on GaN. *Thin Solid Films*, Vol. 515, No. 4, (December 2006) 2111-2115, ISSN: 0040-6090

Martens, K.; Wang, W.F.; Dimoulas, A.; Borghs, G.; Meuris, M.; Groeseneken, G. & Maes, H.,E. (2007). Determining weak Fermi-level pinning in MOS devices by conductance and capacitance analysis and application to GaAs MOS devices. *Solid-State Electronics*, Vol. 51, No. 8, (August 2007) 1101–1108, ISSN: 0038-1101

Matsuzaki, Y.; Hamada, A. & Konagai, M. (2000). Improvement of nanoscale patterning of heavily doped p-type GaAs by atomic force microscope (AFM)-based surface oxidation process. *Journal of Crystal Growth*, Vol. 209, No. 2-3, (February 2000) 509-512, ISSN: 0022-0248

Matsuzaki, K.; Hiramatsu, H.; Nomra, K.; Yanagi, H.; Kamiya, T.; Hirano, M. & Holono, H. (2006). Growth structure, and carrier transport properties of Ga_2O_3 epitaxial film examined for transparent field-effect transistor. *Thin Solid Films*, Vol. 496, No. 1, (February 2006) 37-41, ISSN: 0040-6090

Nakano, Y. & Jimbo, T. (2003). Interface properties of thermally oxidized n-GaN metal–oxide–semiconductor capacitors. *Applied Physics Letters*, Vol. 82, No. 2, (January 2003) 218-220, ISSN: 0003-6951

Nakano a, Y.; Kachi, T. & Jimbo, T. (2003). Electrical properties of thermally oxidized p-GaN metal–oxide–semiconductor diodes. *Applied Physics Letters*, Vol. 82, No. 15, (April 2003) 2443-2245, ISSN: 0003-6951

Nakano b, Y.; Kachi, T. & Jimbo, T. (2003). Inversion behavior in thermally oxidized p-GaN metal–oxide–semiconductor capacitors. *Journal of Vacuum Science and Technology B*, Vol. 21, No. 5 (September/October2003) 2220-2222, ISSN: 1071-1023

Offsey, S. D.; Woodall, J. M., Warren, A. C.; Kirchner, P. D.; Chappell, T. I. & Pettit, G. D. (1986). Unpinned (100) GaAs surfaces in air using photochemistry. *Applied Physics Letters*, Vol. 48, No. 7, (February 1986) 475-477, ISSN: 0003-6951

Orita, M.; Ohta, H.; Hirano, M. & Hosono, H. (2000). Deep-ultraviolet transparent conductive β-Ga_2O_3 thin films. *Applied Physics Letters*, Vol. 77, No. 25, (December 2000) 4166-4168, ISSN: 0003-6951

Orita, M.; Hiramatsu, H.; Ohta, H.; Hirano, M. & Hosono H. (2002). Preparation of highly conductive, deep ultraviolet transparent β-Ga_2O_3 thin film at low deposition temperatures. *Thin Solid Films*, Vol. 411, No. 1, (May 2002) 134–139, ISSN: 0040-6090

Oshima, T.; Okuno, T. & Fujita, S. (2007). Ga_2O_3 Thin Film Growth on c-Plane Sapphire Substrates by Molecular Beam Epitaxy for Deep-Ultraviolet Photodetectors. *Japanese Journal of Applied Physics*, Vol. 46, No. 11, (November 2007) 7217-7220, ISSN: 1347-4065

Pakes, A.; P. Skeldon , G. E. Thompson, J. W. Fraser, S. Moisa, G. I. Sproule, M. J. Graham a& S. B. Newcomb (2003), Anodic oxidation of gallium nitride. *Journal of Materials Science*, Vol. 38, No. 2, (January 2003) 343-349, ISSN: 0022-2461

Passlack, M.; Schubert, E. F.; Hobson, W. S.; Hong, M.; Moriya, N.; Chu, S. N. G.; Konstadinidis, K.; Mannaerts, J. P.; Schnoes, M. L. & Zydzik, G. J. (1995). Ga_2O_3 films for electronic and optoelectronic applications. *Journal of Applied Physics*, Vol. 77, No. 2, (January 1995) 686-693, ISSN: 0021-8979

Passlack, M.; Hong, M.; Mannaerts, J.P.; Opila, R. L.; Chu, S. N. G.; Moriya, N.; Ren, F. & Kwo, J. R. (1997). Low D_{it}, thermodynamically stable Ga_2O_3-GaAs interfaces: fabrication, characterization, and modeling. *Electron Devices, IEEE Transactions on*, Vol. 44, No. 2, (February 1997) 214-225, ISSN: 0018-9383

Passlack, M.; Zurcher, P.; Rajagopalan, K.; Droopad, R.; Abrokwah, J.; Tutt, M.; Park, Y-B.; Johnson, E.; Hartin, O.; Zlotnicka, A.; Fejes, P.; Hill, R.J.W.; Moran, D.A.J.; Li, X.; Zhou, H.; Macintyre, D.; Thoms, S.; Asenov, A.; Kalna, K. & Thayne, I. G. (2007). High mobility III-V MOSFETs for RF and digital applications. *IEEE International Electron Devices Meeting IEDM 2007*, December 2007, pp. 621-624, IEEE, Washington DC, USA

Pearton, S. J.; Zolper, J. C.; Shul, R. J. & Ren, F. (1999), GaN: Processing, defects, and devices, *Journal of the Applied Physics*, Vol. 86, No. 1, (July 1999) 1-78, ISSN: 0021-8979

Peng, L.-H.; Hsu, Y.-C.; Chen, C.-Y.; Ho, J.-K. & Huang, C.-N. (2001). Method of oxidizing nitride material enhanced by illumination with UV light at room temperature, United States Patent, Patent No. 6190508, Feb. 20, 2001

Pucicki, D.; Korbutowicz, R.; Kania, A. & Adamiak, B. (2004). Elaboration of thermal wet oxidation of AlAs/GaAs Distributed Bragg Reflectors. *The Fifth International Conference on Advanced Semiconductor Devices and Microsystems ASDAM 2004*, pp. 179-181, ISBN: 0-7803-8335-7, Smolenice, Slovakia, October 2004, Ed. by J. Osvald, S. Hašcik, NJ, IEEE CNF, (cop. 2004), Piscataway

Readinger, E.D.; Wolter, S.D.; Waltemyer, D.L.; Delucca, J.M.; Mohney, S.E.; Prenitzer, B.I.; Giannuzzi, L.A. & Molnar, R.J. (1999). Wet thermal oxidation of GaN, *Journal of Electronic Materials*, Vol. 28, No. 3, (March 1999) 257-260, ISSN: 0361-5235

Rebien, M.; Henrion, W.; Hong, M.; Mannaerts, J.P. & Fleischer M. (2002). Optical properties of gallium oxide thin films. *Applied Physics Letters*, Vol. 81, No. 2, (July 2002) 250-252, ISSN: 0003-6951

Schwebel, T.; Fleischer, M. & Meixner, H. (2000). A selective, temperature compensated O sensor based on Ga O thin films. *Sensors and Actuators B,* Vol. 65, No. 1-3, (June 2000) 176–180, ISSN: 0925-4005

Shan, F.K.; Liu, G.X.; Lee, W.J.; Lee, G.H.; Kim, I.S. & Shin, B.C. (2005). Structural, electrical, and optical properties of transparent gallium oxide thin films grown by plasma-enhanced atomic layer deposition. *Journal of Applied Physics,* Vol. 98, No. 2, (July 2005) 023504-1-023504-6, ISSN: 0021-8979

Takagi, H.; Kano, G. & Teramoto, I. (1978). Thermal-oxide gate GaAs MOSFET's. *Electron Devices, IEEE Transactions on*, Vol. 25, No. 5, (May 1978) 551-552, ISSN: 0018-9383

Tippins, H. H. (1965). Optical absorption and conductivity in the band edge of β-Ga_2O_3. *Physical Review,* Vol. 140, No. 1A, (October 1965) A316-A319, ISSN: 0031-9007

Tomm, Y.; Reiche, P.; Klimm, D. & Fukuda, T. (2000). Czochralski grown Ga_2O_3 crystals. *Journal of Crystal Growth,* Vol. 220, No. 4, (December 2000) 510-514, ISSN: 0022-0248

Ueda a, N.; Hosono, H.; Waseda, R. & Kawazoe, H. (1997). Synthesis and control of conductivity of ultraviolet transmitting β-Ga_2O_3 single crystals. *Applied Physics Letters,* Vol. 70, No. 26, (June 1997) 3561-3563, ISSN: 0003-6951

Ueda b, N.; Hosono, H.; Waseda, R. & Kawazoe, H. (1997). Anisotropy of electrical and optical properties in β-Ga_2O_3 single crystals. *Applied Physics Letters,* Vol. 71, No. 7, (August 1997) 933-935, ISSN: 0003-6951

Vı́llora, E. G.; Shimamura, K.; Yoshikawa, Y.; Aoki, K. & Ichinose, N. (2004). Large-size β-Ga_2O_3 single crystals and wafers. *Journal of Crystal Growth,* Vol. 270, No. 3-4, (October 2004) 420-426, ISSN: 0022-0248

Wilmsen, C. W. (1985). Oxide/III-V Compound Semiconductor Interfaces - Anodic oxides, in: *Physics and chemistry of III-V compound semiconductor interfaces,* Carl W. Wilmsen (Ed.), Plenum Press 1985, New York and London

Wilmsen, C. W. (1985). Oxide/III-V Compound Semiconductor Interfaces - Plasma-grown oxide, in: *Physics and chemistry of III-V compound semiconductor interfaces,* Carl W. Wilmsen (Ed.), Plenum Press 1985, New York and London

Wilmsen, C. W. (1985). Oxide/III-V Compound Semiconductor Interfaces - Thermal oxides, in: *Physics and chemistry of III-V compound semiconductor interfaces,* Carl W. Wilmsen (Ed.), Plenum Press 1985, New York and London

Wolter, S. D.; Mohney, S. E.; Venugopalan, H.; Wickenden, A. E. & Koleske, D. D. (1998). Kinetic Study of the Oxidation of Gallium Nitride in Dry Air. *Journal of Electrochemical Society,* Vol. 145, No. 2, (February 1998) 629-632, ISSN: 0013-4651

Wolter, S.D.; DeLucca, J.M.; Mohney, S.E.; Kern, R.S.; & Cuo C.P. (2000). An investigation into early stages of oxide growth on gallium nitride. *Thin Solid Films,* Vol. 371, No. 1-2, (August 2000) 153-160, ISSN: 0040-6090

Wu, H.-M.; Lin, J.-Y.; Peng, L.-H.; Lee, C.-M.; Chyi, J.-I. & Chen, E. (2003). Annealing effects on the interfacial properties of GaN MOS prepared by photo-enhanced wet oxidation in International Semiconductor Device Research Symposium 3003, 10-12 Dec. 2003, 2003, 406-407

Xie, X. N.; Chung, H. J.; Sow, C. H. & Wee, A. T. S. (2006). Nanoscale materials patterning and engineering by atomic force microscopy nanolithography, *Materials Science and Engineering R*, Vol. 54, No. 1-2, (November 2006) 1–48, ISSN: 0927-796X

Ye, P.D. (2008). Main determinants for III-V metal-oxide-semiconductor field-effect transistors (invited). *Journal of Vacuum Science and Technology A*, Vol. 26, No. 4, (July 2008) 697–704, ISSN: 0734-2101

Yu, C. F.; Schmidt, M. T.; Podlesnik, D. V.; Yang, E. S. & Osgood, R. M., Jr. (1988). Ultraviolet-light-enhanced reaction of oxygen with gallium arsenide surfaces. *Journal of Vacuum Science and Technology* A, Vol. 6, No. 3, (May 1988) 754-756, ISSN: 0734-2101

Zhou, Y.; Ahyi, C.; Isaacs-Smith, T.; Bozack, M.; Tin, C.-C.; Williams, J.; Park, M.; Cheng, A.-J.; Park, J.-H.; Kim, D.-J.; Wang, D.;. Preble, E.A; Hanser, A. & Evans, K. (2008). Formation, etching and electrical characterization of a thermally grown gallium oxide on the Ga-face of a bulk GaN substrate. *Solid-State Electronics*, Vol. 52, No. 5, (May 2008) 756–764, ISSN: 0038-1101

7

Selective Oxidation on High-Indium-Content InAlAs/InGaAs Metamorphic High-Electron-Mobility Transistors

Yeong-Her Wang [a,*] and Kuan-Wei Lee [b]
[a] *Institute of Microelectronics, Department of Electrical Engineering, Advanced Optoelectronic Technology Center, National Cheng-Kung University, Tainan 701 Taiwan*
[b] *Department of Electronic Engineering, I-Shou University, Kaohsiung County 840 Taiwan*

1. Introduction

Up to now, many efforts have been continuously channeled toward the development of oxidation techniques on the III-V compounds for GaAs-based device application, which include thermal oxidation [1-7], chemical anodization [8-12], photochemical oxidation [13-16], plasma oxidation [17-20], Ga_2O_3 grown by molecular beam epitaxy (MBE) [21-23], Al_2O_3 grown by atomic layer deposition (ALD) [24], oxidized GaAs or InAlAs prepared by ultraviolet and ozone [25-27], and so on. Although the electrical quality of the GaAs-based MOS structures demonstrated to date is not as good as those obtained from the more mature SiO_2/Si system, some of them have yielded promising results for electronic and optoelectronic applications. However, the growth of oxides on the III-V surface is more complex than that on Si. Most of these methods require condensed gases, energy sources (such as excited plasma, electric potential, or optical illumination) or ultrahigh vacuum chamber, and so on, which complicate the oxidation process.
In the past years, a technique named liquid phase oxidation (LPO) [28] on GaAs-based materials operated at low temperature (30 °C to 70 °C) has been proposed and investigated. Much progress has been made to form a high-quality oxide on GaAs, for example, the mechanism and kinetics of oxidation [29], fabrication of GaAs MOSFET [30], pre-treatment and post-oxidation annealing of the oxide [31, 32], and GaAs-based devices [33, 34]. The oxidation takes place through the in-diffusion of oxygen at the semiconductor-oxide interface, where a fresh interface at the original semiconductor surface is achieved. This is an easy, economic, and low-temperature method to grow uniform and smooth native oxide films on GaAs-based materials. Utilizing the electroless technique, neither vacuum, gas condensation equipment, nor an assisting energy source is needed. Meanwhile, the technique has potential advantages for electronic and optical device applications due to its

substantial flexibility in device heterostructure designs and fabrications.

Another purpose of the work is to use the photoresist (PR) or metal as a mask for selective oxide growth on InAlAs with the low-cost, low-temperature LPO method. PR is widely utilized for photolithography processes and can be used as a mask for some device fabrication processes. However, the appearance of inherent problems such as flowing, outgassing, or blistering makes the PR unstable and useless at a high temperature [35]. The pH values of the aqueous oxidation solution for the LPO system range approximately from 5 to 3. Within the temperature and pH range, the PR is very stable. Utilizing the LPO method, the proposed application uses the PR as a stable mask for selective oxide growth on InAlAs.

InAlAs/InGaAs metamorphic high electron mobility transistors (MHEMTs) on GaAs substrates are characterized by high gains and low noise in millimeter-wave applications. They provide promising advantages over the structures grown on InP substrates, since they are less expensive, less fragile, and are available on a large scale. Meanwhile, efforts have been substantially devoted on the improvement of the instability and breakdown voltage. To solve the first problem, InP [36] or InGaP [37] has been used to achieve long-term reliability and to act as an etch-stop layer in a selective-etch recessed-gate process. However, if the InP is used as a Schottky layer, a special structure must be involved to enhance the Schottky barrier height on InP, which may still suffer from the high gate leakage issue. For the second problem, the composite channel [38] or the doped channel [39] has been used to overcome the small bandgap energy of the InGaAs channel. Aside from this, higher aluminum content in the InAlAs Schottky layer also induces a gate leakage issue, which causes the deterioration of device performance, especially when operating at higher bias conditions.

In conventional HEMT device, the undoped Schottky layer was used as the gate insulator, operating in a MIS transistor-like mode [40, 41]. Further improvement in leakage current and breakdown voltage for Schottky gate HEMT can be surmounted by using oxide film as an insulator between the two-dimensional electron gas (2DEG) channel and the gate electrode. The MOS-HEMT not only has the advantages of the MOS structure but also has the high-density, high-mobility, 2DEG channel. In addition, a very low interface trap density is needed in the oxide-semiconductor interface for MOSFET, however, which is different from the 2DEG channel positioned away from the oxide film with a barrier layer for MOS-HEMT in this study. When a negative voltage is applied to the gate, the electrons are depleted from a triangular quantum well. In this case, the vertical electric field points from the channel towards the gate electrode. As a result, some of the holes that are produced during impact ionization can get across the InAlAs barrier layer and are collected easily at the gate electrode without oxide film. Further discussion of impact ionization will follow later in context. When gate bias is made more positive, the bands straighten out and the vertical field drops. When the gate is more positively biased, the electrons are accumulated in a rectangular quantum well [34].

In order to achieve a better performance for InAlAs/InGaAs HEMTs, such as a smaller leakage current and higher breakdown voltage, one has to understand the mechanism of gate leakage and find the optimal device parameters. In this work, a thin InAlAs native oxide layer prepared by means of LPO as the gate dielectric for a 0.65 μm InAlAs/InGaAs MOS-MHEMT application is demonstrated.

2. Experimental

2.1 Liquid phase oxidation

For the LPO method, the most important and fundamental procedure is to prepare the growth solution. First, gallium-ion-containing nitric acid solution is obtained by the sufficient dissolution of high purity (6N) gallium metal in hot (60 °C) and concentrated nitric acid (70%) for more than 8 h and is then diluted with de-ionized (DI) water, ready for use. The second process is the pH adjustment of the solution, which is yet another critical process for the LPO method. The adjustment processes are performed by adding diluted ammonia water solution into the nitric acid solution. The pH value of the solution is usually adjusted within the range of 4.0 to 5.0, found to be the optimum initial pH value for oxidation. Finally, a clear solution is obtained by filtration with a pore size less than 0.1 μm. Figure 1 shows the simple growth system for LPO which consists of a temperature-controller heater and a pH meter. The GaAs-based wafers were first cleaned by organic solvents, followed by polishing and etching to remove the contaminants and residual oxides. These as-received wafers were prepared with minimized defect density before transferring into the LPO system. The oxidation procedure is performed by simply immersing the as-received wafers into the growth solution at a constant temperature. Moreover, in order to ensure the growth of uniform oxide layers, it is necessary to stir the growth solution and monitor the pH value during the oxidation. Without stirring the growth solution, the uniformity of the as-grown oxide will be relatively poor. Using the method described above, the wafers were oxidized at a constant temperature of 50 °C and finally rinsed by DI water and dried in nitrogen.

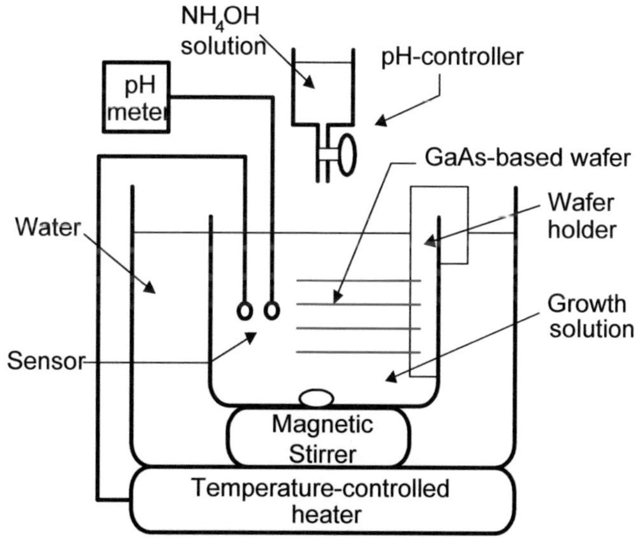

Fig. 1. The LPO system configuration.

2.2 Selective oxidation on InAlAs

The selective oxidation process is schematically illustrated in Fig. 2. After etching the InGaAs capping layer, the PR was coated on the InAlAs layer, and the pattern of which was designed by the photolithographic processes. Then the sample was transferred into the growth solution for oxidation. An oxide layer can be grown only on a bare InAlAs surface that is not covered by PR. Since the oxidation occurs only at the oxide-semiconductor interface, the hetero deposition of films on PR can be avoided. After removing the PR, the final selectively oxidized structure can be obtained. As shown in Fig. 3, a high contrast between InAlAs and oxide layer on the top surface can be seen by the scanning electron microscopy (SEM) image. Similar results can also be observed by using metal masks (e.g., Au/Ge/Ni, Au, etc.) for selective oxidation.

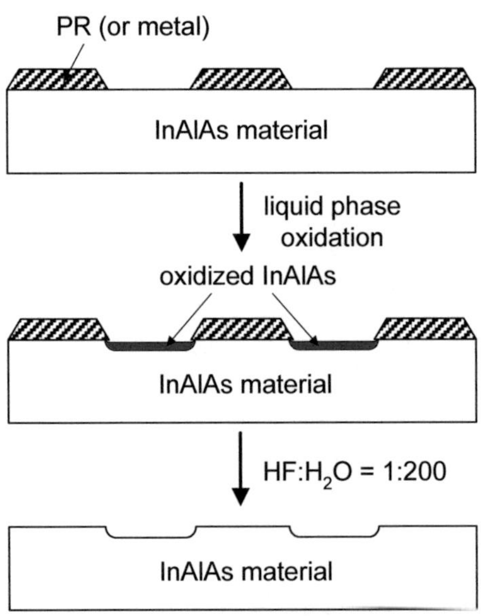

Fig. 2. Cross-sectional view of the proposed selective oxidation procedure on InAlAs material.

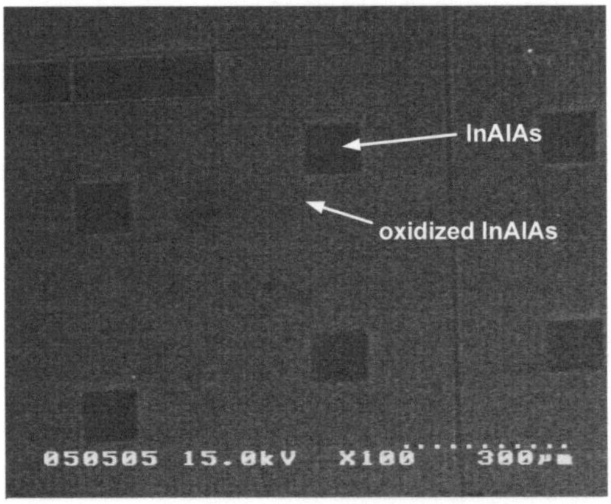

Fig. 3. Example of a top view of a SEM image. The high contrast of the InAlAs and oxide surface can be seen.

In this study, the MHEMT epitaxial structure was grown by metal-organic chemical vapor deposition (MOCVD) on a semi-insulating GaAs substrate as shown in Fig. 4. The measured room- temperature Hall mobility and sheet carrier concentration were 7000 cm²/Vs and 2×10^{12} cm^{-2}, respectively. The MOS-MHEMT fabrication started with mesa isolation by wet etching down to the buffer layer. The ohmic contacts of the Au/Ge/Ni metal were deposited by evaporation, and were then patterned by lift-off processes, followed by rapid thermal annealing. The gate recess was etched by the citric buffer etchant which was composed of the volume ratio of CA:H_2O_2 = 1:1 (CA was made by mixing the monohydrate citric acid and H_2O of 1:1 by weight) [42]. This step also leads a selective sidewall recessing to etch the exposed part of channel layer simultaneously, resulting in a reduction of gate leakage [43], as shown in Fig. 5. Then applying the LPO procedure, the wafer was immersed into the growth solution to generate a gate oxide at 50 °C for 15-30 min, yielding an oxidation rate of about 20 nm/h. After which, the oxide film selectively and simultaneously passivated the walls of the isolated surface. Utilizing the LPO, the proposed application uses the Au/Ge/Ni metal as a mask for selective oxide growth on InAlAs. Finally, the gate metal Au was deposited. The gate dimension and the drain-to-source spacing are 0.65×200 μm² and 3 μm, respectively. The current-voltage (*I-V*) properties were characterized using HP4156B, and the microwave on-wafer measurements were conducted from 0.45-50 GHz in common-source configuration using an Agilent E8364A PNA network analyzer at 300 K.

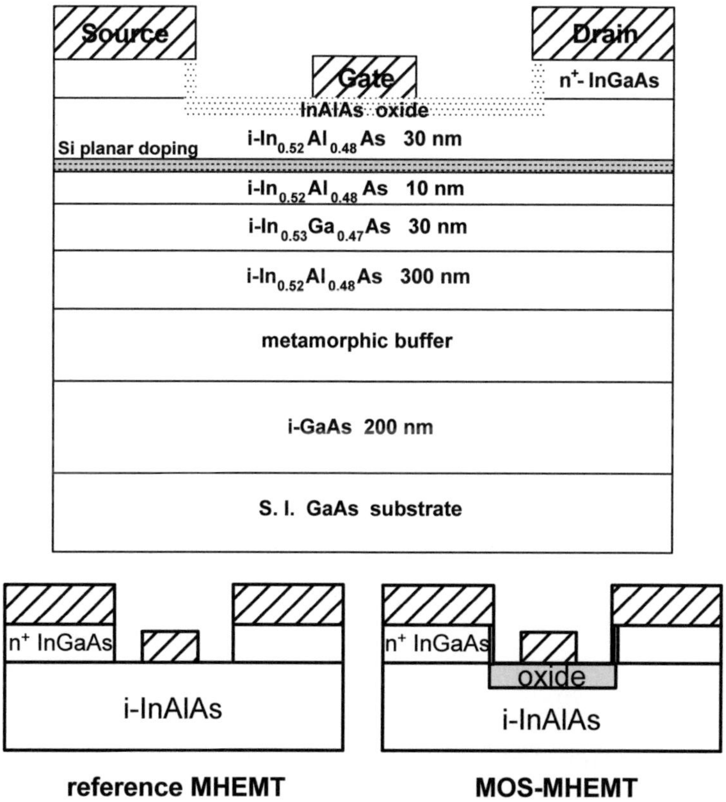

Fig. 4. Cross-sectional view of the studied InAlAs/InGaAs MOS-MHEMT structure.

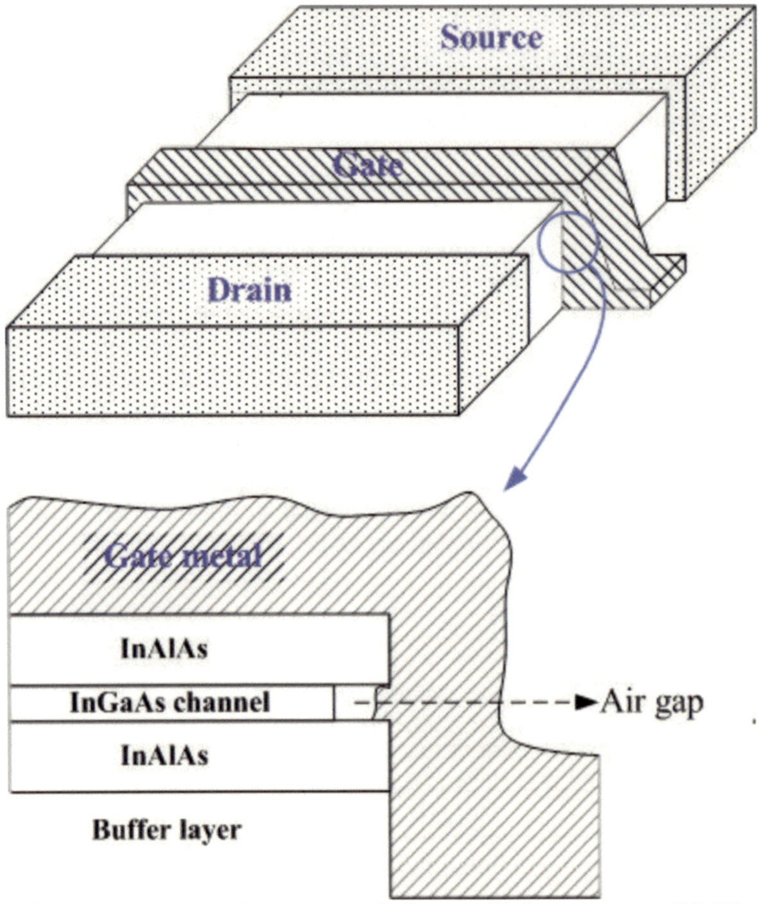

Fig. 5. The schematic cross-section view of the studied structure. (b) The diagram of separation of the gate metal from the channel.

3. Results and discussion

The oxide surface is mainly composed of Al_2O_3, although arsenic oxide and indium oxide may still remain to some extent, as confirmed by the values of the peak energy and energy separations of X-ray photoelectron spectroscopy (XPS) between main core levels (i.e., Al-2p, In-3d and As-3d) in the oxide phases. Figure 6(a)-(c) show the XPS spectra of the surface of the etched samples by the citric buffer etchant before and after oxidation for Al-2p, In-3d, and As-3d core level, respectively. It is especially noteworthy that a stronger Al_2O_3 peak is detected on the Al-2p spectrum before oxidation, and the H_2O_2 plays a dominant role to achieve the selective etching. After oxidation, however, the signals of the Al-2p core level show that the Al oxides and AlAs on the surface are weak and even disappear after 2 h. This implies that the oxidizing materials of Al dissolve in this growth solution during the oxidation process, that is, Al oxide could be removed because etching in growth solution occurs through the repetition of formation and dissolution of oxide. Similar behaviors have also been investigated in AlGaAs material by LPO method [33].

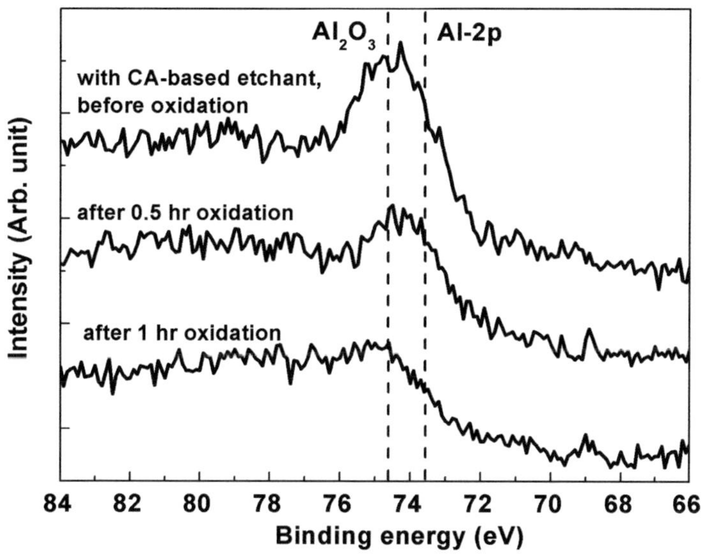
Al_2O_3
Al-2p
with CA-based etchant, before oxidation
after 0.5 hr oxidation
after 1 hr oxidation
Intensity (Arb. unit)
84 82 80 78 76 74 72 70 68 66
Binding energy (eV)

(a)

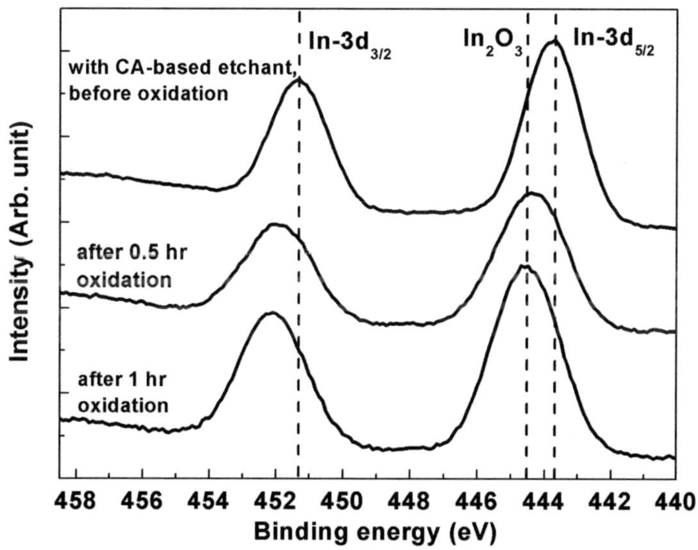
In-$3d_{3/2}$
In_2O_3
In-$3d_{5/2}$
with CA-based etchant, before oxidation
after 0.5 hr oxidation
after 1 hr oxidation
Intensity (Arb. unit)
458 456 454 452 450 448 446 444 442 440
Binding energy (eV)

(b)

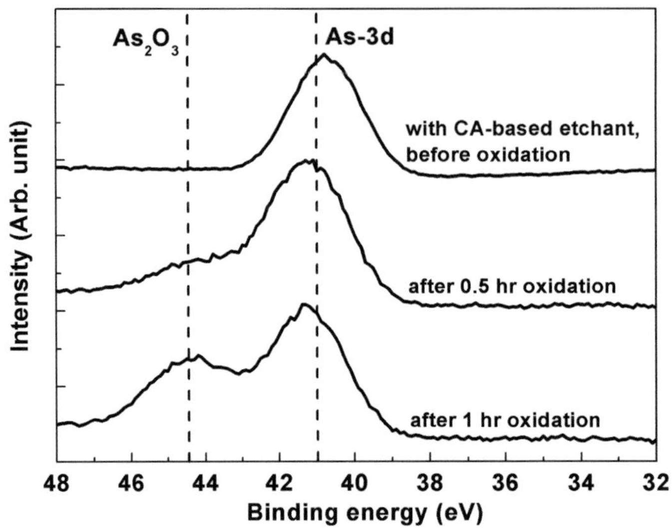

Fig. 6. (a) Al-2p, (b) In-3d, and (c) As-3d XPS spectra of the surface of the etched samples by the citric buffer etchant before and after liquid phase oxidation.

Figure 7 shows the comparison of atomic force microscopy (AFM) 3D image for the InAlAs surface after gate recess by H_3PO_4-based etchant (H_3PO_4:H_2O_2:H_2O = 1:1:30) and CA-based etchant (CA:H_2O_2 = 1:1). Before gate metallization, the root mean square (rms) value of surface roughness is estimated to be 0.72 nm and 0.14 nm by the H_3PO_4-based etchant and the CA-based etchant, respectively. These results are superior to those of previous studies using the same device structure [34]. That is, the device shows excellent etched uniformity by the citric buffer etchant. A good etched surface roughness is very important to achieve high device performance, which is very promising for power and microwave device applications.

On the other hand, the oxide film provides an improvement in the breakdown voltage in terms of the gate leakage current of the MOS-MHEMT, supported by the typical gate-to-drain *I-V* characteristics, as shown in Fig. 8. The reverse gate-to-drain breakdown voltage BV_{GD} (turn-on voltage V_{ON}) of the MOS-MHEMT is as high as -33.7 V (1.5 V), followed by -16.4 V (1.1 V) for the MHEMT with CA-based etchant, and -11.8 V (0.9 V) for the MHEMT with H_3PO_4-based etchant. The BV_{GD} and the V_{ON} are defined as the voltage at which the gate current reaches 1 mA/mm. The higher V_{ON} allows a larger induced current in the InGaAs channel, enhancing the capability of the device power. The gate leakage current can be suppressed by at least more than 4-5 orders of magnitude with an oxide film at V_{GD} = -15 V. The smaller gate leakage current of MOS-MHEMT is owing to the MOS structure and the elimination of sidewall leakage paths that are directly passivated during the oxidation. In other words, the premium BV_{GD} and V_{ON} are attributed to the use of a thin InAlAs oxide film and the gate-recess processes.

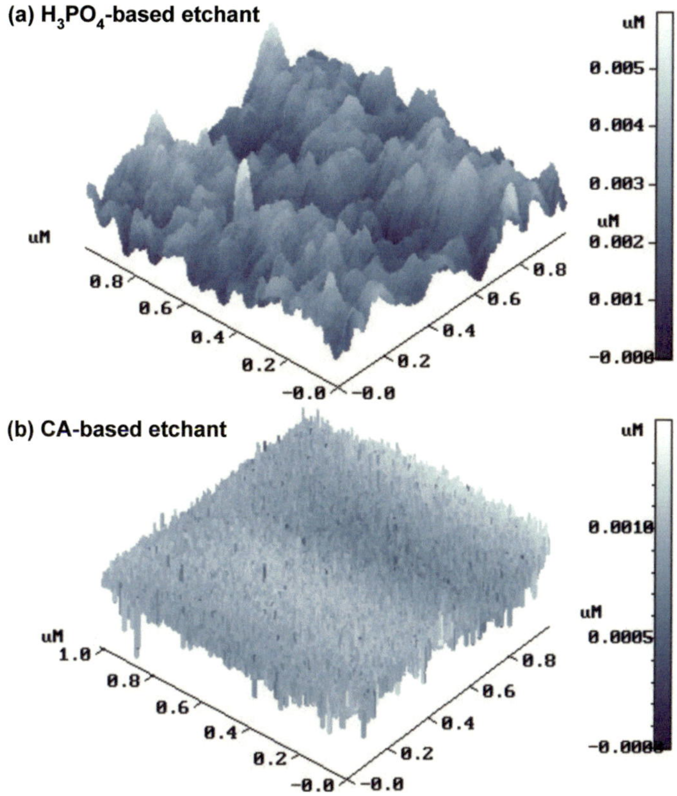

Fig. 7. AFM 3D image of the InAlAs surface after gate recess by (a) H_3PO_4-based and (b) CA-based etchant.

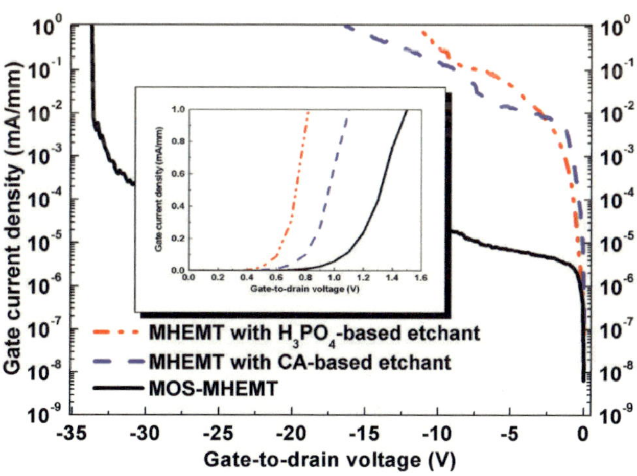

Fig. 8. Comparison of the reversed gate leakage current for MHEMTs with and without an oxide film. The inset shows the comparison of the forward gate-to-drain characteristics.

The measured unity-current-gain cutoff frequency f_T and the maximum oscillation frequency f_{max} are around 41 (22) GHz and 72 (39) GHz at maximum transconductance g_m for MOS-MHEMT (MHEMT with CA-based etchant), respectively, as shown in Fig. 9(a). These improved RF performances show the existence of a smooth and uniform etched surface after selective gate recessing by the CA-based etchant, which is superior to that of the H_3PO_4-based etchant. Furthermore, the enhanced RF performances show the existence of a MOS structure on conventional MHEMT by LPO, which is superior to the reference MHEMT. Fig. 9(b) shows the microwave characteristics versus the gate length for InAlAs/InGaAs MOS-MHEMT with an oxidized InAlAs as gate insulator. The gate-source capacitance Cgs extracted from the S-parameters of the MOS-MHEMT is lower than that of the reference MHEMT at the V_{GS} with maximum g_m. Based on these results, we conjecture that the reason for the increase of cutoff frequency for MOS-MHEMT may be partly attributed to the Cgs.

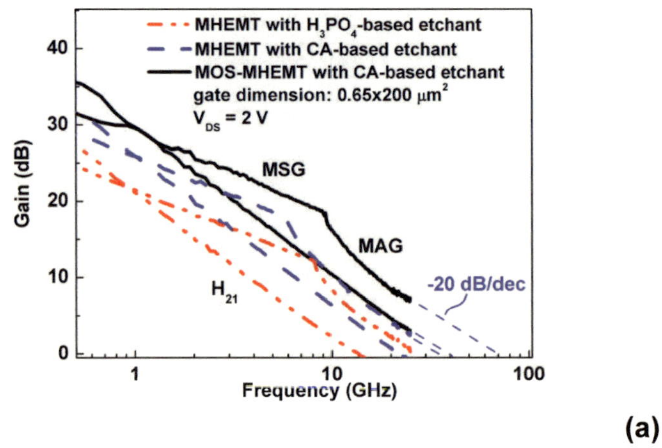

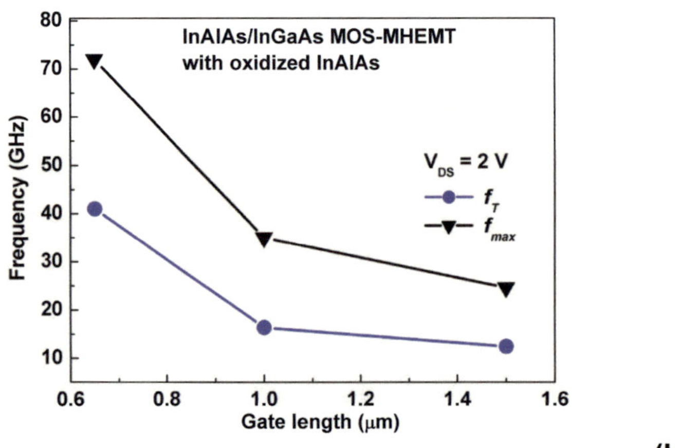

Fig. 9. (a) Comparison of the microwave characteristics at maximum g_m for the MOS-MHEMT and MHEMTs. (b) Microwave characteristics versus gate length for InAlAs/InGaAs MOS-MHEMT with an oxidized InAlAs as gate insulator.

The *I-V* characteristics of the studied MOS-MHEMT is shown in Fig. 10(a). Good pinch-off and saturation characteristics are achieved, though a significant resistance is obtained due to the nonoptimized source/drain contact formation and longer time oxidation. Fig. 10(b) shows the g_m and the drain current density as a function of the V_{GS} at V_{DS} = 2 V for the same device. The maximum g_m and the threshold voltage V_{th} are 104 mS/mm and -0.55 V, respectively. Much higher g_m and drain current density are expected with suitable device structure, such as inserting a Si-planar doping layer to enhance the carrier density.

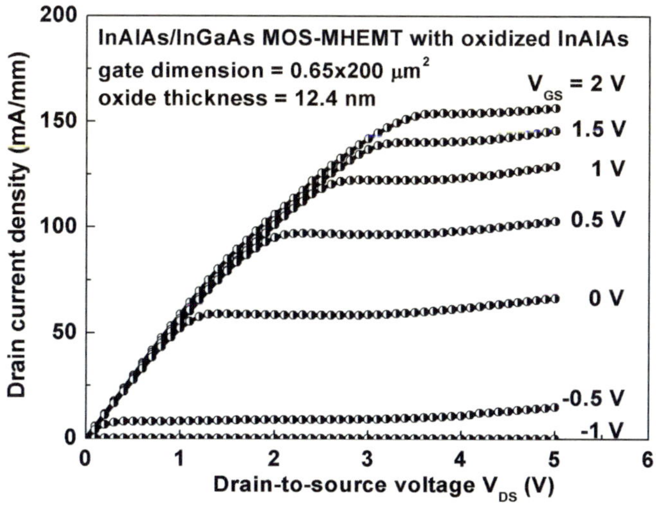

(a)

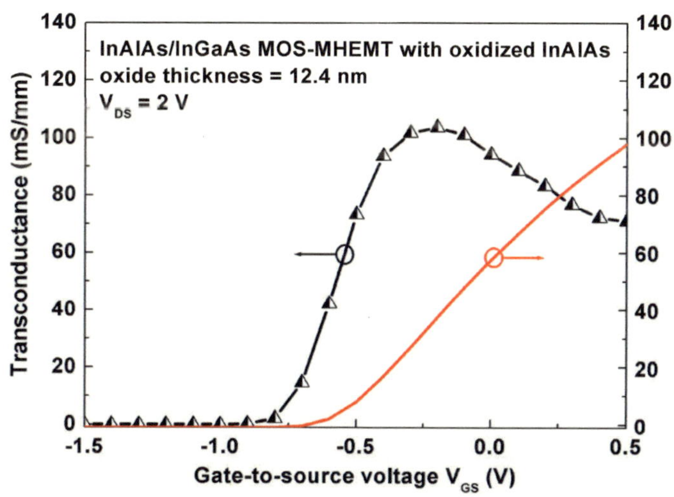

(b)

Fig. 10. (a) *I-V* characteristics of the InAlAs/InGaAs MOS-MHEMT with 12.4-nm-thick oxide. (b) The g_m and the drain current density versus V_{GS} at V_{DS} = 2 V.

Conventional InAlAs/InGaAs MHEMTs drastically suffer from low breakdown voltages due to the enhanced impact ionization effects which occur in the narrow bandgap InGaAs channel. The impact ionization effect is dependent on channel electrical field and the amount of carriers (i.e., drain current). In MOS-MHEMT, however, the electrical field in the gate-to-drain region at a fixed V_{DS} and V_{GS} is smaller than that of the reference MHEMT due to the high barrier height between the gate metal and Schottky layer, resulting in the smaller channel electrical field (i.e., smaller impact ionization effect). In order to investigate the influence of the impact ionization effect, the gate current density as a function of V_{GS} at different V_{DS} is measured. The bell-shaped curve is the typical behavior of the impact ionization effect as shown in Fig. 11 [44, 45]. At pinch-off, i.e., $V_{GS} < V_{th}$, the gate current due to the tunneling effect is also observed. This gate current is very small due to the large diode breakdown voltage for MOS-MHEMT. However, for $V_{GS} > V_{th}$, the gate current is mainly attributed to the impact ionization current which is larger than the tunneling current. Clearly, a remarkable increase in the peak gate current takes place when devices are biased at a higher V_{DS}. Because of the existence of a high electric field in the gate-to-drain region, significant hot electron phenomena occur in the narrow bandgap InGaAs channel. Furthermore, electrons can obtain high energy to generate electron-hole pairs through enhanced impact ionizations, resulting in the easier injection of holes into the gate terminal [46]. However, the peak gate current density for MOS-MHEMT is significantly improved as compared to that of reference MHEMT. Therefore, the electrons and holes generated by the impact ionization are decreased to further reduce the drain and gate current due to the native oxide layer.

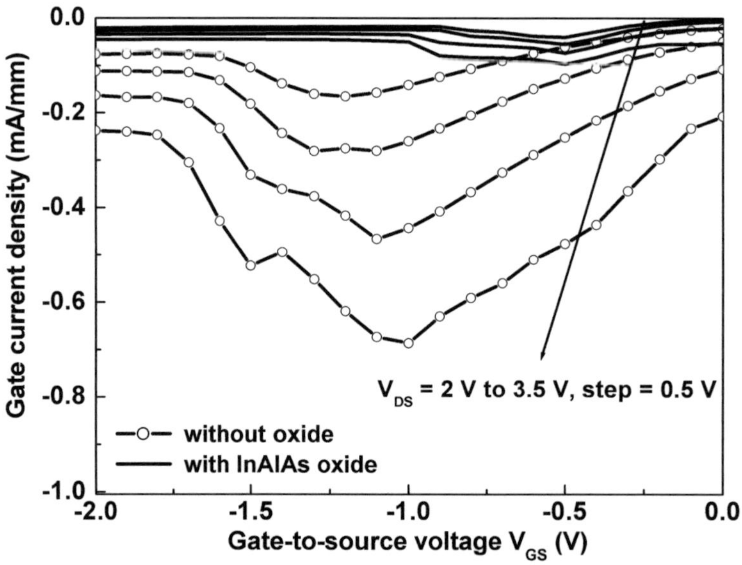

Fig. 11. Comparison of the gate current density versus V_{GS} at various V_{DS} for the MOS-MHEMT and the reference MHEMT.

So far, there have been many reports on InAlAs/InGaAs HEMT device performance, such as dc transfer characteristics, cutoff frequency, noise figure, output power density, and so on. The gate bias is operated above the threshold voltage, i.e., the device is in ON state. C. Jiang *et al.* have investigated the subthreshold current in short-channel GaAs/AlGaAs HEMT with and without a higher potential barrier in the buffer [47]. However, there are very few reports that discuss leakage current in the subthreshold region for InAlAs/InGaAs HEMT. When the gate bias is as low as the pinch-off voltage, some carriers pass through the InGaAs channel. The subthreshold current depends on the device structures, fabrication processes, and epitaxial technology. It determines the ideal OFF state, and impacts the power dissipation and circuit applications, e.g., transmitter/receiver modules. Figure 12 shows the comparison of the subthreshold current density versus the gate bias at V_{DS} = 0.1, 1.1, and 2.1 V for reference MHEMT and MOS-MHEMT. At V_{GS} = V_{th}, both devices have a similar value of drain current. However, in the region of V_{GS} around V_{th}-0.5 V, drain current density of the MOS-MHEMT is about 1000 times smaller than that of the conventional MHEMT. The MOS-MHEMT has a smaller subthreshold swing, ~100 mV/dec, defined as the change in gate voltage needed to reduce the drain current by one decade, than the reference one (250-500 mV/dec) for the same V_{DS}. Here, there is no distinction between the InGaAs channel and buffer layer for both devices. These results clearly suggest that the studied MOS-MHEMT suppresses the subthreshold current due to the reduction of the surface recombination current of the native oxide around the ohmic contact regime, i.e., suppression of the undesirable carriers' injection from the source under OFF state.

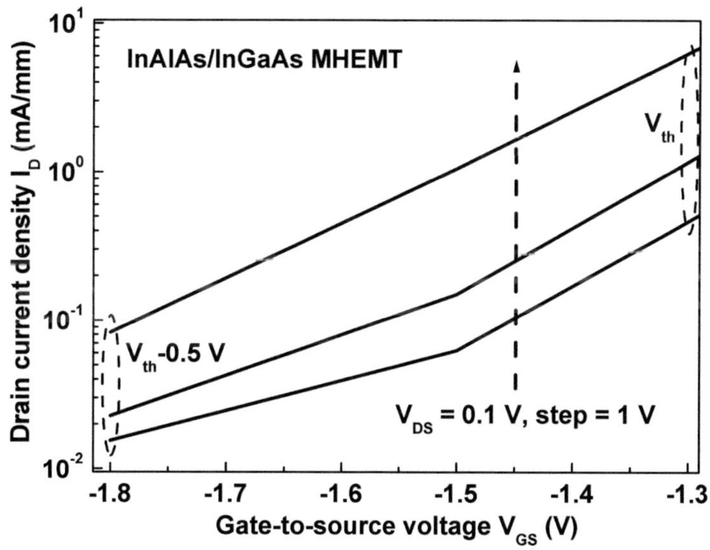

(a)

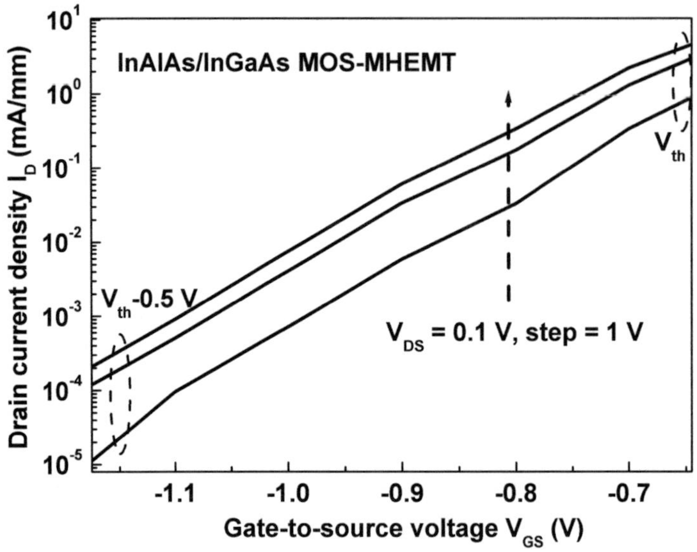

Fig. 12. (a) Subthreshold current density versus gate bias at V_{DS} = 0.1, 1.1, and 2.1 V for a conventional MHEMT. (b) Subthreshold current density versus gate bias at V_{DS} = 0.1, 1.1, and 2.1 V for the MOS-MHEMT.

Group	K.W. Lee, *et al* [48]	K.W. Lee, *et al* [34]	N.C. Paul, *et al* [27]	Y. Sun, *et al* [49]
HEMT Structure	$In_{0.53}Ga_{0.47}As$/ $In_{0.52}Al_{0.48}As$/ $In_{0.53}Ga_{0.47}As$	$In_{0.53}Ga_{0.47}As$/ $In_{0.52}Al_{0.48}As$/ $In_{0.53}Ga_{0.47}As$	$In_{0.53}Ga_{0.47}As$/ $In_{0.52}Al_{0.48}As$/ InP/ $In_{0.52}Al_{0.48}As$/ InGaAs	$In_{0.53}Ga_{0.47}As$/ $In_{0.52}Al_{0.48}As$/ $In_{0.7}Ga_{0.3}As$
Gate oxide	Oxidized InGaAs	Oxidized InAlAs	Oxidized InAlAs	Al_2O_3
Oxidation method	LPO	LPO	UV-ozone	ALD
Temperature (°C)	50	50	100	300
Gate length (μm)	0.65	0.65	1.5	0.26
Maximum V_{GS} (V)	2	1	3	2
Maximum I_{DS} (mA/mm)	390	257	163	115
Maximum g_m (mS/mm)	207	226	200	157
Subthreshold Swing (mV/dec)	-	100	-	200

Table 1. A summary of the DC characteristics of the InAlAs/InGaAs MOS-HEMTs for this study and the previous reported data.

Also, the oxidation of InGaAs and its application to the same device without gate recess were studied [48]. Without gate recessing, the gate oxide is obtained directly by oxidizing the InGaAs capping layer in the LPO system (see Fig. 4). Besides, this process can simplify one mask and grow reliable oxide films as well as sidewall passivation layers simultaneously. Table I summarizes the DC characteristics among the LPO, UV-ozone, and ALD technology on the high-indium-content InAlAs/InGaAs MOS-HEMTs for other groups [27, 49].

4. Conclusion

The oxidized InAlAs as the gate dielectric of InAlAs/InGaAs MOS-MHEMT, which was prepared by LPO at a low temperature before gate metallization, was demonstrated. The selective oxidation between oxide film and PR (or metal) has been studied, and the oxide surface composition for a long oxidation time has been evaluated. The XPS signals of the Al-2p core level indicate that Al and Al-oxides on the oxide surface are weak for a long oxidation time due to the strongly pH-dependent solubility of Al_2O_3. Moreover, as compared to the conventional InAlAs/InGaAs MHEMT, a larger gate bias operation, higher drain-to-source breakdown voltage, lower subthreshold current, lower gate leakage current with the suppressed impact ionization effect, and improved RF performances for the MOS-MHEMT make the proposed economic and low-temperature LPO suitable for device applications. The enhanced high-frequency performances due to the lower Cgs of MOS-MHEMT are expected, exhibiting its promising applications to millimeter-wave integrated circuit designs. Thus, the LPO provides new opportunities to explore many alternative dielectrics to produce gate oxides as well as effective passivation layers on III-V compound semiconductor devices.

5. Acknowledgements

The authors wish to thank Hsien-Chang Lin, Kai-Lin Lee, Chia-Hong Hsieh, Chao-Hsien Tu, and Chih-Chun Hu whose research made this work possible. This work was supported in part by the National Science Council of Taiwan under contract number NSC95-2221-E-006-428-MY3, NSC97-2221-E-214-063, and the MOE Program for Promoting Academic Excellence of Universities.

6. References

[1] H. T. Minden, "Thermal oxidation of GaAs," *J. Electrochem. Soc.*, vol. 109, pp. 733-736, 1962.
[2] D. N. Butcher and B. J. Sealy, "Electrical properties of thermal oxides on GaAs," *Electron. Lett.*, vol. 13, pp. 558-559, 1977.
[3] H. Takagi, G. Kano, and I. Teramoto, "Thermal oxidation of GaAs in arsenic trioxide vapor," *J. Electrochem. Soc.*, vol. 125, pp. 579-581, 1978.
[4] K. Maezawa, T. Mizutani, K. Arai, and F. Yanagawa, "Large transconductance n^+-Ge gate AlGaAs/GaAs MISFET with thin gate insulator," *IEEE Electron Device Lett.*, vol. EDL-7, pp. 454-456, 1986.
[5] N. Basu and K. N. Bhat, "High-pressure thermal oxidation of n-GaAs in an atmosphere of oxygen and water vapor," *J. Appl. Phys.*, vol. 63, pp. 5500-5506, 1988.

[6] T. H. Oh, D. L. Huffaker, L. A. Graham, H. Deng, and D. G. Deppe, "Steam oxidation of GaAs," *Electron. Lett.*, vol. 32, pp. 2024-2026, 1996.
[7] K. D. Choquette, K. M. Geib, C. I. H. Ashby, R. D. Twesten, O. Blum, H. Q. Hou, D. M. Follstaedt, B. E. Hammons, D. Mathes, and R. Hull, "Advances in selective wet oxidation of AlGaAs alloys," *IEEE J. Selected Topic in Quantum Electron.*, vol. 3, pp. 916-926, 1997.
[8] H. Hasegawa, K. E. Forward, and H. L. Hartnagel, "New anodic native oxide of GaAs with improved dielectric and interface properties," *Appl. Phys. Lett.*, vol. 26, pp. 567-569, 1975.
[9] D. J. Coleman, D. W. Shaw, and R. D. Dobrott, "On the mechanism of GaAs anodization," *J. Electrochem. Soc.*, vol. 124, pp. 239-241, 1977.
[10] G. P. Schwartz, J. E. Griffiths, and B. Schwartz, "Interfacial reactions on anodized GaAs," *J. Vac. Sci. Technol.*, vol. 16, pp. 1383-1386, 1979.
[11] M. J. Jeng and J. G. Hwu, "Thin-gate oxides prepared by pure water anodization followed by rapid thermal densification," *IEEE Electron Device Lett.*, vol. 17, pp. 575-577, 1996.
[12] P. Schmuki, G. I. Sproule, J. A. Bardwell, Z. H. Lu, and M. J. Graham, "Thin anodic oxides formed on GaAs in aqueous solutions," *J. Appl. Phys.*, vol. 79, pp. 7303-7311, 1996.
[13] P. A. Bertrand, "The photochemical oxidation of GaAs," *J. Electrochem. Soc.*, vol. 132, pp. 973-976, 1985.
[14] S. D. Offsey, J. M. Woodall, A. C. Warren, D. Kirchner, T. I. Chappell, and G. D. Pettit, "Unpinned (100) GaAs surfaces in air using photochemistry," *Appl. Phys. Lett.*, vol. 48, pp. 475-477, 1986.
[15] T. Sawada, H. Hasegawa, and H. Ohno, "Electronic properties of a photochemical oxide-GaAs interface," *Jpn. J. Appl. Phys.*, vol. 26, pp. L1871-L1873, 1987.
[16] E. Ettedgui, K. T. Park, J. Cao, Y. Gao, and M. W. Ruckman, "Growth of a Cr oxide layer on GaAs(100) by oxidation with condensed water," *J. Appl. Phys.*, vol. 73, pp. 1781-1787, 1993.
[17] T. Sugano, "Oxidation of $GaAs_{1-x}P_x$ surface by oxygen plasma and properties of oxide film," *J. Electrochem. Soc.*, vol. 121, pp. 113-118, 1974.
[18] C. R. Liu and L. Liu, "Measurement and analysis of the infrared absorption band for plasma anodized film on GaAs(100)," *J. Phys. D: Appl. Phys.*, vol. 21, pp. 799-803, 1988.
[19] W. A. Loong and H. L. Chang, "Oxidation of GaAs surface by oxygen plasma and its application as an antireflection layer," *Jpn. J. Appl. Phys.*, vol. 30, pp. L1319-L1320, 1991.
[20] R. Nakamura and H. Ikoma, "Magnetically excited plasma oxidation of GaAs," *Jpn. J. Appl. Phys.*, vol. 35, pp. L8-L11, 1996.
[21] M. Passlack, M. Hong, E. F. Schubert, J. R. Kwo, J. P. Mannaerts, S. N. G. Chu, N. Moriya, and F. A. Thiel, "In situ fabricated Ga_2O_3-GaAs structures with low interface recombination velocity," *Appl. Phys. Lett.*, vol. 66, pp. 625-627, 1995.
[22] F. Ren, J. M. Kuo, M. Hong, W. S. Hobson, J. R. Lothian, J. Lin, H. S. Tsai, J. P. Mannaerts, J. Kwo, S. N. G. Chu, Y. K. Chen, and A. Y. Cho, "$Ga_2O_3(Gd_2O_3)$/InGaAs enhancement-mode n-channel MOSFET's," *IEEE Electron Device Lett.*, vol. 19, pp. 309-311, 1998.

[23] M. Passlack, J. K. Abrokwah, R. Droopad, Z. Yu, C. Overgaard, S. I. Yi, M. Hale, J. Sexton, and A. C. Kummel, "Self-aligned GaAs p-channel enhancement mode MOS heterostructure field-effect transistor," *IEEE Electron Device Lett.*, vol. 23, pp. 508-510, 2002.

[24] P. D. Ye, G. D. Wilk, J. Kwo, B. Yang, H.-J. L. Gossmann, M. Frei, S. N. G. Chu, J. P. Mannaerts, M. Sergent, M. Hong, K. K. Ng, and J. Bude, "GaAs MOSFET with oxide gate dielectric grown by atomic layer deposition," *IEEE Electron Device Lett.*, vol. 24, pp. 209-211, 2003.

[25] R. Driad, Z. H. Lu, S. Laframboise, D. Scansen, W. R. Mckinnon, and S. P. Mcalister, "Reduction of surface recombination in InGaAs/InP heterostructures using UV-irradiation and ozone," *Jpn. J. Appl. Phys.*, vol. 38, pp. 1124-1127, 1999.

[26] K. Iiyama, Y. Kita, Y. Ohta, M. Nasuno, S. Takamiya, K. Higashimine, and N. Ohtsuka, "Fabrication of GaAs MISFET with nm-thin oxidized layer formed by UV and ozone process," *IEEE Trans. Electron Devices*, vol. 49, pp. 1856-1862,, 2002.

[27] N. C. Paul, K. Nakamura, H. Seto, K. Iiyama, and S. Takamiya, "Oxidation of InAlAs and its application to gate insulator of InAlAs/InGaAs metal oxide semiconductor high electron mobility transistor," *Jpn. J. Appl. Phys.*, vol. 44, pp. 1174-1180, 2005.

[28] H. H. Wang, C. J. Huang, Y. H. Wang, and M. P. Houng, "Liquid phase chemical-enhanced oxidation for GaAs operated near room temperature," *Jpn. J. Appl. Phys.*, vol. 37, pp. L67-L70, 1998.

[29] H. H. Wang, J. Y. Wu, Y. H. Wang, and M. P. Houng, "Effects of pH values on the kinetics of liquid phase chemical enhanced oxidation of GaAs," *J. Electrochem. Soc.*, vol. 146, pp. 2328-2332, 1999.

[30] J. Y. Wu, H. H. Wang, Y. H. Wang, and M. P. Houng, "A GaAs MOSFET with a liquid phase oxidized gate," *IEEE Electron Device Lett.*, vol. 20, pp. 18-20, 1999.

[31] D. W. Chou, R. F. Lou, H. H. Wang, Y. H. Wang, and M. P. Houng, "Properties of the GaAs oxide layer grown by the liquid-phase chemical-enhanced technique on the S-passivated GaAs surface," *J. Electronic Materials*, vol. 31, pp. 71-75, 2002.

[32] D. W. Chou, L. T. Wang, H. H. Wang, Y. H. Wang, and M. P. Houng, "Influence of annealing ambient on GaAs oxide prepared by the liquid phase method," *Solid-State Electron.*, vol. 48, pp. 2175-2179, 2004.

[33] K. W. Lee, Y. H. Wang and M. P. Houng, "Liquid phase chemical enhanced oxidation on AlGaAs and its application," *Jpn. J. Appl. Phys.*, vol. 43, pp. 4087-4091, 2004.

[34] K. W. Lee, K. L. Lee, X. Z. Lin, C. H. Tu, and Y. H. Wang, "Improvement of impact ionization effect and subthreshold current in InAlAs/InGaAs metal-oxide-semiconductor metamorphic HEMT with a liquid phase oxidized InAlAs as gate insulator," *IEEE Trans. Electron Devices*, vol. 54, no. 3, pp. 418-424, 2007.

[35] H. H. Wang, Y. H. Wang, and M. P. Houng, "Near-room-temperature selective oxidation on GaAs using photoresist as a mask," *Jpn. J. Appl. Phys.*, vol. 37, pp. L988-L990, 1998.

[36] D. R. Greenberg, J. A. del Alamo, and R. Bhat, "A recessed-gate InAlAs/n^+-InP HFET with an InP etch-stop layer," *IEEE Electron Devices Lett.*, vol. 13, p. 137, 1992.

[37] S. Fujita, T. Noda, C. Nozaki, and Y. Ashizawa, "InGaP/InAlAs HEMT with a strained InGaP Schottky contact layer," *IEEE Electron Devices Lett.*, vol. 14, p. 259, 1993.

[38] T. Enoki, K. Arai, A. Kohzen, and Y. Ishii, "Design and characteristics of InGaAs/InP composite-channel HFET's," *IEEE Trans. Electron Devices*, vol. 42, p. 1413, 1995.

[39] J. A. del Alamo and T. Mizutani, "An $In_{0.52}Al_{0.48}As/n^{+}$-$In_{0.53}Ga_{0.47}As$ MISFET with a modulation-doped channel," *IEEE Electron Devices Lett.*, vol. 10, p. 394, 1989.

[40] H. Kinoshita, Y. Sano, T. Ishida, S. Nishi, M. Akiyama and K. Kaminishi, "A new insulated-gate inverted-structure modulation-doped AlGaAs/GaAs/N-AlGaAs field-effect transistor," *Jpn. J. Appl. Phys.*, vol. 43, p. L836, 1984.

[41] S. Fujita, M. Hirano, and T. Mizutani, "Small-signal characteristics of n^{+}-Ge gate AlGaAs/GaAs MISFET's," *IEEE Electron Device Lett.*, vol. 9, p. 518, 1988.

[42] M. Tong, K. Nummila, A. Ketterson, I. Adesida, C. Caneau, and R. Bhat, "InAlAs/InGaAs/InP MODFET's with uniform threshold voltage obtained by selective wet gate recess," *IEEE Electron Device Lett.*, vol. 13, p. 525, 1992.

[43] S. R. Bahl and J. A. del Alamo, "Elimination of mesa-sidewall gate leakage in InAlAs/InGaAs heterostructures by selective sidewall recessing," *IEEE Electron Device Letters*, vol.13, pp. 195-197, 1992.

[44] G. Meneghesso, T. Grave, M. Manfredi, M. Pavesi, C. Canali, and E. Zanoni, "Analysis of hot carrier transport in AlGaAs/InGaAs pseudomorphic HEMT's by means of electroluminescence," *IEEE Trans. Electron Devices*, vol. 47, p. 2, 2000.

[45] M. Boudrissa, E. Delos, C. Gaquiere, M. Rousseau, Y. Cordier, D. Theron, and J. C. De Jaeger, "Enhancement-mode $Al_{0.66}In_{0.34}As/Ga_{0.67}In_{0.33}As$ metamorphic HEMT: modeling and measurements," *IEEE Trans. Electron Devices*, vol. 48, p. 1037, 2001.

[46] T. Suemitsu, T. Enoki, N. Sano, M. Tomizawa, and Y. Ishii, "An analysis of the kink phenomena in InAlAs/InGaAs HEMT's using two-dimensional device simulation," *IEEE Trans. Electron Devices*, vol. 45, p. 2390, 1998.

[47] C. Jiang, D. C. Tsui, B. J. F. Lin, H. Lee, A. Lepore, and M. Levy, "Subthreshold current in MODFET's of tenth-micrometer gate," *IEEE Electron Device Lett.*, vol. 11, p. 63, 1990.

[48] K. W. Lee, H. C. Lin, K. L. Lee, C. H. Tu, and Y. H. Wang, "Selectively liquid-phase oxidation of InGaAs and application to metal-oxide-semiconductor InAlAs/InGaAs metamorphic HEMT without gate recess," *J. Electrochem. Soc.*, vol. 155, pp. H932-H936, 2008.

[49] Y. Sun, E. W. Kiewra, S. J. Koester, N. Ruiz, A. Callegari, K. E. Fogel, D. K. Sadana, J. Fompeyrine, D. J. Webb, J.-P. Locquet, M. Sousa, R. Germann, K. T. Shiu, and S. R. Forrest, "Enhancement-mode buried-channel $In_{0.7}Ga_{0.3}As/In_{0.52}Al_{0.48}As$ MOSFETs with high-κ gate dielectrics," *IEEE Electron Device Lett.*, vol. 28, pp. 473-475, 2007.

GaN-based metal-oxide-semiconductor devices

Ching-Ting Lee
Institute of Microelectronics, Department of Electrical Engineering,
National Cheng Kung University,
Tainan, Taiwan, Republic of China

1. Introduction

Si-based semiconductors-related technologies are well developed for their applications in electronic devices and integrated circuits (ICs). However, the properties of Si-based semiconductors are seldom being applied in high-frequency and high-power systems. In comparison, the GaN-based semiconductors have also aroused huge interests in recent years owing to their advantages, such as wide and direct energy bandgap, better thermal and chemical stability, and high-electron drift velocity. Therefore, GaN-based semiconductors are widely utilized in electronic devices, including field-effect transistors (FETs) (Khan et al., 1993; Zolper et al., 1996; Klein et al., 1999; Yoshida & Suzuki, 1999; Johnson et al., 2000; Zhang et al., 2000; Jiménez et al., 2002; Binari et al., 2002; Braga et al., 2004; Wallis et al., 2005; Horio et al., 2005; Hong & Kim, 2006; Saripalli et al., 2007) and high-electron mobility transistors (HEMTs) (Egawa et al., 2000; Karmalkar & Mishra, 2001; Vetury et al., 2001; Islam et al., 2002; Koley et al., 2003; Khan et al., 2003; Mizutani et al., 2003; Lu et al., 2003; Meneghesso et al., 2004; Fareed et al., 2005; Inoue et al., 2005; Seo et al., 2008). Among the devices mentioned above, GaN-based HEMTs possess high electron mobility larger than $1000cm^2/Vs$ at room temperature and large operating current due to the formation of two dimensional electron gases (2DEGs). In addition, the GaN-based optoelectronic devices, such as light-emitting diodes (LEDs) (Han et al., 1998; Chang et al., 2004; Fujii et al., 2004; Wierer et al., 2004; Shen et al., 2006; Lee et al., 2006; Chang et al., 2006; Lee et al., 2007; Chen et al., 2007; Chuang et al., 2007), photodetectors (PDs) (Kuksenkov et al., 1998; Walker et al., 1998; Katz et al., 2001; Rumyantsev et al., 2001; Sheu et al., 2002; Seo et al., 2002; Pau et al., 2004; Su et al., 2005; Vardi et al., 2006; Navarro et al., 2009), and laser diodes (LDs) (Nakamura et al., 1996; Nakamura et al., 1996; Saitoh et al., 2003; Suski et al., 2004; Schoedl et al., 2005; Peng et al., 2006; Laino et al., 2007; Braun et al., 2008; Rossetti et al., 2008), have also been investigated and developed extensively for display, lighting, memory system, and communication system in recent years.

It has been reported that utilizing Schottky contact gate could substantially improve the performances of GaN-based metal-semiconductor field-effect transistors (MES-FETs) and metal-semiconductor high-electron mobility transistors (MES-HEMTs) in high-frequency applications (Chumbes et al., 2001; Ohno et al., 2001; Javorka et al., 2002; Okita et al., 2003; Endoh et al., 2004). However, these devices are not suitable for high-power applications because of large gate leakage current, small gate voltage swing (GVS) and small breakdown voltage. To conquer these aforementioned drawbacks, GaN-based metal-oxide-semiconductor field-effect transistors (MOS-FETs) and metal-oxide-semiconductor high-electron mobility transistors (MOS-HEMTs) are proposed by featuring promising structures and their immense potential in high-power and high-frequency applications (Hu et al., 2001; Kao et al., 2005; Simin et al., 2005; Endoh et al., 2006; Higashiwaki et al., 2006). For MOS devices, the gate insulator plays an important role, since the good quality dielectric layers would yield these devices with outstanding performances. Up to now, several dielectrics, such as SiO_2, Pr_2O_3, Si_3N_4, AlN, Ta_2O_5, MgO, $Ga_2O_3(Gd_2O_3)$, Sc_2O_3, Al_2O_3, and stack dielectric materials, etc. (Arulkumaran et al., 1998; Ren et al., 1999; Hong et al., 2000; Therrien et al., 2000; Rumyantsev et al., 2000; Gaffey et al., 2001; Chen et al., 2001; Lay et al., 2001; Kim et al., 2001; Cho et al., 2003; Ma et al., 2003; Mehandru et al., 2003; Bas & Lucovsky, 2004; Irokawa et al., 2004; Cico et al., 2007) have been used in GaN-based MOS devices and the corresponding electrical characteristics have also been reported. In this chapter, after presenting a brief introduction of the working principle of MOS devices, various dielectric film deposition methods are summarized with a particular emphasis in the photoelectrochemical (PEC) oxidation method. At the same time, the different performances of GaN-based MOS-HEMTs with different insulators are also compared and discussed.

2. Operation principle of metal-oxide-semiconductor devices

2.1 The metal-oxide-semiconductor (MOS) diodes

The metal-oxide-semiconductor (MOS) diodes are key parts of MOS transistors. Fig. 1 shows the cross-sectional schematic configuration of conventional MOS diodes. An oxide film as the insulator layer was deposited on the surface of a semiconductor and then the gate electrode was deposited thereafter on the surface of the oxide film. An ohmic metal contact was formed at the bottom surface of the semiconductor.

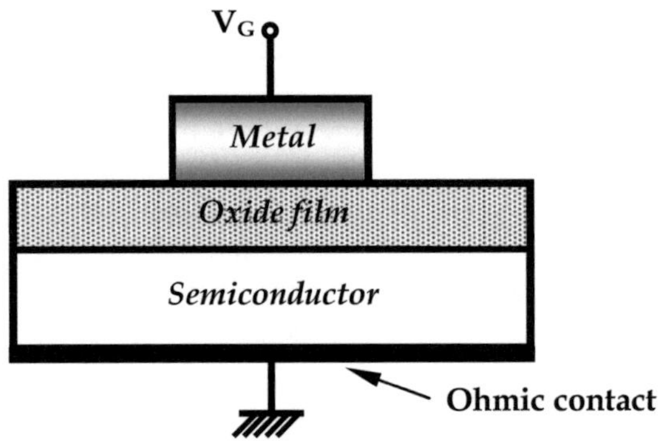

Fig. 1. The cross-sectional schematic configuration of conventional MOS diodes.

Three typical situations are normally occurred at the semiconductor surface, including accumulation, depletion, and inversion, depending on the applied bias between the two electrodes. Fig. 2 shows the energy band diagrams of an ideal MOS diode with n-type semiconductor at different operating conditions (Neaman, 1997). When a positive voltage is applied on the gate electrode of the MOS diode, the free electrons in conduction band of the semiconductor are driven towards the oxide/semiconductor interface, where electrons are accumulated. Accordingly, the energy band at the semiconductor surface is bent downward and the conduction band edge becomes closer to Fermi level. This is so-called accumulation case, as shown in Fig. 2 (a). When a small negative voltage is applied, the electrons are driven away from the oxide/semiconductor interface by the surface electric field which results in the formation of a depletion region. This case is called depletion case, as shown in Fig. 2 (b). When a larger negative bias is applied, a large number of minority carriers (hole) were induced at oxide/semiconductor interface. Therefore, the energy bands bent upward even more so that the intrinsic level at the surface crosses over the Fermi level. The number of holes (minority carriers) at the semiconductor surface is larger than the number of electron (majority carriers), and in which case it is called inversion and shown in Fig. 2 (c). However, the situation is different for GaN-based MOS diodes, within which the inversion layer is hard to form because the GaN-based material is a wide energy gap semiconductor. The reported results showed that the generation rate of the minority carriers (holes) is extremely low at room temperature (Casey et al., 1996; Hashizume et al., 2000). For an n-type GaN-based material with a carriers concentration of $1.2\times10^{17}cm^{-3}$, the generation time of minority carriers is $t \cong 4.2\times10^{18}s$. However, it is impossible to form an inversion layer with an extremely long generation time. When a negative bias is applied, there are an insufficient number of minority carriers present at the interface and the depletion region extends into GaN to maintain electric neutrality. Consequently, the n-type GaN-based MOS diodes show a deep depletion characteristic at negative bias.

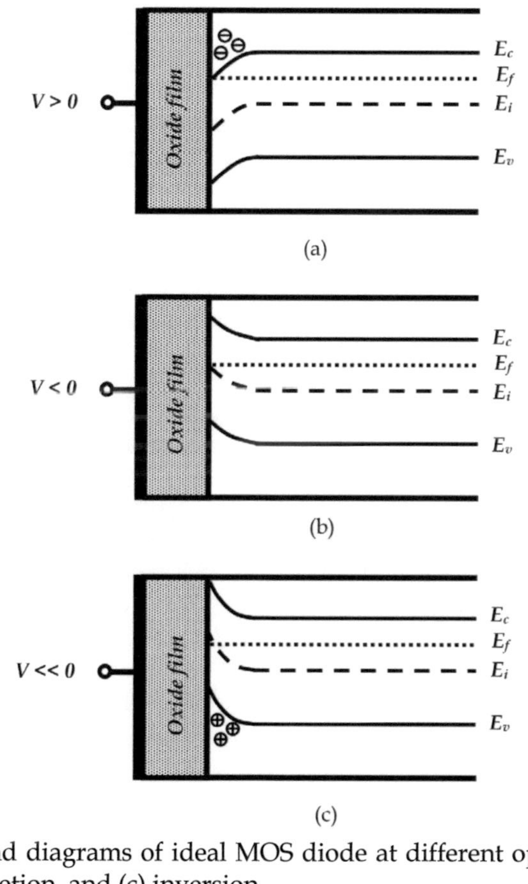

Fig. 2. The energy band diagrams of ideal MOS diode at different operating conditions: (a) accumulation, (b) depletion, and (c) inversion.

2.2 GaN-based metal-oxide-semiconductor field-effect transistors

The GaN-based metal-oxide-semiconductor field-effect transistors (MOS-FETs) is a device which consists of two ohmic electrodes and a MOS diode. Those two ohimc electrodes, located at the two sides of the gate electrode, are connected to GaN-based semiconductor through doped regions. One of the contacts is called the source as it implies that the charge carriers entering the channel originate from this contact, while the other is the drain where the carriers leave the channel. The conductance of the semiconductor near the interface under the gate electrode can be modulated by applying gate bias, as mentioned above for the MOS diode. By applying gate bias, one can modify the effective channel thickness by varying the width of the depletion region, and this in turn varies the output current. If a negative gate voltage is applied on n-channel, the carriers are then depleted from the channel, causing a decrease of the channel conductance; in this case the device behaves as a normally-on (depletion) GaN-based MOS-FETs.

Figure 3 (a) and (b) show the cross-sectional schematic configuration of an n-channel depletion GaN-based MOS-FETs and MES-FETs, respectively. The prior one has better electrical performances due to small gate leakage current, large gate voltage swing and high breakdown field than those of the later one. Fig. 4 shows the depletion layer and the output characteristics of MOS-FETs under various bias conditions (Sze, 2002). As shown in Fig. 4 (a), when gate voltage is zero and drain voltage is small, a small drain current flows from drain to source in the channel and varies linearly with the drain voltage. When the drain voltage increases, the width of depletion layer extends while the cross-sectional area of channel reduces. If the drain voltage is further increased, the depletion layer touches the semi-insulating substrate and the channel became pinch-off, as shown in Fig. 4 (b). The corresponding value of the drain voltage and drain current are called saturation voltage and saturation current, respectively. In Fig. 4 (b), the location P is called the pinch-off point. If the drain voltage increases further, the depletion region widens and pinch-off point then shifts toward the source, as shown in Fig. 4 (c). As the voltage at point P equals to saturation voltage, the current flow in the channel remains fixed. When the drain voltage is larger than the saturation voltage, the drain current becomes independent of the drain voltage. In addition, when a negative voltage is applied to the gate electrode, the electrons are driven away from semiconductor surface, which reduces the electron concentration and forms the depletion layer. Consequently, as the effective channel thickness decreases, and the drain current becomes lower. When the negative gate voltage increases over a certain value, the depletion layer touches the semi-insulating substrate and the channel then operates in a cut-off mode, as shown in Fig. 4 (d). Consequently, the depletion of n-channel depletion-mode MOS-FETs at semiconductor surface can be achieved by applying a sufficient gate bias to inhibit the device current. The GaN-based MES-FETs have similar behaviors. When Schottky gate metal contacts to semiconductors as shown in Fig. 3 (b), the depletion region forms at the metal/semiconductor interface and its width can be controlled by gate bias as mentioned above.

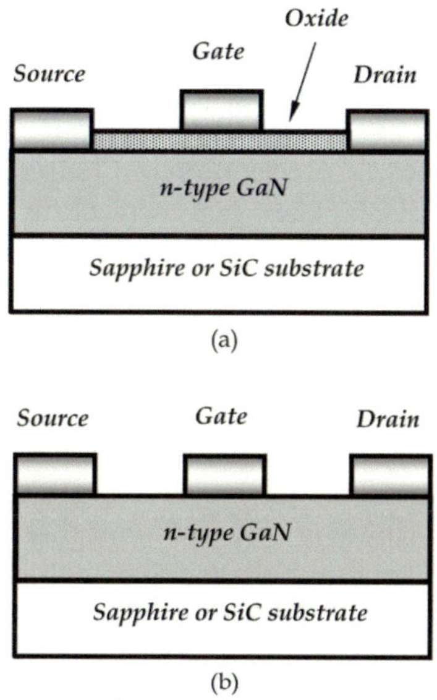

Fig. 3. The cross-sectional schematic of an n-channel depletion (a) MOS-FETs and (b) MES-FETs.

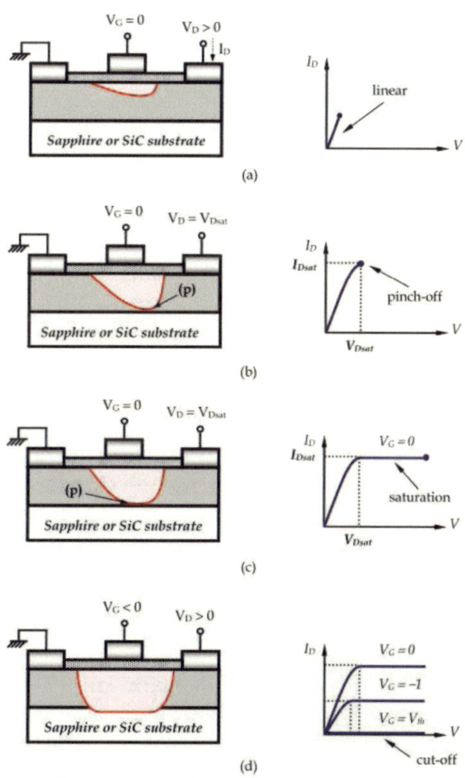

Fig. 4. The depletion layer width and output characteristic of MOS-FETs at various bias conditions. (a) V_G=0 and a small V_D voltage (b) V_G=0 and at pinch-off (c) V_G=0 and at post pinch-off (V_D>V_{Dsat}) (d) a negative V_G voltage and at cut-off case.

2.3 GaN-based metal-oxide-semiconductor high-electron mobility transistors

Metal-oxide-semiconductor high-electron mobility transistors (MOS-HEMTs), namely metal-oxide-semiconductor heterojunction field-effect transistors (MOS-HFETs) (Sze, 2002), are another kind of field effect transistor, which is similar to the MOS-FETs but with its semiconductor layer replaced by a semiconductor heterostructure. Figure 5 shows the energy band diagram of the AlGaN/GaN heterostructure. The band gap of AlGaN (low electron affinity) is wider than the bandgap of GaN (high electron affinity). When they form junctions, both the conduction band and valence band at the AlGaN/GaN interface bend and the conduction band of GaN drops below the Fermi level. Consequently, a triangular potential well is formed at the GaN side of the interface, owing to the conduction band discontinuity. The electrons are confined in this well, and forming a two dimensional electron gas (2DEG). In addition, the spontaneous polarization (Ambacher et al., 1999; Ambacher et al., 2000) and piezoelectric polarization (Smorchkova et al., 1999; Sacconi et al., 2001) effects, which are large in the wurtzite GaN-based semiconductor materials (Johnson et al., 2001; Lu et al., 2001), provide a further improvement in the sheet carrier concentration of the 2DEG, which is typically up to $10^{13}cm^{-2}$. Furthermore, the 2DEG are located in the region of the undoped GaN, where the carrier mobility is high as the scattering effect is reduced.

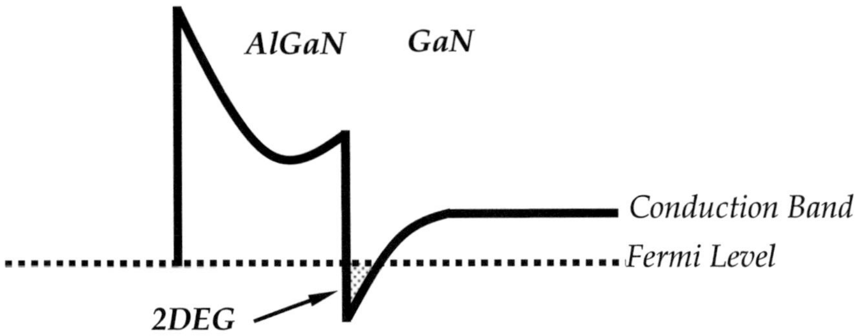

Fig. 5. The energy bandgap diagram of AlGaN/GaN heterostructure.

Figure 6 (a) and (b) shows the schematic configuration of AlGaN/GaN MOS-HEMTs and MES-HEMTs grown on sapphire substrates (Al_2O_3) using metalorganic chemical vapor deposition (MOCVD) system or molecular-beam epitaxy (MBE) system. The carbon-doped GaN layer has high resistance and can confine the carriers transport in the AlGaN/GaN channel (Webb et al., 2001). The electrons transport occurs in the 2DEG channel between two ohmic contacts of source and drain electrode. The insulators are deposited in the drain-source region for surface passivation and gate insulation. The source is usually grounded and the voltage applied between the drain and source is called V_{DS}, while the gate-source voltage is called V_{GS}. The current flowing from source to drain is called I_{DS} and I_{GS} is the current entering the gate through the oxide film. This is the reason why the MOS-HEMTs have small gate leakage current and large breakdown voltage compared to those of MES-HEMTs. The ability of the gate electrode to modulate the source-drain current is expressed as the extrinsic transconductance $g_m=\partial I_{DS}/\partial V_{GS}$. The conductance of the channel varies with the sheet carrier concentration, which is modulated by the gate bias. When the gate bias is positive, more electrons accumulate in the 2DEG channel and the operation current of

AlGaN/GaN MES-HEMTs will be larger. The current decreases with the decrease of the gate bias. The 2DEG channel will be fully depleted and the drain-source current will drop to zero when the gate bias reaches a negative threshold value, which is called cut-off voltage or threshold voltage. This corresponds to the case of which the MOS-HEMTs operating in the depletion mode.

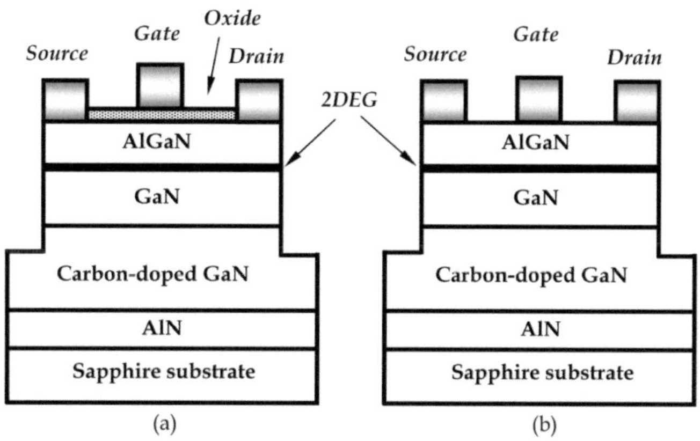

Fig. 6. The schematic configuration of AlGaN/GaN (a) MOS-HEMTs and (b) MES-HEMTs.

3. Obstacle to further improving transistor performances

For GaN-based MES-HEMTs and MOS-HEMTs, there are still some problems to be solved, which limit their performances ultimately. Among them, the self-heating (Gaska et al., 1997; Gaska et al., 1998; Ahmad et al., 2006) and current collapse (Klein et al., 2001; Mittereder et al., 2003; Kuzmik et al., 2004; Zheng et al., 2008) are the most serious effects that degrade the high-frequency and high-power performances of the transistors.

The self-heating effect occurs usually when the transistor operates at high drain-source voltage. Fig. 7 gives a typical example where the drain current decreases at high DC drain-source voltage (Fan et al., 2004). This is related to the fact that the electrons flowing in the device transfer part of their energy to the lattice via trap-related scattering and electron-phonon interaction (Bhapkar et al., 1997), which raises the channel temperature of the devices. This effect is induced by the poor thermal conductivity of the sapphire substrates which are commonly used for fabricating GaN-based transistors. The self-heating effect can be suppressed by using highly thermal conductive SiC substrates in stead of sapphire substrates (Morkoc et al., 1994).

During epitaxial processes, traps exist in GaN buffer layer, AlGaN layer and AlGaN surface. Traps on the semiconductor surface capture electrons and form lots of virtual gates. Those virtual gates deplete the channel and make the drain-source current is smaller than ideal value. In addition, when high drain-source electrical field is applied, some hot electrons transfer from 2DEG channel to adjacent layer with high concentration of traps. Those factors cause the drain-source current decrease and this phenomenon is called the current collapse. The simple schematic configuration of current collapse of transistors is shown in Fig. 8 (Zheng et al., 2008). It can be suppressed by surface passivation and gate insulation by depositing dielectrics on AlGaN surface (Tan et al., 2002; Arulkumaran et al., 2004) or using confinement layer underlying 2DEG channel (Palacios et al., 2006). For obtaining better radio-frequency (rf) performances and power performances, suppressing those two unideal effects is urgent.

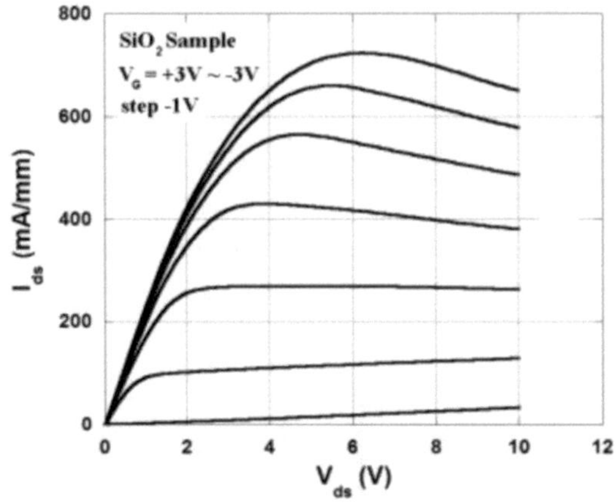

Fig. 7. Self-heating in an AlGaN/GaN MOS-HEMT.

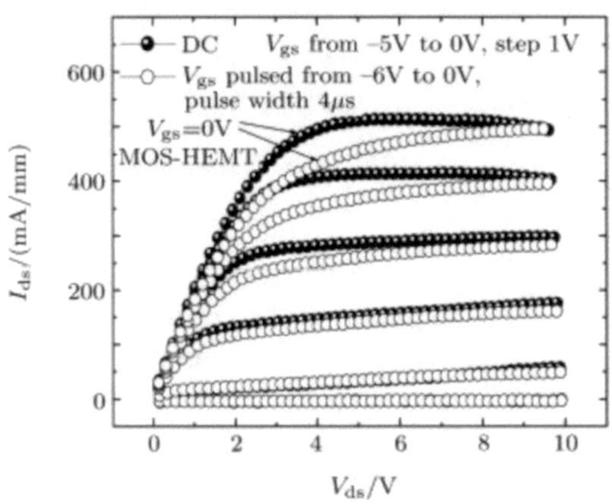

Fig. 8. Current collapse in an Al_2O_3/AlGaN/GaN MOS-HEMT (dashed lines: Before stress; solid lines: After stress).

In addition, the low frequency noises, such as flicker noise, and generation-recombination noise, mean the inevitable disturbances of output signals when devices operated at a low frequency. Fig. 9 shows the spectra of flicker noise and generation-recombination noise. The prior one is proportional to inverse frequency but the later one has Lorentizn distribution. Flicker noise is an important indicator for judging the transistors weather they are suitable for communicating applications or not, because a large level of flicker noise limits the phase noise characteristics and causes the performance degradation of the electronic systems (Balandin et al., 1999). The interface-state densities, defects, and phonons are possible sources of flicker noise. Furthermore, it is a useful tool for confirming crystal quality, and manufacture techniques (Balandin et al., 1999; Balandin et al., 2000). In general, MOS-HEMTs have better behavior in flicker noise than MES-

HEMTs, because the interface-state densities in MOS-HEMTs are well passivated (Vertiatchikh & Eastman, 2003; Chiu et al., 2008). Fabricating transistors with low flicker noise is a vital issue for developing high performance electronic devices and systems.

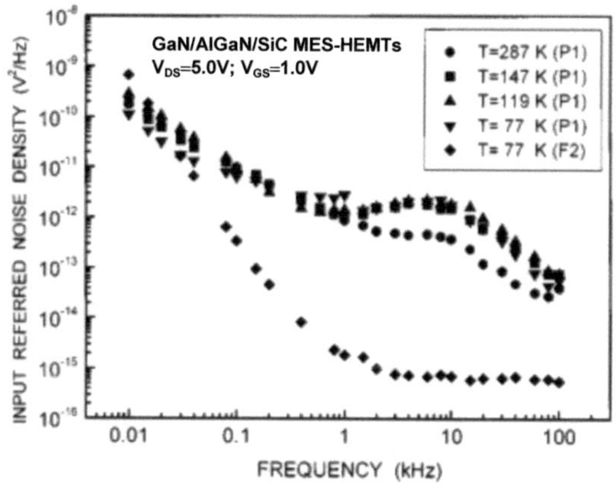

Fig. 9. Low temperature noise characteristics of the GaN MES-HEMTs in the subsaturation regime. Generation-recombination (g-r) bulges are clearly seen in spectra of the doped channel device (P1) at frequency f ≈3–4 kHz.

4. Method of growing gate insulators of GaN-based MOS devices

For fabricating GaN-based MOS devices with excellent performances, an important task is, as with the other MOS devices, to find suitable right methods to form a proper insulator layer on the semiconductor. Many efforts have been focused on the exploration of insulator materials and deposition methods. In most of cases reported in the literature, the insulator layers of the GaN-based MOS devices were deposited externally onto the semiconductor surface by physical vapor deposition (PVD) and chemical vapor deposition (CVD) methods. The main results reported in the literatures will be reviewed first. Then, we will give a more detailed discussion on photoelectrochemical (PEC) oxidation method, with which the oxide layer is formed via a chemical reaction with the semiconductors, instead of relying on an external deposition.

4.1 Electron-beam deposition method

The electron beam evaporation employs an electron-beam emitted by a filament of the electron gun to bombard the target in a high vacuum chamber. The kinetic energy of the electron-beam transforms into the thermal energy upon contact of the target, which in turn melts the target. The e-beam evaporation has good thermal translation efficiency and high evaporation rate. In addition, the current can be controlled, so as to judiciously control the evaporation rate. There are basically no limits to which the materials can be evaporated with the electron-beam evaporation method. The materials can easily be evaporated irrespective of their purity and chemical compounds. The target is placed into a crucible with an adequate cooling through proper arrangement. High thermal melt zone is localized around a position of the target which is bombarded directly by an electron-beam.

Several conventional gate oxides such as $Ga_2O_3(Gd_2O_3)$, SiO_2, TiO_2, AlN, MgO, Ta_2O_5, Pr_2O_3 and stack dielectrics have been deposited using e-beam deposition system for gate insulators of GaN-based MOS devices (Hong et al., 2000; Arulkumaran et al., 2005; Kikuta et al., 2006; Yagi et al., 2006; Chiu et al., 2008). Figure 10 shows the typical capacitance-voltage characteristics of GaN MOS diode with $Ga_2O_3(Gd_2O_3)$ as the gate insulator, measured at various frequencies. The capacitance measured at forward bias is originated by capacitance of oxide films. The smaller capacitance at reverse bias is due to the capacitances in series originated from the oxide layer and the depletion region. The frequency dispersion observed in accumulation mode is attributed to the traps existed in the inner oxide layer. This phenomenon is induced from the Maxwell-Wranger effect (Hippel, 1954). The obtained density state (D_{it}) is less than $10^{11}cm^{-2}eV^{-1}$. The ln(J) v.s. V characteristics of the GaN MOS diodes are shown in Fig. 11. It can be seen that the $Ga_2O_3(Gd_2O_3)$ insulator grown using the electron-beam deposition method exhibits good insulation properties.

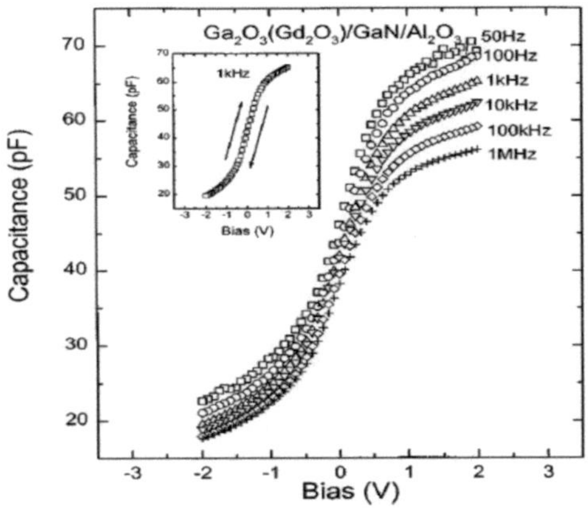

Fig. 10. The capacitance-voltage characteristics of $Ga_2O_3(Gd_2O_3)$/GaN MOS diodes.

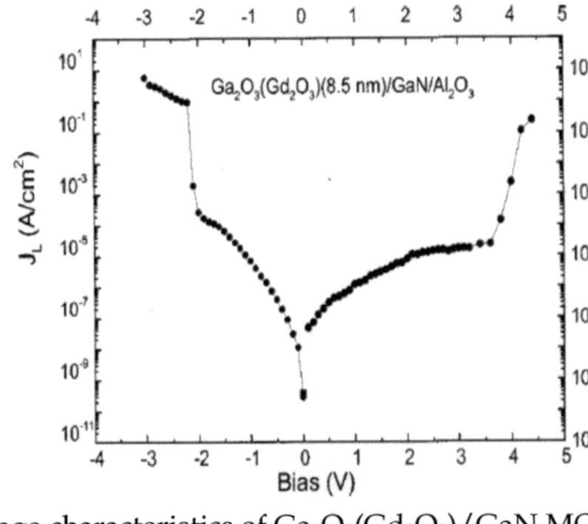

Fig. 11. The current-voltage characteristics of $Ga_2O_3(Gd_2O_3)$/GaN MOS diodes.

The drain-source current-drain-source voltage (I_{DS}-V_{DS}) characteristics of SiO_2/AlGaN/GaN MOS-HEMTs and AlGaN/GaN MES-HEMTs are reported by Arulkumaran and Egawa, et al., and shown in Fig. 12 (a) and (b), respectively (Arulkumaran et al., 2005). It can be seen that the $I_{DS(max)}$ of MOS-HEMTs is about 856mA/mm which is larger than that of MES-HEMTs. Fig. 13 (a) and (b) show the gate leakage current (I_g) as a function of gate-source voltage and the transfer characteristics of MOS-HEMTs and MES-HEMTs, respectively. At reverse voltage of 40V, the I_g of MOS-HEMTs is about three orders smaller than that of MES-HEMTs The $g_{m(max)}$ of MOS-HEMTs and MES-HEMTs are 160mS/mm and 145mS/mm, respectively. Better direct-current (dc) performance can be obtained with the improvement of the extrinsic transconductance of transistors. These results show that the electrical performance can be enhanced through the surface passivation and gate insulation. The similar behaviors of AlGaN/GaN MOS-HEMTs with different gate insulators can be observed in other reports (Kikuta et al., 2006; Yagi et al., 2006).

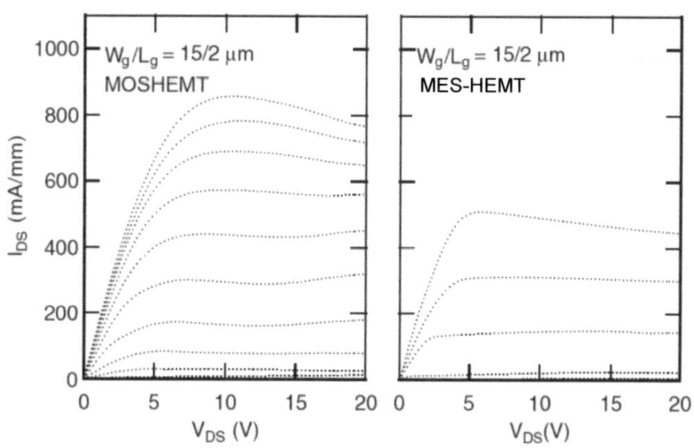

(a) (b)

Fig. 12. The I_{DS}-V_{DS} characteristics of (a) SiO_2/AlGaN/GaN MOS-HEMTs and (b) AlGaN/GaN MES-HEMTs.

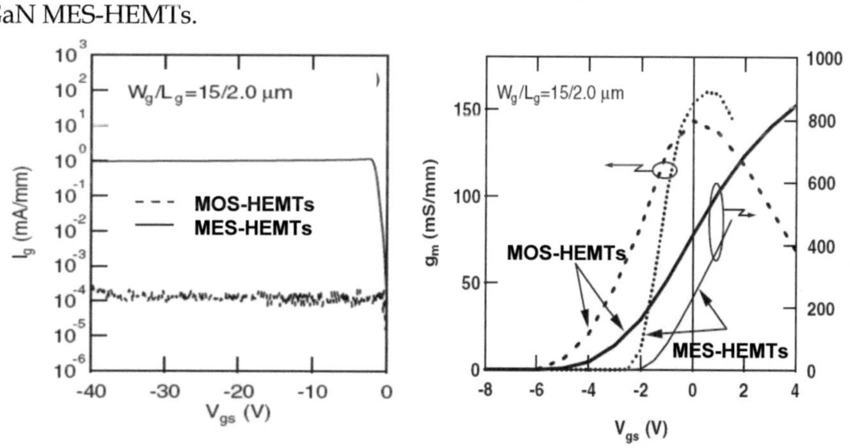

(a) (b)

Fig. 13. The dc electric performance of MOS-HEMTs and MES-HEMTs (a) the gate leakage current as a function of gate-source voltage and (b) the transfer characteristics.

Investigating dielectrics with high permittivity is an interesting issue for better high-frequency and high-performances due to the suppression of gate leakage current and current collapse by applying gate dielectrics in AlGaN/GaN MOS-HEMTs. However, the decrease of the transconductance (Ye et al., 2005) and the large shift of the threshold voltage are obvious expenses. Using dielectrics with high permittivity (high-k) is helpful to solve these problems. A larger dielectric constant could translate to an efficient gate modulation (Liu et al., 2006); thus, a smaller decrease in transconductance and a moderate increase in the threshold voltage could be expected in MOS-HEMTs with high-k gate insulators. The Pr_2O_3 high-k dielectrics (k=30) have also been deposited using an electron-beam evaporator for gate insulation and surface passivation of AlGaN/GaN MOS-HEMTs (Chiu et al., 2008). Fig. 14 shows the flicker noise spectra of MES-HEMTs and MOS-HEMTs, respectively. It can be seen that the level of flicker noise of the MOS-HEMTs is smaller than that of the MES-HEMTs. It means that the lower surface states and gate leakage current is achievable with high-k Pr_2O_3 films grown using an electron-beam evaporator. The microwave-power characteristics of both devices are shown in Fig. 15. The maximum output-power density and PAE are 753mW/mm and 36.8% for MOS-HEMTs at an input power of 15dBm, which are better than those of the MES-HEMTs with corresponding values of 698mW/mm and 32.1%, respectively.

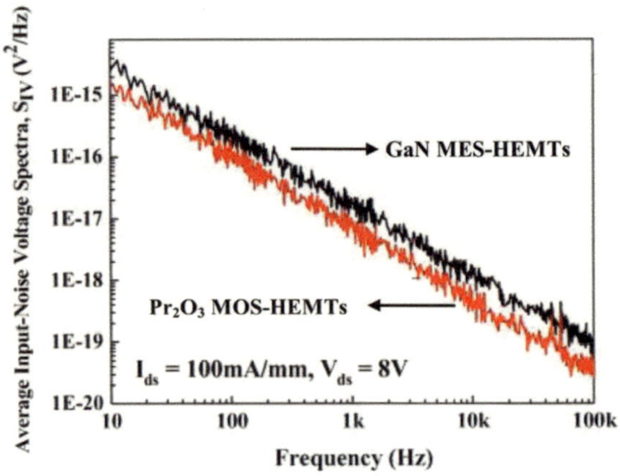

Fig. 14. Flicker noise spectra of MES-HEMTs and MOS-HEMTs, respectively.

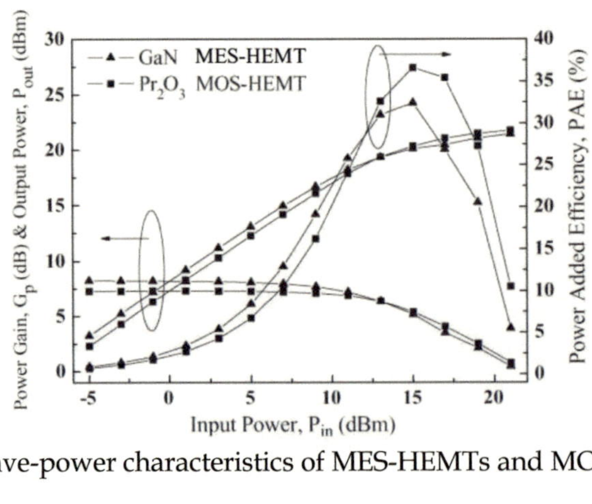

Fig. 15. The mircorwave-power characteristics of MES-HEMTs and MOS-HEMTs.

4.2 Atomic layer deposition (ALD) deposition method

The atomic layer deposition (ALD) deposition is a technique for growing thin films on various substrates with atomic scale precision based on alternate saturated surface reaction in each cycle of deposition. ALD is a chemical gas phase deposition process. The growth of thin film is self-limited and based on surface reaction. The deposition of conformal thin-films onto substrates of varying compositions via sequential surface chemistry enables controllable atomic scale deposition. ALD reaction divides the CVD reaction into two half-reactions. The ALD deposition has several advantages over other techniques, including high quality deposited films, excellent conformity and reproducibility. A variety of thin films can be deposited utilizing ALD deposition with high density and low impurity at a low deposition temperature.

The Al_2O_3 and HfO_2 dielectrics are common insulators grown using ALD system and have been applied in AlGaN/GaN MOS-HEMTs (Park et al., 2004; Ye et al., 2005; Wu et al., 2006; Kim et al., 2007; Medjdoub et al., 2007; Yue et al., 2008; Feng et al., 2009; Chang et al., 2009). The qualities of Al_2O_3 films gauged by several important figures of merit including uniformity, defect density and stoichiometric ratio of the deposited films, are comparably better, when ALD is chosen over the other deposition methods such as sputtering and electron-beam deposition methods. The I_{DS}-V_{DS} and effective electron mobility characteristics of GaN MOS-HEMTs are shown in Fig. 16 (a) and (b), respectively. The $I_{DS(max)}$ of MOS-HEMTs at V_{GS}=6V is 375mA/mm and the off-state breakdown voltage is 145V. In addition, the negative output conductance under high drain bias is due to self-heating effect. The effective carrier mobility (μ_{eff}) of insulator/AlGaN/GaN structure as a function of effective electric field (E_{eff}) is larger compared to the other semiconductors, which implies devices sharing this kind of structure do possess excellent high-frequency and high-power performances.

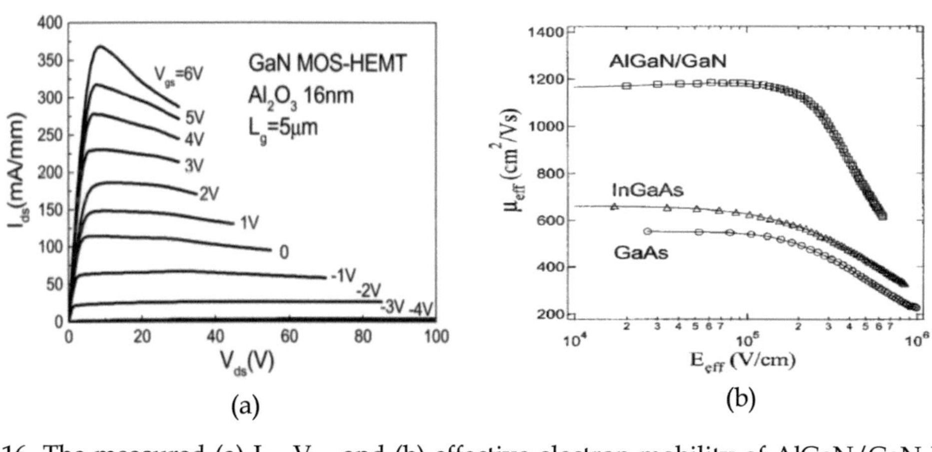

Fig. 16. The measured (a) I_{DS}-V_{DS} and (b) effective electron mobility of AlGaN/GaN MOS-HEMTs.

The HfO_2 insulator is a promising candidate because of its high dielectric constant (20~25) and large bandgap (5.6~5.8eV). Al_2O_3 is also a potential candidate for the above application, which has a good passivation effect and low interface state density. The Al_2O_3 is a better insulator for interfacial passivation layer (IPL) application in high-k gate process of

AlGaN/GaN MOS-HEMTs because it has better chemical and thermal stabilities against AlGaN compared to HfO_2 films (Gusev et al., 2006). The HfO_2/Al_2O_3 stack gate dielectrics deposited using ALD system utilized in AlGaN/GaN MOS-HEMTs are reported by Yue and Hao, et al. (Yue et al., 2008). The $g_{m(max)}$ of MES-HEMTs and MOS-HEMTs are 165mS/mm and 150mS/mm, respectively. The decay of the transconductance is due to the increase of the distance between the channel and Schottky gate of the MOS-HEMT structure. Furthermore, the negative shift in threshold voltage from –4V of MES-HEMTs to –5V of MOS-HEMTs is caused by the same reason. The decrease of $g_{m(max)}$ is only 9% and therefore is better compared with the deterioration degree of transconductance in MOS-HEMTs with low-k gate insulators (Kordos et al., 2005). The gate voltage swing (GVS) of MES-HEMTs and MOS-HEMTs are 2.4V and 1.8V, respectively. MOS-HEMTs not only have better linear operation because of large GVS compared to HEMTs, but also have a smaller intermodulation distortion, a smaller phase noise, and a larger dynamic range. These advantages ultimately render them suitable for practical amplifier applications (Khan et al., 2006). The pulse current-voltage performances of Al_2O_3 passivated MES-HEMTs and stack gate MOS-HEMTs are shown in Fig. 17. The current collapse phenomenon ascribed to the interface state density can not be observed in both devices and therefore it is reasonable to assume the surface states are well passivated in both transistors. Moreover, the unity current gain cut-off frequency and the maximum frequency of oscillation are 12GHz and 34GHz, respectively, as depicted in Fig. 18. According to the results mentioned above, the MOS-HEMTs not only have excellent dc and pulse mode electrical performances than those of passivated MES-HEMTs, but also exhibit outstanding high-frequency properties.

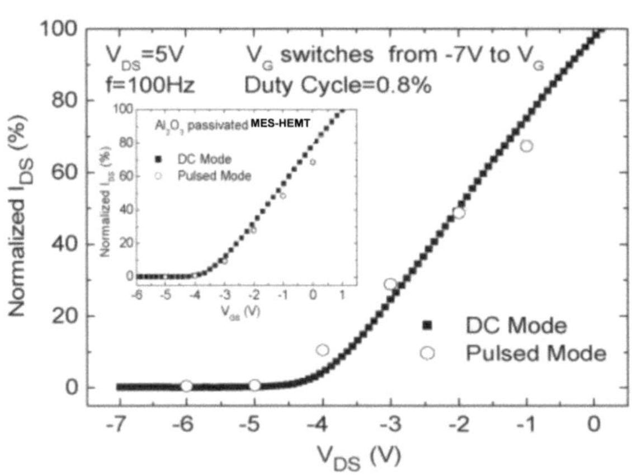

Fig. 17. The pulse current-voltage characteristics of Al_2O_3 passivated MES-HEMTs and stack gate MOS-HEMTs.

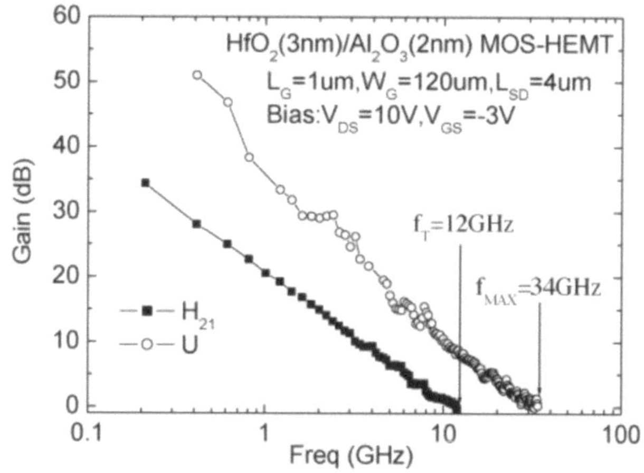

Fig. 18. The short-circuit current gain (H_{21}) and unilateral power gain (U) versus frequency of MOS-HEMTs operated at V_{DS}=10V and V_{GS}=-3V.

4.3 Sputter deposition method

Usually, a sputtering deposition involves two electrical plates in the vacuum chamber and one of them is used to accommodate a target material. When a negative voltage is applied to the target (i.e. cathode) by an external power supply, a number of free electrons are accelerated toward the anode. While on their way to anode these free electrons would expect to hit other gas molecules and then ionize them in the chamber, if the energy of the accelerated electrons is sufficiently high enough. Due to the negative voltage applied on target, the energy of these positive ions attracted toward the target surface depend on the applied voltage. When accelerated positive ions collide with the surface atoms of the target material, their energy are imparted to the target molecules, so that they can be ejected or sputtered out from the surface of the target material. Sputtering of target material is a possible outcome only, when the bombarding ion energy is large enough to disrupt the target molecules. Sometimes the ions bombarding the cathode may knock the electrons out from it, causing them to accelerate under the influence of applied field to enhance a chain of reactions involving the necessary ionization of molecules and the sputtering process. Magnetron sputtering is a powerful and flexible technique and the sputtering process almost places no restriction in the choice of the target materials; these include using a dc power to sputter pure metals and a rf power or pulsed dc power source to sputter semiconductors and isolators.

The performances of GaN-based MOS devices with insulators grown by sputtering technology are demonstrated (Nakano & Kachi, 2003; Liu et al., 2007; Jhin et al., 2008; Shih et al., 2009). Liu and Chor, et al. have reported the performances of AlGaN/GaN MOS-HEMTs with HfO_2 high-k dielectrics deposited using sputtering system (Liu et al., 2006). The I_{DS}-V_{DS} and transfer characteristics are shown in Fig. 19 (a) and (b), respectively. The $I_{DS(max)}$ of 830mA/mm is obtained at V_{DS}=6V. In addition, the self-heating effect does not be found under the high drain-source current operation, because the Si substrate has better thermal conductivity. The GVS and $g_{m(max)}$ of MOS-HEMTs are 2.91V and 115mS/mm, respectively. The negative shift in threshold voltage from -4V of MES-HEMTs to -6V of MOS-HEMTs is caused by the increase in distance between gate and channel. Figure 20 shows the dc and pulse mode measurement of MES-HEMT and MOS-HEMT respectively. The V_{DS} bias is held at 8V and the frequency of pulsed V_{GS} is 100kHz. There is 60% drain current discrepancy between the dc and pulsed modes associated with MES-HEMTs. However, a significant drain current recovery is observed in MOS-HEMTs. This phenomenon clearly indicates that the HfO_2 layer can passivate the traps on AlGaN surface that otherwise would create a depletion region and suppress the current collapse.

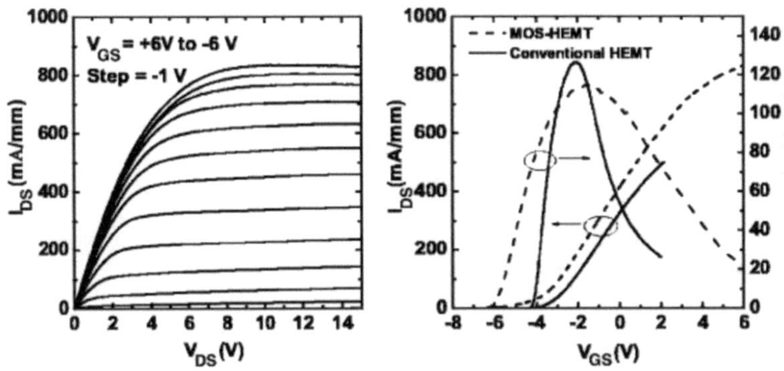

(a) (b)

Fig. 19. The (a) I_{DS}-V_{DS} of MOS-HEMTs and (b) transfer characteristics of MES-HEMTs and MOS-HEMTs.

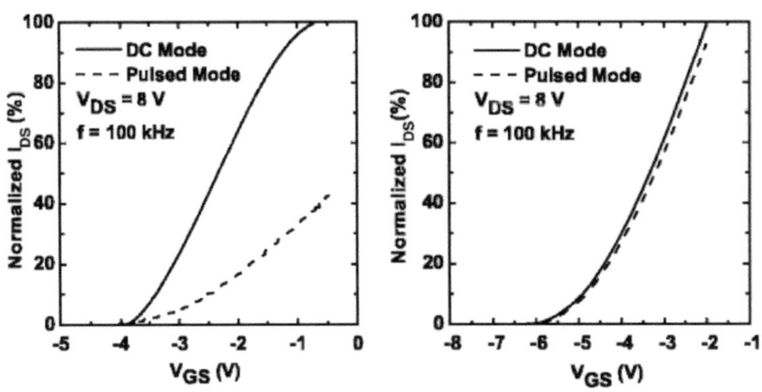

(a) (b)

Fig. 20. The dc and pulse mode measurement of AlGaN/GaN (a) MES-HEMTs and (b) MOS-HEMTs.

The relationship between gate insulation and high-frequency performances are also investigated (Liu et al., 2007). The high-frequency performances of AlGaN/GaN HEMTs with and without HfO_2 gate insulator are shown in Fig. 21 (a) and (b), respectively. The improvement of f_T and f_{max} is attributed to the increase of transconductance.

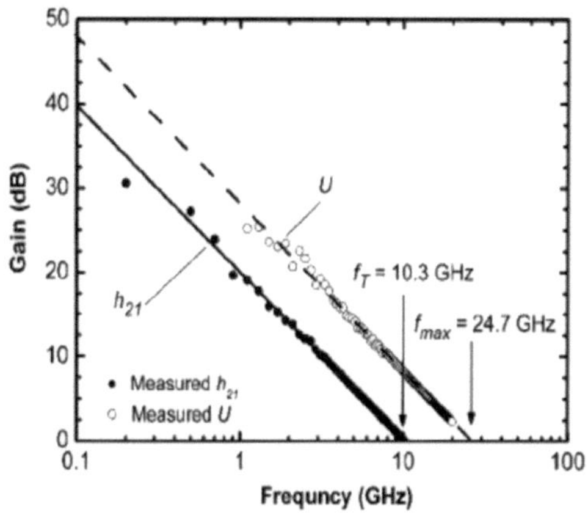

(a)

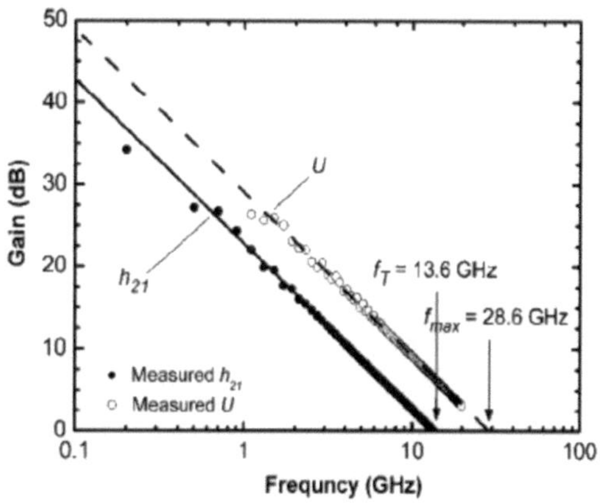

(b)

Fig. 21. The S-parameter measurement of AlGaN/GaN (a) MES-HEMTs and (b) MOS-HEMTs.

4.4 Jet vapor deposition (JVD) method

The jet vapor deposition (JVD) is a novel process for synthesizing wide variety of thin films of metals, semiconductors, and insulators (Ma, 1998). It relies on supersonic jets of a light carrier gas such as helium to transport depositing vapor from the source to the substrate. Because of the separation of the constituent depositing species, and their short transit times, there is very little chance for gas-phase nucleation. In contrast to the conventional CVD silicon nitride, the high-field I-V characteristics of the JVD silicon nitride fit the Fowler-Nordheim (F-N) tunneling theory over 4-5 orders of magnitude in current, but do not fit at all with the Frenkel-Poole (F-P) transport theory. This is consistent with a much lower concentration of electronic traps in the JVD silicon nitride.

The JVD is a novel and interesting technology. The $SiO_2/Si_3N_4/SiO_2$ stack gate insulators of GaN MOS devices grown using JVD system were reported (Gaffey et al., 2001; Ma et al., 2003). Figure 22 (a) shows the measured and theoretical capacitance-voltage characteristics performed at a room temperature and a frequency of 10kHz. The open circle and solid line indicate the measured data and theoretic value of capacitance-voltage (C-V) characteristics, respectively. It can be seen that the measured data are fitted well to theoretic data with out any translation along the voltage axis. This phenomenon indicates that the net density of fixed charges within the insulator is very low. Fig. 22 (b) shows the C-V performances at different frequency at 450°C. The near ideal curve means that the devices work well up to 450°C. The interface-state density (D_{it}) of devices should be low because of weak frequency dependence. The inset shows a magnified view of this C-V curve with respect to a particular voltage region and note that the small frequency dependence is observed.

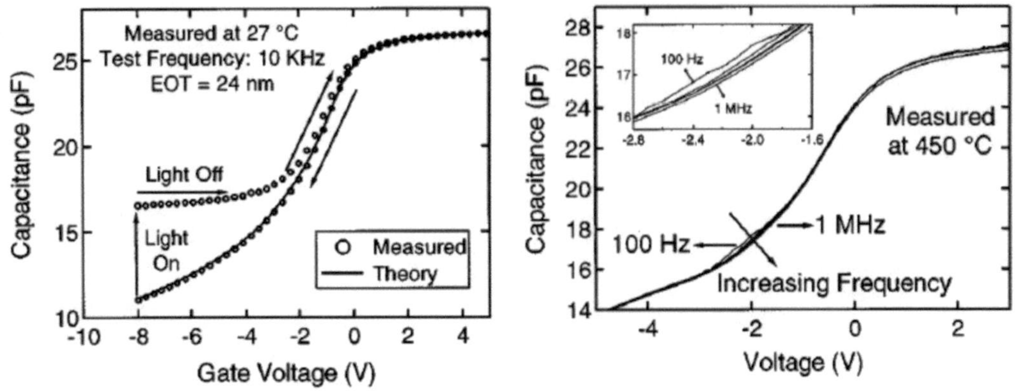

(a) (b)

Fig. 22. (a) The measured and theoretic C-V characteristics of GaN MOS devices at room temperature. (b) The C-V characteristics of GaN MOS devices at varied testing frequency and 450ºC.

The extracted D_{it} values using conductance method (Schroder, 1998) as a function of energy separation from conduction band edge are shown in Fig. 23 (a). Those values are on the order of mid-to-high $10^{11}cm^{-2}eV^{-1}$, which are deemed reasonable. The relationships between gate leakage current (J) and oxide electrical field (E) are measured at 27ºC, 150ºC, 250ºC, 350ºC, and 450ºC and shown in Fig. 23 (b), respectively. The electrical field is defined as V_g/t_{ox}, where V_g is the gate voltage and t_{ox} is the equivalent oxide thickness. It can be seen that the leakage currents are very low and with weak temperature dependence. In addition, an acute breakdown is observed when electrical filed is larger than 12MV/cm over the entire range of temperatures. The observed phenomena demonstrate the realization of the high quality stack gate insulators.

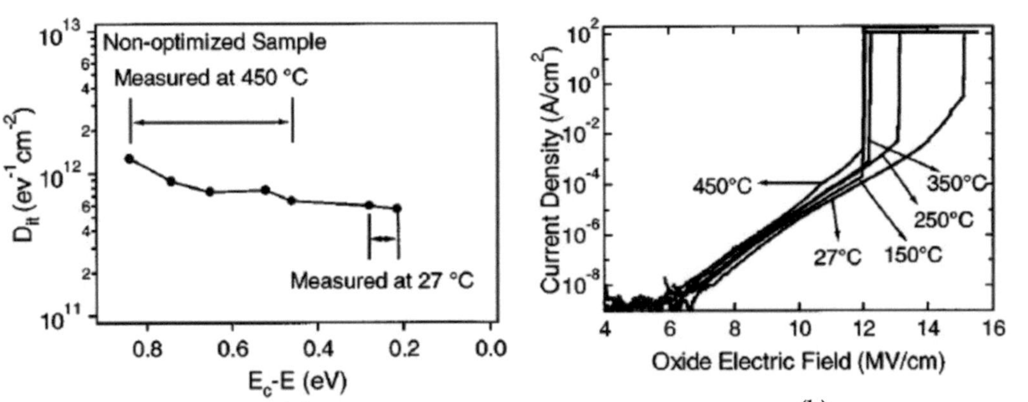

(a) (b)

Fig. 23. (a) The estimated D_{it} values as a function of energy separation from conduction band edge at 450ºC. (b) The gate leakage current density as a function of oxide electrical field of GaN MOS devices measured at various temperatures.

The MOS-FETs using $SiO_2/Si_3N_4/SiO_2$ stack as gate insulators demonstrate reliable performance up to the working temperature as high as 450°C (Ma, 2003). Fig. 24 (a) and (b) shows the I-V characteristics of a typical GaN MOS-FET on 6H-SiC substrate before and after accelerated lifetime test stress. In this experiment, a device was declared "dead" when its gate oxide leakage current exceeded $0.25mA/cm^2$, as measured from an otherwise well-functioned device shown in Fig. 24 (b). The inversion layer electron mobility obtained from this device is around $40cm^2/Vs$ both before and after the accelerated lifetime test. In this case, the effective channel mobility is reduced by the high source and drain contact resistances.

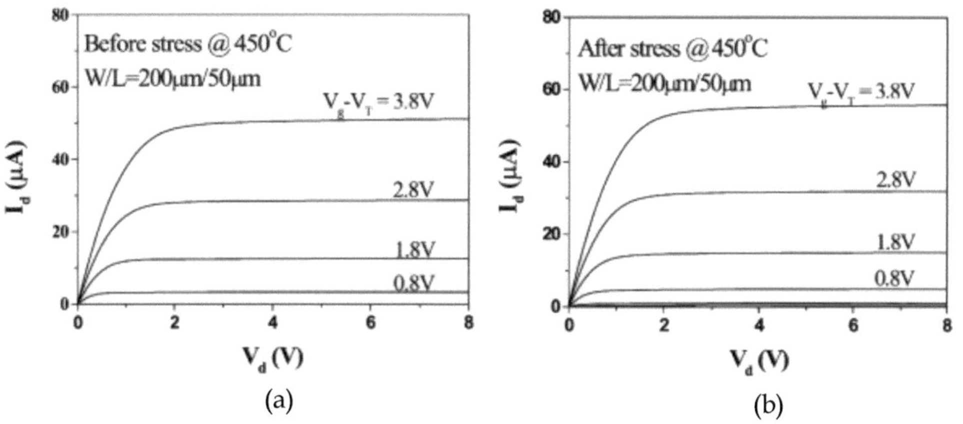

Fig. 24. I–V characteristics of a typical GaN MOS-FETs on 6H-SiC substrate (a) before and (b) after accelerated lifetime test stress.

4.5 Chemical vapor deposition (CVD) method

Traditional CVD deposition systems include atmospheric pressure CVD (APCVD), low-pressure CVD (LPCVD), and plasma-enhances CVD (PECVD). Among the three systems, PECVD appears more advantages due its comparatively low deposition temperature and a good step surface coverage. The power source of a CVD system is usually a radio-frequency (RF) power with frequency of 13.56MHz applied to the vacuum deposition chamber. It induces the gas molecules situated between the parallel electrodes to collide with electrons, and thereby to trigger the dissociation of molecules. Several free radicals, gas ions, electrons, and nuclear gas molecules, called cold plasma, are thus generated due to the increase of internal energy which enhances to induce excitation, ionization, relaxation, and recombination reactions to take place. The un-disturbed RF power provides energy to the gas plasma and the collisions among gas species lead to the chemical reactions. Since the dissociation of the gas plasma provides the energy needed for the chemical vapor deposition directly, the process could therefore take place under a low temperature. It is the most frequently used technique for low temperature deposition.

There are reports about the electrical performances of AlGaN/GaN MOS-HEMTs with gate oxide layers grown using PECVD system (Casey et al., 1996; Arulkumaran et al., 1998; Khan et al., 2000; Simin et al., 2002; Marso et al., 2002; Misstele et al., 2003; Onojima et al., 2007; Higashiwaki et al., 2009). As shown in Fig. 25 (a), the $g_{m(max)}$ of passivated MES-HEMTs is improved compared to the traditional MES-HEMTs (Marso et al., 2002). The threshold voltage of passivated MES-HEMTs is not affected by SiO_2 layer, because the metalized gate makes direct contact with the semiconductor surface of the unpassivated MES-HEMTs as well as that of the passivated MES-HEMTs. The SiO_2 passivation layer of the passivated MES-HEMTs cannot influence the channel in the active region under the gate contact, however, it can reduce the resistance of the source and the drain, thereby improving the dc performances (Higashiwaki et al., 2005). For MOS-HEMTs, the large negative shift of the threshold voltage is observed and it is mainly caused by an increase in gate-to-channel separation due to the presence of the dielectric layer. Besides, in the Fig. 25 (b), the cutoff frequency (f_T) of unpassivated MES-HEMTs, passivated MES-HEMTs and MOS-HEMTs are 16.5GHz, 18.2GHz, and 24.0GHz, respectively.

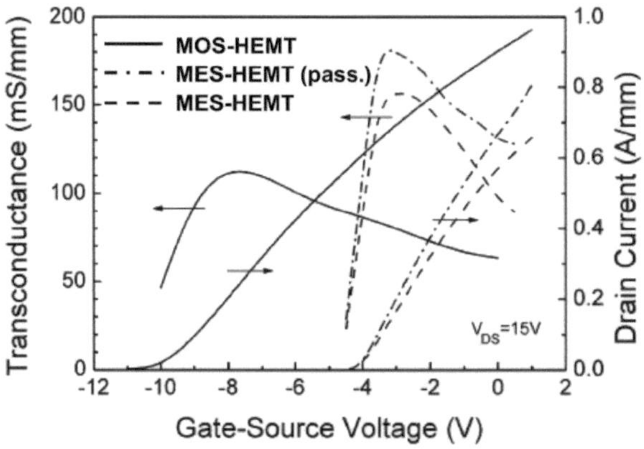

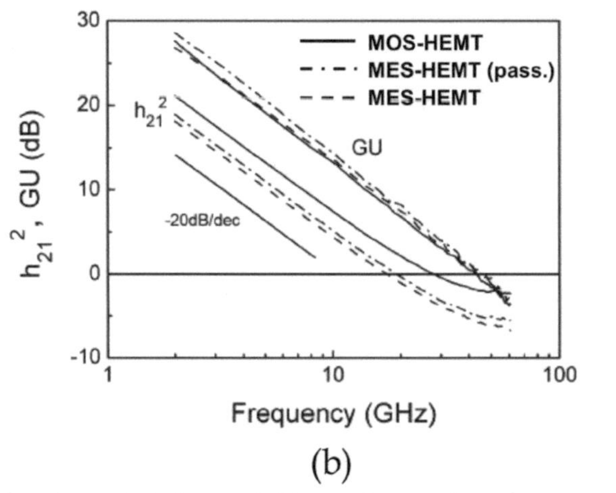

Fig. 25. The (a) transfer characteristics and (b) small-signal performances of traditional MES-HEMTs, passivated MES-HEMTs and MOS-HEMTs.

The power performance of SiO_2/AlGaN/InGaN/GaN MOS-HEMTs are investigated (Simin et al., 2002). Figure 26 shows the power performance and the stability of power performance of MOS-HEMTs. A maximum power of 6.1W/mm in CW mode and 7.5W/mm of in pulse mode can be obtained, when the MOS-HEMTs are operated at V_{DS}=30V and 2GHz. It is worth noting that the power degrades at about 0.5~0.6W/mm in the first 2~3 hours and this stabilization is irreversible.

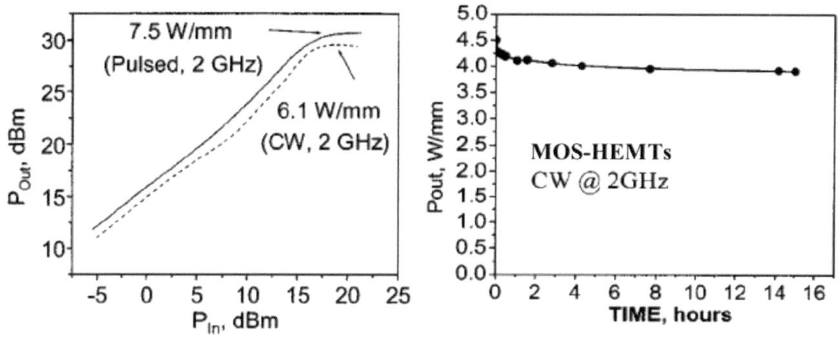

(a) (b)

Fig. 26. The (a) power performances and (b) CW mode power performances of MOS-HEMTs.

In general, the PECVD system is not perfect and there is a concern on the degree of plasma damage or charge-induces damage which would affect the devices during fabrication process. In some cases, such as the passivation of compound semiconductor devices, the plasma damage is serious that it would even affect the throughput of device production. The catalytic chemical vapor deposition (Cat-CVD) method was developed to overcome these problems encountered during the deposition process of PECVD. It is expected that vary low-damage process using the Cat-CVD system can be realized (Wiesmann et al., 1979; Matsumura et al., 1986), which is a low-temperature method being developed without using the plasma. The Cat-CVD method, also known as the hot-wire CVD (HWCVD) method, the deposition gases are decomposed by catalytic cracking reaction with a heated catalyzer placed near the substrate, so that the films are deposited at a low substrate temperatures around 300°C without any plasma. The Cat-CVD apparatus consists mainly of three parts: a gas inlet which feed gas into the low-pressure deposition chamber, an arrangement for gas decomposition via catalytic cracking reaction at the surface of a heated catalyzer, and the substrate for film deposition where decomposed species are transported from the catalyzer. In most cases, a tungsten (W) wire or a W ribbon is used as the catalyzer, because the melting point of W is as high as 3382°C, and also can sustain the temperature as high as 2165°C (Poate et al., 1978) even when the surface of W is converted to silicide by the reaction with SiH_4 gas.

The performances of AlGaN/GaN MOS-HEMTs with SiN_x films for gate insulation and surface passivation grown using Cat-CVD system are demonstrated by Higashiwaki and Mimura, et al. (Endoh et al., 2006; Higashiwaki et al., 2006; Yamashita et al., 2006). The output and transfer characteristics are shown in Fig. 27 (a) and (b), respectively. The $I_{DS(max)}$ and $g_{m(max)}$ are 1.49A/mm at V_{GS}=1V and 402mS/mm at V_{DS}=2V, respectively. A large reduction of the transconductance occurs as the V_{DS} is increased from 2 to 6V, mainly due to the self-heating effect. The off-state gate-drain breakdown voltage is about -18V at a gate leakage current density of 1mA/mm.

The maximum f_T and f_{max} are obtained at different bias voltages, as shown in Fig. 28 (a) and (b). The maximum f_T is 181GHz. The maximum f_{max} is also a very high value, 186GHz or 183GHz obtained from the maximum stable gain (MSG) or unilateral gain (Ug) extrapolation.

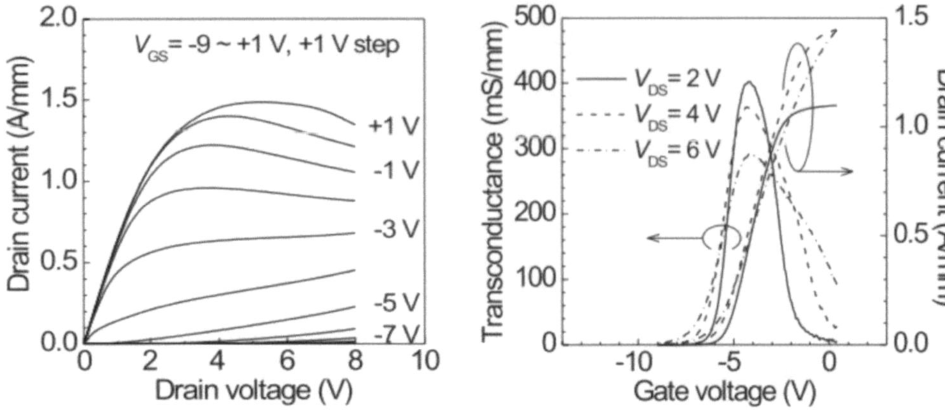

(a) (b)

Fig. 27. The dc performances of MOS-HEMTs (a) output characteristics (b) transfer characteristics.

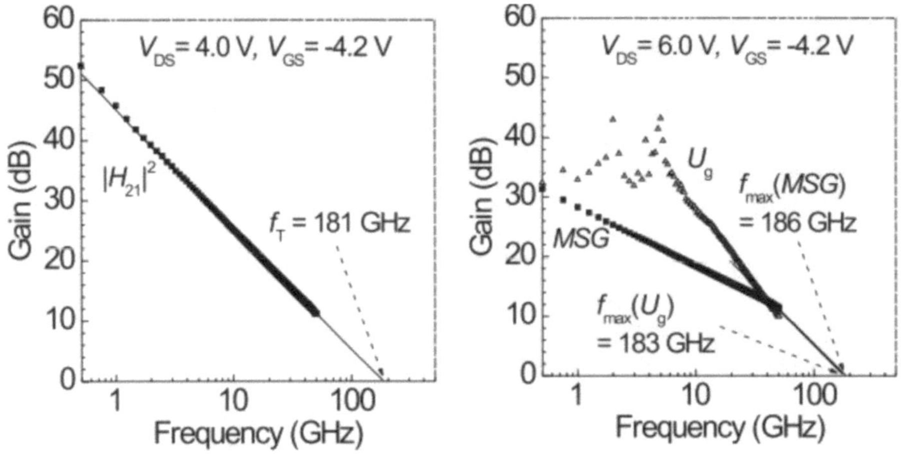

(a) (b)

Fig. 28. The small-signal RF characteristics of MOS-HEMTs with L_G=30nm at V_{GS}=–4.2V, (a) V_{DS}=4.0V and (b)V_{DS}=6.0V.

In recent years, the photochemical vapor deposition (photo-CVD) method has been developed to grow high-quality SiO_2 layers on various MOS devices (Chiou et al., 2003). In this method, a deuterium (D_2) lamp is used as the excitation source. The D_2 lamp emits strong ultraviolet and vacuum ultraviolet radiation, which can effectively decompose SiH_4 and O_2. The O_2 can absorb photons in the wavelength region from 133nm to 175nm and SiH_4 can absorb photons below 147nm (Chang et al., 2003). Therefore, the energy can be directly transferred from the D_2 lamp to the excited Si and O atoms. It has been reported that the quality of oxide layers grown by photo-CVD system is close to that of grown by thermal oxidation. Consequently, the better electrical properties of MOS devices with oxide films

grown using photo-CVD system can be obtained (Wang et al., 2003; Chiou et al., 2003; Chiou et al., 2004; Wang et al., 2005; Liu et al., 2007).
The current-voltage characteristics and transfer performances of MOS-HEMTs, in which the SiO_2 film is grown using photo-CVD method, operated at room temperature and 300°C are shown in Fig. 29 (a) and (b), respectively. It can be found that the values of I_{DS} decrease as the transistors are operated at high temperature. The $I_{DS(max)}$ values of MOS-HEMTs operated at room temperature and 300°C are about 755mA/mm and 323mA/mm, respectively. The $g_{m(max)}$ and GVS of transistors biased at V_{DS}=10V and V_{GS}=-5V at room temperature are 95mS/mm and 8V, respectively. The electrical performance of transistors at 300°C is poor compared to those operated at room temperature due to the decrease of the electron mobility in the 2DEG channel.

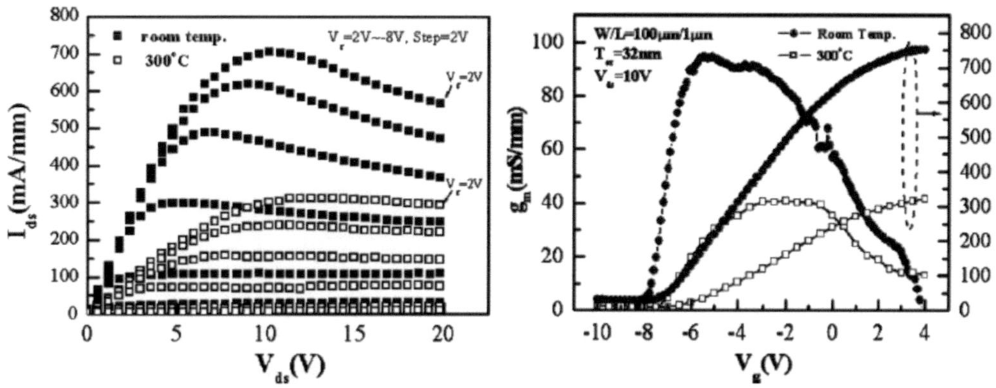

(a) (b)

Fig. 29. The (a) output characteristics and (b) transfer performances of fabricated SiO_2/AlGaN/GaN MOS-HEMTs.

The flicker noise of transistors is an important indicator when they are applied as mixers and oscillators in communication systems. The flicker noise performance of AlGaN/GaN MOS-HEMTs with SiO_2 grown by photo CVD is investigated (Liu et al., 2007). The diagram of noise sources of MOS-HEMTs is shown in Fig. 30. The possible noise sources of transistors are originated from the resistance of gate-region and series resistance of ungated-region. The prior resistance can be modulated by gate voltage while the later remains constant. The normalized noise power spectra of MOS-HEMTs, in which the SiO_2 layer is deposited using Cat-CVD method, operated in saturation region are shown in Fig. 31. It can be seen that the noise are fitted well by 1/f law, when V_{GS} is closed to threshold voltage. However, the exponent of 1/f fitting line increases when gate bias becomes positive. This phenomenon is attributed to the spatial distribution of interfacial traps states (Reimbold et al., 1984).

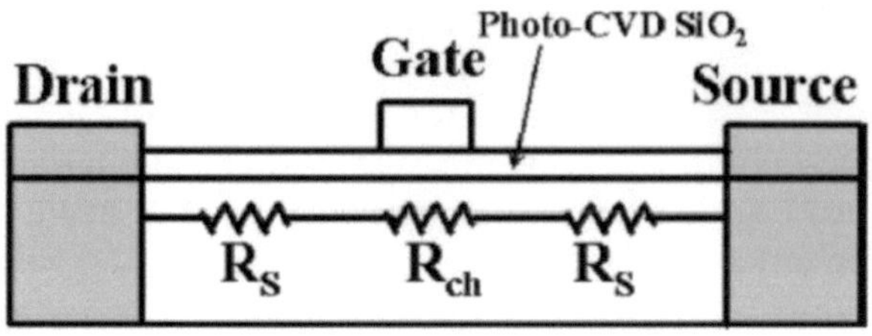

Fig. 30. The diagram of noise sources of MOS-HEMTs.

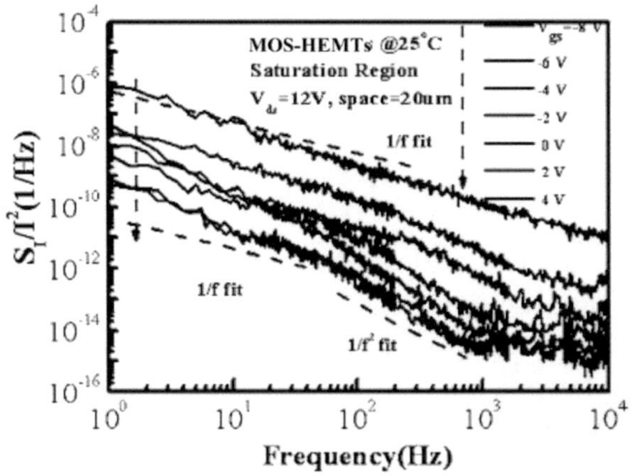

Fig. 31. The normalized noise power spectra of MOS-HEMTs operated in saturation region.

4.6 Metalorganic chemical vapor deposition (MOCVD) method

The metalorganic chemical vapor deposition (MOCVD) is a form of CVD utilizing metalorganic compounds as precursors. Essentially all III-V and II-VI semiconductors and most of their alloys can be successfully grown using MOCVD system, making this possibly the most versatile growth technique for compound semiconductors. Mechanism of MOCVD system growth process in general is governed by thermodynamics, reaction kinetics, flow dynamics, and chemical composition of the gas species in the MOCVD system. Thermodynamics provide the driving force for the reactions, i.e. the equilibrium condition. The kinetic reaction describes the speed of the reaction and dynamic flow control the transport of the reactants to the surface. The chemical composition of the atmosphere determines the reactions involved in the growth, which includes the pre-reactions as well. MOCVD process relies on the vapor transport of the group III alkyls combined with group V hydrides to the heated substrate. At the heated substrate, the molecules pyrolyze to produce the group III and the group V elements needed for formation of the desired III-V semiconductors and subsequent reaction that comes after. Because these precursor molecules are so unstable at the growth temperature and the III-V solid is so stable, the thermodynamic driving force for MOCVD typically is enormous.

Up to now, many works dedicated to study the electrical performance of AlGaN/GaN MOS-HEMTs with dielectrics deposited using MOCVD system are reported (Kawai et al., 1998; Ren et al., 1998; Mehandru et al., 2002; Rai et al., 2006; Pozzovivo et al., 2007; Gregušová et al., 2007; Kordoša et al., 2007; Kuzmik et al., 2008). The performances of

AlGaN/GaN MOS-HEMTs with 4-nm-thick Al_2O_3 gate insulators grown using MOCVD are reported by Gregušová and Stoklas, et al. (Gregušová et al., 2007). Fig.32 (a) and (b) show the static and dynamic output characteristics of fabricated MES-HEMTs and MOS-HEMTs, respectively. The $I_{DS(max)}$ of the MES-HEMTs and MOS-HEMTs are 440mA/mm and 630mA/mm, respectively. The threshold voltages of MES-HEMTs and MOS-HEMTs are -3.14V and -4.26V, respectively. The $g_{m(max)}$ of MES-HEMTs and MOS-HEMTs are 67mS/mm and 116mS/mm, respectively. The larger g_m value of MOS-HEMTs compared to that of MES-HEMTs may be attributed to the incorporation of very thin gate insulators (Gregušová et al., 2007). In Fig. 32 (b), it can be seen that the MOS-HEMTs have less current collapse but the self-heating effect is still found at large drain-source voltage.

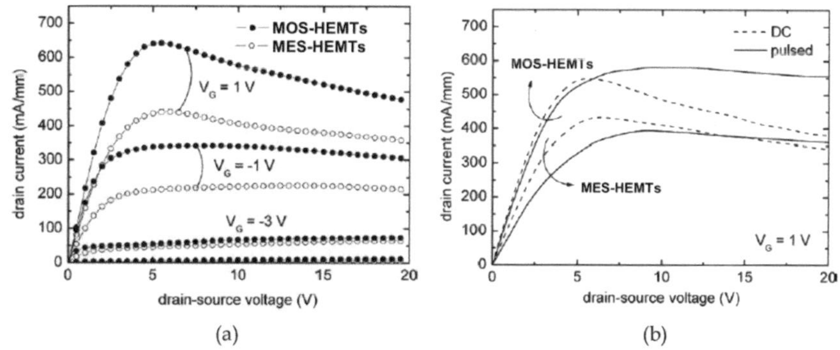

Fig. 32. The (a) static and (b) dynamic output performances of AlGaN/GaN MES-HEMTs and MOS-HEMTs.

The Si_3N_4/AlGaN/GaN MOS-HEMTs with field-plate structure for power application are investigated (Rai et al., 2006). The schematic configuration of MOS-HEMTs and power performances of MES-HEMTs and MOS-HEMTs are shown in Fig. 33 (a) and (b), respectively. When both devices operated at 2GHz and 50V, the output power of MES-HEMTs and MOS-HEMTs are 12 W/mm and 14.32 W/mm, respectively.

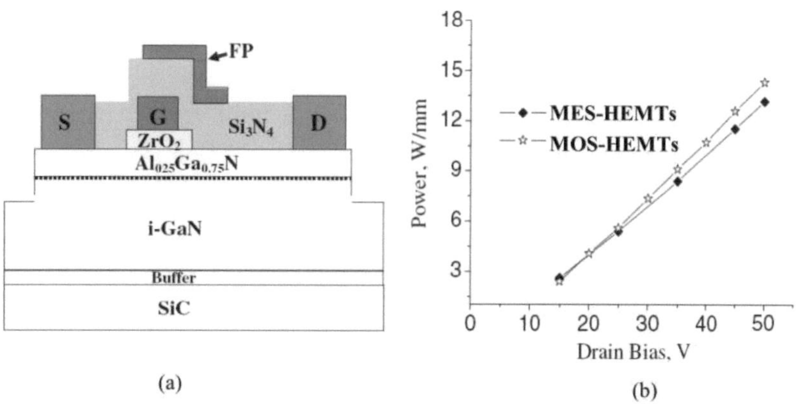

Fig. 33. The (a) schematic configuration of MOS-HEMTs and (b) power performances of AlGaN/GaN MES-HEMTs and MOS-HEMTs.

4.7 Photoelectrochemical (PEC) oxidation method

The dielectrics of AlGaN/GaN MOS-HEMTs are often deposited externally using electron-beam evaporator, sputtering system, MOCVD system, CVD system and others. However, the performances of these MOS devices are still affected by the natures of aforementioned deposition techniques involving particular growth conditions and contaminants inevitably brought to the semiconductor surface. In recent years, a photoelectrochemical (PEC) method has been successfully developed to either etch (Youtsey & Adesida, 1997) or oxidize (Rotter et al., 2000) GaN semiconductor. The PEC oxidation method can oxidize GaN and AlGaN semiconductors directly for gate insulators and surface passivation and it is similar to the formation of SiO_2 film on Si wafer directly using thermal or wet oxidation method. The PEC oxidation technique has many advantages including large growth rate, low cost, plasma damage-free, and oxide/semiconductor interface with few contaminants. Compared with external deposition methods, the PEC oxidation method is able to deliver high quality oxide/semiconductor interface with low interface state density and high insulation layer (Lee et al., 2005). In the PEC experiment, the chemical solution with appropriate pH value and light source with appropriate wavelength are needed. The purpose of using a laser beam with a wavelength shorter than the wavelength corresponding to the energy bandgap of semiconductors is to create more electron-hole pairs for enhancing the PEC oxidation reactions. When n-type GaN semiconductor makes contact with chemical solution, a Schottky contact is formed at the interface owing to the difference in the work function between semiconductor and chemical solution (Lee et al., 2005). The induced holes are moved up toward the interface owing to the built-in electrical field.

Figure 34 illustrates the photoelectrochemical oxidation system. A chemical aqueous solution of H_3PO_4 with a pH value of 3.5 was used as the electrolytic solution. An ampere meter and a pH meter were used to measure the current and pH value, respectively. The work function W_E of electrolytic solution as a function pH value can be expressed as (Finklea, 1988):

$$W_E(\text{eV}) = 4.25 + 0.059 \times \text{pH } \textit{value} \tag{1}$$

Since the pH value of 3.5 was used, the work function of the electrolytic solution is 4.457eV. The work function Ws of n-type GaN is expressed as:

$$W_S(\text{eV}) = \chi + (E_C - E_F) \tag{2}$$

Where χ=4.1eV is the electron affinity of n-type GaN, and E_C-E_F=0.039eV is the energy difference between the conduction band E_C and Fermi level E_F for the n-type GaN with an electron concentration of $5.0\times10^{17}cm^{-3}$. The schematic energy-band diagram for the electrolytic solution/n-type GaN is shown in Fig. 35. A built-in electric field was induced within the depletion region due to the work function difference between the electrolytic solution and the n-type GaN. By using a He-Cd laser with a wavelength of 325nm, electron-hole pairs were generated on the n-type GaN layer. The built-in electric field transported the generated electrons and holes to the n-type GaN and the electrolytic solution/n-type GaN interface, respectively. Since holes were accumulated at the interface, the n-type GaN was oxidized via the following reaction:

$$2GaN + 6h^+ + 3H_2O \rightleftharpoons Ga_2O_3 + 6H^+ + N_2 \tag{3}$$

where h^+ is hole. Not only was the Ga_2O_3 formed, but also the Ga_2O_3 was etched by the electrolytic solution. When the oxidation rate is larger than the etching rate, the Ga_2O_3 layer can be grown directly. Therefore, we can deduce that the growth rate of the Ga_2O_3 depends on the pH value of the electrolytic solution and the intensity of the He-Cd laser. By using a He-Cd laser, the growth rate of the Ga oxide film as a function of the laser intensity is shown in Fig. 36. The dependence of the induced current on the He-Cd laser intensity is also shown in Fig. 36. It can be seen that the growth rate and induced current are almost linearly proportional to the laser intensity. Because the generation rate of electron-hole pairs depends on the laser intensity, the increase in growth rate is attributable to the accumulation of more holes on the interface between the electrolytic solution and the n-type GaN. Oxidation was the process by which a layer of Ga oxide was formed on the n-type GaN surface. The oxidizing species diffused through the already grown Ga oxide to react with the GaN. For a grown Ga oxide thickness t_{ox} etched with KOH solution, we found the Ga oxide surface to be $0.67t_{ox}$ above the original n-type GaN surface, while the Ga oxide/n-type GaN interface was 0.33tox below.

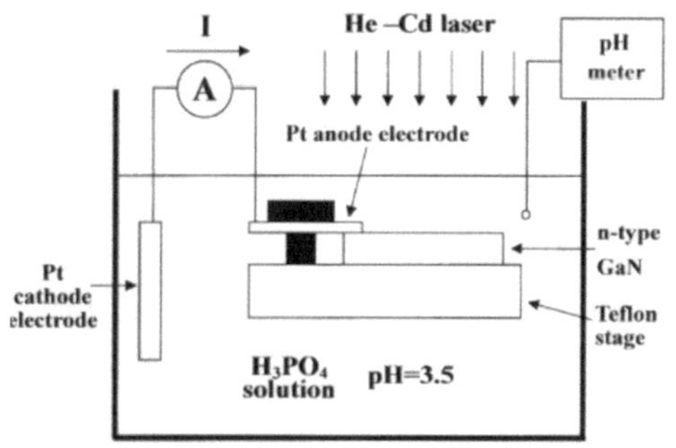

Fig. 34. Photoelectrochemical oxidation system.

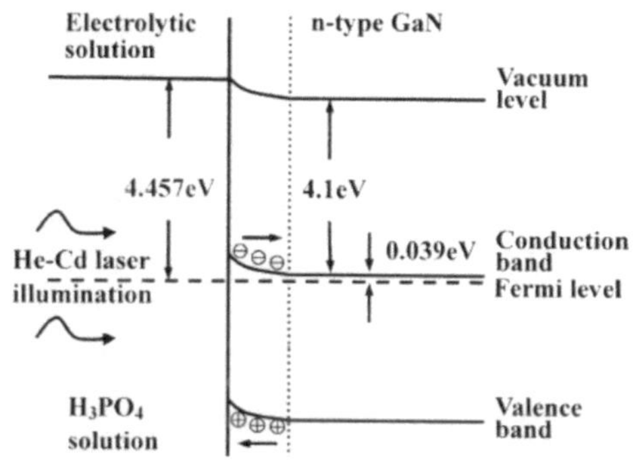

Fig. 35. Schematic energy band structure for H_3PO_4 solution/n-type GaN.

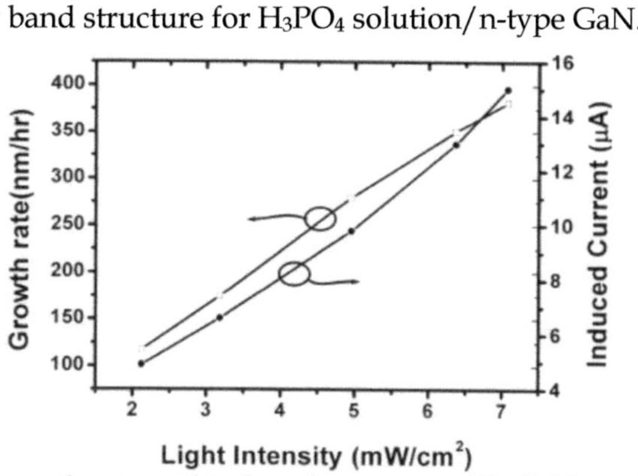

Fig. 36. Dependences of growth rate and induced current on He-Cd laser intensity.

For the p-GaN semiconductors, the schematic energy band diagram for H_3PO_4 solution/p-type GaN is shown in Fig. 37 (a). The flat-band potential energy W_S of p-GaN with a hole concentration of $3\times10^{17}cm^{-3}$ can be calculated from the following equation:

$$W_S(eV)=q\chi + E_g - (E_F - E_V) \quad (4)$$

where $q\chi$=3.3±0.2eV (Wu & Kahn, 1999) and E_g=3.4eV are the electron affinity and the energy bandgap of the p-GaN, respectively. The separation between the Fermi level E_F and the valance band edge E_V of p-GaN is 0.106eV. Owing to the difference between the flat-band potential energy of p-GaN and the work function of the H_3PO_4 solution, a Schottky contact is consequently formed; its schematic energy band diagram is shown in Fig. 37 (a). The holes and electrons ionized in the space-charge layer can move toward the p-GaN layer and H_3PO_4/p-GaN interface, respectively, due to the built-in electric field. These induced holes would move into the bulk layer under the influence of the built-in electrical field if p-GaN is made contact with electrolytic solution (Huang et al., 2009). This may explain why oxidizing p-GaN using traditional PEC oxidation is considered difficult. As shown in Fig. 37 (a), p-GaN is known to have a very large downward band bending in the dark. However, when laser light with enough intensity illuminates the p-GaN surface, the band bending is flattened owing to the surface photovoltage (SPV) effect (Long & Bermudez, 2002) The SPV effect may cause upward band bending to bring a small density of holes up to the surface. To push more holes toward the H_3PO_4/p-GaN interface to effectively oxidize the p-GaN, an applied forward bias is needed to counteract the effect of the built-in electric field. Figure 37 (b) shows the energy band diagram of the bias-assisted H_3PO_4 solution and p-GaN. The grown oxide thickness as a function of forward bias for an oxidation time of 10min is shown in Fig. 38. It is found that the oxidation occurs when the forward bias is larger than 2V, and the oxide thickness is proportional to the net forward bias.

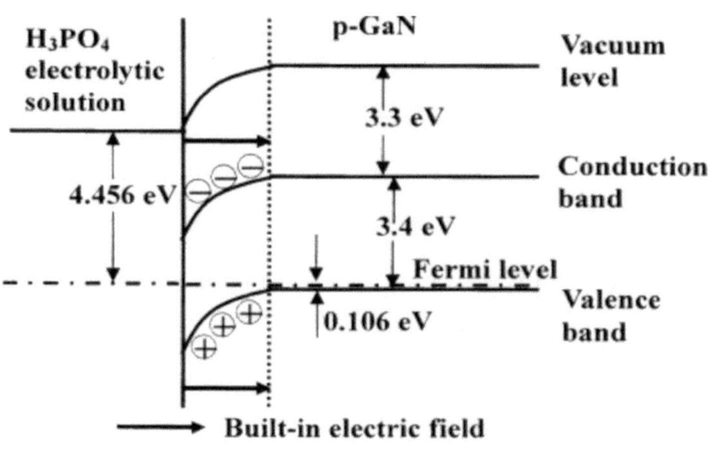

(a)

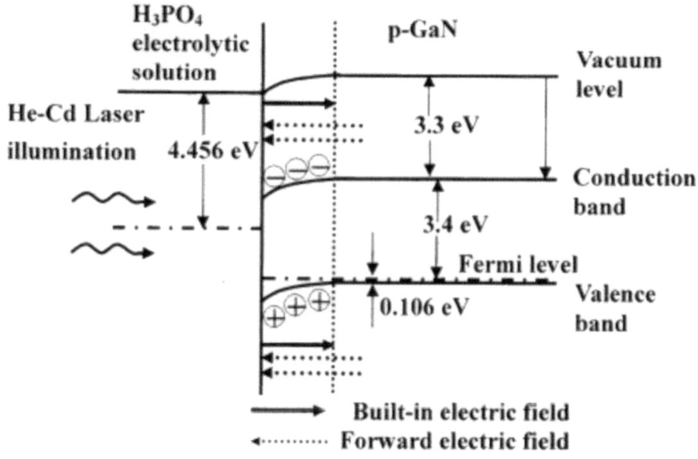

(b)

Fig. 37. Schematic energy band diagram for H_3PO_4 solution/p-type GaN (a) without bias assistance and (b) with bias assistance.

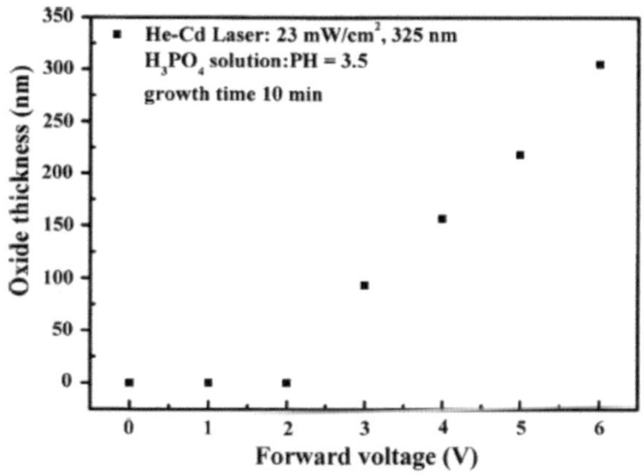

Fig. 38. Grown oxide thickness as a function of forward bias.

It can be found that the etching and oxidative reaction occur concurrently at the interface. Whichever the reaction is dominant could usually be determined using the appropriate PH value of the electrolytic solution in the PEC experiment. Furthermore, these reactions also reveal that the holes play an important role in the PEC oxidation process.

The properties of oxide films grown by oxidizing GaN semiconductor directly using PEC oxidation method is reported by Lee and Chen, et al. (Lee et al., 2005). Figure 39 shows the Ga2$P_{3/2}$ core-level and O1s core-level x-ray photoelectron spectroscopy (XPS) spectra of oxide films grown using PEC oxidation method. It can be seen that the XPS peaks related to Ga-O, O-Ga, and O-P bonds are observed at 1118.9eV, 530.8eV and 532.7eV. We can deduce that the GaN is oxidized successfully and the existence of O-P bonds is caused by H_3PO_4 chemical solution utilized in PEC experiment. The energy-dispersive spectrometer (EDS) measurement is used to analyze the composition of the Ga oxide films and is hereby shown in Fig. 40. The composition of oxide layer is determined to be Ga:O:P=36.8%:62.4%:0.8%. Therefore, the composition of Ga oxide film is determined to be Ga_2O_3. The XRD measurement is utilized to verify the crystalline phases of oxide films annealed at different temperature. The XRD patterns of as grown oxide film and annealed oxide film at various temperatures are shown in Fig. 41. It can be found that the α-Ga_2O_3 (104) corresponded to 33.78% appears after the annealing treatment. According to the results observed in this work, the PEC oxidation method could grow Ga_2O_3 by oxidizing GaN semiconductor directly.

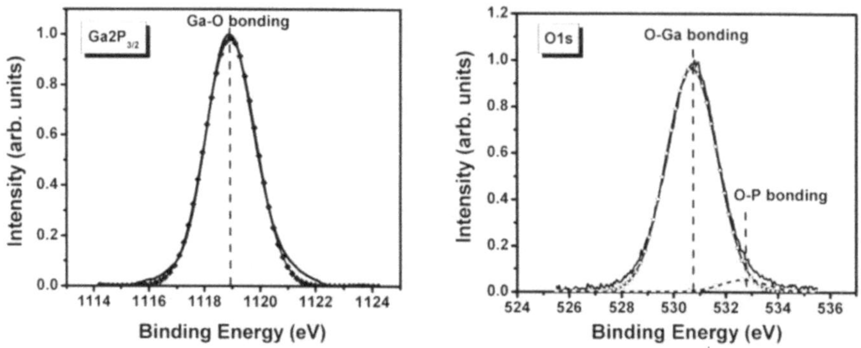

Fig. 39. (a) $Ga2P_{3/2}$ core-level and (b) O1s core-level XPS spectra for grown Ga oxide Films.

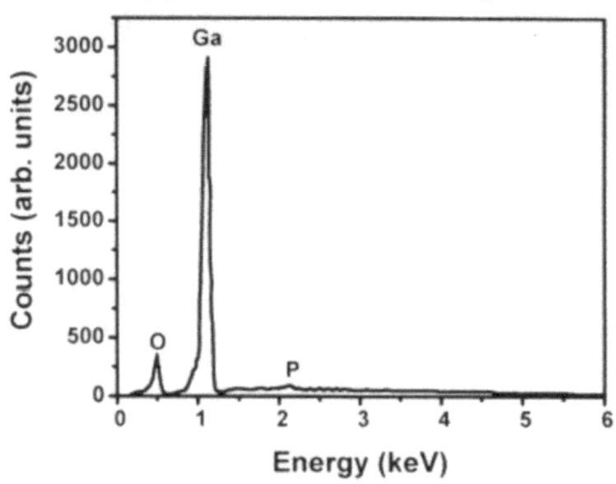

Fig. 40. The EDS spectra of grown Ga oxide films.

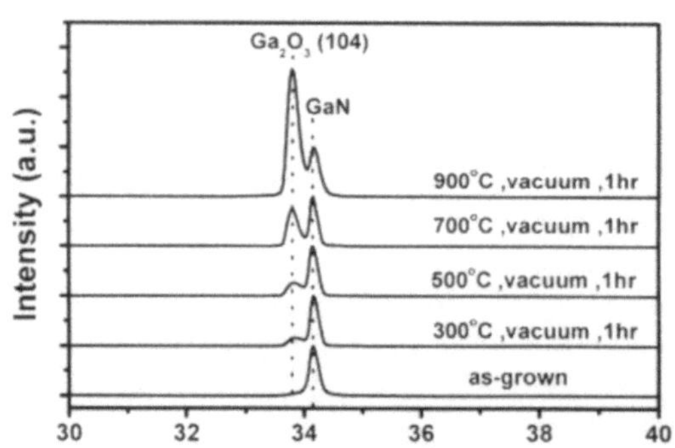

Fig. 41. The Diffraction patterns of as-grown Ga_2O_3 and annealed Ga_2O_3 films measured by x-ray diffraction.

Figure 42 shows the cross-sectional TEM micrograph of the oxidized samples. It is evident that the hexagonal pinholes originated from the defects of the n-type GaN. Using EDS measurement, similar Ga_2O_3 composition was found for the oxidation layers in the flat surface and pit region of the GaN. It is obvious that not only the flat surface of the n-type GaN was oxidized, but that the defect region was also oxidized into Ga oxide.

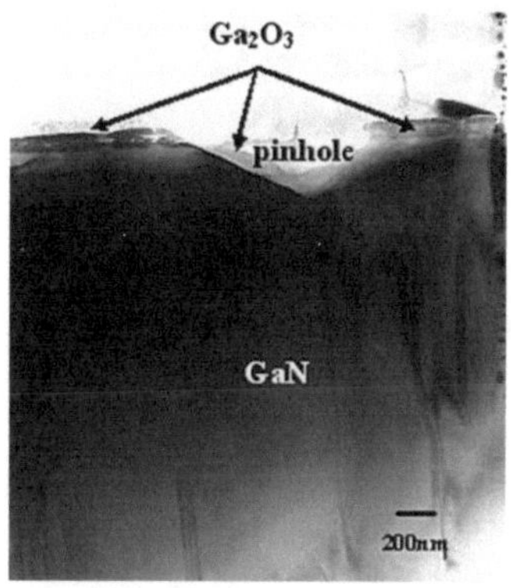

Fig. 42. Cross-sectional TEM micrograph of oxided n-type GaN.

For PEC oxidation of n-AlGaN semiconductors (Huang and Lee, 2007), a He-Cd laser with a power density of 10.0mW/cm^2 and a wavelength of 325nm were used as the light source. An H_3PO_4 chemical solution with a pH value of 3.5 was used as the electrolytic solution. The work function W_E of the H_3PO_4 solution can be calculated as (Finklea, 1988):

$$W_E(\text{eV})=4.25+0.059\times\text{pH value}=4.457\text{eV} \quad (5)$$

The work function W_S of the n-type AlGaN is expressed as:

$$W_S(\text{eV})=3.985+(E_C-E_F) \quad (6)$$

where 3.985eV is the electron affinity of the n-type AlGaN, and $E_C-E_F=0.021$eV is the energy difference between the conduction band E_C and Fermi level E_F of the n-type AlGaN with an electron concentration of 1.2×10^{18}cm^{-3}. By illuminating a He-Cd laser, electron-hole pairs were generated on the surface of the n-type AlGaN layer. The energy band structure for the electrolytic solution and the n-type AlGaN is shown in Fig. 43. An induced built-in electric field resulted in the surface of the n-type AlGaN layer. The generated electrons and holes were transported to the n-type AlGaN layer and the interface between the H_3PO_4 electrolytic solution and the n-type AlGaN by the induced built-in electric field, respectively. Because the generated holes were accumulated at the interface, the surface of the n-type AlGaN layer was gradually oxidized via the following reaction

$$2AlGaN + 12h^+ + 6H_2O \rightleftharpoons Al_2O_3 + Ga_2O_3 + N_2 + 12H^+ \quad (7)$$

where h^+ is holes. Not only was the oxidized film formed, but it was etched by the H_3PO_4 electrolytic solution. The PEC method is a diffuse procedure similar to thermal oxidation of Si in O_2 ambient. Even oxide films were grown; electrolytic solution continuously diffused through the grown oxide film to the interface and gradually formed oxide films. Figure 44 shows the XRD pattern of the grown oxidized film without thermal annealing. The peak of ε-Al_2O_3 (32.63°) and Ga_2O_3 (43.08° and 47.61°) can be found, except for the peaks of Al_2O_3, GaN, and AlGaN which originated from the sapphire substrate, undoped GaN nucleation layer and buffer layer, and AlGaN layer, respectively. Because a 200nm thick Ti metal was deposited on a part of the surface of the sample to assist the oxidation process, a peak of Ti (36.37°) was also found in Fig. 44. Because the oxidized film without thermal annealing can be easily dissolved in developer, acid solution, and alkaloid solution, it is difficult to use in the fabrication process of related devices. When the oxidized film was annealed in O_2 ambient at 700°C for 2h, the thickness was reduced from 400 to about 280nm. To clearly illustrate the oxidized film with the annealing process, the difference in the resulting XRD pattern between the original sample and the annealed oxidized sample is shown in Fig. 45. It can be seen that Ga_2O_3 was transferred to β-Ga_2O_3 (57.56° and 59.23°). Furthermore, the ε-Al_2O_3 crystal phase was transferred to α-Al_2O_3 (52.55°). Both β-Ga_2O_3 and α-Al_2O_3 crystalline phases show a better stability and ability of antietching in developer, acid solution, and alkaloid solution. This phenomenon indicates that the oxidized AlGaN films have a better quality after the annealing treatment. Figure 46 shows the XPS spectra and associated curve-fitting spectra for the core level O1s of the oxidized AlGaN without and with annealing treatment, respectively. As shown in Fig. 46, the binding energies of 530.8, 531.4, and 532.7eV are related with O-Ga, O-Al, and O-P bonds, respectively (Tourtin et al.,). It can be found that the main composition of the oxidized AlGaN film consists of Ga_2O_3 and Al_2O_3. However, O-P existed in the oxidized AlGaN film. The content of P originated from the H_3PO_4 chemical solution. Because the composition related with O-P binding structure was not found in the XRD pattern as shown in Fig. 44, it can be deduced that the content of P is very small. For the XPS spectra of core level O1s for the oxidized AlGaN film annealed in O_2 ambient at 700°C for 2h as shown in Fig. 46, the signal intensity of O-P bonds, O-Ga bonds, and O-Al bonds is enhanced. It can be deduced that more oxygen would bind with P, Ga, and Al atoms and form more PO_x, Ga_2O_3, and Al_2O_3 bonds during the annealing process in O_2 ambient. The components of the oxidized AlGaN layer were measured using SIMS. Figure 47 shows the associated SIMS depth profiles. The source used in SIMS measurement is Cs+. The depth profiles shown in Fig. 47 indicate that the AlGaN was oxidized.

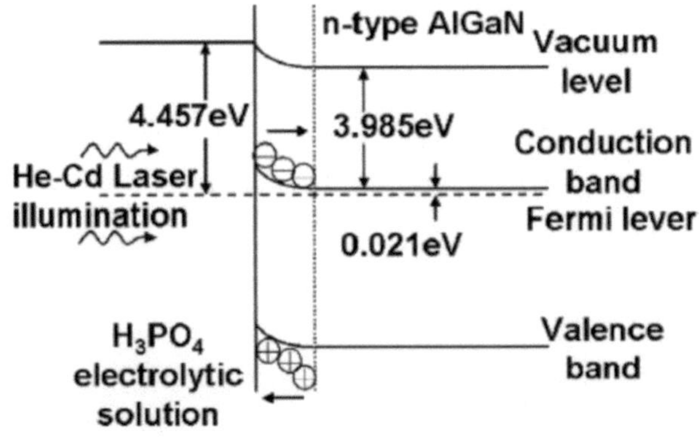

Fig. 43. Schematic energy band structure for H_3PO_4 solution/n-type AlGaN.

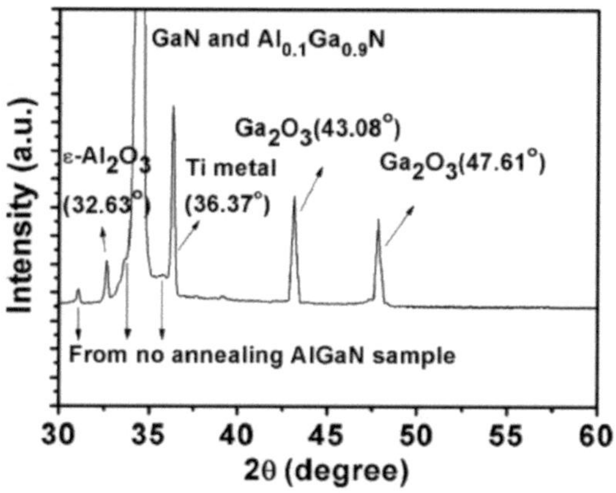

Fig. 44. XRD pattern of an oxidized AlGaN oxide layer without annealing treatment.

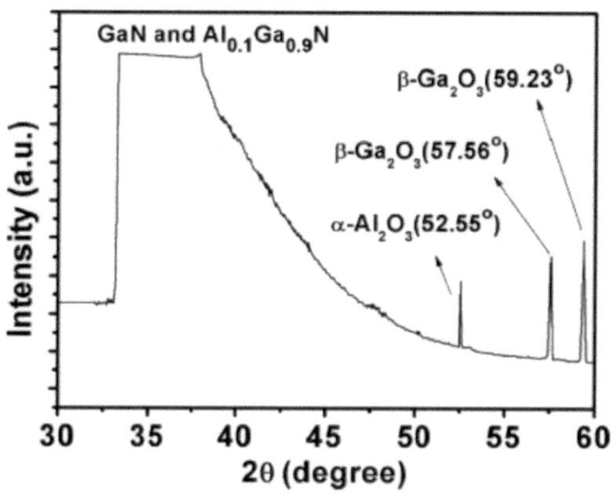

Fig. 45. The comparison of the different XRD patterns of the original sample and an oxidized AlGaN layer annealed at 700°C in O_2 ambient for 2h.

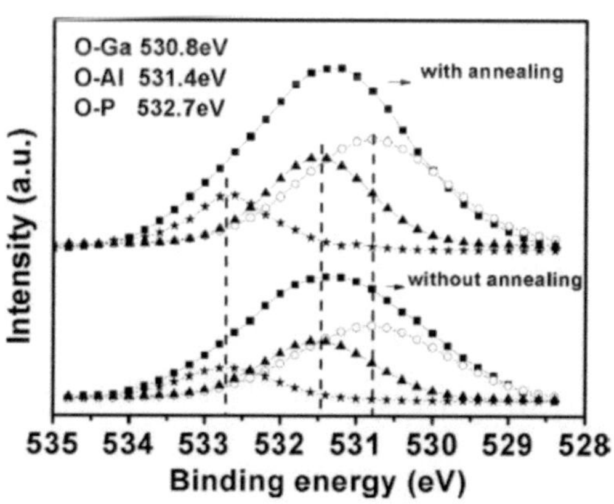

Fig. 46. The XPS spectra of oxidized AlGaN layer with and without annealing treatment.

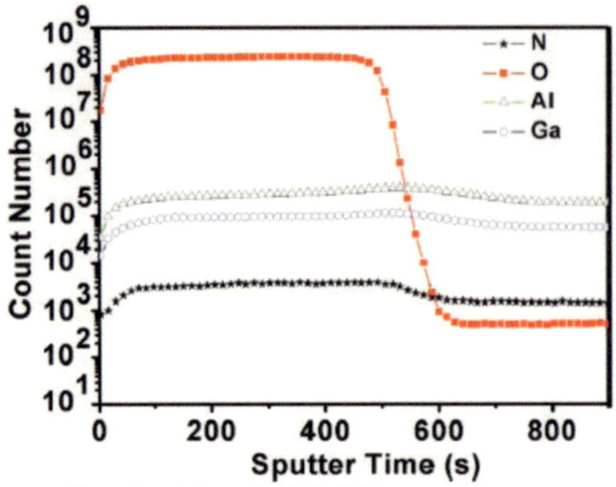

Fig. 47. The SIMS depth profile of oxidized AlGaN layer.

The electrical performance of GaN MOS diodes with as-grown Ga_2O_3 films grown by PEC oxidation method are also demonstrated by Lee and Chen, et al. (Lee et al., 2003). The schematic configuration of GaN MOS diodes is shown in Fig. 48. The inner radius, outer radius, and oxide thickness of gate insulators are 150μm, 400μm, and 100nm, respectively. Figure 49 shows the current-voltage characteristics of MOS diodes. The forward breakdown field and reverse breakdown field of GaN MOS diodes with 100-nm-thick as-grown Ga_2O_3 films are 2.8MV/cm and 5.7MV/cm, respectively. In addition, by applying the forward bias, many electrons are accumulated at the interface. When a reverse bias is applied, there are an insufficient number of holes to be supplied to accumulate at the interface. Since inducing minor carriers with only an applied bias in GaN-based semiconductors is very difficult, hence the deep-depletion phenomenon can easily be observed from the capacitance-voltage characteristics of GaN-based MOS diodes. Figure 50 shows the photoassisted C-V characteristics of the GaN MOS diodes. With forward bias applied, the total capacitances are predominantly attributed to the oxide film. The dielectric constant of 10.6 can be obtained using following equation:

$$C_{ox}=\varepsilon_{ox}\varepsilon_o A/t_{ox} \quad (8)$$

where C_{ox} is the capacitance of oxide film, ε_o is the permittivity in a vacuum, A and t_{ox} are the measured area and thickness of oxide films. The interface state density can be calculated using the following:

$$D_{it}=C_{ox}\Delta V/AqE_g \quad (9)$$

where ΔV is the shift of threshold voltage, and Eg is the energy bandgap of GaN. The interface state density estimated from the above is $2.53\times10^{11}cm^{-2}eV^{-1}$. So, the PEC oxidation method potentially could render high insulating oxide film and oxide/semiconductor interface with low interface state density.

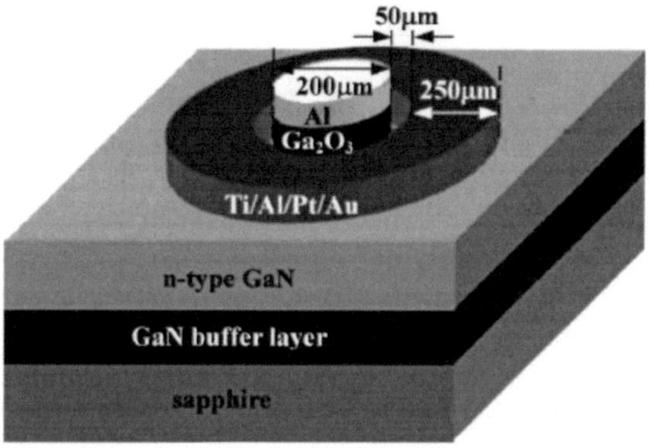

Fig. 48. The schematic configuration of GaN MOS diodes.

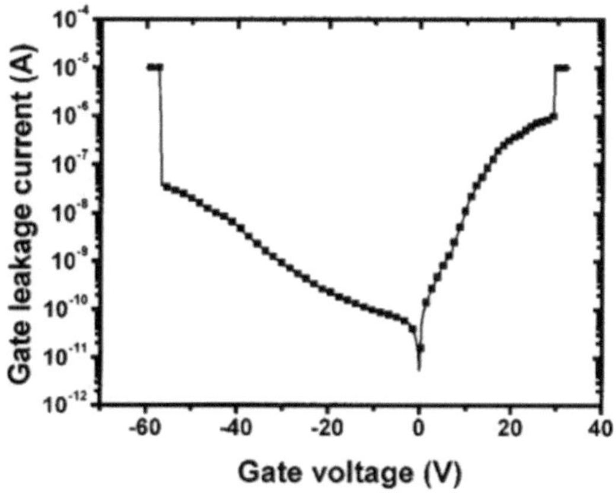

Fig. 49. The current-voltage characteristics of the MOS devices at room temperature.

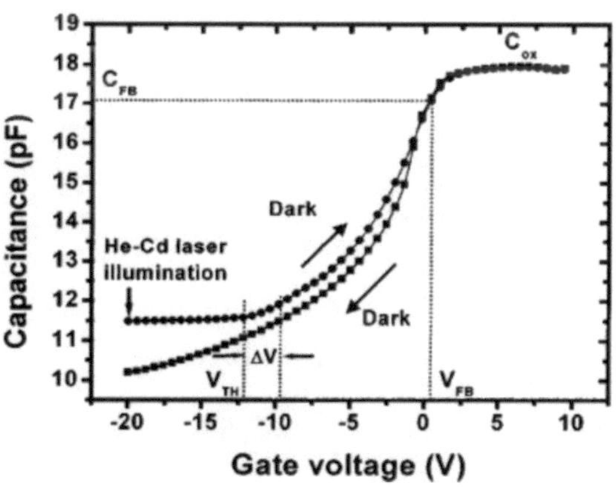

Fig. 50. The photoassisted C-V characteristics of the MOS devices.

The electrical performance of GaN-MOS diodes with SiO_2/Ga_2O_3 stack gate dielectrics is reported by Lee, et al. (Lee et al., 2003). The MOS devices with insulating layer of as-grown Ga_2O_3 (80nm) and SiO_2–as-grown Ga_2O_3 (20–80nm) are referred to as device 1 and device 2, respectively. The devices 3, 4, 5, and 6 are referred to as GaN MOS devices with oxide films of SiO_2 and Ga_2O_3 annealed at 300°C, 500°C, 700°C, and 900°C, respectively. The thickness of as-grown oxide films decreases with an increase in annealing temperature, making the oxide films gradually become denser. Figure 51 shows the cross-sectional schematic configuration of MOS diodes with stack gate oxide films. The current-voltage characteristics of MOS diodes depicted in normal scale and a magnified scale are shown in Fig. 52 and 53 respectively. It can be seen that the forward leakage current is larger than the reverse leakage current for all devices. The breakdown voltages of all MOS devices all behave similarly. When a forward bias is applied on gate, many electrons supplied by n-GaN accumulate at the interface. When a reverse bias is applied, the n-GaN cannot supply a sufficient number of holes to accumulate at the interface. The phenomenon mentioned above can be used to explain why the substantially larger forward leakage current and forward breakdown voltage are obtained compared to the reverse parameters. The interface state density of all MOS diodes can be extracted by using photoassisted capacitance-voltage measurement. The parameters of all GaN MOS diodes are shown in Table 1. The carrier transport mechanism in insulating layer is analyzed using Frenkel-Poole model (Sze, 2002). The Frenkel-Poole emission current density can be expressed as:

$$J(E,T)=BE\exp\{-q\,[\Phi_B-(qE/\pi\varepsilon_i\varepsilon_o)]/kT\} \qquad (10)$$

where B is the proportional constant, Φ_B is the barrier height, ε_i is the dielectric constant of Ga_2O_3, and k is the Boltzmann constant. To estimate the barrier height of the MOS diodes, equation (5) can be rewritten as follows:

$$\ln(J/E)=\ln B-q\,[\Phi_B-(qE/\pi\varepsilon_i\varepsilon_o)^{0.5}]/kT \qquad (11)$$

The values of Φ_B were estimated from the relationship between ln(J/E) and 1/T of the MOS devices operated at -10V and 25°C, 50°C, 75°C, and 100°C, respectively, and are shown in Table 1. It can be seen that the barrier height of oxide films enhances with an increase in annealing temperature of the as-grown Ga_2O_3 oxide layers. Those results indicate that the gate leakage current also decreases with the increase of the annealing temperature.

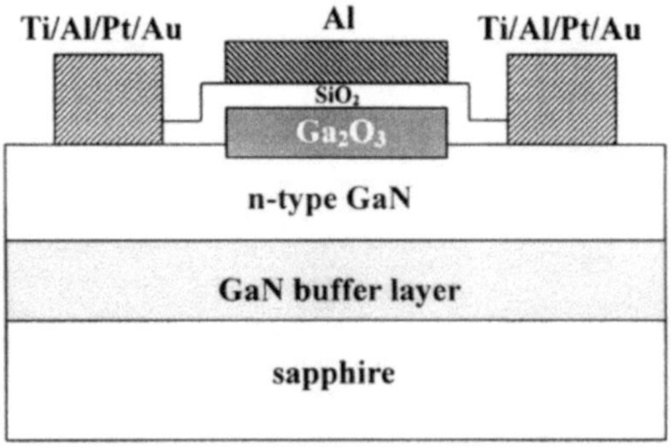

Fig. 51. The cross-sectional schematic configuration of the GaN MOS diode with stack gate insulator.

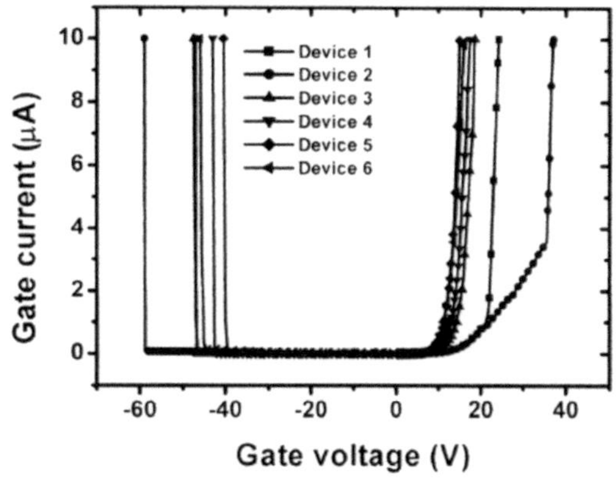

Fig. 52. The current-voltage characteristics of the GaN MOS diodes with stack gate insulators.

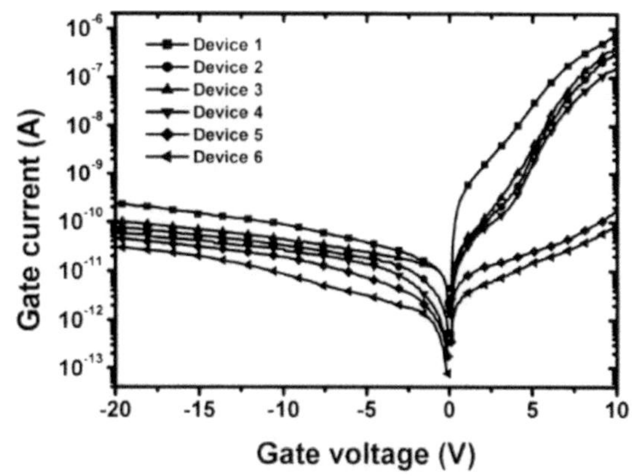

Fig. 53. The current-voltage characteristics of the GaN MOS diodes with stack gate insulators with an magnified current scale.

	Device 1	Device 2	Device 3	Device 4	Device 5	Device 6
Insulating layer thickness (nm)	Ga_2O_3 80	SiO_2/Ga_2O_3 20/80	SiO_2/Ga_2O_3 20/42	SiO_2/Ga_2O_3 20/32.6	SiO_2/Ga_2O_3 20/25.5	SiO_2/Ga_2C_3 20/20.1
Forward breakdown voltage (V)	22.0	35.6	16.2	14.7	12.0	11.7
Forward breakdown field (MV/cm)	2.75	3.56	2.61	2.79	2.64	2.92
Reverse breakdown voltage (V)	47.0	59.0	47.5	43.0	40.5	46.0
Reverse breakdown field (MV/cm)	5.88	5.90	7.66	8.17	8.90	11.50
Barrier height ($q\phi_B$, eV)	0.60	0.63	0.86	0.98	1.09	1.21
Interface state density ($10^{11}cm^{-2}eV^{-1}$)	2.53	2.50	2.42	2.34	2.21	2.00

Table 1. The parameters of the GaN MOS diodes with stack gate insulators.

The schematic configuration of $Al_{0.15}Ga_{0.85}N$ MOS diode with 45-nm-thick annealed oxide film is shown in Fig. 54. The oxide film is annealed at 700°C in O_2 ambient for 2h. The inner radius and outer radius of the structure are 150μm, 400μm, respectively. Fig. 55 shows the current-voltage characteristics of AlGaN MOS diode. The gate leakage current is only 45nA and 69pA at 5V and -15V, respectively. The forward breakdown voltage and reverse breakdown voltage are 2.2MV/cm and 6.6MV/cm, respectively. Those results suggest the excellent insulation of the oxidized AlGaN films is indeed realized after annealing treatment. Furthermore, it can be seen that the forward leakage current is larger than the reverse leakage current. When applying forward bias to the gate electrode, many electrons would accumulate at the oxide/semiconductor interface. On the other hand, when the reverse bias is applied, the AlGaN semiconductor could not supply an enough number of holes to accumulate at the interface, since inducing minor carriers by the applied bias alone in AlGaN semiconductors is very difficult. The interface state density of annealed oxidized AlGaN/AlGaN interface estimated using the photoassisted capacitance-voltage measurement and the C-V characteristics of AlGaN MOS diodes are shown in Fig. 56. The average interface state density is $5.1\times10^{11}cm^{-2}eV^{-1}$.

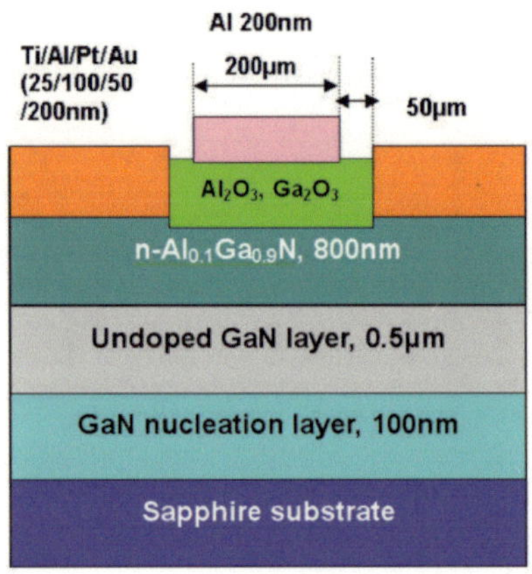

Fig. 54. The schematic configuration of $Al_{0.15}Ga_{0.85}N$ MOS diodes with 45-nm-thick annealed oxide films grown using PEC oxidation method.

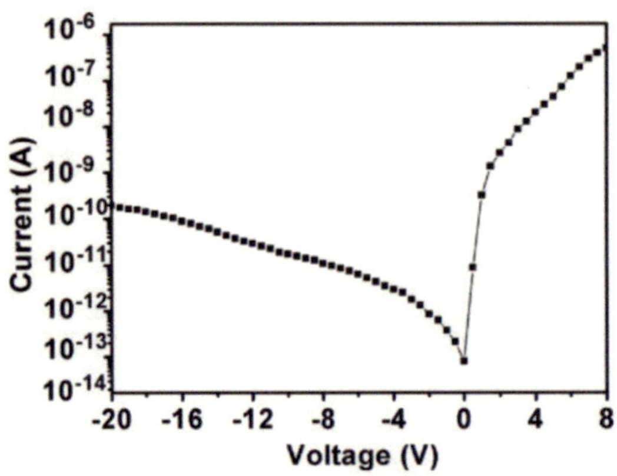

Fig. 55. The current-voltage of $Al_{0.15}Ga_{0.85}N$ MOS diodes.

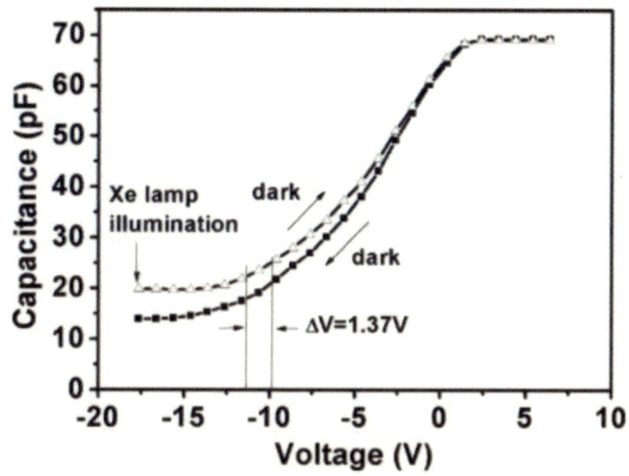

Fig. 56. The photoassisted capacitance-voltage characteristics of $Al_{0.15}Ga_{0.85}N$ MOS diodes.

The performance of AlGaN/GaN MOS-HEMTs with gate oxide films grown using PEC oxidation method are demonstrated by Huang and Yeh, et al. (Huang et al., 2008) for the first time. The schematic configuration of the MOS-HEMTs is shown in Fig. 57. The HEMT structure is consisted of a 100nm $Al_{0.15}Ga_{0.85}N$ layer, a 0.3μm undoped GaN channel, a 1.5μm insulating carbon-doped GaN, and a 20nm AlN nucleation layer, are all grown on a sapphire substrate using molecular-beam epitaxy system (MBE). This HEMT sample shows a sheet resistance of 726Ω/sq., a sheet electron density of $6.93\times10^{12}cm^{-2}$, and Hall mobility of $1240cm^2/Vs$. The gate width, gate length, and the thickness of annealed insulators are 300μm, 3μm, and 45nm, respectively. Fig. 58 shows the output characteristics of AlGaN/GaN MOS-HEMTs. The I_{DSS} at V_{GS}=0V and the threshold voltage are 200mA/mm and -5V, respectively. The transfer characteristics of MOS-HEMTs operated at V_{DS}=10V are shown in Fig. 59. The peak g_m value of 50mS/mm is obtained at V_{GS}=-2.09V. The gate leakage currents at forward gate bias of V_{GS}=10V and reverse gate bias of V_{GS}=-10V are 50pA and 2pA, respectively. It clearly indicates that the forward gate leakage current is larger than the reverse gate leakage current.

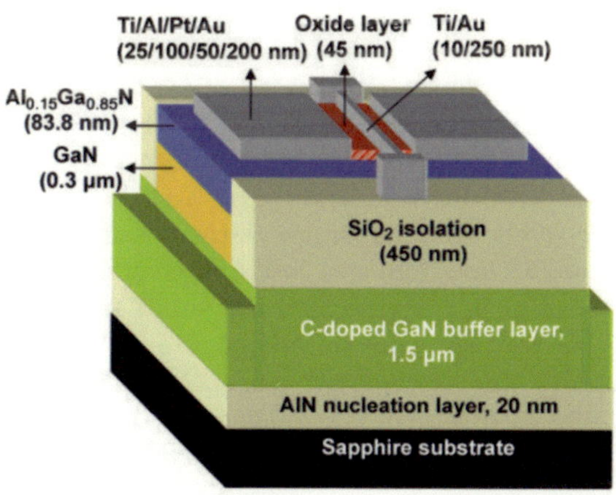

Fig. 57. The cross sectional schematic configuration of $Al_{0.15}Ga_{0.85}N$/GaN MOS-HEMTs.

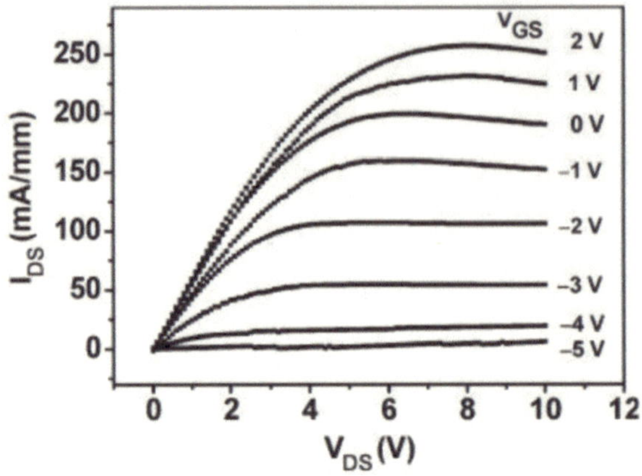

Fig. 58. The output characteristics of $Al_{0.15}Ga_{0.85}N$/GaN MOS-HEMTs.

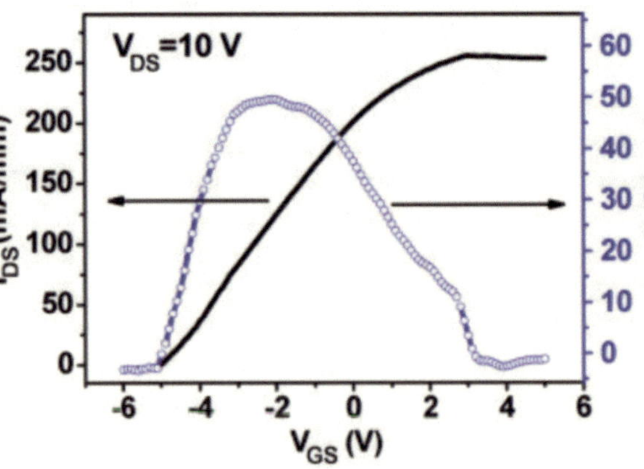

Fig. 59. The transfer characteristics of $Al_{0.15}Ga_{0.85}N$/GaN MOS-HEMTs operated at V_{DS}=10V.

To realize the high-frequency performance of AlGaN/GaN MOS-HEMTs with gate oxide films grown using PEC oxidation method, the transistors with 1-μm-long and 50-μm-wide two-finger gate are also reported (Huang et al., 2008). The schematic configuration of fabricated MOS-HEMTs is shown in Fig. 60. The oxide thickness is 40nm. Fig. 61 shows the dc electrical performances of MOS-EHMTs. The I_{DSS} at V_{GS}=0V and threshold voltage are 580mA/mm and -9V, respectively. The maximum extrinsic transconductance ($g_{m(max)}$) and gate voltage swing (GVS) are 76.72mS/mm and 2.6V, respectively. The gate leakage current is only 102nA and 960nA when V_{GS} are respectively set at -60V and 20V. The forward breakdown voltage and reverse breakdown voltage are 25V and larger than -100V, respectively. Figure 62 shows the short-circuit current gain ($|h_{21}|$) and the maximum available power gain (G_{max}) as a function of frequency derived from S-parameters measured at V_{DS}=10V. The f_T and f_{max} value of AlGaN/GaN MOS-HEMTs are 5.6GHz and 10.6GHz, respectively.

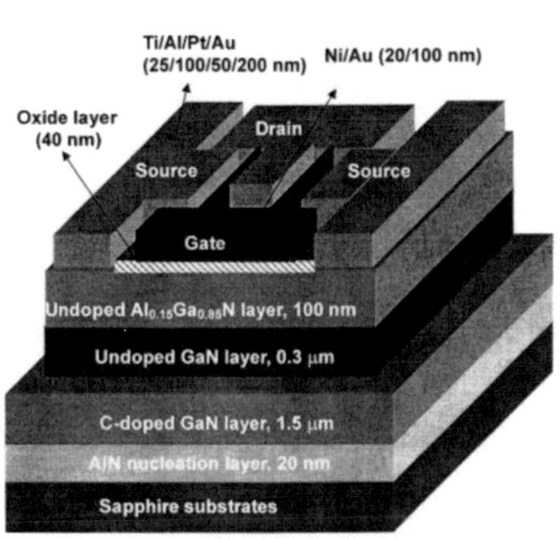

Fig. 60. The cross sectional schematic configuration of $Al_{0.15}Ga_{0.85}N/GaN$ MOS-HEMTs.

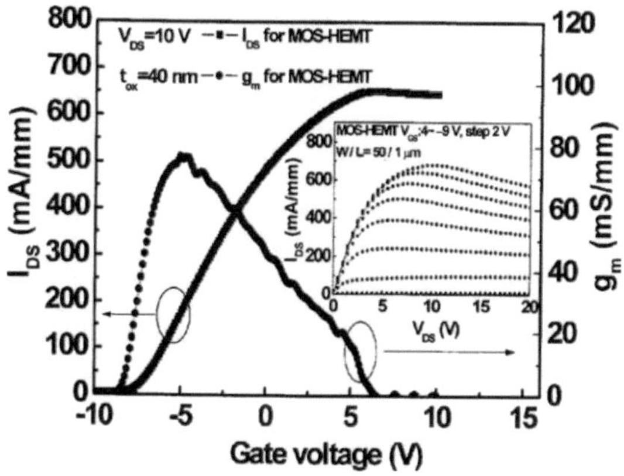

Fig. 61. The dc performances of $Al_{0.15}Ga_{0.85}N/GaN$ MOS-HEMTs.

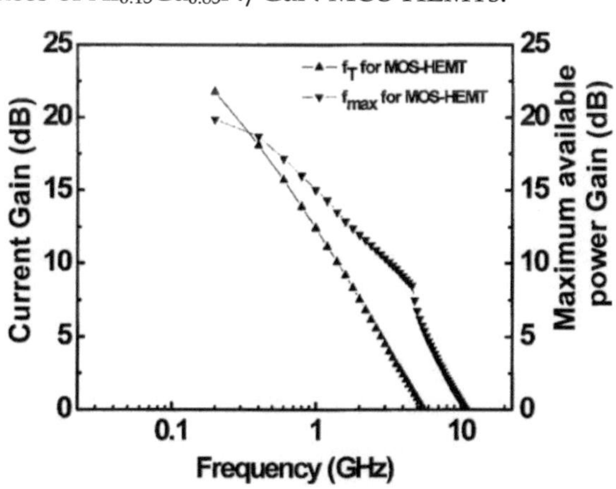

Fig. 62. The short-circuit current gain (h_{21}) and maximum available power gain (G_{max})of $Al_{0.15}Ga_{0.85}N/GaN$ MOS-HEMTs.

When the transistors are applied as oscillators or mixers in communications and electronic systems, the presence of flicker noise or 1/f noise limits the phase noise characteristics, thereby causing the performance degradation of the associated systems. In general, the flicker noises are caused by contact resistance, gate leakage current, and bulk resistance. The contact resistance is only about 18Ω for AlGaN/GaN MOS-HEMTs. The gate leakage current is five orders of magnitude smaller than the drain-source current of the MOS-HEMTs operated at V_{DS}=10V. Therefore, it can be deduced that the bulk noise is the dominant flicker noise source. Figure 63 shows the normalized noise power spectra of the MOS-HEMTs measured in saturation region (V_{DS}=10V) in the frequency range from 4Hz to 10kHz. It can be seen that the flicker noises are proportional to 1/f fitting line.

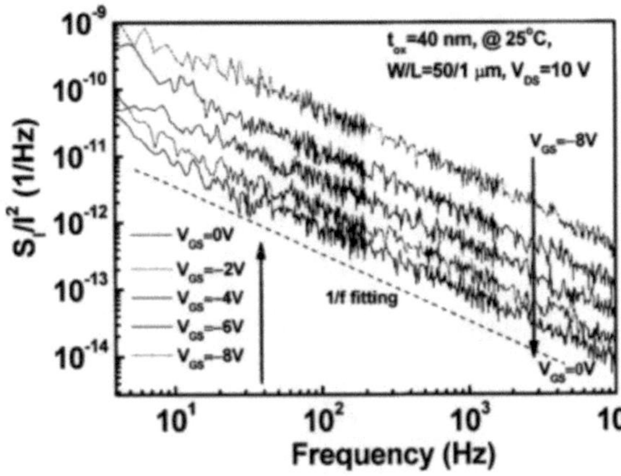

Fig. 63. The normalized noise power spectra of $Al_{0.15}Ga_{0.85}N$/GaN MOS-HEMTs measured at room temperature and V_{DS}=10V.

The mobility fluctuation model (Hooge et al., 1981; Hooge, 1994; Vandamme et al., 1994) is used to analyze the flicker noise performance in this work. The Hooge' s coefficient (α_{ch}) can be calculated from the following equation:

$$\alpha_{ch}=S_I(f) \times f \times N/I^2=[S_I(f) \times f \times (L_g^2/q\times\mu\times R_{ch})]/I^2 \quad (12)$$

where $S_I(f)/I^2$ is the normalized noise power density, f is the frequency, N is the total number of carriers, L_g is the gate length, μ is the carrier mobility, q is the elementary charge, and R_{ch} is the channel resistance. According to the noise spectrum shown in Fig. 63, the 1.25×10^{-3} value of MOS-HEMTs biased at V_{GS}=0V and V_{DS}=10V is obtained at 100Hz and it is comparable to other reported values (Chiou et al., 2006). In Fig. 63, it can also be found that the normalized noise power density increases with the decrease of gate bias. The total low frequency noise (S_{Rt}) between source and drain regions can be expressed as follows:

$$S_{Rt} = S_{Rch} + S_{Rs} \quad (13)$$

where S_{Rch} and S_{Rs} are the noises power densities originated from channel resistance and series resistance in un-gated region, respectively. The total resistance (R_t) of the transistors can be expressed as (Peransin et al., 1990):

$$R_t = R_s + R_{ch} = R_s + l_{gate}\,|V_{off}|\,(Wq\mu n_{ch}V_G) \quad (14)$$

where R_s is the series resistance of un-gated regions, Rch is the resistance of channel, l_{gate} is the gate length (1μm), V_{off} is the pinch-off voltage (-9V), W is the width of the channel (50μm), q is the elementary charge, μ is the mobility of 2DEG (1240cm^2/Vs), n_{ch} is the concentration of 2DEG at $V_G=|V_{off}|$, and $V_G=V_{GS}-V_{off}$ is defined as the effective gate bias. The channel resistance is larger than the series resistance when gate bias is negative (V_{GS}=0V). Therefore, the total flicker noise is dominated by the channel resistance. The normalized noise power density can be expressed as:

$$S_I(f)/I^2=S_{Rt}/R_t^2 = (S_{Rch}+S_{Rs})/(R_{ch}+R_s)^2 \approx S_{Rch}/R_{ch}^2 \quad (15)$$

$$S_I(f)/I^2=\alpha/fN \propto V_G^{-1} \quad (16)$$

Using the equations mentioned above, the reason for observing an increase in the normalized noise power density in response to the decrease of gate bias could then be clearly explained.

The direct-current (dc), radio-frequency (rf) and low frequency noise performances of AlGaN/GaN MOS-HEMTs with gate insulators grown using the PEC oxidation method have been reported in recent years. An oxide/semiconductor interface with low interface state density and good insulation of oxide films can be obtained by oxidizing GaN and AlGaN using the PEC oxidation method directly. Therefore, the PEC oxidation method is a promising technique for fabricating high caliber GaN-based MOS devices and GaN-based integrated circuits in the future.

5. Conclusions

The GaN-based semiconductors are potentially suitable materials for optoelectronic and electronic applications. Among various GaN-based electronic devices investigated, the MOS devices show potentially higher operating frequency, large output power gain, and better thermal stability, in comparison with the well developed Si-based MOS devices. Various deposition methods, which have been used to deposit oxides or insulators on GaN-based semiconductors to fabricate MOS devices, are summarized. The resultant devices demonstrate promising performances. In particular, the newly developed photoelectrochemical (PEC) oxidation method has aroused increasing interests due to its intrinsic advantage in producing high quality oxide/semiconductor interface over the other conventional deposition methods. It is expected that with the progress in dielectric deposition techniques, in combination with improvements in GaN-based materials and device configurations, the GaN-based MOS devices will certainly play an important role in high-frequency and high-power applications such as GaN-based integrated circuits (ICs) and opto-electronic integrated circuits (OEICs) in the future.

The processing techniques for Si-based materials and devices are well developed but unfortunately they are not suitable for fabricating LEDs and LDs. On the other hand, the epitaxial techniques for GaN-based semiconductors are still under development and certainly there are still ample rooms for improvements. Someday, the obstacles presently encountered would be overcome and the electrical properties of GaN-based semiconductors are expected to improve tremendously. There is a great expectation in the future to efficiently combine electronic devices with the optoelectronic devices in order to produce the so-called the optoelectronic integrated circuits (OEICs). Therefore, the most important challenges facing the researchers nowadays are to come up with innovative designs of high performance GaN-based electronic devices before these devices could be anticipated to combine with other GaN-based photonic components to eventually produce the ultimate optoelectronic integrated circuits (OEICs) of high caliber in the future.

6. References

Ahmad, I.; Kasisomayajula, V.; Song, D. Y.; Tian, L.; Berg, J. M. & Holtz, M. (2006). Self-heating in a GaN based heterostructure field effect transistor: ultraviolet and visible raman measurements and simulations. *J. Appl. Phys.*, Vol. 100, No. 11, (Dec. 2006) (113718-1-113718-7), 0021-8979

Ambacher, O.; Smart, J.; Shealy, J. R.; Weimann, N. G.; Chu, K.; Murphy, M.; Schaff, W. J.; Eastman, L. F.; Dimitrov, R.; Wittmer, L.; Stutzmann, M.; Rieger, W. & Hilsenbeck, J. (1999). Two-dimensional electron gases induced by spontaneous and piezoelectric polarization charges in N- and Ga-face AlGaN/GaN heterostructures. *J. Appl. Phys.*, Vol. 85, No. 6, (Mar. 1999) (3222-3233), 0021-8979

Ambacher, O.; Foutz, B.; Smart, J.; Shealy, J. R.; Weimann, N. G.; Chu, K.; Murphy, M.; Sierakowski, A. J.; Schaff, W. J.; Eastman, L. F.; Dimitrov, R.; Mitchell, A. & Stutzmann, M. (2000). Two dimensional electron gases induced by spontaneous and piezoelectric polarization in undoped and doped AlGaN/GaN Heterostructures. *J. Appl. Phys.*, Vol. 87, No. 1, (Jan. 2000) (334-344), 0021-8979

Arulkumaran, S.; Egawa, T.; Ishikawa, H.; Jimbo, T. & Umeno, M. (1998). Investigations of SiO_2/n-GaN and Si_3N_4/n-GaN insulator-semiconductor interfaces with low interface state density. *Appl. Phys. Lett.*, Vol. 73, No. 6, (Aug. 1998) (809-811), 0003-6951

Arulkumaran, S.; Hibino, T.; Egawa, T. & Ishikawa, H. (2004). Current collapse-free i-GaN/AlGaN/GaN high-electron-mobility transistors with and without surface passivation. *Appl. Phys. Lett.*, Vol. 85, No. 23, (Dec. 2004) (5745-5747), 0003-6951

Arulkumaran, S.; Egawa, T. & Ishikawa, H. (2005). Studies of electron beam evaporated SiO_2/AlGaN/GaN metal-oxide-semiconductor high-electron -mobility transistors. *Jpn.J. Appl. Phys.*, Vol. 44, No. 25, (Jun. 2005) (L812-L815), 0021-4922

Balandin, A.; Morozov, S. V.; Cai, S.; Li, R.; Wang, K. L.; Wijeratne, G. & Abernathan, C. R. (1999). Low flicker-noise GaN/AlGaN heterostructure field-effect transistors for microwave communications. *IEEE Trans. Microw. Theory Tech.*, Vol. 47, No. 8, (Aug. 1999) (1413-1417), 0018-9480

Balandin, A.; Morozov, S.; Wijeratne, G.; Cai, S. J.; Li, R.; Li, J.; Wang, K. L.; Viswanathan, C. R. & Dubrovskii, Y. (1999). Effect of channel doping on the low-frequency noise in GaN/AlGaN heterostructure field-effect transistors. *Appl. Phys. Lett.*, Vol. 75, No. 14, (Oct. 1999) (2064-2066), 0003-6951

Balandin, A.; Wang, K. L.; Cai, S.; Li, R.; Viswanathan, C. R.; Wang, E. N. & Wojtowicz, M. (2000). Investigation of flicker noise and deep-levels in GaN/AlGaN transistors. *J. Electron. Mater.*, Vol. 29, No. 3, (Mar. 2000) (297-301), 0361-5235

Balchev, I.; Minkovski, N.; Marinova, T.; Shipochka, M. & Sabotinov, N. (2006). Composition and structure characterization of aluminum after laser ablation. *Mater. Sci. Eng., B*, Vol. 135, No. 2, (Nov. 2006) (108-112), 0921-5107

Bas, C. & Lucovsky, G. (2004). Reductions in Interface defects, D_{it}, by post oxidation plasma-assisted nitridation of GaN-SiO_2 Interfaces in MOS devices. *Surf. Sci.*, Vol. 566-568, No. 5, (Jun. 2004) (356-360), 0039-6028

Bhapkar, U. V. & Shur, M. S. (1997). Monte carlo calculation of velocity-field characteristics of wurtzite GaN. *J. Appl. Phys.*, Vol. 82, No. 4, (Aug. 1997) (1649-1655), 0021-8979

Binari, S. C.; Klein, P. B. & Kazior, T. E. (2002). Trapping effects in GaN and SiC microwave FETs. *Pro. of IEEE,* Vol. 90, No. 6 (June 2002) (1048-4058), 0018-9219

Braga, N.; Mickevicius, R.; Gaska, R.; Shur, M. S.; Khan, M. A. & Simin, G. (2004) Simulation of gate lag and current collapse in gallium nitride field-effect transistors. *Appl. Phys. Lett.*, Vol. 85, No. 20, (Nov. 2004) (4780-4782), 0003-6951

Braun, H.; Solowan, H. M.; Scholz, D.; Meyer, T.; Schwarz, U. T.; Brüninghoff, S.; Lell, A. & Strauß, U. (2008). Lateral and longitudinal mode pattern of broad ridge 405 nm (Al, In)GaN laser diodes. *J. Appl. Phys.*, Vol. 103, No. 7, (Apr. 2008) (073102-1-073102-3), 0021-8979

Casey, J., H. C.; Fountain, G. G.; Alley, R. G.; Keller, B. P. & DenBaars, S. P. (1996). Low interface trap density for remote plasma deposited SiO_2 on n-type GaN. *Appl. Phys. Lett.*, Vol. 68, No. 13, (Mar. 1996) (1850-1852), 0003-6951

Chang, C. S.; Chang, S. J.; Su, Y. K.; Lee, C. T.; Lin, Y. C.; Lai, W. C.; Shei, S. C.; Ke, J. C. & Lo, H. M. (2004). Nitride-based LEDs with textured side walls, *IEEE Photon. Technol. Lett.*, Vol. *16*, No. 3, (Mar. 2004) (750-752), 1041-1135

Chang, S. J.; Su, Y. K.; Chiou, Y. Z.; Chiou, J. R.; Huang, B. R.; Chang, C. S. & Chen, J. F. (2003). Deposition of SiO_2 layers on GaN by photochemical vapor deposition. *J. Electrochem. Soc.*, Vol. 150, No. 2, (Jan. 2003) (C77-C80), 0013-4651

Chang, S. J.; Chang, C. S.; Su, Y. K.; Lee, C. T.; Chen, W. S.; Shen, C. F.; Hsu, Y. P.; Shei, S. C. & Lo, H. M. (2006).Nitride-based flip-chip ITO LEDs. *IEEE Trans. Electron Deices,* Vol. 28, No. 2, (May 2006) (273-277), 1521-3323

Chang, Y. H.; Chiu, H. C.; Chang, W. H.; Kwo, J.; Tsai, C. C.; Hong, J. M. & Hong, M. (2009). GaN metal-oxide-semiconductor diodes with molecular beam epitaxy-Al_2O_3 as a template followed by atomic layer deposition growth. *J. Cryst. Growth,* Vol. 311, No. 7, (Mar. 2009) (2084-2086), 0022-0248

Chen, P.; Wang, W.; Chua, S. J. & Zheng, Y. D. (2001). High-Frequency Capacitance-Voltage Measurement of plasma-enhanced chemical-vapor-deposition-grown SiO_2/n-GaN metal-insulator-semiconductor sturctures. *Appl. Phys. Lett.*, Vol. 79, No. 21, (Nov. 2001) (3530-3532), 0003-6951

Chiou, Y. Z.; Chang, C. S.; Chang, S. J.; Su, Y. K.; Chiou, J. R.; Huang, B. R. & Chen, J. F. (2003). Deposition of SiO_2 Layers on 4H-SiC by photochemical vapor deposition. *J. Vac. Sci. Tcehnol. B,* Vol. 21, No. 1, (Jan. 2003) (329-331), 1071-1023

Chiou, Y. Z.; Chang, S. J.; Su, Y. K.; Wang, C. K.; Lin, T. K. & Huang, B. R. (2003). Photo-CVD SiO_2 Layers on AlGaN and AlGaN-GaN MOSHFET. *IEEE Trans. Electron Devices,* Vol. 50, No. 8, (Aug. 2003) (1748-1752), 0018-9383

Chiou, Y. Z. (2004). Photo-CVD SiO_2 layers on AlGaN/GaN/AlGaN MOS-HEMTs. *J. Electrochem.Soc.*, Vol. 151, No. 12, (Oct. 2004) (G863-G865), 0013-4651

Chiou, Y. Z.; Su, Y. K.; Gong, J.; Chang, S. J. & Wang, C. K. (2006). Noise analysis of nitride-based metal-oxide-semiconductor heterostructure field-effect transistors with photo-chemical vapor deposition SiO_2 gate oxide in the linear and saturation regions. *Jpn. J. Appl. Phys.*, Vol. 45, No. 4B, (Apr. 2006) (3405-3409), 0021-4922

Chiu, H. C.; Yang, C. W.; Lin, Y. H.; Lin, R. M.; Chang, L. B. & Horng, K. Y. (2008). Device characteristics of AlGaN/GaN MOS-HEMTs using high-k praseodymium oxide layer. *IEEE Trans. Electron Deices,* Vol. 55, No. 11, (Nov. 2008) (3305-3309), 0018-9383

Chen, P. S.; Lee, C. S.; Yan, J. T. & Lee, C. T. (2007). Performance improvement and mechanism of chlorine-treated InGaN-GaN light-emitting diodes. *Electrochem. Solid State Lett.,*Vol. 10, No. 6, (Mar. 2007) (1099-0062), H165-H167

Chuang, R. W. ; Wu, R. X.; Lai, L. W.; & Lee, C. T. (2007). ZnO-on-GaN heterojunction light-emitting diode grown by vapor cooling condensation technique. *Appl. Phys. Lett.,* Vol. 91, No. 23, (Dec. 2007) (0003-6951), 231113

Chumbes, E. M.; Smart, J. A.; Prunty, T. & Shealy, J. R. (2001). Microwave Performance of AlGaN/GaN metal insulator semiconductor field effect transistors on sapphire substrate. *IEEE Trans. Electron Deices,* Vol. 48, No.3, (Mar. 2001) (416-419), 0018-9383

Cico, K.; Kuzmik, J.; Gregusova, D.; Stoklas, R.; Lalinsky, T.; Georgakilas, A.; Pogany, D. & Frohlich, K. (2007). Optimization and performance of Al_2O_3/GaN metal-oxide-semiconductor structures. *Microelectron. Reliab.,* Vol. 47, No. 4-5, (Apr./May 2007) (790-793), 0026-2714

Cho, H.; Lee, K. P.; Gila, B. P.; Abernathy, C. R.; Pearton, S. J. and Ren, F. (2003). Gate breakdown characteristics of MgO/GaN MOSFETs. *Solid-State Electron.,* Vol. 47, No. 9, (Sep. 2003) (1597-1600), 0038-1101

Egawa, T.; Ishikawa, H.; Umeno, M. & Jimbo, T.. (2000), Recessed gate AlGaN/GaN modulation-doped field-effect transistors on sapphire, *Appl. Phys. Lett.,* Vol. 76, No. 1, (Jan. 2000) (121-123), 0003-6951

Endoh, A.; Yamashita, Y.; Ikeda, K.; Higashiwaki, M.; Hikosaka, K.; Matsui, T.; Hiyamizu, S. & Mimura, T. (2004). Non-recessed-gate enhancement-mode AlGaN/GaN high electron mobility transistors with high RF performances. *Jpn. J. Appl. Phys.,* Vol. 43, No. 4B, (Apr. 2004) (2255-2258), 0021-4922

Endoh, A.; Yamashita, Y.; Hirose, N.; Hikosaka, K.; Matsui, T.; Hiyamizu, S. & Mimura, T. (2006). High performance AlGaN/GaN metal-insulator-semiconductor high electron mobility transistors fabricated using SiN/SiO_2/SiN triple-layer insulators. *Jpn. J. Appl. Phys.,* Vol. 45, No. 4B, (Apr. 2006) (3364-3367), 0021-4922

Fan, Q.; Leach, J. H.; Wu, M.; Xiao, B.; Gu, X.; Morkoç, H. & Handel, P. H. (2004). Comparison of various gate dielectrics on the performance of AlGaN/GaN HFETs. *Proc. SPIE,* Vol. 6894, No. 2008, (Feb. 2008) (689426-1-689426-8), 0277-786X

Fareed, R. S. Q.; Hu, X.; Tarakji, A.; Deng, J.; Gaska, R. & Shur, M. (2005). High-power AlGaN/InGaN/AlGaN/GaN recessed gate heterostructure field-effect transistors, *Appl. Phys. Lett.,* Vol 86, No. 14, (Apr. 2005) (143512-1-143512-3), 0003-6951

Feng, Q.; Hao, Y. & Yue, Y. Z. (2009). The Reduction of Gate Leakage of AlGaN/GaN metal-insulator-semiconductor high electron mobility transistors by N_2 plasma pretreatment. *Semicond. Sci. Technol.,* Vol. 24, No. 2, (Jan. 2009) (025030-1-025030-6), 0268-1242

Finklea, H. O. (1988). *Semiconductor electrodes,* Elsevier Science, 0444429263, Netherlands, Amsterdam

Fujii, T.; Gao, Y.; Sharma, R.; Hu, E. L.; DenBaars, S. P. & Nakamura, S. (2004). Increase in the extraction efficiency of GaN-based light-emitting diodes via surface roughening, *Appl. Phys. Lett.,* Vol 84, No. 6, (Feb. 2004) (855-857), 0003-6951

Gaffey, B.; Guido, L. J.; Wang, X. W. & Ma, T. P. (2001). High-quality oxide/nitride/oxide gate insulator for GaN MIS structures. *IEEE Trans. Electron Deices,* Vol. 48, No. 3, (Mar. 2001) (458-464), 0018-9383

Gaska, R.; Chen, Q.; Yang, J.; Khan, M. A.; Shur, M. S.; Ping, A. & Adesida, I. (1997). AlGaN-GaN heterostructure FETs with offset gate design. *Electron. Lett.,* Vol. 33, No. 14, (Jul. 1997) (1255-1257), 0013-5194

Gaska, R.; Chen, Q.; Yang, J.; Osinsky, A.; Asif Khan, M. & Shur, M. S. (1997). High-temperature performance of AlGaN/GaN HFETs on SiC substrates. *IEEE Electron Device Lett.,* Vol. 18, No. 10, (Oct, 1997) (492-494), 0741-3106

Gaska, R.; Osinsky, A.; Yang, J. W. & Shur, M. S. (1998). Self-heating in high-power AlGaN-GaN HFETs. *IEEE Electron Device Lett.,* Vol. 19, No. 3, (Mar. 1998) (89-91), 0741-3106

Gregušová, D.; Stoklas, R.; Čičo, K.; Heidelberger, G.; Marso, M.; Novák, J. & Kordoš, P. (2007). Characterization of AlGaN/GaN MOSHFETs with Al_2O_3 as gate oxide. *Phys. Stal. Sol. (c),* Vol. 4, No. 7, (Jun. 2007) (2720-2723), 1862-6351

Gregušová, D.; Stoklas, R.; Cico, K.; Lalinsky, T. & Kordoš, P. (2007). AlGaN/GaN metal-oxide-semiconductor heterostructure field-effect transistors with 4 nm thick Al_2O_3 gate oxide. *Semicond. Sci. Technol.,* Vol. 22, No. 8, (Aug. 2007) (947-951), 0268-1242

Gusev, E. P.; Narayanan, V. & Frank, M. M. (2006). Advanced high-k dielectric stacks with poly Si and metal gates: recent progress and current challenges. *IBM J. Res. Develop.,* Vol. 50, No. 4-5, (Dev. 2006) (387-410), 0018-8646

Han, J.; Crawford, M. H.; Shul, R. J.; Figiel, J. J.; Banas, M.; Zhang, L.; Song, Y. K.; Zhou, H.; & Nurmikko, A. V. (1998). AlGaN/GaN quantum well ultraviolet light emitting diodes, *Appl. Phys. Lett.,* Vol 73, No. 12, (Sept. 1998)(1688-1690), 0003-6951

Higashiwaki, M.; Hirose, N. & Matsui, T. (2005). Cat-CVD SiN-passivated AlGaN-GaN HFETs with thin and high Al composition barrier layers. *IEEE Electron Device Lett.,* Vol. 26, No. 3, (Mar. 2005) (139-141), 0741-3106

Higashiwaki, M.; Mimura, T. and Matsui, T. (2006). AlN/GaN insulated-gate HFETs using Cat-CVD SiN. *IEEE Electron Device Lett.,* Vol. 27, No. 9, (Sep. 2006) (719-721), 0741-3106

Higashiwaki, M.; Mimura, T. & Matsui, T. (2006). 30-nm-gate AlGaN/GaN heterostructure field-effect transistors with a current-gain cutoff frequency of 181GHz. *Jpn. J. Appl. Phys.,* Vol. 45, No. 42, (Oct. 2006) (L1111-L1113), 0021-4922

Higashiwaki, M.; Chen, Z.; Chu, R.; Pei, Y.; Keller, S.; Mishra, U. K.; Hirose, N.; Matsui, T. & Mimura, T. (2009). A comparative study of effects SiN_X deposition method on AlGaN/GaN heterostructure field-effect transistors. *Appl. Phys. Lett.,* Vol. 94, No. 5, (Feb. 2009) (053513-1-053513-3), 0003-6951

Hashizume, T.; Alekseev, E.; Pavlidisb, D.; Boutros, K. S. & Redwing, J. (2000). Capacitance-voltage characterization of AlN/GaN metal-insulator-semiconductor structures grown on sapphire substrate by metalorganic chemical vapor deposition. *J. Appl. Phys.,* Vol. 88, No. 4, (Aug. 2000) (1983-1986), 0021-8979

Hippel, A. R. V. (1954). *Dielectrics and waves,* Wiley, 0890068038, New York

Huang, L. H. & Lee, C. T. (2007). Investigation and analysis of AlGaN MOS devices with an oxidation layer grown using the photoelectrochemical oxidation method. *J. Electrochem. Soc.,* Vol. 154, No. 10, (Apr. 2007) (H862-H866), 0013-4651

Huang, L. H.; Yeh, S. H.; Lee, C. T.; Tang, H.; Bardwell, J. & Webb, J. b. (2008). AlGaN/GaN metal-oxide-semiconductor high-electron mobility transistors using oxide insulator grown by photoelectrochemical oxidation method. *IEEE Electron Device Lett.*, Vol. 29, No. 4, (Apr. 2008) (284-286), 0741-3106

Huang, L. H.; Yeh, S. H. & Lee, C. T. (2008). High frequency and low frequency noise of AlGaN/GaN metal-oxide-semiconductor high-electron mobility transistors with gate insulator grown using photoelectrochemical oxidation method. *Appl. Phys. Lett.*, Vol. 93, No. 4, (Jul. 2008) (043511-1-043511-3), 0003-6951

Huang, L. H.; Kan, K. C. & Lee, C. T. (2009). Analysis of oxidized p-GaN films directly grown using bias-assisted photoelectrochemical method. *J. Electron. Mater.*, Vol. 38, No. 4, (Apr. 2009) (529-532), 0361-5235

Hong, M.; Anselm, K. A.; Kwo, J.; Ng, H. M.; Baillargeon, J. N.; Kortan, A. R.; Mannaerts, J. P.; Cho, A. Y.; Lee, C. M.; Chyi, J. I. & Lay, T. S. (2000). Properties of $Ga_2O_3(Gd_2O_3)$/GaN metal-insulator-semiconductor diodes. *J. Vac. Sci. Tcehnol. B,* Vol. 18, No. 3, (May/Jun. 2000) (1453-1456), 1071-1023

Hong, S. J. & Kim, K. (2006), Low-resistance ohmic contacts for high-power GaN field-effect transistors obtained by selective area growth using plasma-assisted molecular beam epitaxy, *Appl. Phys. Lett.*, Vol. 89, No. 4, (Jul. 2006) (042101-1-042101-3), 0003-6951

Hooge, F. N.; Kleinpenning, T. G. M. & Vandamme, L. K. J. (1981). Experimental Studies on 1/f Noise. *Rep. Prog. Phys.*, Vol. 44, No. 5, (May 1981) (479- 532), 0034-4885

Hooge, F. N. (1994). 1/f noise sources. *IEEE Trans. Electron Deices,* Vol. 41, No. 11, (Nov. 1994) (1926-1935), 0018-9383

Horio, K.; Yonemoto, K.; Takayanagi, H.; & Nakano, H. (2005). Physics-based simulation of buffer-trapping effects on slow current transients and current collapse in GaN field effect transistors. *J. Appl. Phys.*, Vol. 98, No. 12, (Dec. 2005) (124502-1-124502-7), 0021-8979

Hu, X.; Koudymov, A.; Simin, G.; Yang, J.; Khan, M. A.; Tarakji, A.; Shur, M. S. & Gaska, R. (2001). Si_3N_4/AlGaN/GaN-metal-insulator-semiconductor heterostructure feld-effect transistors. *Appl. Phys. Lett.*, Vol. 79, No. 17, (Aug. 2001) (2832-2834), 0003-6951

Inoue, T.; Ando, Y.; Miyamoto, H.; Nakayama, T.; Okamoto, Y.; Hataya, K. & Kuzuhara, M. (2005). 30-GHz-band over 5-W power performance of short-channel AlGaN-GaN heterojunction FETs, *IEEE Trans. Microw. Theory Tech.*, Vol. 53, No. 1, (Jan. 2005) (74-80), 0018-9480

Irokawa, Y.; Nakano, Y.; Ishiko, M. & Kachi, T. (2004). MgO/p-GaN enhancement mode metal-oxide semiconductor field-effect transistors. *Appl. Phys. Lett.*, Vol. 84, No. 15, (Apr. 2004) (2919-2921), 0003-6951

Islam, S. S. & Anwar, A. F. M. (2002). Temperature-dependent nonlinearities in GaN/AlGaN HEMTs, *IEEE Trans. Electron Devices,* Vol. 49, No. 5, (May. 2002) (710-717), 0018-9383

J. M. Poate, K. N. Tu, and J. W. Mayer, *Thin Films, Inter-diffusion and Reactions,* New York: Wiley, 1978.

Javorka, P.; Alam, A.; Wolter, M.; Fox, A.; Marso, M.; Heuken, M.; Lüth, H. & Kordoš, P. (2002). AlGaN/GaN HEMTs on (111) silicon substrates. *IEEE Electron Device Lett.*, Vol. 23, No. 1, (Jan. 2002) (4-6), 0741-3106

Jhin, H. K. J.; Byun, D.; Kim, D. & Adesida, I. (2008). Reactive sputtering of Al_2O_3 on AlGaN/GaN heterostructure field-effect transistor. *Thin Solid Films,* Vol. 516, No. 18, (Jul. 2008) (6483-6486), 0040-6090

Jiménez, A.; Buttari, D.; Jena, D.; Coffie, R.; Heikman, S.; Zhang, N. Q.; Shen, L.; Calleja, E.; Muñoz, E.; Speck, J. & Mishra, U. K. (2002) Effect of P-doped overlayer thickness on RF-dispersion in GaN junction FETs. *IEEE Electron Device Lett.,*Vol. 23, No. 6 (Jun. 2002) (306-308), 0741-3106

Johnson, J. W.; Luo, B.; Ren F.; Gila, B. P.; Krishnamoorthy, W.; Abernathy, C. R. S.; Pearton, J.; Chyi, J. I.; Nee, T. E.; Lee, C. M. & Chuo, C. C. (2000). Gd_2O_3/GaN metal-oxide-semiconductor field-effect transistor. *Appl. Phys. Lett.,* Vol. 77, No. 20, (Nov. 2000) (3230-3232), 0003-6951

Johnson, J. W.; Baca, A. G.; Briggs, R. D.; Shul, R. J.; Wendt, J. R.; Monier, C.; Ren, F.; Pearton, S. J.; Dabiran, A. M.; Wowchack, A. M.; Polley, C. J. & Chow, P. P. (2001). Effect of Gate length on DC performance of AlGaN/GaN HEMTs grown by MBE. *Solid-State Electron.,* Vol. 45, No. 12, (Dec. 2001) (1979-1985), 0038-1101

Kao, C. J.; Chen, M. C.; Tun, C. J.; Chi, G. C.; Sheu, J. K.; Lai, W. C.; Lee, M. L.; Ren, F. & Pearton, S. J. (2005). Comparison of low-temperature GaN, SiO_2, and SiN_x as gate insulators on AlGaN/GaN heterostructure field-effect transistors. *J. Appl. Phys.,* Vol. 98, No. 6, (Sep. 2005) (064506-1-064506-5), 0021-8979

Katz, O.; Garber, V. ; Meyler, B. ; Bahir, G. ; & Salzman, J. (2001). Gain mechanism in GaN schottky ultraviolet detectors. *Appl. Phys. Lett.,* Vol.79, No. 10, (Sep.2001) (0003-6951), 1417 - 1419

Karmalkar, S. & Mishra, U. K. (2001), Enhancement of breakdown voltage in AlGaN/GaN high electron mobility transistors using a field plate, *IEEE Trans. Electron Devices,* Vol. 48, No. 8, (Aug. 2001) (1515-1521), 0018-9383

Kawai, H.; Hara, M.; Nakamura, F.; Asatsuma, T.; Kobayashi, T. & Imanaga, S. (1998). An AlN/GaN insulated gate heterostructure field effect transistor with regrown n^+GaN source and drain contact. *J. Cryst. Growth,* Vol. 189, No. 90, (Oct. 1997) (738-741), 0022-1248

Kawai, H.; Hara, M.; Nakamura, F. & Imanaga, S. (1998). AlN/GaN insulated gate heterostructure FET with regrown n^+GaN Oomic contact. *Electron. Lett.,* Vol. 34, No. 6, (Mar. 1998) (592-593), 0013-5194

Khan, M. A.; Kuznia, J. N.; Bhattarai, A. R. & Olson, D. T. (1993). Metal semiconductor field effect transistor based on single crystal GaN. *Appl. Phys. Lett.,* Vol. 62, No. 15, (Apr. 1993) (1786-1787), 0003-6951

Khan, M. A.; Hu, X.; Sumin, G.; Lunev, A.; Yang, J.; Gaska, R. & Shur, M. S. (2000). AlGaN/GaN metal oxide semiconductor heterostructure field effect transistor. *IEEE Electron Device Lett.,* Vol. 21, No. 2, (Feb. 2000) (63-65), 0741-3106

Khan, M. A.; Hu, X.; Tarakji, A.; Simin, G.; Yang, J.; Gaska, R. & Shur, M. S. (2000). AlGaN/GaN metal-oxide-semiconductor heterostructure field-effect-transistor on SiC substrates. *Appl. Phys. Lett.,* Vol. 77, No. 9, (Aug. 2000) (1339-1341), 0003-6951

Khan, M. A.; Simin, G.; Yang, J.; Zhang, J.; Koudymov, A.; Shur, M. S.; Gaska, R.; Hu, X. & Tarakji, A. (2003). Insulating gate III-N heterostructure field-effect transistors for high-power microwave and switching applications, *IEEE Trans. Microw. Theory Tech.,* Vol. 51, No. 2, (Feb. 2003) (886-893), 0018-9840

Kim, J.; Mehandru, R.; Luo, B.; Ren, F.; Gila, B. P.; Onstine, A. H.; Abernathy, C. R.; Pearton, S. J. & Irokawa, Y. (2002). Inversion behavior in Sc_2O_3/GaN diodes. *Appl. Phys. Lett.*, Vol. 81, No. 2, (Jul. 2002) (373-375), 0003-6951

Kim, D. H.; Kumar, V.; Chen, G.; Dabiran, A. M.; Wowchak, A. M.; Osinsky, A. & Adesida, I. (2007). ALD Al_2O_3 passivated MBE-Grown AlGaN/GaN HEMTs on 6H-SiC. *Electron. Lett.*, Vol. 43, No. 2, (Jan. 2007) (129-130), 0013-5194

Kikuta, D.; Ao, J. P. & Ohno, Y. (2006). Gate leakage and electrical performance of AlGaN/GaN MIS-type HFET with evaporated silicon oxide layer. *Solid-State Electron.*, Vol. 50, No. 3, (Mar. 2006) (316-321), 0038-1101

Klein, P. B.; Klein, J. A.; Freitas, Jr., Binari, S. C. & Wickenden, A. E. (1999) Observation of deep traps responsible for current collapse in GaN metal-semiconductor field-effect transistors *Appl. Phys. Lett.*, Vol. 75, No. 25, (Dec. 1999) (4016-4018), 0003-6951

Klein, P. B.; Binari, S. C.; Ikossi, K.; Wickenden, A. E.; Koleske, D. D. & Henry, R. L. (2001). Current collapse and the role of carbon in AlGaN/GaN high electron mobility transistors grown by metalorganic vapor-phase epitaxy. *Appl. Phys. Lett.*, Vol. 79, No. 21, (Nov. 2001) (3527-3529). 0003-6951

Koley, G.; Tilak, V.; Eastman, L. F. & Spencer, M. G. (2003). Slow transients observed in AlGaN/GaN HFETs effects of SiN_x passivation and UV illumination," *IEEE Trans. Electron Devices*, Vol.50, No. 4, (Apr. 2003) (886-893), 0018-9383

Kordos, P.; Heidelberger, G.; Bernat, J.; Fox, A.; Marso, M. & Luth, H. (2005). High-Power SiO_2/AlGaN/GaN metal-oxide-semiconductor heterostructure field-effect transistors. *Appl. Phys. Lett.*, Vol. 87, No. 14, (Sep. 2005) (143501-1-143501-3), 0003-6951

Kordoša, P.; Gregušová, D.; Stoklas, R.; Čičo, K. & Novák, J. (2007). Improved transport properties of Al_2O_3/AlGaN/GaN metal-oxide-semiconductor heterostructure field-effect transistor. *Appl. Phys. Lett.*, Vol. 90, No. 12, (Mar. 2007) (123513-1-123513-3), 0003-6951

Kuksenkov, D. V.; Temkin, H.; Osinsky, A. ; Gaska, R. & Khan, M. A. (1998). Low-frequency noise and performance of GaN p-n junction photodetectors. *J.Appl.Phys.*,Vol. 83, No. 4, (Feb. 1998) (0021-8979), 2142-2146

Kuzmik, J.; Carlin, J. F.; Gonschorek, M.; Kostopoulos, A.; Konstantinidis, G.; Pozzovivo, G.; Golka, S.; Georgakilas, A.; Grandjean, N.; Strasser, G. & Pogany, D. (2007). Gate-lag and drain-lag effects in (GaN)/InAlN/GaN and InAlN/AlN/GaN HEMTs. *Phys. Stal. Sol. (a)*, Vol. 204, No. 6, (May 2007) (2019-2022), 2019-2022

Kuzmik, J.; Pozzovivo, G.; Abermann, S.; Carlin, J. F.; Gonschorek, M.; Feltin, E.; Grandjean, N.; Bertagnolli, E.; Strasser, G. & Pogany, D. (2008). Technology and performance of InAlN/AlN/GaN HEMTs with gate insulation and current collapse suppression using ZrO_2 or HfO_2. *IEEE Trans. Electron Deices*, Vol. 55, No. 3, (Mar. 2008) (937-941), 0018-9383

Laino, V.; Roemer, F.; Witzigmann, B.; Lauterbach, C.; Schwarz, U. T.; Rumbolz, C.; Schillgalies, M. O.; Furitsch, M.; Lell, A. & Härle, V. (2007). Substrate modes of (Al,In)GaN semiconductor laser diodes on SiC and GaN substrate. *IEEE J. Quantum Electron.*, Vol. 43, No. 1, (Jan. 2007) (16-24), 0018-9197

Lay, T. S.; M. Hong,; Kwo, J.; Mannaerts, J. P.; Hung, W. H. & Huang, D. J. (2001). Energy-band parameters at the GaAs- and GaN-Ga_2O_3(Gd_2O_3) interfaces. *Solid-State Electron*, Vol. 45, No. 9, (Sep. 2001) (1679-1682), 0038-1101

Lee, C. T.; Lee, H. Y. & Chen, H. W. (2003). GaN MOS devices using SiO_2-Ga_2O_3 insulator grown by photoelectrochemical oxidation method. *IEEE Electron Device Lett.*, Vol. 24, No. 2, (Feb. 2003) (54-56), 0741-3106

Lee, C. T.; Chen, H. W. & Lee, H. Y. (2003). Metal-oxide-semiconductor devices using Ga_2O_3 dielectrics on n-type GaN. *Appl. Phys. Lett.*, Vol. 82, No. 24. (Jun. 2003) (4303-4306), 0003-6951

Lee, C. T.; Chen, H. W.; Hwang, F. T. & Lee, H. Y. (2005). Investigation of Ga oxide films directly grown on n-type GaN by photoelectrochemical oxidation using He-Cd laser. *J. Electron. Mater.*, Vol. 34, No. 3, (Mar. 2005) (282-286), 0361-5235

Lee, C. T.; Yang, U. Z.; Lee, C. S. & Chen, P. S. (2006). White light emission of monolithic carbon-implanted InGaN-GaN light-emitting diodes. *IEEE Photon. Technol. Lett.*, Vol. 18, No. 15, (Dec. 2006) (2517-2519), 1041-1135

Lee, H. Y.; Pan, K. H. ; Lin, C. C.; Chang, Y. C.; Kao, F. J. & Lee,C. T. (2007). Current spreading of III-nitride light-emitting diodes using plasma treatment. *J. Vac. Sci. Tcehnol. B, vol.***25**, No.4,(Jul 2007)(1071-1023), 1280 - 1283

Liu, C.; Chor, E. F. & Tan, L. S. (2006). Investigation of HfO_2/AlGaN/GaN metal-oxide-semiconductor high electron mobility transistors. *Appl. Phys. Lett.*, Vol. 88, No. 07, (Apr. 2006) (173504-1-173504-3), 0003-6951

Liu, C.; Chor, E. F. & Tan, L. S. (2007). Enhanced device performance of AlGaN/GaN HEMTs using HfO_2 high-k dielectric for surface passivation and gate oxide. *Semicond. Sci. Technol.*, Vol. 22, No. 5, (Mar. 2007) (522-527), 0268-1242

Liu, C. H.; Lam, K. T.; Chang, S. J.; Wang, C. K. & Sun, Y. S. (2007). Flicker noise of AlGaN/GaN metal-oxide-semiconductor heterostructure field-effect transistor with a photo-CVD SiO_2 layer. *J. Electrochem. Soc.*, Vol. 154, No. 2, (Dec. 2006) (H119-H123), 0013-4651

Lin, Y. J.; Tsai, C. D.; Lyu, Y. T. & Lee, C. T. (2000). X-ray photoelectron spectroscopy study of $(NH_4)_2S_x$-treated Mg-doped GaN layers. *Appl. Phys. Lett.*, Vol. 77, No. 5, (Jul. 2000) (687-689), 0003-6951

Long, J. P. & Bermudez, V. M. (2002). Band bending and photoemission-induced surface photovoltages on clean n- and p-GaN (0001) surfaces. *Phys. Rev. B*, Vol. 66, No. 12, (Sep. 2002) (121308-1-121308 4), 0163-1829

Lu, W.; Yang, J.; Khan, M. A. & Adesida, I. (2001). AlGaN/GaN HEMTs on SiC with over 100GHz f_T and low microwave noise. *IEEE Trans. Electron Deices*, Vol. 48, No. 3, (Mar. 2001) (581-585), 0018–9383

Lu, W.; Kumar, V.; Piner, E. L. & Adesida, I. (2003). DC, RF, and microwave noise performance of AlGaN-GaN field effect transistors dependence of aluminum concentration. *IEEE Trans. Electron Devices*, Vol. 50, No. 4, (Apr. 2003) (1069-1074), 0018-9383

Ma, T. P. (1998). Making Silicon Nitride Film a Viable Gate Dielectric. *IEEE Trans. Electron Deices*, Vol. 45, No. 3, (Mar. 1998) (680-690), 0018-9383

Ma, T. P. (2003). Gate Dielectrics for Si, SiC, and GaN as Synthesized by Jet Vapor Deposition. *Microelectr. J.*, Vol. 34, No. 5-8, (May/Aug. 2003) (363-370), 0026-2692

Marso, M.; Heidelberger, G.; Indlekofer, K. M.; Bernat, J.; Fox, A.; Kordos, P. & Luth, H. (2002). Origin of improved RF performance of AlGaN/GaN MOSHFETs compared to HFETs. *IEEE Electron Device Lett.*, Vol. 53, No. 7, (Jul. 2002) (1517-1523), 0018-9383

Matsumura, H. (1986). Catalytic chemical vapor deposition (CTL-CVD) method producing high quality hydrogenated amorphous silicon. *Jpn. J. Appl. Phys.*, Vol. 25, No. 12, (Nov. 1986) (L949-L951), 0021-4922

Medjdoub, F.; Sarazin, N.; Tordjman, M.; Magis, M.; di Forte-Poisson, M. A.; Knez, M.; Delos, E.; Gaquiére, C.; Delage, S. L. & Kohn, E. (2007). Characteristics of Al_2O_3/AlInN/GaN MOSHEMT. *Electron. Lett.*, Vol. 43, No. 12, (Jun. 2007) (691-692), 0013-5194

Mehandru, R.; Gila, B. P.; Kim, J.; Johnson, J. W.; Lee, K. P.; Luo, B.; Onstine, A. H.; Abernathy, C. R.; Pearton, S. J. & Ren, F. (2002). Electrical characterization of GaN metal oxide semiconductor diode using Sc_2O_3 as the gate oxide. *Electrochem. Solid State Lett.*, Vol. 5, No. 7, (Apr. 2002) (G51-G53), 1099-0062

Mehandru, R.; Luo, B.; Kim, J.; Ren, F.; Gila, B. P.; Onstine, A. H.; Abernathy, C. R.; Pearton, S. J.; Gotthold, D.; Birkhahn, R.; Peres, B.; Fitch, R.; Gillespie, J.; Jenkins, T.; Sewell, J.; Via, D. & Crespo, A. (2003). AlGaN/GaN metal-oxide-semiconductor high electron mobility transistors using Sc_2O_3 as the gate oxide and surface passivation. *Appl. Phys. Lett.*, Vol. 82, No. 15, (Apr. 2003) (2530-2532), 0003-6951

Meneghesso, G.; Verzellesi, G.; Pierobon, R.; Rampazzo, F.; Chini, A.; Mishra, U. K.; Canali, C. & Zanoni, E. (2004). Surface-related drain current dispersion effects in AlGaN-GaN HEMTs, *IEEE Trans. Electron Devices*, Vol. 51, No. 10, (Oct. 2003) (1554 - 1561), 0018-9383

Misstele, D.; Rotter, T.; Horn, A.; Katz, O.; Bougrioua, Z.; Aderhold, J.; Graul, J.; Bahir, G. & Salzman, J. (2003). Incorporation of dielectric layers into the processing of III-nitride-based heterostructure field-effect transistors. *J. Electron. Mater.*, Vol. 32, No. 5, (May 2003) (355-363), 0361-5235

Mittereder, J. A.; Binari, S. C.; Klein, P. B.; Roussos, J. A.; Katzer, D. S.; Storm, D. F.; Koleske, D. D.; Wickenden, A. E. & Henry, R. L. (2003). Current collapse induced in AlGaN/GaN high-electron-mobility transistors by bias stress. *Appl. Phys. Lett.*, Vol. 83, No. 8, (Aug. 2003) (1650-1652), 0003-6951

Mizutani, T.; Ohno, Y.; Akita, M.; Kishimoto, S.; & Maezawa, K. (2003). A study on current collapse in AlGaN/GaN HEMTs induced by bias stress. *IEEE Trans. Electron Devices*, Vol. 5 , No. 10, (Oct. 2003) (2015-2020), 0018-9383

Morkoc, H.; Strite, S.; Gao, G. B.; Lin, M. E.; Sverdlov, B. & Burns, M. (1994). Large-band-gap SiC, III-V nitride, and II-VI ZnSe-based semiconductor device technologies. *J. Appl. Phys.*, Vol. 76, No. 3, (Aug. 1994) (1363-1398), 0021-8979

Neaman, D. A. (1997). *Semiconductor physics and devices: basic principles*, Irwin, 025608405X, Homewood IL

Nakamura, S.; Senoh, M.; Nagahama, S.; Iwasa, N.; Yamada, T.; Matsushita, T.; Kiyoku, H.; & Sugimoto, Y. (1996). InGaN-based multi-quantum-well-structure laser diodes. *Jpn. J. Appl. Phys. Lett.*, Vol. 35, No. 1B, (L74-L76) (Jan. 1996). 0021-4922

Nakamura, S.; Senoh, M.; Nagahama, S. I.; Iwasa, N.; & Yamada, T. (1996). Characteristics of InGaN multi-quantum-well-structure laser diodes, *Appl. Phys. Lett.*, Vol. 68, No.3, (Mar. 1996) (3269-3271), 0003-6951

Nakamura, S.; Senoh, M.; Nagahama, S.; Iwasa, N.; Yamada, T.; Sushita, T. M.; Kiyoku, H.; Sugimoto, Y.; Kozaki, T.; Umemoto, H.; Sano, M.; & Chocho, K. (1998). InGaN/GaN/AlGaN-based laser diodes with modula-tion-doped strained-layer superlattices grown on an epitaxially laterally overgrown gan substrate. *Appl. Phys. Lett.*, Vol. **72**, No. 12, (Nov. 1998) (211-213), 0003-6951

Nakano, Y. & Kachi, T. (2003). Characteristics of SiO_2/n-GaN interfaces with β-Ga_2O_3 interlayers. *Appl. Phys. Lett.*, Vol. 83, No. 21, (Nov. 2003) (4336-4338), 0003-6951

Navarro, A.; Rivera, C.; Pereiro, J.; Muñoz, E.; Imer, B.; DenBaars, S. P.; & Speck, J. S. (2009). High responsivity a-plane GaN-based metal-semiconductor-metal photodetectors for polarization-sensitive applications. *Appl. Phys. Lett.*, Vol. 94, (May 2009) (213512-213514), 0003-6951

Ohno, Y. & Kuzuhara, M. (2001). Applications of GaN-based heterojunction fets for advanced wireless communication. *IEEE Trans. Electron Deices*, Vol. 48, No. 3, (Mar. 2001) (517-523), 0018-9383

Okita, H.; Kaifu, K.; Mita, J.; Yamada, T.; Sano, Y.; Ishikawa, H.; Egawa, T. & Jimbo, T. (2003). High transconductance AlGaN/GaN-HEMT with recessed gate on sapphire substrate. *Phys. Stat. Sol. (a)*, Vol. 200, No. 1, (Nov. 2003) (187-190), 0031-8965

Onojima, N.; Higashiwaki, M.; Suda, J.; Kimoto, T.; Mimura, T. & Mastsui, T. (2007). Reduction in potential barrier height of AlGaN/GaN heterostructures by SiN passivation. *J. Appl. Phys.*, Vol. 101, No. 4, (Feb. 2007) (043703-1-043703-6), 0021-8979

Palacios, T.; Chakraborty, A.; Heikman, S.; Keller, S.; DenBaars, S. P. & Mishra, U. K. (2006). AlGaN/GaN high electron mobility transistors with InGaN back-barriers. *IEEE Electron Device Lett.*, Vol. 27, No. 1, (Jan. 2006) (13-15), 0741-3106

Park, K. Y.; Cho, H. I.; Choi, H. C.; Bae, Y. H.; Lee, C. S.; Lee, J. L. & Lee, J. H. (2004). Device characteristics of AlGaN/GaN MIS-HFET using Al_2O_3-HfO_2 laminated high-k dielectric. *Jpn.J. Appl. Phys.*, Vol. 43, No. 11A, (Oct. 2004) (L1433-L1435), 0021-4922

Pau, J. L.; Rivera, C.; MuĖoz, E.; Calleja, E.; Schühle, U.; Frayssinet, E.; Beaumont, B.; Faurie, J. P.; & Gibart, P. (2004). Response of ultra-low dislocation density gan photodetectors in the near- and vacuum-ultraviolet. *J. Appl. Phys.*, Vol. 95, No. 12, (June 2004) (8275-8279), 0021-8979

Peng, P. C.; Peng, W. R.; Feng, K. M.; Chiou, H. Y.; Chen, J.; Kuo, H. C.; Wang, S. C. & Chi, S. (2006). OCDMA light source using directly modulated Fabry–Pérot laser diode in an external injection scheme. *IEEE Photon. Technol. Lett.*, Vol. 18, No. 9, (May 2006) (1103-1105), 1041-1135

Peransin, J. M.; Vignaud, P.; Rigaud, D. & Vandamme, L. K. J. (1990). 1/f noise in MODFETs at low drain bias. *IEEE Trans. Electron Deices*, Vol. 37, No. 10, (Oct. 1990) (2250-2253), 0018-9383

Poate , J. M.; Tu, K. N. & Mayer, J. W. (1978). *Thin films, inter-diffusion and reactions*, Wiley, 0471022381, New York

Pozzovivo, G.; Kuzmik, J.; Golka, S.; Schrenk, W.; Strasser, G.; Pogany, D.; Čičo, K.; Ťapajna, M.; Fröhlich, K.; Carlin, J. F.; Gonschorek, M.; Feltin, E. & Grandjean, N. (2007). Gate insulation and drain current saturation mechanism in InAlN/GaN metal-oxide-semiconductor high-electron-mobility transistors. *Appl. Phys. Lett.*, Vol. 91, No. 4, (Jul. 2007) (043509-1-043509-3), 0003-6951

Rai, S.; Adivarahan, V.; Tipirneni, N.; Koudymov, A.; Yang, J.; Simin, G. & Khan, M. A. (2006). Low threshold-14W/mm ZrO_2/AlGaN/GaN metal-oxide-semiconductor heterostructure field effect transistors. *Jpn. J. Appl. Phys.*, Vol. 45, No. 6A, (Jun. 2006) (4985-4987), 0021-4922

Reimbold, G. (1984). Modified 1/f trapping noise theory and experiments in MOS transistors biased from weak to strong inversion-influence of interface states. *IEEE Trans. Electron Deices*, Vol. 31, No. 9, (Sep. 1984) (1190-1198), 0018-9383

Ren, F.; Abernathy, C. R.; Mackenzie, J. D.; Gila, B. P.; Pearton, S. J.; Hong, M.; Marcus, M. A.; Schurman, M. J.; Baca, A. G. & Shul, R. J. (1998). Demonstration of GaN MIS diodes by using AlN and $Ga_2O_3(Gd_2O_3)$ as dielectrics. *Solid-State Electron.*, Vol. 42, No. 12, (Dec. 1998) (2177-2181), 0038-1101

Ren, F.; Pearton, S. J.; Abernathy, C. R.; Baca, A.; Cheng, P.; Shul, R. J.; Chu, S. N. G.; Hong, M.; Schurman, M. J. & Lothian, J. R. (1999). GaN metal oxide semiconductor field effect transistors. *Solid-State Electron.*, Vol. 43, No. 9. (Sep. 1999) (1817-1820), 0038-1101

Rotter, T.; Mistele, D.; Stemmer, J.; Fedler, F.; Aderhold, J. & Graul, J. (2000). Photoinduced oxide film formation on n-type GaN surfaces using alkaline solution. *Appl. Phys. Lett.*, Vol. 76, No. 26, (Jun. 2000) (3923-3925), 0003-6951

Rossetti, M.; Smeeton, T. M.; Tan, W. S.; Kauer, M.; Hooper, S. E.; Heffernan, J.; Xiu, H. & Humphreys, C. J. (2008). Degradation of InGaN/GaN Laser diodes analyzed by microphotoluminescence and microelectroluminescence Mappings. *Appl. Phys. Lett.*, Vol. 92, No. 15, (Apr. 2008) (151110-1- 151110-3), 0003-6951

Rumyantsev, S. L.; Pala, N.; Shur, M. S.; Gaska, R.; Levinshtein, M. E.; Khan, M. A.; Simin, G.; Hu, X. & Yang, J. (2000). Effect of gate leakage current on noise properties of AlGaN/GaN field effect transistors. *J. Appl. Phys.*, Vol. 88, No. 11, (Dec. 2000) (6726-6730), 0021-8979

Rumyantsev, S. L.; Pala, N.; Shur, M. S.; Gaska, R.; & Levinshtein, M. E. (2001). Low-frequency noise in $Al_{0.4}Ga_{0.6}N$-based schottky barrier photodetectors, *Appl. Phys. Lett.*, Vol. 79, No. 6, (May 2001) (866-868), 0003-6951

Sacconi, F.; Carlo, A. D.; Lugli, P. & Morkoc, H. (2001). Spontaneous and piezoelectric polarization effects on the output characteristics of AlGaN/GaN heterojunction modulation doped FETs. *IEEE Trans. Electron Deices*, Vol. 48, No. 3, (Mar. 2001) (450-457), 0018-9383

Saitoh, T.; Kumagai, M.; Wang, H.; Tawara, T.; Nishida, T.; Akasaka, T.; & Kobayashi, N. (2003). Highly reflective distributed bragg reflectors using a deeply etched semiconductor/air grating for InGaN/GaN laser diodes. *Appl. Phys. Lett.*, Vol.82, No. 23, (Jun. 2003) (4426-4429). 0003-6951

Saripalli, Y. N.; Pei, L.; Biggerstaff,T.; Ramachandran, S.; Duscher, G. J.; Johnson, M. A. L.; Zeng, C.; Dandu, K.; Jin, Y. & D. Barlag, W. (2007), Transmission electron microscopy studies of regrown GaN ohmic contacts on patterned substrates for metal oxide semiconductor field effect transistor applications, *Appl. Phys. Lett.*, Vol. 90, No. 20, (May 2007) (204106-1-204106-3), 0003-6951

Schoedl, T.; Schwarza, U. T.; Kümmler, V.; Furitsch, M.; Leber, A.; Miler, A.; Lell, A. & Härle, V. (2005). Facet degradation of GaN heterostructure laser diodes. *J. Appl. Phys.*, Vol. 97, No. 12, (Jun. 2005) (123102-1-123102-8), 0021-8979

Schroder, D. K. (1998). *Semiconductor material and device characterization,* Wiley, 0471241393, New York

Seo, S.; Lee, K. K.; Kang, Sangbeom; Huang, S.; Doolittle, W. A.; Jokerst, N. M.; Brown, A. S. & Brooke, M. A. (2002). The heterogenous integration of GaN thin-film metal-semiconductor-metal photodetectors onto silicon. *IEEE Photon. Technol. Lett.,* Vol. 14, No. 2, (Feb. 2002) (185-187), 1041-1135

Shen, C. F.; Chang, S. J.; Ko, T. K.; Kuo, C. T.; Shei, S. C.; Chen, W. S.; Lee, C. T.; Chang C. S.,; & Chiou, Y. Z. (2006). Nitride-based light emitting diodes with textured sidewalls and pillar waveguides, *IEEE Photon. Technol. Lett.,* Vol 18, No. 23, (Dec. 2006) (2517-2519), 1041-1135

Shih, C. F.; Li, W. M.; Shu, S. C.; Hsiao, C. Y. & Hung, K. T. (2009). Electrical Properties of Al/HfO_2/n-GaN Prepared by Reactive Sputtering Method. *Jpn.J. Appl. Phys.,* Vol. 48, No. 2, (Feb. 2009) (020224-1-020224-3), 0021-4922

Shih, C. F.; Hung, K. T.; Hsiao, C. Y.; Shu, S. C. & Li, W. M. (2009). Investigations of GaN Metal-oxide-semiconductor Capacitors With Sputtered HfO_2 Gate Dielectrics. *J. Alloy. Compd,* Vol. 480, No. 2, (Jul. 2009) (541-546), 0925-8388

Seo, H. C.; Chapman, P.; Cho, H. I.; Lee, J. H. & Kim, K. (2008). Ti-based nonalloyed ohmic contacts for $Al_{0.15}Ga_{0.85}N$/GaN high electron mobility transistors using regrown n^+-GaN by plasma assisted molecular beam epitaxy, *Appl. Phys. Lett.,* Vol 93, No. 10, (Sept. 2008) (102102-1-102102-3), 0003-6961

Sheu, J. K.; Lee, M. L.; Yeh, L. S.; Kao, C. J.; Tun, C. J.; Chen, M. G.; Chi, G. C.; Chang, S. J.; Su, Y. K. & Lee, C. T. (2002). Planar GaN n-p photodetectors formed by Si implantation into p-GaN. *Appl. Phys. Lett.,* Vol. 81, No. 22, (Dec. 2006) (4263-4265), 0003-6951

Simin, G.; Koudymou, A.; Fatima, H.; Zhang, J.; Yang, J.; Khan, M. A.; Hu, X.; Tarakji, A.; Gaska, R. & Shur, M. S. (2002). SiO_2/AlGaN/InGaN/GaN MOSDHFETs. *IEEE Electron Device Lett.,* Vol. 23, No. 8, (Aug. 2002) (458-460), 0741-3106

Simin, G.; Adivarahan, V.; Yang, J.; Koudymov, A.; Rai, S. & Khan, M. A. (2005). Stable 20W/mm AlGaN-GaN MOSHFET. *Electron. Lett.,* Vol. 41, No. 13, (Jun. 2005) (774-775), 0013-5194

Smorchkova, I. P.; Elsass, C. R.; Ibbetson, J. P.; Vetury, R.; Heying, B.; Fini, P.; Haus, E.; DenBaars, S. P.; Speck, J. S. & Mishra, U. K. (1999). Polarization-induced charge and electron mobility in AlGaN/GaN heterostructures grown by plasma-assisted molecular-beam epitaxy. *J. Appl. Phys.,* Vol. 86, No. 8, (Oct. 1999) (4520-4526), 0021-8979

Su, Y. K.; Chang, P. C.; Chen, C. H.; Chang, S. J.; Yu, C. L.; Lee, C. T.; Lee, H. Y.; Gong, J.; Chen, P. C.; & Wang, C. H. (2005). Nitride-based MSM UV photodetectors with photo-chemical annealing schottky contacts. *Solid-State Electron.,* Vol. 49, No. 3, (Mar. 2005) (459-463), 0038-1101

Suski, T.; Franssen, G.; Perlin, P.; Bohdan, R.; Bercha, A.; Adamiec, P.; Dybala, F.; Trzeciakowski, W.; Prystawko, P.; Leszczyński, M.; Grzegory, I.; & Porowski, S. (2004). A pressure-tuned blue-violet InGaN/GaN laser diode grown on bulk gan crystal. *Appl. Phys. Lett.,* Vol. 62, No. 23, (Feb. 2004) (1236-1238), 0003-6951

Sze, S. M. (2002). *Semiconductor Devices-Physics & Technology,* Wiley, 0471333727, New York

Tan, W. S.; Houston, P. A.; Parbrook, P. J.; Hill, G. & Airey, R. J. (2002). Comparison of different surface passivation dielectrics in AlGaN/GaN heterostructure field-effect transistors. *J. Phys. D: Appl. Phys.*, Vol. 35, No. 7, (Mar. 2002) (595-598), 0022-3727

Therrien, R.; Lucovsky, G. & Davis, R. (2000). Charge Redistribution at GaN-Ga_2O_3 Interfaces: a Microscopic Mechanism for Low Defect Density Interfaces in Remote-Plasma-Processed MOS Devices Prepared on Polar GaN Faces. *Appl. Phys. Lett.*, Vol. 166, No. 1, (Oct. 2000) (513-519), 0169-4332

Tourtin, F.; Armand, P.; Ibanez, A.; Tourillon, G. & Philippot, E. (1998). Gallium phosphate thin solid films: structural and chemical determination of the oxygen surroundings by XANES and XPS. *Thin Solid Films*, Vol. 322, No. 1-2, (Jun. 1998) (85-92), 0040-6090

Vandamme, L. K. J.; Li, X. & Rigaud, D. (1994). 1/f noise in MOS devices, mobility or number fluctuations. *IEEE Trans. Electron Deices*, Vol. 41, No. 11, (Nov. 1994) (1936-1945), 0018-9383

Vardi, A.; Akopian, N.; Bahir, G.; Doyennette, L.; Tchernycheva, M.; Nevou, L.; Julien, F. H.; Guillot, F. & Monroy, E. (2006). Room temperature demonstration of GaN/AlN quantum dot intraband infrared photodetector at fiber-optics communication wavelength. *Appl. Phys. Lett.*, Vol. 88, No. 14, (April 2006) (143101-1-143101-3), 0003-6951

Vertiatchikh, A. V. & Eastman, L. F. (2003). Effect of The Surface and Barrier Defects on The AlGaN/GaN HEMT Low-Frequency Noise Performance. *IEEE Electron Device Lett.*, Vol. 24, No. 9, (Sep. 2003) (535-537), 0741-3106

Vetury, R.; Zhang, N. Q.; Keller, S. & Mishra, U. K. (2001), The impact of surface states on the DC and RF characteristics of AlGaN/GaN HFETs, *IEEE Trans. Electron Devices*, Vol. 48, No. 3, (Mar. 2001) (560-566), 0018-9383

Walker, D.; Saxler, A.; Kung, P. ; Zhang, X. ; Hamilton, M. ; Diaz, J. & Razeghi, M. (1998). Visible blind GaN p-i-n photodiodes. *Appl. Phys. Lett.*, Vol. 72, No. 25 (Jun. 1998) (3303-3305), 0003-6951

Wallis, D. J.; Balmer, R. S.; Keir, A. M. & Martin, T. (2005). Z-contrast imaging of AlN exclusion layers in GaN field-effect transistors. *Appl. Phys. Lett.*, Vol. 87, No. 4, (July 2005) (042101-1-042101-3), 0003-6951

Wang, C. K.; Chiou, Y. Z.; Chang, S. J.; Su, Y. K.; Huang, B. R.; Lin, T. K. & Chen, S. C. (2003). AlGaN/GaN metal-oxide-semiconductor heterostructure field-effect transistor with photo-chemical-vapor Deposition SiO_2 gate oxide. *J. Electron. Mater.*, Vol. 32, No. 5, (May 2003) (407-410), 0361-5235

Wang, C. K.; Chang, S. J.; Su, Y. K.; Chiou, Y. Z.; Kuo, C. H.; Chang, C. S.; Lin, T. K.; Ko, T. K. & Tang, J. J. (2005). High temperature performance and low frequency noise characteristics of AlGaN/GaN/AlGaN double heterostructure metal-oxide-semiconductor heterostructure field-effect-transistors with photochemical vapor deposition SiO_2 layer. *Jpn. J. Appl. Phys.*, Vol. 44, No. 4B, (Apr. 2005) (2458-2461), 0021-4922

Wang, C. K.; Chuang, R. W.; Chang, S. J.; Su, Y. K.; Wei, S. C.; Lin, T. K.; Ko, T. K.; Chiou, Y. Z. & Tang, J. J. (2005). High temperature and high frequency characteristics of AlGaN/GaN MOS-HFETs with photochemical vapor deposition SiO_2 layer. *Mater. Sci. Eng. B*, Vol. 119, No. 1, (May. 2005) (25-28), 0921-5107

Webb, J. B.; Tang, H.; Bardwell, J. A.; Moisa, S.; Peters, C. & MacElwee, T. (2001). Defect reduction in GaN epilayers and HFET structures grown on (0001) sapphire by ammonia MBE. *J. Cryst. Growth,* Vol. 230, No. 3-4, (Sep. 2001) (584-589), 0022-0248

Wierer, J. J.; Krames, M. R.; Epler, J. E.; Gardner, N. F.; Craford, M. G.; Wendt, J. R.; Simmons, J. A.; & Sigalas, M. M. (2004). InGaN/GaN quantum-well heterostructure light-emitting diodes employing photonic crystal structures, *Appl. Phys. Lett.,* Vol 84, No. 19, (May. 2004) (3885-3887), 0003-6951

Wiesmann, H.; Ghosh, A. K.; McMahon, T. & Strongin, M. (1979). A-Si:H produced by high-temperature thermal decomposition of silane. *J. Appl. Phys.,* Vol. 50, No. 5, (May 1979) (3752-3754), 0021-8979

Wu, C. I. & Kahn, A. (1999). Electronic states and effective negative electron affinity at cesiated p-GaN surfaces. *J. Appl. Phys.,* Vol. 86, No. 6, (Sep. 1999) (3209-3212), 0021-8979

Wu, Y. Q.; Ye, P. D.; Wilk, G. D. & Yang, B. (2006). GaN metal-oxide-semiconductor field-effect-transistor with atomic layer deposited Al_2O_3 as gate dielectric. *Mater. Sci. Eng. B-Solid State Mater. Adv. Technol.,* Vol. 135, No. 3, (Dec. 2006) (282-284), 0021-5107

Yagi, S.; Shimizu, M.; Inada, M.; Yamamoto, Y.; Piao, G.; Okumura, H.; Yano, Y.; Akutsu, N. & Ohashi, H. (2006). High breakdown voltage AlGaN/GaN MIS-HEMT with SiN and TiO_2 gate insulator. *Solid-State Electron.,* Vol. 50, No. 6, (1057-1601) (Jun. 2006), 0038-1101

Yamashita, Y.; Endoh, A.; Hirose, N.; Hikosaka, K.; Matsui, T.; Hiyamizu, S. & Mimura, T. (2006). Effect of bottom SiN thickness for AlGaN/GaN metal-insulator-semiconductor high electron mobility transistors using SiN/SiO_2/SiN triple-layer insulators. *Jpn. J. Appl. Phys.,* Vol. 45, No. 26, (Jun. 2006) (L666-L668), 0021-4922

Ye, P. D.; Yang, B.; Ng, K. K. & Bude, J. (2005). GaN metal-oxide-semiconductor high-electron-mobility-transistor with atomic layer deposited Al_2O_3 as gate dielectric. *Appl. Phys. Lett.,* Vol. 86, No. 6, (Jan. 2005) (063501-1-063501-3), 0003-6951

Yoshida, S. & Suzuki, J. (1999) High-temperature reliability of GaN metal semiconductor field-effect transistor and bipolar junction transistor. *J. Appl. Phys.,*Vol. 85, No. 11, (Jun. 1999) (7931-7934), 0021-8979

Youtsey, C. & Adesida, I. (1997). Highly anisotropic photoenhanced wet etching of n-type GaN. *Appl. Phys. Lett.,* Vol. 71, No. 15, (Oct. 1997) (2151-2153), 0003-6951

Yue, Y. Z.; Hao, Y.; Zhang, J. C.; Ni, J. Y.; Mao, W.; Feng, Q. & Liu, L. J. (2008). AlGaN/GaN MOS-HEMT with HfO_2 dielectric and Al_2O_3 interfacial passivation layer grown by atomic layer deposition. *IEEE Electron Device Lett.,* Vol. 29, No. 8, (Aug. 2008) (838-840), 0741-3106

Zhang, L.; Lester, L. F.; Baca, A. G.; Shul, R. J.; Chang, P. C.; Willison, C. G.; Mishra, U. K.; Denbaars, S. P. & Zolper, J. C. (2000) Epitaxially-grown GaN junction field effect transistors. *IEEE Trans. Electron Devices,* Vol. 47, No.3 (Mar. 2000) (507-511), 0018-9383

Zheng, Y. Y.; Yue, H.; Cheng, Z. J.; Qian, F.; Yu, N. J. & Hua, M. X. (2008). A study on Al_2O_3 passivation in GaN MOS-HEMT by pulsed stress. *Chin. Phys.,* Vol. 17, No. 4, (Apr. 2008) (1405-1409), 1674-1056

Zolper, J. C.; Shul, R. J.; Baca, A. G.; Wilson, R. G.; Pearton, S. J.; and Stall, R. A. (1996). Ion-implanted GaN junction field effect transistor. *Appl. Phys. Lett.*, Vol. 68, No. 16, (Apr. 1996) (2273-2275), 0003-6951

Concepts of Optimizing Power Semiconductor Devices Using Novel Nano-Structure for Low Losses

Ye, Hua[1] and Haldar, Pradeep[2]
[1]*Microsoft Corporation*
[2]*College of Nanoscale Science and Engineering*
State University of New York at Albany
USA

1. Introduction

In the chapter, the authors discuss two new concepts of optimizing power devices that directly addressing the limitations of current IGBT (Insulated Gate Bipolor Transistors) and SJ (Superjunction) MOSFET technologies.

Power MOSFETs and IGBTs are the two main competing power semiconductor devices for switching electric power in electrical power conversion systems at mid-voltage ratings. Power MOSFETs conduct current as soon as a forward bias voltage is applied between the drain and the source electrodes; however, as the blocking voltage capability increases, the on-resistance of conventional power MOSFETs increases proportionally to the second order of its blocking voltage (Hu, 1979). In order to overcome the limitation of conventional power MOSFET, IGBT is introduced. Unlike conventional power MOSFET, the forward voltage drop of IGBTs does not follow a second order dependence on blocking voltage because the conductivity of the voltage blocking drift layer can modulated by carrier injection during forward bias. However, IGBTs cannot carry any significant current until the external bias surpasses an internal barrier voltage (heel voltage). This distinction, among other considerations, makes the selection of power semiconductor switches a trade-off between MOSFETs and IGBTs. For instance, paralleling IGBTs will not reduce the heel voltage.

Another technology to address the limitation of conventional power MOSFET is SJ MOSFET that employs the charge compensation concept have been significantly researched in an effort to break the "silicon limit" and led to growing commercialization (Coe, 1988; Chen, 1993; Fujihira, 1997; Shenoy, et al., 1999; Deboy, et al., 1998). These devices use an alternating p and n charge compensation structure to replace the planar voltage-blocking drift layer in the conventional power MOSFET, where the n-columns can be much more heavily doped than the planar drift layer, leading to significant reduction in specific on-resistance. The breakdown voltage of an SJ MOSFET is proportional to the depth of the p and n columns. At the same time, reducing the widths of the alternating p and n columns leads to higher allowable doping levels and thus smaller on-resistance (Fujihira, 1997). However, fabricating the SJ structure with increasing depths of p and n columns and decreasing column sizes leads to increasing process difficulties. In addition, the criticality of match the doping levels in the p and n regions with their widths on the breakdown voltage further increases the process difficulties (Shenoy, et al., 1999). State-of-the-art fabrication techniques -- such as high-energy implantation, multi-epitaxial growth, and trench-filling -- have been demonstrated to be only sufficient to create low to mid voltage range (<1000V) devices (Deboy, et al., 1998; Miura, et al., 2005; von Borany, et al., 2004; Rub, et al., 2004; Onishi, et al., 2002; Minato, et al., 2000; Rochefort, et al., 2002; Saito, et al., 2005; Liang, et al., 2001; Chen & Liang, 2007; Gan, et al., 2001;).

The first concept discussed in this chapter is a proposal of a mid-to-high voltage power switch that utilizes reverse band-to-band tunneling and an avalanche injection mechanism called Tunnelling Junction Enhanced MOSFET (TJE-MOSFET) (Ye & Haldar, 2008). This device is predicted to have the best properties of both power MOSFETs and IGBTs (Insulated Gate Bipolar Transistors) - the two main competing power semiconductor technologies at mid-voltage (500-1000V) ratings. The structure and the operating mechanism of the TJE-MOSFET are described. The proposed novel device operates in a way similar to an IGBT; however, due to the inclusion of a nano-structured band-to-band tunneling junction, the internal barrier voltage for forward conduction is much smaller than that in an IGBT. Numerical simulation suggests that, at the same current level, the forward voltage drop of the TJE-MOSFET is much smaller than that of an IGBT. Compared to power MOSFETs, the new device has a lower forward voltage drop even at very low current levels.

The second concept is a novel SJ MOSFET fabrication process based on porous silicon formation (Ye & Haldar, 2008). The voltage blocking SJ structure is directly created within the lightly doped thin silicon wafer instead of growing the costly thick epitaxial layer. The charge compensating structures are created by etching the structured macro-pores, followed by passivating the walls and filling the pores with oppositely charged poly-silicon. The effects of charge imbalance and the thickness of the passivation layer are studied by physics-based numerical device simulations. It is found that even with some amount of charge imbalance, the proposed method can still produce high-voltage MOSFETs with much better performance than existing technology. A thick oxide layer between the p and n columns is found to be helpful in alleviating the JFET (Junction-Field-Effect Transitor) effect when the doping concentrations in the p and n columns are low in comparison with a conventional SJ structure. The inclusion of an oxide layer between the p and n columns is found to help increase the device efficiency in addition to its ability to prevent dopant interdiffusion.

2. Tunnelling Junction Enhanced MOSFET (TJE-MOSFET)

2.1 Background

A band-to-band tunneling junction diode working in the forward bias regime has been widely used in a variety of the applications such as switching, oscillation, and amplification by taking advantage of its negative resistance characteristics. Reverse-biased tunneling has received much less attention until recently. A few attempts of taking advantage of reverse band-to-band tunneling breakdown in order to create a new family of transistors that aims at replacing the today's CMOS technology have been reported recently (Aydin, et al., 2004). In addition, reverse band-to-band tunneling is also found to be important in CMOS at room temperature for dopant concentrations above $5\times10^{17}cm^{-3}$, which presents a limit to scaling of future CMOS technology (Solomon, et al, 2004). Solomon et al. (Solomon, et al, 2004) have studied ion-implanted p/n junction diodes with doping levels up to 10^{20} cm^{-3} by measuring current-voltage characteristics in both forward and reverse bias conditions. Their measurements show that for a highly doped p/n junction diode, very high current densities are achieved at very low reverse bias voltage, which is dominated by band-to-band tunneling. They conclude that the higher the junction doping concentration, the smaller the effective tunneling distance, resulting in higher tunneling current densities.

In this section, a novel power switch is proposed, which utilizes a reverse biased nanoscale band-to-band tunneling structure in order to reduce the forward voltage drop during conduction. The device structure and the operating mechanism are described. The proposed TJE-MOSFET operates in a way similar to an IGBT. However, by taking advantage of a reverse-biased band-to-band tunneling junction, the internal barrier voltage for forward conduction is much smaller than that of an IGBT. Numerical simulation suggests that, at the same current level, the forward voltage drop of the TJE-MOSFET is much smaller than that of an IGBT. Compared to power MOSFETs (conventional as well as the superjunction MSOFETs), the TJE-MOSFET has a much lower forward voltage drop even at very low current levels.

2.2 Structure and Operation Mechanism of the Device

The structure of the TJE-MOSFET is very similar to that of a power MOSFET or IGBT as shown in Figure 1(a-c), where they all share a similar gate structure. They all feature a lightly-doped n- drift layer which is used to block the high voltage during the OFF-state when the junction between this layer and the p-base layer (J2) is reverse biased. The differences are at the back side of the devices. Compared to power MOSFETs and IGBTs, the TJE-MOSFET features a unique sharp (abrupt) and highly doped p++/n++ junction J1. The doping levels in the p++ and n++ layer are on the orders of 3×10^{19} to 1×10^{21} cm^{-3}. The p++ layer has to be very thin with thickness on the order of several to several tens of nanometers. An optional n layer several microns thick with mid-level doping can be included as a minority carrier injection buffer layer and/or field-stop layer if a punch-through design is desired. The operation of the device is similar to a power MOSFET or IGBT in that the ON and the OFF states of the device are controlled by altering the bias voltage at the gate electrode.

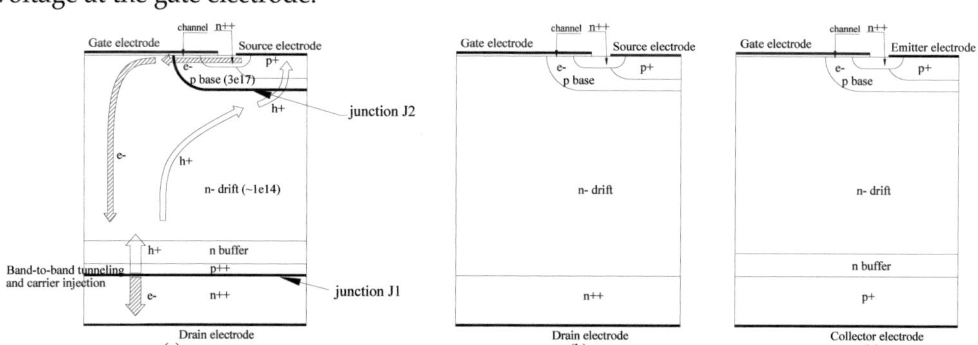

Fig. 1 (a) Structure of the device; (b) power MOSFET; (c) IGBT

During the forward conduction mode or switch-on, the channels near the gate oxide in the p-base region are created by applying a positive gate-to-source bias voltage above the gate threshold voltage. The drain electrode is positively biased. This makes the highly-doped p++/n++ junction (J1) reverse-biased. Due to the extremely high doping concentration on both sides of J1, the conduction band edge on the n++ side of J1 overlaps with the valance band edge on the p++ side as shown in Figure 2(a). As junction J1 is reverse biased, electrons are allowed to tunnel from the filled valance band states below the Fermi level E_{fp} on the p++ side to the empty conduction band states above the Fermi level E_{fn} on the n++ side. At the same time, holes are left over on the p++ side. As the reverse bias voltage increases, E_{fn} continues to move down with respect to E_{fp}, leaving more filled states on the p++ side and more empty states on the n++ side; therefore, the tunneling of electrons increases. This process can also be viewed as the injection of holes from the n++ side into the p++ side at the junction J1. Since the electric field across the junction J1 is very high, the electrons and holes created by the tunneling are accelerated by the field to gain more energy. Thus a carrier multiplication process is followed by an impact ionization mechanism to create more electron-hole pairs. The electrons drift toward the drain electrode and the holes drift into the n- drift region and then diffuse toward the p-base region. This process can be viewed as avalanche injection of holes into the n- drift region from the reverse-biased junction J1. The purpose of the n buffer layer right above the p++ layer is to control the injection of holes and acts as a field stopper. As the channel exists in the forward conduction mode, electrons flow from the n+ source region into the n- drift region and recombine with the injected holes. The remaining holes that diffuse near the p-base region are collected in the p-base region and then drift toward the source electrode on top of the p-base region. The hole and electron current components during the conduction mode are shown in Figure 1(a). Due to the high-level injection of holes into the n- drift region, the concentration of electrons in the n- drift region becomes much higher than its doping concentration in order to maintain charge neutrality. This phenomenon is called conductivity modulation and is well understood in the operation of bipolar junction transistors, IGBTs, thyristors, etc. Due to conductivity modulation, the forward voltage drop during conduction becomes very small despite low doping levels in the n drift layer.

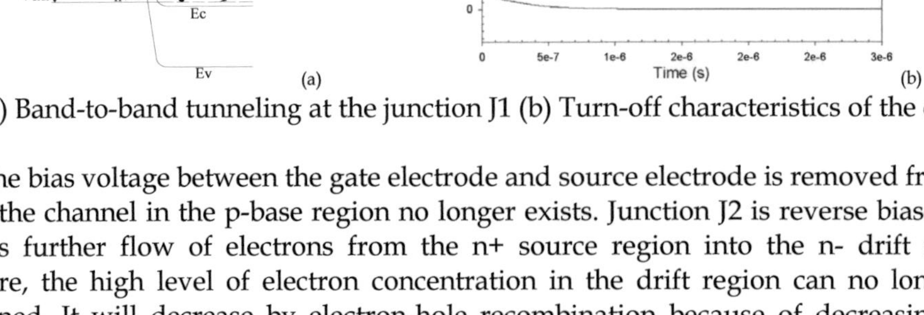

Fig. 2 (a) Band-to-band tunneling at the junction J1 (b) Turn-off characteristics of the device

When the bias voltage between the gate electrode and source electrode is removed from the device, the channel in the p-base region no longer exists. Junction J2 is reverse biased and prevents further flow of electrons from the n+ source region into the n- drift region. Therefore, the high level of electron concentration in the drift region can no longer be maintained. It will decrease by electron-hole recombination because of decreasing hole concentration. As the carrier concentration decreases in the drift region, the voltage will gradually build up at the reverse biased junction J2, and this junction will sustain all the applied OFF-state voltage. The decrease of the forward current follows a similar pattern to the turn-off operation of IGBTs. As the gate voltage reduces below the gate threshold voltage, the electron current component will suddenly decrease to zero leading to a sharp drop of total current. However, current continues to flow through the device due to the high hole concentration in the n- drift region. This current gradually decreases as the hole concentration in the n- drift region gradually decreases by electron-hole recombination. The turn-off curve is illustrated in Figure 2(b) as obtained from numerical simulation that is described in the next section.

Unlike an IGBT, where high-level injection occurs only when the applied voltage across the p/n junction near the collector electrode increases above the internal barrier of the junction (0.7V at room temperature), high-level injection in the TJE-MOSFET can happen at much smaller forward bias. Numerical simulations suggest that a much smaller forward voltage drop can be realized in the device when compared to an IGBT with the same current density level. Simulations also suggest that the forward voltage drop decreases with increasing doping concentration at the p++/n++ junction J1.

2.2 Numerical Simulation and Discussion

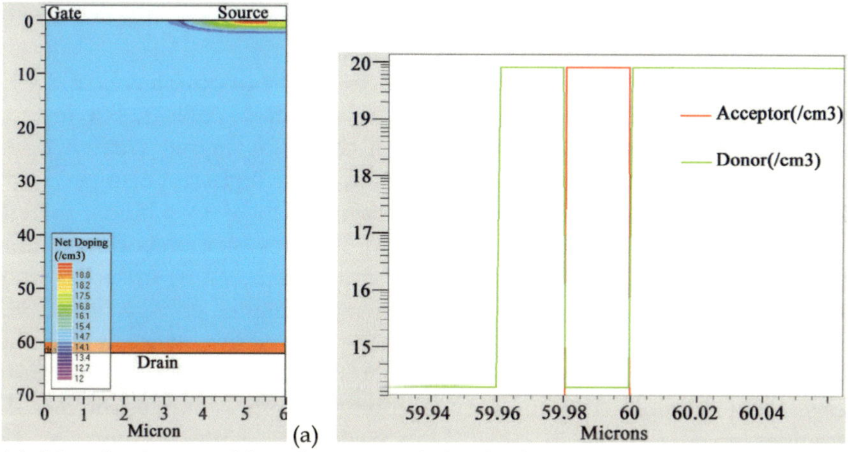

Fig. 3 (a) Net doping profile schematic of the half unit cell of the simulated device (b) Doping concentrations near the p++/n++ junction

Numerical simulations were carried out to evaluate the potential of the TJE-MOSFET concept. A Silvaco Atlas device simulator was used in the analysis. Fig. 3 shows the geometry and doping concentration profile of the simulated half unit cell. A 20 nm thick p++ layer ($8\times10^{19}cm^{-3}$) is created above the n++ substrate ($8\times10^{19}cm^{-3}$). Another 20nm thick n+ layer is created above the p++ layer for the purpose of controlling the injection efficiency. Fig. 4(a) shows the band diagram of the TJE-MOSFET near the p++/n++ junction at equilibrium.

Overlap of the valance and conduction bands is clearly seen in this figure. Fig. 4(b) shows the carrier concentration within the device during the ON-state with a drain bias of 1V. It clearly shows that both the hole and electron concentrations are much higher than the doping concentration in the region, a phenomenon called conductivity modulation.

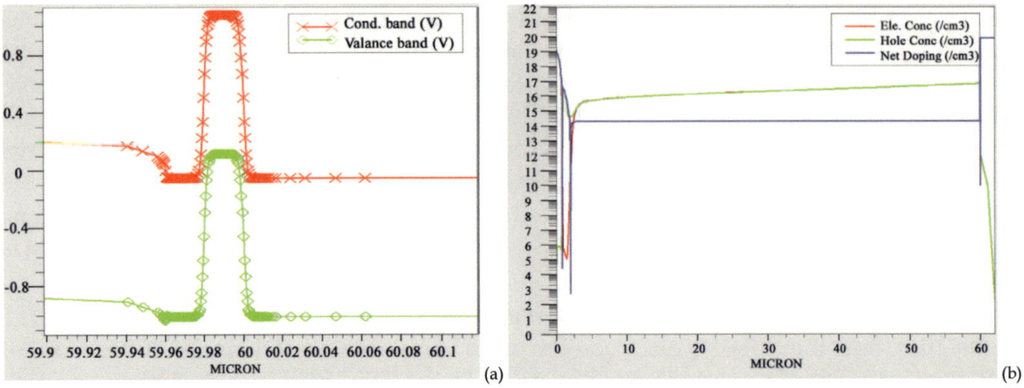

Fig. 4 (a) Band energy diagram near the p++/n++ junction at equilibrium (b) Carrier concentration during conduction

Fig.5(a) shows the I-V characteristics of the TJE-MOSFET vs. other devices (i.e., MOSFETs and IGBTs) with the same n- drift thickness and doping level. The major advantages of the TJE-MOSFET are its superior conduction characteristics compared with those of the existing power devices. Normally, power MOSFETs are used in low-voltage and low current density applications while IGBTs are used in high-voltage and high current density applications. The TJE-MOSFET is very competitive in both applications.

As shown in Fig.5(a), the proposed the device (with p++/n++ doping levels of 8×10^{19}) can carry much higher current density than conventional power MOSFETs. For instance, at a forward voltage drop of 1V, the TJE-MOSFET can carry 25x higher current density. At higher voltage drops, the current density can be significantly higher. It also performs better than a SJ MOSFET with a 2.5μm pillar width at a forward voltage drop higher than 0.9V. As described in the Introduction section, the SJ MOSFET requires accurately doped alternating p and n pillars, the on-resistance of an SJ MOSFET can be reduced by orders of magnitude compared to the conventional power MOSFET. However, higher voltage SJ MOSFETs are not particularly easy to fabricate. For a 1000V-rated Super-junction device, a pillar height of 60μm is needed. Creating such narrow and deep pillars which have exactly opposite doping concentrations is very difficult using current semiconductor processing technology. The TJE-MOSFET provides an alternative to create high performance power switches. The TJE-MOSFET can share most of the processing techniques with conventional power MOSFETs or IGBTs. However, additional steps are required to create the sharp and heavily-doped p++/n++ metallurgical junction, which is challenging. The high thermal budgets of conventional epitaxy methods for creating the n-drift layer and subsequent steps would definitely alter the earlier created p++ and n++ layers; therefore, novel processing techniques will be needed to address this challenge. For instance, a lightly-doped thin wafer might be used as the voltage blocking layer instead of an epitaxially created n- drift layer. Low-temperature epitaxial processes such as MBE might be required to create the p++ and n++ layers.

Compared to IGBTs, the TJE-MOSFETs have superior forward conduction characteristics. At the same current density, the forward voltage drop of the TJE-MOSFET is much smaller than that of an IGBT. Furthermore, the TJE-MOSFET has no heel voltage as seen in an IGBT. Fig.5(a) clearly shows that the TJE-MOSFET can carry $50A/cm^2$ of current density at a forward voltage drop of 0.7V while the current density of an IGBT is negligible at this

voltage. This means that the TJE-MOSFET can carry current right after a bias is applied much like a MOSFET. Therefore, the TJE-MOSFET is more suitable than IGBTs in lower current density applications. This also enables the parallelization of the TJE-MOSFET in order to further improve the conduction characteristics. A trade-off between the forward voltage drop and turn-off time is considered when designing an IGBT. Since the TJE-MOSFET has a much lower forward voltage drop than an IGBT at the same current density, there should be more flexibility to optimize between conduction loss and switching loss.
Furthermore, the simulations also suggest that the current density of the TJE-MOSFET can be further improved by increasing the doping levels in the p++/n++ junction as shown in Fig.5(b). This is due to the fact that band-to-band tunneling current is exponentially proportional to the inverse of tunneling distance at a reverse-biased junction. The improvement should only be restricted by the highest doping levels that can be reached in these junctions.

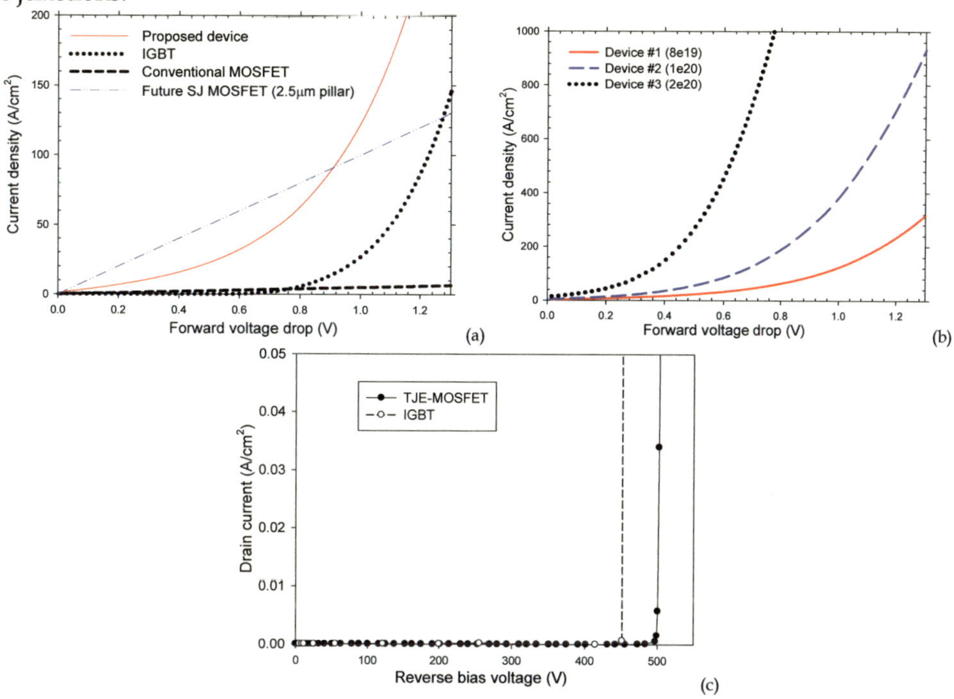

Fig.5 (a) I-V characteristics of the TJE-MOSFET vs. other power devices with same geometry (b) I-V Characteristics comparison between the TJE-MOSFETs with various doping concentrations at the p++/n++ junctions (c) Reverse characteristics comparison between the THE-MOSFET and IGBT

This indicates that the TJE-MOSFET has the potential to outperform SJ MOSFETs at very low current density levels in terms of conduction characteristics. Fig.5(c) shows the breakdown characteristics of the TJE-MOSFET, which is very similar to the IGBT with similar geometry and n- drift layer doping levels. It should be noted that the drift layer doping levels in neither the TJE-MOSFET nor the IGBT were optimized and is taken a value of 2×10^{14} cm^{-3}.

3. Novel High Voltage SJ MOSFET Based on Porous Silicon Formation

3.1 Device Structure and Fabrication Process

A schematic of the trench gate type SJ MOSFET half unit cell based on porous silicon formation is shown in Figure 6(a). A lightly doped (10^{15}-$10^{16}cm^{-3}$ depending on the targeted pore size and pitch) n-type thin wafer (120-200 micron thick depending on voltage ratings) is electrochemically etched to form deep macro-pores with small pore size and pitch though most of the wafer thickness, leaving only a thin layer un-etched at the back side of the wafer. Photo-assisted electro-chemical etching of silicon to form high-aspect-ratio microstructures has been under development since the 1970s and has advanced to reach the level of submicron precision (Coe, 1988; Chen, 1993).

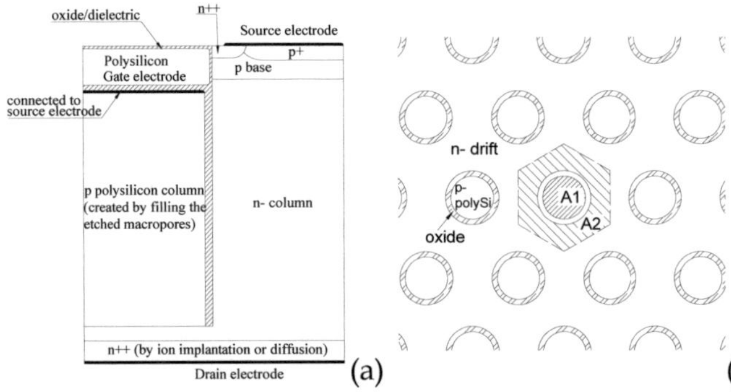

(a) (b)

Fig. 6. (a) Schematic of the half unit-cell of the porous silicon based SJ; (b) Schematic of the circular cell in hexagonal array topological design

This technique has been widely used in fabricating silicon-based photonic devices and electronic devices, as well as micro-electro-mechanical systems (MEMS) (Theunissen, 1972; Kleimann, et al., 2005; Charlton & Parker, 1997; Lehmann, et al., 1996; Wang, et al., 2003). The pattern of the macro-pores is controlled by photolithography based on the cell topological design (such as the simple linear trench design or the circular cell in hexagonal array design as shown in Figure 6(b)). The pore size is controlled by the applied current density and illumination intensity during etching. An n++ layer on the back side can then be created by ion implantation or other standard doping method for the drain electrode. The macro-pores are deposited with a thin conformal coating of oxide/dielectric layer using standard microelectronic processes such as thermal oxidation or oxide-nitride-oxide (ONO) deposition. The oxide/dielectric on the bottom of the macro-pores is then anisotropically etched, leaving the oxide/dielectric layer only on the side walls of the macro-pores. Doped p-type poly-silicon is deposited in the macro-pores to form the p region with the doping level matching their sizes and pitches as well as the doping of the wafer. The poly-silicon doping can be *in situ* doping or use a sequence of conformal poly deposition, tilted implantation, poly refill and drive-in. These processes are sufficient for an aspect ratio as large as 100. The p-type poly-silicon and the reminder of the n-type silicon between the pores form the alternating p and n columns with a thin layer of oxide/dielectric layer on the side walls. This oxide/dielectric layer prevents interdiffusion during the subsequent processing steps. After the formation of the SJ structure and possibly a chemical-mechanical polishing step, conventional ways of creating the power MOSFET structures including the gate and source electrodes can be employed. This process can fabricate SJ structures with a column size of a couple of microns and depth up to a couple of hundred microns, with the potential to create SJ devices at voltages above 2000V.

3.2 Device Simulation Description

We have used physically-based device simulation software Synopsys Sentaurus Device to perform the simulations. Fermi-Dirac statistics is employed in the carrier density calculation. For low field bulk mobility, Klaassen's unified mobility model (Klaassen, 1992; Klaassen, 1992) is used, which considers the effects of lattice scattering, impurity scattering, carrier-carrier scattering, and impurity clustering effects at high concentration. The mobility degradation within the inversion layers is considered by employing Lombardi's model (Lombardi, et al., 1988) that accounts for the effects of surface acoustic phonon scattering and surface roughness scattering. Canali's model (Canali, et al., 1975) is used to account for the carrier velocity saturation in high electric fields. Doping-dependent Shockley-Read-Hall recombination and Auger recombination are considered. The breakdown of the device is simulated by employing Lackner's avalanche generation model (Lackner, 1991).
The device under consideration has a cell structure as shown in Figure 6(a) and uses the topological design of a circular cell in a hexagonal array (Figure 6(b)). In order to satisfy the charge compensation requirement, the size of the poly-silicon p columns and pitch of the p columns are designed by

$$p \cdot A_1 = n \cdot A_2 , \tag{1}$$

where p is the doping density in the p-type poly-silicon column and n is the wafer doping density; A_1 and A_2 are the areas of the p-type poly-silicon and area of n-type silicon surrounding the p-poly, respectively, as shown in Figure 6(b). The depths of the p and n columns are taken as $160\mu m$; p-base depth is $2.7\mu m$ and n+ source depth of $0.8\mu m$, giving a channel length of $1.9\mu m$. The gate oxide thickness is 50nm. Gaussian profiles are assumed for p-base and n+ source with the peak concentrations of $3\times10^{17} cm^{-3}$ and $10^{19} cm^{-3}$, respectively. The thickness of the n-layer underneath the p and n columns is $5\mu m$, and the n+ drain doping is $10^{19} cm^{-3}$. In the baseline half unit-cell, the width of the p-poly-silicon is $1\mu m$, which is equivalent to a p-column diameter of $2\mu m$. The thickness of the oxide layer (between the p and n columns) and n-column width are $0.1\mu m$ and $0.692\mu m$, respectively. This makes the p-column center-to-center distance to be $3.6\mu m$. The area ratio between the poly-silicon p-column and the n-type silicon surround it, A_1/A_2, is then ½. The doping level of the wafer is $9\times10^{15} cm^{-3}$ and the doping level of the p-poly-silicon column is $1.8\times10^{16} cm^{-3}$. In order to account for the cell topology of a circular cell in a hexagonal array, cylindrical symmetry is assumed in the two-dimensional device simulations for the half unit cell.

3.3 Simulation Results and Discussion

Figure 7 shows the simulated breakdown and I-V characteristics of the baseline device. The device has a breakdown voltage of 2490V and a specific on-resistance of $20 m\Omega\cdot cm^2$. The specific on-resistance is calculated at the drain voltage of 0.5V by first evaluating the total current within a 1 cm^2 device area. At this voltage, the silicon limit for conventional power MOSFETs is $2600 m\Omega\cdot cm^2$ by way of comparison.

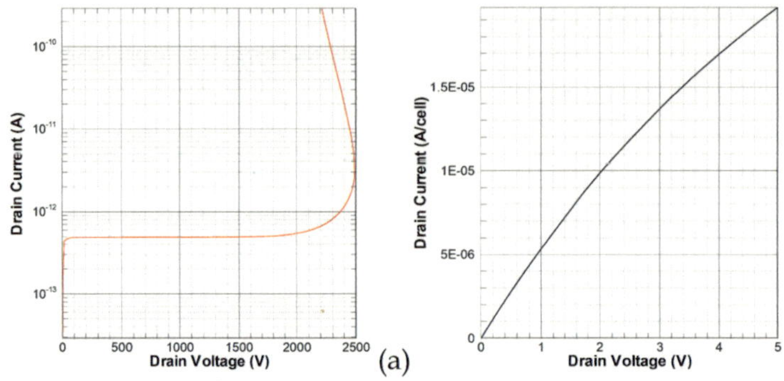

Fig. 7. Simulation results for baseline cell: (a) Breakdown characteristics; (b) I-V characteristic at Vg=15V.

As aforementioned, the charges in the p and n columns need to be perfectly matched in order to reach the highest breakdown voltage at certain n- and p-column doping levels. By introducing an oxide layer between the p and n columns, interdiffusion between the two regions during deposition and subsequent processing steps can be eliminated. Even so, controlling the doping in the p-poly-silicon to exactly satisfy the charge balancing is very challenging. It is useful to see how the charge imbalance in the p-poly-silicon would affect the cell breakdown voltage. Simulations were first conducted to evaluate the breakdown voltage variations with respect to the doping levels in the n-type silicon wafer while the doping in the p-poly-silicon column is perfectly matched. The charge imbalance scenarios, where the doping concentrations in the poly-silicon are taken as 5% or 10% above or below the perfect matching levels, are then simulated. It can be seen in Figure 8(a), when a perfect doping match is assumed, variations of doping in the wafer contribute little to the values of the breakdown voltage (around 2500V for a column depth of 160 μm) until a critical level is reached. Above this level, the breakdown voltage decreases drastically due to the fact that the p and n columns can no longer be fully depleted during reverse biasing. On the other hand, if the doping level in the p-poly-silicon does not match the wafer, the breakdown voltage decreases drastically as the doping in the wafer increases. A 10% imbalance leads to a bigger reduction in breakdown voltage than a 5% imbalance at the same wafer doping level. This result indicates process control for doping the p-poly-silicon is critical in creating an ultra-high voltage MOSFET with ultra-low on-resistance. The doping imbalance is more tolerable (less reduction in breakdown voltage) as the doping concentration in the wafer gets lower.

However, lowering the doping concentration in the wafer leads to higher on-resistance (as shown in Figure 8(b)). Consequently, if a certain level of charge imbalance in the p-poly-silicon is inevitable, the doping concentration in the wafer has to be chosen such that a trade-off between the breakdown voltage and the specific on-resistance can be reached. Even with a small amount of charge imbalance, a properly designed device can still have far better performance than a conventional one. For instance, starting from a wafer with a doping level of $10^{15} cm^{-3}$ and assuming a 5% charge imbalance in the p poly-silicon, a MOSFET with a breakdown voltage of 2100V and specific on-resistance of 410 mΩ·cm^2 can be produced. This is much better than the silicon limit of 1700 mΩ·cm^2 at this voltage for a conventional power MOSFET. Figure 3(c) shows the figures of merit (V_b^2/R_{on}A) for these scenarios.

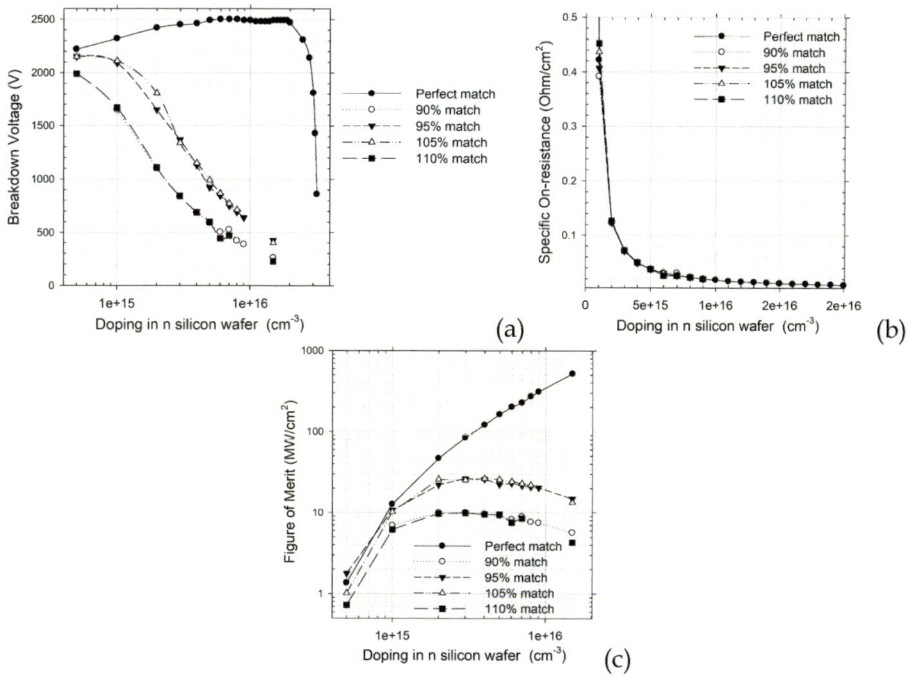

Fig.3. (8) Breakdown voltage, (b) specific on-resistance, and (c) figure of merit vs. doping concentration of silicon wafer for various charge imbalance conditions

It shows that a figure of merit of more than 500 MW/cm^2 is possible if perfect control of the p-poly-doping can be achieved. A small amount of charge imbalance greatly reduces the figure of merit. However, with a 5% charge imbalance, figures of merit between 10 and 30 MW/cm^2 are achievable for wafer doping levels between 10^{15} to 10^{16} cm^{-3}, which covers breakdown voltages between 2100 and 1000 V. This is still much better than the figures of merit of silicon limits for conventional power MOSFETs, which are between 2.5 and 4 MW/cm^2 at these voltage ratings.

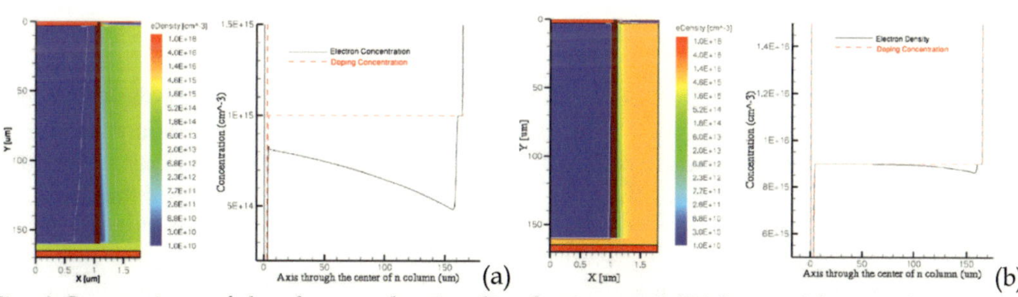

Fig. 4 Comparison of the electron density distributions at 0.5V forward bias in the unit cell and along the vertical axis through the center of n column between (a) wafer doping of $10^{15} cm^{-3}$ and (b) wafer doping of $9\times10^{15} cm^{-3}$. The oxide thickness between p and n column is $0.1\mu m$.

It is shown in Figure 3(b) that the specific on-resistance increases with decreasing wafer doping levels and it does not follow a linear relationship. Examination of the electron density under forward-bias condition reveals that the nonlinear increase of the specific on-resistance at lower wafer doping levels is due to the much more severe depletion in the n-column, or JFET (Junction Field-Effect Transistor) effect. For instance, Figure 4(a) and (b) show the comparison of the electron density distributions at a forward-bias of 0.5V for the device with the wafer doping levels of $10^{15} cm^{-3}$ and $9\times10^{15} cm^{-3}$ (baseline device), respectively. The figure on the left shows the electron distribution over the half-unit cell. The white line denotes the boundary of the depletion region. It can be seen that there is a depletion region in the n-column along the oxide interface in both cases. The width of the depletion region for a wafer doping of $10^{15} cm^{-3}$ is much larger than that for a wafer doping of $9\times10^{15} cm^{-3}$. The figure on the right shows both the electron density and doping concentration distributions along the center line of the n-column. It is clearly shown that the electron density in the n-column is significantly lower than the doping concentration for the case of low wafer doping of $10^{15} cm^{-3}$; on the other hand, the electron density is kept at the doping level for the case of high wafer doping of $9\times10^{15} cm^{-3}$. This explains why the on-resistance of the device with a wafer doping of 10^{15} cm^{-3} is 21 times larger than the one with a wafer doping of $9\times10^{15} cm^{-3}$, while the doping level in the n-column is only 9 times smaller.

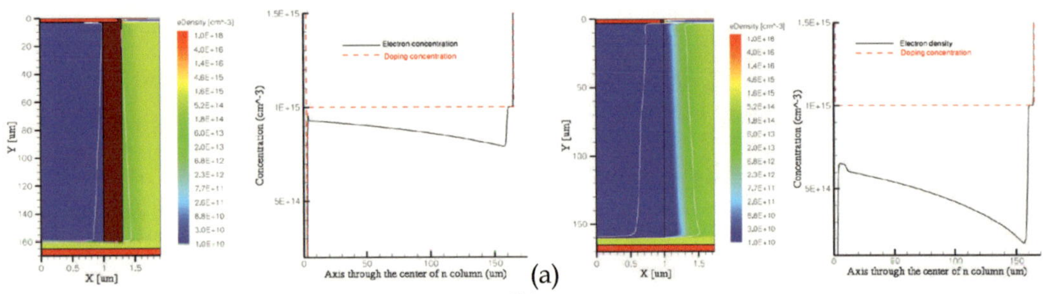

(a)

(b)

Fig. 5. Electron density distributions at 0.5V forward bias in the unit cell and along the vertical axis through the center of n column for: (a) the case where the oxide thickness between p and n column is 0.3 μm, and (b) the case of conventional SJ structure (no oxide between p and n columns). The wafer doping level is $10^{15} cm^{-3}$.

It can be said that the JFET effect is not a concern if high enough doping concentration is used in the n and p columns. However, as aforementioned, if a small amount of charge imbalance is inevitable, we have to reduce the wafer doping level in order to retain the high breakdown voltage. In this case, the JFET effect greatly reduces the on-resistance of the device. One way to partially alleviate this problem is to increase the thickness of the oxide layer between the p and n columns. As the oxide thickness increases, the electrical field near the oxide interface becomes smaller so that the depletion of the carriers in the nearby regions becomes smaller. For example, Figure 5(a) shows the electron density distribution for the device with a wafer doping level of $10^{15} cm^{-3}$ and an oxide thickness of 0.3 μm. Compared to Figure 4(a), it is clearly shown that by increasing the oxide thickness from 0.1 μm to 0.3 μm, the depletion width is reduced and the electron density during forward-bias is increased. This leads to lower specific on-resistance from 422 to 289 mΩ·cm^2, a 30% reduction. We have also simulated the case of a conventional SJ MOSFET, where the oxide layer is nonexistent. It turns out that, at the same wafer doping level of $10^{15}cm^{-3}$, the JFET effect is much more severe as shown in Figure 5(b); the specific on-resistance turns out to be 1084 mΩ·cm², almost 4 times that of the device with a 0.3 μm oxide layer. Further simulations on device breakdown have shown that adding an oxide layer between the p and n columns has almost no effect on the breakdown voltage. This clearly shows that putting an oxide layer between the p and n columns not only helps to prevent interdiffusion between the columns during processing but increases the performance of the device as well by reducing the JFET effect at lower wafer doping levels. This is important when charge imbalance is expected and wafer doping is purposely reduced to maintain the device breakdown voltage.

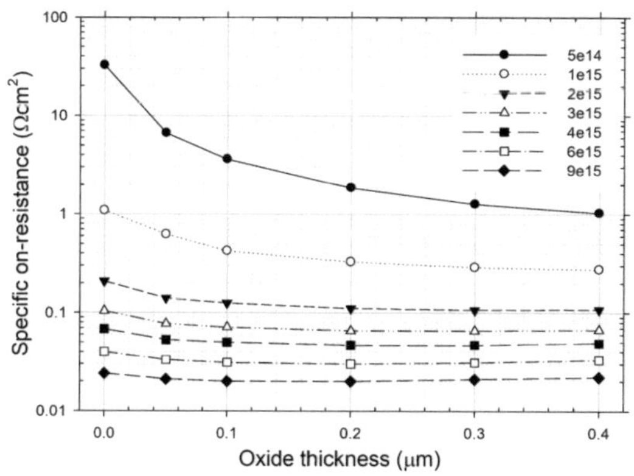

Fig. 6 On-resistance vs. oxide thickness for various wafer doping levels

Although a thicker oxide layer helps to reduce JFET effect, it does not carry any current like the p-column; therefore, it creates a "dead" region in the device unit cell. This means the oxide thickness needs to be optimized for each wafer doping level so that the lowest on-resistance can be achieved. Figure 6 shows how the oxide thickness affects the specific on-resistance of the device at various wafer doping levels. If a lower wafer doping level is to be used, a larger width of oxide thickness would be needed. As the wafer doping level increases toward its superjunction limit, the JFET effect becomes less of a concern and a thinner oxide layer gives better on-resistance.

4. Conclusions and Future Directions

This chapter describes a novel concept for a mid-to-high voltage power semiconductor switch that utilizes band-to-band tunneling and an avalanche injection mechanism. Numerical simulations suggest the TJE-MOSFET has better forward conduction characteristics than both IGBTs and power MOSFETs. The TJE-MOSFET can be very competitive to MOSFETs and IGBTs in mid power range applications.

A promising method of making high-voltage SJ MOSFETs without costly epitaxial growth is also described in this chapter. The charge compensating structures are created by etching the structured macro-pores, followed by passivating the walls and filling the pores with oppositely-charged poly-silicon. The effects of charge imbalance and the thickness of the passivation layer are studied by physically-based numerical device simulations. It is found that even with a small amount of charge imbalance, the proposed method can still produce high-voltage MOSFETs with much better performance than existing technology. A thick oxide layer between the p and n columns is found to be helpful of alleviating the JFET effects when the doping concentrations in the p and n columns are low. In comparison with a conventional superjunction structure, the inclusion of an oxide layer between p and n columns is found to help increase the device efficiency in addition to its ability to prevent dopant interdiffusion.

5. References

Aydin, C., Zaslavsky, A., Luryi, S., Cristoloveanu, S., Mariolle, D., Fraboulet, D., and Deleonibus, S. (2004). Lateral interband tunneling transistor in silicon-on-insulator. *Applied Physics Letters*. vol. 84, no. 10, pp. 1780-1782.

Canali, C., Majni, G., Minder, R., and Ottaviani, G. (1975). Electron and Hole Drift Velocity Measurements in Silicon and Their Empirical Relation to Electric Field and Temperature. *IEEE Transactions on Electron Devices*. vol. 22, no. 11, pp. 1045-1047.

Charlton, M.D.B. and Parker, G.J. (1997). Fabrication of High Aspect Ratio Silicon Microstructures by Anodic Etching. *Journal of Micromechanics and Microengineering*. vol. 7, pp.155-158.

Chen, X. (1993)Semiconductor power devices with alternating conductivity type high-voltage breakdown regions. *US Patent 5216275*.

Chen, Y., and Liang, Y. C. (2007). Design of Gradient Oxide-Bypassed Superjunction Power MOSFETE Devices. *IEEE Transactions on Power Electronics*, vol. 22, no. 4, pp. 1303-1310.

Coe, D. J. (1998). High voltage semiconductor device. *US Patent 4754310*.

Deboy, G., März, M., Stengle, J.-P., Strack, H., Tihanyi, J., and Weber, H. (1998). A new generation of high voltage MOSFETs breaks the limit line of silicon. *Technical Digest of IEDM 98*, Dec. 1998, pp. 683-685.

Fujihira, T. (1997). Theory of Semiconductor Superjunction Devices. *Japan Journal of Applied Physcis*, vol. 36, no.10, pp. 6254-6262.

Gan, K. P., Liang, Y. C., Samudra, G. S., and Yong, L. (2001). A Simple Technology for Superjunction Device Fabrication: Polyflanked VDMOSFET. *IEEE Electron Device Letters*, vol. 22, no. 8, pp. 407-409.

Hu, C. (1979). Optimum doping profile for minimum ohmic resistance and high-breakdown voltage. *IEEE Transactions on Electron Devices*, vol. 26, no. 3, pp. 243-244.

Klaassen, D. B. M. (1992). A unified mobility model for device simulation--I. Model equations and concentration dependence. *Solid-State Electron*. vol. 35, no. 7, pp. 953-959.

Klaassen, D. B. M. (1992). A unified mobility model for device simulation--II. Temperature dependence of carrier mobility and lifetime. *Solid-State Electron*. vol. 35, no. 7, pp. 961-967.

Kleimann, P., Badel, X., and Linnros, J. (2005). Toward the Formation of Three-Dimensional Nanostructures by Electrochemical Etching of Silicon. *Applied Physics Letters*. vol. 86, pp.183108.

Lackner, T. (1991). Avalanche Multiplication in Semiconductors: A Modification of Chynoweth's Law. *Solid-State Electron*. vol. 34, no. 1, pp. 33-42.

Lehmann, V., Honlein, W., Reisinger, H., Spitzer, A., Wendt, H., and Willer, J. (1996). A Novel Capacitor Technology Based on Porous Silicon. *Thin Solid Films*. vol.276, pp.138-142.

Liang, Y. C., Gan, K. P., and Samudra, G. S. (2001). Oxide-bypassed VDMOS (OBVDMOS): An alternative to superjunction high voltage MOS power devices. *IEEE Electron Device Letters*. vol. 22, no. 8, pp. 407-409.

Lombardi, C., Manzini, S., Saporito, A., and Vanzi, M. (1988). A physically based mobility model for numerical simulation of nonplanar devices. *Computer-Aided Design of Integrated Circuits and Systems, IEEE Transactions on*. vol. 7, no. 11, pp. 1164-1171.

Minato, T., Nitta, T., Uenisi, A., Yano, M., Harada, M., and Hine, S. (2000). Which is cooler, Trench or Multi-Epitaxy. in *Proc.of the 12th International Symposium on Power Semiconductor Devices and ICs, Toulous, France*: pp. 73-76, 2000.

Miura, Y., Ninomiya, H., and Kobayashi, K. (2005). High Performance Superjunction UMOSFETs with Split P-Colums Fabricated by Multi-Ion-Implantations. in *Proc.of the 17th International Symposium on Power Semiconductor Devices & IC's*, Santa Barbara, CA: pp. 1-4.

Onishi, Y., Iwamoto, S., Sato, T., Nagaoka, T., Ueno, K., and Fujihira, T. (2002). 24 m□cm2 680 V silicon superjunction MOSFET. in *Proc.of the 14th International Symposium on Power Semiconductor Devices and ICs*, pp. 241-244.

Rochefort, C., van Dalen, R., Duhayon, N., and Vandervorst, W. (2002). Manufacturing of high aspect-ratio p-n junctions using vapor phase doping for application in multi-Resurf devices. in *Proc.of the 14th International Symposium on Power Semiconductor Devices and ICs*, pp. 237-240, 2002.

Rub, M., Bar, M., Deboy, G., Niedernostheide, F. J., Schmitt, M., Schulze, H. J., and Willmeroth, A. (2004). 550V Superjunction 3.9mm2 Transistor Formed by 25 MeV Masked Boron Implantation. in *Proc.of the 16th International Symposium on Power Semiconductor Devices & IC's*, Kitakyushu, Japan: pp. 445-458.

Shenoy, P.M., Bhalla, A., and Dolny, G.M. (1999). Analysis of the Effect of Charge Imbalance on the Static and Dynamic Characteristics of the Super Junction MOSFET. *Proceedings of ISPSD'99*, May 1999, pp. 99-102.

Solomon, P. M., Frank, D. J., D'Emic, C., Dokumaci, O., Ronsheim, P., and Haensch, W. E., Universal tunneling behavior in technologically relevant P/N junction diodes. *Journal of Applied Physics*. vol. 95, no. 10, pp. 5800-5812.

Theunissen, M.J.J. (1972). Etch Channel Formation during Anodic Dissolution of N-Type Silicon in Aqueous Hydrofluoric Acid. *Solid-State Science and Technology*. vol.119, no. 3, pp.351-360.

von Borany, J., Friedrich, M., Rüb, M., Deboy, G., Butschke, J., and Letzkus, F. (2004). Application of Ultra-High Energy Boron Implantation for Superjunction Power (CoolMOS TM) Devices. *Nuclear Inst. and Methods in Physics Research, B*, vol. 237, no. 1-2, pp. 62-67.

W. Saito, I. Omura, S. Aida, S. Koduki, M. Izumisawa, H. Yoshioka, and T. Ogura, "Over 1000V Semi-Superjunction MOSFET with Ultra-Low On-Resistance Blow the Si-Limit," in Proc.of 17th International Symposium on Power Semiconductor Device & IC's, Santa Barbara, CA: 2005.

Wang, L., Nichelatti, A., Schellevis, H., de Boer, C., Visser, C., Nguyen, T.N., and Sarro, P.M. (2003). High Aspect Ratio Through-Wafer Interconnections for 3D-Microsystems. *Proc. of IEEE MEMS-03 Kyoto*, pp.634-637.

Ye, H. and Haldar, P., (2008). A MOS Gated Power Semiconductor Switch Using Band-to-Band Tunneling and Avalanche Injection Mechanism. *IEEE Trans. On Electron Devices*. Vol.55, No.6, pp.1524-1528.

Ye, H. and Haldar, P. (2008). Optimization of the Porous Silicon Based Superjunction Power MOSFET. *IEEE Trans. On Electron Devices.* Vol.55, No.8, pp.2246-2251.

10

The Critical Feedback Level in Nanostructure-Based Semiconductor Lasers

F. Grillot [(1)(2)], N. A. Naderi [(2)], M. Pochet [(2)], C.-Y. Lin [(2)] and L. F. Lester [(2)]
(1) Université Européenne de Bretagne, CNRS, Laboratoire Foton, INSA de Rennes, 20 Avenue des Buttes de Coësmes, 35708 Rennes, Cedex 7, France
(2) Center for High Technology Materials, The University of New Mexico, 1313 Goddard SE, Albuquerque, United States

1. Introduction

The extension of optical networks to local residential subscribers requires the development of extremely low-cost laser transmitter sources (Tohmori et al., 1997). While wafer fabrication allows for large-scale production, which drastically reduces the cost per laser, packaging remains a cost bottleneck, as it is not supported by parallel processing. Cost reduction must therefore be based on packaging simplification, such as flip-chip bonding and direct coupling of the laser into the fiber (Fernier et al., 1998). One of the characteristic problems of semiconductor lasers that complicates packaging is their sensitivity to external optical feedback. Fiber optic communication systems can be limited by unwanted optical feedback arising at fiber facets and junctions (Clarke, 1991), (Kitaoka et al., 1996). Although Faraday isolators are used extensively to reduce back reflections by as much as 60-dB, the elimination of the optical isolator remains a big challenge and is desirable because it leads to simplified packaging and will greatly reduce cost (Grillot et al., 2002). To understand the conditions under which the isolator can be eliminated, it is instructive to briefly review the physics of feedback in semiconductor lasers. This investigation naturally leads to highlighting the particular advantages of nanostructures in altering or improving the feedback resistance of the device.

The performance of a semiconductor laser can be strongly altered by external optical feedback. During early research, the importance of the distance between the laser facet and an external mirror reflector was pointed out in determining the nature of the semiconductor laser's response to optical feedback (Hirota & Suematsu, 1979). Even small reflections in the percent range were found to affect the laser stability dramatically. Although external optical feedback can be considered as a source of instability in many situations, it also can produce several beneficial effects that can improve laser performance. At the extremes of very weak and very strong optical feedback, linewidth narrowing and noise suppression can occur. This advantage, along with the large gain bandwidth of the semiconductor laser, can produce a highly tunable, narrow linewidth source that attracts many applications in spectroscopy, metrology and telecommunications (Kane & Shore, 2005).

Five distinct regimes based on spectral observation were reported for 1.55-µm distributed feedback (DFB) semiconductor lasers (Tkach & Chraplyvy, 1986). At the lowest feedback level, regime I, the laser operates on a single external cavity mode that emerges from the solitary laser mode. Depending on the phase of the feedback, the laser linewidth can be narrowed or broadened. Within regime II, the mode appears to split into two modes arising from rapid mode hopping. Noise-induced hopping between two external cavity modes is the underlying reason for this behavior. The transition to regime II has been shown to correspond to multiple solutions to the steady state equation that determines the frequency of the laser. This condition is satisfied when the parameter X_F is equal to unity in the following expression:

$$X_F = K_F \tau_e \sqrt{1+\alpha_H^2} \tag{1}$$

where τ_e is the external cavity roundtrip time while $K_F = (2C/\tau_i)\sqrt{R_p}$ is denoted as the feedback parameter. R_P is the feedback ratio and is defined as $R_P = P_1/P_0$ (with P_1 the power returned to the facet and P_0 the emitted power), C is the coupling coefficient from the laser facet to the external cavity, and τ_i is the internal roundtrip time within the laser cavity. The so-called linewidth enhancement factor (α_H-factor) corresponding to the coupling between the phase and the amplitude of the electric field is usually defined as follows:

$$\alpha_H = -\frac{4\pi}{\lambda}\frac{dn/dN}{dg/dN} = -\frac{4\pi\Gamma}{\lambda}\frac{dn/dN}{dG_{net}/dN} \tag{2}$$

where g is the material gain. The α_H-factor depends on the ratio of the evolution of the refractive index n with the carrier density N to that of the differential gain dg/dn. Γ is the optical confinement factor and $G_{net}=\Gamma g-\alpha_i$ is the net modal gain where α_i is the internal loss coefficient. The α_H-factor is used to distinguish the behavior of semiconductor lasers with respect to other types of lasers (Henry, 1982), and influences several fundamental aspects of semiconductor lasers, such as the linewidth (Su et al., 2004) and the laser behavior under optical feedback (Su et al., 2003). In regime III the laser re-stabilizes in a single external cavity mode (the lowest linewidth mode) with constant power. As the feedback level is further increased, the impact of optical feedback becomes independent of the external cavity length and the laser undergoes a transition to a chaotic state characterized by coherence collapse (CC) and denoted as regime IV (Lenstra et al., 1985). Coherence collapse or critical feedback is the common name given to describe the irregular dynamics occurring when the laser is operated above threshold, and has been greatly studied over the last twenty years. This regime has been described as co-existing chaotic attractors (Mork et al., 1988) and as an important source of noise (Mork et al., 1988), (Tromborg & Mork, 1990). The main consequence of the coherence collapse regime consists of a drastic collapse of the laser's coherence time leading to an enhancement of the laser's linewidth up to several gigahertz. For lasers used as an optical transmitter, the coherence collapse has been experimentally (Grillot et al., 2002) and theoretically (Clarke, 1991) demonstrated to cause a strong degradation in the bit error rate (BER). As the feedback level is further increased, the laser enters regime V, which is characterized by single-mode, constant intensity and narrow

linewidth operation. This regime can only be reached when laser diodes with antireflection coated facets are used.
The purpose of this chapter is to show both theoretically and experimentally that the variations of the above-threshold α_H-factor negatively impact the feedback level for which the coherence collapse regime occurs. The coherence collapse threshold is explored in nanostructure based semiconductor lasers, which in the context of this work encompasses quantum dot (QD), quantum dash (QDash), and quantum well (QW) structures. It is particularly well-known that QD and QDash based semiconductor lasers have attracted a lot of interest in the last decade owing to their expected remarkable properties arising from charge carrier confinement in the three space dimensions (Arakawa & Sakaki, 1982). Indeed, low threshold current densities and high material gain (Bimberg et al., 1997), (Liu et al., 1999), temperature insensitivity (Deppe et al., 2002), and reduced α_H-factor at the lasing wavelength (Saito et al., 2000), (Martinez et al., 2005) have been reported. This latter property combined with a high damping factor (O'Brien et al., 2003), (Erneux et al., 2008) was found to be of utmost importance because it should increase the tolerance to optical feedback in these devices (Azouigui et al., 2007), (Su et al., 2003) and may also offer potential advantages for direct modulation without transmission dispersion penalty.
However, in some cases, the above-threshold α_H-factor was experimentally found to be much larger as compared to traditional QW lasers ([a]Grillot et al., 2008), (Dagens et al., 2005). Such an enhancement is naturally not beneficial in practice for many reasons (Henry, 1982) and it provokes a rapid collapse of the laser's coherence time. Thus, considering the energy level contributions such as the ground state (GS) and the excited state (ES), this chapter shows that the analytical relation giving the onset of the critical feedback level can be rewritten. The carrier filling from the ES is found to produce an additional term, which accelerates the route to chaos. Also depending on how the above-threshold α_H-factor behaves, the critical feedback level can exhibit two different trends with output power. Thus, the influence of the ES coupled to the emphasized non-linear effects makes such devices more sensitive to optical feedback causing larger variations in the onset of the coherence collapse. These results highlight that the control of the α_H-factor has to be considered as a significant input for the realization of feedback-resistant lasers. It is also pointed out that the prediction of the onset of the coherence collapse remains an important feature for all applications requiring a low noise level or a proper control of the laser coherence.

2. Predicting the onset of the critical feedback regime

Different analytical models giving the onset of the coherence collapse have been derived over the last twenty years. This section aims to provide an overview of the most relevant relations estimating this critical feedback level.

2.1 Optical Power Transfer Function Analysis

This model was derived from the microwave modulation characteristics of laser diodes (Helms & Petermann, 1989). Starting from the conventional rate equations for a single-mode laser diode with optical feedback evaluated through small-signal analysis, it was shown that the modulation transfer function of a feedback laser can be expressed as:

$$H_K(j\omega_m)=(1-K(j\omega_m))\frac{H(j\omega_m)}{1-K(j\omega_m)H(j\omega_m)} \tag{3}$$

where ω_m is the modulation frequency. In (3), $H(j\omega_m)$ is the normalized transfer function of the laser diode in the absence of optical feedback given by:

$$H(j\omega_m)=\frac{1}{\left(j\frac{\omega_m}{\omega_r}\right)^2+\left(j\frac{\omega_m}{\omega_d}\right)+1} \tag{4}$$

ω_r and ω_d are the relaxation and damping resonance frequencies, respectively. Equation (3) is derived for the minimum linewidth mode due to its high degree of stability over other cavity modes (Shunk & Petermann, 1988). The modification of the solitary laser modulation response due to weak feedback and α_H>1 is found to be (Helms & Petermann, 1989):

$$K(j\omega_m)=j\frac{K_F}{\omega_m}\sqrt{1+\alpha_H^2}\left(1-e^{-j\omega_m\tau_e}\right) \tag{5}$$

Examination of (3) reveals that the system will be unstable due to a small perturbation if an unstable pole occurs. The existence of such a pole does not necessarily mean that coherence collapse occurs but the minimum feedback level at which an unstable pole occurs does correspond to the onset of the coherence collapse. As a result, considering the case of a long external cavity (e.g. $\omega_r\tau_e$>>1) as well as $\omega_m \approx \omega_r$ and $\omega_r << \omega_d$, the unstable pole occurs in (3) when $K(j\omega_m)H(j\omega_m)=1$. This condition leads to an analytical expression for the onset of coherence collapse expressed by:

$$R_{Pc}=\left(\gamma\frac{\tau_i}{4C}\right)^2 f(\alpha_H) \tag{6}$$

with $\gamma=\omega_r^2/\omega_d$ the damping factor in rad/sec and $f(\alpha_H)=1/\left(1+\alpha_H^2\right)$. Expression (6) depends only on the solitary laser response and the α_H-factor. There is no explicit dependence on the external cavity length since a long-cavity asymptotic assumption was used in the derivation. For the case of a short external cavity (e.g. $\omega_r\tau_e$<<1), the onset of the coherence collapse is found to be dependent on the external cavity length as demonstrated in (Shunk & Petermann, 1989). In validating equation (6) against numerical simulations of an external cavity laser diode, it was shown that the expression deviated from the numerical results for low α_H-factor values. Consequently, the expression for the critical feedback level was found empirically driven by the same expression but instead $f(\alpha_H)$ is replaced by $g(\alpha_H)$, which is expressed as: $g(\alpha_H)=(1+\alpha_H^2)/\alpha_H^4$. This relation theoretically predicts that coherence collapse does not occur if $\alpha_H \to 0$, as explained through Henry and Kazarinov's ellipse whose eccentricity decreases with the α_H-factor (Henry & Kazarinov, 1986). For the case where α_H=0, all the fixed points describing the stability of the system (*modes* and *antimodes*) are located on a circle around the solitary laser mode. Under this condition, the minimum gain mode and the minimum linewidth mode are the same and do not compete with each other and thereby coherence collapse cannot occur (Henry & Kazarinov, 1986). Despite a general agreement that the critical feedback level may be strongly up-shifted for low α_H-factors ([a]Grillot et al., 2008), the complete suppression of the critical feedback level has never been reported. This last condition has been particularly expected for QD and QDash lasers because of their remarkable properties arising from charge carrier confinement in three spatial dimensions (Arakawa & Sakaki, 1982) and lower α_H-factors (Bimberg et al., 1997), (Newell et al., 1999).

2.2 External Cavity Mode Analysis

Based on the Lang and Kobayashi rate equations (Lang & Kobayashi, 1980) in the presence of optical feedback, the variation of the angular frequency induced by external optical feedback can be written as:

$$\Delta\omega\tau_i = (\omega - \omega_0)\tau_i = -2C\sqrt{R_P}\sqrt{1+\alpha_H^2}\sin\left(\varphi + \tan^{-1}(\alpha_H)\right) \qquad (7)$$

with φ the phase of the feedback wave and ω_0 the pulsation of the free-running laser. (Henry & Kazarinov, 1986) demonstrated that the onset of the coherence collapse occurs when the feedback parameter K_F, given in (1), is comparable to the relaxation frequency ω_r. Thus, based on (7), the onset of the coherence collapse can be determined when the frequency shift is maximized ($\Delta\omega_{\max}\tau_i = 2C\sqrt{R_{PC}}\sqrt{1+\alpha_H^2}$) for a certain critical feedback level. A strong increase in the optical power spectrum at $\omega_0 \pm \omega_r$ marks the feedback level defining the transition to the coherence collapse regime when $\Delta\omega_{\max} \geq \omega_r$. As a result, the onset of coherence collapse can be calculated via the following relationship (Binder & Cormarck, 1989):

$$R_{Pc} = \left(\omega_r \frac{\tau_i}{2C}\right)^2 f(\alpha_H) \qquad (8)$$

Although this equation is based on a model considering only the steady-state solutions under optical feedback, it was shown to agree with predictions made in (Shunk & Petermann, 1988) and (Henry & Kazarinov, 1986). Compared with (6), this new expression does not include the damping rate, which has been found to be much larger in QD and QDash based devices compared to QW ones (O'Brien et al. 2003), ([b]Grillot et al., 2008).
Another expression similar to (8) but with $(2)^{1/2}$ instead of a 2 in the numerator was derived by analyzing the stability of the oscillation condition solutions for a laser under optical feedback (Mork et al., 1992). Coherence collapse is seen as a chaotic attractor and that chaos is reached for increasing feedback levels through a quasi-periodic route interrupted by frequency locking. For a long external cavity ($\omega_r\tau_e$>>1), this new expression provides an approximation to the value of the feedback parameter K_F at which instability sets in.
Because the two approaches discussed in this section are highly similar, (8) can be rewritten as:

$$R_{P_c} = \left(\omega_r \frac{\tau_i}{pC} \right)^2 f(\alpha_H) \tag{9}$$

with p=2 or $(2)^{1/2}$ depending on the model under consideration.

2.3 Mode Competition Theory

The generation mechanism for excess intensity noise due to optical feedback has been analyzed theoretically and experimentally in (Yamada & Suhara, 1990). Modal rate equations under the weakly coupled condition with external optical feedback were derived which account for the mode competition phenomena found in Distributed Feedback (DFB) and Fabry-Perot (FP) lasers. The critical optical feedback level was calculated by confirming the build-up of external cavity modes and expressed as follows (Yamada & Suhura, 1990), (Suhura & Yamada, 1993):

$$R_{P_c} = 22R \left[\frac{nL}{l(1-R)} \right]^2 = \frac{11}{2C^2} \left(\frac{nL}{l} \right)^2 \tag{10}$$

with L and l the lengths of the laser cavity and the external cavity, respectively, n the refractive index and R the reflectivity of the emitting facet. The critical feedback level was found to be inversely proportional to the distance squared (l^2) to the reflection point. This relation, which holds only for FP lasers, was used to analyze the generation of the excess noise on the basis of mode competition theory and calculated the minimum noise level in the presence of external optical feedback. Calculations have shown that external cavity modes build up above a critical level of external optical feedback and result in excess noise due to the competition between external cavity modes and lasing modes. This method does not account for variations in the bias current, limiting its applicability for devices that have characteristic parameters that vary as the bias current is changed.

3. Recent advances in nanostructure-based semiconductor lasers

Low cost, directly modulated lasers will play a major role in the next generation telecommunication links (Local and Metropolitan Area Networks) for uncooled and isolator-free applications. Materials grown on an InP substrate have generally been the reference systems for 1.55-µm applications. Actual laser sources which are based on the InGaAsP/InP QW active material exhibit typical threshold currents lower than 10mA and a characteristic temperature T_0 of ~ 60 K (20-80 °C) (Thijs et al., 1994), (Phillips et al., 1999). More recently, GaInAlAs/InP QW lasers have shown an improved T_0 of ~ 90 K from 20 to 120 °C owing to a higher conduction band offset. Lasers realized in this material system allowed data transmission over a 40 km fiber span at 10Gbps up to 80°C (Stegmuller et al., 1993), (Makino et al., 2005). Temperature-insensitive lasers fabricated on GaAs are very attractive for low-cost high volume production due to a reduced cost of GaAs wafers and a mature technology. Two approaches have been intensively investigated in the last decade namely InAs QDs and GaInAsN QW emitting in the 1.3-µm window. GaInAsN(Sb) alloys grown on GaAs substrates extend the emission wavelength further into the 1.55-µm band.

This material system offers high electron confinement, a prerequisite for high T_0, and a high differential gain. A recent result of GaInAsNSb/GaAs QWs emitting at 1.55-µm shows a relatively low threshold current density of 579 A/cm² with propagation losses of ~ 4.8 cm^{-1} (Bank et al., 2006). However, due to a lower material gain, typical threshold currents of ridge waveguide lasers amount to 60mA (Gupta et al., 2006). The less mature material of GaInAsNSb development obviously needs further improvements. Many efforts have been devoted to the GaAs-based QD material system for emission in the 1.3-µm band, owing to a better material maturity (Bimberg et al., 1997), (Liu et al., 1999), (Deppe et al., 2002), (Saito et al., 2000), (Martinez et al., 2005), which allowed the demonstration of temperature insensitive 10Gbit/s transmission up to 85°C (Gerschütz et al., 2006), (Dagens et al., 2006). However for long-haul telecomm applications, lasers emission at 1.55-µm is mandatory. Two different approaches can match this application: metamorphic QDs grown on GaAs and QDs grown on InP substrates. Laser emission up to 1.45-µm has recently been demonstrated using metamorphic InAs/GaAs QD layer. The typical threshold current density equals ~ 1-1.5 kA/cm², and the characteristic temperature is ~65K (20-85 °C) (Karachinsky et al., 2005). More recently, coupled QD-QW tunnel injection lasers present a very low threshold current density of 63 A/cm² and an interesting 3-dB modulation bandwidth of ~ 8 GHz (Mi et al., 2006). But further extension of the emission wavelength at 1.55-µm on GaAs remains an issue. So far, only QD-based active layers grown on InP have allowed emission in the telecommunication window. Growth of QDs on (100) substrates has already been demonstrated using Molecular Beam Epitaxy (MBE) (Kim et al., 2004), or Molecular Organic Vapor Phase Epitaxy (MOVPE) (Franke et al., 2001). Elongated dots, or so-called QDashs have also been obtained by MBE (Wang et al., 2001), (Deubert et al., 2005) which have led to the demonstration of high performance lasers (Lelarge et al., 2007). An alternative approach is the use of the InP(311)B wafer orientation, which has also enabled demonstration of 3D confined nanostructures with a QD density as high as 10^{11} cm^{-2} (Caroff et al., 2005) with as well as low chirp of a 2.5Gbps directly-modulated single mode waveguide laser emitting at 1.6-µm (Saito et al., 2001). As an example, fig. 1 represents 1×1-µm² atomic force microscopy (AFM) images of QDs and QDashs grown on InP(311)B and InP(001) substrates, respectively. Let us note that very recently, dynamic properties of InAs/InP(311B) QD lasers emitting on the GS at 1.52-µm have been reported. The α_H-factor is found to be as low as 1.8 at the gain peak just below threshold and increases to about 7 above threshold but remains constant with the current. The rather high value is attributed to band filling of the thick wetting layer high degeneracy states. The sole emission from the GS at high current and at a high temperature of 75°C as well as a distinct relaxation oscillation peak in the frequency modulation response indicate the absence of phonon relaxation bottleneck originating from an ES (Martinez et al., 2008), (cGrillot et al., 2008).

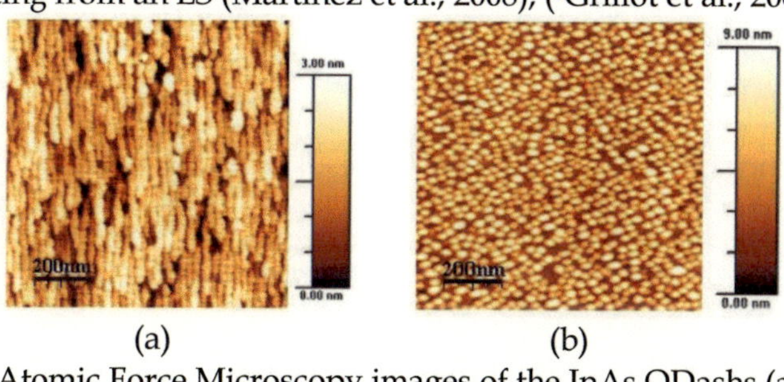

(a) (b)

Fig. 1. (a) 1x1-µm² Atomic Force Microscopy images of the InAs QDashs (a) and (b) QDs.

4. Results and discussion

This section gives the experimental results on both the static and the dynamic characteristics of the semiconductor lasers under study.

4.1 Device description

The device structure shown in fig. 2 was grown on an n^+-InP substrate. The active region is a dot-in-well (DWELL) structure consisting of 5 layers of InAs QDashs embedded in compressively-strained $Al_{0.20}Ga_{0.16}In_{0.64}As$ QWs which are separated by 30-nm un-doped tensile-strained $Al_{0.28}Ga_{0.22}In_{0.50}As$ spacers. Each side of the active region is surrounded by 105-nm waveguide layers of lattice-matched $Al_{0.30}Ga_{0.18}In_{0.52}As$. The p-cladding layer is step-doped AlInAs with a thickness of 1.5-µm to reduce free carrier loss and the n-cladding is a 500-nm thick layer of AlInAs. The laser structure is capped with a 100-nm InGaAs layer. Processing consisted of patterning a four-micron wide ridge waveguide with a 500-µm cleaved cavity length.

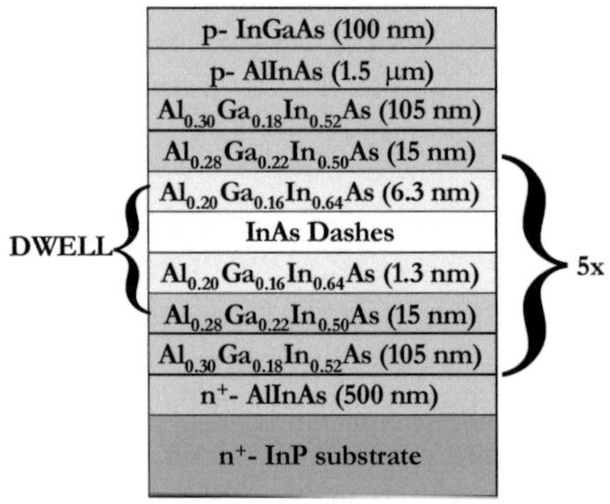

Fig. 2. Layer structure of the InAs/InP QDash Fabry-Perot (FP) device fabricated with the DWELL technology

The threshold current leading to a GS-emission is ~45mA and the external differential efficiency is about 0.2W/A. Beyond a pump current of ~100mA, ES lasing emission occurs. Fig. 3 shows the light-current characteristic measured at room temperature. As observed in fig. 3, the onset of ES lasing leads to a kink in the light-current characteristics as well as a modification of the slope efficiency (Veselinov et al., 2007)

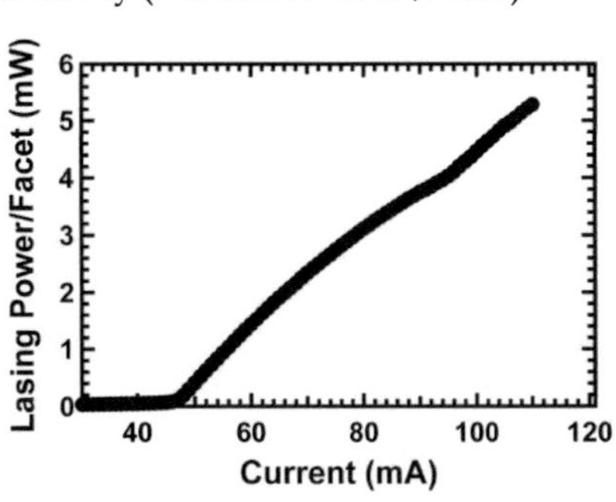

Fig. 3. The light current characteristic of the QDash FP laser under study

4.2 Effective gain compression

Conventional small-signal analysis of the semiconductor laser rate equations leads to a damped oscillator solution that is characterized by a relaxation frequency and an associated damping rate. To account for saturation of the optical gain generated by the semiconductor media with the photon density in the cavity, it is common to include a so-called gain compression term as well (Coldren & Corzine, 1995). Measuring the frequency response as a function of the output power is a common method to evaluate gain compression in semiconductor lasers. In the case of the QD laser, it has been shown that effects of gain compression are more important than those measured on QW devices (Su et al., 2005), (Su & Lester, 2005). In order to explain this phenomenon, a modified nonlinear gain coefficient has been introduced leading to a new expression for the relaxation frequency under strong gain saturation (Su & Lester, 2005):

$$f_r^2 = \frac{v_g a S}{4\pi^2 \tau_p (1+\varepsilon_S S)} \approx \frac{v_g a_0 S}{4\pi^2 \tau_p (1+\varepsilon_{Seff} S)} \tag{11}$$

with v_g being the group velocity, a the differential gain, a_0 the differential gain at threshold (unsaturated value), S the photon density, τ_p the photon lifetime, ε_S the gain compression factor related to the photon density and ε_{Seff} the effective gain compression factor which is defined as:

$$\varepsilon_{Seff} = \varepsilon_S \frac{1}{1-\frac{g_{th}}{g_{\max}}} \tag{12}$$

where g_{th} is the gain at threshold and g_{max} is the maximum gain for GS-lasing. Equation (12) indicates that the gain compression is enhanced due to gain saturation by a factor of $g_{max}/(g_{max}-g_{th})$. In Fig. 4 the evolution of the normalized gain compression $\varepsilon_{Seff}/\varepsilon_S$ is plotted as a function of the ratio g_{max}/g_{th}. This shows that the higher the ratio g_{max}/g_{th} the lower the effects of gain compression. If $g_{max}>>g_{th}$ the graph tends to an asymptote such that $\varepsilon_{Seff}/\varepsilon_S \rightarrow 1$. For cases where $g_{max}\rightarrow g_{th}$, gain compression effects are strengthened: the ratio increases drastically and can be extremely large if not enough gain is provided within the structure ($g_{max} \approx g_{th}$). As an example, for the QD laser under study, $g_{max}/g_{th} \approx 2$ meaning that the effects of gain compression are doubled causing critical degradation to the laser bandwidth.

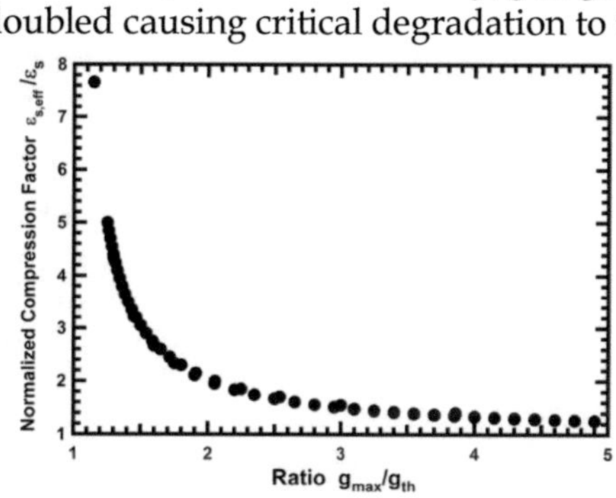

Fig. 4. Normalized compression factor as a function of g_{max}/g_{th}

Applying this same theory to the case of the QDash laser, the square of the measured resonance frequency is plotted in fig. 5 as a function of the output power, which is linked to the photon density through the relation $P = h\nu V v_g \alpha_m S$ with $h\nu$ the energy per photon, V the cavity volume and $\alpha_m v_g$ the energy loss through the mirrors, α_m being the mirror loss. The experimental dependence of the relaxation oscillation frequency shows a deviation from the expected proportionality given by expression (11) (case with $\varepsilon_S = 0$) on the square root of the optical output power. Thus, the experimental trend depicted in fig. 5 for the QDash laser is modelled via the following relation (Su & Lester, 2005):

$$f_r^2 = \frac{AP}{1+\dfrac{P}{P_{sat}}} = \frac{AP}{1+\varepsilon_P P} \tag{13}$$

The curve-fit based on equation (13) is used to express the gain compression in terms of a saturation power such that $\varepsilon_S S = \varepsilon_P P = P/P_{sat}$ with ε_P the gain compression coefficient related to the output power P. The value of P_{sat} is indicative of the level of output power where nonlinear effects start to be significant. For the QDash device under test, the curve-fit leads to a $P_{sat} < 17$mW and a gain compression coefficient of approximately $\varepsilon_P = 1/P_{sat} \approx 0.06$ mW^{-1}. The maximum of the resonance frequency can be directly deduced from the curve-fitting as $\Omega_r = (AP_{sat})^{1/2}$ and was expected to be ~7.6GHz. Taking into account the facet reflectivity as well as the modal volume of the laser, the order of magnitude for the gain compression factor ε_S is in the range of 5×10^{-15} cm^3 to 1×10^{-16}cm^3 which is much larger than the typical values measured on conventional QW lasers (typically around 10^{-17} cm^3) (Petermann, 1988).

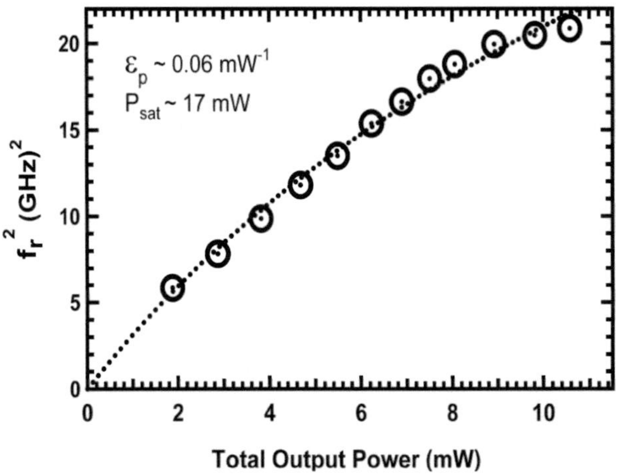

Fig. 5. The square of the resonance frequency versus the output power (open circles)

In fig. 6, the evolution of the damping rate against the relaxation frequency squared leads to a K-factor of 0.45ns as well as an effective carrier lifetime of γ_N^{-1}=0.16ns. The maximum intrinsic modulation bandwidth $f_{\max} = 2\pi\sqrt{2}/K$ is 19.7GHz. This f_{max} is never actually achieved in the QDash laser because of the aforementioned gain compression and the short effective carrier lifetime.

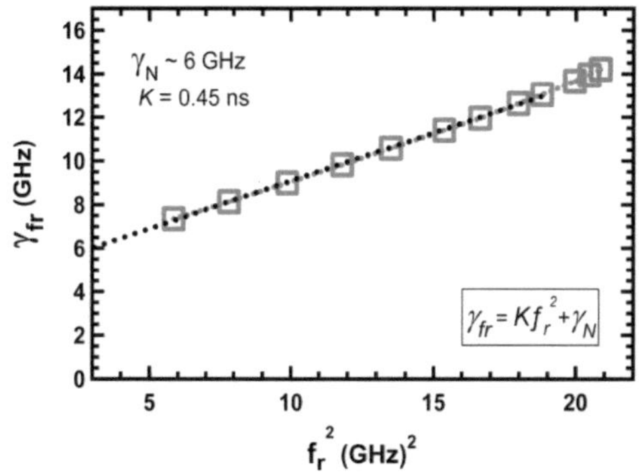

Fig. 6. The damping factor versus the square of the relaxation frequency

4.3 On the above-threshold α_H -factor

The above-threshold GS α_H-factor was measured using the injection locking (IL) technique, which is based on the asymmetry of the stable locking region over a range of detuning on both the positive and negative side of the locked mode (Liu et al., 2001). Using the IL technique, the α_H-factor can be determined using the following relationship:

$$\alpha_H = \sqrt{\left(\frac{\Delta\lambda_+}{\Delta\lambda_-}\right)^2 - 1} \tag{14}$$

where $\Delta\lambda = \Delta\lambda_{master} - \Delta\lambda_{slave}$, and $\Delta\lambda_{+/-}$ reflects the master's wavelength being either positively or negatively detuned with respect to the slave's wavelength. The ratio of $\Delta\lambda_+/\Delta\lambda_-$ should theoretically remain the same for any value of side mode suppression ratio (SMSR), which was kept at 35-dB for this measurement. The measured GS α_H-factor as a function of bias current is depicted in fig. 7. It was observed that the GS α_H-factor increased from ~1.0 to ~14 as the bias current was increased from the threshold value to 105mA. This enhancement is mostly attributed to the plasma effect as well as to the carrier filling of the non-lasing states (Wei & Chan, 2005), which results in a differential gain reduction above threshold. This strong degradation of the GS α_H-factor with the bias current produces a significant variation in the feedback sensitivity of the laser.

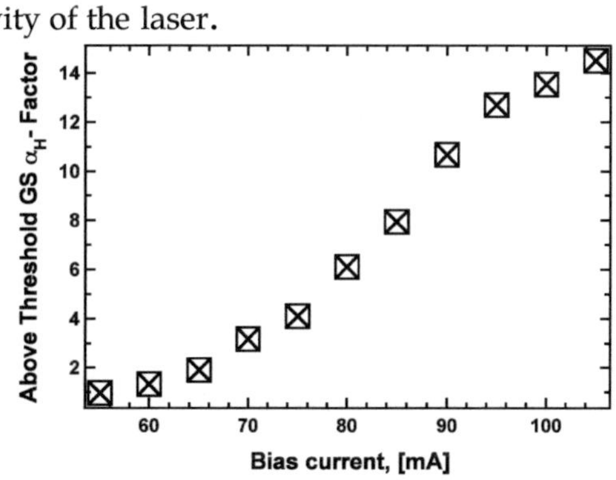

Fig. 7. The above threshold GS α_H-factor of the QDash laser versus the bias current measured by the injection-locking method.

On one hand, in QW lasers, which are made from a homogeneously broadened gain medium, the carrier density and distribution are clamped at threshold. As a result, the change of the α_H-factor is due to the decrease of the differential gain from gain compression and can be expressed as (Agrawal, 1990):

$$\alpha_H = \alpha_0(1+\varepsilon_P P) \tag{15}$$

where α_0 is the linewidth enhancement factor at threshold. Since the carrier distribution is clamped, α_0 itself does not change as the output power increases. As an example, fig. 8 shows the measured α_H-factor versus the output power for a 300-µm-long AR/HR coated DFB laser made from six compressively-strained QW layers. The threshold current is ~8mA at room temperature for the QW DFB device. Black squares correspond to experimental data. As described by equation (15), the effective αH-factor linearly increases with the output power to about 4.3 at 10mW. By curved-fitting the data in fig. 7, the α_H-factor at threshold is found to be around 4 while the gain compression coefficient equals ~3x10^{-2} mW^{-1}. Compared to QD or QDash lasers, such a value of the gain compression coefficient is much smaller leading to a higher saturation power, which lowers the enhancement of the effective α_H-factor over the range of output power. It is worthwhile noting that modifying the laser's rate equations and including the effects of intraband relaxation, (15) can be reexpressed as follows (Agrawal, 1990):

$$\alpha_H = \alpha_0(1+\varepsilon_P P) + \beta\varepsilon_P P\,\frac{(1+\varepsilon_P P)}{(2+\varepsilon_P P)} \tag{16}$$

with β the parameter related to the slope of the linear gain which controls the nonlinear phase change. The situation for which β=0 corresponds to an oscillation purely located at the gain peak. For most cases, the second part of (16) usually remains small enough to be neglected.

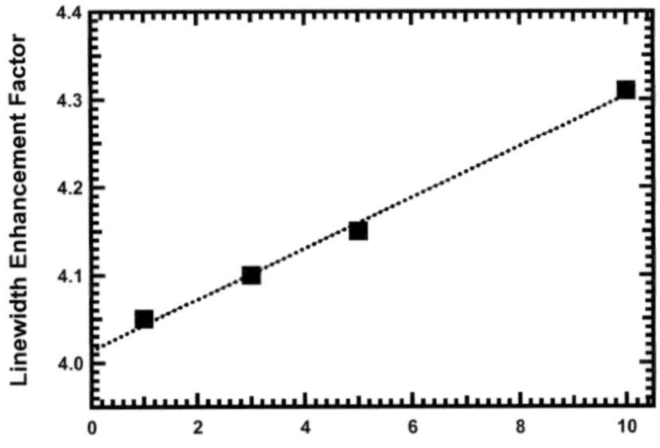

Fig. 8. The effective linewidth enhancement factor α_H as a function of the output power for the QW DFB laser.

On the other hand, in QD or QDash lasers, the carrier density and distribution are not clearly clamped at threshold. As a consequence of this fact, the lasing wavelength can switch from GS to ES as the current injection increases meaning that a carrier accumulation occurs in the ES even though lasing in the GS is still occurring. The filling of the ES inevitably increases the α_H-factor of the GS and introduces an additional dependence with the injected current. Thus taking into account the gain variation at the GS and at the ES, the index change at the GS wavelength can be written as follows:

$$\delta n = \sum_{k=GS,ES} \alpha_k \delta g_k \tag{17}$$

with k being the index of summation for GS and ES respectively. Equation (2) leads to:

$$\delta n = \left(\alpha_{ES} \frac{a_{ES}}{a_{GS}} + \alpha_{GS} \right) \delta g_{GS} = \alpha_H \delta g \tag{18}$$

In equation (18), δg and δn are the changes of the gain and refractive index at the GS, respectively, αH is the linewidth enhancement factor actually measured in the device and evaluated at the GS-wavelength, a_{ES} and a_{GS} are the differential gain values at the ES and at the GS, respectively, α_{ES} describes the change of the GS index caused by the ES gain and α_{GS} is related to the GS index change caused by the GS gain variation. When the laser operates above threshold, α_{GS} keeps increasing with $\alpha_{GS}(1+\varepsilon_P P)$ as previously shown for the case of QW devices.

The gain saturation in a QD media can be described by the following equation (Su et al., 2005):

$$g_{GS} = g_{\max} \left[1 - e^{-\ln(2)\left(\frac{N}{Ntr} - 1 \right)} \right] \tag{19}$$

with N the carrier density and N_{tr} the transparency carrier density. When the laser operates above threshold, the differential gain for the GS lasing is defined as follows:

$$a_{GS} = \frac{dg_{GS}}{dN} = \frac{\ln(2)}{N_{tr}} \left(g_{\max} - g_{GS} \right) \tag{20}$$

with $g_{GS}=g_{th}(1+\varepsilon_P P)$ the uncompressed material gain increasing with the output power. Equation (19) leads to:

$$a_{GS} = a_0 \left(1 - \frac{g_{th}}{g_{\max} - g_{th}} \varepsilon_P P \right) = a_0 \left(1 - \frac{g_{th}}{g_{\max} - g_{th}} \varepsilon_S S \right) \tag{21}$$

with a_0 the differential gain at threshold. Then using equations (15), (18) and (21), the linewidth enhancement factor can be analytically written as:

$$\alpha_H(P)=\alpha_1(1+\varepsilon_P P)+\frac{\alpha_0}{1-\frac{g_{th}}{g_{\max}-g_{th}}\varepsilon_P P} \tag{22}$$

with $\alpha_1 \equiv \alpha_{GS}$ and $\alpha_0=\alpha_{ES}(a_{ES}/a_0)$. The first term in (22) denotes the gain compression effect at the GS (similar to QW lasers) while the second is the contribution of the carrier filling from the ES that is related to the gain saturation in the GS. For the case of strong gain saturation or lasing on the peak of the GS gain, equation (21) can be reduced to:

$$\alpha_H(P)=\frac{\alpha_0}{1-\frac{g_{th}}{g_{\max}-g_{th}}\varepsilon_P P} \tag{23}$$

In fig. 9, the normalized linewidth enhancement factor α_H/α_0 is calculated through equation (23) and represented in the (X,Y) plane with $X = P/P_{sat}$ and $Y = g_{max}/g_{th}$. This graph serves as a stability map and simply shows that a larger maximum gain is absolutely required for a lower and stable α_H / α_0 ratio. For instance let us consider the situation for which $g_{max} = 3g_{th}$: at low output powers i.e, $P < P_{sat}$, the normalized α_H-factor remains constant ($\alpha_H/\alpha_0 \sim 3$) since the gain compression is negligible. As the output power approaches and goes beyond P_{sat}, the ratio α_H / α_0 is increased. Gain compression effects lead to an enhancement of the normalized α_H-factor, which can go up to 10 for $P \approx 2P_{sat}$ level of injection for which the ES occurs.

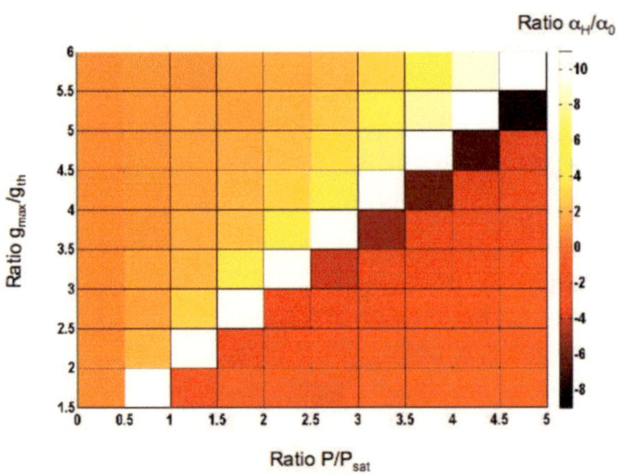

Fig. 9. Stability map based for the normalized linewidth enhancement factor α_H/α_0 in the (P/P_{sat}, g_{max}/g_{th}) plane.

Assuming that $g_{max} = 5g_{th}$, fig. 9 shows that the effects of gain compression are significantly attenuated since the ratio α_H / α_0 remains relatively constant over a wider range of output power. The level at which gain compression starts being critical is now shifted to $P \approx 3P_{sat}$ instead of $P \approx P_{sat}$. It is also important to note that at a certain level of injection, the normalized GS α_H-factor can even become negative. This effect has been experimentally reported in (Dagens et al., 2005) and occurs when the GS gain collapses, e.g when ES lasing occurs.

In fig. 10, the calculated GS α_H-factor (black dots) of the QD-laser from (Dagens et al., 2005) is depicted as a function of the bias current. Red stars superimposed correspond to data measurements from (Dagens et al., 2005) which have been obtained via the AM/FM technique. This method consists of an interferometric method in which the output optical signal from the laser operated under small-signal direct modulation is filtered in a 0.2nm resolution monochromator and sent in a tunable Mach-Zehnder interferometer. From separate measurements on opposite slopes of the interferometer transfer function, phase and amplitude deviations are extracted against the modulating frequency, in the 50MHz to 20GHz range (Sorin et al., 1992). The α_H-factor is given by the phase to amplitude response ratio at the highest frequency within the limits of the device modulation bandwidth. Fig. 10 shows a qualitative agreement between the calculated values and the values experimentally obtained. As expected, the GS α_H-factor increases with the injected current due to the filling of the excited states as well as carrier filling of the non-lasing states (higher lying energy levels such as the wetting layer). Although the α_H-factor is lowered at lower output powers, its increase with bias current stays relatively limited as long as the bias current remains lower than 150mA, e.g. such that $P<P_{sat}$. Beyond P_{sat}, compression effects become significant, and the α_H-factor reaches a maximum of 57 at 200mA before collapsing to negative values. As previously mentioned, the collapse in the α_H-factor is attributed to the occurrence of the ES as well as to the complete filling of the available GS states. In other words, as the ES stimulated emission requires more carriers, it affects the carrier density in the GS, which is significantly reduced. As a result, the GS α_H-factor variations from 57 down to -30 may be explained through a modification of the carrier dynamics such as the carrier transport time including the capture into the GS. This last parameter affects the modulation properties of high-speed lasers via a modification of the differential gain. These results are of significant importance because they show that the α_H-factor can be controlled by properly choosing the ratio g_{max}/g_{th}: the lower g_{th}, the higher g_{max}, and the smaller the linewidth enhancement factor. A high maximum gain can be obtained by optimizing the number of QD layers in the laser structure while gain at threshold is directly linked to the internal and mirror losses. Both g_{th} and g_{max} should be considered simultaneously so as to properly design a laser with a high differential gain and limited gain compression effects. The g_{max}/g_{th} ratio is definitely the key-point in order to obtain a lower α_H-factor for direct modulation in QD and QDash lasers.

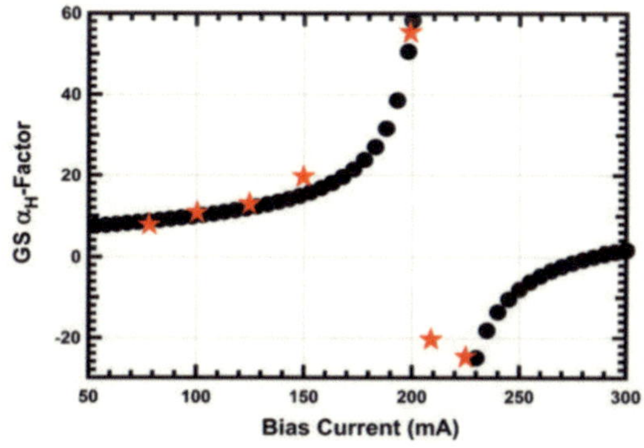

Fig. 10. Calculated GS α_H-factor for a QD laser versus the bias current (black dots). Superimposed red stars correspond to experimental data from (Dagens et al., 2005)

5. Optical feedback sensitivity

This sections aims to investigate the laser's feedback sensitivity by using different analytical models. Also the impact of the α_H-factor on the feedback degradation is carefully studied.

5.1 Description of the optical feedback loop

The experimental apparatus to measure the coherence collapse threshold is depicted in fig. 11. The setup core consists of a 50/50 4-port optical fiber coupler. Emitted light is injected into port 1 using a single-mode lensed fiber in order to avoid excess uncontrolled feedback. The optical feedback is generated using a high-reflectivity dielectric-coated fiber (> 95%) located at port 2. The feedback level is controlled via a variable attenuator and its value is determined by measuring the optical power at port 4 (back reflection monitoring). The effect of the optical feedback is analyzed at port 3 via a 10pm resolution optical spectrum analyzer (OSA). A polarization controller is used to make the feedback beam's polarization identical to that of the emitted wave in order to maximize the feedback effects. The roundtrip time between the laser and the external reflector is ~30ns. As a consequence, the long external cavity condition mentioned in the previous section $\omega_r \tau_e$>> 1 is fulfilled.

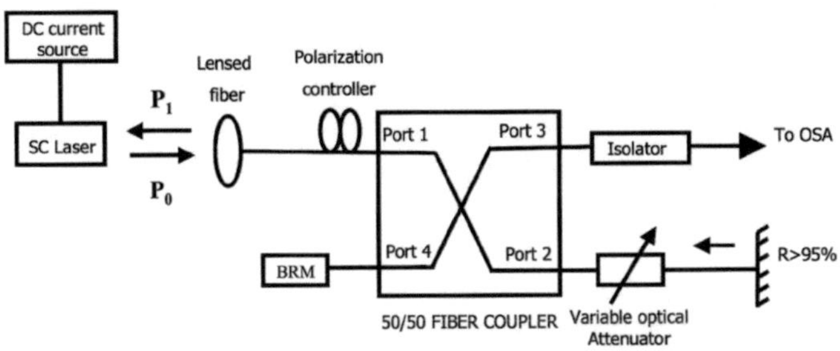

Fig. 11. Schematic diagram of the experimental apparatus for the feedback measurements

The long external cavity condition means that the coherence collapse regime does not depend on the feedback phase nor the external cavity length. Thus, in order to improve the accuracy of the measurements at low output powers, an erbium-doped-fibre-amplifier (EDFA) was used with a narrow band filter to eliminate the noise. The EDFA is positioned between the laser facet and the polarization controller (not shown in fig. 11). As already stated in section 1, the amount of injected feedback into the laser is defined as the ratio $R_{PdB} = 10\log(P_1/P_0)$ where P_1 is the power returned to the facet and P_0 the emitted one. The amount of reflected light that effectively returns into the laser can then be expressed as follows (Su et al., 2003):

$$R_{P\,dB} = P_{BRM} - P_0 + C \quad (24)$$

where P_{BRM} is the optical power measured at port 4, C is the optical coupling loss of the device to the fiber which was estimated to be about -4dB and kept constant during the entire experiment. The device is epoxy-mounted on a heat sink and the temperature is controlled at 20^{0}C. The onset of the coherence collapse was determined by monitoring the laser spectra and noting when the linewidth begins to significantly broaden as shown in (Grillot et al., 2002), (Tkach & Chraplyvy, 1986). As an example, fig. 12 shows the measured optical spectra of a 1.5-µm QD DFB laser. The spectral broadening caused by the optical feedback at coherence collapse level, can significantly degrades the capacity of the high-speed communication systems.

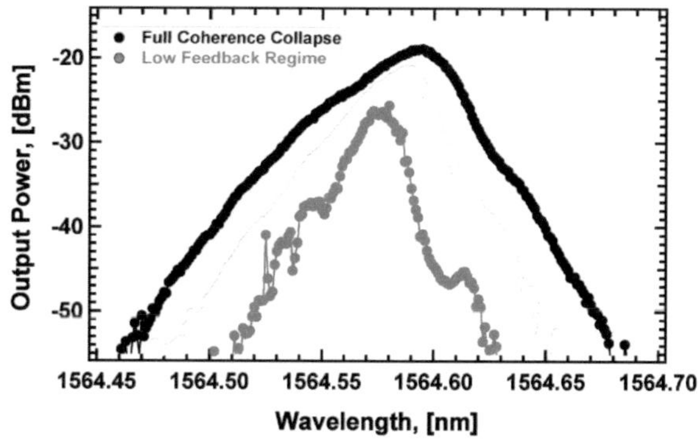

Fig. 12. Optical spectra of a 1.5-µm QD DFB laser. The solid black line corresponds to the fully developed coherence collapse

5.2 Evaluation of the critical feedback level

Based on (6) & (9), a strong degradation of the a_H-factor with the bias current should produce a significant variation in the laser's feedback sensitivity. In fig. 13, the measured onset of the coherence collapse is shown (black squares) for the QDash FP laser depicted in fig. 2 as a function of the bias current at room temperature Note that the dashed line in fig. 13 is added for visual help only. The feedback sensitivity of the laser is found to vary by ~20-dB over the range of examined current levels as the α_H-factor increases at higher bias currents. In order to compare the experimental data with theoretical models previously described, the onset of coherence collapse is calculated by substituting the measured relaxation frequency, damping factor and α_H factor values directly into (6) and (9). Assuming a laser with cleaved facets, the coupling coefficient from the facet to the external cavity $C=(1-R)/2\sqrt{R}$ is calculated to be 0.6 and the internal round trip time in the laser cavity is about ~10ps. As shown in fig. 13, the best agreement with experimental data over the range of current is found with (9) for both values of p. The discrepancy between (9) is 3-dB which corresponds to the factor 2 as described in section 2.3. Such a difference remains within the experimental resolution of +/- 3-dB (see error bars in fig. 13). Using (6) leads to a larger discrepancy, whose minimum value is ~11-dB at 65mA. It is worth noting that for α_H-factors approaching unity (below 60mA), the critical feedback level saturates for all four models considered. This saturation is generated by the function $f(\alpha_H)$, which converges to 1 as α_H gets smaller. Experimentally, the trend does not saturate at this level of bias current since the resistance to optical feedback keeps increasing, demonstrating that the critical feedback level can be up-shifted for lower α_H-factors (Cohen & Lenstra, 1991).

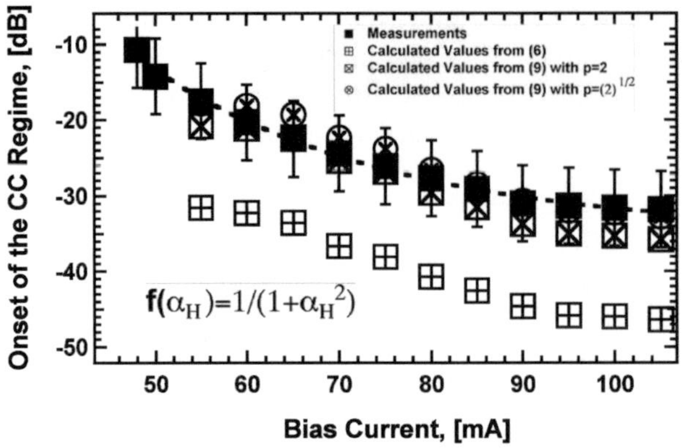

Fig. 13. Coherence collapse threshold as a function of the bias current for the QDash FP laser under study. Dashed line is added for visual help only.

In order to account for the α_H-factor approaching unity, the empirical function $g(\alpha_H)$ described in section 2.1 has been included in (6), and the results are depicted in fig. 14. Note that the dashed and solid lines in fig. 14 are added for visual help only. The calculated critical feedback level is now up-shifted for the lower values of the α_H-factor. At low bias currents, the measured values are found to be in a better agreement with calculations. Although (6) does not match the quantitative values in fig. 14, it qualitatively reproduces the up-shifting observed for small α_H-factors. This effect can be explained through variation of the α_H-factor, which changes $g(\alpha_H)$ by a factor of 500. Thus, at low bias currents, the feedback sensitivity is mostly driven by the $g(\alpha_H)$ function and not variations in the damping factor. Despite the fact that (6) was derived empirically under the assumption of weak optical feedback similar to a more complete analysis based on the Lang and Kobayashi phase equation (Alsing et al., 1996), (Erneux et al., 1996), it is found to exhibit a better accuracy for ultra-low α_H-factors. Thus, the discrepancy between the experimental data and theoretical prediction is decreased from 14-dB to 7-dB at 55mA. When extrapolating the dotted line in fig. 14 to 45mA, the calculated values will be very close to the experimental data. According to the mode competition based method given by expression (10), a critical feedback level of 58-dB is calculated using an external cavity length of 5m. This value is lower than the minimum value calculated with (6), which is about 45-dB. This feedback level corresponds to a critical level at which the external cavity modes start building-up but do not really correspond to the full coherence collapse regime.

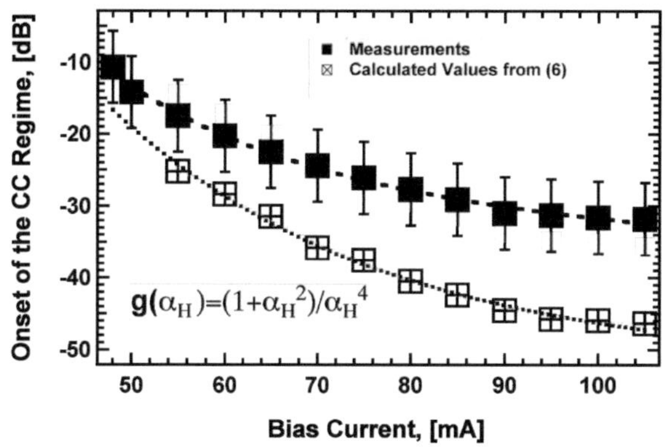

Fig. 14. Coherence collapse threshold as a function of the bias current for the QDash FP laser under study. Dashed and dotted lines are added for visual help only.

Fig. 15 shows the measured coherence collapse threshold as a function of the bias current for the QW laser studied in section 4.3. An increase in the critical feedback level is found to range between 36-dB to 27-dB when the current increases from 12mA to 70mA. In that situation, the onset of the coherence collapse follows a conventional trend (Azouigui et al, 2007), (Azouigui et al, 2009) driven by variations of the relaxation frequency.

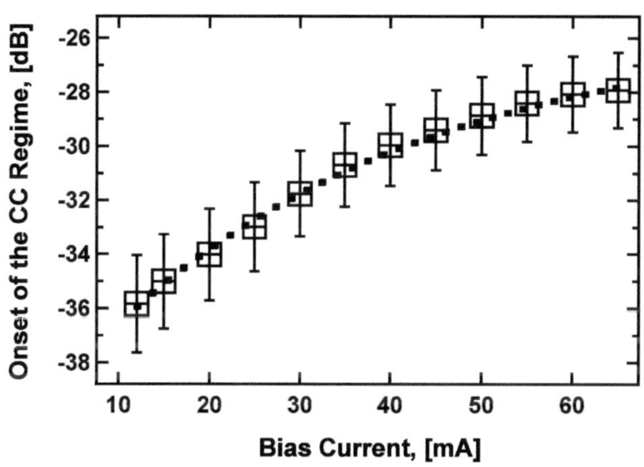

Fig. 15. Coherence collapse threshold as a function of the bias current for the QW DFB laser. The dotted line was added for visual help only.

5.3 Role of the ES in the feedback degradation

In QD or QDash lasers, it has been shown in section 4.3 that the α*H*-factor evaluated at the GS wavlength can be written as:

$$\alpha_H(P)=\alpha_{GS}(P,P_{sat})+\alpha_{ES}(P,P_{sat}) \tag{25}$$

In (24), the first term denotes the gain compression effect at the GS while the second term represents the contribution of the carrier filling from the ES In the presence of a strong gain compression or when the second term in (25) dominates, the α_H-factor follows a non-linear trend above the laser threshold as previously shown. Based on the Lang and Kobayashi rate equations in the presence of optical feedback, it has been shown that an accurate way to calculate the onset of the coherence collapse regime is given by (9). Considering also expression (25), the mutual contributions of the GS and the ES can be considered together so as to re-write the critical feedback level for a QDash laser as follows:

$$R_{P_c}(dB) = R_{Pc0}(dB) + 10\log\left(\frac{1}{1+\frac{\alpha_{ES}(P,P_{sat})\left[\alpha_{ES}(P,P_{sat})+2\alpha_{GS}(P,P_{sat})\right]}{1+\alpha_{GS}^2(P,P_{sat})}}\right) \quad (26)$$

with,

$$R_{Pc0} = \left(2\pi f_r \frac{\tau_i}{pC}\right)^2 \frac{1}{1+\alpha_{GS}^2(P,P_{sat})} \quad (27)$$

The amount R_{Pc0} denotes the contribution of the GS only towards the change in the onset of the coherence collapse. The second term in (25) needs to be considered when the above threshold α_H-factor is dependant on contributions from both the GS (α_{GS}) and the ES (α_{ES}). Expression (26) goes a step further in the analytical description of the onset of the critical feedback level since it includes the additional dependence related to the ES itself.

Fig. 16 shows the calculated GS and ES contributions to the onset of the coherence collapse (with $p=2^{1/2}$) as a function of the bias current for the QDash FP laser depicted in fig. 4. In the calculations, an internal roundtrip time of 10ps and a coupling coefficient $C=(1-R)/2\sqrt{R}$ ~0.6 (for an as-cleaved laser) is considered. The saturation power P_{sat} is close to 17mW, the ratio g_{max}/g_{th} is about 1.5 while coefficients α_0 and α_1 are treated as fitting parameters and are such that $\alpha_0 << 1$ and α_1~2. Solid lines in fig. 16 are used for guiding the eyes only. On one hand, when plotting only the contribution related to the gain compression at the GS (labeled [1]) given by (26), the critical feedback level is found to increase with the bias current As the laser's relaxation frequency is power dependent, such a variation is naturally expected. On the other hand, when considering only the contribution taking into account the carrier filling from the ES (labeled [2]) in (25), an opposite trend is observed. This contribution can be seen as a significant perturbation that results in a shift in the overall coherence collapse threshold. Thus, when both the GS and ES contributions are considered in the overall coherence collapse threshold, the calculated coherence collapse threshold is found to decrease with bias current (black solid line). Let us emphasize that these calculated values are in a good agreement with experimental data (black squares) except at low bias current for which a saturation is theoretically predicted around 23-dB. This discrepancy can be attributed to the fact that the amplitude of the optical feedback gets too large and does not match the low feedback assumption. As a conclusion, the overall experimental trend depicted in fig. 16 appears unconventional since it does not match the relaxation frequency variations even at low bias current levels for which the coherence collapse is up-shifted. This different behavior is specific to QDash structures in which the non-linear effects associated with the ES can be much more emphasized. This phenomenon can make nano-structured lasers more sensitive to optical feedback, which results in larger variations in the onset of the coherence collapse compared to that of the QW devices.

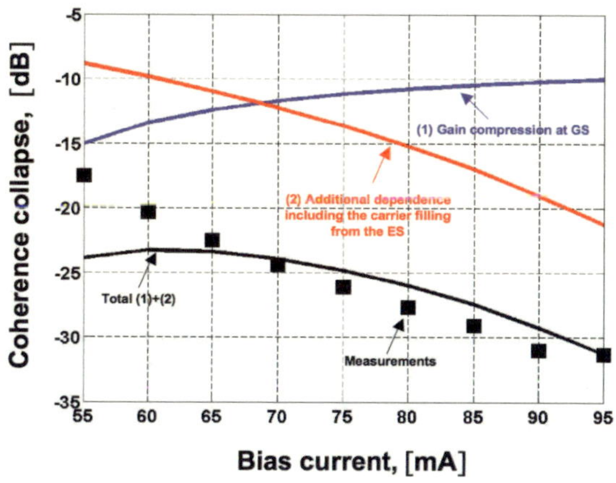

Fig. 16. Coherence collapse threshold as a function of the bias current including the contributions of the GS only, the ES only, both the GS and the ES and comparison with the measured data (black squares).

Fig. 17 shows the measured coherence collapse thresholds as a function of the α_H-factor for both the QDash FP laser (circles) and the QW DFB (squares). This figure illustrates how the route to chaos may change in a semiconductor laser; indeed depending on how the above-threshold α_H-factor behaves, the sensitivity to the coherence collapse may be improved or degraded. In regards to the QW device, the sensitivity to optical feedback is improved when increasing the current. This behavior, which has previously been observed (Azouigui et al., 2007), (Azouigui et al., 2009) is attributed to α_H-factor variations directly related to the relaxation frequency. Thus, the α_H-factor increases quite linearly above the laser's threshold and it remains mostly driven by the gain compression at the GS through the first term of equation (25) such that $\alpha_{GS} >> \alpha_{ES}$. Regarding the QDash device, the result shows a different situation: the resistance to optical feedback is substantially degraded with increasing bias current. This is due to the fact that the α_H-factor variations in the QDash FP laser are much more emphasized since for instance $\alpha_{GS} < \alpha_{ES}$ meaning that the carrier filling from the ES needs to be considered in order to explain the non-linear increase of the GS abovethreshold αH-factor. As a consequence, the critical feedback level does not follow the relaxation frequency variations since the coherence collapse is found to be up-shifted when decreasing the bias current level. Such behaviors can mostly occur in nano-structured lasers in which the influence of the ES coupled to the non-linear effects are emphasized. This phenomenon makes QD and QDash lasers more sensitive to optical feedback, thus the feedback sensitivity can be very different from a laser to another, which results in larger variations in the onset of the coherence collapse as compared to QW devices. At the wavelength of 1.5-μm, the best feedback sensitivity was found to be 24-dB for a 205-μm Cleaved/HR QD DFB laser (Azouigui et al., 2007). This result can be explained by the combination of the cleaved facet that lowers the feedback sensitivity, a higher bias current (> 100mA) as well as smaller α_H-factor (α_H~4.5). For 1.3-μm wavelength range, higher tolerances to optical feedback have been reported in QD lasers with coherence collapse thresholds as high as 14-dB (Su et al., 2004) and 8-dB (O'Brien et al., 2003). Let us stress that making such a comparison with a QW FP laser instead of a QW DFB would not change the conclusion since on QW based structures the increase of the α_H-factor with current is limited as compared to QDash devices.

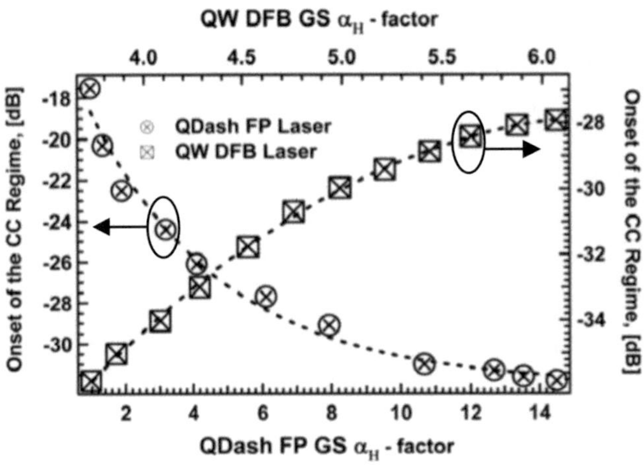

Fig. 17. Coherence collapse thresholds as a function of the α_H-factor for the QW laser (square markers) and for the QDash FP laser (circular markers) under study. Dashed lines are added for visual help.

6. Conclusions

The onset of the coherence collapse regime has been investigated experimentally and theoretically in 1.5-μm nanostructured semiconductor lasers. The prediction of the critical feedback regime has been conducted through different analytical models. Models based on the laser transfer function and on the mode competition analysis have been found to underestimate the onset of the critical feedback level. The models based on external cavity mode stability analysis have been found to be in good agreement with the experimental data. Although the model based on the transfer function has been widely used in the field of optical feedback, it does not systematically yield a strong agreement for nanostructure-based semiconductor devices except for the cases where an ultra low α_H-factor is considered. For QDash lasers, calculations are in agreement with the experiments that demonstrate that the ES filling produces an additional term, which accelerates the route to chaos. This contribution can be seen as a perturbation that reduces the overall coherence collapse threshold. Depending on the variations of the α_H-factor above threshold, the feedback resistance can be improved or deteriorated from one laser to another. The design of QDash lasers with no ES, reduced gain compression effects, and a lower, quasi-constant α_H-factor remains a big challenge. Recently a promising result was achieved using a 1.5-µm InAs/InP(311B) semiconductor laser with truly 3D-confined quantum dots (Martinez et al., 2008) (cGrillot et al., 2008). The laser characteristics exhibited a relatively constant α_H-factor as well as no significant ES emission over a wide range of current. These results highlight that the control of the α_H-factor has to be considered as a significant input for the realization of feedback-resistant lasers. Also, the prediction of the onset of the coherence collapse remains an important feature for all applications requiring a low noise level or a proper control of the laser's coherence.

7. References

Agrawal, G. P., Effect of Gain and Index Nonlinearities on Single-Mode dynamics in Semiconductor Lasers, *IEEE J. Quantum Electron.*, Vol. 26, 11, 1901-1909, 1990.

Alsing, P.M.; Kovanis, V.; Gavrielides, A. & Erneux, T., Lang and Kobayashi phase equation, *Phys. Rev. A*, Vol. 53, 4429-4434, 1996.

Arakawa, Y. & Sakaki, H., Multidimensional quantum well laser and temperature dependence of its threshold current, *Appl. Phys. Lett.*, Vol. 40, 939-941, 1982.

Azouigui, S.; Dagens, B.; Lelarge, F.; Provost, J.G.; Accard, A.; Grillot, F.; Martinez, A.; Zou, Q. & Ramdane, A., Tolerance to Optical Feedback of 10Gbps Quantum-Dash based Lasers emitting at 1.55-µm, *IEEE Photon. Technol. Lett*, Vol. 19, 15, pp. 1181-1183, 2007.

Azouigui, S.; Dagens, B.; Lelarge, F.; Provost, J.-G.; Make, D.; Le Gouezigou, O.; Accard, A.; Martinez, A.; Merghem, K.; Grillot, F.; Dehaese, O.; Piron, R.; Loualiche, S.; Q. Zou and Ramdane, A., Optical Feedback Tolerance of Quantum-Dotand Quantum-Dash-Based Semiconductor Lasers Operating at 1.55 µm, IEEE J. of Select. Topics in Quantum Electron., Vol. 15, 764-773, 2009.

Bank, S.R.; Bae, H.P.; Yuen, H.B.; Wistey, M.A.; Goddard, L.L. & Harris, J.S., Jr., Room-temperature continuous-wave 1.55-µm GaInNAsSb laser on GaAs, *Electron. Lett.*, Vol. 42, 156-157, 2006.

Bimberg, D.; Kirstaedter, N.; Ledentsov, N.N.; Alferov, Zh.I.; Kop'ev, P.S. & Ustinov, V.M., InGaAs-GaAs quantum-dot lasers, *IEEE J. of Select. Topics in Quantum Electron.*, Vol. 3, 196-205, 1997.

Binder, J.O. & Cormarck, G.D., Mode selection and stability of a semiconductor laser with weak optical feedback, *IEEE J. Quantum Electron.*, Vol. 25, 11, 2255-2259, 1989.

Caroff, P.; Paranthoën, C.; Platz, C.; Dehaese, O.; Bertru, N.; Folliot, H.; Labbé, C.; Piron, R.; Homeyer, E.; Le Corre, A.; & Loualiche, S., High-gain and low-threshold InAs quantum-dot lasers on InP, *Appl. Phys. Lett*, Vol. 87, 243107, 2005.

Clarke, R.B., The effects of reflections on the system performance of intensity modulated laser diodes", *J. Lightwave. Tech*, 9, 741-749, 1991.

Cohen, J.S. & Lenstra, D., The Critical Amount of Optical Feedback for Coherence Collapse in Semiconductor Lasers, *IEEE J. Quantum Electron.*, Vol. 27, 10-12, 1991.

Coldren, L.A. & Corzine, S.W., Diode Lasers and Photonic Integrated Circuits, John Wiley & Sons, Inc., New York, 1995.

Dagens, B.; Markus, A; Chen, J. X.; Provost, J.-G.; Make, D.; Le Gouezigou, O.; Landreau, J.; Fiore, A. & Thedrez, B., Giant linewidth enhancement factor and purely frequency modulated emission from quantum dot laser, Vol. 41, 6, 323-324, 2005.

Dagens, B.; Bertran-Pardo, O.; Fischer, M.; Gerschütz, F.; Koeth, J.; Krestnikov, I.; Kovsh, A.; Le Gouezigou, O. & Make, D., Uncooled Directly Modulated Quantum Dot Laser 10Gb/s Transmission at 1.3μm, with Constant Operation Parameters, European Conference on Optical Communication, Th 4.5.7, 2006.

Deppe, D.G.; Huang, H. & Shchekin, O.B., Modulation characteristics of quantum-dot lasers: the influence of p-type doping and the electronic density of states on obtaining high speed, *IEEE J. Quantum Electronics*, Vol. 38, 1587-1593, 2002.

Deubert, S.; Somers, A.; Kaiser, W.; Schwertberger, R.; Reithmaier, J.P. & Forchel, A., InP-based quantum dash lasers for wide gain bandwidth applications, *J. Cryst. Growth*, Vol. 278, 346-350, 2005.

Erneux, T.; Kovanis, V.; Gavrielides, A. & Alsing, P.M., Mechanism for period-doubling bifurcation in a semiconductor laser subject to optical injection , *Phys. Rev. A*, Vol. 53, 4372-4380, 1996.

Erneux, T.; Viktorov, E.A.; Mandel, P.; Azouigui, S. And Ramdane A., Relaxation characteristics of quantum-dash-based semiconductor lasers, Appl. Phys. Lett. 95, 231107, 2009

Fernier, B.; Adams, K.; Artigue, C.; Barrou, T.; Göth, A.; Grard, E.; Jörg, W.; Keller, D.; Lafragette, J.L.; Lestra, A.; Pagnod, P.; Rabaron, S.; Rainsant, J.M.; Scherb, J.; Toullier, D.; Tregoat, D. & and Rehm, W., 1.3 μm low cost plastic module for 622Mbit/s transmission at 85°C, in Proc. ECOC'98, Vol. 1, 445-446, 1998.

Franke, D.; Harde, P.; Boettcher, J.; Moehrle, M.; Sigmund, A. & Kuenzel, H., Improved emission wavelength reproducibility of InP-based all MOVPE grown 1.55 μm quantum dot lasers, International Conference on Indium Phosphide and Related Materials, FrB1.4, 2007.

Gerschütz, F.; Fischer, M.; Koeth, J.; Chacinski, M.; Schatz, R.; Kjebon, O.; Kovsh, A.; Krestinkov, I. & Forchel, A., Temperature insensitive 1.3-μm InGaAs/GaAs quantum dot distributed feedback lasers for 10Gbit/s transmission over 21 km, *Electron. Lett.*, Vol. 42, 1457-1458, 2006.

Grillot, F.; Thedrez, B.; Py, J.; Gauthier-Lafaye, O.; Voiriot, V. & Lafragette, J.L., 2.5 Gbit/s transmission characteristics of 1.3μm DFB lasers with external optical feedback, *IEEE Photon. Technol. Lett*, Vol. 14, 101-103, 2002.

Grillot, F.; Thedrez, B.; Voiriot, V. & Lafragette, J.L., Coherence collapse threshold of 1.3-μm semiconductor DFB lasers, *IEEE Photon. Technol. Lett.*, Vol. 1, 1-3, 2003.

[a]Grillot, F. ; Naderi, N.A. ; Pochet, M. ; Lin, C.-Y. & Lester, L.F., Variation of the Feedback Sensitivity in a 1.55μm InAs/InP Quantum-Dash Fabry-Perot Semiconductor Laser, *Appl. Phys. Lett.*, 13, 191108, 2008.

[b]Grillot, F. ; Dagens, B. ; Provost, J.G. ; Su, H. & Lester, L.F., Gain compression and above-threshold linewidth enhancement factor in 1.3-μm InAs/GaAs quantum dot lasers, *IEEE J. Quantum Electron.*, Vol. 44, 10, 946-951, 2008.

[c]Grillot, F.; Martinez, A.; Merghem. K.; Provost, J.G.; Alexandre, F.; Piron, R.; Dehaese, O.; Loualiche, S.; Lester, L.F. & Ramdane, A., Stable above-threshold linewidth enhancement factor in a 1.52-μm InAs/InP (311B) quantum dot laser, The 21st Annual Meeting of the IEEE LEOS, 535 – 536, 2008.

Gupta, J.A.; Barrios, P.J.; Caballero, J.A.; Poitras, D.; Aers, G.C.; Pakulski, G. & Wu, X., Gain and lifetime of GaInNAsSb narrow ridge waveguide laser diodes in continuous-wave operation at 1.56-µm, *Appl. Phys. Lett*, Vol. 89, 151119, 2006.

Helms, J. & and Petermann, K., Microwave modulation of semiconductor lasers with optical feedback, *Electron. Lett.*, 25, 1369-1371, 1989.

Helms, J. & Petermann, K., A simple analytic expression for the stable operation range of laser diodes with optical feedback, *IEEE J. Quantum Electron.*, Vol. 26, 833-836, 1990.

Henry, C.H., Theory of the linewidth of semiconductor lasers, *IEEE J. Quantum Electron.*, 18, 259-264, 1982.

Henry, C.H. & Kazarinov, R.F., Instabilities of semiconductor lasers due to optical feedback from distant reflectors, *IEEE J. Quantum Electron.*, Vol. 22, 294-301, 1986.

Hirota, O. & Suematsu, Y., Noise properties of injection lasers due to reflected waves, *IEEE J. of Quantum Electron.*,15, 142-149, 1979.

D. M. Kane and K. A. Shore, Unlocking dynamical diversity, Wiley, 2005.

Karachinsky, L.Ya.; Kettler, T.; Gordeev, N.Yu.; Novikov, I.I.; Maximov, M.V.; Shernyakov, Yu.M.; Kryzhanovskaya, N.V.; Zhukov, A.E.; Semenova, E.S.; Vasil'ev, A.P.; Ustinov, V.M.; Ledentsov, N.N.; Kovsh, A.R.; Shchukin, V.A.; Mikhrin, S.S.; Lochmann, A.; Schulz, O.; Reissmann, L. & Bimberg, D., High-power singlemode CW operation of 1.5 lm-range quantum dot GaAs-based laser, *Electron. Lett.*, Vol. 41, 8-9, 2005.

Kim, J.S.; Lee, J.H.; Hong, S.U.; Han, W.S.; Kwack, H.; Lee, C.W. & OH, D.K., Room-temperature operation of InP-based InAs quantum dot laser, *IEEE Photon. Technol. Lett*, Vol. 16, 1607-1609, 2004.

Kitaoka, Y.; Sato, H.; Mizuuchi, K.; Yamamoto, K. & Kato, M., Intensity noise of laser diode with optical feedback, *IEEE J. of Quantum Electron.*, Vol. 32, 822-828, 1996.

Lang, R. & Kobayashi, K., External optical feedback effects on semiconductor injection laser properties, *IEEE J. Quantum Electron.*, Vol. QE-16, 347-355, 1980.

Lelarge, F.; Dagens, B.; Renaudier, J.; Brenot, R.; Accard, A.; Van Dijk, F.; Make, D.; Le Gouezigou, O.; Provost, J.G.; Poingt, F.; Landreau, J.; Drisse, O.; Derouin, E.; Rousseau, B.; Pommereau, F. & Duan, G.H., Recent Advances on InAs/InP Quantum Dash Based Semiconductor Lasers and Optical Amplifiers Operating at 1.55-µm, *IEEE J. of Selected Topics in Quantum Electronics.*, Vol. 13, 111-124, 2007.

Lenstra, D.; Verbeek, B.H. & Den Boef, A.J., Coherence collapse in single-mode semiconductor lasers due to optical feedback, *IEEE J. Quantum Electron.*, Vol. QE-21, 674-679, 1985.

Li, Y.; Naderi, N.A. ; Kovanis, V. & Lester, L.F., Modulation Response of an Injection-Locked 1550 nm Quantum Dash Semiconductor Laser, The 20th Annual Meeting of the IEEE LEOS, 498-499, 2007.

Liu, G.T.; Stintz, A.; Li, H.; Malloy, K.J. & Lester, L.F., Extremely low room-temperature threshold current density diode lasers using InAs dots in $In_{0.15}Ga_{0.85}As$ quantum well, *Electron. Lett.*, Vol. 35, 1163-1165, 1999.

Liu, G.; Jin, X. & Chuang, S.L., Measurement of linewidth enhancement factor of semiconductor lasers using an injection-locking technique, *IEEE Photon. Technol. Lett.*, Vol.13, 430 – 432, 2001.

Makino, S.; Shinoda, K.; Kitatani, T.; Tsuchiya, T. & Aoki, M., Wide temperature range (0 to 85°C), 40-km SMF transmission of a 1.55 µm, 10-Gbit/s InGaAlAs electroabsorption modulator integrated DFB laser, Technical Digest OFC, Vol. 6, 6-11, postdealine paper PDP14, 2005.

Martinez, A.; Lemaître, A.; Merghem, K.; Ferlazzo, L.; Dupuis, C.; Ramdane, A.; Provost, J.G.; Dagens, B.; Le Gouezigou, O. & Gauthier-Lafaye, O., Static and dynamic measurements of the α-factor of five-quantum-dot-layer single-mode lasers emitting at 1.3-µm on GaAs, *Appl. Phys. Lett*, Vol. 86, 211115, 2005.

Martinez, A.; Merghem, K.; Bouchoule, S.; Moreau, G.; Ramdane, A.; Provost, J.-G.; Alexandre, F.; Grillot, F.; Dehaese, O.; Piron, R. & Loualiche, S., Dynamic properties of InAs/InP(311B) quantum dot Fabry-Perot lasers emitting at 1.52-µm, *Appl. Phys. Lett.*, Vol. 93, 021101, 2008.

Mi, Z.; Bhattacharya, P. & Yang, J., Growth and characteristics of ultralow threshold 1.45 m metamorphic InAs tunnel injection quantum dot lasers on GaAs, *Appl. Phys. Lett.*, Vol. 89, 153109, 2006.

Mork, J.; Tromborg, B. & Mark, J., Chaos in semiconductor lasers with optical feedback: theory and experiment, *IEEE J. Quantum Electron.*, Vol. QE-28, 11, 93-108, 1992.

Mork, J.; Tromborg, B. & Christiansen, P.L., Bistability and low-frequency fluctuations in semiconductor lasers with optical feedback: a theoretical analysis, *IEEE J. Quantum Electron.* Vol. 24, 123-133, 1988.

O'Brien, D.; Hegarty, S.P.; Huyet, G.; McInerney, J.G. ; Kettler, T.; Laemmlin, M. ; Bimberg, D.; Ustinov, V.M.; Zhukov, A.E. ; Sikhrin, S.S. & Kovsh, A.R., Feedback sensitivity of 1.3µm InAs/GaAs quantum dot lasers, *Electron. Lett.*, 39, 1819-1820, 2003.

Petermann, K., Laser Diode Modulation and Noise, Kluwer Academic Publishers, 1988.

Petermann, K., External Optical Feedback Phenomena in Semiconductor Lasers, *IEEE J. of Select. Topics in Quantum Electronics*, Vol. 1, 480-489, 1995.

Phillips, A.F.; Sweeney, S.J.; Adams, A.R. & Thijs, P.J.A., Temperature dependence of 1.3- and 1.5-µm compressively strained InGaAs(P) MQW semiconductors lasers, *IEEE J. of Select. Topics in Quantum Electronics*, Vol. 5, 401-412, 1999.

Saito, H.; Nishi, K.; Kamei, A. & Sugou, S., Low chirp observed in directly quantum dot lasers, *IEEE Photon. Technol. Lett*, Vol. 12, 1298-1300, 2000.

Saito, H.; Nishi, K. & S. Sugou, Low chirp operation in 1.6-µm quantum dot laser under 2.5GHz direct modulation, *Electron. Letts.*, 37, 1293-1295, 2001.

Shunk, N. & Petermann, K., Numerical analysis of the feedback regimes for a single-mode semiconductor laser with external feedback, *IEEE J. Quantum Electron.*, Vol. 24, 1242-1247, 1988.

Shunk, N. & Petermann, K., Stability analysis for laser diodes with short external cavities, *IEEE Photon. Technol. Lett.*, Vol.1, 49-51, 1989.

Sorin, W. V.; Chang, K. W.; Conrad, G. A. & Hernday, P. R., Frequency domain analysis of an optical FM discriminator, *J. Lightwave Technol.*, Vol. 10, 787–793, 1992.

Stegmuller, B.; Borchert, B. & Gessner, R., 1.57-µm Strained-Layer Quantum-Well GaInAlAs Ridge-Waveguide Laser Diodes with High Temperature (130°C) and Ultrahigh-speed (17 GHz) Performance, *IEEE Photon. Technol. Lett.*, Vol. 5, 597-599, 1993.

Su, H.; Zhang, L.; Gray, A.L.; Wang, R.; Newell, T.C.; Malloy, K.J. & and Lester, L.F., High external feedback resistance of laterally loss-coupled distributed feedback quantum dot semiconductor lasers, *IEEE Photon. Technol. Lett.*, Vol.15, 1504-1506, 2003.

Su, H.; Zhang, L.; Gray, A.L.; Wang, R.; Newell, T.C.; Malloy, K.J. & Lester, L.F., Linewidth Study of InAs-InGaAs quantum dot distributed feedback lasers, *IEEE Photon. Technol. Lett.*, Vol.16, 2206-2208, 2004.

Su, H.; Zhang, L.; Gray, A.L.; Wang, R.; Varangis, P.M. & Lester, L.F., Gain compression coefficient and above-threshold linewidth enhancement factor in InAs/GaAs quantum dot DFB lasers, *Proceeding SPIE* 5722-11, 2005.

Su, H. & and Lester, L.F., Dynamic properties of quantum dot distributed feedback lasers: high speed linewidth and chirp, *J. Phys. D: Appl. Phys.*, Vol. 38, 2005.

Suhara, M. & and Yamada, M., Analysis of excess noise due to external optical feedback in DFB semiconductor lasers on the basis of the mode competition theory, *IEICE Trans. Electron.*, Vol. EC76-C, 6, 1007-1017, 1993.

Thijs, P. J. A; Tiemeijer, L.F.; Binsma, J.J.M & and Van Dongen, T., Progress in longwavelength strained-layer InGaAs(P) quantum-well semiconductor lasers and amplifiers, *IEEE J. Quantum Electron.*, Vol. 30, 477–499, 1994.

Tkach, R.W. & Chraplyvy, A.R., Regimes of feedback effects in 1.5-μm distributed feedback lasers, *J. Lightwave. Tech.*, Vol. LT-4, 1655-1661, 1986.

Tohmori, Y.; Itaya, Y. & Toba, H., 1.3 μm spot-size converter integrated laser diodes (SS-LDs) for access network applications, The 10th Annual Meeting of the IEEE LEOS, TuQ1, 270-271,1997.

Tromborg, B. & Mork, J., Non-linear injection locking dynamics and the onset of coherence collapse in external cavity lasers, *IEEE J. Quantum Electron.*, Vol. 26, 642-654, 1990.

Veselinov, K; Grillot, F.; Cornet, C.; Even, J.; Bekiarski, A.; Gioannini, M. & Loualiche, S., Analysis of the Double Laser Emission Occurring in 1.55μm InAs/InP (113)B Quantum-Dot Lasers, *IEEE J. Quantum Electron*, Vol. 43, 810-816, 2007.

Wang, R.H.; Stintz, A.; Varangis, P.M.; Newell, T.C.; Li, H.; Malloy, K.J. & Lester, L.F., Room-temperature operation of InAs quantum-dash lasers on InP (100), *IEEE Photon. Technol. Lett.*, Vol. 13, 767-769, 2001.

Wei, J. H. & Chan, K. S., A theoretical analysis of quantum dash structures, *J. Applied Phys.*, 97, 123524, 2005.

Yamada, M. & Suhara, M., Analysis of excess noise due to external optical feedback in semiconductor lasers on mode competition theory, *IEICE Trans. Electron.*, Vol. EC73-C, 77-82, 1990.

Acknowledgments

Authors are grateful to Professor Jean-Claude Simon, Head of FOTON for the financial support as well as for fruitful discussions. Authors acknowledge Professor Thomas Erneux from Brussels University and Professor Vassilios Kovanis from the US Air Force for fruitful discussions.

This work was supported in part by the Air Force Research Laboratory. The views expressed in this book chapter are those of the authors and do not reflect the official policy or positions of the United States Air Force, Department of Defense, or the US government.

11

Electron transport effect on optical response of quantum-cascade structures

Mykhailo Klymenko and Oleksiy Shulika
Kharkov National University of Radio Electronics
Ukraine

Igor Sukhoivanov
University of Guanajuato
Mexico

1. Introduction

The quantum-cascade laser is an unique source of the THz laser radiation operated in continuous-wave and pulse regimes [Gmachl et al (2001)]. History of these lasers counts more than ten years. However, many aspects of the carrier transport and interaction with light field are still unclear. Very important question concerning physics of the quantum-cascade structures (QCS) is the following: which kind of transport, coherent or incoherent, is prevailed in QCS? There were many discussions about the problem, and several attempts to estimate kind of transport were successful especially [Iotti et al (2001)],[Weber et al (2009)]. The answer on this question depends on conditions of QCS operation. For example, the coherent electron transport is of interest in the non-equilibrium regime at femtosecond and picosecond time intervals. The incoherent transport is prevalent at the high excitation level in the stationary quasi-equilibrium regime. In both cases, the electron transport influence on optical properties of the device. In this connection, the development of the theory for coherent and incoherent electron transport regimes, included many-body effects and light-matter interactions in QCS, is of actual interest.

In this chapter, we provide modeling of optical and transport properties of QCS uncovering influence of the electron transport on optical characteristics. Lasing, light absorption and spontaneous emission in QCS are accompanied and affected by many complicated transport processes such as electron diffusion, drift, tunneling, recombination, generation, capture and escape mediated by electron-electron, electron-phonon and electron-photon scattering events [Piprek (2005)]. Most of these effects can be treated within the quasi-equilibrium approximation. However, the approximation is not valid at ultrashort time intervals which are of interest nowadays due to rapid development of the femtosecond spectroscopy for semiconductor nanostructures [Rulliere (2005)]. Other area demanding consideration of ultra-fast non-equilibrium processes is THz emitting of QCS in the pulse regime; that is under rapid development currently due to promising applications in fundamental and applied science [Lee (2009)].

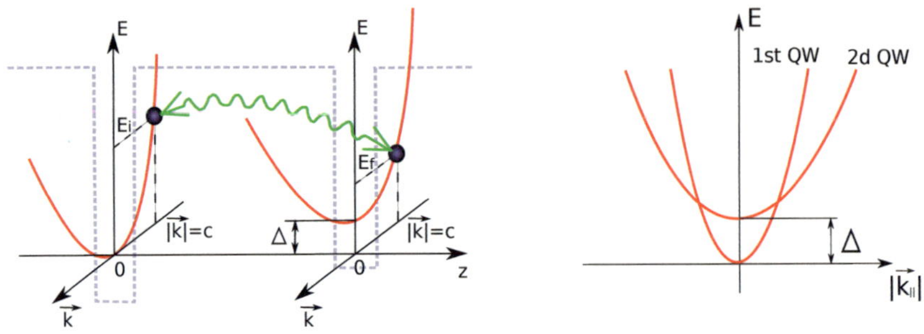

(a) (b)

Fig. 1. Coherent electron transport between two quantum wells with subband structure via tunneling

To predict output optical characteristics of the quantum-cascade lasers, it is necessary to simulate dynamics of electrons, holes and photons in the non-equilibrium state. The peculiarity of QCS is that the system is related to the open quantum systems. Moreover, the structure is characterized by pronounced non-equilibrium regime of operation. There are several approaches to modelling the transport in such a system. One of them is based on non-equilibrium Green's functions [Lee et al (2002)], another one is based on the density matrix theory [Iotti et al (2001)]. These two are not the only approaches to modeling transport in QCS. For example, rate equations are widely used in connection to this problem [Vukmirović (2005)]. However, these two methods are most rigorous and controllable. They can be realized at various levels of approximation and allow to estimate approximation error. In ideal case, they do not require any fitting parameter and give results *ab initio*. Our consideration in this chapter is based on the density matrix theory. As a result, we will derive kinetic equations describing dynamics of carriers, polarization and inter-quantum-well tunneling currents for non-equilibrium regime of the operation, and then, we discuss main features of transport and optical properties of QCS.

2. Coherent and incoherent transport

2.1 Coherent transport

In this section, we will represent theoretical instruments proper for high accuracy modeling of the electron transport in QCS. Up to date, many efforts have been made and much progress has been achieved in modeling of the electron transport in QCS. Especially, it concerns stationary operating regime. Recently, advance in modeling of femtosecond optical response of QCS has been reached [Iotti et al (2001)], [Weber et al (2009)]. Most successful approaches to electron transport modeling have been realized applying the density matrix theory [Meier (2007)]. This theory is especially suitable for the large open quantum systems with many-body interactions. Therefore, we apply exactly the density matrix theory to realize systematic treatment of the coherent and incoherent electron transport in semiconductor nanostructures.

Here, the simplest model heterostructure consisting of two interacting quantum wells is considered to make statements compact and clear. The sketch of energy levels for such a structures is shown in Fig. 1.

Electron states in each quantum well are characterized by single-band structure when quantum wells are uncoupled. The band structure, shown in Fig. 1(b), can be analytically expressed as:

$$\varepsilon_{1,k} = E_1 + \frac{\hbar^2 k^2}{2m_1}, \tag{1}$$

$$\varepsilon_{2,k} = E_2 + \frac{\hbar^2 k^2}{2m_2} = E_1 + \Delta + \frac{\hbar^2 k^2}{2m_2}. \tag{2}$$

The electron in each quantum well is characterized by continuous energy spectra and has states $|1,k\rangle$ in one quantum well and $|2,k\rangle$ in another one. The number in the ket vector corresponds to the subband kind and the letter is the in-plane electron wave vector limited by the 1st Brillouin zone (axial approximation is applied [Haug (2004)]). States should satisfy completeness conditions [Meier (2007)]:

$$\hat{1} = \sum_j \sum_k |j,k\rangle \langle k,j|. \tag{3}$$

In practice, each quantum well and barrier layer can be made of different semiconductor materials. This means that each quantum well can be characterized by own width, depth and effective masses. That is why, we consider subband dispersion curves which are shifted relative each other by some value Δ and characterized by different curvature.

It is well known that, if one turns on the interaction between quantum wells, the quantum states of whole system are changed and each state of the system is splitted into bound and antibound state. That is the picture for the stationary regime. However, the case of interest is the time evolution of the system that is prepared in some non-stationary state in the initial time moment. Even though quantum system is prepared in some defined stationary state, following artificial modification of the system (by measurement event for example) can change the energy spectrum and the ininial state is not stationary anymore. This leads to nontrivial dynamical evolution of observables. Considered here open quantum systems are interacted with environment that leads to modifications of its parameters and dynamical evolution of observables. Therefore, we focus on the case when the sysatem is prepared in the stationary state of noninteracted quantum wells with following turning on of interactions.

The Hamiltonian for considered model system can be represented in the form:

$$\hat{H} = \hat{H}_1 + \hat{H}_2 + \hat{H}_{int}, \tag{4}$$

here: $\hat{H}_1$ is Hamiltonian for the first quantum well, $\hat{H}_2$ is Hamiltonian for the second quantum well and $\hat{H}_{int}$ describes the interaction between quantum wells.

Acting by the unity operator from left and right sides on the Hamiltonian, one gets:

$$\hat{H} = \hat{1} \cdot \hat{H} \cdot \hat{1} = \sum_k \left(\varepsilon_{1,k} |1,k\rangle \langle k,1| + \varepsilon_{2,k} |2,k\rangle \langle k,2| \right) + \sum_{i \neq j} \sum_k h_{ij} |i,k\rangle \langle k,j|, \tag{5}$$

where: h_{ij} is the coupling coefficient describing the intensity of interactions between quantum wells.

As far as considered structure is the open quantum system, we use trusted instrument from the quantum statistical physics that is the density operator:

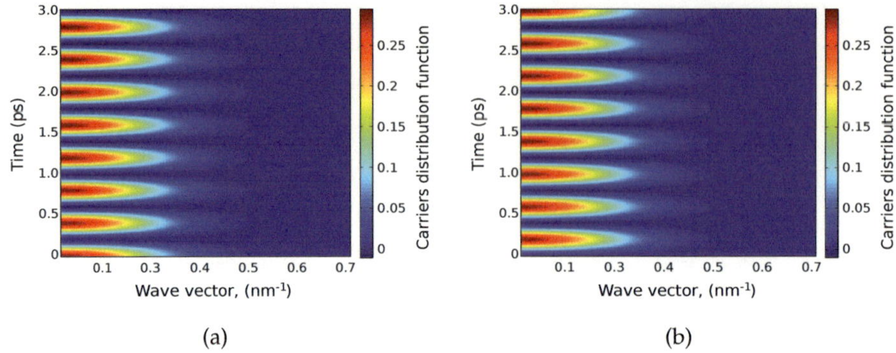

Fig. 2. Time-dependent electron distribution function for a) the left quantum well and b) right quantum well having the same band structure

$$\hat{\rho} = |t\rangle\langle t|. \tag{6}$$

The density operator can be represented in the matrix form using defined system of basis functions. As an example, we build matrix representation of the density operator using basis $|j,k\rangle$ defined above:

$$\rho_k = \langle k,i|\hat{\rho}|j,k\rangle = \begin{pmatrix} \rho_{11} & \rho_{12} \\ \rho_{21} & \rho_{22} \end{pmatrix} = \begin{pmatrix} |\langle k,1|t\rangle|^2 & \langle k,1|t\rangle\langle t|2,k\rangle \\ \langle k,2|t\rangle\langle t|1,k\rangle & |\langle k,2|t\rangle|^2 \end{pmatrix}. \tag{7}$$

Diagonal elements describe the probability of finding the electron at the time t in some defined energy band. Nondiagonal elements corresponds to some kind of correlations which give probability of the particle transition between states at the time t. The non-diagonal matrix elements are related to microscopic polarization or currents. In turn, the polarization is directly related to the electrical current according to classical electrodynamics as well as quantum one.

Time evolution of the density operator is defined by the Liouville-von Neumann equation [Meier (2007)]:

$$i\hbar\frac{\partial\rho}{\partial t} = [H,\hat{\rho}]_{+}. \tag{8}$$

In the Heisenberg representation, this equation is coincided with the Heisenberg equation for time-dependent operators. Eq. 8 can be written related to each element of the density matrix using quantum-mechanical averaging with the basis defined above . Resulting system of equations reads:

$$i\hbar\frac{\partial\rho_{11,k}}{\partial t} = h_{12,k}\left(\rho_{21,k} - \rho_{12,k}\right), \tag{9}$$

$$i\hbar\frac{\partial\rho_{22,k}}{\partial t} = -h_{12,k}\left(\rho_{21,k} - \rho_{12,k}\right), \tag{10}$$

$$i\hbar\frac{\partial\rho_{12,k}}{\partial t} = \left(\varepsilon_{2,k} - \varepsilon_{1,k}\right)\rho_{21,k} + h_{12,k}\cdot\left(\rho_{22,k} - \rho_{11,k}\right) \tag{11}$$

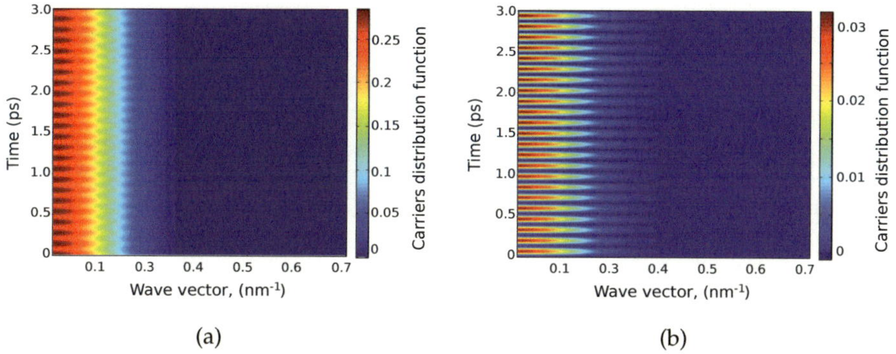

Fig. 3. Time-dependent electron distribution function for a) the left quantum well and b) right quantum well having bands shifted on 30 meV relative to each other

These are ordinary differential equations also known as kinetic equations. The number of equations is equal $3N_k$, where N_k is the number of discretization points in k-space. Eqs. (9)-(11) are written for some defined point k in the Brillouin zone. To analyze electron transport in our simple model system, we should solve this system of equations analytically or numerically. Here, we choose the second way to show general approach to such a mathematical problem. The fourth order Runge-Kutta method is applied to solve the problem. This method is stable and accurate enough to satisfy our requirements on CPU time and computational accuracy. As far as we deal with first order ordinary differential equations, the initial condition should be added. We assume that, at the initial time, all electrons are located in the 1st noninteracting quantum well with some defined distribution function. Initial distribution is chosen to be the Fermi-Dirac distribution with some defined temperature and Fermi level ($T = 300$ K and $E_f = \varepsilon_{1,0}$). Solutions of kinetic equations are time dependencies of microcurrents or polarizations and electron distribution functions for each band and each value of the in-plane wave vector k. At the initial time, interaction between quantum wells is turning on that is reflected in the coupling coefficient:

$$h_{12,k} = \begin{cases} 0, & t < 0; \\ const, & t \geq 0. \end{cases} \tag{12}$$

Let us consider first the effect of band structure on electron transport. Solving of the equations for two identical subbands, one obtains the result shown in Fig.2. Electrons oscillate between quantum wells through the barrier. The frequency of oscillations is determined only by the coupling coefficient. At some instant times, all electrons totaly depopulate the band in a quantum well transiting to another one.

In the case, when bands have the same shape and are shifted relative to each other, the electron distribution function is characterized by the time dependence shown in Fig. 3. The mismatch of energy levels leads to decreasing of electrons amount passing through the barrier. Most of particles do not leave the state occupied at the initial time moment. In particular case represented in Fig. 3, band mismatch is equal 30 meV and the maximal fraction of passed particles amounts 10 %.

Also, changes of the oscillation frequency is observed. Thus, the frequency of oscillations is dependent on the band mismatch as well as coupling coefficient. Explicit dependence could be

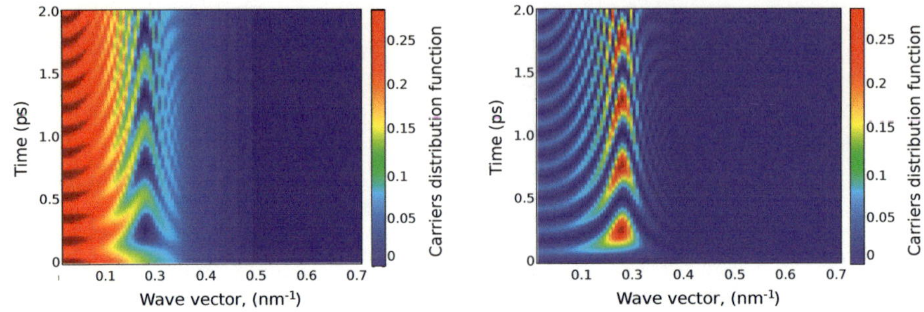

(a) (b)

Fig. 4. Time-dependent electron distribution function for a) the left quantum well and b) right quantum well having bands with different effective masses

derived from equations (9)-(11). This can be realized by Fourier transformation of equations with following algebraic manipulations. The result reads:

$$\omega = \frac{1}{2\hbar}\sqrt{\Delta^2 + 4h_{12,k}}, \tag{13}$$

here Δ is the band mismatch shown in Fig.1.

Another approach to obtaining this result is solving of the stationary Schrödinger equation for coupled quantum wells. The oscillation frequency is proportional to splitting of energy levels caused by resonant tunneling [Meier (2007)].
As far as each quantum well can be characterized by own width, depth and effective masses, we will provide investigation of the coherent electron transport between bands with different curvature of dispersion dependencies. As an example, let us consider two bands shown in Fig. 1(b) having parameters $E_1 = 0.03$, $E_2 = 0$, $m_1 = 5m$ and $m_2 = m$.
Band dispersion curves are crossed at the point $k = 0.27nm^{-1}$. The most part of electrons are propagated with this non-zero in-plane wave vector and oscillation frequency is dependent on the electron in-plane wave vector (see Fig. 4(b)). The lowest frequency corresponds to minimal gap between bands. That is in agreement with formula (13).

2.2 Incoherent transport

All cases of electron transport considered above are related to the coherent electron transport due to any decoherence effect has not been included in the consideration yet. Decoherence can be caused by scattering events leading to relaxation into the stationary equilibrium state. Thus, one should include additional term in the Hamiltonian (4) describing scatterings.

$$\hat{H} = \hat{H}_1 + \hat{H}_2 + \hat{H}_{int} + \hat{H}_{scatt}. \tag{14}$$

It is necessary to note that the single-particle formalism used above is not applicable directly to scattering processes, because such processes are essentially many-body effects. However, some approximation can conserve the problem be single-particle. For example, one can apply the mean-field approximation to the many-body problem. This approach is often used in

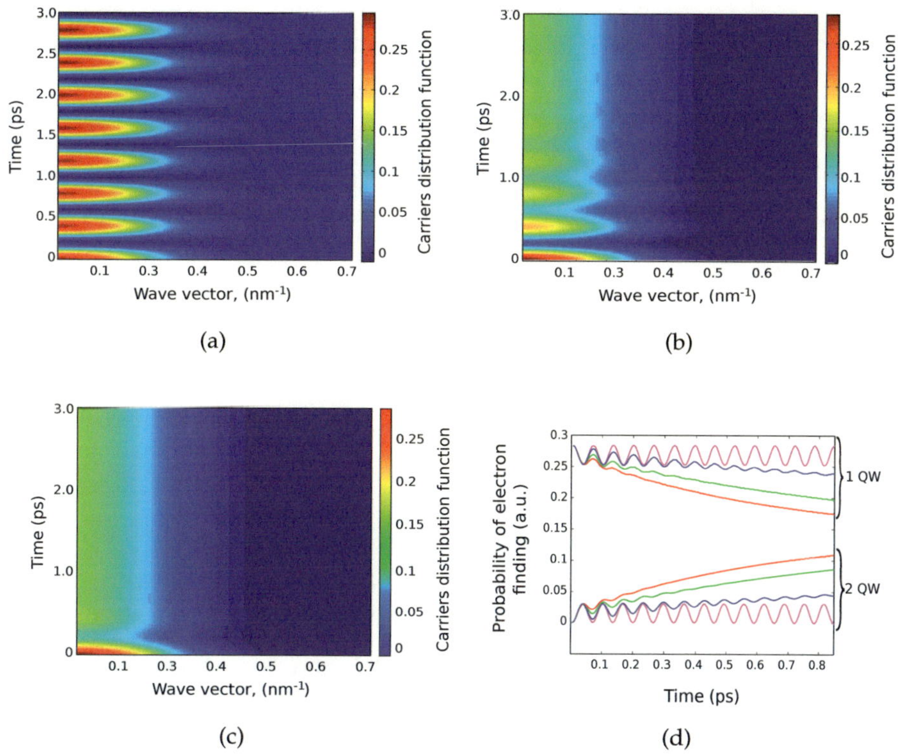

Fig. 5. Time evolution of the electron distribution function for a) $\gamma = 10ps$, b) $\gamma = 1.5ps$ and c) $\gamma = 0.5ps$ and d) $k = 0$, band mismatch $\Delta = 0.03eV$ and dephasing times taken from the previous case.

connection with phenomenological relaxation and dephasing times describing influence of many-body effects on single-particle equations.

Formally, effect of the scattering term in (14) can be represented in Eqs. (9)-(11) by additional terms at the right side of kinetic equations.

$$i\hbar\frac{\partial\rho_{11}}{\partial t} = h_{12}\left(\rho_{21} - \rho_{12}\right) + \left.\frac{\partial\rho_{11}}{\partial t}\right|_{scatt}, \tag{15}$$

$$i\hbar\frac{\partial\rho_{22}}{\partial t} = -h_{12}\left(\rho_{21} - \rho_{12}\right) + \left.\frac{\partial\rho_{22}}{\partial t}\right|_{scatt}, \tag{16}$$

$$i\hbar\frac{\partial\rho_{12}}{\partial t} = \left(\varepsilon_2 - \varepsilon_1\right)\rho_{21} + h_{12}\left(\rho_{22} - \rho_{11}\right) + \left.\frac{\partial\rho_{12}}{\partial t}\right|_{scatt}, \tag{17}$$

These additional terms can be computed *ab initio* using many-body theory and a set of approximations which will be represented in the next section. Here, we use phenomenological relaxation and dephasing times to investigate many-body effects. In this case, Eqs. (15)-(17) are modified as follows:

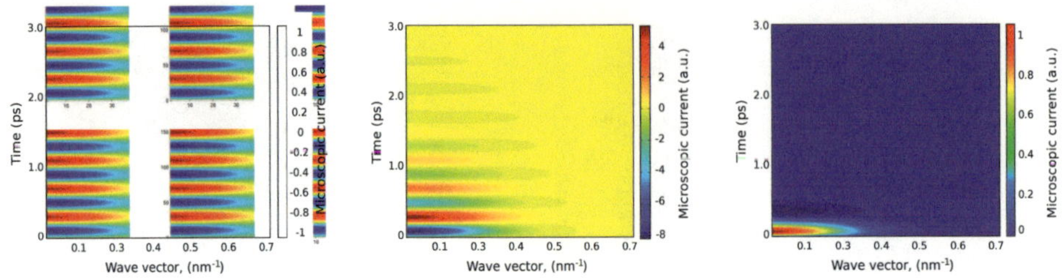

(a) (b) (c)

Fig. 6. Microscopic currents for a) $\gamma = 10ps$, b) $\gamma = 1.5ps$ and c) $\gamma = 0.5ps$

$$i\hbar\frac{\partial\rho_{11}}{\partial t} = h_{12}\left(\rho_{21} - \rho_{12}\right) - \frac{i(\rho_{11} - f_1)}{\tau}, \tag{18}$$

$$i\hbar\frac{\partial\rho_{22}}{\partial t} = -h_{12}\left(\rho_{21} - \rho_{12}\right) - \frac{i(\rho_{22} - f_2)}{\tau}, \tag{19}$$

$$i\hbar\frac{\partial\rho_{12}}{\partial t} = \left(\varepsilon_2 - \varepsilon_1\right)\rho_{21} + h_{12}\left(\rho_{22} - \rho_{11}\right) - \frac{i\rho_{12}}{\gamma}. \tag{20}$$

here γ is the dephasing time, τ is the relaxation time, f_1 and f_2 are stacionary electron distributions in each quantum well.

Relaxation and dephasing times can be determined from experimental data (optical pump-probe experiments [Vu (2006)]). If the total number of electrons is time-independent, the relaxation times tend to infinity and the last terms in Eq.(18) and (19) can be neglected. This is the case considered in this section. Thus, we will investigate the effect of dephasing only. Time dependencies of electron distribution function are shown in Fig. 5 for different dephasing times. Corresponding microscopic currents are shown in Fig. 6.
Microscopic currents reflects the probability of electron transition from one quantum well to another. Dephasing leads to decay of oscillations and becoming of the stationary distribution of electrons. If dephasing is absent (dephasing time is very high), the electron transport is pure coherent (see Fig. 5(a)). In Fig. 5(c), another limit case is shown when the dephasing time is very small. In this case, the transient process require few time and stationary regime becomes very fast. The transient process occurs because the quantum system is prepared in non-stationary state at the initial time, and it tends to the stationary state. Scattering events allow energy quanta exchange between particles leading to the relaxation into the stationary state. As follows from Fig. 6(c), particle exchange between quantum wells occurs at short time interval when the dephasing time is great.
The evolution of the electron distribution function for zero in-plane wave vector (k=0) is shown in Fig. 5(d) for the case when bands are shifted relative each other by 30 meV. In this case, dephasing times are the same as in previous examples. Dephasing leads to leveling of electron concentration in each quantum well and becoming of the stationary state. Oscillations of the electron distribution function are decayed with increasing of the dephasing time. The result of dephasing absence is endless oscillations of the electron distribution function and asymmetrical population of subbands.

3. Optical response

3.1 Model structure and its Hamiltonian

As we did in the previous section, we introduce here the new model structure that reflects main effect in the QCS and is still simple enough for modeling and analysis. Effects of interest are light-matter interactions together with transport processes in the structure.

In this section, we focus our attention on optical processes in the QCS with vertical transitions [Faist (1995)]. Term "vertical transitions" means that photon assisted tunneling is excluded from the consideration. The QCS have N optical-sensitive active regions which interact with each other via exchanging of electrons through injectors. The Hamiltonian of the system reads:

$$H = \sum_{j=1}^{N} H_j + \sum_{j=1}^{N-1} H_{j,j+1} + H_L + H_R + H_{L,1} + H_{N,R}. \tag{21}$$

The first sum in RS contains electron kinetic energy, electron scattering and light-matter interaction terms for all active regions. The second sum describes electron transport through injectors. The term H_L and H_R corresponds to the energy of regions terminated considered planar structure at the both sides. Finally, terms $H_{L,1}$ and $H_{N,R}$ describe exchanging of electrons between the structure and terminal regions. Such expression of Hamiltonian is quite natural if all optical transitions appear inside active regions. This is our case because the QCS with vertical transitions is under consideration [Faist (1995)]. The terminal regions shown in Fig. 7(c) as circles is implemented artificially. These regions include the whole rest space of the system except some considered region been of interest. Necessity of such regions is caused by influence of environment in the open quantum system. Modeling of all periods of QCSs requires much computational resources. So, the second reason of terminal regions application is the approximation allowing to consider dynamical behavior of electrons only in one or several periods. In this case, whole rest structure is assumed to be in the quasi-equilibrium state, and it is contained in the terminal regions. We use approximation that the terminal regions are characterized by some kind of stationary distribution function. We call them bathes in analogy to statistical mechanics. Alternative approach is application of periodic boundary conditions [Lee et al (2002)].

In this section, we consider only one period of the QCS. Corresponding model structure is shown in Fig. 7(c). It contains only one active region surrounded by two injectors and two terminal regions. The Hamiltonian for the model structure consist of five terms in the simplest case, when many-body effects are not considered:

$$H = H_L + H_R + H_a + H_{La} + H_{aR}, \tag{22}$$

where H_L is the energy of the left reservoir, H_R is the energy of the right reservoir, H_a is the energy of the active region, H_{La} and H_{aR} describe transitions between reservoirs and the active region.

In the active regions, light-matter interactions proceed involving electron-phonon, electron-electron and electron-impurity scatterings. Thus, we should include into consideration many-body effects to simulate the optical response of the semiconductor media correctly. The approximations that all many-body effects appear in the active region is applied. In this case, the Hamiltonian for the active region reads:

$$H_a = H_{kin} + H_I + H_{ph-el} + H_{el-el} + H_{el-imp}, \tag{23}$$

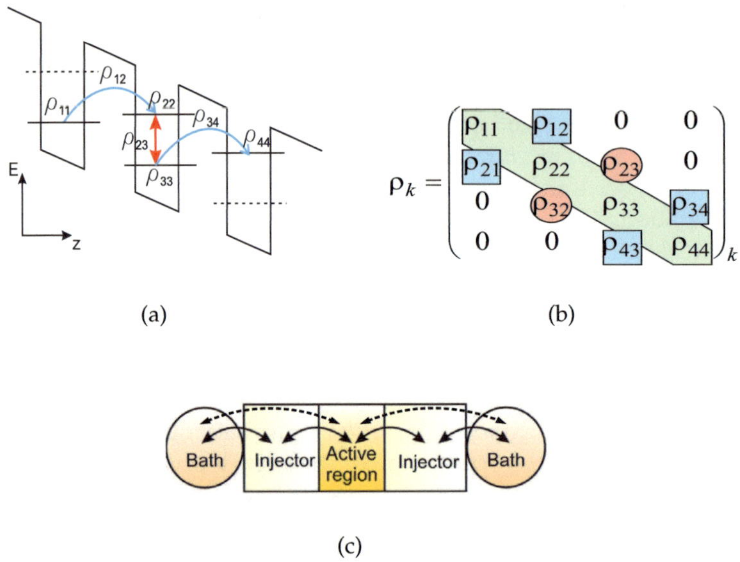

Fig. 7. Electron transitions (a), density matrix (b) and configuration (c) for the model structure

where: H_{kin} is the kinetic energy term; H_I is the light-matter interactions term; H_{ph-el} is the photonelectron scatterings term; H_{el-el} is the electron-electron interactions term; H_{el-imp} is the electron impurities scatterings term.

In this chapter, we consider electron-electron interactions at the Hartree-Fock level of approximations. All other interactions are taken into account phenomenologically via the dephasing time. In the frame of many body theory, each term in (22) and (23) is represented as a product of field operators. They could be expanded in some set of single-particle basis functions. Expansion coefficients are creation/annihilation operators. Thus, if the basis is known, the problem can be formulated in terms of creation/anihilation operators:

$$H_{kin} = \sum_{i,k} \varepsilon_{i,k} a^{\dagger}_{i,k} a_{i,k} \quad , \tag{24}$$

$$H_L = \sum_{\kappa_1} \varepsilon_{\kappa_1} L^{\dagger}_{\kappa_1} L_{\kappa_1} \quad , \tag{25}$$

$$H_R = \sum_{\kappa_2} \varepsilon_{\kappa_2} R^{\dagger}_{\kappa_2} R_{\kappa_2} \quad , \tag{26}$$

$$H_{La} = \sum_{i,k,\kappa_1} \left(h_{i,k,\kappa_1} L^{\dagger}_{\kappa_1} a_{i,k} + h_{\kappa_1,i,k} a^{\dagger}_{i,k} L_{\kappa_1} \right), \tag{27}$$

$$H_{aR} = \sum_{i,k,\kappa_2} \left(h_{i,k,\kappa_2} R^{\dagger}_{\kappa_2} a_{i,k} + h_{\kappa_2,i,k} a^{\dagger}_{i,k} R_{\kappa_2} \right), \tag{28}$$

$$H_I = \sum_{k} \left(d_{12} a^{\dagger}_{1,k} a_{2,k} + d^{*}_{21} a^{\dagger}_{2,k} a_{1,k} \right), \tag{29}$$

$$H_{el-el} = \sum_{\substack{i,j,i',j',\\ k,k',q\neq 0}} V_q^{i,j,i',j'} a^{\dagger}_{j',k'+q} a^{\dagger}_{i',k-q} a_{i,k} a_{j,k'}. \tag{30}$$

here:

$a^{\dagger}_{i,k}$	is the creation operator for the in-plane wave vector k and subband i in the active region
$a_{i,k}$	is the annihilation operator for the in-plane wave vector k and subband i in the active region
$L^{\dagger}_{\kappa_1}$	is the creation operator for the in-plane wave vector $\kappa 1$ in the left bath
L_{κ_1}	is the annihilation operator for the in-plane wave vector $\kappa 1$ in the left bath
$R^{\dagger}_{\kappa_2}$	is the creation operator for the in-plane wave vector $\kappa 2$ in the right bath
R_{κ_2}	is the annihilation operator for the in-plane wave vector $\kappa 2$ in the right bath
d_{12}	is the dipole matrix element
h_{i,k,κ_1}	is the coupling coefficient between the active region and left bath
h_{i,k,κ_2}	is the coupling coefficient between the active region and right bath
$V_q^{i,j,i',j'}$	is the Coulomb potential
k	is the in-plane wave vector for the active region
κ_1	is the in-plane wave vector for the left bath
κ_2	is the in-plane wave vector for the right bath
q	is the wave vector $q = \lvert k - k' \rvert$
i, j, i', j'	are subband indexes for the active region, $i, j, i', j' = 1, 2$

In the active region, we assume presence of only two subbands while bathes are characterized by single bands. Therefore, states in the active region have the quantum number, additional to wave vector, which is subband index $i = 1, 2$. Coupling coefficients defines properties of the transition regions between the active region and bathes. Such a transition region can be single injection barrier separating the active region and injector. Also, the whole injector can be considered as an effective barrier. The width for such a barrier is dependent on the energy and momentum of propagated particles. This approximation can be applied if electrons propagate through the injector in the ballistic transport regime (without inelastic scattering). The transmission dependence on the electron energy and momentum have been computed in [Klymenko et al (2008)] for layered structures in the ballistic limit.
The density matrix elements can be represented using creation and annihilation operators:

$$\rho_{ij,k} = \langle a^{\dagger}_{i,k} a_{j,k} \rangle. \tag{31}$$

The structure of the density matrix is represented in Fig. 7(a) and 7(b). Matrix elements at the main diagonal are probabilities of electron finding at some defined state. In other words, these elements are electron distribution functions for subbands in the active region and bathes. Elements at upper and lower subdiagonals describe transitions between subbands. The density matrix has tridiagonal structure due to the chain configuration of the transitions. It means that electron can not transit from one bath to another one avoiding the active region. That is undoubtedly an approximation and the probability of such an even exists. However, the approximation is good enough that is proved by computations of probabilities for these transitions. Squares in Fig. 7(b) indicate density matrix elements corresponding to the transitions between the active region and bathes. Circles correspond to transitions between subbands within the active region. Hereafter, non-zero density matrix elements are expressed in terms of creation/anihilation operators:

$$P_k = \langle a^{\dagger}_{2,k} a_{1,k} \rangle, \tag{32}$$

$$n_{i,k} = \langle a^{\dagger}_{i,k} a_{i,k} \rangle, \tag{33}$$

$$n^L_{\kappa 1} = \langle L^{\dagger}_{\kappa 1} L_{\kappa 1} \rangle, \tag{34}$$

$$n^R_{\kappa 2} = \langle R^{\dagger}_{\kappa 2} R_{\kappa 2} \rangle, \tag{35}$$

$$J_{\kappa 1,i,k} = \langle L^{\dagger}_{\kappa 1} a_{i,k} \rangle, \tag{36}$$

$$J_{\kappa 2,i,k} = \langle R^{\dagger}_{\kappa 2} a_{i,k} \rangle. \tag{37}$$

In consecutive order, these are the microscopic polarization, electron distribution function in the active region, electron distribution function in the left and right bath respectively, and microscopic polarizations caused by currents from the left bath to the active region and from the active region to the right bath.
To obtain information about the time evolution of any operator product or density matrix element, one should write and then solve the system of Heisenberg equations.

$$-i\hbar \frac{P_k}{dt} = \langle \left[H, a^{\dagger}_{2,k} a_{1,k} \right] \rangle, \tag{38}$$

$$-i\hbar \frac{dn_{i,k}}{dt} = \langle \left[H, a^{\dagger}_{i,k} a_{i,k} \right] \rangle, \tag{39}$$

$$-i\hbar \frac{n^L_{\kappa 1}}{dt} = \langle \left[H, L^{\dagger}_{\kappa 1} L_{\kappa 1} \right] \rangle, \tag{40}$$

$$-i\hbar \frac{n^R_{\kappa 1}}{dt} = \langle \left[H, R^{\dagger}_{\kappa 1} R_{\kappa 1} \right] \rangle, \tag{41}$$

$$-i\hbar \frac{J_{\kappa 1,i,k}}{dt} = \langle \left[H, L^{\dagger}_{\kappa 1} a_{i,k} \right] \rangle, \tag{42}$$

$$-i\hbar \frac{J_{\kappa 2,i,k}}{dt} = \langle \left[H, R^{\dagger}_{\kappa 2} a_{i,k} \right] \rangle. \tag{43}$$

3.2 Kinetic equations

After evolution of commutators in (38)-(43), one gets following equations:

$$\frac{\partial P_k}{\partial t} = -i \left(e_{2,k} - e_{1,k} \right) P_k - i \left(n_{2,k} - n_{1,k} \right) \omega_{R,k} + \left. \frac{\partial P_k}{\partial t} \right|_{scatt}, \tag{44}$$

$$\frac{\partial n_{2,k}}{\partial t} = -2Im \left(\omega_{R,k} P^*_k \right) + 2Im \left(h_{i,k,\kappa 1} J_{\kappa 1,i,k} \right) + \left. \frac{\partial n_{2,k}}{\partial t} \right|_{scatt}, \tag{45}$$

$$\frac{\partial n_{1,k}}{\partial t} = -2Im \left(\omega_{R,k} P_k \right) + 2Im \left(h_{i,k,\kappa 2} J_{\kappa 2,i,k} \right) + \left. \frac{\partial n_{1,k}}{\partial t} \right|_{scatt}, \tag{46}$$

$$\frac{J_{\kappa 1,i,k}}{dt} = -i \left(e_{L,k} - e_{2,k} \right) J_{\kappa 1,i,k} - \frac{i h_{i,k,\kappa 1}}{\hbar} \left(n^L_{\kappa 1} - n_{2,k} \right) + \left. \frac{\partial J_{\kappa 1,i,k}}{\partial t} \right|_{scatt}, \tag{47}$$

$$\frac{J_{\kappa 2,i,k}}{dt} = -i \left(e_{1,k} - e_{R,k} \right) J_{\kappa 2,i,k} - \frac{i h_{i,k,\kappa 2}}{\hbar} \left(n_{1,k} - n^R_{\kappa 2} \right) + \left. \frac{\partial J_{\kappa 2,i,k}}{\partial t} \right|_{scatt}, \tag{48}$$

$$n^L_{\kappa 1} = f^L, \tag{49}$$

$$n^R_{\kappa 2} = f^R. \tag{50}$$

$$e_{i,k} = \frac{\varepsilon_{i,k}}{\hbar} - \frac{1}{\hbar}\sum_{k' \neq k} V^{iiii}_{|k'-k|} n_{i,k'} \tag{51}$$

$$\omega_{R,k} = \frac{d_{12}E(z,t)}{\hbar} + \frac{1}{\hbar}\sum_{k' \neq k} V^{iiii}_{|k'-k|} P_{k'} \tag{52}$$

here $e_{i,k}$ is the renormalized transition frequency; $\omega_{R,k}$ is the renormalized Rabi frequency; $e_{R,k} = \varepsilon_{R,k}/\hbar$ and $e_{L,k} = \varepsilon_{L,k}/\hbar$

Equations (49) and (50) reflect approximation of the stationary carrier distribution in bathes. Thus, the kinetic equation is not necessary, and Fermi-Dirac distribution functions can be uses for the approximation. The expressions (51) reflects the renormalization of the transition frequency due to exchange interactions. Also, electron-electron interactions lead to the renormalization of the Rabi frequency represented by Eq. (52). Equations (44)-(46) have the form similar to the semiconductor Bloch equations [Haug (2004)]. Dissimilarities lie in additional terms describing electron transport between the active region and bathes. Additional equations are appeared to provide self-consistent treatment of the electron transport.
As in the previous section, we use the fourth order Runge-Kutta method to solve the problem numericaly [Chow (1999)].

3.3 Band structure, single-particle optical response in quasi-equilibrium

Inclusion of the strain effects in the consideration leads to strong modification of the electron dispersion as well.
Band structures of both interband and intersubband heterostructures are schematically shown in Fig.8. The heterostructures of both kinds have additional subband structure inside the allowed bands. In the interband structures the optical radiation is a result of electron transitions from the conduction subband to the valence subband. As a result, the minimal quantum of the energy is limited by the band gap of the quantum-well material. Curvatures of the bands involved in the transition have very different magnitudes and, what is more important, different senses of curvature. It results in the joint density of states which is stepped one in this case.
Optical transitions in the quantum-cascade heterostructures occur between subbands within an allowed band (see Fig. 8(b)). In contrast to the interband heterostructures, the subband structure is governed by the conduction band offset and width of the qauntum well layer. Minimal transition energy is not limited by the fundamental band gap and can be tailored by a material composition of the quantum well and the thickness of the quantum-well layer. Therefore, quantum-cascade structures are widely used to achieve lasing in THz range. The charge carriers inside the band are characterized by the effective mass The curvature of dispersion curves is almost the same, and their senses of curvature are coincided. It results in the narrow joint density of states, Fig.8(b). Although difference in the curvature of the dispersion curves can be small, it has great influence on the optical characteristics of the quantum-cascade structures. We have examined three cases when subbands with different curvatures are involved in the optical transition. They are shown schematically on Fig.9, where E_{f1} and E_{f2} are quasi-Fermi levels for corresponding subband.
Different relations between effective masses for subbands leads to different absorption spectra. When $m_1 > m_2$ we have $\hbar\omega|_{k=0} > \hbar\omega|_{k\neq 0}$. On the contrary, we have $\hbar\omega|_{k=0} < \hbar\omega|_{k\neq 0}$

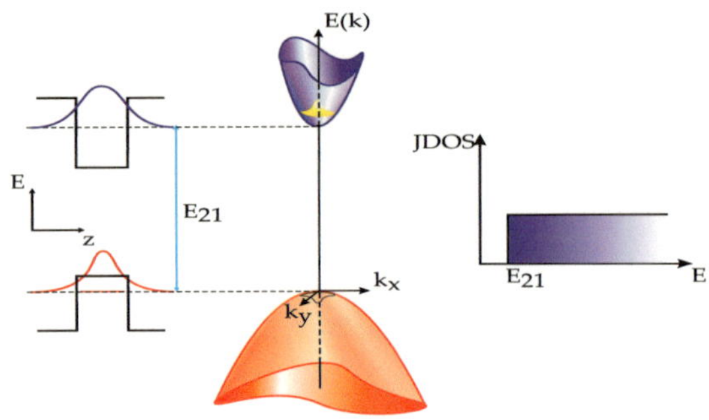

(a)

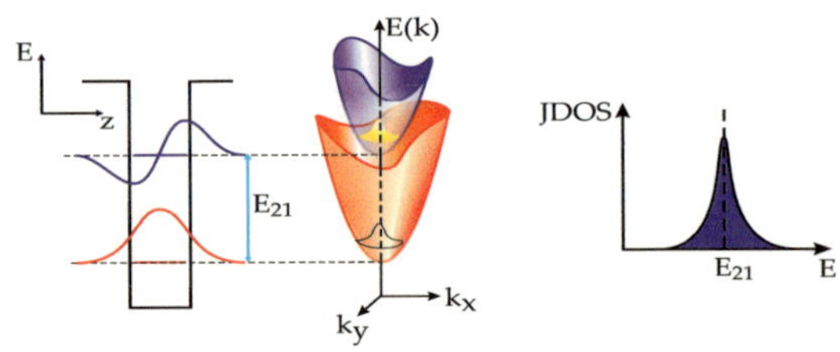

(b)

Fig. 8. Sketches of the band diagrams, band structures and joint DOS for two cases of interband and intersubband transitions.

when $m_1 < m_2$. And, in the case of equal effective masses, one gets $\hbar\omega|_{k=0} = \hbar\omega|_{k\neq0}$. Fig. 10 contains calculated single-particle absorption spectra. Vertical line indicates the energy of intersubband transition E_{12} at the center of the Brillouin zone without renormalization , i.e. $E_{12} = E_1|_{k=0} - E_2|_{k=0}$. Two important features are observed. Depending on the relation between the effective masses in the subbands, maximum of the absorption get red- or blue-shifted relative to the case of the equal effective masses. The value of the shift is about 20 meV, what is very important in the THz range. Difference of effective masses leads to additional broadening of the absorption spectrum and decreasing of its maximum comparing with the case when effective masses are equal. Thus, the band structure with energy-dependent effective mass affects strongly on optical response of QCS.

3.4 Many-body effects within the Hartree-Fock approximation

In this section, we take quick look at many-body effects in the QCS at the Hartree-Fock level of approximations. At this level of approximations, electron-electron interaction effects are described in the frame of the mean-field approximation when only exchange interactions and Rabi frequency renormalization are taking into account. Fig. 11 contains computed absorption spectra for the quasi-equilibrium regime. Three cases have been considered: single-particle

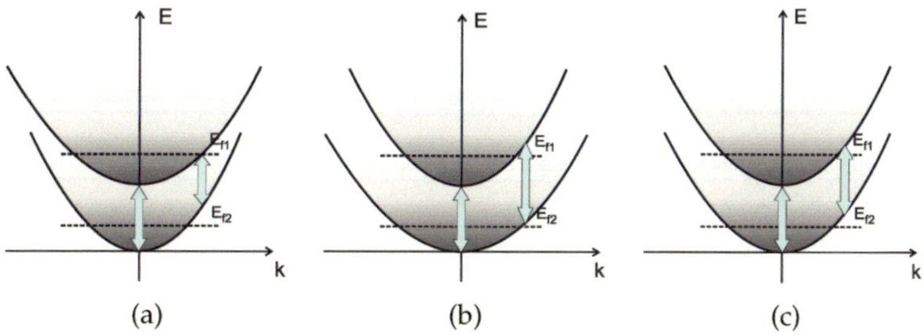

Fig. 9. Sketches of the band structures for various combinations of the effective masses in two subbands involved in radiation transitions: a) $m_1 > m_2$; b) $m_1 < m_2$; c) $m_1 = m_2$.

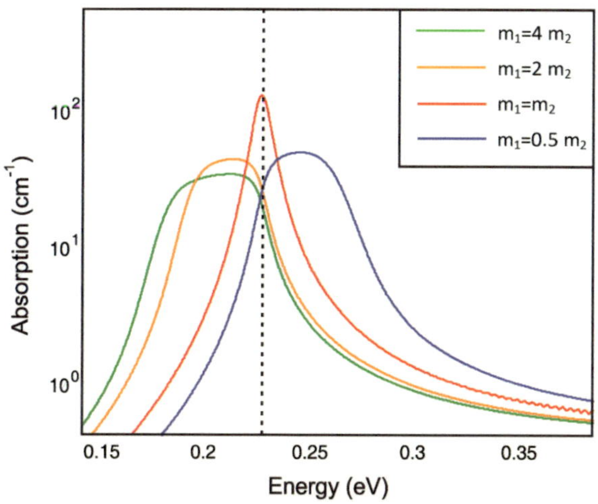

Fig. 10. Single-particle absorption spectra for various combinations of the effective mass in two subbands involved into radiation transitions.

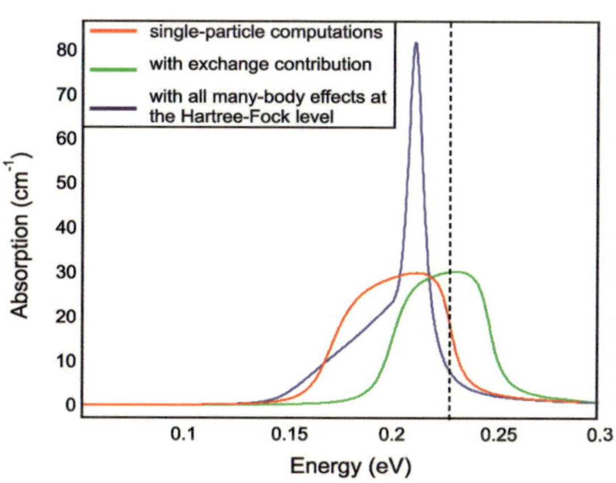

Fig. 11. Many-body effects in the optical absorption spectrum

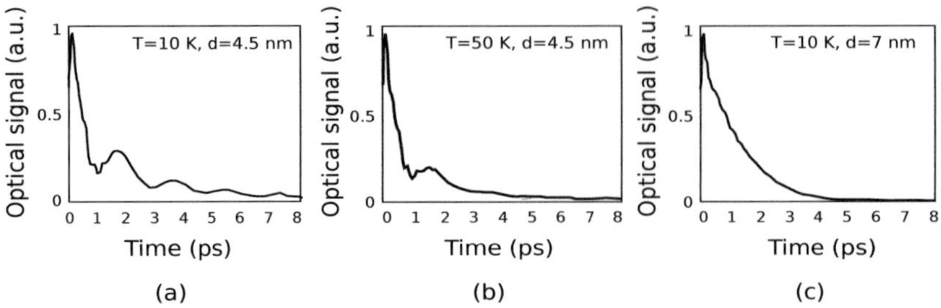

Fig. 12. Optical signals in pump-probe experiments. Adapted from [Weber et al (2009)].

optical response, effect of transition energy renormalization due to the exchange contribution and all many-body effects at the Hartree-Fock level of approximation including Rabi frequency renormalization. All these cases are attended by dephasing treated phenomenologically. Presented results are evidence of high importance of many-body effects which lead to dramatical changes in absorption spectra. In Fig. 11, the dashed line marks energy gap between subbands at the center of Brillouin zone ($k = 0$).

The exchange energy term causes shifting of the absorption spectra into high energies. Contribution of the exchange energy term leads to decreasing of energy for electrons populating subbands. Energy reduction for each subband is proportional to its electron population. Therefore, transition energy is increased if a lower subband contains more carriers comparing with higher one. In the opposite case, when higher subband is more populated, the transition energy is decreased. Both cases have been reported in papers [Mi (2005)] for the first case and [Pereira (2004)] for the second one). That is the distinguished feature of intersubband transitions. Energy of interband transitions is always decreased if the exchange contribution is taking into account. Energy of intersubband transitions can be shifted in any directions depending on subbands populations.

Hartree-Fock approximation includes the Rabi frequency renormalization represented in the polarization equation (44). Joint action of the exchange contribution and Rabi frequency renormalization on the spectrum are marked by the blue line in Fig. 11. As follows from results, Rabi frequency renormalization (also known as depolarization) leads to the occurrence of a narrow peak in the absorption spectrum. The frequency corresponding to this peak is the frequency of optically excited coherent collective oscillations in the electron plasma. Such plasma colective oscillations are called the intersubband plasmons [Mi (2005)]. Theory of coupled photon and intersubband plasmon was developed in [Pereira (2007)], and this theory gives rise of new quasiparticle titled antipolariton.

3.5 Electron transport effect

The effects of the coherent transport can be observed in pump-probe experiments at the femtosecond and picosecond time intervals. The pump-probe experiment consists in propagation through the investigated media of two optical pulses shifted in time relative each other. First pump pulse is characterized by high intensity, and it excites optically-active media. The second pulse reads changes in the media undergoing optical absorption or gain. More details about pump-probe techniques can be found in [Weber et al (2009)]. Fig. 12 contains results of pump-probe optical experiments reported in [Weber et al (2009)]. The pump pulse have the shape of the Gaussian function.

Each subfigure corresponds to defined parameters which are the temperature and width of the injection barrier in the QCS. Oscillations of the optical response signal at low temperature and barrier's width is caused by coherent electron transport between active region and injector through the injection barrier. The decay of oscillations with increasing of temperature is effect of many-body interactions. Scatterings leads to destroying of the coherence via dephasing. Represented data also reflects the effect of injection barrier width on electron transport. As have been mentioned above, the coherent electron transport is strongly dependent on the interaction between quantum wells defined by parameters of the potential barrier. As far as the width of barrier is increased, the interaction between quantum wells is decreased and, therefore, the frequency of oscillations is decreased.

4. Conclusions

In this chapter, we have considered influence of the electron transport on the optical properties of quantum-cascade structures. The electron transport can be treated as evolution of the electron distribution function in time and space. On the one hand, optical processes are strongly dependent on this function, and, on the other hand, they cause changes of the distribution function due to radiative transitions of charge carriers. Therefore, transport and optical processes are strongly coupled via the electron distribution function. This situation is common for all semiconductor structures. However, the case of QCS has many particularities connected with intersubband transitions and tunneling coupling of the active regions in neighboring cascades. At very short time intervals, electrons coherently pass from one active region to another through injector. Depending on injectors width and structure, carriers can propagate through whole injector without inelastic scatterings. In the oposite case, electron from the active region makes coherent transitions to some energy level in the injector. Thus, it has been shown that the coherent transport influence optical chacteristics at the time interval been of order up to one picosecond. This result is confirmed by experimental data.
Our consideration is based on the density matrix theory. This approach is appropriate for equilibrium case as well as for non-equilibrium one and open quantum systems. We have derived kinetic equations describing dynamics of the electron distribution function, polarization and tunneling microcurrents.
The single-particle band structure influences strongly the shape of optical absorption spectra. Consideration of the position- and energy-dependent effective mass increases acuracy of obtained results.
Many-body effects are relevant for all operational regimes of QCS. They determine the inhomogeneous broadening of spectral characteristics and their peaks position at the energy scale. The temperature dependence of optical characteristics is caused by many-body effects.
It is necessary to provide future investigations of the interference between electron transport and optical processes including in the consideration many-body interactions in injectors and correlations of electrons through several periods.

5. References

Gmachl, C.; Capasso, F.; Sivco, D.L.; Cho, A.Y. (2001) Recent progress in quantum cascade lasers and applications. *Rep. Prog. Phys.*, 64, 11, November 2001, 1533-1601, ISSN
Iotti, R.C.; Rossi, F. (2001) Nature of charge transport in quantum cascade lasers. *Phys. Rev. Lett.*, 87, 14, October 2001, 146603-1-4, ISSN

Weber, C.; Wacker, A.; Knorr A. (2009) Density-matrix theory of the optical dynamics and transport in quantum cascade structures: The role of coherence. *Phys. Rev. B*, 79, 2009, 165322-1-14, ISSN

Optoelectronic devices: advaced simulation and analysis, Piprek J. (Ed.), Springer, ISBN 0-387-22659-1, New York

Femtosecond laser pulses, Rulliere C. (Ed.), Springer, ISBN 0-387-01769-0, New York

Lee, Y.-S. (2009) *Principles of Terahertz Science and Technology*, Springer, ISBN 978-0-387-09539-4, New York

Lee, S.-C.; Wacker, A. (2002) Nonequilibrium GreenâĂŹs function theory for transport and gain properties of quantum cascade structures. *Phys. Rev. B*, 66, 2002, 245314-1-18, ISSN

Vukmirović, N., Jovanović, V.C.; Indjin, D.; Ikonić, Z.; Harrison, I.; Milanović, V. (2005) Optically pumped terahertz laser based on intersubbnad transitions in a GaN/AlGaN double quantum well. *J. Appl. Phys.*, 97, 2005, 103106-1-5, ISSN

Meier, T.; Thomas, P.; Koch, S.W. (2007) *Coherent Semiconductor Optics*, Springer, ISBN 10-3-540-32554-9, Berlin

Haug, H.; Koch, S.W. (2004) *Quantum theory of the optical and electronic properties of semiconductors*, World Scientific Publishing, ISBN 981-238-609-2, Danvers

Vu, Q.T.; Haug, H.; Koch, S.W.; (2006) Relaxation and dephasing quantum kinetics for a quantum dot in a optically excited quantum wells. *Phys. Rev. B*, 73, 2006, 205317-1-8, ISSN

Faist, J., Capasso, F.; Sirtori, C.; Sivco, D.L.; Hutchinson, A.L.; Cho, A.Y. (1995) Vertial transitions quantum cascade laser with Bragg confined excited state. *Appl. Phys. Lett.*, 66, 05, January 1995, 538-540, ISSN

Klymenko, M.V.; Safonov, I.M., Shulika, O.V., Sukhoivanov, I.A. (2008) Ballistic transport in semiconductor superlattices with non-zero in-plane wave vector. *Physica Stat. Solidi B*, 245, 8, June 2008, 1598-1603, ISSN

Chow, W.W.; Koch, S.W. (1999) *Semiconductor lasers: fundamentals*, Springer, ISBN 3-540-66166-1, Berlin

Mi, X.W.; Cao, J.C.; Zhang, C.; Meng, F.B. (2005) Effects of collective excitations on the quantum well intersubband absorption. *J. Appl. Phys.*, 98, 2005, 103530-1-5, ISSN

Pereira, M.F., Lee, S.-C.; Wacker, A. (2004) Controling many-body effects in the midinfrared gain and terahertz absorption of quantum cascade laser structures. *Phys. Rev. B*, 69, 20, 2004, 205310-1-7, ISSN

Pereira, M.F., (2007) Intersubband antipolaritons: microscopic approach. *Phys. Rev. B*, 75, 19, 2007, 195301-1-5, ISSN

12

Preparation of transparent conductive AZO thin films for solar cells

Vladimir Tvarozek[1], Pavol Sutta[2], Sona Flickyngerova[1], Ivan Novotny[1], Pavol Gaspierik[1], Marie Netrvalova[2] and Erik Vavrinsky[1]
[1]*Department of Microelectronics, Slovak University of Technology in Bratislava*
[2]*Department of Materials & Technology, West Bohemian University, Plzen*
[1]*Slovakia,* [2]*Czech Republic*

1. Introduction

Transparent conducting oxides (TCOs) based on ZnO are promising for application in thin-film solar photovoltaic cells (PVCs) and various optoelectronic devices (Minami, 2005). Desired parameters of ZnO and doped ZnO:Al (AZO) thin films are given by their role in superstrate configuration of tandem Si solar cell (Zeman, 2007): the light enters the cell through the glass substrate where two pin absorber thin-film structures are placed between two TCO layers with back metal contact. The upper front contact AZO layer should fulfill several important requirements: high transparency in VIS/near IR solar spectrum; high electrical conductivity; suitable surface texture in order to enhance light scattering and absorption inside the cell; high chemical stability and adhesion to silicon. Moreover, bottom ZnO interlayer between Si and metal (usually Ag) contact is acting as barier and adhesion layer as well as optical matching layer to Ag back contact to improve its reflection of radiation, particularly in near IR region (Dadamseh et al., 2008). Optimization of the front contact TCO has proven to be crucial for getting the high cell efficiency (Berginski et al., 2008).

RF sputtering is owning several advantages in comparison with the other physical and chemical deposition methods: a low-temperature ion-assisted deposition of metals, semiconductors, insulators, the before/post deposition modification of substrate/thin - film surface by ions on the micro-/nano- level; change of deposition rate in wide range (0,1 to 10 nm/s); to control further parameters which are important for thin film growth (substrate temperature, plasma density, composition of working gas, ion bombardment of film during deposition). In addition there is a significant contribution of secondary electron bombardment to the atomic scale heating of the film when it is prepared by the RF diode sputtering.

Therefore RF sputtering of AZO films from ceramic target is often used to get the best their electrical and optical properties. An influence of different technological parameters was investigated: partial pressure of oxygen (Tsui & Hirohashi, 2000), substrate temperature (Fu & Zhuang, 2004), (Ali, 2006), (Berginski et al., 2008), substrate bias voltage (Ma & Hao, 2002), (Lim & Kim, 2006), post-deposition annealing (Fang at al., 2002), (Oh et al., 2007), (Berginski et al., 2008), surface-texturing by chemical etching (Kluth & Rech, 1999), (Berginski et al., 2008) or ion-sputter etching (Flickyngerova, et al. 2009). The complex study and an optimization of various deposition parameters were done by using in-line AC magnetron sputtering system with Zn/Al compound targets (Sittinger et al., 2006).

In general, sputter deposition is determined by complex processes proceeded: (a) at the target bombarded by energetic ions, (b) in the low-temperature plasma, (c) on the surface of substrate and growing film. In general, thin film growth is influenced by the kinetic energy of coating species on the substrate - in addition to substrate temperature a total energy flux is acting to the substrate and growing thin film. It depends mainly on the amount and the energy of: (i) sputtered coating species, (ii) energetic neutral working gas atoms (neutralized and reflected at the target), (iii) energetic secondary electrons emitted from the target, (iv) negative ions coming from the working gas plasma or target, (v) ions bombarding the substrate in bias or reactive mode. These effects can cause significant changes in the crystallic structure, surface morphology and chemical stoichiometry of sputtered thin films, i.e. they can modify their electrical and optical properties. The existence of high-energy particles bombarding the film during both the planar diode and the planar magnetron sputtering of ZnO was confirmed (Tominaga et al., 1982). It was found from energy analyses that the high-energy neutral oxygen atoms should be taken into account above working pressures 1.3 Pa and negative oxygen ions accelerated at the target becomes important at pressures in the range of 0.1 Pa. The negative ion resputtering by oxygen ions during sputtering of ZnO:Al thin films has caused extended defects in the film crystalline structure (interstitials, lattice expansion, grain boundaries) - it was responsible for the degradation of electrical properties of these films (Kluth et al., 2003), (Rieth & Holloway, 2004). Thornton's microstructural model developed for sputtered metal thin films (Thorton & Hoffman, 1989) they modified for magnetron sputtered ZnO at low-/medium-/high-pressure regions (0.04 - 4 Pa) and they discussed the correlation of sputter parameters (sputter gas pressure and substrate temperature) to structural and electrical properties of thin film. These results and next ones obtained also later (Kluth et al., 2006) showed a strong dependence of ZnO:Al thin film properties on sputter gas pressure and oxygen content in working gas.

Structural models based on Thornton's assumptions are well satisfied in the technological approach of sputtering of metals. In the parameter „Ar working gas pressure" he implicitly included collisions between the sputtered and Ar atoms at elevated pressures causing the deposited atoms to arrive at the substrate in randomized directions that promote oblique coating. Therefore to use more physical approach, in addition to substrate temperature Ts, we introduced a total energy flux density E_Φ [W/m^2] affecting to the substrate and the growing thin film (Fig. 1). A total energy flux density, by other words power density E_Φ,, can be expressed by microscopic quantities known from the kinetic theory of gases, low-temperature plasma physics and the models of sputtering processes. It can be also estimated by macroscopic sputtering parameters like supply RF power, deposition rate, average DC voltage induced on target, flow or pressure of working gases, substrate bias voltage or power (Tvarozek et al., 2007). The substrate temperature is normalized to the melting temperature T_m of sputtered material, T_s / T_m. The substrate temperatures are usually very far from melting point of ZnO (T_m = 1975°C) during the sputtering that's why we found useful to express T_s/T_m in logarithmic scale. The ratio of the total energy flux density E_Φ and its minimum value $E_{\Phi min}$ specified by the sputtering mode and the geometrical arrangement of the sputtering system is $E_\Phi / E_{\Phi min}$. Optimal conditions for deposition of semiconductor oxides and nitrides (ITO, TiN, ZnO, ZnO:N, ZnO:Al, ZnO:Ga, ZnO:Sc) in our diode sputtering system corresponded to the relative total energy flux density $E_{\Phi} / E_{\Phi min}$ in the range of $4 \div 7$, $E_{\Phi min} \sim 1 \times 10^4 \, W/m^2$, (Fig. 1, dashed lines).

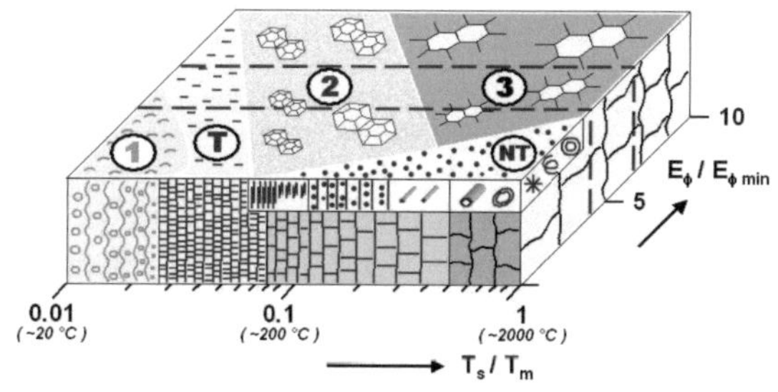

Fig. 1. Crystalline structure zone model of sputtered ZnO thin films: Zone 1 - porous structure of tapered amorphous or crystalline nanograins separated by voids, Zone T - dense polycrystalline structure of fibrous and nanocrystalline grains, Zone 2 - columnar grain structure, Zone 3 - single-crystal micrograin structure, Zone NT - nanostructures and nanoelements.

The aim of present work has been to find correlations among the technological parameters (power density, substrate temperature and post-deposition annealing) and structural / electrical / optical properties of AZO thin films. In the beginning to accelerate our investigation of desirable thin film properties we used the RF diode sputtering where one can get continual changes of thin film thickness (of composition also) in one deposition run.

2. Modelling and simulation

Computer simulations have proved to be an indispensable tool for obtaining a better understanding of solar photovoltaic cells (PVC) performance and for determining trends for optimizing material parameters and solar cell structures. We focused on the simulations of both the parasitic effect in real bulk PVCs and progressive thin film solar PVCs, based on amorphous silicon and transparent conductive layers of ZnO, ZnO:Al.
Sputtering is an important technique for deposition of both multicomponent thin films for solar applications as well as multilayer coatings with only few nanometers thin layers (so-called superlattices) which exhibit superior hardness, high wear, corrosion resistance and thermal stability (Panjan, 2007). Sputter deposition is attractive particularly in industrial applications due to the need of high deposition rates and uniform coverage over large areas. Therefore it is desirable to know what influence has the sputter system arrangement on spatial distribution of sputtered particles on the top of substrate (so-called deposition profile), i.e. on homogeneity of growing film properties.

2.1 Electric properties of PVC

The most important electric parameters, which are used to characterize the quality of PVC, are defined: the short-circuit current I_{SC} (the current through the solar cell when the voltage

across the solar cell is zero), the open-circuit voltage V_{OC} (the maximum voltage available from a solar cell, at zero current), the fill factor FF (indicating how far the product $I_{SC}V_{OC}$ is from the power delivered by the PVC) and the conversion efficiency (η).
The conversion efficiency is defined as the ratio of the photovoltaically generated electric output of the cell to the radiation power falling on it P_{in}:

$$\eta = \frac{I_m V_m}{P_{in}} = FF\frac{I_{SC}V_{OC}}{E \times A} = FF\frac{J_{SC}V_{OC}}{E}, \tag{1}$$

where FF is the fill factor of PVC $I_m V_m / I_{SC} V_{OC}$ (or area ratio A / B in Fig. 2), E is value of irradiance and J_{SC} is the short-circuit current density I_{SC}/A. The values of V_m and I_m are the co-ordinates for maximal power point (he designates the optimal operating point of PVC), and can be estimated from the open circuit voltage and short circuit current: $V_m \sim (0.75\text{-}0.9) \times V_{OC}$, $I_m \sim (0.85\text{-}0.95) \times I_{SC}$ (Goetzberger & Hoffmann, 2005). Efficiency is measured under standard test conditions (temperature of PVC 25°C, irradiance 1000 Wm^{-2}, air mass 1.5).

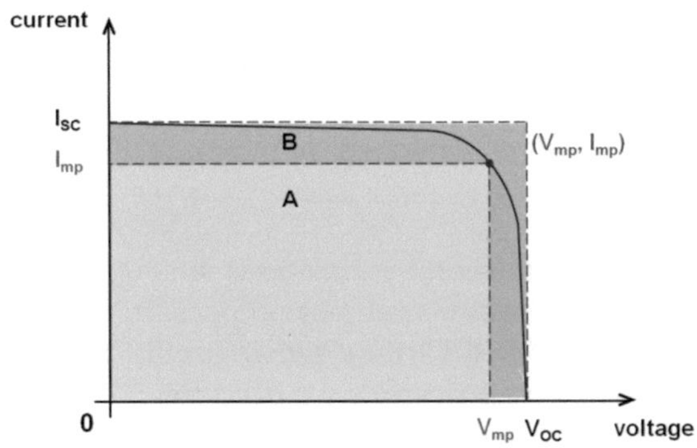

Fig. 2. C-V and P-V (dash line) characteristics of illuminated solar cell

2.2 PSPICE model of bulk solar cell the 1st generation

The model of simple 1st generation PVC (i.e. *p-n* junction represented as bulk silicon diode of large-area), following equivalent circuit diagram (Fig. 3) by PSpice software (PSpice A/D Circuit Simulator, 2009) for analysis of electronic circuits and their simulations, was created.
By means of the two-diode model we achieved the better description of PVC. Diode D_1 is representing the carrier injection current I_{INJ} and diode D_2 the recombination current I_R. The values of saturation current densities and ideality factors for this diodes are different: J_{s1} = 1e-12 Acm^{-2}, J_{s2} = 1e-8 Acm^{-2}, $m_1 = 1$ (ideal diode), $m_2 = 2$. We define the size of PVC area A = 100 cm^2. The other components are of resistive nature, a parallel (or shunt) resistance R_P and the series resistance R_S.
For obtaining high efficiency of PVC, the parallel parasitic resistance R_P (described loss currents at the edges of the solar cell and surface inhomogeneities) should be as high as possible and the series resistance R_S (the resistance through the wafer, the resistance of the back surface contact and the contact grid on the front surface) as a low as possible, ideally it's a deal: $R_S = 0$, $R_P \rightarrow \infty$. The values of parasitic resistances depend on PVC size, consequently also from area.

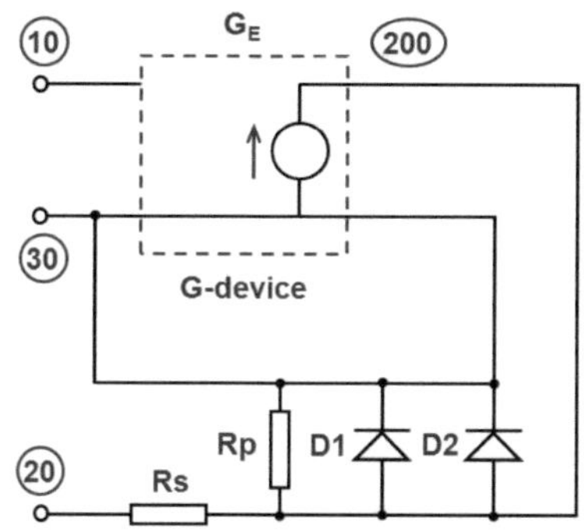

Fig. 3. Equivalent circuit of real bulk PVC modelled in PSpice

In PSpice is the value of the short circuit current I_{SC} assigned to a voltage-controlled current source (G-device, Fig.2) and is given by:

$$G_E = \frac{J_{SC}A}{1000}E \tag{2}$$

We considered that the value of short-circuit current density J_{SC} is given at standard test conditions.

The effect of parasitic resistances R_S and R_P on C-V characteristic is shown in the Fig.4.

As can be seen, for parasitic serial resistance R_S (Fig.4a) the values of the short-circuit current and of the fill factor (it follows too efficiency) can be expressively reduced. At the high values of R_S occur the big reduction of the short-circuit current value I_{SC}. The open-circuit voltage is independent of the series resistance. The product $R_S.I_{SC}$ should never have been greater than 25 mV in praxis (outside temperature 25 °C).

The parallel resistance also degrades the performance of the PVC, Fig.4b.

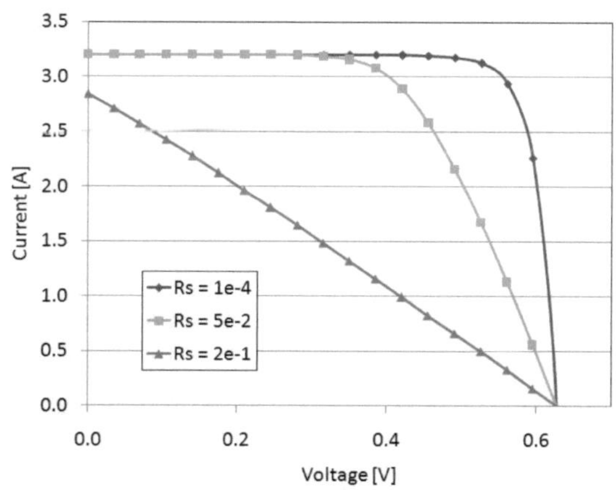

a)

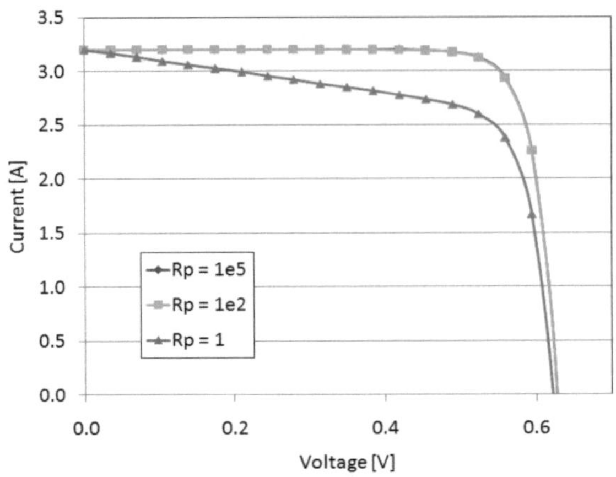

b)

Fig. 4. C-V characteristics of bulk PVC structure (illumination for AM1.5) for modification of serial (a) and parallel resistance (b)

Small values of the parallel resistance heavily degrade the fill factor (i.e. efficiency). Also are the value of open-circuit voltage reduced, the short-circuit current is independent of the parallel resistance.

The concrete values of parasitic resistances, that we used by the simulation with PSpice are:

- serial parasitic resistance R_S: 1e-4, 5e-2, 2e-1 Ω,
- parallel parasitic resistance R_P: 1e5, 1e2, 1 Ω.

Selected parameters of illuminated PVC with the parasitic resistances R_S and R_P is shown in the Tab.1. The parameters I_{SC} and V_{OC} are assigned from the graph, parameters FF and η are calculated by (eq. 1).

R_S (Ω)	I_{SC} (A)	V_{OC} (V)	FF (-)	η (%)
1e-4	3.200	0.628	0.825	16.58
5e-2	3.200	0.628	0.604	12.14
2e-1	2.844	0.628	0.261	4.66
R_P (Ω)	**I_{SC} (A)**	**V_{OC} (V)**	**FF (-)**	**η (%)**
1e5	3.2	0.628	0.825	16.58
1e2	3.2	0.628	0.824	16.56
1	3.2	0.622	0.686	13.65

Table 1. Selected parameters of bulk PVC structure (illumination for AM1.5)

2.3 ASA model of thin film solar cell the 2nd generation

Progressive solar PVC, 2nd and 3rd generation with higher efficiency of 20÷40%, are formed in thin film structures, predominantly based on amorphous silicon (a-Si:H, *p-i-n* junction) as the absorber material and transparent conducting oxide (TCO) semiconductors for transparent electrodes, e.g. single junction *p-i-n* a-Si PVC structure "glass/TCO/a-Si:H (*p-i-n*)/TCO/Ag or Al (reflective back contact)" or tandem solar cell structure "glass/TCO/a-Si:H (*p-i-n*)/µc-Si:H (*p-i-n*)/ TCO/Ag or Al (reflective back contact)" (Zeman, 2007). For the simulation of the thin film PVC we have used the ASA program, developed at Delft University of Technology (Zeman et al., 2005), which is designed for the simulation of multilayered heterojunction device structures.

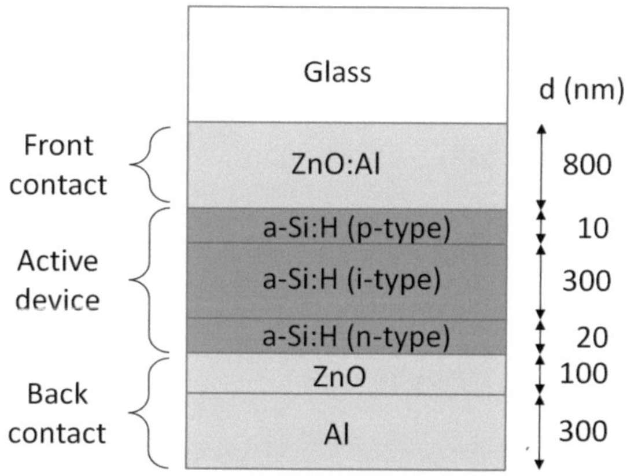

Fig. 5. Superstrate PVC configuration of single junction (*p-i-n*) structure

We focused for "superstrate" configuration of thin-film solar PVC: Glass/ZnO:Al/a-Si:H (*p-i-n*)/ZnO/Al (reflective back contact). Schematic structure of a single junction a-Si:H PVC is shown in Fig. 5. The active device consists of three layers: a *p*-type a-Si:H layer, an intrinsic a-Si:H layer and an n type a-Si:H layer. This layers form a *p-i-n* single junction. The doped layers set up an internal electric field across the intrinsic a-Si:H layer and establish low loss ohmic electrical contacts between the a-Si:H part of the PVC and the external electrodes (Zeman, 2007).

The thickness of the *i*-region should be optimized for maximum current generation. In practice is limit the *i*-region thickness to around 0.5 µm (Nelson, 2003).

Transparent conducting oxides based on ZnO are promising for application in thin-film solar photovoltaic cells. The upper front contact Zno:Al layer should fulfil several important requirements: high transparency in VIS/near IR solar spectrum; high electrical conductivity; suitable surface texture in order to enhance light scattering and absorption inside the cell; high chemical stability and adhesion to silicon. Moreover, bottom ZnO interlayer between Si and metal (usually Ag) contact is acting as barier and adhesion layer as well as optical matching layer to Ag back contact to improve its reflection of radiation, particularly in near IR region (Dagamseh et al., 2008). Optimization of the front contact TCO has proven to be crucial for the high cell efficienty (Berginski et al., 2008).

Computer simulations for single junction a-Si:H PVC structure (Fig. 4) we compile in ASA software. The thicknesses of particular layers are show in Fig. 4. All important electric properties are set direct in the C-V characteristic for illuminated *p-i-n* PVC structure (Fig.6). Also in this case are the parameters I_{SC} and V_{OC} assigned from the graph, parameters *FF* and η are calculated by (eq. 1).

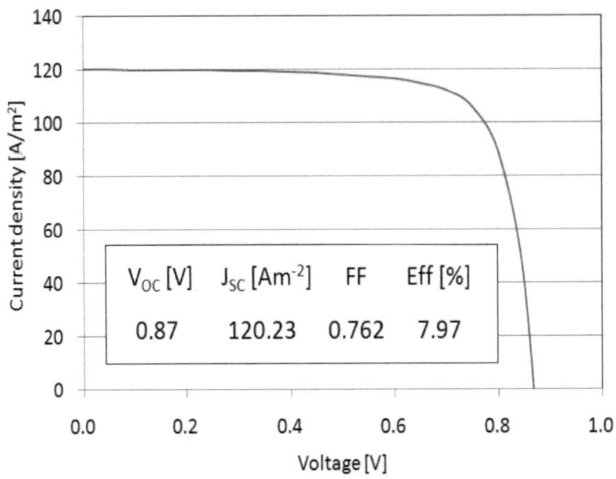

Fig. 6. C-V characteristics of TF PVC structure (illumination for AM1.5) for the thicknesses of ZnO:Al layer 800 nm and ZnO layer 100 nm

For the simulations the next parameters of TCO layers were used (obtained experimentally at the wavelength 500 nm):

1) ZnO:Al:

- refractive index $n = 2.675$
- absorption coefficient $\alpha = 1.71e5\ m^{-1}$

2) ZnO:

- refractive index $n = 2.052$
- absorption coefficient $\alpha = 7.55e5\ m^{-1}$

Reflection and quantum efficiency of selected thin film PVC structures for different thicknesses of ZnO layers a shown on Fig. 7 and Fig. 8.

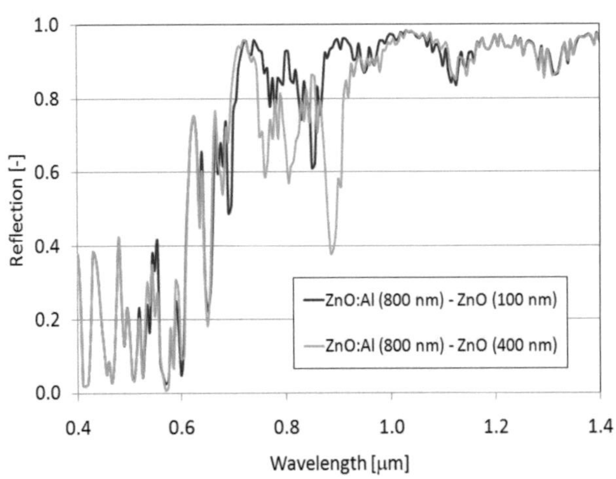

Fig. 7. Reflection of TF PVC structure for different thicknesses of ZnO layers

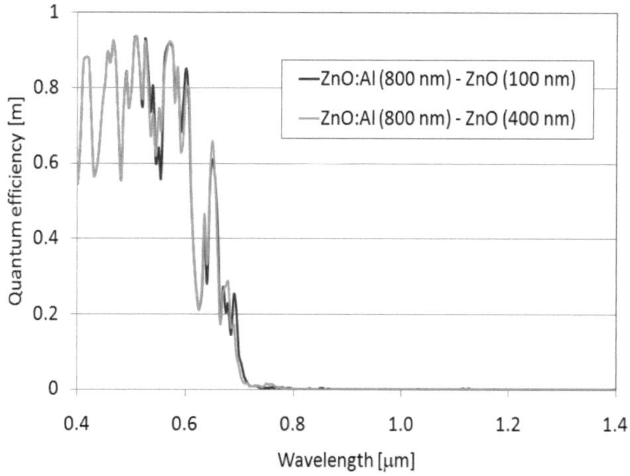

Fig. 8. Quantum efficiency of TF PVC structure for different thicknesses of ZnO layers

Reflection in the near IR region has a tendency to increase with the decrease of thickness of front ZnO layer. From dependences of quantum efficiency on wavelength result the fact, that the simple TF PVC structure shows lower effectiveness in the near IR region.

2.4 Sputtering

Computer simulations of magnetron sputtering corresponding to multi-source and multi-fold substrate rotation facilities have been progressively improved [Rother, 1999], [Jehn, 1999]. These calculations consider over-cosine distribution of sputter-deposited fluxes on the substrate area A_s , $dJ/dA_s \sim cos^n\,\Theta$, $1 < n \leq 2$, which reflects the kinetic energy losses of sputtered particles due to their collisions with working gas atoms.

We have applied the model and simulations of the spatial distribution of sputtered particles (Tvarozek et al., 1982) - deposition profile - in our diode system consisting of the plan-parallel arrangement of target and substrate in the distance of D (Fig. 9) with these idealized assumptions:

(a) Material of the target is emitted uniformly from the target area A_t;

(b) Angular distribution of the intensity of particles rejected from the target (sputtered particle flux J in the direction given by an angle Θ to the normal of surface) conforms with the Knudsen cosine law (Kaminsky, 1965)

$$J = J_0 \cos \Theta \tag{3}$$

or with its slight modifications, J_0 is the sputtered flux perpendicular to the target surface;

(c) Scattering of particles on the way to substrate is neglected;

(d) Accommodation coefficient of particles condensed on the substrate is equal 1 (re-sputtering is not included).

Then the flux of material sputtered from the target surface element dA_t to the substrate area element dA_s in the point P is given

$$d^2J = (J_0/\pi L^2)\cos^2\Theta\, dA_t\, dA_s, \tag{4}$$

where Θ is an angle between dA_t - dA_s path and the normal of target/substrate planes ($dA_t = dA_s \cos\Theta$). By integrating of Eq. (4) over the target area A_t we can get the deposition profile $G(P)$ in the arbitrary point P on the substrate

$$G(P) = (\pi/J_0)\, dJ/dA_s = \int_{A_t} \cos^2\Theta / L^2\, dA \tag{5}$$

It is useful to normalize the deposition profile $G_N(P)$ towards the maximal value of $G(0)$ which is usually the beginning of the co-ordinal system across the substrate

$$G_N(P) = G(P) / G(0) \tag{6}$$

For computing and simulations of deposition profiles we used Agilent Visual Engineering Environment (Agilent VEE Pro, 2009) - graphical language programming environment optimized for the use with electronic instruments, that provides a quick path to measurement and analysis.

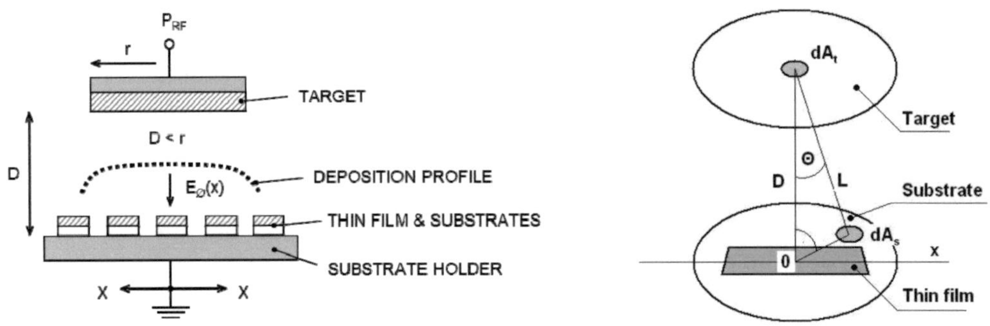

Fig. 9. Arrangement of the diode sputtering system

Simulated deposition profiles were compared with the lateral radial distribution of thin film thickness over the substrate. RF diode sputtering was performed at the working Ar gas pressure p_{Ar} = 1.3 Pa and Corning glass substrates were placed on different radial positions under the target of a diameter of 152.4 mm (ZnO+2 wt. % Al_2O_3) or of 76.2 mm (ZnO+2 wt. % Ga_2O_3 and ZnO+2 wt. % Sc_2O_3). In this case, simulations and experiments confirmed that an intensity of particles emitted from target approximately conforms to the Knudsen cosine law (Fig. 10 a). Changes of thin film thicknesses were less than 5% in the central substrate region of diameter ≤ ½ of target diameter. In lateral distances larger than target radius, the thin film thickness decreased down to 1/4 of the maximal value.

The detail comparison of various theoretical and experimental deposition profiles has shown that the sputtered fluxes from target follow slightly over-cosine angular distribution during the diode sputtering at pressures below 1 Pa (Fig. 10 b). Sputtered particle flux J in the direction given by an angle Θ towards normal of target surface can be described by equation

$$J = J_0 \, (c_1 \cos \Theta + c_2 \cos^2 \Theta) \tag{7}$$

where the coefficients were estimated from experimental deposition profiles (approximately $c_2 \approx 0.1\, c_1$).
Using the RF diode sputtering in the low-pressure region ($p \leq 1.3$ Pa), the mean free path of sputtered particles (~ 10^{-2} m) is comparable with the distance of target - substrate and therefore we can assume "collision-less" regime, particularly for high energetic particles passed through RF discharge. Neutral sputtered particles and energetic species (ions neutralized at the target and reflected from it) sputtered from the target conform to the Knudsen cosine law slightly modified by over-cosine dependence and their spatial distribution on the substrate (deposition profile) is described by power cosine dependence (Eq. 5).

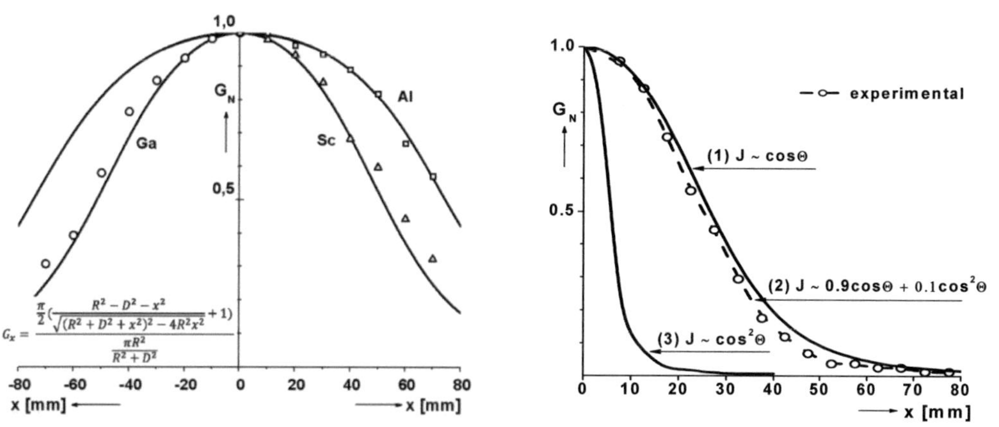

Fig. 10. Experimental and theoretical deposition profiles of: (a) ZnO:Al target with diameter of 152.4 mm and ZnO:Ga, ZnO:Sc targets with diameter of 76.2 mm; (b) target in diameter of 50.8 mm and comparison of deposition profiles calculated on the base of the Knudsen cosine law (curve 1) and the over-cosine angular dependence of sputtered particles (curves 2 and 3) with the experimental deposition profile

This "directional" sputtering can be exploited for preparation of strong textured thin films, e.g. c-axis inclined ZnO (Link, 2006) applicable in solar cells and in various optoelectronic devices, but also in an engineering of the nanoscale morphology, e.g. for preparation of a new class of optical nanomaterials - sculptured thin films - consisting of shaped, parallel, identical nanowires generally grown by physical vapor deposition techniques (Lakhtakia, 2008), (Jen, 2009).

3. Technology

We prepared ZnO:Al thin films in a planar RF sputtering diode system Perkin Elmer 2400/8L, using a ceramic target (ZnO+2 wt.% Al_2O_3) in Ar working gas at constant pressure of 1.3 Pa. Films with thickness from 560 nm to 800 nm, depended on the sputtering power density and deposition time, were deposited on Corning glass 7059 substrates. Three technology parameters (chosen to tailor the physical properties of the AZO films) were investigated: the sputtering RF power density E_Φ (1.1 ÷ 4.4 x 10^4 W/m²), the substrate temperature T_s (RT, 100 ÷ 300°C) and the post-deposition annealing at temperatures T_a (200 ÷ 400°C) in the forming gas (80% N_2 + 20% H_2) for 1 hour.
The real structure characterization (crystallite size and their crystallographic orientation, micro-strains and biaxial lattice stresses) of the films was investigated by X-ray diffraction (XRD) analysis. The XRD patterns were carried out using an automatic powder X-ray diffractometer AXS Bruker D8 equipped with a position sensitive area detector Histar. Cobalt Kα radiation (λ = 0.179 nm) was used as an X-rays source. Six strongest diffraction lines in the range of 2ϑ - diffraction angle from 30° to 80° of a ZnO powder are used as reference patterns. The position, height, integrated intensity and FWHM are the main four parameters that characterize the diffraction lines. The broadening of a diffraction line is a result of a real material structure, where the size of the crystallites and the micro-strains are the most important contributors to the broadening of the line. In this study we used a procedure utilizing an integral breadth of a diffraction line (Langford, 1978), (Delhez et al., 1982). Equation 8 characterizes the integral breadth β that includes two parameters namely the height and the integrated intensity:

$$\beta = \frac{I_{\text{int}}}{I_0} \tag{8}$$

where I_{int} is the integrated intensity (area below the line) and I_0 is the maximal intensity of the diffraction line. In general, the instrumental resolution of the equipment has also to be taken into account in order to obtain a physical (depended only on the properties of the matter) component of the broadening of the diffraction line. Furthermore, the physical component of the integral breadth of the diffraction line is a convolution of Cauchy and Gaussian components and so it is necessary to do de-convolution into a Cauchy part β_C^f and a Gaussian part β_G^f before the main real structure parameters are carried out. The Cauchy and Gaussian parts of the integral breadth of the line represent the size of the crystallites and the micro-strains, respectively. The average crystallite size and micro-strains are determined using Equations 9 and 10, respectively (Delhez et al., 1982):

$$\langle D \rangle = \frac{\lambda}{\beta_C^f \cos\vartheta} \tag{9}$$

where <*D*> is the average crystallite size in the direction perpendicular to the diffracting lattice planes, λ is the X-ray wavelength used and ϑ is the Bragg's angle;

$$\langle \varepsilon \rangle = \frac{\beta_G^f}{4\,\text{tg}\,\vartheta} \tag{10}$$

where <ε> is the average micro-strain in the diffracting volume. Biaxial lattice stresses were calculated from a shift of the most intensive diffraction line (002) according to the equation (Sutta et al., 1982)

$$\sigma_1 + \sigma_2 = -\frac{E}{\mu} \cdot \frac{d - d_o}{d_o} \tag{11}$$

where d_0 is the strain free reference lattice spacing, d is the lattice spacing obtained from the experiment, E is Young's modulus and μ is Poisson's ratio. Preferred orientation of crystallites (texture) in a certain direction can be approximately characterized by the Harris texture index (Okolo et al., 2005)

$$T_i = \frac{n \cdot I_i / R_i}{\sum_{j=1}^{n} I_j / R_j} \tag{12}$$

where n is the number of investigated diffraction lines, I_i is the observed intensity and R_i is the corresponding intensity of the sample with randomly oriented crystallites (this value can be calculated from the theory if the structure of material is known or it can be picked up from the standard diffraction data files). In our case the R_i values were calculated from the theory. The maximal value that the Harris texture index can reach is equal to the number of lines taken into the calculation. For samples with completely randomly oriented crystallites the Harris texture index is equal to one.

The grain size was observed by transmission electron microscopy (TEM) by JOEL Electron Microscope. The thickness of AZO thin films was measured by DEKTAK 150, their electrical resistivity was obtained by Van der Pauw method and optical transmittance was determined by spectrometer Specord 210 and Avantes AVASPEC Fiber Optic Spectrometer. The optical transmittance in the UV spectrum region (blue-shift of the absorption edge) gives the information about the width of optical band-gap. The absorption edge for direct inter-band transitions is given by

$$\alpha h\nu - A(h\nu - E_g)^{1/2} \tag{13}$$

where A is a constant for a direct transition, a is the optical absorption coefficient, which is given from dividing absorbance by film thickness, and $h\nu$ is the photon energy. The direct band-gap of materials was obtained by plotting and extrapolation of $(ah\nu)^2$ vs. $h\nu$.

4. Results and discussion

4.1 Deposition profile and properties of sputtered AZO thin films

AZO thin films were sputtered in the planar diode sputtering system (Fig. 9) at power density 4 times higher than the value of minimal energy flux density (~ 1 x 10^4 W/m^2). Corning glass substrates with dimension of 10x10 mm were placed on different positions under the target, ZnO+2 wt. % Al_2O_3, with diameter of 152 mm. We have applied the model of spatial distribution of sputtered particles in our diode system (see Section 2.4) to get continual changes of thickness of thin films in one deposition run from 700 nm in the center to 300 nm at the border of substrate holder (Fig. 10 a). All sputtered AZO films were polycrystalline with the columnar structure and a very strong texture in the [001] direction perpendicular to the surface, what was confirmed by XRD, TEM and SEM analyses (Fig. 11, Fig.12). Properties of AZO thin films have been changed in lateral direction across the substrate holder. In the central region we can expect the highest power density (corresponding to deposition profile) what caused also an increase of mean grain size of 50 nm in comparison with value of 20 nm in peripheral position (Fig. 11 a).

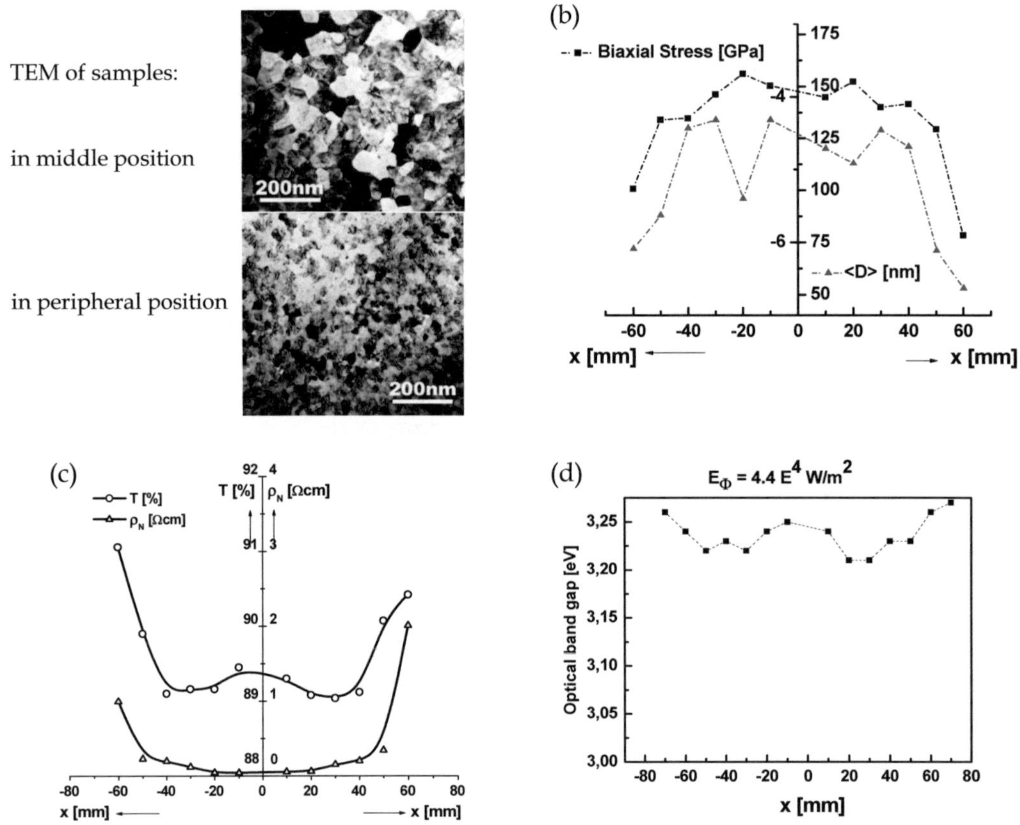

Fig. 11. Changes of properties of AZO thin films with the substrate position: (a) grain size observed by TEM, (b) lattice biaxial stresses $\sigma_1+\sigma_2$ and size of crystallites <*D*> evaluate from XRD patterns, (c) transmission *T* and resistivity ρ_N, (d) optical band gap E_{go}

Evaluation of XRD patterns also showed the tendency mentioned previously (Fig. 11 b): the average size of crystallites <*D*> (regions of coherent X-ray scattering) changed from 54 nm to 136 nm (in central region) and compressive lattice biaxial stresses $\sigma_1+\sigma_2$ increased from -3.8 GPa to -6.4 GPa (in peripheral position). Resistivity of AZO films placed in middle substrate region was in the range of 10^{-2} Ωcm and gradually towards the side of holder, has increased up to 3 Ωcm (Fig. 11 c). Remarkable increase of resistivity with position may be explained by the variation of particular fluxes of sputtered particles, dominantly by change of mutual ratios of sputtered ZnO, Al and O fluxes. Particularly, the bombardment of growing film by negative O ions generates large amount of oxygen intersticials that act as trap for free electrons, thereby increasing the resistivity. Transmission of films sputtered in the central region was about 89 % and it increased to 91 % at the edge of substrate holder. The course of the optical band-gap dependence on substrate position is copying the transmittance dependence (Fig. 11 d).

For the explanation of previous mentioned phenomena, both effects the negative oxygen ion bombardment, or resputtering respectively, of growing film and the local atomic-scale heating of film surface by energetic secondary electrons, should be taken in the consideration. We assume, that it is a difference of the spatial distribution between energetic negative oxygen ions and secondary electrons, coming from the target and accelerating at it. Even the both fluxes have obtained the same kinetic energy (in order of 10 - 100 eV), the mean velocity of electrons and ions differs of 10^3-times because another masses. Therefore in the case of RF diode sputtering, the distribution of negative oxygen ions sputtered from the target will be more uniform across its area (in comparison with sputtered neutral particles) since the flux of negative oxygen ions is collimated by the induced electric field of target. The distribution of secondary electrons bombarding the growing film will follow the deposition profile because the induced target voltage has no significant influence on their angular emission due to small electron mass and high velocity. Then we can expect a higher local atomic-scale heating of the growing film in the central region.

We can conclude, that our arrangement provides sputtering of AZO thin films with relatively homogenous properties in the central region of diameter of 50 mm.

4.2 Influence of technological parameters on structural properties

Based on previous results we used the central region for sputtering and the effect of three technological parameters on structure of AZO thin films has been investigated: power density, substrate temperature and post-deposition annealing.

An increase of both sputtering power density up to 4,4.10 4 W/m2 and substrate temperature to 200°C as well as post-deposition annealing provided highly textured films with a preferential c-axis [001] orientation (Fig. 13, 15, 17). Typical columnar structure of AZO films and their surface morphology are shown in Fig. 12.

(a) (b)

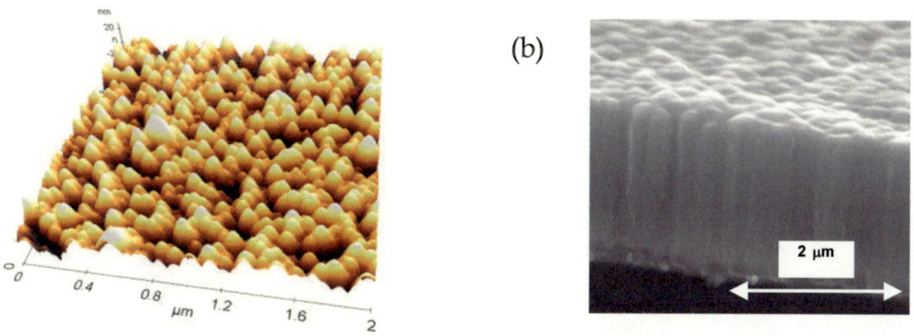

Fig. 12. The cross-section of typical polycrystalline AZO thin film with columnar structure in [001] direction perpendicular to the substrate and a hillock surface morphology

Significant asymmetry of (002) diffraction line when room substrate temperature was used indicates that there is a region with heterogeneous structure at the substrate - film interface (Fig. 13 a). Further energy delivered to the growing film by heating the substrate from RT to

200°C improved the crystallinity of the films (Fig. 15).The asymmetry of (002) line as well as heterogeneous regions completely diminished at higher substrate temperatures and small shifts from the reference line position were observed, which indicates only unimportant lattice strains (Fig. 15 b). The widths of (002) lines became narrower and dimensions of crystallites growing were from 60 to 120 nm for films deposited at RT and to more than 200 nm for those deposited at higher substrate temperatures. The widths of azimuthal line profiles also decreased from 15 to 3.5° with increasing energy delivered to the growing film during the deposition. This indicates lower declination of individual crystallites from the normal to the substrate (stronger texture). At the highest substrate temperature (300°C) also the other diffraction lines appeared, which is caused by more randomly oriented crystallites of the film. Other authors report the same changes with the temperature (Fu et al., 2003). In this case the film loosed its anisotropy. These phenomena induced by power density and temperature during deposition indirectly confirmed by changes of the surface morphology of films (Fig. 14, Fig. 16).

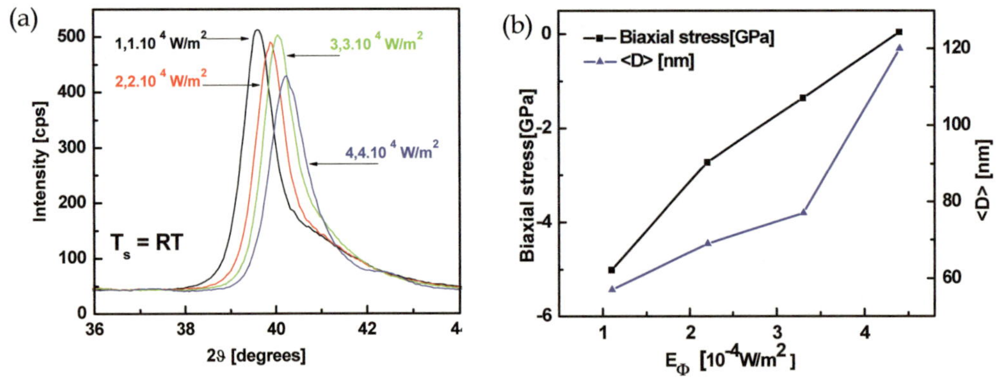

Fig. 13. AZO films sputtered at different power densities, $E_{\Phi} = 1,1 - 4,4.10^{4}\,W/m^2$: (a) XRD patterns, (b) dependences of biaxial lattice stress and crystallite size on power density

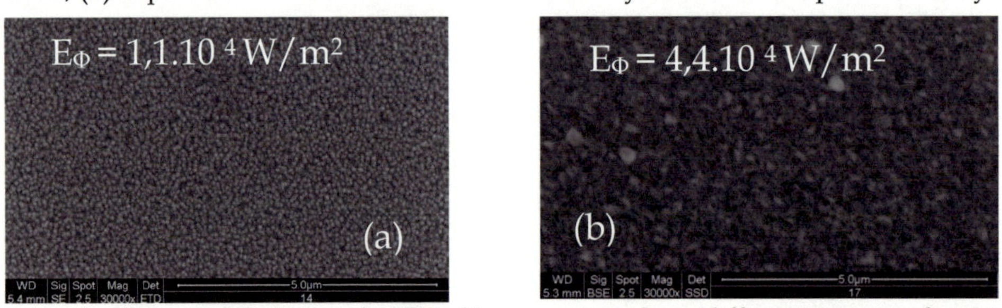

Fig. 14. Surface morphology of AZO films sputtered at different power densities: (a) $E_{\Phi} = 1,1.10^{4}\,W/m^2$, (b) $E_{\Phi} = 4,4.10^{4}\,W/m^2$

The shift up of the 2ϑ with increasing RF power, as well as substrate and annealing temperatures, is a result of the increase of Al^{3+} substituents (Al^{3+} that substitute for Zn^{2+} in the ZnO lattice) and a reduction of the interplanar distance, which changes the lattice distortion in AZO films from compressive to tensile lattice stresses. An increase of the power density and temperature during growth are resulting in the larger grains (growth from 60 to more than 200 nm) and better crystalline structure (no line asymmetry). In our case the maximal value of the Harris texture index can reach the number 3 (3 lines of ZnO can be taken into the calculation in the range of 2ϑ = 36 - 44°). Value of H_i = 2,97 has confirmed the best crystallite texture (002) of films sputtered at the substrate temperature of 200 °C.

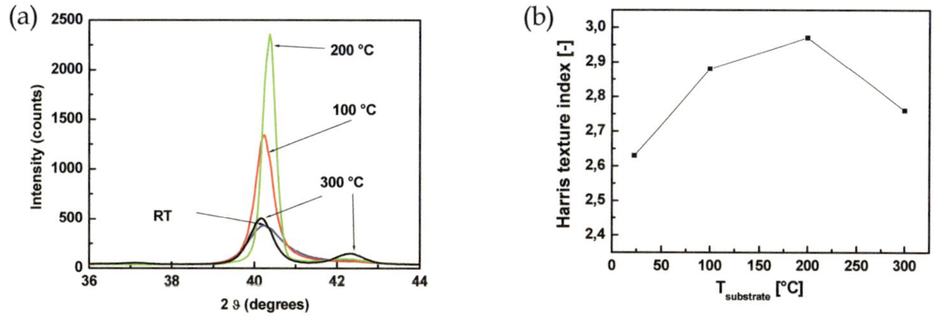

Fig. 15. AZO films sputtered at different substrate temperatures and constant power density, $E_\Phi = 4{,}4.10^4\ W/m^2$: (a) RTG patterns, (b) Harris texture index

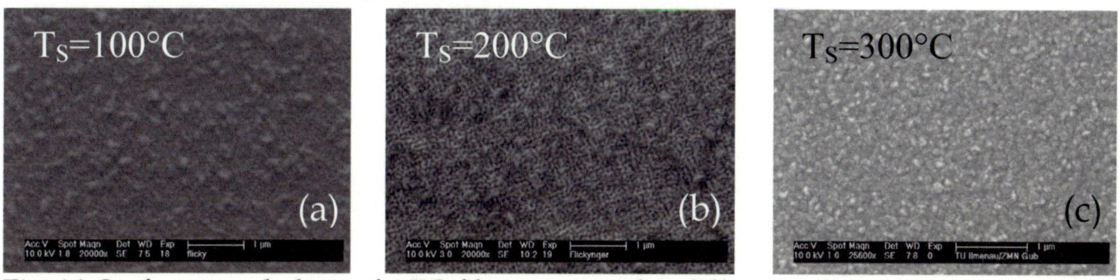

Fig. 16. Surface morphology of AZO films sputtered at different substrate temperatures T_S

Post-deposition annealing in the forming gas at 400 °C diminished an asymmetry of (002) diffraction line (Fig. 17).

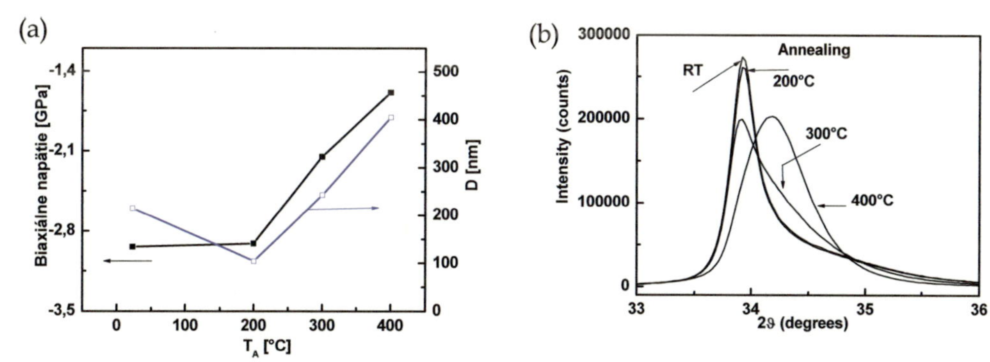

Fig. 17. Post-deposition annealing (in forming gas at T_A = 200 - 400 °C) of AZO films sputtered at room temperature of substrate and power density $E_\Phi = 4{,}4.10^4\ W/m^2$: (a) RTG patterns, (b) dependences of biaxial lattice stress and crystallite size on annealing temperature

The average size of crystallites was ~ 100 nm at T_A = 200°C and it increased to ~ 400 nm by the growth of annealing temperature up to T_A = 400 °C (Fig. 17 b). Post-deposition annealing reduced biaxial lattice stresses due to elimination of grain boundary defects. Both effects can be explained by recrystallization and phase transformation.

4.3 Influence of technological parameters on electrical and optical properties

The substrate temperature increase during sputtering caused the decrease of sheet resistance from 300 Ω/square to 50 Ω/square with a minimum 36 Ω/square at T_s = 200°C (Fig. 18 a). Sputter power density has changed considerable R_s from 1137 Ω/square at $E_\Phi = 1.1.10^4$ W/m² to 300 Ω/square at $E_\Phi = 4.4.10^4$ W/m² (Fig. 18 b). After annealing (over the T_A > 200°C) the decrease of the sheet resistance to 10.4 Ω/square was observed (Fig. 18 c). The lowering of the sheet resistance by temperature (during or after deposition) was caused by both the improvement of crystalline structure and the increase of free electron concentration particularly, what has been supported by optical measurements.

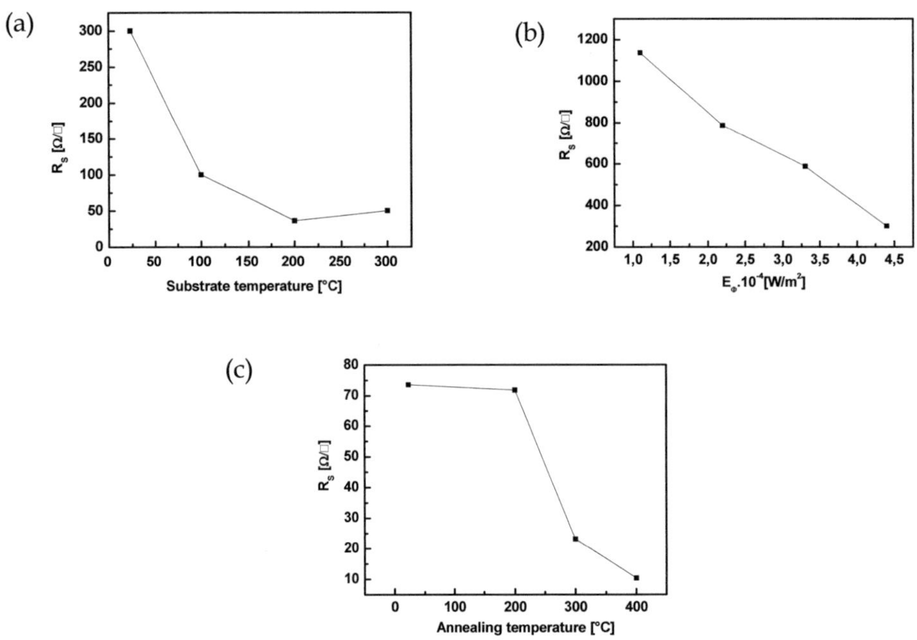

Fig. 18. The effect of: (a) substrate temperature, (b) power density and (c) annealing temperature on the sheet resistance

AZO thin films sputtered at a higher substrate temperature showed the "blue shift" in optical spectrum (Fig. 19), so-called the Burstein – Moss effect, which is dependent on the dopant concentration and the effective mass state density (Sernelius et al. 1988). The widening of the optical band-gap with the substrate temperature is originated by the increase of the electron concentration caused by Al doping (Liu et al. 2006), (Moon et al. 2006). No significant influence of the substrate temperature on the transmittance was observed (Fig. 20).

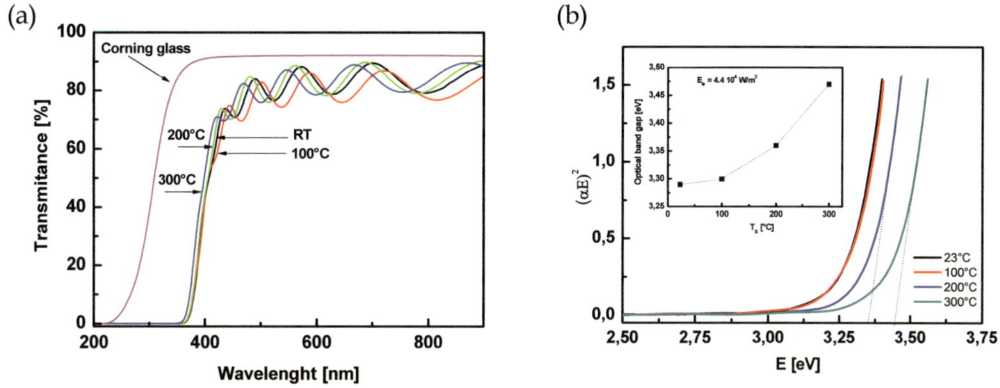

Fig. 19. The influence of the substrate temperature on: (a) the transmittance spectrum with glass substrate, (b) the blue–shift of the absorption edge and the optical band-gap width was obtained by plotting and extrapolation of $(ah\nu)^2$ vs. $h\nu$.

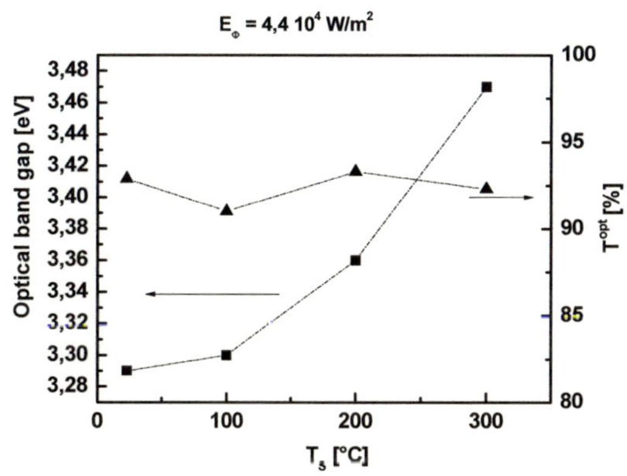

Fig. 20. The effect of the substrate temperature on the transmittance for AZO layer and the optical band-gap width

The dependence of transmittance on the power density showed a minimum at a middle powers and the highest value of 93 % was reached at power density of $E_\Phi = 4.4.10^4$ W/m^2 (Fig. 21). The post-deposition annealing had a positive effect on the increase of transmittance accompanied with the decrease of the optical band-gap width (Fig. 22).

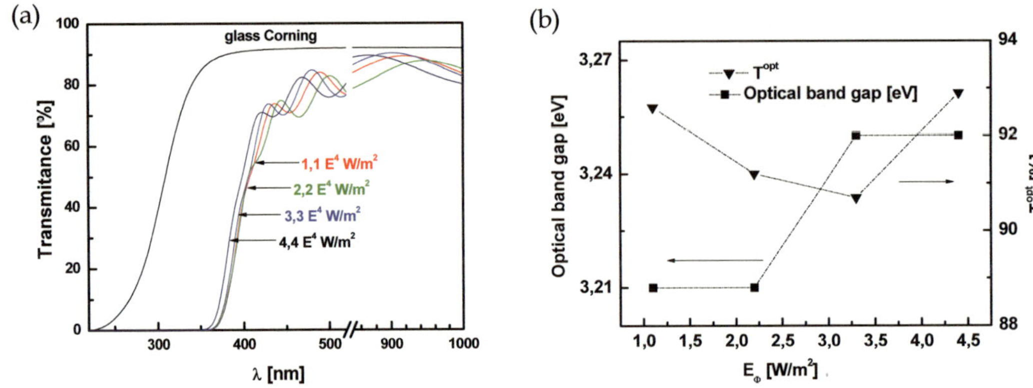

Fig. 21. The influence of the power density on: (a) the transmittance spectrum with glass substrate, (b) the transmittance only for AZO layer and the optical band-gap width

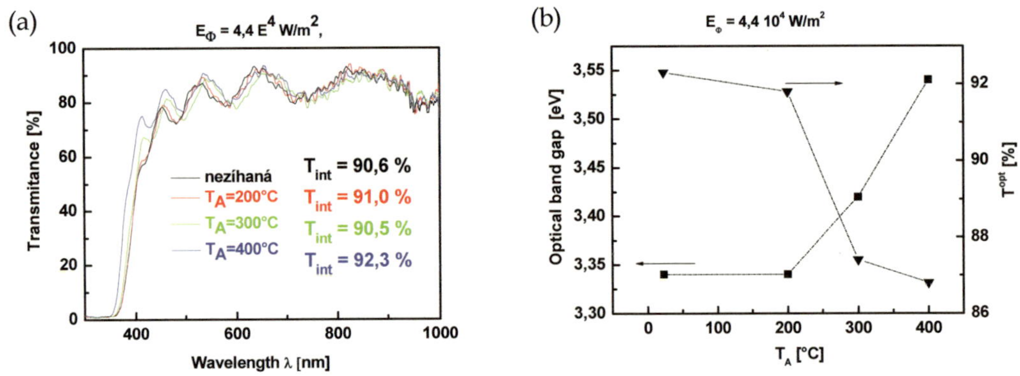

Fig. 22. The effect of the annealing temperature on: (a) the transmittance spectrum with glass substrate, (b) the transmittance only for AZO layer and the optical band-gap width

4.3 Summary

An influence of substrate temperature, power density, post-deposition annealing temperature on sheet resistance and transparency of sputtered AZO thin films are summarized in Table 1.

T_s [°C]	R_s [Ω]	T^{opt} [%]	E_Φ. 10^{-4} [W/m²]	R_s [Ω]	T^{opt} [%]	T_A [°C]	T^{opt} [%]	R_s [Ω]
RT	300	92.9	1.1	1137	92.6	0	92.3	73.6
100	100	91.0	2.2	787	91.2	200	91.8	71.8
200	36	93.3	3.3	587	90.7	300	87.4	23.2
300	50	92.3	4.4	300	92.9	400	86.8	10.4

Tab. 1. The effect of substrate temperature T_s, power density E_Φ, post-deposition annealing temperature T_A on sheet resistance R_s and transmittance T^{opt} of sputtered AZO thin films

There is an inherent trade-off between how high transparent and low resistive TCO layer is, which is represented by the figure of merit: $F = T_{opt}/R_s$, where T_{opt} (%) is the transmittance for AZO layer and R_s (Ω/square) is the sheet resistance. The higher figure of merit ($F > 1\% / \Omega$) there is the better the performance of the TCO thin film. The substrate temperature increase during sputtering caused an improvement of the figure of merit to the value of $F = 2.6\ \%/\Omega$ at 200 °C (Fig. 23 a). It showed a low value of 8 x 10^{-2} %/Ω at the initial power density $E_\Phi = 1.1 \times 10^4\ W/m^2$ which became greater with growth of power density (Fig. 23 b). The greatest effect on the figure of merit exhibited the post-deposition annealing what increased of it up to value of F =8.9 %/Ω at T_A = 400 °C (Fig. 23 c).

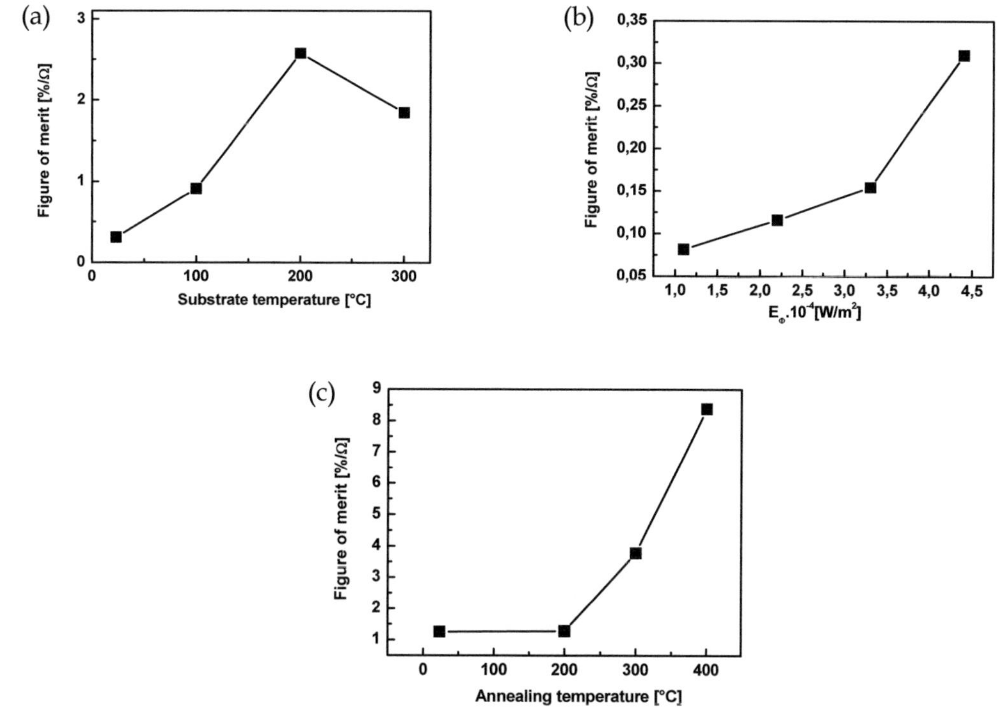

Fig. 23. Influence of substrate temperature (a), power density (b) and annealing temperature (c) on the figure of merit

5. Conclusion

AZO thin films sputtered at power density of 4.4 x $10^4\ W/m^2$, substrate temperature of 200°C and annealed in the forming gas at 400°C showed a highly (002) oriented crystalline structure with the larger grains. Optimal substrate temperature could not surpass 200°C and on the other side, an annealing temperature should be higher than 200°C. Desired properties of AZO thin films for solar cell application - the high figure of merit ($F \geq 8\ \%/\Omega$), low sheet resistance ($R_s \leq 10\ \Omega$/square) and the highest transmittance ($T_{opt} \geq 86\%$) were obtained in highly oriented AZO films prepared at high values of power density, optimal substrate and annealing temperatures. The next experiments are focusing on the incorporation of AZO films into the real thin film solar cell, on the evaluation of their behaviour and on the enhancement of light scattering by AZO film using chemical or ion etching modification of their surface.

6. Acknowledgments

Presented work was supported by the MSMT Czech Republic project 1M06031.

7. References

Agilent VEE Pro, http://www.home.agilent.com/agilent/home.jspx, 2009

Ali, A. I. (2006). Growth and Characterization of ZnO:Al Thin Film Using RF Sputtering for Transparent Conducting Oxide, *Journal of the Korean Physical Society*, Vol., 49, (December 2006), pp. 652-656

Berginski, M.; Hupkes, J.; Reetz, W.; Rech, B. & Wuttig, M. (2008). Recent development on surface-textured ZnO:Al films prepared by sputtering for thin-film solar cell application, *Thin Solid Films*, Vol., 516, (October 2007), pp. 5836-5841

Dagamseh, A. M. K.; Vet, B.; Tichelaar, F. D.; Sutta, P. & Zeman, M. (2008). ZnO:Al films prepared by rf magnetron sputtering as back reflectors in thin-film silicon solar cells, *Thin Solid Films*, doi:10.1016/j.tsf2008.05.009

Delhez, R.; Keijser, T. H. & Mittemeijer (1982). Determination of crystallite size and lattice distortions through X-ray diffraction line profile analysis, *Fres. Z. Anal. Chem.*, Vol., 312, pp. 1-16

Fang G. J.; Li, D. & Yao, B. L. (2002). Influence of post-deposition annealing on the properties of transparent conductive nanocrystalline ZAO thin films prepared by RF magnetron sputtering with highly conductive ceramic target, *Thin Solid Films*, Vol., 418, pp. 156–162

Flickyngerova, S.; Netrvalova, M.; Prusakova L.; Novotny, I.; Sutta, P. & Tvarozek, V. (2009). Modification of AZO thin-film properties by annealing and ion etching, Vacuum, doi:10.1016/j.vacuum.2009.04.006

Fu E.G.; Zhuang D.M.; Zhang, G.; Yang, W. & Zhao, M. (2003). *Appl. Surf. Science*, Vol., 217, pp. 88-92

Fu E.G. & Zhuang D.M. (2004). Properties of transparent conductive ZnO:Al thin films prepared by magnetron sputtering, *Microelectronics Journal*, Vol., 35, pp. 383–387

Goetzberger, A. & Hoffmann, V.U. (2005). Photovoltaic Solar Energy Generation, Springer Verlag, 3-540-23676-7, Berlin

Jehn, H. A. & Rother, B. (1999). Homogeneity of multi-component PVD hard coatings deposited by multi-source arrangements, *Surface and Coatings Technology*, Vol., 112, pp. 103–107

Jen, Y-J.; Lakhtakia, A.; Yu, Ch-W. & Lin, Ch-T. (2009). Vapor-deposited thin films with negative real refractive index in the visible regime, *Optics Express*, Vol., 17, Issue 10, pp. 7784-7789, doi:10.1364/OE.17.007784

Kaminsky M. (1965). Atomic and ionic impact phenomena on metal surfaces, Springler Verlag, Berlin Heideberg

Kluth, O. & Rech: B (1999). Texture etched ZnO:Al coated glass substrates for silicon based thin Film cells, *Thin Solid Films*, Vol., 351, pp. 247-253

Kluth, O.; Schöpe, G.; Hüpkes, J.; Agashe, C.; Müller, J. & Rech B. (2003). Modified Thornton model for magnetron sputtered zinc oxide: film structure and etching behaviour, *Thin Solid Films*, Vol., 442, pp. 80–85

Lakhtakia, A.; Demirel, M.C.; Horn, M.W. & Xu, J. (2008). Six Emerging Directions in Sculptured Thin-Film Research, *Advances in Solid State Physics*, Vol., 46, pp. 295–307

Lim, D.G. & Kim D.H (2006). Improved electrical properties of ZnO:Al transparent conducting oxide films using a substrate bias, *Superlattices and Microstructures*, Vol., 39, pp. 107–114

Link, M.; Schreiter, M.; Weber, J.; Gabl, R.; Pitzer, D.; Primig, R. & Wersing W. (2006). c-axis inclined ZnO films for shear-wave transducers deposited by reactive sputtering using an additional blind, *J. Vac. Sci. Technol.*, Vol., A 24(2), pp. 218-222

Langford J. I. (1978). A rapid method for analyzing the breadths of diffraction and spectral lines using the Voigt function, *J. Appl. Cryst.*, Vol., **11,** pp. 10-14

Liu, Y.; Zhao, L. & Lian, J. (2006). Al-doped ZnO films by pulsed laser deposition at room temperature, Vacuum, Vol., 81, pp. 18–21

Ma H.L. & Hao, X. T. (2002). Bias voltage dependence of properties for ZnO:Al films deposited on flexible substrate, *Surface and Coatings Technology*, Vol., 161, pp. 58–61

Minami, T. (2005). Transparent Conducting oxide semiconductors for transparent electrodes, *Semicond. Sci. Technol.*, Vol., 20, pp. 35-44

Moon ,Y. K.; Kim, S. H. & Park J. W. (2006). The influence of substrate temperature on the properties of aluminum-doped zinc oxide thin films deposited by DC magnetron sputtering, *J Mater Sci: Mater Electron,* Vol., 17, pp. 973–977

Nelson, J. (2003). The Physics of Solar Cells, Imperial College Press, 1-86094-349-7, London

Oh, B. Y.; Jeong, M.Ch. & Myoung, J. M. (2007). Stabilization in electrical characteristics of hydrogen-annealed ZnO:Al films, *Applied Surface Science* Vol., 253, pp. 7157–7161

Okolo, B.; Lamparter, P.; Welzel, U.; Wagner, T.; Mittemeijer, E.J. (2005). The effect of deposition parameters and substrate surface condition on texture, morphology and stress in magnetron-sputter-deposited Cu thin films, *Thin Solid Films* Vol., 474, pp. 50- 63

Panjan, M.; Cekada, M.; Panjan, P.; Zalar, A. & Peterman, T. (2008). Sputtering simulation of multilayer coatings in industrial PVD system with three-fold rotation, *Vacuum*, Vol., 82, pp.158–161

PSpice A/D Circuit Simulator, http://www.ema-eda.com/products/orcad/pspice.aspx, 2009

Rieth, L. W. & Holloway, P. H. (2004). Influence. , J. Vac. Sci. Technol. A , Vol., 22 (No. 1), pp. 20-29

Rother, B.; Ebersbach, G. & Gabriel, H. M. (1999). Substrate-rotation systems and productivity of industrial PVD processes, *Surface and Coatings Technology*, Vol., 116–119, pp. 694–698

Sernelius, B. E. (1988). Band gap tailoring of ZnO by means of heavy Al doping, *Physical Review* B, Vol., 37, number 17, pp. 10244-10248

Sittinger, V.; Ruske, F.; Werner W.; Szyszka B.; Rech, B.; Hupkes, J.; Schope, G. & Stiebig, H. (2006). ZnO:Al films deposited by in-line reactive AC magnetron sputtering for a-Si:H thin film solar cells, Thin Solid Films, 496, pp. 16 - 25

Sutta, P. & Jackuliak, Q. (1998). Macro-stress Formation in Thin Films and its Determination by X-ray Diffraction, In: Proc. of the 2nd International Conference on Advanced Semiconductor Devices and Microsystems, *ASDAM'98*, pp. 227-230

Thorton, J. A. & Hoffman, D. W. (1989). Stress-related effects in thin films, *Thin Solid Films*, Vol., 171, pp. 5-31

Tominaga, K.; Iwamura, S.; Shintani Y. & Tada, O. (1982). Energy Analysis of high - energy Neutral Atoms in the Sputtering of ZnO and BaTiO3, Jpn. J. Appl. Phys., Vol., 21, (No. 5), pp. 688-695

Tvarozek, V.; Harman, R. & Jesenakova, V. (1982). Simultaneous sputtering of several materials from composite target, *Jemna mechanika a optika*, Vol. XXVII, No. 4, pp. 89-94

Tvarozek, V.; Novotny, I.; Sutta, P.; Flickyngerova, S.; Shtereva K. & Vavrinsky E. (2007). Influence of sputtering parameters on crystalline structure of ZnO thin films, *Thin Solid Films*, Vol., 515, pp. 8756-8760

Zeman, M. (2007). Advanced Amorphous Silicon Solar Cell Technolgies, In: *Thin film solar cells: fabrication, characterization and applications*, Poortmans, J. & Arkhipov, V., pp. 173-236, John Wiley & Sons, 978-0-470-09126-5, New York

Zeman, M.; Van den Heuvel, J.; Kroon, M.; Willemen, J.; Pieters, B. & Krč, J. (2005). ASA, Users Manual,Version 5.0, Delft University of Technology, The Netherlands

13

Role of rare-earth elements in the technology of III-V semiconductors prepared by liquid phase epitaxy

Jan Grym, Olga Procházková, Jiří Zavadil and Karel Žďánský
Institute of Photonics and Electronics
Academy of Sciences CR, v.v.i.
Czech Republic

1. Introduction

First applications of rare-earth (RE) elements in semiconductor technology are rooted in radiation tolerance improvements of silicon solar cells and purification of GaP crystals. The idea was later adopted in the technology of germanium and compound semiconductors. Since the 1980's, considerable attention has been directed towards REs applications in III-V compounds both for epitaxial films and bulk crystals (Zakharenkov et al., 1997).

The uniqueness of REs arises from the fact that the lowest-energy electrons are not spatially the outermost electrons of the ion, and thus have a limited direct interaction with the ion's environment. The shielding of the 4f electrons by the outer filled shells of 5p and 5s electrons prevents the 4f electrons from directly participating in bonding (Thiel et al., 2002). The RE ions maintain much of the character exhibited by a free ion. This non-bonding property of the 4f electrons is responsible for the well-known chemical similarity of different REs. Since transitions between the electronic states of the shielded 4f electrons give rise to spectrally narrow electronic transitions, materials containing REs exhibit unique optical properties. By careful selection of the appropriate ion, intense, narrow-band emission can be gained across much of the visible region and into the near-infrared (Kenyon, 2002). Inspired by the striking results accomplished in the field of optical amplifiers and lasers based on RE-doped fibres (Simpson, 2001), substantial research activity has been recently carried out on RE-doped semiconductor materials for optoelectronics (Klik et al., 2001).

In most cases, however, achieving effective doping of III-V compounds by REs during growth from the liquid phase has proven difficult; the high chemical reactivity and the low solid solubility are the main restrictions on introducing RE atoms into the crystal lattices (Kozanecki & Groetzschel, 1990). On the other hand, the enhanced chemical affinity of REs towards most species of the shallow impurities leads to the formation of insoluble aggregates in the melt. Under suitable growth conditions, these aggregates are rejected by the growth front and are not incorporated into the grown layer: gettering of impurities takes place. Especially Si and main group-six elements acting as shallow donors in III-V semiconductors are effectively gettered due to REs high affinity towards them (Wu et al., 1992). Removal of detrimental impurities is of vital importance in applications such as PIN photodiodes (Ho et al., 1995) or nuclear particle detector structures (Procházková et al., 2005a), where high electron and hole drift velocities are appreciated.

1.1 Main Objectives

Recently, we have performed a unique study of the impact of REs (Tb, Dy, Pr, Tm, Er, Gd, Nd, Lu, Ce) and their oxides (PrO_x, TbO_x, Tm_2O_3, Gd_2O_3, Eu_2O_3) on the properties of InP layers (Procházková et al., 2002; Procházková et al., 2005a; Grym et al., 2009). This study was motivated by the lack of systematic research in the field of liquid phase epitaxy (LPE) grown III-V semiconductors from RE treated melts. REs open the door for the preparation of high purity III-V layers without extended baking of the melts or other complicated and time consuming methods. In this chapter we cover the following topics:

- Short introduction to LPE.
- Discussion of the behaviour of REs in the liquid and solid phase during LPE, their incorporation and gettering.
- Comparison of the behaviour of different RE species in the growth process of InP layers, their structural, electrical, and optical properties. InP has been chosen as a simple binary system to perform this investigation.
- Preparation of p-type InP layers, which have not been systematically investigated by other groups. Detailed description of the gettering phenomenon will be given together with the explanation of the conductivity conversion from n to p-type.
- Application of REs in the device technology: light emitting diodes (LEDs) based on InGaAsP/InP double heterostructure for near infrared spectroscopy.

2. Present Status

Even though numerous papers on the gettering effect of particular REs in III-V semiconductors have been published within the last three decades, no systematic study of the whole set of REs in a given III-V system has been carried out. The details of the relationship between the growth conditions, possible incorporation mechanisms, and the purifying phenomena have not been established yet.

LPE is a mature technology, which has been used in the production of III-V compound semiconductor devices for more than 40 years (Nelson, 1963); it triggered pioneering work of a vast number of semiconductor devices including LEDs, laser diodes, infrared detectors, or heterojunction bipolar transistors. LPE is capable of producing high quality layers, taking place close to thermodynamic equilibrium, with a superior luminescence efficiency and minority carrier lifetime. To emphasize several unique advantages of LPE, at least the following should be listed: (i) high growth rates; (ii) a wide range of available dopants making LPE an excellent tool for the investigation of fundamental doping studies; (iii) the low point defect densities; (iv) no toxic precursors; (v) low equipment and operating costs. Plenty of achievements ranking LPE first in the world are summarized in the review paper of Kuphal (Kuphal, 1991). However, in recent years, LPE has fallen into disfavour, especially in device applications that require large-area uniformity, extremely thin layers, abrupt composition control, and smooth interfaces. Superlattices, quantum wells, strained layers, or nonisoperiodic structures with a high lattice mismatch, all of these are grown by molecular beam epitaxy (MBE) or metal organic vapour phase deposition (MOVPE) (Capper & Mauk, 2007). LPE has nearly disappeared from universities so that the know-how exists in the industry only and papers on LPE are scarce. Still, lots of niches in semiconductor technology remain to be served by LPE. We believe that LPE growth from RE treated melts is one of them.

LPE growth is typically carried out from supersaturated solutions composed of source materials in a graphite boat. The boat is placed in a quartz reactor tube in the atmosphere of high purity hydrogen. There are several sources of impurities that may be introduced into the grown layer (Dhar, 2005):

- Source materials and chemicals to clean them.
- Parts of the graphite boat being in contact with the growth solution.
- Contaminants deposited on the inner wall of the quartz reactor tube. These contaminants can be transported to the solution by the ambient gas during high temperature growth.
- The carrier hydrogen gas itself.
- Tools and containers for storing, handling, and cleaning the substrate and source materials.

Several procedures help to prevent these impurities to be incorporated into the layer being grown. The materials for growth are of a high purity level. At present, indium is available at 6N or even 7N purity, REs typically at 3N but recently some of their oxides up to 4N+ purity. These materials, before loading into the growth boat, are thoroughly cleaned to remove the contaminated surface. The graphite boat is made of ultra high purity graphite with low porosity. The reactor tube is made of high quality quartz and the inside wall is periodically cleaned and baked-out at high temperatures. High-purity hydrogen generator or palladium diffusion cell is used to guarantee high purity hydrogen flow. Typical LPE InP layers grown under these conditions posses electron concentrations above 10^{17} cm^{-3} at room temperature. In addition to the above self-evident precautions, there are several methods to suppress residual impurity concentrations (Rhee & Bhattacharya, 1983; Kumar et al., 1995):

- Prolonged baking of the growth solution above the growth temperature. A long bake-out under the dry hydrogen atmosphere leads to the removal of volatile impurities such as Zn, Mg, Cd, Te, and Se from the In melt by the evaporation. However, S is only partly removed due to the formation of In-S compounds and Si remains due to its low vapour pressure.
- Introduction of controlled amounts of H_2O in the growth ambient. Si is oxidized and thus prevented from being incorporated into the epitaxial layer in the electrically active form. However, this method can lead to inferior surface morphology and creation of oxygen-related traps.
- Extended prebaking of the melt can be alternatively realized in high vacuum generally leading to suppressed S concentrations.
- Other improvements including growth in PH_3 atmosphere or use of dummy substrates as the source material.
- And finally, the addition of REs acting as effective gettering agents.

A brief review of REs studied in connection with III-V semiconductors prepared by LPE follows. Emphasis is put on InP and InP-based compounds. The review is sorted by individual REs. Among the REs investigated in InP, only ytterbium atoms occupy exclusively one type of the lattice site in InP. The Yb impurity in InP was proved to be incorporated as a cubic Yb^{3+} ($4f^{13}$) centre on cation site (In) by Rutherford backscattering spectroscopy (Kozanecki & Groetzschel, 1990). This means that its luminescent properties are independent of the growth and doping techniques.

It is not surprising that Yb was probably the most intensively studied RE in the context of III-V compounds. In 1981, Zakharenkov reported Yb-related luminescence band in LPE grown InP (Zakharenkov et al., 1981). Further studies of RE activated luminescence in Yb and Er implanted InP, GaP, and GaAs were performed by Ennen (Ennen et al., 1983). LPE InP:Yb layers were prepared by Korber group (Korber et al., 1986). High doping levels and high growth temperatures were applied to increase Yb solubility. Employing low concentration of Yb in the melt, its gettering effect was demonstrated and high purity samples were prepared. The same group fabricated a light-emitting diode based on InP:Yb LPE layer showing intense emission at 1000 nm due to the intracentre transition of Yb^{3+} ions. (Haydl et al., 1985). Later, excitation and decay mechanisms of the Yb^{3+} in InP LPE layers were studied (Korber & Hangleiter, 1988). Nakagome confirmed incorporation of Yb in LPE InP layers by SIMS. Only a negligible portion of Yb was uniformly dispersed, most of Yb was embedded as micro-particles of Yb oxides and phosphides (Nakagome et al., 1987). He also observed deterioration of the surface morphology at higher Yb concentrations and growth temperatures. Kozanecki studied lattice location and optical activity of Yb in III-V compounds (Kozanecki & Groetzschel, 1990). He proves rather exceptional behaviour of Yb in InP consisting in relatively easy substitution of In by Yb. He states that this behaviour is related to similar ionic radii between Yb^{3+} and In^{3+} minimizing the elastic strain energy generated by the impurity, and the partially covalent Yb–P bonding. Novotný showed gettering effect of Yb in InP LPE layers (Novotný et al., 1999). The PL spectra of the studied samples were markedly narrowed and Yb^{3+} sharp intracentre transitions occurred. Different concentrations of Yb led to the preparation of both n-and p-type conductivity layers. Recently published paper (Krukovsky et al., 2004) deals with growth of GaAs prepared from Yb treated melts and demonstrates its gettering effect.

Optoelectronic materials doped with erbium atoms have received extensive attention due to their impact on optical communication systems operating at 1540 nm. Luminescent properties of erbium in III-V semiconductors were summarized in a review paper of Zavada (Zavada & Zhang, 1995). More recent review of rare-earth doped materials for optoelectronics can be found in the paper of Kenyon (Kenyon, 2002). Investigation of Er doping of InP prepared by LPE was performed by Chatterjee (Chatterjee & Haigh, 1990). Prevention of erbium oxide and hydride formation to suppress development of erbium precipitates is discussed in detail. Together with a vast number of papers on Er doped semiconductors, several papers also discuss Er gettering properties. Wu examined effect of Er admixture on structural, electrical, and optical properties of InGaAsP grown by LPE. He reports significantly diminished carrier concentrations (3×10^{15} cm^{-3}) and a mirror-like surface morphology up to certain Er concentration limit (Wu et al., 1992). This work is further extended by PL studies of these samples (Chiu et al., 1993). Other paper of Wu reports on preparation of very high purity InP by LPE using Er gettering (Wu & Chiu, 1993). High quality of the layers is demonstrated by narrowing of the PL peaks and by the Hall effect measurements resulting in lowered electron concentrations to 5×10^{14} cm^{-3} when introducing an optimum amount of Er into the growth solution. Ho and Wu took advantage of the high purification efficiency in the fabrication of a PIN mesa photodiode, where the GaInAs absorbing layer was prepared from Er treated melts (Ho et al., 1995). In 1996, Gao gave a detailed survey on the preparation of InGaAs using Yb, Gd, and Er treated melts. Free carrier concentration reaches 1×10^{14} cm^{-3}. However, this extremely low concentration is attributed to a large degree of compensation.

Further investigations were performed on Ho and Nd treated InP and GaInAsP LPE layers (Procházková et al., 1997; Procházková et al., 1999). A high donor gettering effciency was demonstrated. Detailed studies of the gettering effect of n-type InP layers were performed by Zavadil (Zavadil et al., 1999) and Žďánský (Žďánský et al., 1999). Žďánský determined donor and acceptor concentrations from temperature variation of resistivity and Hall coefficient, and room temperature capacitance-voltage measurements. Two types of donors and an acceptor were taken into account.

Lee prepared Nd-doped AlGaAs by LPE (Lee et al., 1996) in order to apply these layers in Nd:AlGaAs lasers or LEDs with wavelength 0.91, 1.08 and 1.35 μm. He reports mirror-like surface morphologies up to 0.4 wt% of Nd in the growth solution and uniform distribution of Nd in AlGaAs layers as well as effective gettering of residual impurities. However, higher amounts of Nd in the growth melt lead to surface roughening with many defect sites, Nd forms microparticles and segregates.

Kovalenko observed n → p conductivity conversion at 0.1 wt% of Gd admixture on the GaAs LPE layers and a decreased electron concentration 2×10^{15} cm^{-3} (Kovalenko et al., 1993). A further increase of Gd concentration above 0.1 wt% slightly increases the hole concentration. The author suggests that Gd is not incorporated into the GaAs layers. The more recent paper of Gao (Gao et al., 1999) reports the growth of very pure InAs by introducing Gd into the growth melt. Gao stresses that LPE growth occurs at thermodynamic equilibrium, and in comparison with MBE or MOVPE, the resulting crystalline perfection is superior with few defects. The electron concentration is reduced to 6×10^{15} cm^{-3} when optimum Gd concentration is added to the growth melt. While the surface of conventionally grown InAs layers is mirror-like, even a small admixture of Gd (10^{-6} mol%) leads to deterioration of the surface morphology. The deterioration of the surface morphology is assigned to the formation of precipitates and their nodules distributed throughout the melt. In order to suppress the number of nodules deposited in the layer, a new boat design, containing two recesses, is proposed. The supplementary recess is used for a sacrificial substrate on which nodules from the melt are deposited.

Kumar reported on the role of Dy in LPE growth of InP (Kumar & Bose, 1992). He attributes the gettering effect of donor impurities to the formation of stable silicides (Dy_3Si_5 and $DySi_2$), sulfides (Dy_2S_3) and tellurides (Dy_2Te_3), which do not dissolve in the indium melt. All layers are of n-type conductivity and the electron concentration is decreased to 4×10^{15} cm^{-3}. Another paper of Misprint in ligature due to oxygen in LPE grown InGaAs with Dy admixture (Kumar et al., 1995). He further states that Dy gettering not only results in decreased carrier concentration and increased mobility but also better morphology and lower etch pit density is achieved.

Reports on the gettering properties of Pr are quite scarce. Pr was studied in GaAs, InGaAs, and InP by Jiang (Jiang, 1993). He correlates a linewidth narrowing of PL spectra with an improved crystalline quality due to Pr presence in the growth melt.

REEs in the semiconductor technology have been thoroughly investigated since the last quarter of the 20th century also in Russia. Studies concerning the use of rare-earth elements in the liquid-phase epitaxy of the InP, InGaAsP, InGaAs, and GaP compounds and with the fabrication of various optoelectronic and microelectronic devices and structures based on these compounds are summarized in two review articles (Gorelenok et al., 1995; Gorelenok et al., 2003).
Reports on RE oxide admixtures in the growth technology of semiconductors are limited to praseodymium oxide (Novák et al., 1989). Gettering properties of PrO_2 in InGaAs grown by LPE were described by Novák (Novák et al., 1991). When PrO_2 is directly added to the growth melt, layers of both conductivity types are grown. While at low PrO_2 concentrations n-type layers are prepared, higher PrO_2 concentrations lead to the growth of p-type layers with hole concentrations in the range of 2×10^{15} cm^{-3} to 2×10^{16} cm^{-3}. Transport properties of these p-type layers were examined in detail by Kourkoutas (Kourkoutas et al., 1991). Finally, studies of incorporation of Pr into the lattice of InGaAs were performed at high PrO_2 concentrations in the growth melt (Novák et al., 1993). Pr is incorporated in the form of inactive complexes. These complexes can be activated by thermal annealing. The activation occurs solely in a thin layer near the surface.

3. Experimental

A conventional sliding boat system was available for the growth of InP and InGaAsP layers by LPE. InP epitaxial layers were prepared by the supercooling technique on (100)-oriented substrates with RE or RE oxide addition to the melt. The role of growth conditions, particularly (i) the growth temperature, (ii) the cooling rate, (iii) the growth time, and (iv) the method of the growth melt preparation were investigated together with varying RE content in the melt. The initial growth temperature was altered from 600 to 660 °C with the initial supercooling of 5 to 10 °C and the cooling rate of 0.1 to 0.7 °C/min. The growth was terminated after 15 to 30 minutes. The layer thickness varied from 4 to 15 µm. Relatively thick layers were prepared due to their intended application in radiation detectors. To suppress the great affinity of REs, especially with respect to oxygen and hydrogen, it was necessary to prevent the reactive metallic RE to come into contact with the surrounding ambient at the stage before the growth. The LPE process was realized in two cycles. In the first cycle, required amounts of In and undoped polycrystalline InP were homogenized at the temperature of 700 °C for one hour in the Pd-purified hydrogen ambient. The system was cooled, and in the second cycle, pieces of RE were mechanically embedded into the melt to form the growth solution. A polished single crystal (100)-oriented semi-insulating InP:Fe or n-type InP:Sn substrate was placed in the moving part of the boat. The substrate was covered by an InP slide to suppress its thermal decomposition. The temperature was again raised to 700 °C and held constant for one hour. The system was then cooled down to the growth temperature. Just prior to growth, the substrate was etched in situ by passing the substrate bellow a pure In or undersaturated In-InP melt.
The supersaturation of the solution cannot be evaluated precisely. During growth, refractory compounds of phosphorus with REs (pnictides) are formed in the liquid phase (Nakagome et al., 1987). These compounds are insoluble in indium. The effective concentration of phosphorus is diminished and so is the supersaturation (Gorelenok et al., 2003). This supplementary (negative) supersaturation may vary with RE concentration in the growth

solution. Since the growth is usually performed from only slightly supersaturated solutions, this effect must be taken into consideration, especially when growing multilayer structures in order to avoid etching of the previous layer (Astles, 1990).
Structural defects were revealed by several chemical etchants. Optical microscopy with Nomarski differential interference contrast was employed to study the surface morphology and the structural defect density. Scanning Electron Microscopy (SEM) served to trace the substrate-layer interface and the layer thickness after chemical etching. Estimates of the electrical properties on the contactless samples were gained from capacitance-voltage (C-V) measurements performed with the mercury probe at room temperature. In the probe, a smaller area circular Schottky contact with the diameter of 0.3 mm and a concentric larger area annulus Schottky contact with the outer diameter of 3 mm are formed under the pressure of 20 torr. Capacitance is monitored by a bridge with the test frequency of 1 MHz. The samples prepared on SI substrates were further characterized by the temperature dependent Hall effect measurement using a home made computer controlled apparatus with high impedance inputs and a switch box in van der Pauw configuration. The current source and current sink can be individually applied to any sample contact. The error voltages are eliminated by taking eight d.c. measurements of the Hall voltage at each temperature with two directions of the magnetic field. The set-up is equipped with a closed-cycle helium cryogenic system for the temperature range 6—320 K or with a liquid nitrogen cryostat for the temperature range 80—450 K. Photoluminescence (PL) spectra were taken at various temperatures and various levels of excitation power. The low temperature measurements were performed in order to gain information on the impurity and defect states, since the thermal energy is low enough and a variety of transitions can be resolved. The experimental set-up consists of an optical cryostat, a monochromator and a detection part. The optical cryostat is based on a closed cycle helium refrigeration system and automatic temperature controller that enables measurements in the interval of 4-300 K. Photoluminescence spectra are analyzed by 1 m focal length monochromator coupled with liquid nitrogen-cooled high purity Ge detection system and/or thermoelectrically cooled GaAs photomultiplier in the spectral range 400—1700 nm. The excitation was provided by the He-Ne and Ar ion laser. The excitation densities varied in the range of 0.1—600 mW/cm^2 using suitable neutral density filters.

4. Results and Discussion

4.1 Structure and Surface Morphology

Most optoelectronic devices malfunction with the presence of dislocations and other structural defects. These defects cause rapid recombination of holes with electrons without conversion of their available energy into photons; nonradiative recombination arises, uselessly heating the crystal (Queisser & Haller, 1988). The number of crystallographic defects can be decreased by the optimization of the growth technique (Procházková & Zavadil, 1999). The etch pit method is an effective way to easily measure the dislocation density (Nishikawa et al., 1989). The dependence of the InP layer surface morphology and defect density on the individual REs and their concentrations was traced.
The surface morphology of most layers grown with a small addition (several tenths of weight percent) of REs was desirably smooth and mirror-like with a minimum of surface droplets. For higher concentrations, the layers become imperfect with many defect sites on

the surface. The InP layer-substrate interface—revealed on the cleaved edge by chemical etching—was flat and free of inclusions. In general, the effect of individual REs on the surface morphology, dislocation density and interface quality was similar and only slightly varied due to different solubility of REs in the growth solution. This is in contrast with the studies of Nd and Yb addition prior to the optimisation of REs addition into the growth melt. In the case of Nd admixture, the surface morphology was very rough with isolated areas associated with the growth melt droplets even at relatively low concentrations exceeding 0.1 wt% (Procházková et al., 1999).

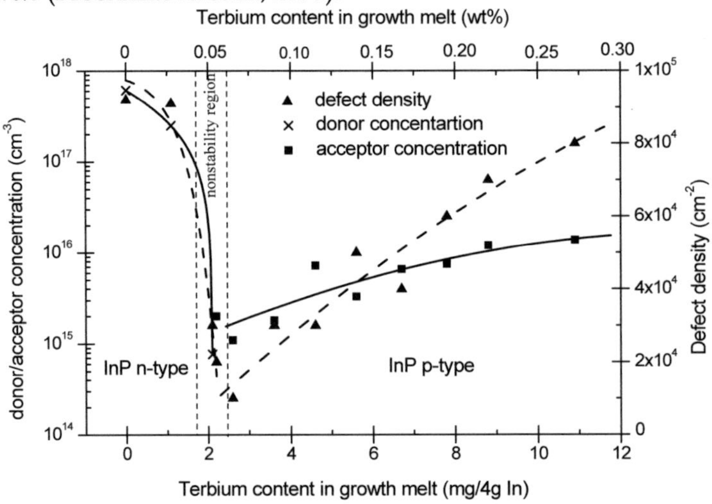

Fig. 1. Dependence of the donor/acceptor concentration of InP layer together with the density of structural defects on Tb content in the growth melt.

The layer thickness exhibited dependence not only on the temperature and the supercooling regime but also on the presence of individual REs in the melt. Again, Nd and Yb admixtures led to markedly decreased growth rates, while the other REs showed only subtle effect on the growth process. Obviously, RE oxides were employed at higher concentrations up to several weight percent—owing to their lower reactivity as compared to elemental REs—without observable deterioration of the surface morphology. The etch pit density for growth from Tb-treated melts together with impurity concentrations are depicted in Fig. 1.

4.2 Electrical and Optical Properties

Firstly, REs will be divided into several groups according to their behaviour during the growth process of InP layers and their impact on electrical and optical properties of these layers. Some general observations valid for these groups of REs will be given. Thereafter, specific behaviour of particular REs will be discussed one after another.

The expected gettering effect has been observed for all REs. However, their purifying efficiency varied considerably for individual RE species. The admixture of certain REs causes not only substantial reduction of residual shallow impurities but also conversion of electrical conductivity from n to p type with one exception, that of Lu maintaining n-type conductivity even at relatively high Lu concentration reaching the solubility limit in In. Among the studied REs, only Ce was incorporated into the InP lattice.

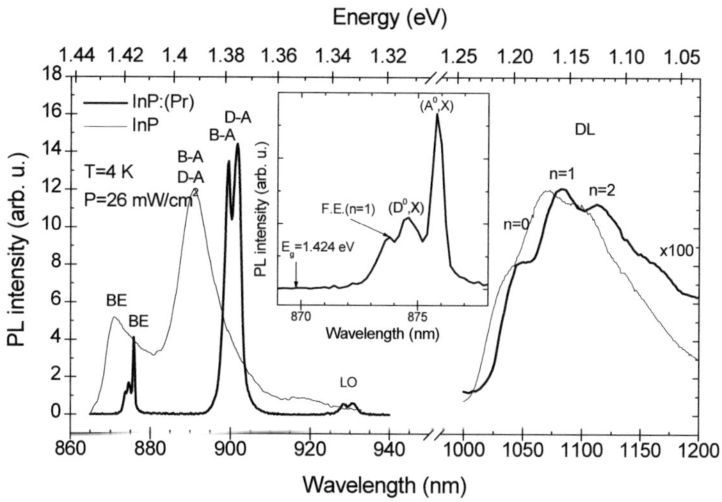

Fig. 2. Comparison of PL spectra of conventionally grown InP and p-type InP:(Pr) (N_A=3x10^{14} cm^{-3}, Pr concentration 0.3 wt%). Magnified excitonic band is shown in the inset.

A typical dependence of the shallow impurity concentration on RE content in the growth melt for Tb, Dy, Tm, Pr, and Gd (group I) is shown in Fig. 1. for the case of Tb admixture.
Introduction of REs to the growth solution results in simultaneous gettering of shallow impurities. Donor impurities are preferentially gettered (Wu & Chiu, 1993). This is in accord with the well known high affinity of REs towards Si and group VI elements (Gschneidner, 1978). This preferential gettering leads to the conductivity conversion from n- to p-type. Further increase of RE addition results in moderately elevated acceptor concentrations. We claim that there are two mechanisms behind this elevation. First, new acceptor species are introduced into the growth solution with REs. The 3N purity of REs, which is currently the highest purity available on the market, is much lower than that of 6N In and InP source materials. At low concentrations of REs, the gettering effect remains virtually undisturbed. However, when RE concentration exceeds the amount necessary for the removal of all donor species, the inadvertent introduction of impurities with RE admixture to the growth solution takes place and results in elevated acceptor concentrations in the grown layers. Second, RE pnictides - particularly compounds of P and RE - are formed in the growth solution. Consequently, the stoichiometric ratio of In and P is altered at the growth interface so that the generation of P vacancies is favored (Ennen et al., 1983). The increased number of vacancies results in increased p-type activity of amphoteric impurities (Žďánský et al., 2001). Very similar behaviour could be observed for all REs oxides (group II). Clearly, for oxides, the concentration at which the conductivity conversion occurs is shifted towards higher values.
The PL spectra show fine features with narrow peaks supporting the results of C-V measurements. Typical PL spectra comparing layers grown with and without RE (Pr) admixture are shown in Fig. 2. The observed radiative transitions in studied InP samples could be grouped into three categories: band-edge (BE) transitions at about 1.418 eV (875 nm), shallow impurity related transitions at 1.38 eV (900 nm), and deep-level transitions at 1.14 eV (1090 nm) (Pearsall, 2000). There is a free space in the final line of this page.

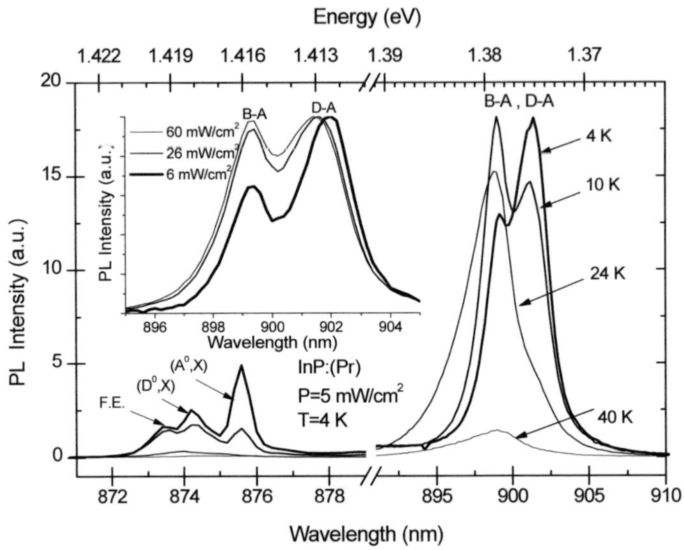

Fig. 3. Temperature dependence of NBE part of the PL spectra of p-type InP:(Pr), (N_A=3x10^{14} cm^{-3}, Pr concentration 0.3 wt%) with the inset depicting D-A peak shift with increasing excitation power.

superlinear behaviour with increasing excitation power and results from the decay of excitons. Transitions due to free exciton (FE) and excitons bound to the neutral donor (D^0,X) or the neutral acceptor (A^0,X) are well resolved. Transitions described as B-A and D-A are related with shallow acceptors and correspond to conduction band-acceptor and donor-acceptor pair transitions, respectively as revealed by the examination of their temperature dependence (see Fig. 3). The band of the lowest energy is rapidly quenched around 25-30 K and is thus assigned to D-A transitions. The other sub-band quenches around 70 K and is thus assigned to B-A transitions (Swaminathan et al., 1985). The peak LO is a phonon replica of the band related to shallow impurities and its position is in accordance with the known value of 43 meV for LO phonon. The peak related to shallow impurities is an unresolved convolution of (B-A) and (D-A) transitions in the case of Ce, Lu, and zero admixtures while separate peaks are well resolved at low excitation power on samples prepared with group I and group II admixtures. The D-A peak shifts with increasing excitation power (see the inset in Fig. 3) since more carriers are generated and the average distance between the donor and the acceptor undergoing the transition decreases (Hsu et al., 1994). The smaller average distance of pairs involved causes the observed blue shift of the D-A transition. The position of the D-A around 1.375 eV corresponds fairly well with the ionization energy of carbon acceptor in InP reported by Skromme (Skromme et al., 1984). Carbon probably originates from the graphite sliding boat.

The long-wavelength part of the spectra is usually dominated by Mn related band consisting of three partly resolved peaks at 1.184 eV (n=0), 1.145 eV (n=1), and 1.107 eV (n=2), which are interpreted as a zero phonon line, and one, and two phonon replicas, respectively (Fig. 2). This characteristic band, observed in majority of samples, whether rare-earth treated or not, is attributed to the recombination of free or loosely bound electrons with holes bound to the Mn acceptor occupying an In site.

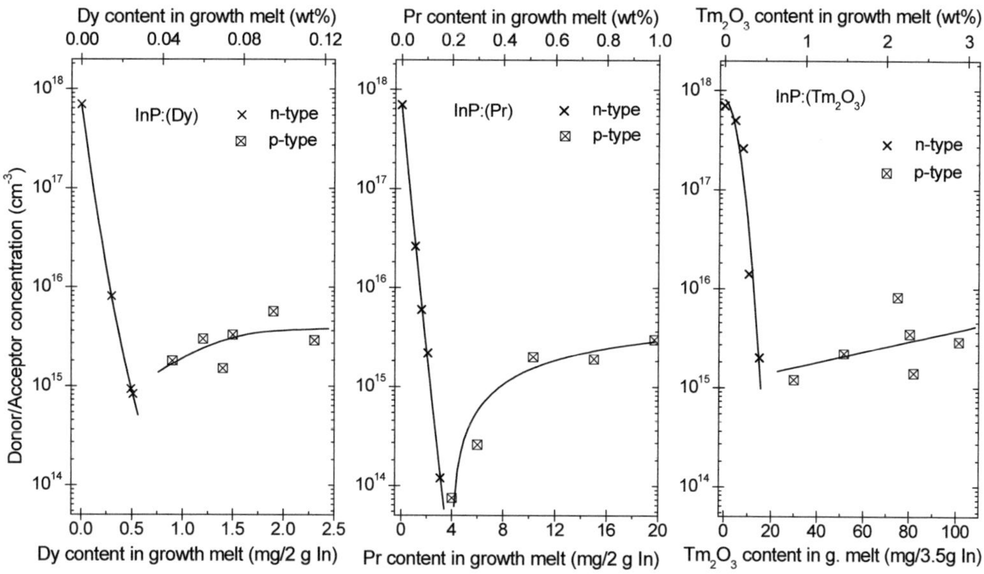

Fig. 4. Dependence of the donor/acceptor concentration of InP layer on Dy (left), Pr (middle), and Tm_2O_3 (right) content in the growth melt.

The deep-level (DL) luminescence is strongest for conventionally grown layers (zero RE admixture). Its intensity decreases with increasing RE admixture up to some certain limit when the highest purity layers are grown and typical conductivity change occurs. Further RE increment does not have significant impact on the DL part of the spectra. These observations indicate that REs may act as scavenging agents for deep levels in addition to shallow donor gettering.

Now, specific behaviour of the individual RE elements and their oxides will be discussed.

4.2.1 Terbium, Dysprosium, and Praseodymium

The shallow impurity concentration as a function of Tb concentration in the growth melt (Fig. 1) was already discussed when describing general behaviour of REs. Tb concentrations in the melt above 0.05 wt% lead to the change of the conductivity type n →p. A similar behaviour can be observed for Pr and Dy additions (Fig. 4). Only the concentration at which the conductivity change occurs is shifted towards lower (Dy) and higher (Pr) values. This is due to their different reactivity towards impurity species and their different solubility in the growth melt. Preparation of high purity p-type layers is one of the goals of our studies. From that point of view, Dy with conversion around 0.03 wt% seems to be a good candidate for the growth of p-type layers. Recall that higher concentrations of REs may deteriorate structural properties of the layers. On the other hand, Pr shows better purification efficiency, even though the n→p conversion takes place at higher RE concentration around 0.2 wt%.

The PL spectra of n and p-type InP layers prepared with the addition of Tb, Dy, and Pr were qualitatively similar in the NBE region (see Fig. 2 and Fig. 3). However, spectra are different in the spectral region above 1000 nm, where the deep level related luminescence dominates.

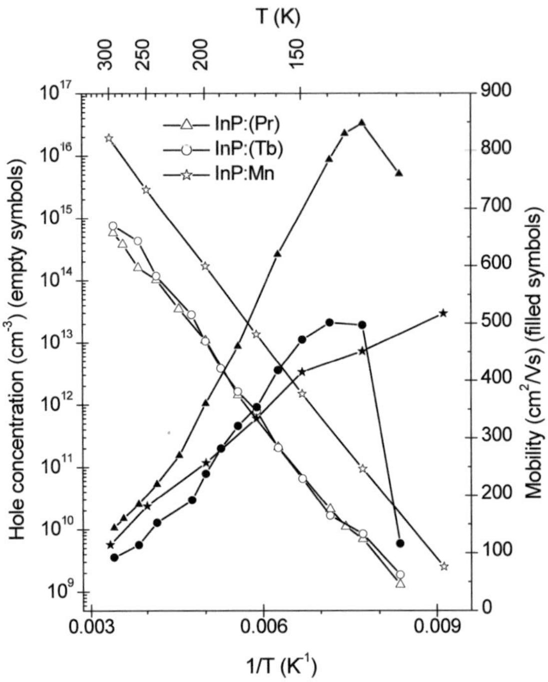

Fig. 5. Temperature dependence of the hole concentration and mobility of p-type InP:(Pr) and InP:(Tb) layers prepared on InP:Fe substrate. The curves of the same quantities of InP:Mn are shown for comparison.

To explain differences in the deep level luminescence and its origin, let us first take advantage of the Hall measurements. Fig. 5 shows curves of the hole concentration p and the hole mobility μ_p as a function of reciprocal temperature for InP layers grown with the admixture of Pr and Tb. Data for Mn doped InP bulk crystal are also shown for comparison. The logarithmic plots of the hole concentration show straight lines in the range of several decades and are nearly identical for two different samples. Sample InP:(Tb) was prepared from the melt containing 0.07 wt% of Tb and sample InP:(Pr) with 0.25 wt% of Pr. Very similar behaviour could be observed for the layers grown with Dy. The binding energy of the dominant acceptor determined from the slope of the straight lines is equal to 0.22 eV. This value is close to the binding energy of Ge acceptor (210±20 meV) and to the binding energy of Mn acceptor (230 meV) (Žd'ánský et al., 2002).

We already know that deep level parts of the PL spectra of the layers grown with Tb and Dy, and most of those grown with Pr admixture are dominated by the Mn band formed by the group of three peaks interpreted as a zero phonon line and its one- and two-phonon replicas (Fig. 2). Different behaviour of some of the layers grown from Pr treated melts is demonstrated in Fig. 6., where low-temperature PL spectra of Pr and Tb treated spectra are plotted for wavelength exceeding 1000 nm. The peak at 1.1952 eV is close to the estimated position of the no-phonon line (1.215 eV) of the band-Ge acceptor transitions (Žd'ánský et al., 2001). To sum up, the dominant acceptor responsible for n→p conductivity conversion in InP layers grown with Tb and Dy was identified as Mn, while for some samples grown with Pr it was identified as Ge. Both Ge and Mn are residual contaminants in undoped InP and probably become dominant electrical impurity due to the distinct preferential gettering of the individual REs. Secondary ion mass spectroscopy measurements are currently under way to properly resolve this behaviour.

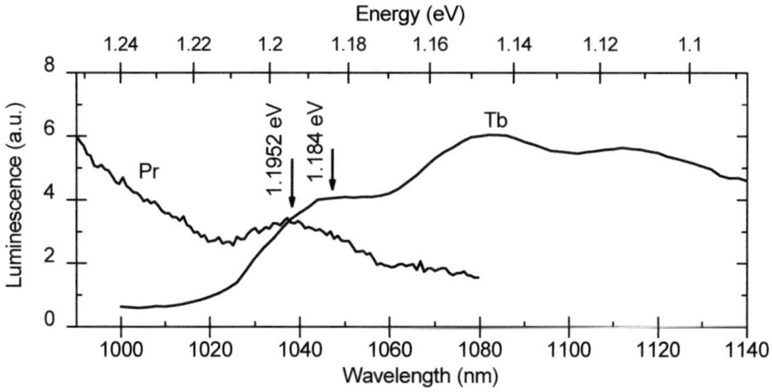

Fig. 6. InP prepared with Tb admixture showing a characterictic Mn band, and with Pr admixture with a peak at 1.1952 eV due to Ge.

Fine spectral features of the PL spectra, low donor and acceptor concentrations, and high mobilities indicate that Pr addition to the growth melts results in high purity p-type InP layers. The RT hole mobility of the InP:(Pr) sample of 144 cm^2/Vs is close to the value theoretically expected for pure p-type InP (Kranzer, 1974). Maximum mobility of 848 cm^2/Vs is reached at 130 K. This value is slightly smaller than low concentration Zn, Cd, and Mg doped p-type InP (Kuphal, 1981).

4.2.2 Thulium

PL spectra of Tm treated samples show fine features in the excitonic and shallow levels related band (

Fig. 7). Measurement at low levels of excitation power enables us to distinguish three subbands within a part of the spectra associated with the shallow levels. Inspection of their temperature dependence revealed that the band of the lowest energy is rapidly quenched around 25 K and is assigned to donor-acceptor pair transitions. The other two subbands quench around 60 K and are due to conduction band-acceptor transitions. Recall similar behaviour of temperature dependence for Pr sample (Fig. 3). The two B-A subbands are tentatively assigned to Cd and Zn (Pearsall, 2000).

Sample Tm-14 (0.09 wt% of Tm) was grown on a semi-insulating (SI) InP:Fe substrate and instead of n→p conductivity change, transition to SI state was observed. Similarly, all other samples with different Tm concentrations were converted to SI state. For that reason, electrical properties were evaluated only by using mercury probe on the samples prepared on InP:Sn substrates (Table 1). Notice that certain admixtures (around 0.08 wt%) lead to the preparation of high purity samples. The conductivity change occurs at relatively uncertain value of Tm addition. The transition to semi-insulating state related to the diffusion of Fe into adjacent InP layer is an undesirable effect. Fast out-diffusion of iron doped SI InP substrates enhanced by the presence of zinc, cadmium, and beryllium was reported and possible mechanisms of iron-acceptor exchange and acceptor interstitial leakage was proposed (Kazmierski et al., 1992). Iron redistribution is extremely fast in adjacent p-type material, while no significant diffusion into n-type material occurs.

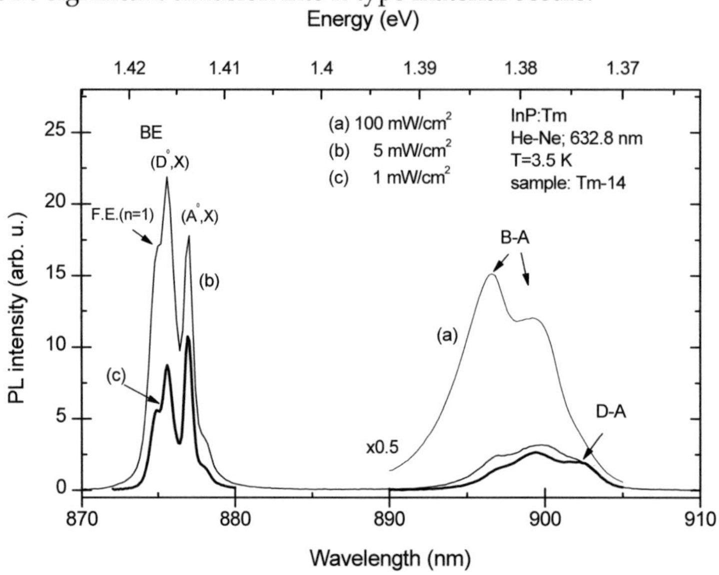

Fig. 7. NBE part of the PL spectra of p-type InP prepared with Tm admixture showing fine spectral features.

The experiments with different REs reveal the indeterminateness in the behaviour of iron in contact with p-type layers prepared from the solutions with REs admixtures. As for the layers prepared with Tb, Dy, Pr, and Ce, no iron diffusion has been observed, while samples prepared with Tm and also Gd admixtures were converted to the SI state. On the contrary, the behaviour of typical p-type dopants (Zn, Cd, Be) was universal (Kazmierski et al., 1992). This leads to the suggestion that only Tm and Gd act in the same way as a typical p-type dopant. The proposed mechanism involves an acceptor insterstitial out diffusion from the p-type material into the SI substrate. The vacancy equilibrium is shifted in p-type layer leading to the excess of the phosphorus vacancy concentration, and thus, enables a higher incorporation ratio of the substitutional Fe (Procházková et al., 2005b). The iron redistribution can be well suppressed by inserting a thin buffer layer of undoped InP. The absence of significant iron diffusion through the buffer layer was validated by C-V measurements and PL spectroscopy. However, insertion of the buffer layer does not solve the problem of the Hall effect measurement requiring SI base.

4.2.3 Gadolinium

Gd treated samples show extremely low concentration at which the n → p conversion occurs (see Table 2). This observation is a proof of a high Gd reactivity (Zakharenkov et al., 1997). The PL spectra show some interesting features in the NBE part leading to the suggestion that donor impurities are gettered extremely successfully. On the other hand, high acceptor concentrations in p-type layers with relatively low admixtures are probably caused by the introduction of new impurities to the melt with Gd. However, these suggestions must be proved by further studies. As stated above, Gd admixture leads to conversion to SI state when preparing layers on InP:Fe substrate.

sample	REC (mg)	REC (wt%)	CT	N_D or N_A (cm^{-3})	sample	REC (mg)	REC (wt%)	CT	N_D or N_A (cm^{-3})
Tm-2	1.1	0.031	n	3.0E+16	Gd-10	0.3	0.008	n	2.0E+15
Tm-15	2.7	0.077	n	1.0E+15	Gd-16	0.4	0.011	p	3.5E+15
Tm-7	2.8	0.080	n	7.4E+14	Gd-12	0.5	0.014	p	7.7E+15
Tm-9	2.8	0.080	p	8.9E+14	Gd-11	1.0	0.028	p	6.3E+15
Tm-36	2.9	0.083	p	4.2E+15	Gd-14	2.1	0.059	p	1.2E+16
Tm-6	3.2	0.091	p	1.0E+15	Gd-15	5.0	0.141	p	2.2E+16

Table 1. Donor (acceptor) concentrations of the samples prepared with Tm and Gd admixture. REC stands for RE concentration, CT for conductivity type.

4.2.4 Cerium

As stated in the introductory part of this section, Ce is the only element among the studied ones, which was incorporated into the InP lattice. In our earlier studies of Yb admixture (not included in this review) it was shown that Yb was incorporated into the InP lattice, layers exhibit n→p conductivity conversion and Yb itself was identified as the dominant acceptor responsible for the conductivity crossover (Žďánský et al., 2002). To give evidence of this statement, let us first look at the left panel of Fig. 8 showing the acceptor concentration as a function of Ce content in the melt. All layers, even those prepared at very low Ce concentration, are of p-type conductivity. When increasing the Ce concentration, the acceptor concentration in the layers is also increased. This may imply that Ce behaves as an acceptor itself. It was recently shown that dominant acceptors responsible for the n→p conductivity conversion of the LPE InP:Yb layers are Yb^{3+} ions incorporated into the InP lattice as a cubic Yb^{3+} center on the cation site (In) (Žďánský et al., 2002). Typical temperature dependence of the hole concentration and mobility derived from the Hall measurements for p-type InP layers doped with Ce is shown in the right panel of Fig. 8. The data for Yb doped LPE layers and Zn doped bulk crystals are given for comparison. Similar behaviour of the layers doped with Yb and Ce can be clearly seen. Hall measurements presented in Fig. 8 have been performed in the range from room temperature to about 35 K. Below this threshold temperature, the conductivity of Ce and Yb doped samples changes very slowly so that the temperature equilibrium cannot be reached within hours. A slow decay of conductivity is initiated at about 60 K. The samples doped with Yb and Ce reach a metastable state and results of Hall measurements are not reliable any more. Introducing Ce and Yb into the InP lattice, their atoms are exposed to a strong electron-lattice interaction due to their large atomic radii and due to a large difference between electronegativities of Yb, Ce, and In (Zavadil et al., 2007). A potential barrier raised by this interaction is supposed to be responsible for the observed metastability of electrical conductivity (Procházková et al., 2005c). The curves of the hole concentration reveal straight lines at low temperatures giving the binding energy of Ce equal to 31 meV.

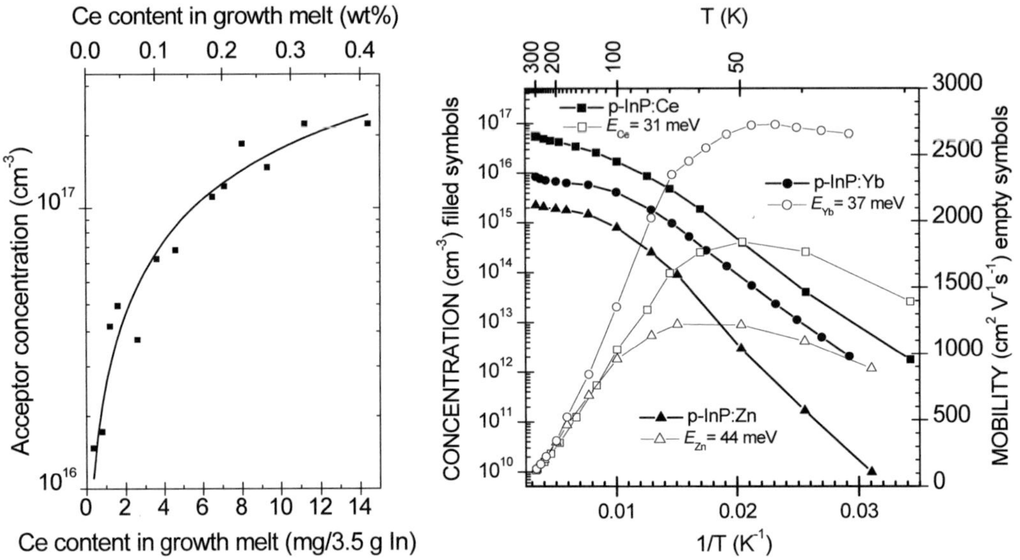

Fig. 8. Dependance of the acceptor concentration on Ce content in the growth melt (left panel) and temperature dependence of the hole concentration and mobility of p-type InP layers doped with Ce, Yb, and Zn (right panel).

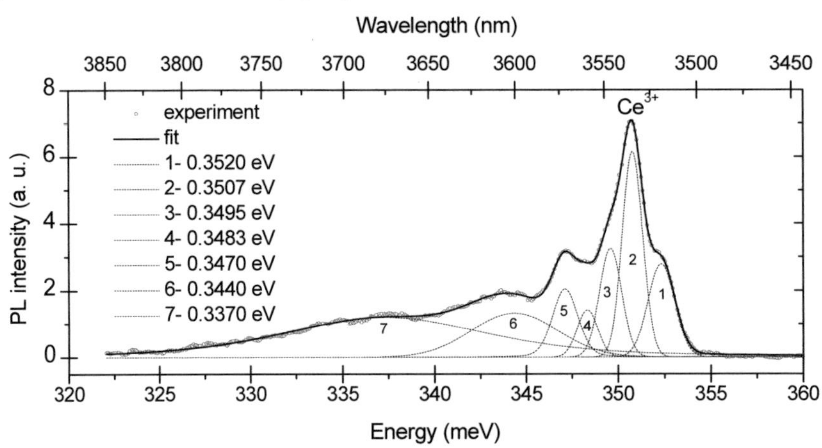

Fig. 9. PL spectrum of the InP layer prepared with Ce admixture.

Among the other REs, Ce and Yb have relatively simple energy level scheme of 4f shell with only one excited state. Their position within the lanthanide series can be viewed as complementary; Ce^{3+} possesses one 4f electron, while Yb^{3+} possesses thirteen 4f electrons. The PL spectrum of InP:Ce (0.2 wt%) layer measured in the mid-infra-red range is shown in Fig. 9. The sharp inner shell $2F^{7/2} \rightarrow 2F^{5/2}$ transition of Ce^{3+} ion at 3534 nm (0.3507 eV) together with a fine structure caused by crystal field splitting is clearly seen. This is a direct proof of Ce^{3+} ions being incorporated into the InP lattice.

4.2.5 Lutetium

Samples prepared with with the admixture of Lu were of n-type conductivity at any Lu concentration. A list of results of C-V measurements for selected samples follows: sample LuI-1 (0.014 wt%) with $N_D=3.5\times10^{17}$ cm^{-3}; sample LuI-2 (0.049 wt%) with $N_D=1.7\times10^{17}$ cm^{-3}; sample LuI-4 (0.108 wt%) with $N_D=3.4\times10^{16}$ cm^{-3}; and sample LuI-6 (0.173 wt%) with $N_D=8.8\times10^{15}$ cm^{-3}. Higher Lu concentration could not be applied due to the deterioration of the surface morphology.

Let us briefly sum up the effect of the studied REs on electrical and optical properties of the InP layers. Dy produced the conductivity conversion from n→p at the lowest concentration. Pr and Tm addition resulted in high purity samples. Typical Pr treated samples exhibited high mobilities. The fine spectral features in the NBE part of the spectra were described on Pr and Tm treated samples. The dominant impurity in the case of p-type layers were determined as Ge or Mn for Pr treated samples, Mn for Tb and Dy samples, and Ce itself for Ce samples. The temperature dependent Hall effect measurement could not be performed on the samples with Tm and Gd admixture due to their conversion to SI state caused by Fe out-diffusion from the InP:Fe substrate. Finally, Lu did not exhibit as large an impact as the other elements and its addition always resulted in the growth of n-type layers.

REs show high oxidation potential and therefore must be handled carefully in the atmosphere with an oxygen content (Korber et al., 1986). The insertion of partially oxidized REs into the growth solution may cause difficulties. REs are inert towards hydrogen atmosphere. However, this holds only for room temperature. At the temperature above 300 °C, the pure REs absorb hydrogen instantly and create very stable hydrides. The extent of the hydride formation is practically limited by the previous oxidation of the pure RE metal (Novák et al., 1991). It follows that the gettering effect is extremely sensitive to the RE oxide and hydride formation. The alteration of this ratio from run to run causes the deterioration of the reproducibility of the epitaxial growth process. Taking into account certain level of uncertainty in the behavior of the elemental REs, from now on, the role of RE oxides will be discussed.

4.2.6 Thulium oxide

The standard two-period cycle for the sample preparation did not lead to a substantial improvement of the layer quality. Namely, its background concentrations were not decreased significantly. Thus, a four-period cycle of the sample preparation was proposed. During the first two periods, solution of In and Tm_2O_3 was prepared and homogenized at the temperature of 730 °C. In the third period, InP was added and shortly homogenized again. The last period was a standard growth period as described in section describing experimental procedures. Very similar impact of Tm_2O_3 in comparison with elemental thulium was observed. The expected lower reactivity of the oxide results in much larger value of concentration (around 0.5 wt%) at which the conductivity crossover is observed (Fig. 4). PL spectra allowed to distinguish the B-A transitions from D-A pair ones, and enabled identification of exciton related transitions. Similarly to Tm treated samples, threesubbands are typically resolved in the shallow levels related part of the spectra (left panel of Fig. 10).

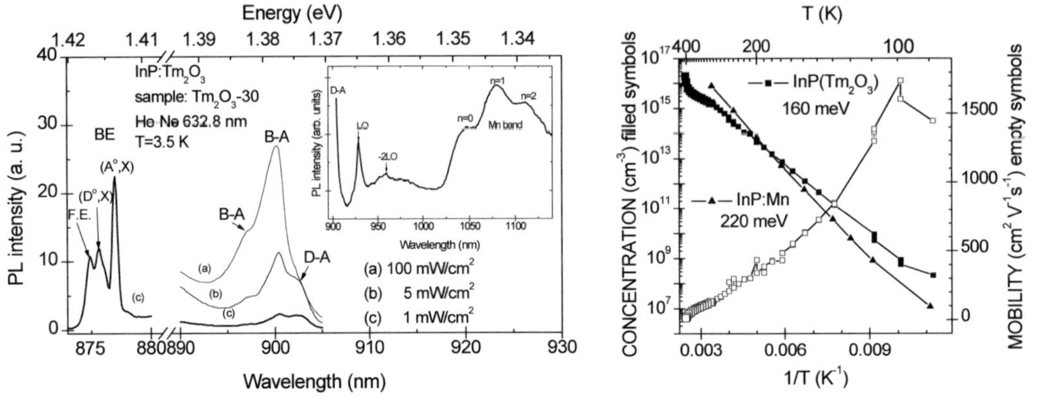

Fig. 10. NBE part of the PL spectra of p-type InP prepared with Tm_2O_3 admixture showing fine spectral features with the inset depicting its long wavelength part (left panel) and Hole concentration and mobility of p-type InP layers grown with Tm_2O_3 admixture (right panel).

Hole concentration and mobility of the layers are given in the right panel of Fig. 10 together with corresponding data measured on InP:Mn sample for comparison. The binding energy of the dominant acceptor responsible for n→p conductivity conversion was determined as 160 meV. Low temperature PL spectra (see the inset in the left panel of Fig. 10) exhibit a broad band at 960 nm (1.29 eV) that fairly well corresponds with the above mentioned energy. The origin of this impurity is not clear. The sample with Tm_2O_3 (Tm_2O_3-30, 2.3 wt%) is characterized by the following values of free carrier concentration and mobility: $p(300)=1.7\times10^{15}\,cm^{-3}$, $\mu(300)=127\,cm^2V^{-1}s^{-1}$. Maximum mobility $\mu(100)=1764\,cm^2V^{-1}s^{-1}$ is reached at the temperature of 100 K.

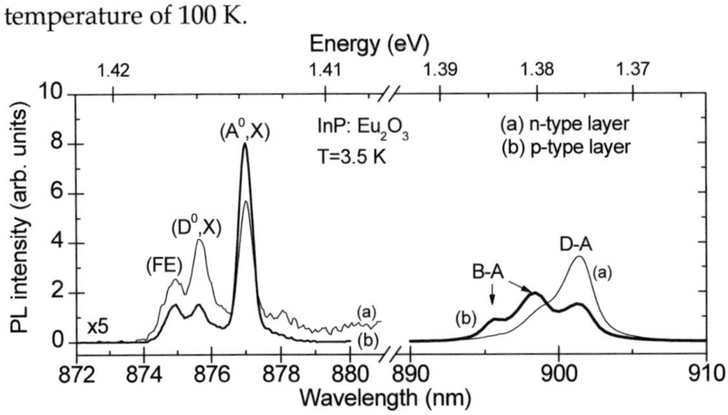

Fig. 11. NBE part of the PL spectra of both n-type and p-type InP prepared with Eu_2O_3 admixture

4.2.7 Europium oxide

Europium oxide Europium oxide appears to have a very high purification effect. Unfortunately there is no strict level of Eu_2O_3 at which the conductivity change takes place. The choice of substrate seems to play an important role. Two n-type layers of very low donor concentration were grown on InP:Fe substrate:

sample EuO-6 (0.39 wt%) with N_D=2.5×10^{14} cm^{-3}

sample EuO-15 (1.13 wt%) with N_D=1.2×10^{14} cm^{-3}

The corresponding layers prepared on InP:Sn substrate are of both n-and p-type:

sample EuO-7 (0.49 wt%, n-type) with N_D =1.1×10^{15} cm^{-3}

sample EuO-19 (1.49 wt%, p-type) with N_A =2.6×10^{15} cm^{-3}

A further series of experiments has to be performed to clarify the behaviour of Eu_2O_3 and its relation to the substrate properties. The PL spectra show fine features with narrow peaks supporting the results of C-V measurements. The NBE part of the spectra clearly demonstrates differences between n-type and p-type layers.

4.2.8 Other oxides

To round off the enumeration of RE oxides, PrO_x, Gd_2O_3, and TbO_x admixtures have to be mentioned. These samples were characterized by the C-V measurements. The data are summarized in Table 2 (asterisk means different conditions of the sample preparation). PrO_x samples were further characterized by PL measurements. The behavior of PrO_x did not markedly differ from the behavior of elemental Pr. Evidently, the conductivity conversion arose at higher concentration around 1 wt%. Admixture of Gd_2O_3 and TbO_x always resulted in preparation of n-type layers. This was probably caused by their low solubility in the In-based melt. The last sample TbOI-6* was prepared by a four-period cycle similar to the earlier mentioned Tm_2O_3 layers. Moreover, the temperature of homogenization was increased to 800 °C. To briefly sum up RE oxides show some promising advantages over the elemental REs. However, the process of their introduction into the melt and its subsequent homogenization must be solved in order to result in repetitive preparation of high quality layers.

sample	REC (mg)	REC (wt%)	CT	N_D or N_A (cm^{-3})	sample	REC (mg)	REC (wt%)	CT	N_D or N_A (cm^{-3})
Gd_2O_3-1	1.9	0.09	n	2.6E+17	Nd_2O_3-0	0.0	0.00	n	2.8E+17
Gd_2O_3-2	3.1	0.15	n	1.7E+17	Nd_2O_3-4	2.5	0.12	n	7.1E+14
Gd_2O_3-3	15.1	0.74	n	8.0E+16	Nd_2O_3-3	5.0	0.24	p	3.4E+15
Gd_2O_3-4	30.0	1.46	n	9.3E+16	Nd_2O_3-5	7.0	0.34	p	3.0E+15
Gd_2O_3-5	60.0	2.88	n	1.4E+16	Nd_2O_3-2	10.0	0.49	p	3.2E+15
Gd_2O_3-8*	30.2	0.85	n	6.8E+15	Nd_2O_3-6	15.0	0.73	p	3.0E+15
PrO_x-8	5.0	0.25	n	5.7E+15	Nd_2O_3-7	45.0	2.17	p	3.9E+15
PrO_x-2	15.8	0.79	n	2.2E+15	TbO_x-1	25.0	1.25	n	6.6E+17
PrO_x-1	24.5	1.23	p	2.8E+15	TbO_x-2	50.8	2.54	n	2.3E+17
PrO_x-6	30.0	1.50	p	2.0E+15	TbO_x-3	97.6	4.88	n	1.3E+16
PrO_x-5	59.5	2.98	p	3.8E+15	TbO_x-6*	49.3	2.46	n	7.8E+15

Table 2. Donor (acceptor) concentrations of the samples prepared with different RE oxide admixtures. REC stands for RE concentration, CT for conductivity type.

4.3 Towards the Application
In this section we try to demonstrate the application of the gettering phenomena in device concepts. InP-based radiation detectors and double heterostructure LEDs in the near infrared region were selected as examples.

4.3.1 Radiation Detectors
While the mainstream of research effort in the area of semiconductor technology is nowadays aimed at producing nanometer-thin structures, there remain niches, such as fabrication of radiation detector structures, where relatively thick layers are valued and required.
InP-based materials have various advantages, such as bandgap energy suitable for light emitters in the long-wavelength region, an extremely high saturation velocity of electrons suitable for the active channel in high-power and high-speed electronic devices, as well as high thermal conductivity and high threshold of optical catastrophic degradation (Wada & Hasegawa, 1999). On the other hand, InP-based materials are less mature in wafer- and device-processing technologies such as surface passivation, metal-semiconductor gate technology or ohmic contact preparation. New studies and developments are required in various fields of technology to fully exploit the intrinsic advantages of InP.
InP single crystals are promising materials for the preparation of radiation detectors operating at room temperature. The room temperature operation is possible due to sufficiently wide bandgap energy Eg=1.35 eV at room temperature and high mass density ρ=4.8 g/cm^3. The high value of atomic number of In (Z_{In}=49) predicates high stopping power, which is proportional to Z^n ($n \approx 4$-5) for a high energy photon (Pelfer, 2001). Preparation of thick high-purity InP layers would be appreciated to exploit the above mentioned potential in radiation detectors. High purity and homogeneity of the layers is essential to attain large charge collection efficiency. Conventionally prepared bulk and epitaxial InP crystals are of n-type conductivity due to intrinsic donor impurities. Intentional doping with shallow acceptors is generally the way to prepare p-type material. In this chapter, we have demonstrated that the application of certain concentration of appropriate RE leads to the preparation of high purity InP of both conductivity types without intentional doping.
High purity InP layers of both n- and p- type conductivity with carrier concentration diminished to 10^{14} cm^{-3} thanks to Pr additions to the growth melts were used to fabricate detector structures with p-n junction. High purity n-type and the p-type layers were subsequently grown on Sn doped n-type substrate. Each of the layers was 10 µm thick. Metal contacts were deposited by vacuum evaporation: Au-Be/Cr/Au contacts on p-type InP and AuGeNi contacts on n-type InP. The structures were diced into chips of 500x500 µm. Spectra of α-particles from the ^{241}Am source of 5.48 MeV were measured at room temperature in vacuum with negative voltage on the irradiated (p-type InP) side. The highest achieved CCE was 40 % with FWHM energy resolution of 8 % (Procházková et al., 2009).

4.3.2 Double Heterostructure LEDs
InP itself had not received much interest until it was used as a substrate for InGaAsP-based LEDs and lasers (Hsieh et al., 1976) and InGaAs transport devices (Takeda et al., 1976).

Lattice matched InGaAsP to InP substrate can cover the energy range from 0.75 eV to 1.35 eV at RT. This energy range is important in silica-fibre-based optical communications and infrared spectroscopy (Rakovics et al., 2002). We demonstrate the purifying effect of Pr on the preparation of InGaAsP layers and InGaAsP/InP double heterostructure LEDs by liquid phase epitaxy. Background impurities in the active region, especially at high concentrations, introduce defects that act as recombination centres. High background concentration may also lead to carrier spill-over into one of the confinement regions and decrease the radiative efficiency.

Four-layer structures were grown consisting of an n-type buffer layer (7 μm), Pr-treated InGaAsP active layer (400 nm), p-type InP:Zn confinement layer (2 μm), and p-type InGaAsP:Zn contact layer. Ohmic contacts were deposited by vacuum evaporation: Au-Be/Cr/Au contacts on p-type InGaAsP and AuGeNi contacts on n-type InP. Electrical and optical properties were evaluated on the chips with the size of 500×500 μm, PL spectra were measured on the samples without p-type confinement and contact layer.

Narrowing of the PL peak at RT was observed already when 0.1 wt% Pr admixture was applied. The FWHM dropped from 48 meV for a conventionally grown layer to 42 meV for the layer grown with 0.1 wt% of Pr. Narrowing of the peak is caused by the elimination of a donor band, which - in the case of heavily doped material - extends into the conduction band and gives rise to the peak broadening and shifting towards lower energy (Kasap, 2001). The RT electroluminescence spectral bandwidth measured at 10 mA was 57 meV. Typical external quantum yield of the fabricated LEDs was 1.7 % and integrated optical power was 1.8 mW, both measured at RT and electrical current of 100 mA.

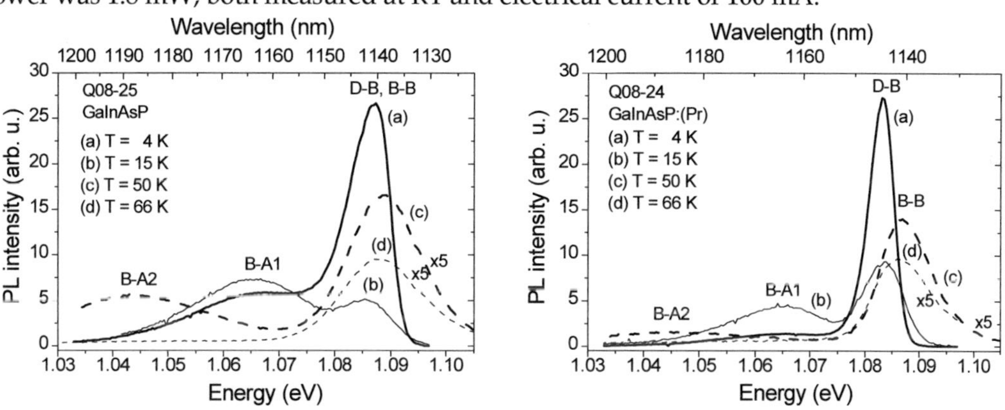

Fig. 12. Low temperature PL spectra of InGaAsP (left panel) and InGaAsP:(Pr) (right panel) are shown for temperatures 4, 15, 50 and 66 K by curves (a), (b), (c) and (d), respectively.

InGaAsP layers with different amounts of Pr admixture, corresponding to impurity concentrations of 1.6×10^{17} cm^{-3} (0 wt% of Pr), 1.2×10^{17} cm^{-3} (0.15 wt% of Pr), 9.2×10^{16} cm^{-3} (0.2 wt% of Pr), and 5.4×10^{15} cm^{-3} (0.5 wt% of Pr), were further investigated by low temperature PL spectroscopy. PL spectra of conventionally grown InGaAsP layer (i.e. without Pr addition) measured at four temperatures are shown in the left panel of Fig. 12. Temperature dependence of PL spectra was investigated in the temperature range 4—70 K in order to deduce contributions from various radiative transitions. The dominant band (D-B, B-B) at the lowest temperature of 4 K is at 1.084 eV with the halfwidth of 10 meV. The contributions from free or bound excitons, band to band (B-B) and shallow donor to valence band (D-B) transitions are not resolved. With increasing temperature, in the range 4—70 K, the band is broadened and slightly shifted but individual contributions are still not resolved. Transitions due to shallow acceptors with binding energies (23 meV and 45 meV) could also be seen for temperatures bellow 70 K. When the temperature is increased above 70 K, only the broad band due to band to band transition is seen and is gradually shifted to lower energy with increasing temperature. This band is centred at 1.03 eV (1200 nm) at 280 K, and has the halfwidth of 42 meV.

All samples prepared with Pr admixture exhibit a distinct narrowing of PL spectra and D-B transition could be resolved from B-B transitions in the temperature range 4—15 K. In view of the fact that the donor binding energy E_D~4.5 eV is very small, the donor related band is resolved only for temperatures below 20 K in samples purified by Pr addition. Typical low temperature PL spectra of InGaAsP:(Pr) samples are shown in the right panel of Fig. 12. At the lowest temperature, the spectrum is dominated by the narrow band (D-B) with the halfwidth of 5 meV, which is due to donor to valence band transitions. When the temperature is increased above 15 K, but not exceeding 70 K, the band is shifted by about 5 meV to higher energy (B-B) and is dominated by band to band transitions.

By the inspection of Fig. 12 we can see that considerable suppression of acceptor related radiative transitions takes place in Pr treated samples. This suggests effective gettering of acceptor impurities.

5. Conclusions

High purity InP layers were prepared by liquid phase epitaxy. Their growth was performed from the melts containing, besides essential components, also REs (Tb, Dy, Pr, Tm, Gd, Lu, Ce) and some of their oxides (PrO_x, Tm_2O_3, Gd_2O_3, Eu_2O_3, and TbO_x). The content of individual RE species was systematically altered to investigate their impact on physical properties of the grown layers.

The expected gettering effect was observed for all REs. However, their purifying efficiency varied considerably for individual RE species. Introducing REs to the growth solution, simultaneous gettering of shallow impurities takes place. InP layers with shallow impurity concentration decreased by up to four orders of magnitude were grown. Donor impurities are preferentially gettered due to the high affinity of REs towards Si and group VI elements. The preferential gettering leads to conductivity conversion from n- to p-type when increasing RE concentration in the growth melt. Very similar behaviour could be observed for RE oxides. Due to their lower reactivity, the concentration at which the conductivity conversion occurred, was shifted towards higher values. In the case of Lu, Gd_2O_3, and TbO_x addition, all layers maintained n-type conductivity even at relatively high concentrations reaching its solubility limit in the growth melt. Among the studied REs, only Ce was conclusively incorporated into the InP lattice.

To emphasize the finest results, the addition of 0.2 wt% of Pr led to the following values of the hole concentration and mobility at room temperature: $p(300)=8\times10^{13}$ cm^{-3} and $\mu(300)=144$ cm^{2}V^{-1}s^{-1}. The samples showing low concentrations of residual impurities were characterized by PL spectroscopy. The observed significant narrowing of PL curves and the corresponding appearance of fine spectral features are typical characteristics of pure materials, low in defects.

With the help of temperature dependent Hall measurements and PL spectroscopy, the nature of the dominant acceptor in p-type InP was recognized. The dominant acceptor in InP layers grown with Tb, Dy, and Pr was identified as Mn. In some samples grown with Pr, Ge was a dominant acceptor. Samples prepared on InP:Fe substrates with Tm and Gd admixtures were converted to semi-insulating state. This conversion is attributed to the out-diffusion of iron from the InP:Fe substrate.

The surface morphology of the layers grown with a small addition of RE (several tenths of weight percent for elemental REs and several percent for their oxides) was desirably smooth and mirror-like with a minimum of surface droplets. Optimized admixtures of REs resulted in lowered dislocation densities.

Preliminary studies of the application of the unique properties of REs were performed. Pr has been chosen for these investigations since its admixtures (i) provoke conductivity conversion at relatively low concentrations in the melt, (ii) do not influence the growth rate, and (iii) under appropriate growth conditions lead to lowered dislocation densities and good surface morphology (Grym et al., 2009). Applications in semiconductor radiation detectors and double heterostructure LEDs were demonstrated.

This work was supported by the projects #102/08/P617 and #102/09/1037 of the Czech Science Foundation. The authors thank M. Hamplová, Z. Podvalová, J. Zelinka, and V. Malina for technical assistance, P. Gladkov for the PL measurements of Ce doped samples, and R. Yatskiv for the measurements of the detection performance of the structures with p-n junction.

6. References

Astles, M.G. (1990). Liquid--Phase Epitaxial Growth of III-V Compound Semiconductor Materials and their Device Applications, IOP Publishing Ltd

Capper, P. & Mauk, M., Eds. (2007). Liquid Phase Epitaxy of Electronic, Optical and Optoelectronic Materials, John Wiley & Sons, ISBN: 978-0-470-85290-3

Chatterjee, A.K. & Haigh, J. (1990). Investigation of erbium doping of InP and (Ga,In)(As,P) layers grown by LPE, Journal of Crystal Growth, Vol. 106, pp. 537-542

Chiu, C.M.; Wu, M.C. & Chang, C.C. (1993). Photoluminescence of undoped and Er-doped InGaAsP layers grown by liquid phase epitaxy, Solid State Electronics, Vol. 36, pp. 1101-1106

Dhar, S. (2005). Growth of high purity semiconductor epitaxial layers by liquid phase pitaxy and their characterization, Bulletin of Materials Science, Vol. 28, pp. 349--353

Ennen, H., et al. (1983). Rare earth activated luminescence in InP, GaP and GaAs, Journal of Crystal Growth, Vol. 64, pp. 165-168

Gao, H.; Krier, A. & Sherstnev, V. (1999). High quality InAs grown by liquid phase epitaxy using gadolinium gettering, Semiconductor Science and Technology, Vol. 14, pp. 441-445

Gorelenok, A.; Kamanin, A. & Shmidt, N. (1995). Rare-earth elements in the technology of InP, InGaAsP and devices based on these semiconductor compounds, Microelectronics Journal, Vol. 26, No. 7, pp. 705-723

Gorelenok, A.; Kamanin, A. & Shmidt, N. (2003). Rare-earth elements in the technology of III–V compounds and devices based on these compounds, Semiconductors, Vol. 37, No. 8, pp. 894-914

Grym, J., et al. (2009). LPE growth of InP layers from rare-earth treated melts for radiation detector structures, Materials Science and Engineering: B, pp. In Press, Corrected Proof, doi:10.1016/j.mseb.2009.03.004

Gschneidner, K.A., Ed. (1978). Handbook on the Physics and Chemistry of Rare-earth, North-Holland Publishing Company, ISBN: 978-0-444-82178-2

Haydl, W.H., et al. (1985). Ytterbium-doped InP light-emitting diode at 1000 nm, Applied Physics Letters, Vol. 46, pp. 870-872

Ho, W.J.; Wu, M.C. & Tu, Y.K. (1995). High responsivity GaInAs PIN photodiode by using Erbium gettering, IEEE Transactions on Electron Devices, Vol. 42, pp. 639-644

Hsieh, J.J.; Rossi, J.A. & Donnelly, J.P. (1976). Room-temperature cw operation of GaInAsP/InP double-heterostructure diode lasers emitting at 1.1 um, Applied Physics Letters, Vol. 28, No. 12, pp. 709-711

Hsu, J.K., et al. (1994). Photoluminescence studies of interstitial Zn in InP due to rapid thermal annealing, Journal of Vacuum Science & Technology B: Microelectronics and Nanometer Structures, Vol. 12, No. 3, pp. 1416-1418

Jiang, G.C. (1993). Gettering properties of praseodymium in GaAs, InGaAs and InP grown by liquid phase epitaxy, Crystal Research and Technology, Vol. 31, pp. 365-371

Kasap, S.O. (2001). Optoelectronics and Photonics: Principles and Practices, Prentice Hall ISBN: 0201610876

Kazmierski, C.; Robein, D. & Gao, Y. (1992). Universal iron behavior in Zn-, Cd- and Be-doped InP, Journal of Crystal Growth, Vol. 116, pp. 75-80

Kenyon, A.J. (2002). Recent developments in rare-earth doped materials for optoelectronics, Progress in Quantum Electronics, Vol. 26, pp. 225-284

Klik, M., et al. (2001). Optically induced Auger recombination of Yb3+ in p-type InP, Physica B, Vol. 308-310, pp. 884-887

Korber, W., et al. (1986). Rare earth ions in LPE III-V semiconductors, Journal of Crystal Growth, Vol. 79, pp. 741-744

Korber, W. & Hangleiter, A. (1988). Excitation and decay mechanisms of the intra-4f luminescence of Yb3+ in epitaxial InP:Yb layers, Applied Physics Letters, Vol. 52, pp. 114-116

Kourkoutas, C.D., et al. (1991). Transport properties of praseodymium doped p-type InGaAs layers, Solid State Communications, Vol. 78, pp. 543-546

Kovalenko, V.; Krasnov, V. & Malyshev, V. (1993). Gettering of donor impurities by Gd in GaAs, Semiconductor Science and Technology, Vol. 8, pp. 1755-1757

Kozanecki, A. & Groetzschel, R. (1990). Lattice location and optical activity of Yb in III-V semiconducting compounds, Journal of Applied Physics, Vol. 68, pp. 517-522

Kranzer, D. (1974). Mobility of holes of zinc-blende III-V and II-VI compounds, Physica status solidi (a), Vol. 26, No. 1, pp. 11-52, ISSN: 1521-396X

Krukovsky, S.I., et al. (2004). Effect of rare earth addition on GaAs-based layers grown by liquid phase epitaxy, The European Physical Journal: Applied Physics, Vol. 27, pp. 177-79

Kumar, A. & Bose, D.N. (1992). Liquid phase epitaxy growth of high purity InP using rare earth dysprosium gettering, Materials Science and Engineering B, Vol. 12, pp. 389-392

Kumar, A.; Pal, D. & Bose, D. (1995). Liquid pahse epitaxy growth of InGaAs with rare-earth gettering: characterization and deep level transient spectroscopy studies, Journal of Electronic Materials, Vol. 24, pp. 833-840
Kuphal, E. (1981). Preparation and characterization of LPE InP, Journal of Crystal Growth, Vol. 54, pp. 117-126
Kuphal, E. (1991). Liquid phase epitaxy, Applied Physics A: Materials Science & Processing, Vol. 52, No. 6, pp. 380-409
Lee, C.T.; Yeh, J.H. & Lyu, Y.T. (1996). Characterization of Nd-doped AlGaAs grwon by liquid phase epitaxy, Journal of Crystal Growth, Vol. 163, pp. 343-347
Nakagome, H.; Takahei, K. & Homma, Y. (1987). Liquid phase epitaxy and characterization of rare-earth-ion (Yb, Er) doped InP, Journal of Crystal Growth, Vol. 85, pp. 345-356
Nelson, H. (1963). Epitaxial growth from the liquid state and its application to the fabrication of tunnel and laser diodes, RCA Review, Vol. 24, pp. 603
Nishikawa, H., et al. (1989). Etch Pit Observation of InP, Japanese Journal of Applied Physics, Vol. 28, No. 5, pp. 941-942
Novák, J., et al. (1989). Growth and properties of low-doped InGaAs LPE layers using rare-earth oxides, Journal of Crystal Growth, Vol. 96, pp. 645-648
Novák, J.; Hasenhorl, S. & Kuliffayová, M. (1991). Gettering properties of PrO_2 in $In_{0.53}Ga_{0.47}As$ LPE growth, Journal of Crystal Growth, Vol. 110, pp. 862-866
Novák, J.; Hasenhorl, S. & Malacký, L. (1993). Large activation of praseodymium in InGaAs, Semiconductor Science and Technology, Vol. 8, pp. 747-749
Novotný, J., et al. (1999). Preparation and properties of Er and Yb doped InP-based semiconductor compounds, Czechoslovak Journal of Physics, Vol. 49, No. 5, pp. 757-763
Pearsall, T.P., Ed. (2000). Properties, Processing and Applications of Indium Phosphide, INSPEC, The Institution of Electrical Engineers, London, UK, ISBN: 085296 949-x
Pelfer, P.G. (2001). Present status and perspectives of the radiation detectors based on InP materials, Nuclear Instruments and Methods in Physics Research A, Vol. 458, pp. 400-5
Procházková, O., et al. (1997). Effect of holmium addition during LPE growth on the properties of InP and GaInAsP layers, Czechoslovak Journal of Physics, Vol. 47, No. 7, pp. 685-691
Procházková, O., et al. (1999). Effect of rare-earth addition on liquid phase epitaxial InP-based semiconductor layers, Materials Science and Engineering, Vol. B66, pp. 63-66
Procházková, O. & Zavadil, J. (1999). Rare earth elements in semiconductors technology - Part I., Science Foundation in China, Vol. 7, No. 2, pp. 44-47, ISSN: 1005-0841
Procházková, O., et al. (2002). Preparation of InP-based semiconductor materials with low density of defects: effect of Nd, Ho and Tb addition, Materials Science and Engineering B, Vol. 91-92, pp. 407-411
Procházková, O., et al. (2005a). Preparation of p-type InP layers for detection of radiation, Journal of Crystal Growth, Vol. 275, No. 1-2, pp. e959-e963
Procházková, O.; Zavadil, J. & Žďánsky, K. (2005b). Effect of Fe inter-diffusion on properties of InP layers grown with addition of RE elements, Physica Status Solidi C, Vol. 4, pp. 1269-1274
Procházková, O., et al. (2009). Impact of Pr on properties of InP based layers for light sources and detectors, Physica Status Solidi (c), pp. In print

Queisser, H.J. & Haller, E.E. (1988). Defects in Semiconductors: Some Fatal, Some Vital, Science, Vol. 281, No. 5379, pp. 945-950

Rakovics, V., et al. (2002). Spectral characteristics of InGaAsP/InP infrared emitting diodes grown by LPE, Materials Science and Engineering B, Vol. 91-92, pp. 491-494

Rhee, K. & Bhattacharya, K. (1983). Impurities and Traps in Liquid Phase Epitaxial InP in relation to melt prebaking, Journal of Electrochemical Society: Solid State Science and Technology, Vol. 130, pp. 700-703

Simpson, J.R. (2001). Rare Earth Doped Fiber Fabrication: Techniques and Physical Properties, In Rare Earth Doped Fiber Lasers and Amplifiers. Digonnet, M.J.F. (Ed), 1, CRC, ISBN: 978-0824704582

Skromme, B.J., et al. (1984). Identification of the residual acceptors in undoped high purity InP, Applied Physics Letters, Vol. 44, No. 3, pp. 319-321

Swaminathan, V.; Donnelly, V.M. & Long, J. (1985). A photoluminescence study of Cd-related centers in InP, Journal of Applied Physics, Vol. 58, No. 12, pp. 4565-4572

Takeda, Y., et al. (1976). Electron mobility and energy gap of In0.53Ga0.47As on InP substrate, Vol. 47, No. 12, pp. 5405-5408

Thiel, C.W.; Sun, Y. & Cone, R.L. (2002). Progress in relating rare-earth ion 4f and 5d energy levels to host bands in optical materials for hole burning, quantum information and phosphors, Journal of Modern Optics, Vol. 49, pp. 2399-2411

Wada, O. & Hasegawa, H., Eds. (1999). InP-based Materials and Devices: Physics and Technology, John Wiley & sons, ISBN: 0-471-18191-9

Wu, M.C., et al. (1992). Erbium doping in InGaAsP grown by liquid phase epitaxy, Journal of Applied Physics, Vol. 71, pp. 456-460

Wu, M.C. & Chiu, C. (1993). Very high purity InP layer grown by liquid-phase epitaxy using erbium gettering, Journal of Applied Physics, Vol. 73, pp. 468

Zakharenkov, L.F., et al. (1981). Fizika i Tekhnika Poluprovodnikov (Soviet Physics-Semiconductors), Vol. 15, pp. 1631-1636

Zakharenkov, L.F., et al. (1997). Rare-Earth Elements in the Technology of III-V Compounds, In Semiconductor Technology - Processing and Novel Fabrication Techniques. Levinshtein, M.E. & Shur, M.S. (Eds), 91-130, John Wiley&Sons

Zavada, J.M. & Zhang, D. (1995). Luminescence properties of erbium in III-V compound semiconductors, Solid State Electronics, Vol. 38, pp. 1285-1293

Zavadil, J.; Procházková, O. & Žďánský, K. (1999). Rare Earth Elements in Semiconductors. Characterization -- part II. Proceedings of Proceedings of Advanced Materials and Devices for Optoelectronics, 48-51, Chinese Academy of Science, Beijing, China

Zavadil, J., et al. (2007). Ce and Yb doped InP layers grown for radiation detection, physica status solidi (c), Vol. 4, No. 4, pp. 1444-1447, ISSN: 1610-1642

Žďánský, K.; Procházková, O. & Nohavica, D. (1999). Temperature dependence of electron Hall concentration in n-type InP and n-type InGaP. Proceedings of Proceedings of Advanced Materials and Devices for Optoelectronics, 25-43, Chinese Academy of Science, Beijing, China

Žďánský, K., et al. (2001). P-type InP grown by LPE with rare earths: not intentional Ge acceptor doping, Materials Science and Engineering B, Vol. 80, pp. 10-13

Žďánský, K., et al. (2002). P-type InP grown by liquid phase epitaxy from melts with rare earth admixtures, Materials Science and Engineering B, Vol. 91-92, pp. 38-42

14

Applications of saturable absorption-based nonlinear vertical-cavity semiconductor devices for all-optical signal processing

Claudio Porzi[1], Mircea Guina[2], Nicola Calabretta[3], Antonella Bogoni and Luca Poti'[3]
[1] *Scuola Superiore Sant'Anna, Pisa Italy.*
[2] *Optoelectronics Research Centre, Tampere University of Technology, Tampere Finland*
[3]*COBRA Research Institute, Eindhoven University of Technology, Eindhoven The Netherlands.*
[3] *Consorzio Nazionale Interuniversitario per le Telecomunicazioni, Pisa Italy*

1. Introduction

Modern ultrahigh bandwidth optical communication networks create a demand for devices with ever higher performance, conjugating high operation speed, reduced fabrication costs, compactness, and advanced functionality. Passive nonlinear devices based on semiconductor quantum wells (QWs) are promising candidates for many all-optical signal processing (AOSP) applications, that will be described in this chapter. A semiconductor quantum well is formed by a thin layer (typical thickness is ~10 nm) of a semiconductor material placed between two other layers of a second semiconductor material, the middle layer having a smaller band-gap energy than the external layers. Carriers are thus confined in one dimension within the smaller band-gap layer, leading to important modifications of the electronic and optical properties (e.g. higher and discrete density of states for carriers compared to bulk structure). Such QWs are usually arranged in a periodic fashion, with the basic structure alternating several times to form a multiple quantum well (MQW) heterostucture whose optical features can be controlled by the number of QWs. Semiconductor QWs can be fabricated with precise control of layers thickness using standard epitaxial growth techniques like molecular beam epitaxy (MBE) or metal-organic chemical vapour deposition (MOCVD). For applications in the field of optical communications, the material band-gap should fit the desired operating wavelength. By using compounds with different compositions, the energy gap can be adjusted. The III-V compounds (i.e., elements from column III and V of the periodic table) enable operating wavelengths in the three fiber-optic communication bands at 0.85 µm, 1.3 µm and 1.55 µm to be used. The use of QWs has brought many benefits for the development of a large variety of optical telecom devices, such as laser diodes (LDs), semiconductor optical amplifiers (SOAs), electroabsorption modulators (EAMs) and saturable absorbers (SAs). Generally speaking, it is possible to differentiate these devices into two classes, *active* and *passive*. Passive devices differ from active devices in that carriers are not injected into the material (typically by applying an electrical current) to provide optical gain like in LDs or SOAs. Passive devices are therefore usually simpler to fabricate, and require lower overall operating powers than active devices. Typically, passive MQW-based devices exploit the change in light absorption provided by the semiconductor structure under an applied electrical or optical control signal. Change of light absorption within MQWs in the presence of an electric field is due to the quantum-confined Stark effect (QCSE) (as opposed to the weaker Franz-Keldysh effect in bulk materials). On the other hand, the phenomenon responsible for the saturation of absorption under an applied optical field in quantum-confined structures is known as phase-space-filling (in bulk materials one usually refers to the band-filling effect). The optical saturation of absorption is a consequence of the fact that the available states for the electrons and holes are filled by the free carriers generated by the applied optical field. Thus, for low values of incident intensity, photons experience (relatively) high absorption losses, whereas at high intensity values the absorption is saturated, and the material becomes, ideally, transparent. The QCSE is the basis of operation for EAMs, whereas the optically induced saturation of absorption is exploited in SAs. When the absorber is coupled with a mirror on one side (typically a high-reflectivity semiconductor Bragg mirror grown on a substrate), these devices are referred to as semiconductor saturable absorber mirrors (SESAMs), when they are used for mode-locking of ultrafast lasers (Keller et al., 1996), or asymmetric Fabry-Pérot (AFP) etalons/modulators when they are used for signal processing applications (Loka et Smith, 1998). The non-linear phenomenon of saturable absorption can be greatly enhanced by placing the semiconductor MQWs at the antinodes of the vertical Fabry-Pérot cavity formed between the high-reflective semiconductor mirror and the top surface mirror; typically, a vertical-cavity architecture is realized by growing the absorbing layer on a high reflectivity (~ 100%) bottom semiconductor Bragg mirror and finishing the structure with a partially reflective (< 100%) semiconductor or dielectric mirror. Vertical-cavity technology offers a simplification of the device fabrication, versatility in coupling with optical fibres, increased functionality, polarization independence, and most importantly, low switching powers and a high switching contrast. Indeed, due to the cavity effect, the energy that must be applied to the device in order to saturate the nonlinear medium is much lower than the saturating energy of the bare SA, resulting in a decreased "effective" saturation power. Furthermore, the cavity can be designed in such a way that the optical field reflected from the cavity can be totally cancelled when the gate is in the *OFF* state, allowing the normal SA characteristic to be enhanced, and high *ON/OFF* contrast ratio *(CR)* values to be achieved. This concept has been exploited to realize high-contrast EAMs (Yan et al., 1991) and for nonlinear optical processing, in applications for noise suppression for signal regeneration (Isomaki et al, 2003; Massoubre et al., 2006), all-optical switching (Takahashi et al., 1996; Loka et Smith, 1998) and wavelength conversion (Burr et al., 2003; Akiyama et al., 1998). However, the design parameters of the nonlinear AFP cavity can also be tailored to provide a specific target value of the device reflectivity for a given value of the input signal energy which is not necessarily associated with a low-energy photon flux. This allows the application field of nonlinear vertical-cavity semiconductor devices for AOSP to be widened.

In this chapter we review a device engineering method we have developed to provide high functionality of passive nonlinear vertical-cavity devices exploiting saturable absorption in semiconductor MQWs. Section 2 is concerned with the design and fabrication of the vertical-cavity nonlinear gate. The advantages of the proposed approach for all optical gating in terms of enhanced throughput, increased optical signal-to-noise ratio (OSNR), and high switching *CR* are then discussed in Section 3. Subsequently, a specific application in a wavelength conversion experiment is presented in Section 4, whereas in Section 5 and Section 6 we show how this approach enables complex functions for all optical networking and different logic operations, respectively, to be implemented which would not otherwise be possible with standard SAs. Analytical and numerical modelling of devices will be presented throughout the different sections; the models can be exploited for device optimization and design trade-offs. Finally, these results will be summarized in Section 7.

2. Device Structure

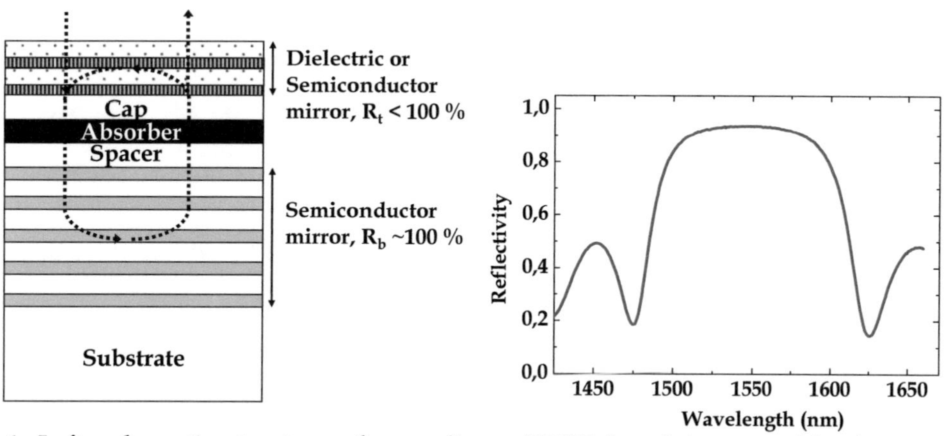

Fig. 1. Left: schematic structure of a nonlinear VCSG based on saturable absorption in MQWs. Right: typical spectral reflectivity of the bottom DBR mirror.

The typical structure of a nonlinear vertical-cavity device based on saturable absorption in semiconductor MQWs is shown in Fig. 1 (left). We will refer to this generic structure throughout the text as a vertical-cavity semiconductor gate (VCSG). The structure consists of an AFP cavity formed between a bottom mirror, with a reflectivity R_b, approaching 100%, and a top mirror with lower reflectivity, R_t. The mirrors can be fabricated based on either semiconductor or dielectric materials or, in the case of the highly reflective bottom mirror, with a thin layer of silver or gold. Saturable losses in the cavity are usually provided by semiconductor QWs (quantum dots or any other medium exhibiting intensity-dependent saturable losses could be used as well). For the experiments presented in the next sections all the samples were grown by solid-source MBE on n-type InP (100) substrate. The samples comprised of a Burstein-Moss shifted distributed Bragg reflector (DBR) with 19.5 pairs of $\lambda/4$ n^+-$Ga_{0.47}In_{0.53}As$/InP, an InP spacer layer, the active region, and a 50-nm InP cap layer (Xiang et al., 2001). The length of the spacer layer varied for different samples depending on the specific target application. The active regions were always comprised of four groups of seven 9-nm-thick $Ga_{0.47}In_{0.53}As$ QWs and 10-nm-thick InP barriers. Each group of QWs has been centered at the antinodes of the 3λ/2 cavity defined between the semiconductor DBR and top surface of the gate. The DBR reflectivity was over 96% in the wavelength range from 1.525 to 1.610 nm. The measured DBR reflectivity spectrum is also shown in Fig 1 (right). In order to decrease the absorption recovery time, the samples were irradiated with I^+ ions with a dose of $5x10^{11}$ cm^{-2} and an energy of 12-MeV.

3. Device Operation

The reflectivity of the generic structure shown in Fig. 1 (left) can be calculated by using the standard formula for an FP cavity incorporating losses (Siegman, 1984). If the resonator has a loss per pass of *(1-exp[-αd]),* α [m^{-1}] being the absorption coefficient of the MQW section, and d [m] its physical length, the reflectivity of the structure in Fig. 1 is given by:

$$R = \frac{\left[\sqrt{R_t} - \sqrt{R_b}\exp(-\alpha d)\right]^2 + 4\sqrt{R_t}\sqrt{R_b}\exp(-\alpha d)\sin^2(\phi_{sp})}{\left[1 - \sqrt{R_t}\sqrt{R_b}\exp(-\alpha d)\right]^2 + 4\sqrt{R_t}\sqrt{R_b}\exp(-\alpha d)\sin^2(\phi_{sp})} \tag{1}$$

Where ϕ_{sp} is the single-pass phase change of a wave travelling in the structure. The typical spectral reflectivity of an uncoated (i.e. without the top mirror) sample is shown in Fig. 2. The side oscillations that can be observed in the figure are due to the effect of the bottom DBR mirror whereas the dip in the central part of the spectrum corresponds to the FP resonance created between the bottom mirror and the Fresnel reflection at the air-semiconductor interface. At resonance, the value of ϕ_{sp} is $m\pi$, where m is an integer.

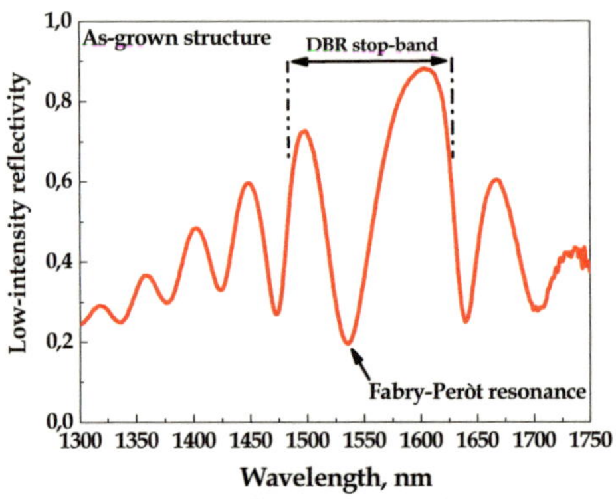

Fig. 2. Low-intensity spectral reflectivity of an uncoated As-grown sample.

The minimum value of reflectivity R_{res} is calculated from (1) as:

$$R_{res} = \left(\frac{\sqrt{R_t} - \sqrt{R_b}\exp(-\alpha \cdot d)}{1 - \sqrt{R_t R_b}\exp(-\alpha \cdot d)}\right)^2 \tag{2}$$

From the last expression it can be seen that the reflected wave is zero if the top dielectric/semiconductor mirror has a reflectivity satisfying the following condition:

$$R_t = R_b e^{-2\alpha d} \tag{3}$$

The condition expressed by (3) is usually called impedance-matching (IM), and has been widely exploited to realize high-contrast electro absorption modulators or all-optical gates (Yan et al., 1991; Loka et Smith, 1998). By taking into account also the non-saturable part of absorption, α_{ns}, the dependence of α on the energy E_{int} impinging on the medium can be expressed as:

$$\alpha = \alpha_{ns} + \alpha_{sat} = \alpha_{ns} + \frac{\alpha_0}{1 + E_{\text{int}} / E_{sat}}. \tag{4}$$

In (4), α_0 is the unsaturated small-signal absorption coefficient, i.e. the value of the saturable part of absorption, α_{sat}, for applied energies much weaker than the medium saturation energy, E_{sat}. On the other hand, the non-saturable part of the absorption, α_{ns}, which can, for example, be caused by the imperfection of the semiconductor structure, does not change with incident energy.

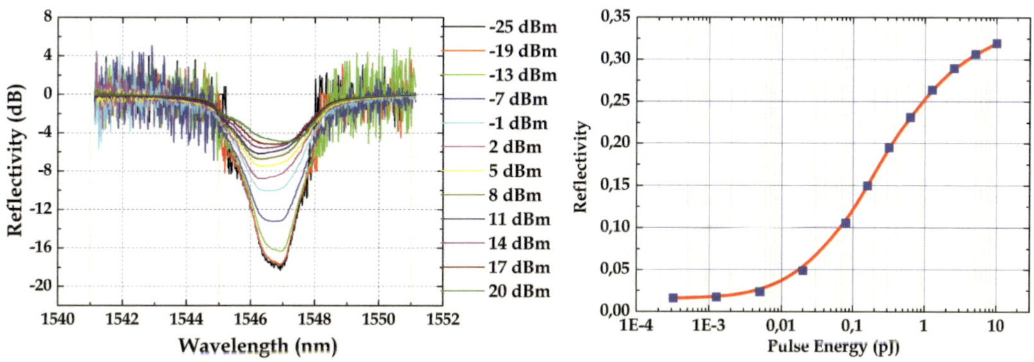

Fig. 3. Left: VCSG spectral reflectivity for different power values of an input pulse tuned at resonance. Right: corresponding nonlinear reflectivity at resonant wavelength.

By substituting (4) into (3) a more general expression for the IM condition can be derived:

$$R_t = R_b e^{-2[\alpha_{ns} d + \alpha_0 d / (1 + E/E_{sat})]}. \tag{5}$$

In many practical cases, the top mirror reflectivity is designed to satisfy IM condition for low values of the applied optical field, that is for:

$$R_t = R_b e^{-2[(\alpha_{ns} + \alpha_0) d]}. \tag{5}$$

Under this condition, the typical SA characteristic (i.e. low throughput at low input intensity and high throughput at high input intensity) is strongly enhanced. Measured spectral reflectivity for a VCSG designed to meet the IM condition at low values of input energies and the corresponding nonlinear reflectivity curve at resonance are shown in Fig. 3. By exploiting the dependence of the roundtrip power transmission on the energy incident on the absorber via the nonlinear absorption coefficient α (see eq. 3), one could partially or even totally reverse the nonlinear characteristic of Fig. 3 (right). It can indeed be deduced from (5) that, for a given R_t, the IM condition can be satisfied for a particular value of the signal energy, provided that R_t is higher than the roundtrip power transmission associated with unsaturated absorption, i.e. for:

$$R_t > R_b e^{-2[(\alpha_{ns}+\alpha_0)d]}. \tag{7}$$

When (7) holds, the cavity impedance can still be matched by using an input optical field that partially saturates the absorption to satisfy the condition given by (3), with α being the general, intensity-dependent expression of (4). In order to distinguish it from the usual case in which the IM is achieved for weak values of the applied field, we will refer hereafter to a structure designed to satisfy the generic condition given by (7) as an *impedance-detuned* VCSG.

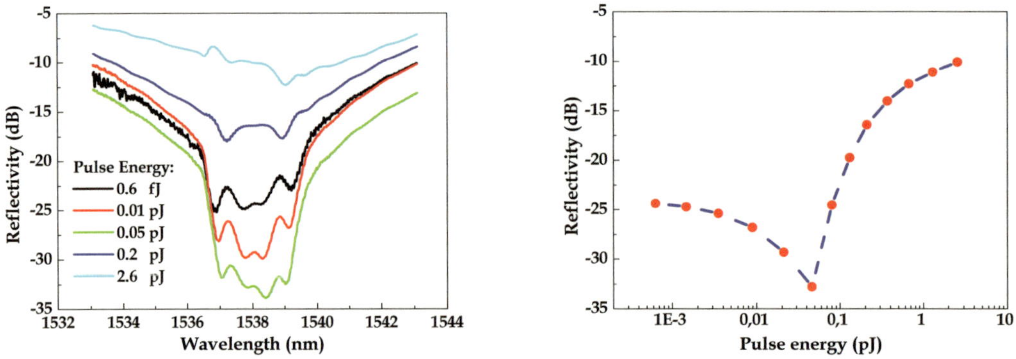

Fig. 4. Left: impedance-detuned VCSG spectral reflectivity for different values of energies of an input pulse tuned at resonance. Right: corresponding nonlinear reflectivity at resonance.

The typical spectral response and the relative characteristic at resonance of an impedance-detuned VCSG for different energy values of the input signal are shown in Fig. 4. In order to obtain a nonlinear characteristic as that of Fig. 4 (right), the low-intensity reflectivity spectrum for an as-grown (i.e., uncoated) structure was first measured. A top dielectric coating consisting of two layers of SiO_2/TiO_2 was then deposited on the sample, providing a reflectivity of ~ 53%, a value that allowed detuning (from the zero energy point) of the IM energy. The reflectivity spectra of Fig. 4 (left) had a minimum value at the resonance of about -33 dB for an applied pulse energy of -0.05 pJ. Due to detuned impedance, the reflectivity for the lowest input energy (0.6 *fJ)* is about 10 dB higher. When the power of the signal was increased above 0.05 pJ, the saturation of the MQWs absorption increased, resulting in the increased reflectivity. Fig. 4 (right) further clarifies the dependence between the minimum reflectivity at cavity resonance and the input energy. The advantage arising from the use of the *impedance-detuned* design is explained in the following. In many practical applications, the reflectivity exhibited by an applied probe signal incident on the VCSG is changed by an intense optical pump field (see, for example, the scheme of Fig 5 showing a typical wavelength conversion set-up). In order to understand the benefits of the proposed design for practical applications, we will use as figures of merit for the switching capability of the device the *ON/OFF CR* and the eye-opening (*EO*) of the reflected probe signal. The *CR* of a modulated signal is defined as the ratio between its high and low levels; with reference to Fig. (5), the *CR* is then given by:

$$CR = \frac{P_{level1}^{\lambda_2}}{P_{level0}^{\lambda_2}} = \frac{P_{CW}^{\lambda_2} R_{ON}}{P_{CW}^{\lambda_2} R_{OFF}} = \frac{R_{ON}}{R_{OFF}} \tag{9}$$

where R_{ON} and R_{OFF} are the reflectivity values of the gate in the *ON* and *OFF* states, respectively. The converted signal *EO* is given by:

$$EO = P_{level1}^{\lambda_2} - P_{level0}^{\lambda_2} = P_{CW}^{\lambda_2}(R_{ON} - R_{OFF}) \tag{10}$$

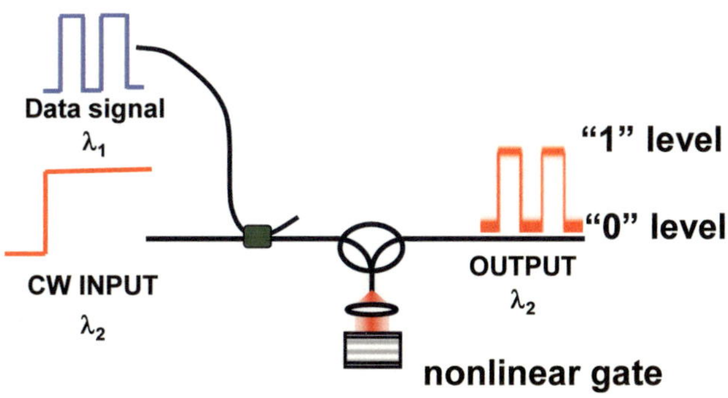

Fig. 5. Schematic of wavelength conversion with VCSG.

The output signal *CR* takes into account the amount of modulation transfer from the data signal onto the CW probe, and a high *CR* (say, at least 10 dB) ensures resilience of the signal to transmission impairments when it has to be launched through an optical transmission channel. On the other hand, the *EO* is related to the power throughput of the gate. A fair throughput is important in order to have a good optical signal-to-noise ratio (OSNR) at the gate output, which makes subsequent optical amplification of the signal without significant degradation possible. Also, a high throughput allows the noise characteristic of the detected signal to be improved at the receiver side. In general, both the *CR* and *EO* (which are related to each other, as can be easily seen from (9) and (10)) have an effect on the properties of the system in terms of the photodetected output signal noise. In digital communications, the systems performance is conveniently expressed in terms of error probability and, in the Gaussian approximation, the receiver sensitivity required to achieve a given bit error rate (BER) depends on the so called Q-factor, given by:

$$Q = \frac{I_1 - I_0}{\delta_1 + \delta_0} \tag{11}$$

where I_1 (I_0) and δ_1 (δ_0) are the mean and standard deviation of the receiver output for a 1 (0) bit, respectively. Clearly, $I_1 - I_0$, is directly proportional to the *EO*. From (11) it can also be seen that decreasing the *CR* leads to an increase of the optical power required at the receiver to attain a given BER (Agrawal, 2002). In general, the receiver performance can be improved by maximizing both the *CR* and the *EO* of the nonlinear gate output signal. In Fig. 6, two possible operating conditions of an optical modulator based of a nonlinear VCSG designed to satisfy the IM condition in the low energy regime are presented.

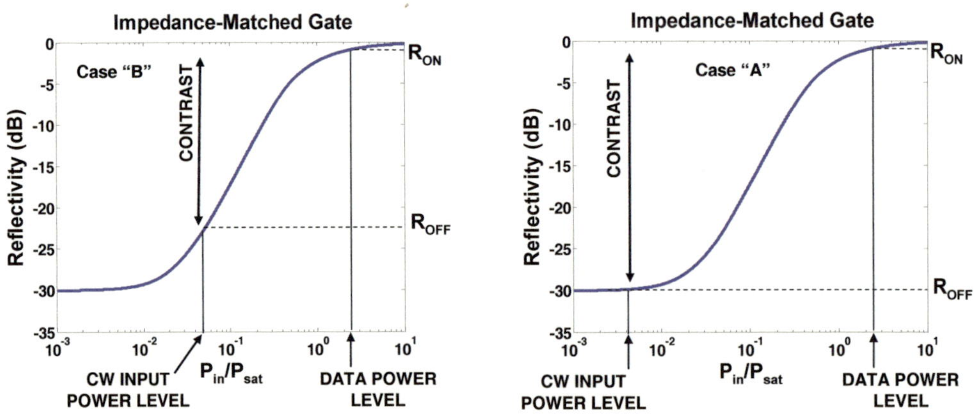

Fig. 6. Left: high CR and limited throughput can be obtained with this configuration. Right: this configuration corresponds to high throughput and low CR of the gate.

In case "A", the CW probe signal is in the low power regime (well below the gate saturation power P_{sat}) whereas the pump signal level is high enough to turn the gate in the *ON* state. A fair *ON/OFF* CR can be achieved with this configuration; however, it is also clear that the power throughput, $P_{CW} \cdot R_{ON}$, and consequently the *EO*, are limited by the low probe signal power level. One could attempt to overcome this limit by increasing the gate saturation power P_{sat}, but clearly this solution would not be beneficial in terms of the required switching power. On the other hand, for case "B" of Fig. 6, the CW probe signal is partially saturating the absorber in the nonlinear gate. This condition describes the situation of an increased throughput (with respect to case "A") at the expense of a reduced *CR*. A trade-off between these two parameters should then be found when operating with the (low-energy) IM design. However, by exploiting the impedance-detuned design it is possible to maximize both parameters, as shown in Fig. 7 (left). By setting the CW probe power to the level that gives the minimum value of reflectivity, a large CR together with an increased throughput (with respect to the case "A" of Fig. 6) can be obtained for the same saturation power. Both figures of merit can then be optimized, regardless of the switching power value, which can hence be kept as low as possible in designing the device (Porzi et al., 2005). As has already been discussed, an impedance-detuned device can be achieved by coating the as-grown structure of Fig. 1 (left) with a top mirror having a reflectivity higher than the roundtrip power transmission value associate with unsaturated absorption, $R_b exp(-2\alpha_0 d)$. The effect of increasing the top mirror reflectivity is shown in Fig. 7 (right); for larger detuning of the impedance, the value of R_{ON} decreases as well due to the unavoidable presence of non-saturable losses, leading to higher values of the gate insertion losses, $(1-R_{ON})$. However, the possibility to shape the nonlinear characteristic is of particular interest for broadening the application fields of nonlinear VCSGs.

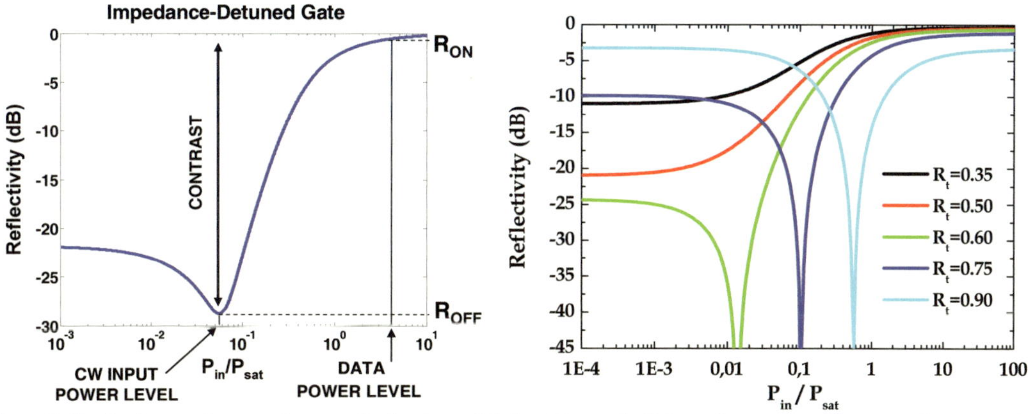

Fig. 7. Left: impedance detuned design: high *CR* and *EO* values can be simultaneously accomplished. Right: nonlinear VCSG characteristic at the resonant wavelength for different values of top mirror reflectivity R_t (other parameters: R_b=100%, $\alpha_0 d$=0.28, $\alpha_{ns}d$=0.01).

4. Broadband wavelength conversion with impedance detuned VCSG

A straightforward application of the MQWs-based nonlinear VCSG is in wavelength conversion, which refers to the modulation transfer from a data signal at a given wavelength onto a CW or clock probe signal at a different wavelength. Wavelength conversion is a key technique for switching and routing in next generation optical communication networks (Brackett et al., 1993), leading to increased network flexibility, throughput, capacity, and efficient solutions to alleviate congestion at network nodes. In order to effectively exploit such a technique in practical applications the quality of the converted signal must be suitable for long distance transmission, which requires a large dynamic *CR* in the converted output. In addition, efficient operation requires low optical input powers for the control signal, and a high power in the converted output. Cross-absorption saturation in semiconductor absorbers have been previously proposed (Burr et al., 2003; Akiyama et al., 1998) to perform wavelength conversion, but their performances were limited due to either poor conversion range or output signal *CR*. The proposed impedance-detuned design allows high switching contrasts and fair gate throughputs to be obtained. Furthermore, this technology allows correct operation over a broad band of input data (i.e. the signal to be converted) wavelengths (Porzi et al., 2006).

4.1 Quasi-static broadband characterization

The reflectivity of a VCSG shows a spectrally dependent behaviour due to its resonant structure. Nevertheless, a locally generated probe signal can be tuned close to a resonant wavelength, whereas the absorption in the QWs due to the broadband excitonic interaction in the proximity of the material band-gap can be saturated by a data pump field tuned away from the resonant wavelength. As a consequence, the reflectivity of the probe signal will be modified as well. In order to investigate the degradation of the gate performance as a function of the optical detuning of pump from resonance, a time-resolved measurement of the *ON/OFF CR* of the wavelength-converted signal for different values of pump-probe detuning was performed. The measurement was carried out in a *quasi-static* condition, by using pump pulses much longer than the carrier recombination time in the absorber medium at a repetition frequency much lower than the absorber recovery rate, to allow complete recovery of absorption between two subsequent pulses. In particular, 8 ns-long pump pulses, at the repetition rate of 10 kHz were applied to the VCSG. The wavelength of the probe signal was set to ~ 1536 nm, where the etalon exhibited a resonance. The wavelength of the pump beam could be varied to provide up to 25 nm of detuning from the cavity resonance.

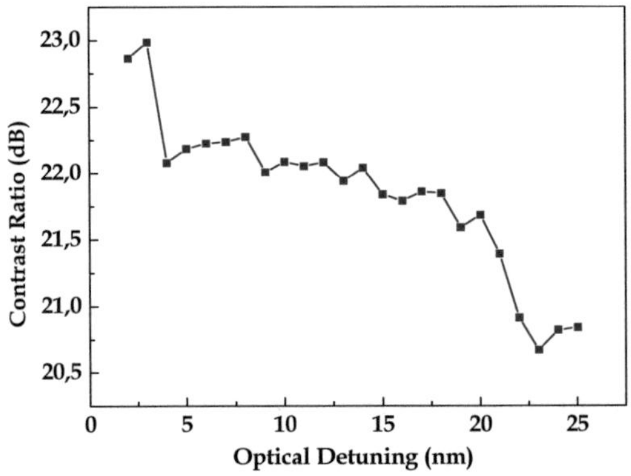

Fig. 8. Probe output extinction ratio as a function of pump detuning.

The power of the probe beam at the VCSG input was set at about -2 dBm in order to match the condition of minimum reflectivity in the absence of pump light. The average pump power was set to 0 dBm at the input of the VCSG for all the values of pump wavelength. The nonlinear characteristic at the resonant wavelength for the device used in the experiment is that of Fig. 4 (right), exhibiting an impedance-detuned feature. The reflected probe signal was photodetected and sent to an oscilloscope; maximum and minimum voltages (V_{max} and V_{min}, respectively) of the detected signal for different pump-probe detuning were stored to calculate the switched output *ON/OFF CR* (in dB), defined as $10log_{10}(V_{max}/V_{min})$. The measured values are shown in Fig. 8 as a function of pump-probe optical detuning. As expected, a maximum *CR* of ~ 23 dB was achieved for low pump-probe detuning (i.e., when the pump light wavelength was close to the cavity resonance), since in this case the pump field distribution is enhanced inside the resonator. However, the device exhibits a rather flat response over a wide range of pump wavelengths and the CR degradation remains below ~ 2.5 dB over 25 nm of pump detuning from resonance. The absolute value for the *CR* in this experiment is set by the value of pump light switching energy (which corresponded to a pump intensity on the absorber as low as ~ 1.5 kW/cm^2).

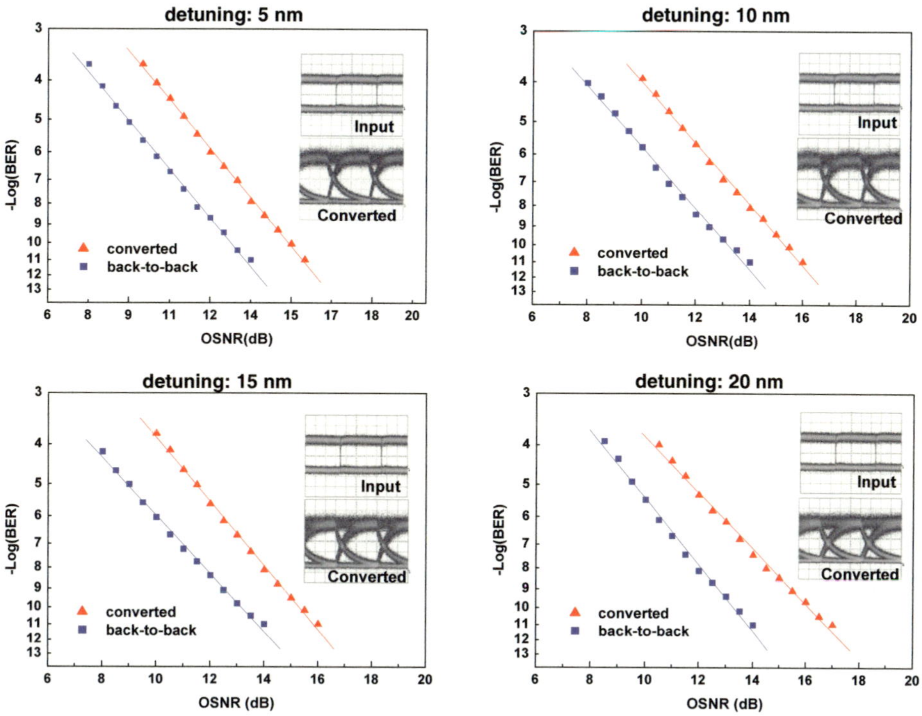

Fig. 9. BERvsOSNR for the converted probe light (triangles) and the input data (circles) for different pump-probe optical detuning.

4.2 Wavelength conversion experimentt

For the wavelength conversion experiment the pump light was modulated by means of a Mach-Zehnder (MZ) modulator driven by a 2^{31}-1 Pseudo-Random Bit Sequence (PRBS). The probe CW signal was set at resonance with a power level satisfying the IM condition, as in the previous case. After selecting the reflected probe light from the gate by means of optical filters, a standard optically amplified receiver was used in order to perform BER measurements of the converted data as a function of OSNR. Also, a wide-bandwidth (50 GHz) sampling oscilloscope was used to monitor the eye diagram of the wavelength-converted data. In this experiment the data repetition rate was limited by the carrier recombination time in the absorber, which was previously measured with a pump-probe technique to be ~500 ps. Nevertheless, error-free operation was observed for data rates up to 2 Gb/s. However, in order to investigate the performance of the VCSG-based wavelength converter in terms of conversion bandwidth, and avoid degradation of the converted data due to carrier recombination effects, the repetition rate of the input data was set at 622 Mb/s. In the past years, several reports of semiconductor saturable absorbers with a response below 1 ps have been reported (Delpon et al., 1998; Takahashi et al. 1996). Thus, by increasing the ion-irradiation dose, the operation speed of the VCSG-based wavelength converter could be easily increased up to the multi-GHz regime. The results of BER vs. OSNR measurements for values of pump-probe detuning of 5, 10, 15 and 20 nm are shown in Fig. 9, together with the relative pump data back-to-back BER curves. The OSNR penalty, with respect to the back-to-back values, is limited to ~ 2.5 dB for all the measurements. The

VCSG input pump powers in the experiments ranged from ~ 6 dBm, at the lowest values of pump-probe detuning, up to ~ 8 dBm for the highest value of detuning. These values corresponded to an optimized eye diagram for the converter output data. With these values of pump input powers, the corresponding converted average power at the VCSS output ranged from ~ -18 dBm to ~ -16 dBm. The converted output power was limited by the pump power levels and by the butt-coupling losses for the back-reflected light from the VCSG. The probe output power when pump was switched off was ~ -35 dBm.

5. Signal processing with VCSG for all-optical packet switching

In this section, self-induced effects in VCSG are discussed, i.e. the case in which the nonlinear reflectivity changes are induced by the signal itself, rather than by an additional control beam as in the case of wavelength conversion. This feature can be useful to realize several functionalities which are required in packet switched networks. In particular, the implementation of an optical header extractor and of an optical seed pulse extractor will be discussed in this section. Optical packet switching (OPS) and optical burst switching (OBS) offer fast dynamic allocation of optical channels, with bandwidth occupation only when the data are actually transmitted, leading to an efficient usage of the available resources. For this scope, an optical header has to be inserted before the data, which is sent in the form of payload packets; the optical header contains information on payload destination. The core functionality of an optical switch is to selectively transmit packets from a particular input port to a particular output port, according to the header routing information. Thus, extraction of the packet header is required at the input interface of the packet switch. Several all-optical header extraction (AO-HE) and processing techniques have been considered and demonstrated in the recent years. The header information could be transported either by using an in-band header section having the same wavelength as the payload data, by using different wavelengths for the header and the payload section, or by using an orthogonal modulation format for the header. In the case of in-band transmission, the header can either have the same modulation format of the payload data or a different one. Among the different techniques for coding an in-band transmitted header, pulse position coding (PPC) is attractive due to its simple implementation. In PPC an optical header is uniquely defined by the time distance between two on-off-keying (OOK) modulated pulses. PPC allows simple header recognition by using a self-correlation technique which does not need generation of header reference patterns in the header recognition block (Calabretta et al., 2001). Another useful functionality for OPS networks is all-optical seed pulse extraction (AO-SPE) from the incoming packets, which is needed for synchronization of different inputs of a switch node in time-slotted operation. By exploiting self-induced effects in nonlinear optical devices, asynchronous AO-HE and AO-SPE can be achieved. SOAs have been widely exploited to implement AO-HE and AO-SPE (Calabretta et al., 2004; Vegas Olmos et al., 2005; Huang et al., 2005). We demonstrate in this section both AO-HE and AO-SPE functionalities by exploiting the flexibility in designing the nonlinear characteristic of a VCSG (Porzi et al., 2007). Self-induced effects in VCSG provide important benefits compared with other all-optical techniques in terms of efficiency, power consumption and polarization dependency. For example, the VCSG preserves the compactness and low power advantages of the SOA, but in addition it is passive and polarization independent.

5.1 Operation principle

As discussed in the previous sections, the nonlinear VCSG can be designed to have an inverse saturable absorber characteristic (i.e. high reflectivity at low input energy and low reflectivity at high input energy) for input energy values below the IM energy (see for instance Fig. 4 (right)). For this impedance-detuned design, when low input energies are used, ranging between zero and IM energy point, the gate reflectivity decreases with increasing energy. By exploiting the nonlinear characteristic at low energy of an impedance detuned VCSG, it is then possible to extract short return-to-zero (RZ) pulses synchronous with the leading edge of an incoming non-return-to-zero (NRZ) optical pulse, as explained in the following. Before the pulse arrival, the absorption in the MQWs is unsaturated and the gate is in the high-reflectivity state. Thus, when an optical NRZ pulse with appropriate peak power enters the gate, the low energy leading edge of the pulse is back reflected from the gate, whereas the following part of the pulse rapidly biases the gate in the IM condition and is strongly suppressed. This effect represents the basic principle we exploited to realize AO-HE and AO-SPE for optical packets with PPC optical headers.

5.2 Device modelling

A VCSG, working at its resonant wavelength, can be modeled as a two-level system with a single time constant, τ_s, which is the recovery time. The reflected output power $P_{ref}(t)$ is directly related to the input signal $P_{in}(t)$ as shown in (2); the absorption coefficient $\alpha(t)$ is time- and input power-dependent. If the round-trip time of the cavity is much lower than the time-scale associated with input power variations, as in most practical applications, the time dependence of the absorption can be described using the following nonlinear rate equation (Kartner et al., 1996):

$$\frac{\partial \alpha(t)}{\partial t} = \frac{\alpha_0 - \alpha(t)}{\tau_s} - \frac{\xi(t) P_{in}(t)}{E_{sat}} \alpha(t) \tag{12}$$

Where the ξ parameter describes the ratio between the energies of the optical field within the cavity and the input optical fields and accounts for the FP resonant effects. At resonance, its value is given by (Brovelli et al., 1995):

$$\xi = \frac{1 - Rt}{(1 - \sqrt{R_t R_b \exp[-\alpha d]})^2} . \tag{13}$$

In a weakly saturated absorption regime, small variations of α do not affect the ξ parameter, which can be considered constant, and (12) has an analytical solution in the form:

$$\alpha(t) = \alpha_0 \exp\left[-\frac{1}{\tau_s} \int_0^t \left(1 + \frac{P(\tau)\tau_s}{E_{sat}} \right) d\tau \right] \cdot \left(1 + \frac{1}{\tau_s} \int_0^t \exp\left[\frac{1}{\tau_s} \int_0^t \left(1 + \frac{P(\tau)\tau_s}{E_{sat}} \right) d\tau \right] \right) . \tag{14}$$

where *P(t)* is the optical internal power. In the strongly saturated absorption regime, the previous approximation quickly fails, and can be used only to obtain a rough prediction for device dynamics; Numerical methods have to be used in order to obtain an accurate description for this case.

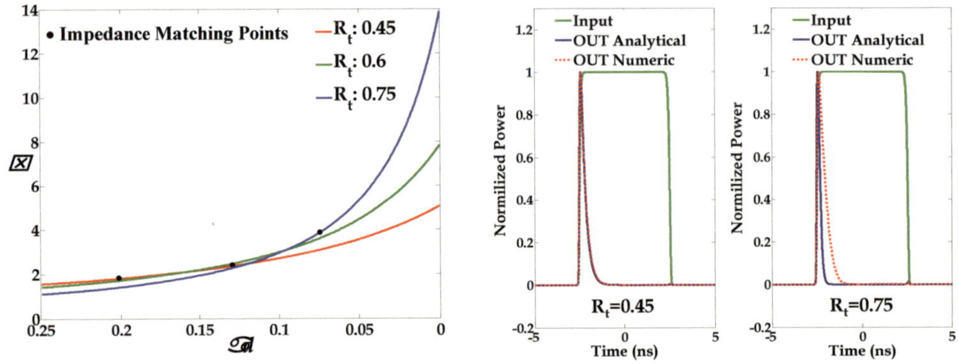

Fig. 10. Left: ξ as a function of the product *ad* for three different R_t values. IM conditions are indicated using black dots. Right: Numerical model vs. analytical solutions when an optical gate is used as input signal, for two different values of R_t. Other parameters: $\alpha_0 d = 0.25$, $E_{sat} = 1.5$ pJ, $\tau_s = 0.5$ ns.

In order to better understand the analytical approximation, ξ is plotted in Fig. 10 (left) as a function of the product αd for three different R_t values. In the same picture IM conditions are highlighted. For the proposed application, when a gate pulse is launched through the device, the αd product varies between its minimum and the IM point. At the same time, the ξ parameter exhibits a small variation, that increases as R_t gets higher. Thus, one pulse is generated by applying an optical gate to the VCSG. Fig. 10 (right) illustrates this situation, calculated using both numerical integration and analytical solution for two different values of the top mirror reflectivity. As discussed, the analytical solution diverges from the numerical one as the value of R_t is increased.

5.3 All-Optical header extraction with VCSG

The experimental set-up that that was employed for demonstrating AO-HE with optical packets encompassing a PPC header section is shown in Fig. 11. The CW laser source, matching the VCSG's resonant wavelength, was modulated by an amplitude modulator to produce the data packets. The packet format is also shown in the same figure (bottom). The header section consisted of two NRZ pulses at 2.5 Gbit/s (label A), separated by a sequence of alternating '0' and '1' bits (label B) at the same bit rate as the data payload (12 Gb/s), with the exception of a sequence of '0s' (label C) placed in front of the second header pulse. The 1.2 ns long sequence of '0s' was introduced to ensure complete recovery of absorption before the second header pulse entered the VCSG. The header was uniquely defined by the position of the second pulse (Calabretta et al., 2004). The payload was Manchester encoded to guarantee that the average signal power was constant, regardless of the specific bit pattern. The guard time between the packets was 1.2 ns.

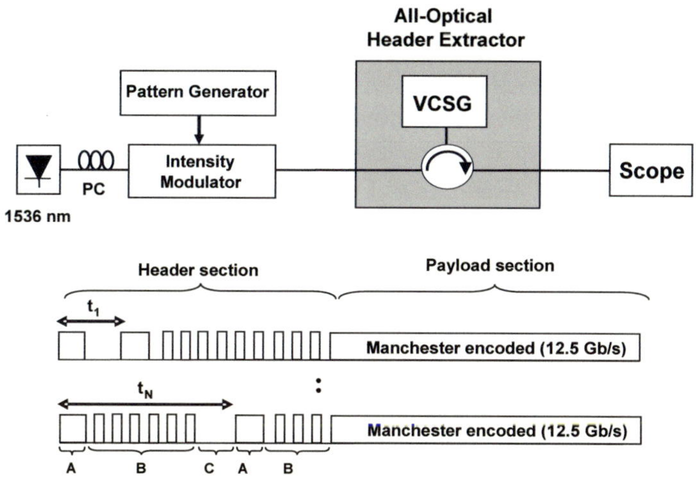

Fig. 11. The set-up and format of the packets used in the AO-HE experiment.

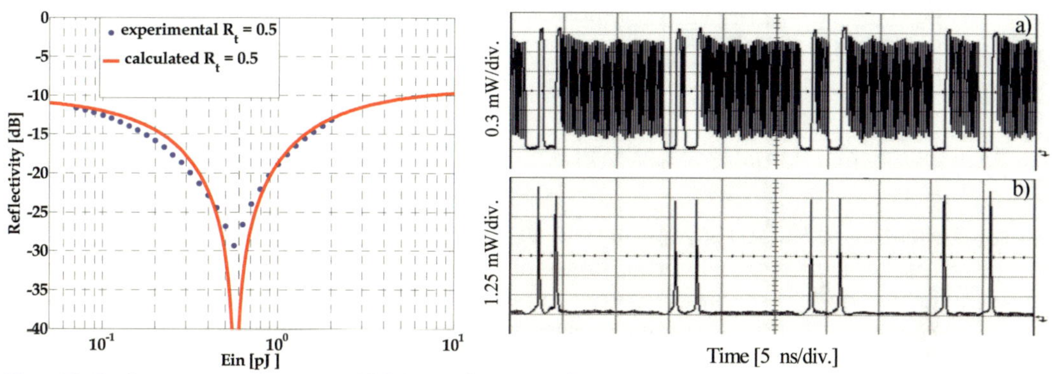

Fig. 12. Left: measured static VCSG nonlinear reflectivity (dots) and fitting with theoretical model (continuous). Right: Experimental results. a): Input data packets. b): Extracted headers at VCSG output

It should be noted that the payload data rate is not limited by the VCSG, and could be extended to higher data rates, provided that the absorber recovery time is sufficiently longer than the bit period. Indeed, the only limitation is that the absorber recovery time should be faster than the guard time between packets and slower than the data period in the payload. It is then expected that even shorter packet guard times would be sufficient at data rates exceeding 12.5 Gb/s. Fig 12 (left) reveals the measurement of the static nonlinear reflectivity of the VCSG at the resonant wavelength. The impedance matching was achieved for a pulse energy of ~0.5 pJ, where the reflectivity is 18 dB lower than the reflectivity for low energy signals. This value sets the limit *CR* of the device in the dynamic operation. A static model, obtained by inserting the steady-state solution $d\alpha/dt$=0 of (12) in (2), was used to fit the experimental data and extract the absorber parameters. The fitting, also shown in the figure, corresponds to E_{sat} = 1.3 pJ, $\alpha_0 d$ = 0.25, and non-saturable round-trip losses $\alpha_{ns}d$ = 0.08. The value of τ_s was determined with a pump-probe measurement to be ~500 ps. In the experiment, four optical packets with different headers (see Fig. 12a, right) were sent to the

VCSG. The output of the VCSG is shown in Fig 12b (right), clearly revealing the two extracted pulses synchronous with the leading edges of the two header pulses in each packet. The payloads immediately following the pulses in the second header were almost completely suppressed, since they encountered the VCSG in the impedance matched state. The *CR* between the two pulses and the suppressed payload was higher than 14 dB for each extracted header pulse. The average input power at the VCSG was about 0 dBm. The gate efficiency (the ratio between input header power and extracted pulse peak power) was calculated to be ~6 %. The effect of top mirror reflectivity on the gate reflectivity at low-energy can be seen in Fig. 7 (right), showing that higher R_t values result in higher values of the low-energy reflectivity, which in turns would increase the efficiency of the VCSG-based AO-HE. However, from the static characteristics it can be also seen that the energy required for matching the impedance of the gate would also increase with R_t.

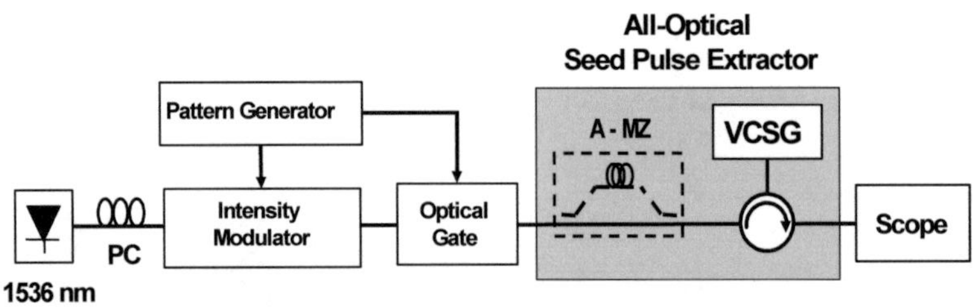

Fig. 13. Experimental setup for the AO-SPE with VCSG.

5.4 All-Optical Seed Pulse Extraction with VCSG

In optical packet switched networks, phase alignment between each packet at an optical packet switching node is required in order to perform packet routing (Sakamoto et al, 2002). Self-synchronization is one of the methods used to achieve packet phase alignment. With self-synchronization, a single pulse seed synchronous with the beginning of the packet is selected from an incoming data pattern. The seed pulse can be used for generating a local clock for packet synchronization, and for driving the subsequent node subsystems (Xia et al., 1999; Huanget al., 2005). An all-optical seed pulse extractor based on self-induced nonlinear reflectivity in a VCSG is described in this section. The experimental set-up that was employed is shown in Fig. 13. Again, the CW laser source at 1536 nm, matching the resonant wavelength of the VCSG, was modulated by an intensity modulator to produce data packets. The modulator was driven by a pattern generator at 12.5 Gbit/s. The optical gate was used to provide an appropriate guard time between the packets for complete recovery of the MQWs absorption. The packet format was the same as in Fig 11. With respect to the previous experiment, an asymmetric Mach-Zehnder (A-MZ) interferometer was inserted, having a delay between the two arms equal to the sequence of "0" bits before the second header (label C); the delay was set to be ~1.2 ns. The output of the A-MZ consisted of the sum of the data packet and its delayed replica. In this way the '0s' sequence was no longer present at the VCSG input. Thus, only the leading edge of the first NRZ header produced a pulse synchronized with the beginning of the packet, while the rest of the packet was suppressed because it experienced IM reflectivity. Fig. 14 shows the input/output waveforms of the AO-SPE. The measured output signal CR was 12 dB.

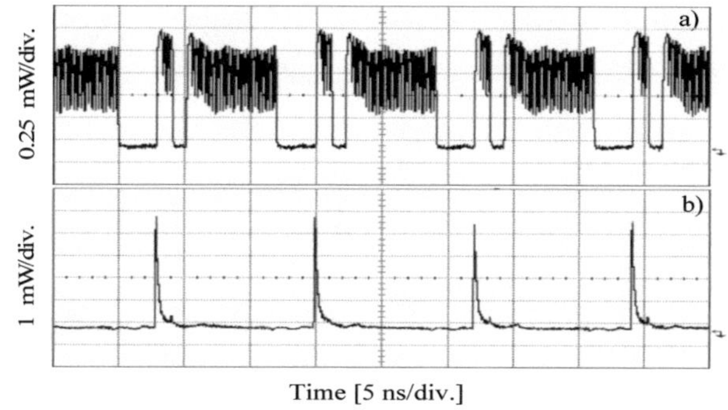

Fig. 14. VCSG input (a), and output (b) traces for the experiment of Fig. 13.

6. All-Optical NAND/NOR Logic Gates based on VCSG

Logical operations performed in the optical domain are required to enable ultra-fast signal processing in all-optical networks. All-optical logic gates can be used to perform many functions in optical packet-switched network. These functions include header recognition and/or modification, packet contention handling, data encoding/decoding, and realization of half- and full-adders. Up to now, semiconductor SAs based on MQWs have been only used to implement ultra-fast *AND* functions (Takahashi et al. 1996, Loka et Smith, 1998), due to their typical nonlinear behaviour, exhibiting low throughput at low powers and high throughput at high powers. An inverse saturable absorption mechanism would allow the use of VCSGs to realize *NAND/NOR* logical operations (Porzi et al., 2008).These functions are particularly important because any logical operations can be realized using only *NAND* or *NOR* operators.

6.1 Operation principle and device optimization

From (6) and (2) it can be seen that, if the VCSG's top mirror reflectivity R_t satisfies the condition:

$$R_t = R_b e^{-2\alpha_{ns} d} \quad , \tag{15}$$

the minimum reflectivity (i.e. the IM) is achieved at high input energy (E>> E_{sat}), as can be easily verified from (5) and (2). In this case, at the resonant wavelength, the VCSG exhibits an inverse saturable absorber characteristic. The simulated behaviour of a VCSG designed to meet the IM condition at high input energies is shown in Fig 15 (left).

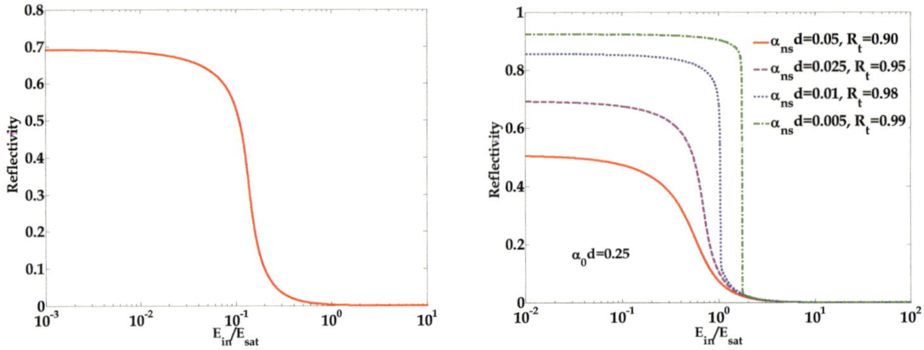

Fig 15. Left: Simulated nonlinear reflectivity at resonance wavelength for a VCSG designed to satisfy the IM condition for totally saturated absorption. Simulation parameters: $\alpha_0 d=0.25$, $\alpha_{ns}d=0.025$, $R_b=1$, and $R_t=0.95$. Right: different plots corresponding to different values of $\alpha_{ns}d$.

In order to implement effective *NAND/NOR* logical operations, it would be desirable to have a step-like characteristic with a sharp transition between the *ON* and the *OFF* states. The well-known phenomenon of field enhancement (at resonance) in a Fabry-Pérot cavity can be exploited to enhance the nonlinear reflectivity change and obtain a steep *ON/OFF* transition. At resonance, the intensity distribution inside the structure can be much higher (or lower) than the input energy, depending on the working conditions. The ratio between the energy inside the cavity and the energy incident on it at the resonant wavelength is given by the parameter ξ, introduced in eq (13). Large changes of ξ for small changes of the input energy enable steep transitions between the *ON* and *OFF* states to be realized. Since the *OFF* state corresponds to the case in which only the non-saturable losses are present in the cavity, the internal energy enhancement factor ξ should exhibit a large increase when α approaches α_{ns}. For $E>>E_{sat}$,

$$\xi_{E\to\infty} = \frac{(1-R_t)}{\left(1-\sqrt{R_t R_b}\,e^{-2\alpha_{ns}d}\right)^2} \tag{16}$$

If the top mirror is optimized for impedance matching at high powers, then the previous expression becomes:

$$\xi_{E\to\infty}^{IM} = \frac{1}{1-R_b e^{-2\alpha_{ns}d}} \tag{17}$$

And, for R_b=1 and $\alpha_{ns}\to 0$ then $\xi\to\infty$. Indeed, in the limit of $\alpha_{ns}\to 0$ and $\alpha_0\to 0$ it can be seen from eq. (15) that we are dealing with the case of an ideal, lossless, symmetric Fabry-Pérot resonator, which has infinite finesse. In practical cases, small values of α_{ns} enable to high values of the internal field enhancement factor to be achieved when the absorber is saturated. In this way, the dynamic input energy range for which the gate changes its operating state is also reduced, giving rise to a step-like characteristic as a function of the input energy. Fig 15 (right) reveals the effect of decreasing α_{ns} (for a fixed value of α_0), on the nonlinear reflectivity characteristic measured at the resonant wavelength. The corresponding values of R_t (given by (15)) are also shown. It can be seen that very sharp transitions correspond to very high values of R_t. Furthermore, in the high-finesse limit, optical bistability may appear, preventing correct operation of the device. The operation principle of the NAND/NOR gates is briefly explained in the following. If a CW probe field is tuned at the cavity resonance, with optical power P_{pb} partially saturating the MQWs absorption, it can initially experience a high value of reflectivity. If two pump signals are applied to the device, and if the single pump power P_{pmp} is not enough to switch the gate in the *OFF* state, whereas twice the pump power is, the reflected probe power represents the *NOR* of the input pump signals. On the other hand, if the power of a single pump signal is enough to turn the gate in the *OFF* state, *NAND* between the two pumps is retrieved by filtering the reflected probe power.

6.2 Simulation results

In order to realize *NAND/NOR* operation with the reverse SA characteristic, a pump-probe configuration is required in which two data pump beams would affect the reflectivity experienced by the probe signal, providing the result of the desired logical operations. For practical applications, where transparency of the device with respect to input pump signals is desirable, only the probe signal is tuned at a cavity resonance, whereas the wavelength of the two pumps could be tuned away from resonance. A spectral analysis of the gate operation is then required. The dependence on the single-pass phase for the reflectivity of the VCSG is given by (1), whereas the wavelength-dependent internal energy enhancement factor is given by:

$$\xi = \frac{(1-R_t)}{\left[1-\sqrt{R_t}\sqrt{R_b}\exp(-\alpha d)\right]^2 + 4\sqrt{R_t}\sqrt{R_b}\exp(-\alpha d)\sin^2(\phi_{sp})} \tag{18}$$

By using eqs (1) and (20), and the power dependence of α expressed by eq. (4), it is possible to calculate the spectral reflectivity of the VCSG for different values of both the input pump and probe energy and wavelength. The calculated spectral reflectivity of the nonlinear gate under different input power conditions is shown in Fig. 16 (left) and Fig. 16 (right) for the cases of *NAND* and *NOR* operation, respectively. In the *NAND* case, for suitable values of the input probe and pump powers, it can be seen that the reflectivity is always high when either the probe field or the probe and a single pump pulse are applied to the gate. On the other hand, if twice the pump power is applied to the gate, the reflectivity at the resonant wavelength is drastically reduced. In the *NOR* case, one single pump pulse contains enough energy to turn the gate to the *OFF* state. It can also be seen that the resonance width changes drastically when passing from the *ON* to the *OFF* state. The resonance width in the *OFF* state is inversely proportional to the steepness of the nonlinear characteristic, thus a trade-off between resonance bandwidth and the *ON/OFF* contrast ratio has to be considered in designing the device. Except for the pump powers, the parameters used to calculate the spectral reflectivity were the same for the examples shown in the two figures. Thus, for a suitable bias level of the probe beam, the two functions can be obtained in the same device by changing the pump power. In an alternative arrangement, the biasing probe power can be changed, while keeping the pump power constant, to switch between the two logical operations.

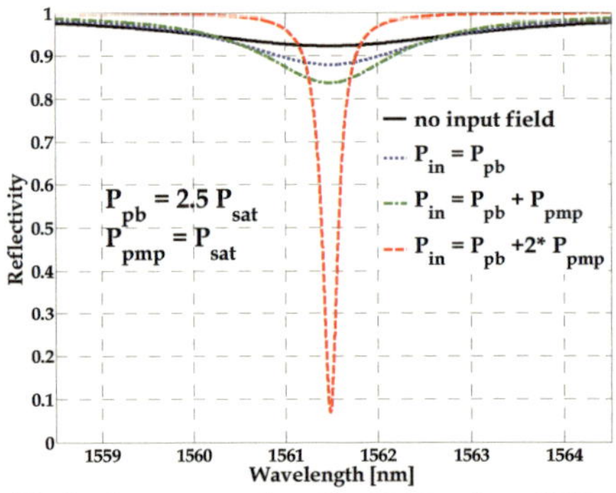

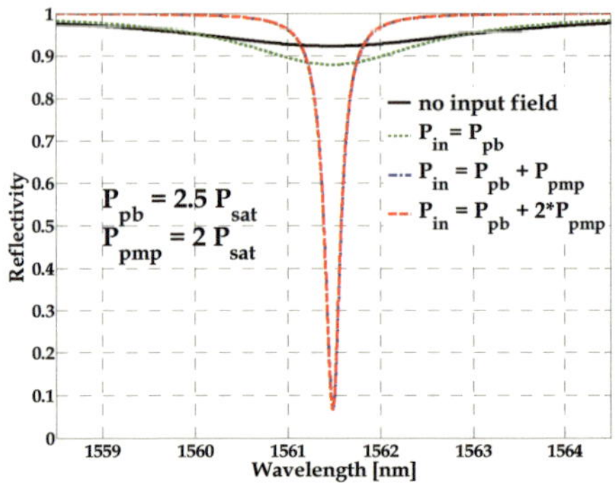

Fig. 16. Left: spectral reflectivity for different input powers for *NAND* gate. Right: spectral reflectivity for different input powers for *NOR* gate. Other simulation parameters: $\alpha_0 d$=0.25, $\alpha_{ns} d$=0.005, and 1 nm pump detuning from resonance.

6.3 Effects of cavity parameters

In a previous section it has been shown that low values of $\alpha_{ns} d$ are desirable for good operation. Here, also the effects of $\alpha_0 d$ on the device performances are investigated. In our simulations we first calculated the nonlinear reflectivity at the resonant wavelength to assess an appropriate value for the input probe power, and then we evaluated the reflectivity variation experienced by the probe field as a function of pump power, for a given pump detuning from resonance. The simulations were performed by means of the nonlinear spectral model previously introduced. In the simulations, the value of R_b was assumed to be 100% and that of R_t was set to the value satisfying the condition given by (15). Pump detuning from resonance was 1 nm. The results of the simulations are shown in Fig 17. Each pair of figures represents the nonlinear reflectivity at resonance, induced by the probe (upper plots) and the pump field (lower plots), for a fixed value of probe power. Each figure corresponds to a fixed value of $\alpha_0 d$ for three different values of $\alpha_{ns} d$. As previously discussed, the steepest characteristic is obtained for the smallest value of $\alpha_{ns} d$. However, very small values of the non-saturable losses $\alpha_{ns} d$ also lead to unpractical values of top mirror reflectivity (> 0.99). Furthermore, as will be clear later on in this section, this situation also makes bistability more likely to occur. On the other hand, increasing $\alpha_0 d$ leads to good extinction ratio values even in the presence of moderate values of $\alpha_{ns} d$, and increases the gate efficiency. However, high values of $\alpha_0 d$ could be difficult to obtain in practice, and would also increase the saturation power (hence preventing low-power operation) and increase the non-saturable losses. A trade-off between ideal device parameters and practical implementation should hence be considered. In the following, only small or moderate values of $\alpha_0 d$ will be considered, which are more likely to be realized in a practical device with low values of non-saturable losses. The results we have obtained can be improved by using higher values of $\alpha_0 d$.

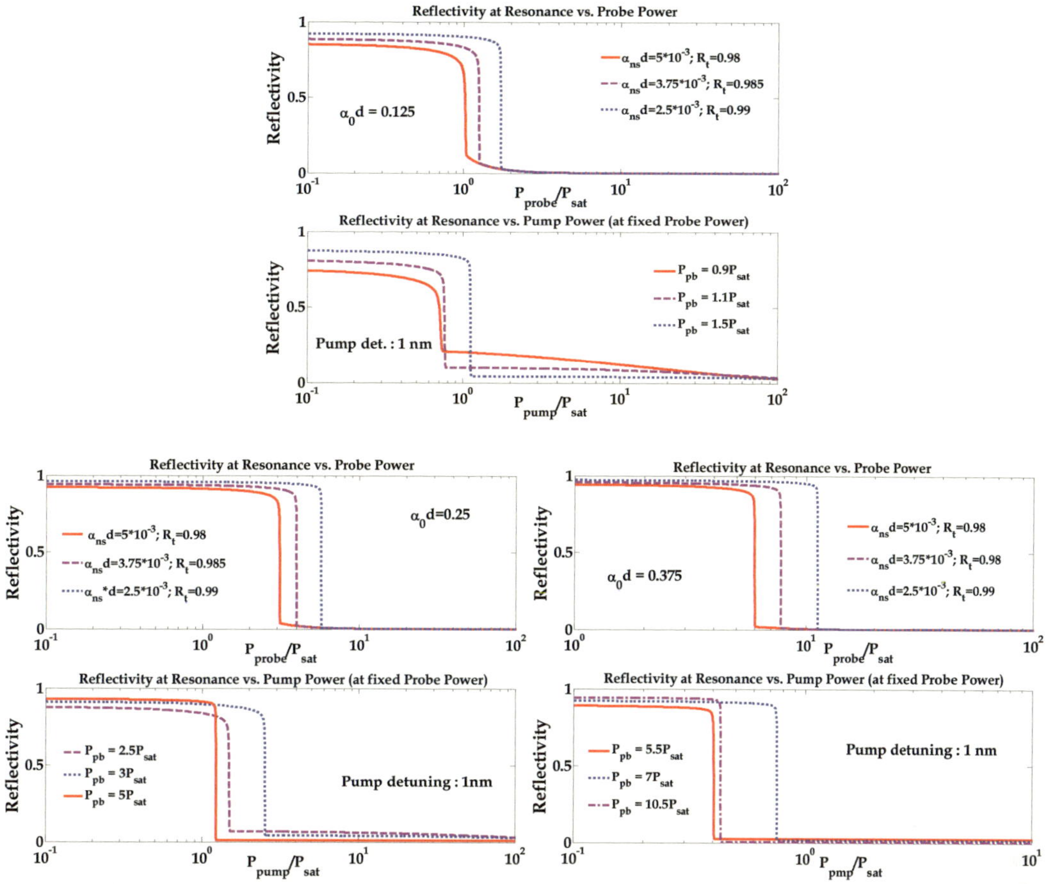

Fig. 17. Effect of $\alpha_{ns}d$ parameter for three different values of $\alpha_0 d$.

6.4 Bistability Analysis

Under particular conditions, bistable operation may appear in the device, meaning that when the pump field is switched off the device could remain in the OFF state rather than going back to its initial ON state. Although it could be useful in other system applications, this situation should be avoided for logical *NAND/NOR* operations. An analysis of bistability can be made by using a procedure similar to that described in (Garmire, 1989), in which the nonlinear effect was only associated with refractive index changes, rather than absorption. Here, for the sake of simplicity, we neglect the nonlinear index change (assumed to be small with respect to the effect of absorption), and consider only the effects of absorption saturation. The input-reflected (I_{in}-I_{ref}) light characteristic at resonance for different values of $\alpha_{ns}d$ and fixed $\alpha_0 d$ are shown in Fig. 18 (left). It can be seen that for the lower value of $\alpha_{ns}d$, the device exhibits bistable behaviour. Thus, bistability sets a limit on the ideal minimum value of internal non-saturable losses.

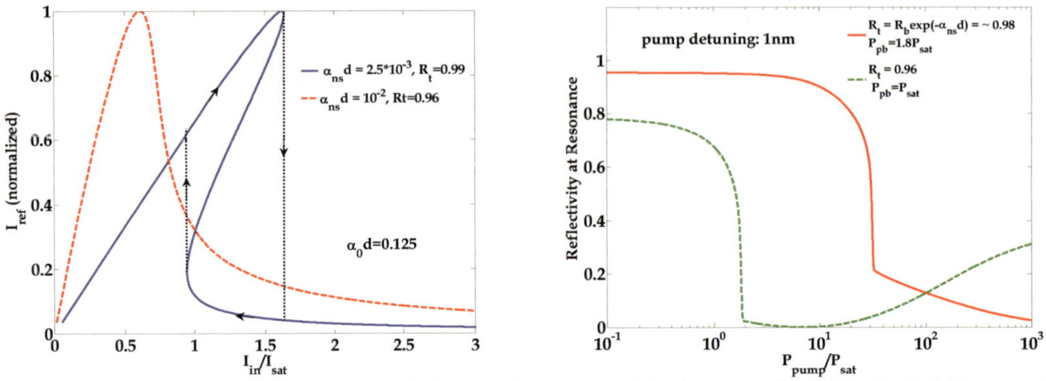

Fig. 18. Left: I_{in}-I_{ref} characteristics of the nonlinear VCSG showing bistable (solid line) and non-bistable (dashed line) behavior for two different values of $\alpha_{ns}d$, and fixed $\alpha_0 d$. Right: Nonlinear reflectivity vs. P_{pmp} for a probe power outside the hysteresis region for $R_t=exp(-a_{ns}d)$ (solid line), and $R_t=0.97exp(-a_{ns}d)$ (dashed line).

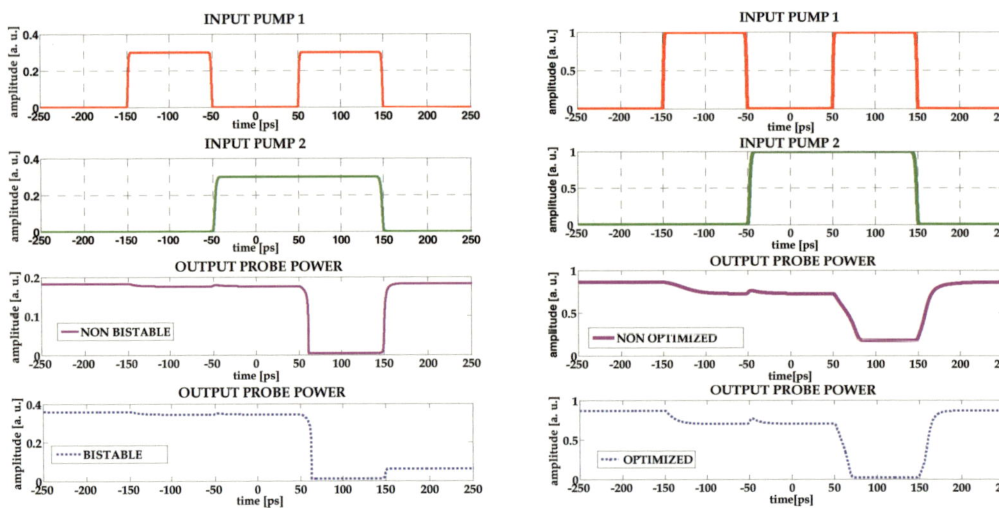

Fig. 19. Left: dynamic NAND operation with NRZ pulses at 10 Gb/s for pump and probe signals tuned at two different resonances ($\alpha_{ns}d$= 0.005, $\alpha_0 d$ = 0.25, $P_{pb}=P_{sat}$); the output probe power for non bistable (solid line) and bistable (dashed lines) cases, are shown. Right: dynamic *NAND* operation relative to the static characteristics of Fig 18; output probe power for non optimized (solid line) and optimized (dashed lines) cases, are shown

Occurrence of bistability can be avoided in two ways, as explained below. The first one is to design a cavity which is inherently non-bistable, like the one associated with the dashed curve in the Fig. 18 (left). The second one is to bias the gate with a proper value of input probe power which lies outside the hysteresis region of the solid curve in Fig. 18 (left). Both these solutions affect the quality of the output signal. As an example, Fig. 18 (right, solid line), shows the reflectivity at a resonance as a function of P_{pmp} for a probe power sufficiently low to avoid bistability, a pump wavelength detuned 1 nm apart from the resonance, and a top mirror reflectivity $R_t=exp(-a_{ns}d)$. It can be seen that the smooth slope of the reflectivity in the proximity of highly saturated absorption (high pump power) leads to a poor output probe *CR*. This suggests an improved operation by matching the gate impedance for a value of absorption corresponding to the applied pump power. That is, a top mirror reflectivity slightly lower than $exp(-a_{ns}d)$. The effect is shown in Figure 18 (right, dashed line), for a top mirror reflectivity R_t= 96%, corresponding to $0.97 \cdot exp(-a_{ns}d)$. The increased *CR* is attained at the expenses of slightly reduced gate efficiency. Lowering the top mirror reflectivity also helps to further prevent bistability, by reducing cavity finesse. The required switching power is reduced too.

6.5 Dynamic Operation

The dynamic model expressed by (12) can be extended to take into account pump and probe optical fields with a generic detuning from resonance:

$$\frac{d\alpha}{dt}=\frac{\alpha_0-\alpha}{\tau_s}-\frac{(1-R_t)\cdot P_{pb}\cdot\alpha}{E_{sat}\left[\left(1-\sqrt{R_t}\cdot\sqrt{R_b}\cdot e^{-(\alpha+\alpha_{ns})d}\right)^2+4\sqrt{R_t}\cdot\sqrt{R_b}e^{-(\alpha+\alpha_{ns})d}\sin^2(\phi_{pb})\right]}-\frac{(1-R_t)\cdot P_{pmp}\cdot\alpha}{E_{sat}\left[\left(1-\sqrt{R_t}\cdot\sqrt{R_b}\cdot e^{-(\alpha+\alpha_{ns})d}\right)^2+4\sqrt{R_t}\cdot\sqrt{R_b}e^{-(\alpha+\alpha_{ns})d}\sin^2(\phi_{pmp})\right]} \quad (19)$$

where ϕ_{pb} and ϕ_{pmp} are the single-pass phases associated with probe and pump wavelengths, respectively; and τ_s is the carrier recombination time in the MQWs. In the following we have assumed τ_s=5 ps. Let us first consider the simple case in which both the pump and the probe fields are tuned in proximity of two cavity resonances. This case correspond to an optimal condition in which a steep transition can be preserved, even for a probe biasing power away from the nonlinear characteristic transition edge to prevent bistability, thanks to the enhancement factor experienced by the pump field in the resonator. For this situation, the results of dynamic simulations are shown in Fig. 19 (right), for the case of NRZ input pump data at 10 Gb/s. For comparison, the dynamic behavior in the case of bistable operation is also shown in the figure (dashed line). However, in practical applications transparency of the operation to the data wavelength is a desirable condition. As previously discussed, hysteresis can be avoided by setting the bias power not too close to the transition edge. Fig. 19 (left, solid line) shows dynamic *NAND* operation for the case relative to Fig. 18 (right, solid line). As expected, although the operation is preserved, the *CR* of the output signal is degraded due to a non ideal characteristic of the gate. The case relative to the static analysis of Fig. 18 (right, dashed lines), where the gate impedance was matched for a partially saturated absorption coefficient corresponding to the value inferred by two data pulses, was also investigated in the dynamic simulation, and the results are shown in Fig. 19 (right, dashed line). It can be seen that *NAND* operation with high *CR* for the filtered probe light can be obtained for this case. Implementation of *NOR* operation is straightforward. In this case there is less constraint on the steepness of the probe nonlinear reflectivity characteristic induced by the pump field. For *NOR* operation, indeed even a relatively smooth characteristic can be tolerated, provided that each pump pulse has enough power to switch the gate to the *OFF* state. However, low-power operation is always desirable, and we can take advantage of the previous considerations to implement efficient, low-power, *NOR* gate with reverse-SA VCSG. As an example, Fig. 20 shows *NOR* operation of the VCSG relative to the case of Fig 18 (right, dashed line) with a pump power ranging within the low-reflectivity region of the figure.

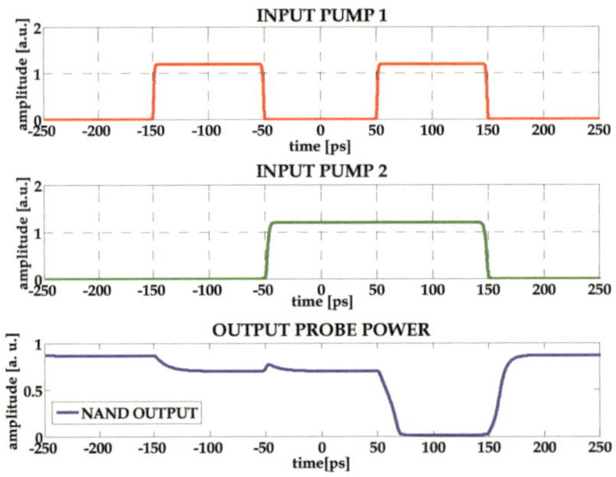

Fig. 20. *NOR* operation with VCSG parameters as in Fig 18 (right, dashed line).

7. Conclusion

In this chapter we have given detailed information on the design, fabrication, and operation of VCSGs based on saturable absorption in semiconductor quantum-wells. Different applications for optical communication subsystems have been discussed. These applications include, but are not limited to, wavelength conversion/optical gating, advanced signal processing functionalities for all-optical networks, and all-optical logical operations. The versatility in tailoring the nonlinear characteristics of the VCSG for a particular application has been discussed using numerical and analytical models.

8. References

Agrawal G. P. (2002), *Fiber-Optic Communication Systems 3rd Ed.*, John Wiley & Sons Inc, 0471215716, New York

Akiyama, T.; Tsuchiya, M.; Kamiya, T. (1998). Sub-pJ operation of broadband asymmetric Fabry-Pérot all-optical gate with coupled cavity structure. *Applied Physics Letters*, Vol. 72, No. 13, (Mar. 1998) pp. 1545-1547, 0003-6951

Brackett C. A.; Acampora A. S.; Sweitzer J.; Tangonan G.; Smith M. T.; Lennon W.; Wang K.-C.; Hobbs R. H. (1993). A scalable multiwavelength multihop optical network: A proposal for research on all-optical networks. *Journal of Lightwave Technology*, Vol. 11, No. 5/6 (May/June 1993) pp. 736-753, 0733-8724

Brovelli L.; Keller U.; and Chiu T. (1995). Design and operation of antiresonant Fabry-Pérot saturable semiconductor absorbers for mode-locked solid-state lasers. *Journal of the Optical Society of America B-Optical Physics*, Vol. 12, No. 2, (Feb. 1995) pp. 311-320, 0740-3224

Burr E. P.; Pantouvaki M.; Seeds A. J.; Gwilliam R. M.; Pinches S. M.; Button C. C. (2003). Wavelength conversion of 1.53-μm-wavelength picosecond pulses in an ion-implanted multiple-quantum-well all-optical switch. *Optics letters*, Vol. 28, No. 6 (Mar. 2003) pp. 483-485, 0146-9592

Calabretta N.; Liu Y.; Waardt D. H.; Hill M. T.; Khoe G. D.; Dorren H. J. S. (2001). Multiple-output all-optical header processing technique based on two-pulse correlation principle. *Electronics Letters*, Vol. 37, No. 20, (Sept 2001) pp. 1238–1240, 0013-5194

Calabretta N.; Liu Y.; Huijskens F. M.; Hill M. T.; De Waardt H.; Khoe G. D.; Dorren H. J. S. (2004). Optical signal processing based on self-induced polarization rotation in a semiconductor optical amplifier. *Journal Of Lightwave Technology*, Vol. 22, No. 2, (Feb. 2004) pp. 372-381, 0733-8724

Delpon E. L.; Oudar J. L.; Bouché N.; Raj R.; Shen A.; Stelmakh N.; Lourtioz J. M. (1998). Ultrafast excitonic saturable absorption in ion-implanted InGaAs/InAlAs multiple quantum wells. *Appied Physics Leters*, Vol. 72, No. 7, (Feb. 1998) pp.759–761, 0003-6951

Garmire E. (1998). Criteria for optical bistability in a lossy saturating Fabry-Pérot. *Journal of Quantum Electronics*, Vol. 25, No. 3, (March 1989) pp. 289 - 295, 0018-9197

Huang X.; Ye P.; Zahng M.; Wang L. (2005). A Novel Self-Synchronization Scheme for All-optical Packet Networks. *IEEE Photonics Technology Letters*, Vol. 17, No. 3, (March. 2005) pp. 645-647, 1041-1135

Isomaki, A.; Vainionpaa, A. M.; Lyytikainen, J.; Okhotnikov, O. G. (2003). Semiconductor mirror for optical noise suppression and dynamic dispersion compensation. *Journal of Quantum Electronics*, Vol. 39, No. 11, (Nov. 2003) pp. 1481-1485, 0018-9197

Kartner F. X., Jung I. D., Keller U. (1996) Soliton Mode-Locking with Saturable Absorbers, *Journal Of Selected Topics in Quantum Electronics*, Vol. 2, No. 3, (Sept. 1996.) pp. 540-556, 1077-260X

Keller, U.; Weingarten, K. J.; Kartner, F. X.; Kopf, D.; Braun, B.; Jung, I. D.; Fluck, R.; Honninger, C.; Matuschek, N.; Aus der Au, J. (1996). Semiconductor saturable absorber mirrors (SESAM's) for femtosecond to nanosecond pulse generation in solid-state lasers. *Selected Topics in Quantum Electronics*, Vol. 2, No. 3, (Sep. 1996) pp. 435-453 1077-260X

Loka H. S.; Smith P. W. E. (1998). Ultrafast All-Optical Switching in an Asymmetric Fabry–Pérot Device Using Low-Temperature-Grown GaAs. *IEEE Photonics Technology Letters*, vol. 10, No. 12, (Dec. 1998) pp. 269-271, 1733-1735, 1041-1135

Massoubre D.; Fatome J.; Pitois S.; Decobert J.; Landreau J.; Millot G.; and Oudar J. L. (2006). All-aptical exctintion ratio enhancement of a 160 GHz pulse train using a saturable absorber vertical microcavity. *Optics letters*, Vol. 31, no. 4, (Feb. 2006) pp. 537-539, 0146-9592

Porzi C.; Isomaki A., Guina M.; and Okothnikov O. (2005). Impedance-detuned high-contrast vertical cavity semiconductor switch, *Proceeding OFC 2005.*, pp. OThM5, 1-55752-783-0, Anaheim Convention Center, Anaheim, California

Porzi C.; Bogoni A.; Poti L.; Guina M.; and Okhotnikov O. (2006). Characterization and operation of vertical cavity semiconductor all-optical broadband wavelength converter, *Proceeding SPIE Integrated Optics, Silicon Photonics, and Photonic Integrated Circuits*, pp. Z1830-Z1830, 0-8194-6239-X, Strasbourg, France

Porzi C.; Calabretta N.; Guina M.; Okhotnikov O.; Bogoni A.; and Potì L. (2007). All-Optical Processing for Pulse Position Coded Header in Packet Switched Optical Networks Using Vertical Cavity Semiconductor Gates. *Selected Topics in Quantum Electronics,* Vol. 13, No. 5, (Sept.-Oct. 2007) pp 1579–1588, 1077-260X

Porzi C.; Guina M. ; Bogoni A. ; Poti, L. (2008). All-Optical nand/nor Logic Gates Based on Semiconductor Saturable Absorber Etalons. *Selected Topics in Quantum Electronics,* Vol. 14, No. 3, (May-June 2008) pp 927–937, 1077-260X

Siegman A. E. (1986), *Lasers*, University Science, 0935702113, Mill Valley, California

Sakamoto T.; Okada A.; Hirayama M.; Sakai Y.; Moriwaki O.; Ogawa I.; Sato R.; Noguchi K.; Matsuoka M. (2002). Demonstration of an optical packet synchronizer for an optical packet switch. *Proceeding OFC 2002,* pp. 762-763, 1-5572-700-8, Anaheim Convention Center, Anaheim, California

Takahashi R.; Kawamura Y.; Iwamura H. (1996). Ultrafast 1.55 mm all-optical switching using low-temperature-grown multiple quantum wells. *Applied. Phyics Letters,* Vol. 68, No. 2, (Jan. 1996) pp. 153–155, 0003-6951

Vegas Olmos J. J.; Tafur Monroy I.; Turkieveia J. P.; Liu Y.; Koonen A. M. (2006). Self-Controlled All-Optical Label and Payload Separator or Variable Length Burst in a Time-Serial IM/DPSK Scheme. *IEEE Photonics Technology Letters,* vol. 17, No. 8, (Aug. 2005) pp. 1692-1694, 1041-1135

Xia T. J.; Kao Y. H.; Liang Y.; Lou J. W.; Ahn K. H.; Boyraz O.; Nowak G. A.; Said A. A.; Slam M. N. (1999). Novel Self- Synchronization Scheme for High-Speed Packet TDM Networks. *IEEE Photonics Technology Letters*, vol. 11, No. 2, (Feb. 1999) pp. 269-271, 1041-1135

Xiang N.; Okhotnikov O.; Vainionpaa A.; Guina M.; Pessa M. (2001). Broadband semiconductor saturable absorber mirror at 1.55. *Electronics. Letters*, vol. 37, No. 6, (Mar. 2001) pp. 374-375, 0013-5194

Yan, R. H.; Simes, R. J.; Coldren, L. A. (1991). Surface-normal electroabsorption reflection modulators using asymmetric Fabry-Pérot structures. *IEEE Journal of Quantum Electronics,* Vol. 27, No. 7, (June 1991) pp. 1922 - 1931, 0018-9197

15

All-optical flip-flops based on semiconductor technologies

Antonella Bogoni[1], Gianluca Berrettini[2], Paolo Ghelfi[1],
Antonio Malacarne[2], Gianluca Meloni[1], Luca Potì[1] and Jing Wang[3]
[1]Consorzio Nazionale Interuniversitario per le Telecomunicazioni (CNIT), Pisa Italy
[2] Scuola Superiore Sant'Anna, Pisa Italy
[3]Department of Electronic Engineering, Tsinghua University, Beijing China

1. Introduction

Optical technologies represent the main bet for future communication systems. Among the others, digital subsystems for optical processing are of great interest thanks to their intrinsic properties in terms of bandwidth, transparency, immunity to the electromagnetic interference, cost, power consumption, as well as robustness in hostile environment. Key basic functions are represented by logic gate, logic function, flip-flop memories, optical random access memories, etc.. Research in this field is in its very early stages even if some interesting techniques have been already theoretically addressed and experimentally demonstrated. Here we review the state of the art for all-optical flip-flop based on semiconductor technologies: best result will be highlighted in terms of transition speed, switching energy, complexity and power consumption; we will then discuss some new achievement we have recently reached.

All-optical packet switching seems to be the most promising way to take advantage of fiber bandwidth to increase routers forwarding capacity, being able to achieve very high data rate operations. All-optical flip-flops have been widely investigated mainly because they can be exploited in all-optical packet switches, where switching, routing and forwarding are directly carried out in the optical domain. Some examples concerning optical packet switches are shown in (Dorren et al., 2003; Liu et al., 2005; Bogoni et al., 2007; Herrera et al., 2007), where an optical flip-flop stores the switch control information and drives the switching operation. Former solutions for all-optical flip-flops have been demonstrated exploiting discrete devices (Dorren et al., 2003) or Erbium-doped fiber properties (Malacarne et al., 2007) which suffer from slow switching times and high set/reset input powers. Several integrated or integrable solutions (Hill et al., 2004; Liu et al., 2006) present a switching energy in the fJ range and switching times of tens of ps at the expenses of poor contrast ratios. On the other hand in (Hill et al., 2005) an integrated scheme exhibiting a very high contrast ratio value but with transition times in the ns range is reported. In any case a

trade off between contrast ratio and edges speed must be found as a function of the flip-flop application. Micro-resonators-based bistable element has been demonstrated (Van et al., 2002) presenting high optical operating power, pJ switching energies and microsecond switching times, theoretically reducible down to the order of tens of ps. Making a comparison with electronics, recent large-scale integration (LSI) circuits (Keyes, 2001) show switching energies of 1fJ even though with slower switching speeds. In (Dorren et al., 2003), a solution based on coupled ring lasers is proposed. This solution offers a certain number of advantages: it can provide high contrast ratios between states; there is no difference in the mechanisms for switching from state 1 to state 2 and vice-versa, making symmetric set and reset operations; it presents a large input light wavelength range and a controllable switching threshold. Moreover, considering an integrated version of this kind of flip-flop, through numerical analysis a switching energy in fJ range has been demonstrated.
Here we will describe the above mentioned solutions underlining the main benefits, drawback, limitation and perspectives. We will then present our activities on clocked flip-flops, and an example of their use in an all-optical counter. Finally, we will present an SOA-based flip-flop which is able to switch with very short rising and falling edges, and we use it in a realistic switching operation. Integrability of our solutions is also discussed.

2. State of the art

One of the simplest way that was originally proposed to implement an optical flip-flop includes two coupled lasers (Hill et al., 2001), as depicted in Fig. 1 (a). The system can have two stable states. In state 1, light from laser 1 suppresses lasing in laser 2. In this state, the optical flip-flop memory emits CW light at wavelength λ_1. Conversely, in state 2, light from laser 2 suppresses lasing in laser 1, and the optical flip-flop memory emits CW light at wavelength λ_2. To change states, lasing in the dominant laser can be inhibited by injecting external light with a different wavelength and opportune power. The output pulse of an optical header processor can be used to set the optical flip-flop memory into the desired wavelength. From the theory it also follows that laser driving currents and coupling coefficient determines the required switching light power.
This flip-flop has also been implemented in a ring configuration based on Semiconductor Optical Amplifiers (SOA), as shown in Fig. 1 (b) (Dorren et al., 2003). Two SOAs act as the lasers gain media. Fabry–Pérot filters (FPF) with a bandwidth of 0.18nm have been used as wavelength selective elements. Optical pulses were used to set and reset the flip-flop. The optical spectrum of the flip-flops' output states is shown in Fig. 2.
The switching time between the two lasing modes is inversely proportional to the length of the laser cavities. Thus, in order to allow switching times in the range of picoseconds, an integrated solution has to be adopted. This was realized in (Hill et al., 2004), where a photonic flip-flop based on two coupled micro-ring lasers with dimensions of 20x40 μm^2 was reported, exhibiting a switching time of 18ps and a switching energy of a few fJ.
The micro-ring lasers were fabricated in active areas of the integrated circuit containing bulk 1.55nm bandgap InGaAsP in the light guiding layer. Separate electrical contacts allowed each laser's wavelength to be individually tuned by adjusting the laser current. Passive waveguides connected the micro-ring lasers to the integrated circuit edges (Fig. 3). Micro-ring lasers typically have two inherent lasing modes; laser light traveling in the clockwise (CW) direction, and laser light in the anticlockwise (ACW) direction.

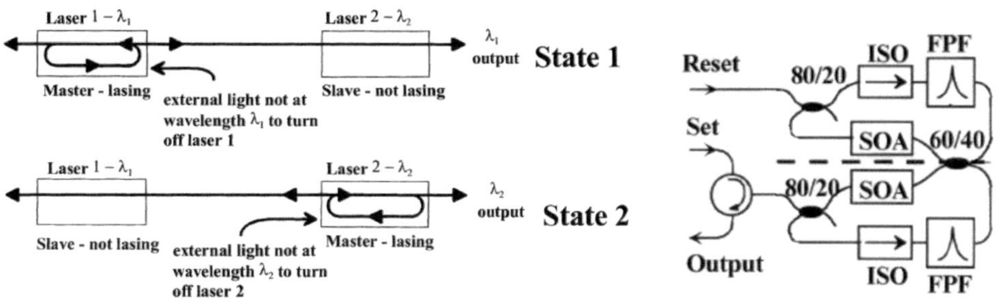

Fig. 1. (a): Arrangement of two coupled identical lasing cavities forming a flip-flop, showing the two possible states. (b): Implementation of the optical flip-flop memory

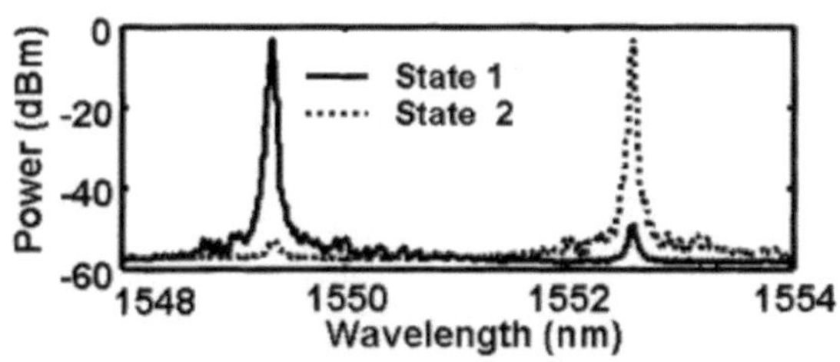

Fig. 2. Spectral output of two states of the optical flip-flop memory.

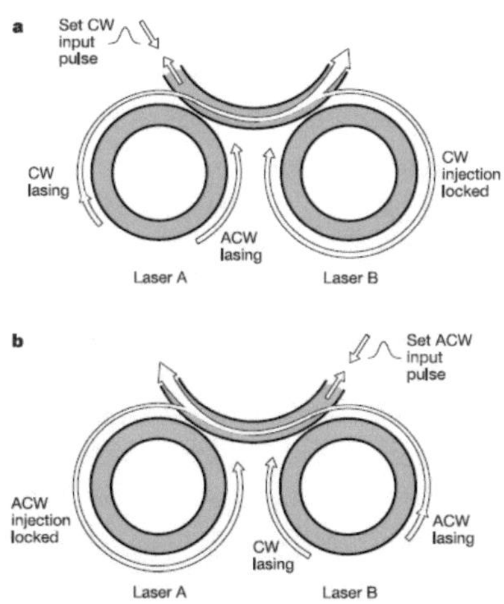

Fig. 3. Two micro-ring lasers coupled via a waveguide to form an optical flip-flop.

In state A, CW light from laser A is injected via the waveguide into laser B. The light from laser A will undergo significant resonant amplification in laser B if the resonant frequencies of the two laser cavities are close. This injected light competes with the laser B self-oscillations for available power from the laser gain medium. If sufficient light is injected into laser B, then the laser B gain will be decreased below threshold. This extinguishes the laser B self-oscillation, and laser A captures or injection-locks (Buczek et al., 1971) laser B, forcing light to circulate only in the CW direction. To set the system in one state or another, light

close to the lasing wavelength and polarization can be injected into the waveguide connecting the lasers. This light will set both lasers simultaneously lasing in either the CW or ACW direction. The different states can be distinguished by the different power levels at the two outputs. The power level at the output associated with the locked laser will be three times that of the other output. Additionally, the lasing wavelengths of the lasers may be different, allowing the states to be distinguished by the wavelength of the light output.
Another scheme recently proposed (Malacarne et al., 2007) exploits absorption and fluorescence of few meters of erbium–ytterbium (Er–Yb)-doped fiber. This solution suffers from slow switching times and high set/reset input powers, and since it doesn't exploit semiconductor devices, it will not be studied in depth here.
In (Liu et al., 2006) a solution that offer the advantage of being fully packaged, was presented. It is based on an hybrid integrated circuit consisting of two coupled Mach-Zehnder interferometers (MZIs), each having one SOA in one arm. The schematic of the circuit is shown in Fig. 4.

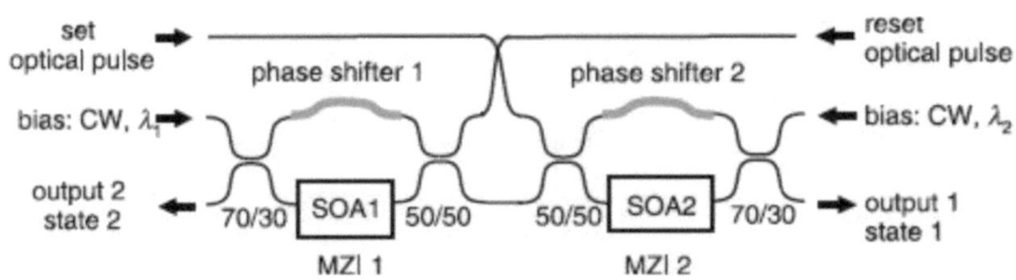

Fig. 4. Schematic diagram of optical flip-flop memory proposed in (Liu et al., 2006).

Each MZI (MZI 1 and MZI 2 in the figure) has an SOA in one arm. A laser emits a continuous-wave (CW) bias light at wavelength λ_1 that is fed into MZI 1. The MZI 1 output is sent into MZI 2, which has the same structure, but biased by a CW light with a different wavelength, λ_2. The system has two possible states: in state 1, the MZI 1 output suppresses output from MZI 2, so λ_1 dominates the output; in state 2, the MZI 2 output suppresses output from MZI 1, and then λ_2 is dominant. When the CW light with λ_1 is injected into MZI 1, MZI 1 is biased in such a way that the light out of MZI 1 goes mostly into the low branch of the 50/50 coupler output. This light then flows into MZI 2 via the 50/50 coupler in MZI 2, and affects the gain and phase shift for light propagating through it. The MZI 1 light perturbs the SOA 2 properties so that the CW bias 2 light (λ_2) propagating through SOA 2 and phase shifter 2 goes mostly into the top output of the 50/50 coupler in MZI 2. Then the CW bias 2 light (λ_1) does not travel into the MZI 1, and does not affect the properties of SOA 1. Actually, the MZI 1 output suppresses output from MZI 2. The states of the system can be switched by sending a light pulse (via Set or Reset port) into the MZI that is currently dominant. This light will switch the MZI output away from suppressing the other MZI, allowing the other MZI then to become dominant.
An optical flip-flop based on two-mode bistability in a multimode interference bistable laser diode (MMI-BLD) has also been reported (Takenaka et al., 2005). A schematic view of the MMI-BLD is shown in Fig. 5 (a). All waveguides including the 2x2 MMI coupler consist of active materials. Saturable absorbers are located at the end of the output ports to obtain hysteresis. The 2x2 MMI is designed as a cross coupler, so that only two cross-coupled lasing modes can exist as illustrated in the insets of Fig. 5 (a). Two-mode bistability between

these two lasing modes will occur due to cross gain saturation and the saturable absorbers if the injection current is within the hysteresis loop (Takenaka & Nakano, 2003).

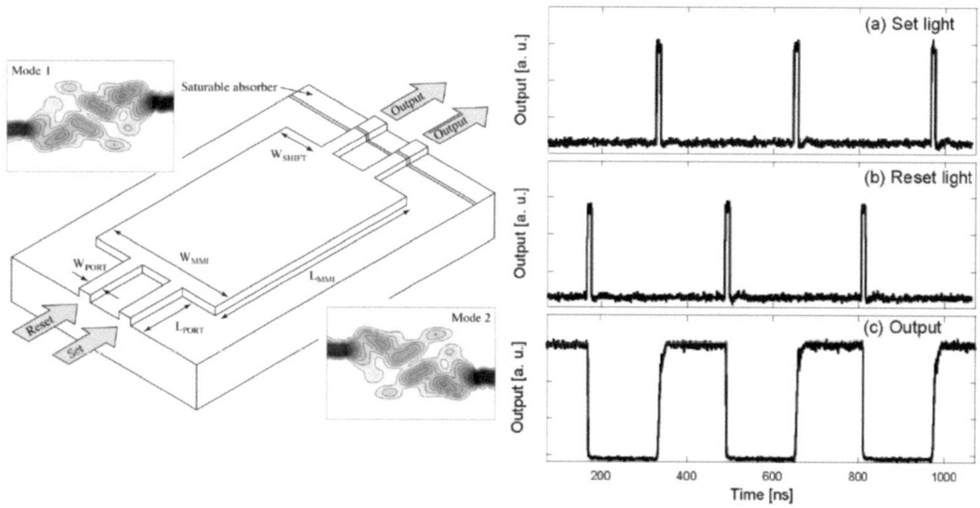

(a) (b)

Fig. 5. (a): Schematic view of the MMI-BLD. Two cross-coupled lasing modes are illustrated in the insets. (b): All-optical flip-flop operation of the MMI-BLD.

A set signal injected into the set port saturates the absorption to Mode 1, causing Mode1 to start lasing. At the same time, cross-gain saturation and the absorption to Mode 2 by the saturable absorber suppress Mode 2. In a similar manner, a reset signal switches the lasing mode from Mode 1 to Mode 2. Therefore, all-optical flip-flop operation is achievable with the MMI-BLD, because external light injection to each input port will select the mode to lase. The corresponding operation, showing the optical power at one of the waveguide output when set and reset pulses are applied is depicted in Fig. 5 (b).

In (Huybrechts et al., 2008) a single DFB laser diode has been used to realize a flip flop. A DFB laser injected with CW light shows two different stable states: one in which the laser is lasing and another one where it is switched off. When the laser is lasing, the gain will be clamped and relatively small. Therefore, the injected light experiences only a small amplification and has almost no influence on the laser light. In the second state, the laser is switched off and the injected light experiences a high amplification. This results in a rising power progression throughout the cavity and therefore a non-uniform distribution of the carriers, known as spatial hole burning. This will affect the refractive index, leading to a distortion of the Bragg reflections in the laser diode. The losses inside the cavity will become higher and the threshold for lasing will rise. Eventually the laser will stay switched off. The two states are equally possible for a range of input powers of the injected light and this gives a bistability in the lasing power (Fig. 6 (a)). This bistability can be exploited to obtain flip-flop operation by injecting short optical pulses: a pulse injected at the same side as the CW light will move the DFB laser out of the hysteresis curve and will switch off the laser; to switch the laser on again, a pulse is injected from the other side, since this will restore the uniformity of the carrier distribution. In the experiment, the set and reset pulses were obtained from an ultra-short pulse source generating 7ps-long pulses. The obtained results are depicted in Fig. 6 (b). The set-pulses have an energy of 75fJ and the reset-pulses 190 fJ. The repetition rate is 1.25GHz and the switch-on time is 75ps. An almost immediate switch-off time of 20ps has been obtained, which corresponds with the resolution of the optical scope.

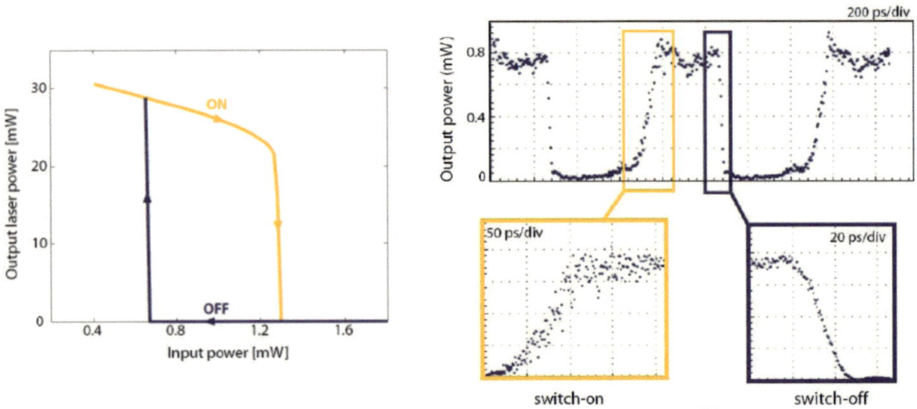

(a) (b)

Fig. 6. (a): Bistability of an injected DFB laser as a function of the injected power. (b): Results.

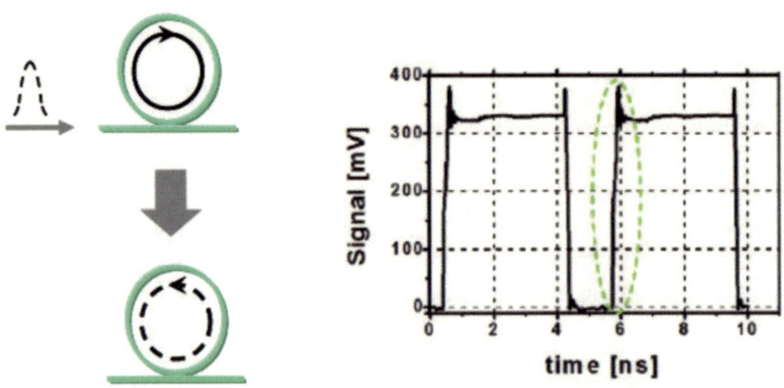

(a) (b)

Fig. 7. (a): Operation principle of the monolithic semiconductor ring laser. (b): Results.

As discussed previously, integrable solutions are preferred since they would allow high-density packaging, with the possibility of reducing costs, power consumption, and operation speed. To achieve these results, researchers are investigating novel technologies in order to reduce as much as possible device dimensions. A possible solution towards this direction is the use of a monolithic semiconductor micro-ring laser (Trita et al., 2009) which shows an intrinsic and robust directional bistability between its CW and ACW propagating modes. If the ring laser is correctly set, injecting a laser pulse in one direction makes the laser emit in that direction (Fig. 7 (a)). Experiments show a switching time of about 20ps for both rising and falling edges, with set/reset pulses of 5ps and 150fJ energy.

Another promising technology is nano-photonics, exploited in the realization of photonic crystals (PCs) and quantum dots (QDs). By combining these technologies one could take advantage of both the band-gap effect and the highly dispersive property of PCs, and the high-density of state and high nonlinear property of QDs.

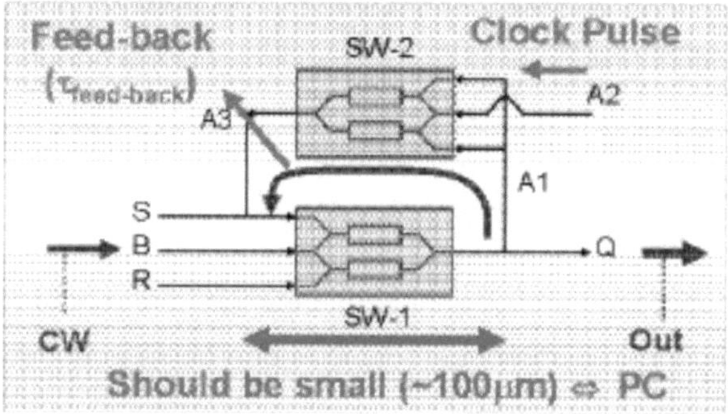

Fig. 8. Schematic diagram of the PC-FF.

A Mach Zehnder-type all-optical flip-flop developed by combining GaAs-based two-dimensional photonic crystal (2DPC) slab waveguides and InAs-based optical nonlinear QDs has been proposed in (Azakawa, 2007). The photonic crystal-based flip-flop (PC-FF) schematic is shown in Fig. 8, and is based on two photonic-crystal-based Symmetric Mach Zehnder (PC-SMZ) switches. The principle of the PC-SMZ is based on the time-differential phase modulation caused by the nonlinear-induced refractive index change in one arm of the two interferometers. 2DPC waveguides are composed of single missing line defects, while nonlinear-induced phase shift arms are selectively embedded with QDs. The mechanism of the third-order nonlinear property is an absorption saturation of the QD caused by a control (pump) pulse. A resultant refractive index change produces a phase shift for the signal (probe) pulse. A wavelength of the control pulse is set to the absorption peak of the QD, while a wavelength of the signal pulse is set in the high transmission range in the 2DPC waveguide with the QD. A single PC-SMZ switch would operate as a pseudo-flip-flop, meaning that the on-state is limited by the carrier relaxation time in the nonlinear material (~ 100ps in the experiment). In order to change the pseudo FF into the normal FF operation, the scheme of Fig. 8 was proposed. An output signal of the PC-SMZ impinges into an optical AND element (which is another PC-SMZ switch) via a feedback loop, where another input pulse, i.e., a clock pulse impinges. An output of the AND element is combined to the set pulse, as shown in the figure. The clock pulse serves as a refresh pulse to expand the on-state period against the relaxation of the carrier, while the feedback signal restricts the clock pulse to be controlled by the set and reset pulses. The feasibility of this idea has been verified only by computer simulation.

3. Flip-flops based on coupled SOA ring lasers: advantages and limitations

In order to investigate advantages and drawbacks of SOA-based solution we consider the setup shown in Fig. 9. The flip-flop consists of two coupled ring lasers emitting at two different wavelengths (λ_1=1550nm and λ_2=1560nm). In each ring, an SOA acts as the gain element, a 0.25nm band-pass filter (BPF) is used to as select the wavelength, and an isolator makes the light propagation unidirectional. Both the SOAs are polarization insensitive

Multi-Quantum Well (MQW) structures with a small-signal gain of 31dB, saturation power of 13dBm and Amplified Spontaneous Emission (ASE) noise peak at 1547nm.

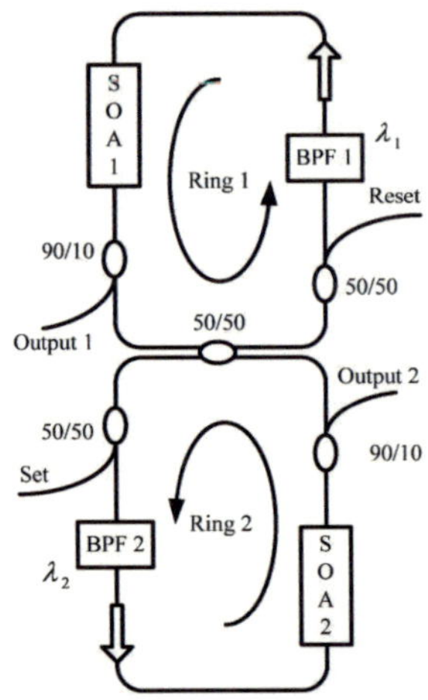

Fig. 9. Experimental Setup of the all-optical flip-flop based on SOAs.

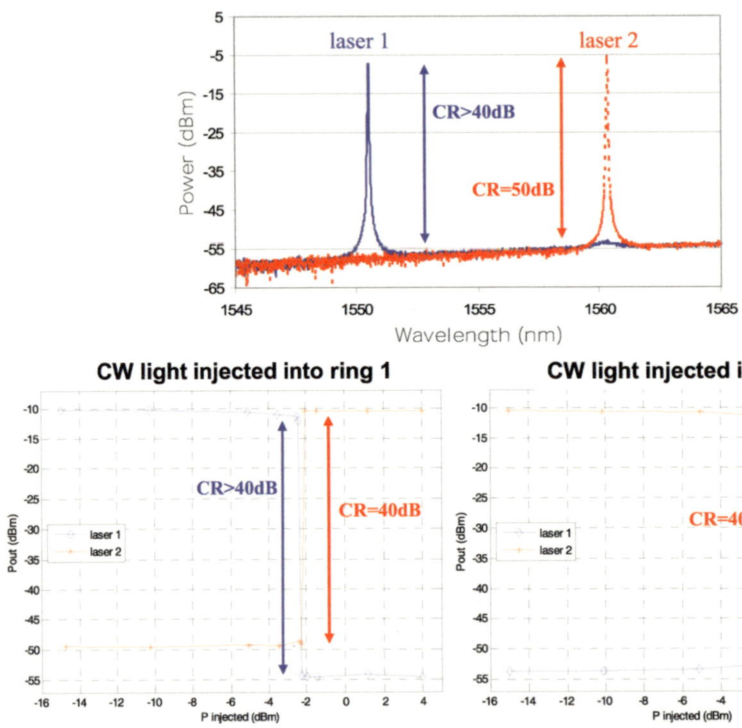

Fig. 10. Top: optical spectra of the two states; Bottom: output power of lasers versus input power injected into cavity 1 (left) and into cavity 2 (right).

The system can have two states. In "state 1", light from ring 1 suppresses lasing in ring 2, reaching cavity 2 through the 50/50 coupler and saturating the SOA 2 gain. In this state, the optical flip-flop output 1 emits CW light at wavelength λ_1.In "state 2" light from ring 2 suppresses lasing in ring 1 (saturating SOA 1 gain), and output 2 emits CW light at wavelength λ_2. To dynamically change state, lasing in the dominant cavity can be switched off by injecting external pulsed light with a wavelength different from λ_1 and λ_2 (λ_{IN}=1554.5nm). In Fig. 10 experimental measurements of the two states optical spectra are investigated and a graph of the output power of both the ring lasers, versus the CW input power injected into each cavity is reported. The output contrast ratios are higher than 40dB.

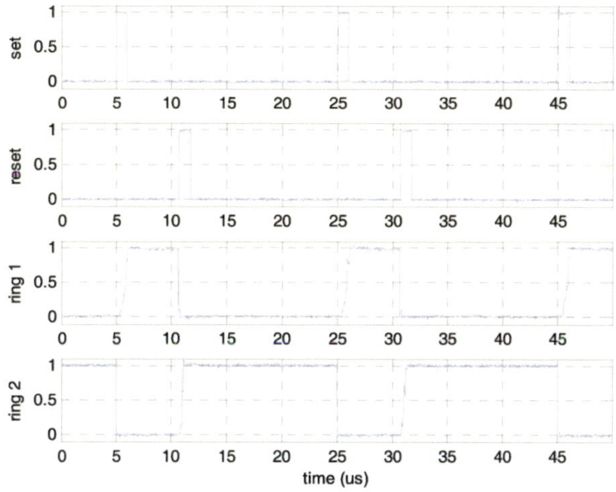

Fig. 11. Experimental results of the all-optical flip-flop output.

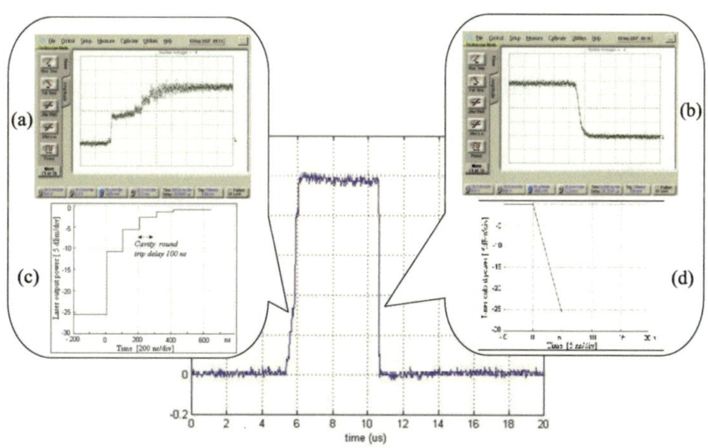

Fig. 12. Measured (a)-(b) and simulated (c)-(d) behavior of the flip-flop output edges.

By injecting two regular sequences of pulses into the set and reset ports, we demonstrate the dynamic flip-flop operation shown in Fig. 11. We experimentally observed that the flip-flop falling time only depends on the edge time of control pulses (5ns in this section), while the rising time is determined by the cavity length and by the length of the fiber between the two SOAs. In our setup, each ring has a cavity length of 20m corresponding to a round-trip time of about 100ns. Experimental measurements (Fig. 12 (a)) show that the building-up process of one state takes place step by step and each step corresponds to a cavity round-trip time equal to 100ns. The total rising edge behavior lasts several hundreds of ns. The experimental falling edge behavior is shown in Fig. 12 (b), with a transition time of 5ns, equal to the input pulse edge.

Dynamics behavior of the two SOA-based coupled lasing cavities has been analyzed through simulations as well, whose details can be found in (Barman et al., 2007). Assuming the same parameters of the experimental setup (cavity length and cavity loss, injected pulses edge time and average power), as can be observed in Fig. 12 (c)-(d), simulation results for rising and falling edges are in good agreement with experimental measurements, confirming the step behavior of the rising edge and at the same time a falling edge as fast as the input pulse edge. We also simulated an integrated version of this flip-flop, considering 2mm cavity length and 0.5mm SOA length. Results predict 12ps falling time and ~40ps rising time with injected input pulsewidth of 12ps and pulse energy of 15.6fJ, comparable with the results of one of the latest optical flip-flop integrated version (Hill et al., 2004).

4. SOA-based clocked flip-flops

Most of the all-optical flip-flops proposed in literature are non-clocked devices, whose output changes immediately following the set/reset signals, thus they are also referred to as Set-Reset (SR) latch. As a digital device that temporarily memorizes the past input signal and processes it with current inputs, optical flip-flop is expected to be synchronized with a system clock, and to work in a timely programmed mode. Moreover, in some complicated optical computing applications such as optical shift registers or counters, various types of clocked flip-flops are necessary, such as SR, D, T, and JK flip-flops.

Starting from the basic structure defined in the previous paragraph, here we show clocked all-optical flip-flops including SR, D, T, and JK types, exploiting also AND logic gates based on nonlinear effects in SOA (Wang et al., 2009, a).

4.1 Clocked SR flip-flop

The characteristic table of the set/reset (SR) flip-flop is shown in Fig. 13 (a). If S=R=0, the flip-flop remains at its previous state; if S=1 R=0, it is set to "state 1"; if S=0 R=1, it is set to "state 0". S=R=1 is forbidden since the flip-flop is unstable in this case. The setup of clocked SR flip-flop is shown in Fig. 13 (b): it consists of two AND gates and one SR latch. "AND 1" and "AND 2" perform AND function between the clock pulse and S and R, respectively. The outputs of "AND 1" and "AND 2" are connected to the "Set" and "Reset" ports of the latch respectively. The operation principle of this clocked flip-flop is shown in Fig. 13 (c): when a clock pulse comes, if S=R=0 it can not pass through either "AND 1" or "AND 2", so "Set" and "Reset" ports receive no pulse and the latch maintains its previous state (Q_{next}=Q); if S=1 R=0, the clock pulse can pass through "AND 1" but is blocked by "AND 2", so only "Set" receives a pulse and the latch is set to "state 1" (Q_{next}=1); if S=0 R=1, the clock pulse can pass through "AND 2" but is blocked by "AND 1", so the latch is set to "state 0" (Q_{next}=0). S=R=1 is forbidden since the latch is unstable when "Set" and "Reset" receive pulses simultaneously. The flip-flop is clocked because it only changes state when a clock pulse comes, according to the S and R values at that time. S and R values at any other time are ignored.

S	R	Q_{next}	Comment
0	0	Q	Hold state
1	0	1	Set
0	1	0	Reset
1	1	N/A	Forbidden

(a)

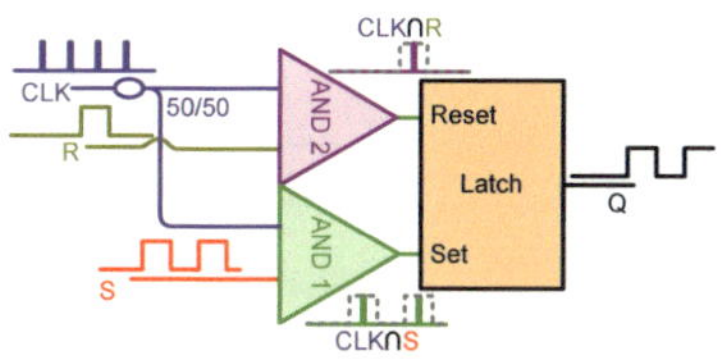

(b)

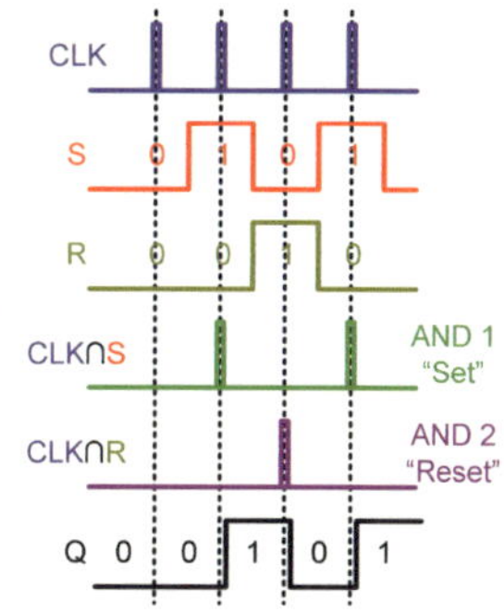

(c)

Fig. 13. Clocked SR flip-flop: (a) characteristic table; (b) logic circuits; (c) working principle.

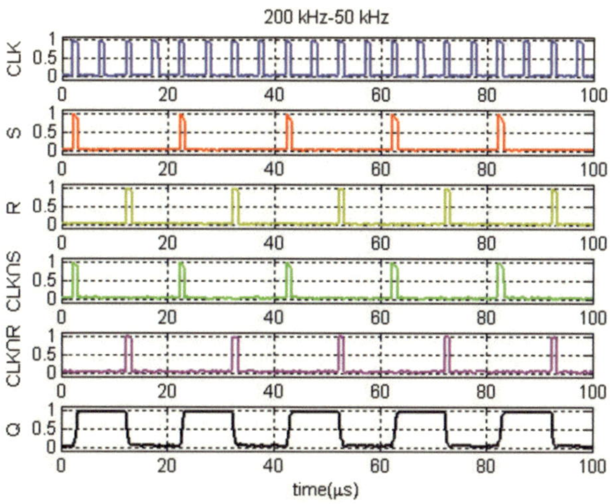

Fig. 14. Clocked SR flip-flop operation.

In Fig. 14 the experimental operation of the clocked SR flip-flop is reported. The clock pulse has a repetition rate of 200kHz with a pulse-width of 1μs. S and R signals also have a pulse-width of 1μs but at a repetition rate of 50kHz, synchronized with the clock. The wavelengths of clock, S and R are λ_{CLK}=1554.1nm, λ_S=1552.5nm and λ_R=1550.5nm respectively, and the outputs of "AND 1" and "AND 2" are at $\lambda_1=2\lambda_S-\lambda_{CLK}$=1550.9nm and $\lambda_2=2\lambda_R-\lambda_{CLK}$=1546.9nm. The flip-flop only responses to the S and R values when a clock pulse comes, but ignores the S and R at any other time, in agreement with Fig. 13 (c).

4.2 Clocked D flip-flop

The characteristic table of D flip-flop is shown in Fig. 15 (a). D represents the data signal. If D=0, the flip-flop is set to "state 0"; if D=1, the flip-flop is set to "state 1". The setup of clocked D flip-flop is shown in Fig. 15 (b): "AND 1" gate performs AND function between the clock pulse and D, whereas "AND 2" performs AND function between clock and inverted D. The operation principle of D flip-flop is shown in Fig. 15 (c): when a clock pulse comes, if D=1 it can pass through "AND 1" but is blocked by "AND 2", so only "Set" port receives a pulse and the latch is set to "state 1" (Q=1); similarly if D=0 the clock pulse can pass through "AND 2" but is blocked by "AND 1", only "Reset" receives a pulse and the latch is set to "state 0" (Q=0). The flip-flop is clocked because it only changes state when a clock pulse comes, according to the D values at that time, but ignores D at any other time.

D	Q_{next}	Comment
0	0	Reset
1	1	Set

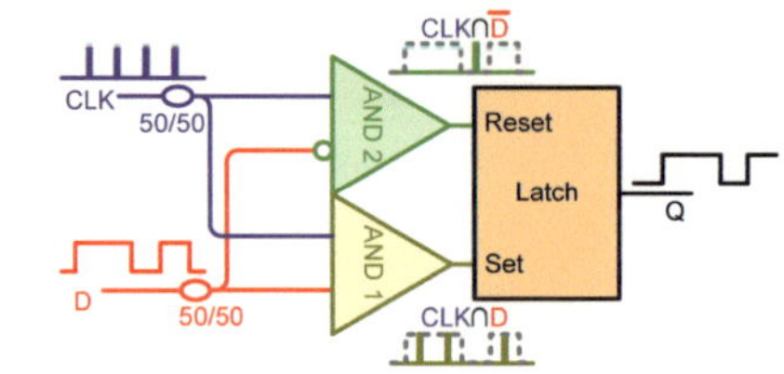

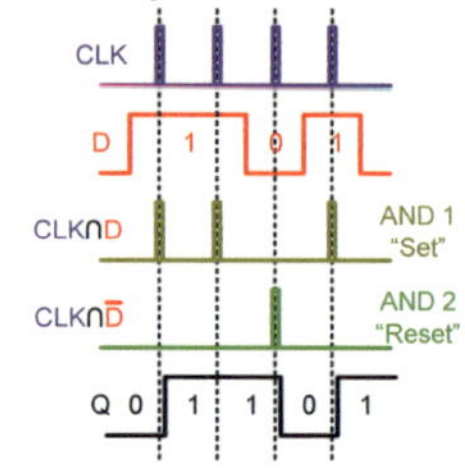

(a) (b) (c)

Fig. 15. Clocked D flip-flop: (a) characteristic table; (b) logic circuits; (c) working principle.

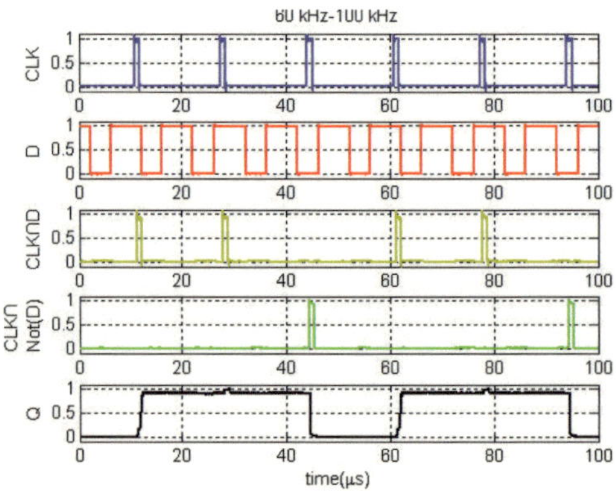

Fig. 16. Clocked D flip-flop operation.

In Fig. 16 clocked D type flip-flop operation is experimentally demonstrated. The clock pulse has a repetition rate of 60kHz with a pulsewidth of 1μs; whereas D has a repetition rate of 100kHz with a pulsewidth of 6μs. The wavelength of clock and D are λ_{CLK}=1554.1nm and λ_D=1552.5nm respectively, so the output of "AND 1" is at $\lambda_1=2\lambda_D-\lambda_{CLK}$=1550.9nm and the output of "AND 2" is at $\lambda_2=\lambda_{CLK}$=1554.1nm, the same with the clock pulse. The flip-flop only responses to the D values when clock pulses come, and therefore is clocked.

4.3 Clocked T flip-flop

The characteristic table of T flip-flop is shown in Fig. 17 (a). T represents the toggling signal. If T=0, the flip-flop maintains its previous state; if T=1, the flip-flop changes its state. The setup of clocked T flip-flop is shown in Fig. 17 (b). Different from SR and D flip-flops, in T flip-flop, the next state is not determined by external control signals, such as S, R, and D, but depends on the previous state, so feedback of output Q is used in T flip-flop to carry out the toggling operation. "AND 1" performs AND function between the clock pulse and T; whereas "AND 2" performs AND between the output of "AND 1" and the feedback output Q. "AND 3" carries out AND function between output of "AND 1" and inverted Q. The operation principle of T flip-flop is shown in Fig. 17 (c): when a clock pulse comes, if T=0 it is blocked by "AND 1", neither "Set" nor "Reset" receives pulse, and the latch remains at its previous state. If T=1, the clock pulse can pass through "AND 1"; then, if Q=1 it can pass through "AND 2" but is blocked by "AND 3", so only "Reset" receives a pulse and the latch toggles to "state 0" (Q=0); if Q=0 the clock pulse can pass through "AND 3" but is blocked by "AND 2", only "Set" receives a pulse and the latch toggles to "state 1" (Q=1). In this way, the flip-flop is triggered by the clock pulse, changing its state if T=1, or maintaining its state if T=0.

T	Q_{next}	Comment
0	Q	Hold state
1	$\overline{Q}$	Toggle

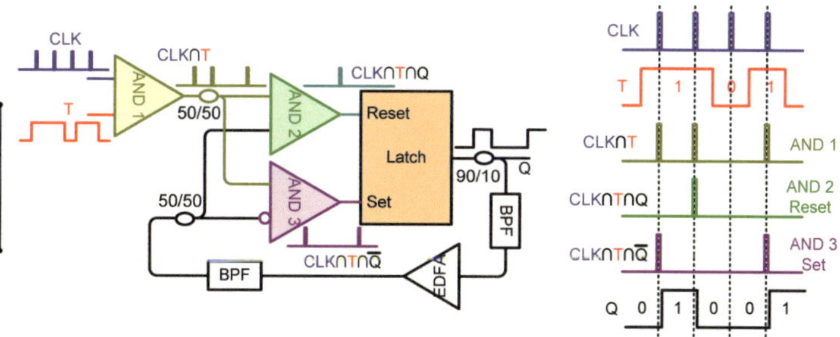

(a) (b) (c)

Fig. 17. Clocked T flip-flop: (a) characteristic Table; (b) logic circuits; (c) working principle.

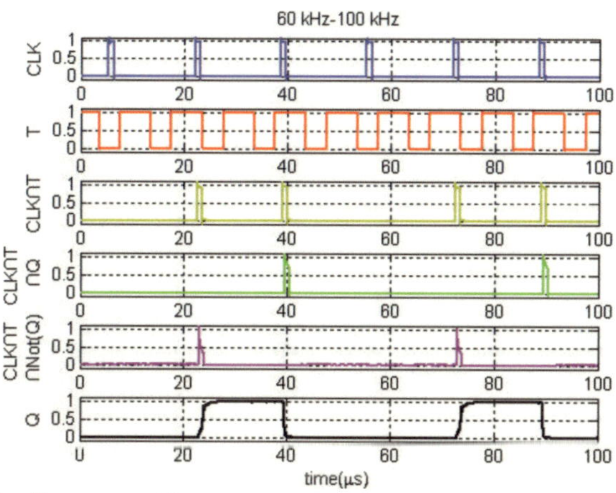

Fig. 18. Clocked T flip-flop operation.

In Fig. 18 clocked T flip-flop operation is experimentally demonstrated. The clock pulse has a repetition rate of 60kHz with a pulse-width of 1μs; whereas T has a repetition rate of 100kHz with a pulse-width of 6μs. The wavelength of clock pulse and T are λ_{CLK}=1554.1nm and λ_T=1552.5nm respectively, so the output of "AND 1" is at $\lambda_1=2\lambda_T-\lambda_{CLK}$=1550.9nm. The flip-flop output, Q, has a wavelength of λ_Q=1549.3nm, so the output of "AND 2" is at $\lambda_2=2\lambda_Q-\lambda_1$=1547.7nm and the output of "AND 3" is at $\lambda_3=\lambda_1$=1550.9nm, the same with the output of "AND 1". The flip-flop is clocked since the state toggling is only triggered when a clock pulse comes and T=1.

4.4 Clocked JK flip-flop

The characteristic table of JK flip-flop is shown in Fig. 19(a), which could be considered as a combination of SR flip-flop and T flip-flop. Like SR flip-flop, J and K signals are also used as set and reset signals: J=K=0 makes the flip-flop maintain its previous state; J=1 K=0 sets it to "state 1"; and J=0 K=1 sets it "state 0". However, in SR flip-flop, S=R=1 is forbidden, but in JK flip-flop, J=K=1 is allowed and the flip-flop toggles its state in this condition, like a T flip-flop.

J	K	Q_{next}	Comment
0	0	Q	Hold state
1	0	1	Set
0	1	0	Reset
1	1	$\overline{Q}$	Toggle

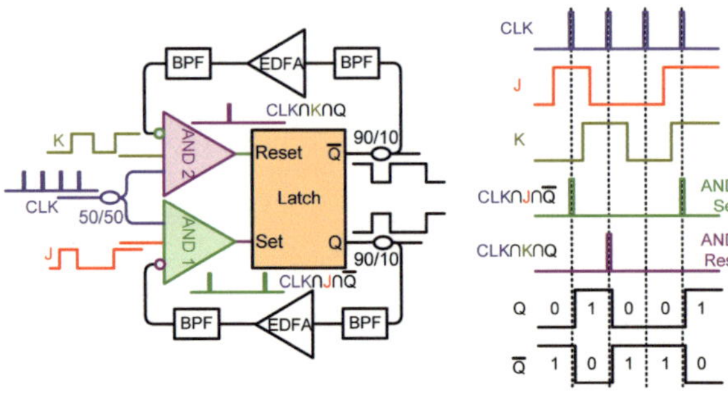

(a) (b) (c)

Fig. 19. Clocked JK flip-flop: (a) characteristic table; (b) logic circuits; (c) working principle.

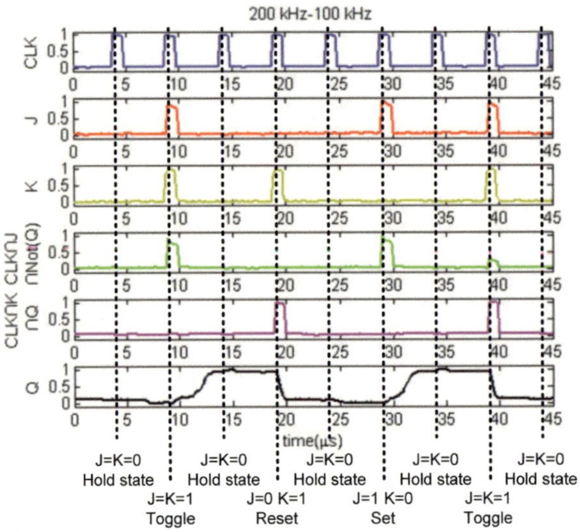

Fig. 20. Clocked JK flip-flop operation.

The setup of clocked JK flip-flop is shown in Fig. 19(b). The two complementary outputs of two ring lasers of SR latch are used as Q and inverted Q respectively. "AND 1" carries out AND function between the clock, J, and inverted Q; whereas "AND 2" carries out AND between the clock, K, and Q. Similar to SR flip-flop, the JK flip-flop can be set and reset by external signals, so CLK∩J and CLK∩K are partially carried out in two AND gates. However, the JK flip-flop can toggle its state like a T flip-flop, so the feedback of Q at previous state must also be taken into account in the two AND gates. When a clock pulse comes, if J=K=0 it can not pass through "AND 1" and "AND 2", so neither "Set" nor "Reset" receives a pulse, and the latch remains at its previous state. If J=1 K=0, the clock pulse is

blocked by "AND 2", but in "AND 1" there are two possible cases. If Q=1 the clock pulse is blocked, so "Set" receives no pulse and the latch will remain at "state 1"; otherwise if Q=0 the clock pulse can pass through "AND 1", and the latch will be set to "state 1". So in the case of J=1 K=0, the flip-flop will be set to "state 1" no matter in which state it was. Similarly, if J=0 K=1, the clock pulse is blocked by "AND 1". But for "AND 2", if Q=1 the clock pulse can pass through, so the latch will be set to "state 0", otherwise if Q=1 the clock pulse is blocked and the latch will stay in "state 0". So the flip-flop will be set to "state 0" no matter in which state it was. Finally, if J=K=1 we also have to consider two cases of Q. If Q=1, the clock pulse is blocked by "AND 1" but can pass through "AND 2", so the latch is set to "state 0"; otherwise, the clock pulse can pass through "AND 1" but is blocked by "AND 2", and the latch is set to "state 1". In both two cases, the flip-flop changes its state, which is called state toggling.
In Fig. 20 clocked JK flip-flop operation is experimentally demonstrated. The clock pulse has a repetition rate of 200kHz and a pulsewidth of 1μs. J and K both quasi-periodic pulse trains, with repetition rate of 100kHz and pulsewidth of 1μs, synchronized with the clock. However, in order to realize all four cases of J=K=0, J=1 K=0, J=0 K=1, and J=K=1, in every 4 periods (40μs) of J and K, there is one pulse missed, as shown in Fig.12. It could be observed that the JK flip-flop operation has a good agreement with Fig. 19(c). The wavelengths of clock, J, and K are λ_{CLK}=1554.1nm, λ_J=1552.5nm and λ_K=1550.5nm respectively, and the wavelength of Q is λ_Q=1549.3nm, so the output of "AND 1" is at $\lambda_1=2\lambda_J-\lambda_{CLK}$=1550.9nm and the output of "AND 2" is at $\lambda_2=2\lambda_K-\lambda_{CLK}$=1546.9nm.

4.5 Three-state flip-flop

Together with clocked flip-flops, another interesting evolution of the basic flip-flop shown in paragraph 3 is the upgrade to multi-state flip-flop. A multi-state memory could in fact extend a 1×2 optical switch to a larger dimension of 1×N, depending on the number of states of the memory.
The setup of the three-state optical memory is shown in Fig. 21 (Wang et al., 2008, a), which consists of three coupled SOA fiber ring lasers operating at three different wavelengths. The memory has three states. In "state 1", only ring 1 is lasing, whereas ring 2 and ring 3 are suppressed; the output light of SOA 1 is split by coupler A into two portions: one portion passes through Path 1 (the dashed red line) and then saturates SOA 3, making ring 3 suppressed; the other portion passes through Path 2 (the dashed green line) and then saturates SOA 2, making ring 2 suppressed. In "state 1", the optical memory emits a CW light at the wavelength of λ_1 from output 1 port. Similarly, in "state 2", only ring 2 is lasing, and the memory emits a CW light at λ_2. Finally in "state 3", only ring 3 is lasing.
To dynamically change the state, three setting couplers are inserted into the ring cavities, each corresponding to a particular state. One pulse injected into set 1 port is split to saturate SOA 3 and SOA 2, and it could not reach SOA 1. Thus ring 2 and ring 3 are both suppressed while ring 1 could lase; the memory is set to "state 1". Similarly for set 2 and set 3.

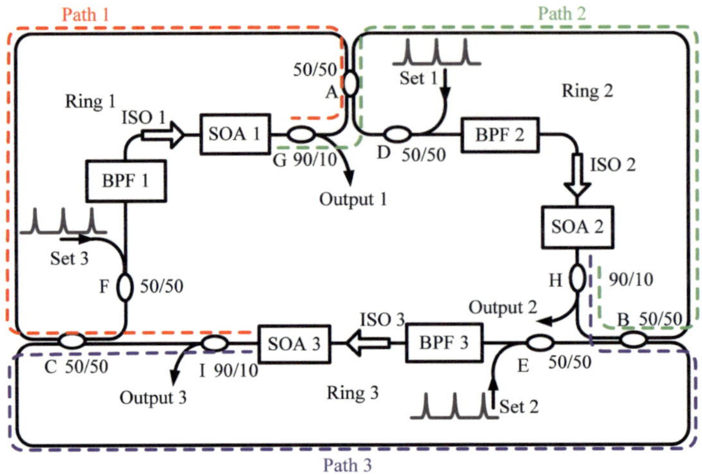

Fig. 21. Experimental setup of three-state all-optical memory

The experiments has shown an "on-off" extinction ratio of 40 dB for each state. The required switching energy is in the order of 12 to 19nJ, depending on the wavelength chosen for the set pulses. In the exploited set-up the ring length of the three cavities is about 42m, giving a rise time of about 210ns, while falling time can be as low as 20ps. Of course, photonic integration will reduce the rise time down to 40ps as well, making GHz switching possible. By coupling N ring lasers, the scheme could be scaled up to N-state, in which output light of one SOA saturates N-1 other SOAs, requiring higher optical power for stable flip-flop operation. Moreover, N(N-1)/2 couplers would be used to couple N ring lasers together and the cavity length would also be increased. Photonic integration or hybrid integration would be useful to reduce both the cavity loss and the cavity length, and make high optical power and fast switching speed possible.

5. Latch-based all-optical counter

An extremely interesting and promising application of clocked flip-flops is the all-optical counter. As a key component in both areas of optical computing and communication, all-optical binary counter can be used as a finite-state machine in optical computing and can also be used for header recognizing and payload processing in optical packet switching networks. Nevertheless, there are few papers related to all-optical counter (Poustie et al., 2000; Benner et al., 1990; Feuerstein et al., 1991). In (Poustie et al., 2000) an all-optical binary counter based on terahertz optical asymmetric demultiplexer (TOAD) switching gate was demonstrated, which is however not integrable due to the nonlinear fiber loop mirrors in the TOADs. In (Benner et al., 1990; Feuerstein et al., 1991) a counter is presented but it requires optical-to-electrical conversion in the coupler switches. Furthermore, in these reported schemes, due to the lack of optical latch or other memory element, the storage of optical bit is realized by fiber loop memory, which requires precise synchronization of the arrival time of optical pulses and makes the counting speed fixed, depending on the fiber length in the loop memory.

Extending the setup of the above mentioned T flip-flop, we have demonstrated the first SR latch based all-optical binary counter (Wang et al., 2009,b), which is able to work at different counting speeds without the necessity of any reconfiguration or re-synchronization. The SR latch is used for optical bit storage, to memorize the accumulated number of input pulses and to carry out binary modulo-2 addition between the accumulated number and new input pulses. The AND logic gate is used for binary carry signal generation when the input and stored bit are both "1". We also presented two-bit binary counting operation as well as 1/2 and 1/4 all-optical frequency division at different frequencies, and Q-factor measurement is performed to evaluate the signal degradation and confirm the cascadability of the scheme. Finally, the operation speed limitation of clocked flip-flop and the counter is investigated.

The setup of optical counter is shown in Fig. 22 (a), which consists of two cascaded stages. Carry 1 signal from stage 1 is used as the input of stage 2. The latches' output, Q_2Q_1, represent the output of the counter.

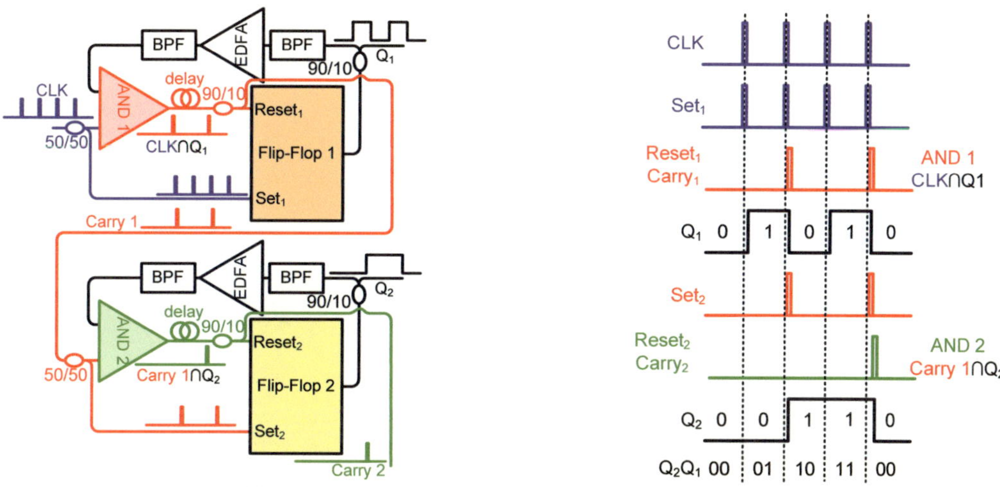

Fig. 22. All-optical binary counter: (a) logic circuits; (b) working principle.

The working principle of the counter is shown in Fig. 22 (b). At first, both latch 1 and latch 2 are in "state 0", Q_2Q_1=00. When the first clock pulse comes, it injects into "Set_1" directly, but since Q_1=0, it can not pass through "AND 1", so only "Set_1" receives a pulse and latch 1 is set to "state 1", Q_2Q_1=01. When the 2nd clock pulse comes, since Q_1=1 it can pass through "AND 1" and reach both "Set_1" and "$Reset_1$" ports. However, due to the fiber delay line, "$Reset_1$" receives the pulse later than "Set_1", so latch 1 is then set to "state 0". The output pulse of "AND 1" is used as "Carry 1" and is injected into stage 2. Since Q2=0, "Carry 1" pulse can not pass through "AND 2", so only "Set_2" receives the pulse and latch 2 is set to "state 1". Now we have Q_2Q_1=10. When the third pulse comes, it is blocked by "AND 1" since Q1=0, so latch 1 is set to "state 1", Q_2Q_1=11. Finally, when the 4th pulse comes, since Q_1=1 it can pass through "AND 1" and reach "$Reset_1$". Due to the fiber delay, "Reset1" receives a pulse later so latch 1 is set to "state 0". Then the output "Carry 1" pulse from "AND 1" injects into stage 2, passes through "AND 2" and sets latch 2 to "state 0". Now the counter returns to the initial state, Q_2Q_1=00, and the "Carry 2" pulse from "AND 2" can be used as the input of next stage. In each stage, the SR latch is used as a memory element to carry out binary modulo-2 addition and store the current state of the counter; whereas the AND gate is used to generate carry pulse when a clock pulse injects into a stage that has already been in "state 1". Different from the schemes proposed in (Poustie et al., 2000; Benner et al., 1990; Feuerstein et al., 1991), whose bit storage is implemented by fiber loop memory and has a fixed counting speed determined by the fiber length, the counter shown in Fig. 22 utilizes SR latches to memorize its current state, and can work at different counting speeds without the necessity of any reconfiguration or re-synchronization.

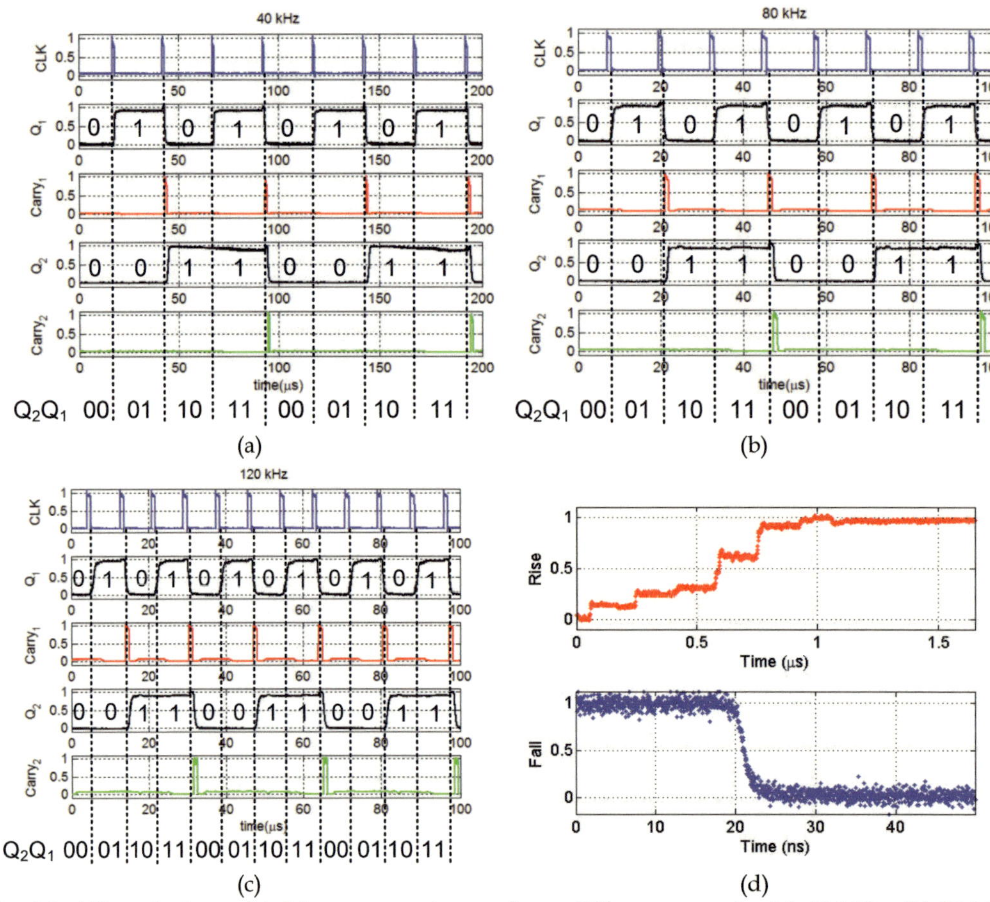

Fig. 23. All-optical two-bit binary counting at three different speeds: (a) 40 kHz; (b) 80 kHz; (c) 120 kHz. (d): transition time of SR latch

Referring to (Wang et al., 2009, b) for all the details of the experiment, Fig. 23 demonstrates that the counter can work at three different counting speeds, 40 kHz, 80 kHz, and 120 kHz without any reconfiguration. It is observed that each time a clock pulse comes, Q_2Q_1 adds 1, from 00 to 01, 10, 11, and finally returns to 00, and when Q_i (i=1,2) changes from 1 to 0 a carry pulse is generated, having a good agreement with Fig. 22 (b). Q_1 and "Carry 1" have a repetition rate 1/2 of the clock; whereas Q_2 and "Carry 2" have a repetition rate 1/4 of the clock. The counter can therefore be used as an all-optical frequency divider.

In principle, by cascading n counter stages it is possible to demonstrate n-bit binary counter which can count from 0 to 2^n-1. However, the cascadability of this scheme is limited by the signal degradation of carry pulses, which mainly comes from the accumulated ASE noise of SOA. To evaluate the signal degradation of carry pulses quantitatively, Q-factor measurement has been carried out. The Q-factors of Q_1 and Q_2 are 16.1 and 19.9 respectively, only determined by the properties of two latches. The Q-factor of "Carry 1" is 17.0, while exploiting an ASE pedestal suppression technique, we obtained "Carry 2" pulse with Q-factor of 15.0, only slightly lower than "Carry 1". These values confirm the good cascadability of this scheme.
In the experiment the operation speed is limited to hundreds of kHz. Since the AND gates are all based on nonlinear effects in the SOA, which have very fast dynamics, the operation speed limitation is mainly due to the switching-on of the SR latch, reported also in Fig. 23 (d). This time depends on the cavity length of fiber ring lasers, and in our setup each ring is about 40m due to the discrete fiber pigtailed implementation. Again, photonic integration is a feasible solution to reduce the cavity length to the range of millimeters, shortening the transition time to <100 ps, and making GHz operation speed possible.

6. Ultra-fast SOA-based all-optical flip-flop

An all-optical flip-flop based on two coupled ring lasers presents a fast falling edge (as fast as the input pulse rising edge), but a slow rising edge (several round-trip times), which mainly limits the flip-flop operating speed for optical packet switching. In this paragraph, using two SOA-based optical NOT logic gates and two identical slow flip-flops, we obtain an optical flip-flop with ultra-fast transition times for both rising and falling edges (Malacarne et al., 2008). The experimental setup is shown in Fig. 24, while the operating principle is described in Fig. 25. Flip-flop 1 is controlled by reset and assistant pulses whereas flip-flop 2 is controlled by assistant and set pulses. Exploiting a 10GHz pattern generator we produce a 16ps-edge pulsed sequence with a pulse-width of 1μs and a repetition rate of 50KHz. Such a wide pulse has been set in order to maintain the gain saturation level into the ring laser to be quenched for several round trip time, allowing to reach a lasing steady condition. The reset pulse is delayed by 10μs (T_{d1}) with respect to the set pulse whereas the assistant pulse is delayed by 15μs (T_{d1}+T_{d2}) with respect to the set pulse. As shown in Fig. 25, a set pulse is firstly injected into ring 3 switching off signal B. Secondly, a reset pulse is injected into ring 1 switching off signal A. Then two assistant pulses are injected into ring 2 and ring 4 simultaneously. They switch off ring 2 and ring 4, switching on ring 1 and ring 3 respectively. Consequently, signals A and B are switched on at the same time. As pointed out above, both signals A and B have a fast falling edge, but a slow rising edge. Exploiting the optical NOT logic gate 1, signal A is inverted in order to obtain signal C, which therefore presents a fast rising edge and a slow falling edge. Since signals A and B are switched on by two assistant pulses simultaneously, the slow falling edge of signal C is almost synchronized with the slow rising edge of signal B, and when they are added together, the slow edges compensate each other in terms of intensity profile. This way, signal D (the sum of signals B and C) has a fast rising edge due to signal C and a fast falling edge coming from signal B. The wavelengths of signals A, B and C are 1550nm, 1558.2nm and 1557.4nm respectively, thus signal D is made of two different wavelengths, as highlighted in Fig. 24, and a tunable filter with -3dB bandwidth of 4.5nm is used to filter and equalize these two wavelength components. Using NOT logic gate 2, we invert signal D and thus obtain signal E, at the same time convert it to one single wavelength λ_E=1560nm. Signal E is switched on and off by the set and reset pulses respectively, showing fast rising and falling edges.

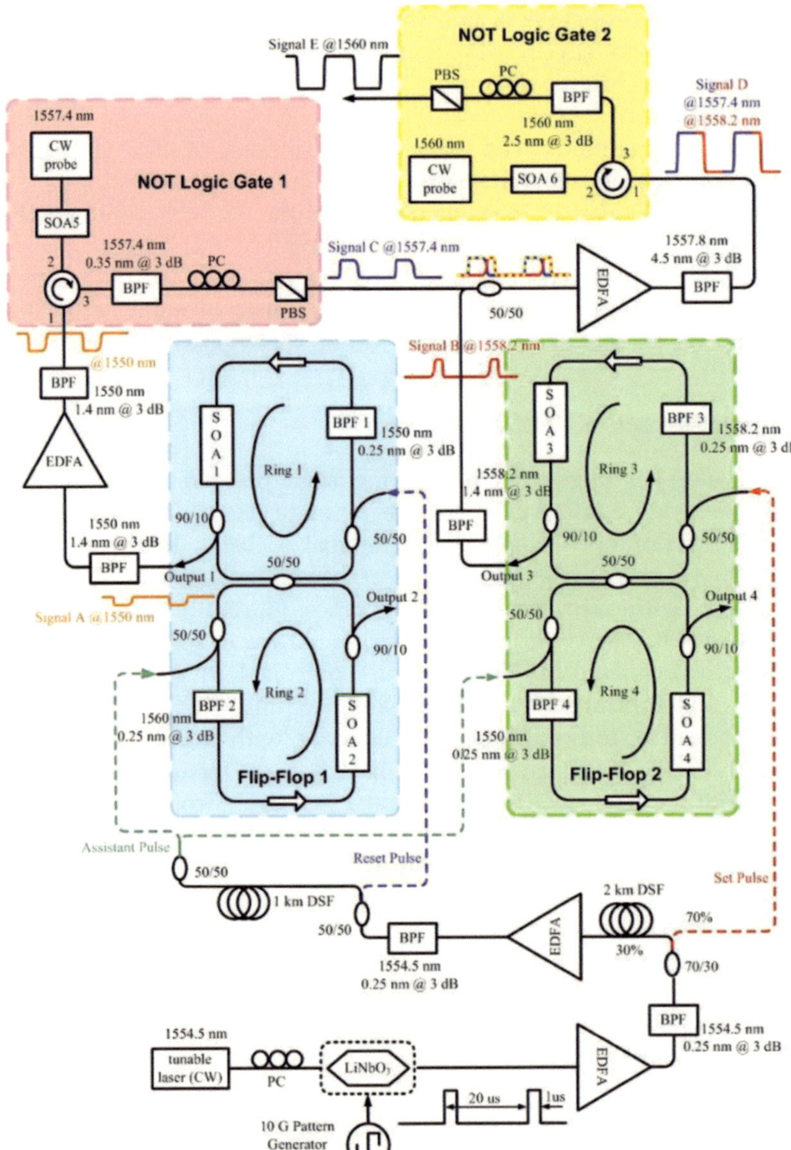

Fig. 24. Experimental setup of the ultra-fast all-optical flip-flop. PC: polarization state controller. Signal A is inverted by NOT logic gate 1 obtaining signal C and added with signal B. Signal D (B+C) is inverted by NOT logic gate 2 obtaining signal E.

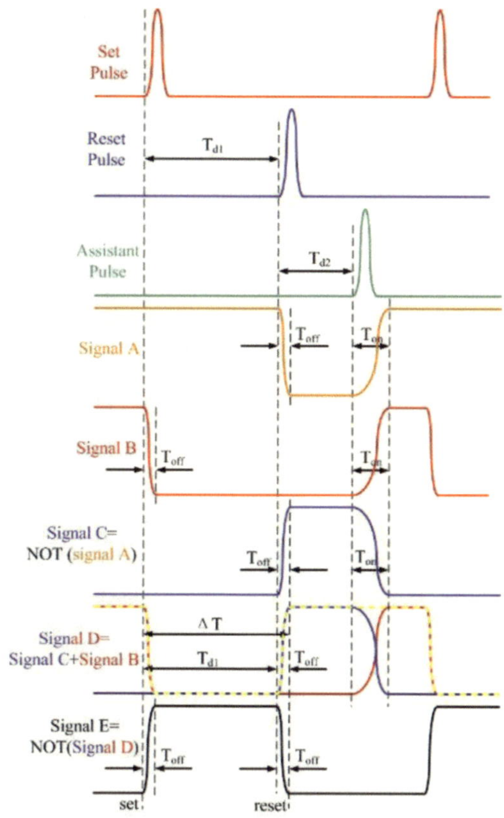

Fig. 25. Working principle of the ultra-fast all-optical flip-flop. Signal A, B, C, D and E. T_{d1}: delay between set and reset pulses; T_{d2}: delay between reset and assistant pulses; T_{on}: rising time; T_{off}: falling time; $\Delta T=T_{d1}+T_{off}$.

The optical NOT logic gates are implemented exploiting cross gain modulation (XGM) in SOAs. Concerning NOT logic gate 1, in SOA 5 a CW probe light counter-propagates with respect to signal A. The gain of SOA 5 is modulated by the intensity profile of signal A through XGM. In particular, when signal A has a low input power, the gain provided by SOA 5 for the CW probe will be high, whereas when signal A has a high power the CW probe will experience a lower gain. Ultimately the CW probe undergoes the gain variations obtaining the inversion of signal A, i.e. signal C.

Signals from A to E are shown in Fig. 26. Since the slow edges of signals B and C do not have a linear behavior, their sum gives rise to a residual peak during the high level of signal D. After NOT logic gate 2 this dynamic is suppressed because of the gain saturation level of SOA 6. CW probe power injected into SOA 6 has been set in order to optimize its saturation level (as CW probe injected into SOA 5). Exploiting input set and reset pulsewidths of 1μs with edge time of 16ps, signal E presents rising and falling times of 18.8ps and 21.9ps respectively, as shown in Fig. 26 (b) and (c) (measured with a total bandwidth of 53GHz), preserving a contrast ratio of 17.5dB. It is possible to obtain a higher contrast ratio just decreasing the CW probe signal powers in SOA 5 and SOA 6, reducing their gain saturation level, with the drawback of slower switching times (Berrettini, 2006, a). Moreover, integrated coupled ring lasers would experience a round trip time in the ps range (instead of 100ns as in our experiment), allowing to use an injected pulsewidth in the ps range too.

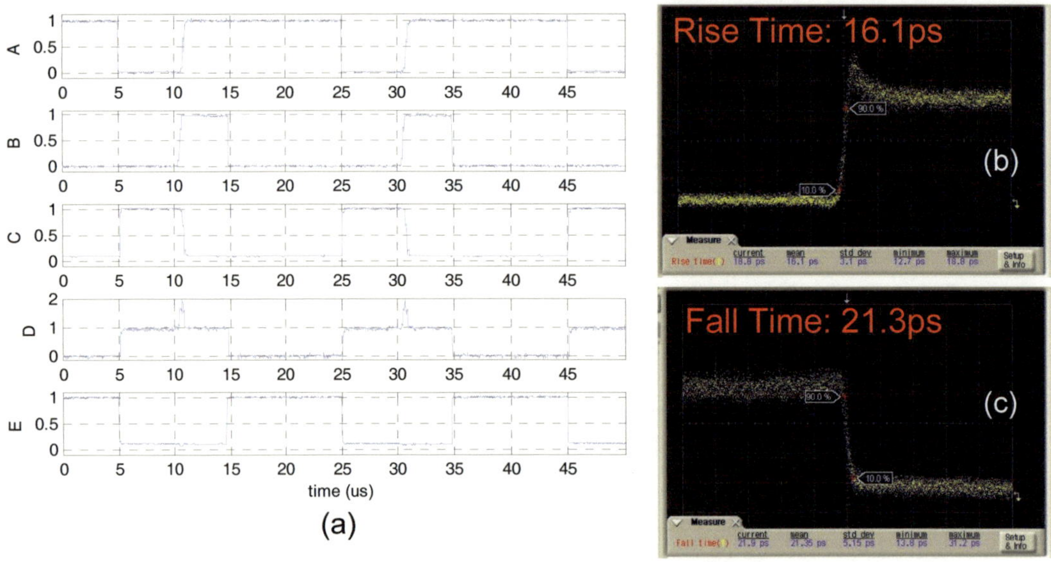

Fig. 26. (a): Signal A, B, C, D and E. Signal C = NOT (signal A); signal D=signal B + signal C; signal E = NOT (signal D). (b)-(c): Signal E rising (a) and falling (b) edges.

7. 10Gb/s switching operation with no bit loss exploiting the ultra-fast all-optical flip-flop

Fast dynamics (rising and falling times of 20ps) and high extinction ratio (17.5dB) make the ultra-fast all-optical flip-flop suitable to be exploited to control a 2×2 SOA-based all-optical switch (Berrettini, 2006, b).

The experimental setup is shown in Fig. 27. The switching operation is based on XGM effect in two different SOAs. Depending on the high or low intensity level of the control signal (pump), in one SOA the gain is strongly reduced while the other SOA is not saturated. The two input signals are generated by splitting a single 10Gb/s Non-Return-to-Zero (NRZ) continuous data stream. The stream is generated by modulating a CW laser at λ_{IN}=1550nm by means of a Mach Zehnder modulator driven by a 10Gb/s pattern generator running in (2^{31}-1)-long PRBS mode. At the same time the ultra-fast flip-flop output is used as pump signal of the optical switch and controls the switch state (BAR or CROSS). The inverted pump signal needed for switching operation is obtained within the optical switch block through signal inversion by means of XGM in an SOA. The data streams average power at the switch inputs are set to -7dBm, while the high pump level is 11.5dBm. We have chosen continuous data streams instead of packet traffic to demonstrate and point out that it is possible to obtain a switching operation without any bit loss, exploiting the 20ps-fast dynamics of the flip-flop. Indeed, as can be observed in Fig. 28, we can confirm a fast switching operation (faster than the 10Gb/s single bit edge), connecting only input 1 (disconnecting input 2) of the switch and visualizing output 1 on a sampling oscilloscope, switching the output data signal on and off within one bit time.

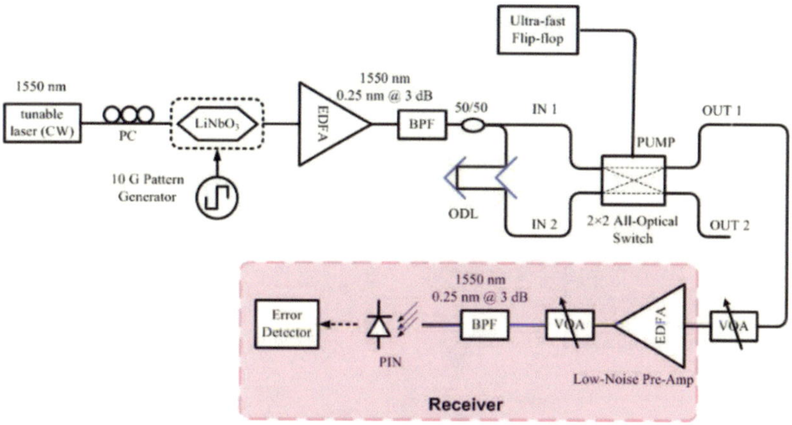

Fig. 27. All-optical switching operation experimental setup using a 2×2 SOA-based optical switch controlled by the ultra-fast all-optical flip-flop.

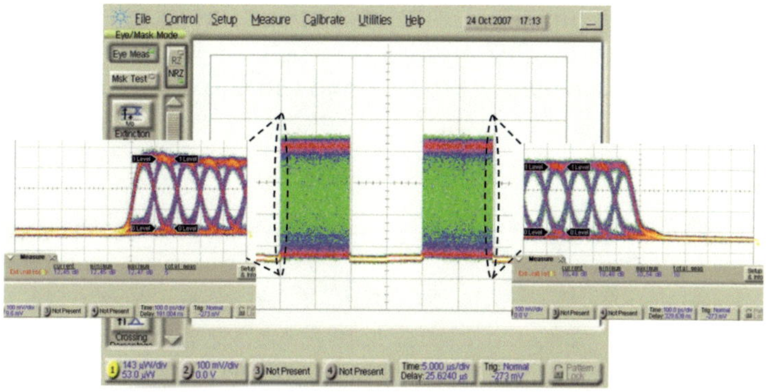

Fig. 28. Output 1 of the 2×2 all-optical switch, when just input 1 is connected (input 2 is disconnected). Insets shows the fast switching-on and switching-off transitions.

Contrast ratio between switched on and switched off signal is about 14dB. This way we avoid any distorted transition bit between switched on and switched off output signals, and vice-versa. Connecting both inputs 1 and 2 of the switch, high or low intensity level of the input pump signal sets the switch in BAR or CROSS state. During BAR state, input 1 of the switch is routed to output 1 (and input 2 is routed to output 2), while during CROSS state input 2 is routed to output 1 (and input 1 is routed to output 2). Fig. 29 (left) shows both input data eye-diagrams and output 1 eye-diagrams in BAR and CROSS configurations, measured by a wide-band photodiode and a sampling oscilloscope. As it can be noticed, the output signal is not affected by pattern effects, showing clearly open eye-diagrams, confirming the effectiveness of the scheme.
Right: (right) shows the BER measurements at output 1 of the switch, in both BAR and CROSS configurations. The used receiver is composed by an optical pre-amplifier with 5dB

noise figure, followed by a VOA, a BPF and a photo-receiver, whose input power is kept constant (by means of the VOA) at -16.7dBm in order to avoid thermal noise. As shown in Right:, making a comparison with the back-to-back case, the maximum penalty at BER=10^{-9} is about 1dB, making the switch driven by the ultra-fast all-optical flip-flop suitable for cascaded schemes.

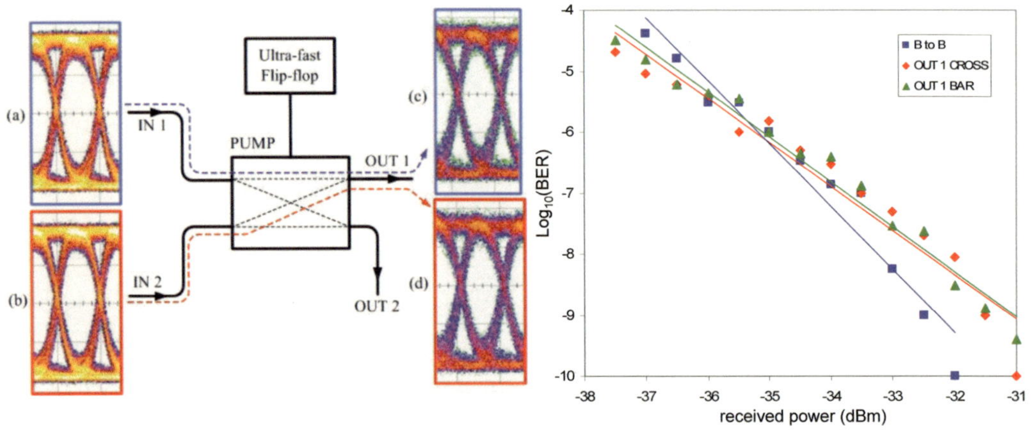

Fig. 29. Left: eye-diagrams of inputs 1 and 2 data frames (a)-(b) and output 1 in BAR (c) and CROSS (d) configurations of the 2×2 all-optical switch. Right: BER curves in the back-to-back (B to B) case and at switch output 1 in BAR and CROSS configurations.

8. References

Asakawa, K. (2007). PC-SMZ-Based All-Optical Flip-Flop Switch: PC-FF, *Proceedings ICTON 2007*, paper Mo.C2.6.

Barman, A.D.; Debnath, S.; Scaffardi, M.; Poti, L.; Bogoni, A. (2007), Modelling and implementation of photonic digital subsystem for bit comparison, *Proceedings of Photonics in Switching*, USA, San Francisco, TuB2.4.

Benner, A.F.; Bowman, J.; Erkkila, T. (1990). Digital optical counter using directional coupler switches, *Applied Optics*, vol. 30, n. 29, 4179-4189.

Berrettini, G.; Simi, A.; Malacarne, A.; Bogoni, A.; Poti, L. (2006). Ultrafast Integrable and Reconfigurable XNOR, AND, NOR, and NOT Photonic Logic Gate, *IEEE Photon. Technol. Lett.*, vol. 18, no. 8, pp. 917-919.

Berrettini, G.; Lauri, E.; Ghelfi, P.; Bogoni, A.; Potì L. (2006). Ultra-Fast Integrable 2x2 All-Optical Switch, *Proceedings of ECOC 2006*, France, Cannes , We3.P.102.

Bogoni, A.; Andriolli, N.; Scaffardi, M.; Berrettini, G.; Meloni, G.; Malacarne, A.; Porzi, C.; Castoldi, P.; Poti, L. (2007). A Synchronous All-optical 160 Gb/s Photonic Interconnection Network, *Proceedings of Optical Fiber Communication Conference*, USA, Anaheim , March 25-29.

Buczek, C.J.; Freiberg, R.J.; Skolnick, M.L. (1971). CO_2 regenerative ring power amplifiers, *J. Appl. Phys.*, vol. 42, pp. 3133–3137.

Dorren, H.J.S.; Hill, M.T.; Liu, Y.; Calabretta, N.; Srivatsa, A.; Huijskens, F.M.; de Waardt, H.; Khoe G.D.; (2003). Optical packet switching and buffering by using all-optical signal processing methods, *IEEE J. Lightwave Technol.*, vol. 21, n. 1, pp. 2-12.

Feuerstein, R.J.; Soukup, T.; Heuring, V.P. (1991). 100-MHz optical counter that uses directional coupler switches, *Optics Letters*, vol. 16, n. 20, 1599-1601.

Herrera, J.; Tangdiongga, E.; Liu, Y.; Hill, M.T.; McDougall, R.; Poustie, A.; Maxwell, G.; Ramos, F.; Marti, J.; de Waardt, H.; Khoe, G.D.; Koonen, A.M.J.; Dorren, H.J.S. (2007). 160-Gb/s All-Optical Packet-Switching With In-Band Filter-Based Label Extraction and a Hybrid-Integrated Optical Flip-Flop, *IEEE Photon. Technol. Lett.*, vol. 19, no. 13, pp 990-992.

Hill, M.T.; de Waardt, H.; Khoe, G.D.; Dorren, H.J.S. (2001), All optical flip-flop based on coupled laser diodes, *IEEE J. Quantum Electron.*, vol. 37, no. 3, pp. 405-413.

Hill, M.Y.; Dorren, H.J.S; de Vries, T.; Leijtens, X.J.M; den Besten, J.H.; Smalbrugge, B.; Oei, Y.S.; Binsma, H.; Khoe, G.D.; Smit, M.K. (2004) A fast low-power optical memory based on coupled micro-ring lasers, *Nature*, vol. 432, pp. 206-208.

Hill, M.T.; de Vries , T. ; Dorren, H.J.S. ; Leijtens, X.J.M.; van Zantvoort, J.H.C.; den Besten, J.H.; Smalbrugge, E.; Oei, Y.S.; Binsma, J.J.M.; Khoe, G.D.; Smit, M.K. (2005). Integrated Two-State AWG-Based Multiwavelength Laser, *IEEE Photon. Technol. Lett.*, vol. 17, no. 5, pp. 956-958.

Huybrechts, K.; Morthier, G.; Baets, R. (2008). Fast all-optical flip-flop based on a single distributed feedback laser diode, *Optics Express*, v. 16, n. 15, pp. 11405-11410.

Keyes, R.W. (2001). Fundamental limits of silicon technology, *Proceedings of IEEE*, vol.89, n. 3, pp. 227-239.

Liu, Y.; Hill, M.T.; Calabretta, N.; de Waardt, H.; Khoe, G.D.; Dorren, H.J.S. (2003) Three-State All-Optical Memory Based on Coupled Ring Lasers, *IEEE Photon. Technol. Lett.*, vol. 15, no. 10, pp. 1461-1463.

Liu, Y.; Tangdiongga, E.; Hill, M.T.; van Zantvoort, J.H.C.; Smalbrugge, E.; de Vries, T.; Binsma, H.; Oei, Y.S.; Leijtens, Y.S.; Smit, M.K.; Khoe, M.K.; Dorren H.J.S. (2005). All-Optical Switching of 80 Gb/s Data Packets using a Wavelength Converter Controlled by a Monolithically Integrated Optical Flip-Flop", *Proceedings of ECOC 2005*, vol. 6, pp. 27-28, 25-29.

Liu, Y.; McDougall, R.;Hill, M.T.; Maxwell, G.; Zhang, S.; Harmon, R.; Huijskens, F.M.; Rivers, L.; Dorren, H.J.S.; Poustie, A. (2006). Packaged and hybrid integrated all-optical flip-flop memory", *Electronics Letters*, vol. 42, no. 24, 1399-1400.

Malacarne, A.; Bogoni, A.; Potì, L. (2007). Erbium–Ytterbium-Doped Fiber-Based Optical Flip-Flop", *IEEE Photon. Technol. Lett.*, vol. 19, no. 12, pp. 904-906.

Malacarne, A.; Wang, J.; Zhang, Y.; Barman, A.D.; Berrettini, G.; Potì, L.; Bogoni, A. (2008). 20 ps-Transition Time All-Optical SOA-based Flip-Flop used for Photonic 10 Gb/s Switching Operation without any Bit Loss, *IEEE Photon. Technol. Lett.*, vol. 14, no. 3, pp. 803-815.

Poustie, A.; Manning, R.J.; Kelly, A.E. (2000). All-optical binary counter, *Optics Express*, vol. 6, n. 3, 69-74.

Ramos, F.; Kehayas, E.; Martinez, J.M.; Clavero, R.; Marti, J.; Stampoulidis, L.; Tsiokos, D.; Avramopoulos, H.; Zhang, J.; Holm-Nielsen, P.V.; Chi, N.; Jeppesen, P.; Yan, N.; Tafur Monroy, I.; Koonen, A.M.J.; Hill, M.T.; Liu, Y.; Dorren, H.J.S.; Van Caenegem, R.; Colle, D.; Pickavet, M.; Rip ti, B. (2005). IST-LASAGNE: towards all-optical label swapping employing optical logic gates and optical flip-flops, *IEEE J. Lightwave Technol.*, vol. 23, n. 10, pp. 2993-3011.

Takenaka M., Nakano, Y. (2003). Multimode interference bistable laser diode, *IEEE Photon. Technol. Lett.*, vol. 15, no. 8, pp. 1035–1037.

Takenaka, M.; Raburn, M.; Nakano, Y. (2005). All-Optical Flip-Flop Multimode Interference Bistable Laser Diode, *IEEE Photon. Technol. Lett.*, vol. 17, no. 5, pp. 968-970.

Trita, A.; Latorre Vidal, M.J.; Zanola, M.; Javaloyes, J.; Mezosi, G.; Bragheri, F.; Cristiani, I.; Sorel, M.; Scirè, A.; Balle, S.; Giuliani, G. (2009). Commutazione tutto-ottica ultraveloce in un flip-flop set-reset monolitico basato su laser ad anello, *Proceedings of Fotonica 2009*, Pisa, paper B3_1.

Van, V.; Ibrahim, T.A.; Absil, P.P.; Johnson, F.G.; Grover, R.; Ho, P.-T. (2002). Optical signal processing using nonlinear semiconductor microring resonators, *IEEE J. Select. Top. Quant. Electron.*, vol. 8, n. 3, pp. 705-713.

Wang, J.; Zhang, Y.; Malacarne, A.; Potì, L.; Bogoni, A. (2008). SOA Fiber Ring Laser-Based Three-State Optical Memory, *IEEE Photon. Technol. Lett.*, vol. 20, n. 20, 1697-1699.

Wang, J.; Berrettini, G.; Meloni, G.; Potì, L.; Bogoni, A. (2009). All-Optical clocked D type flip-flop exploiting SOA-based optical SR latch and logic gates, *accepted for publication at Photonics in Switching 2009*, ThI 1-6, Pisa, Italy.

Wang, J.; Meloni, G.; Berrettini, G.; Potì, L.; Bogoni, A. (2009). All-Optical Counter Based on Optical Flip-Flop and Optical AND Gate, *accepted for publication at ECOC 2009*.

16

Germanium Photodetector Technologies for Optical Communication Applications

Kah-Wee Ang, Guo-Qiang Lo and Dim-Lee Kwong
*Institute of Microelectronics, Agency for Science, Technology & Research (A*STAR)*
Singapore

1. Introduction

Converging computing and communication capabilities on single ship platform is becoming increasingly important to keep up with the performance roadmap known as Moore's Law (Kimerling et al., 2006). Today, data transmissions at a bit-rate of 10 Gb/s over long distance makes photonic interconnection an easier approach to implement than electrical interconnect. At this data rate, the conventional copper solution has begun to encounter extreme challenges related to power consumption and reach (Wada et al., 2002). Moreover, the growing issues in electro-magnetic interference, signal cross-talk, and heavier weight make it an inferior approach for high bandwidth applications (Gunn, 2006). To keep up with the scaling of interconnect bandwidth, an alternative solution makes use of optical interconnect technology to meet the ever increasing bit rate requirement of data communication. Over the past decades, conventional optical components were typically made of exotic III-V compound materials such as gallium-arsenide (GaAs) and indium-phosphide (InP) due to their excellent light emission and absorption properties. Unfortunately, compound-semiconductor devices are generally too complicated to process and costly to implement in optical interconnects.

In search for a cost-effective solution, Si photonic emerges to hold great promise for its inexpensive material and its compatibility with current complementary metal-oxide-semiconductor (CMOS) process technology (Soref, 2006). However to make silicon photonic communication a reality, several key technological challenges have to be addressed. Inferior optical properties of Si have thus far the major show-stopper to preclude the development of a key active photonic component needed to perform optical to electrical encoding. Very recently, germanium (Ge) has attracted growing interest for the realization of high performance photodetector due to its favourable absorption coefficient (Hartmann et al., 2004). However, Ge can be a challenging material to integrate in a CMOS environment for its low thermal budget constraint and its large lattice mismatch of ~4.2% with Si (Luan et al., 1999). High defect densities seen in the Ge-on-Silicon-on-Insulator (Ge-on-SOI) epitaxial film could induce unfavourable carrier recombination process that would degrade the detector's quantum efficiency.

In this chapter, the current development of optical detection technologies on silicon photonics platform is reviewed. The discussion first begins with the development of Ge-on-

SOI hetero-epitaxy process technology. The approach, based on the low temperature pseudo-graded silicon-germanium buffer engineering, is adopted to relieve the large lattice mismatch of ~4.2% between the two heterostructure materials. This enables a high quality Ge epitaxy film with low threading dislocation density to be grown on Si. The subsequent sections of this chapter deal with the state-of-the-art Ge photodetector technologies. We begin with the discussion on the designs of evanescent-coupled Ge *p-i-n* photodetector featuring an integrated SOI micro-waveguide. Performance metrics in these detectors in terms of dark current, responsivity, and bandwidth are evaluated. The mechanism accountable for the leakage generation in such device and its dependence on the applied electric field strength are elucidated. Factors limiting the detector speed performance and guidelines to enable bandwidth scaling are also discussed.

This chapter also aims to discuss the demonstration of Schottky barrier engineered Ge photodetector featuring metal-semiconductor-metal (MSM) configuration. The problem and mechanism responsible for the generation of high leakage current in such a detector are dealt with. Novel concepts adopted to address this issue through Schottky barrier modulation are presented. The approaches are based on bandgap engineering as well as Fermi level de-pinning by segregating valence mending adsorbate near the metal/germanium interface. The recent technological breakthrough in employing all Group-IV based materials to realize high gain-bandwidth product Ge/Si avalanche photodetector (APD) is presented next. The fabrication process and the design of Ge/Si APD featuring separate-absorption-charge-multiplication (SACM) configuration are discussed. We conclude the chapter with a summary providing the readers with the comparative views on the performance metrics of the various Ge-based photodetector schemes.

2. Hetero-Epitaxy of Germanium on Silicon

The key challenge to high quality germanium (Ge) epitaxy growth on silicon (Si) rests with the huge lattice mismatch between the two heterostructure materials. The existence of ~4.2% lattice mismatch strain has been shown to give rise to two major issues: (1) high densities of threading dislocations and (2) rough surface morphology due to 3D Stranski-Krastanov (SK) growth. Both of these defects present much concerns for the generation of high leakage current which would compromise the efficiency of a photodetector. Strategies proposed in the literature to overcome these challenges vary to a large extent. One most intuitive approach is to grow a silicon-germanium (SiGe) layer by compositionally grading its Ge concentration up to 100%. Using low-energy plasma enhanced chemical vapour deposition, Oh et al. (2002) showed that for every 10% increase in the Ge mole fraction, a linearly graded SiGe buffer of ~1μm thickness is required. This results in a need to grow a relatively thick SiGe buffer layer of 10μm before a low dislocation density Ge epilayer can be deposited, which imposes much difficulties for process integration.

In an effort to further reduce the thickness of these active layers, Huang et al. (2004) proposed another approach based on the optimization of two thin SiGe buffer layers with varying Ge concentration. In such approach, a 0.6μm thick $Si_{0.45}Ge_{0.55}$ buffer was first grown and then followed by an intermediate $Si_{0.35}Ge_{0.65}$ buffer with a thickness of 0.4μm. An in-situ annealing for 15 min at 750°C was subsequently performed to further reduce the dislocation density before the growth of a 2.5μm thick Ge epilayer at a process temperature of 400°C. Through this approach, it allows the threading dislocations to be trapped at the hetero-interfaces. This enables a significant reduction in the dislocation density of the as-grown Ge epilayer, thereby improving the detector's dark current performance.

In yet another approach, Colace et al. (1999) proposed a direct hetero-epitaxy growth of Ge on Si through the use of a low temperature thin SiGe buffer layer (a few 10nm). The insertion of such thin buffer avoids the occurrence of 3D SK growth, and allows the misfit dislocations to be concentrated at the hetero-interfaces. However such approach requires a cyclic annealing process to be carried out at both high and low temperature (900°C/780°C) to reduce the threading dislocation density within the Ge active film. Using a similar cyclic thermal annealing approach, Luan et al. (1999) had also demonstrated a significant improvement in both the surface roughness and the dislocation density. When combined with the selective area growth, an average threading dislocation density as low as 2.3×10^6 cm^{-2} was achieved. However, the needs for a high temperature post-epitaxy Ge anneal with long cycle time present a major concern for CMOS implementation.

In this work, selective epitaxial growth of Ge on silicon-on-insulator (SOI) was performed using an ultra-high vacuum chemical vapor deposition (Ang et al., 2010) reactor. Unlike the conventional approaches, a thin pseudo-graded SiGe buffer with a thickness of ~20nm is proposed in this study to relieve the large lattice mismatch stress between the two heterostructure materials (Fig. 1). The Ge mole fraction within the SiGe buffer is compositionally graded from 10% to ~50%. The precursor gases used for the SiGe growth comprise of diluted germane (GeH_4) and pure disilane (Si_2H_6). A relatively thin Ge seed layer of ~30nm is subsequently grown on the SiGe buffer at a process temperature of 370°C. The use of a low temperature growth is intended to suppress adatoms migration on Si and thus prevents the formation of 3D SK growth, which allows a flat Ge surface morphology to be achieved. Upon obtaining a smooth Ge seed layer, the epitaxy process temperature is then increased to ~550°C to facilitate faster epitaxy growth to obtain the desired Ge thickness. Using this approach, high quality Ge epilayer with a thickness of up to ~2μm has been demonstrated, along with the achievement of threading dislocation density as low as $\sim 10^7$ cm^{-2} without undergoing any high temperature cyclical thermal annealing step.

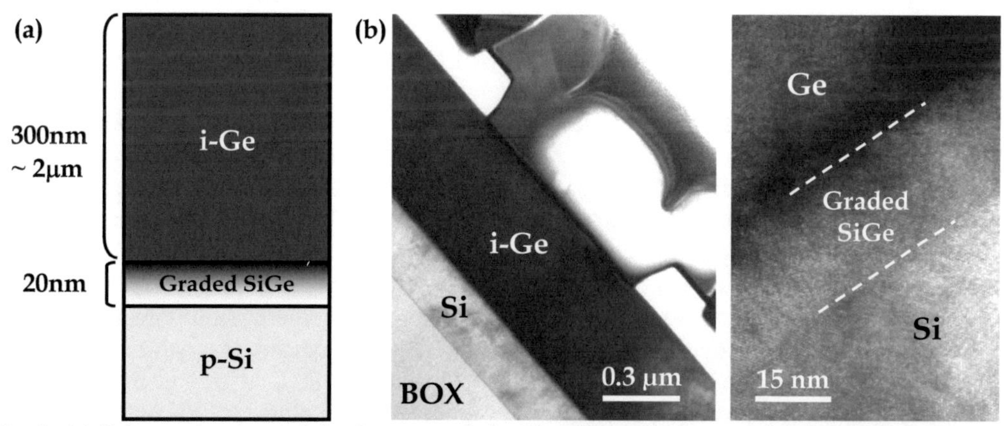

Fig. 1. (a) Schematic view of the layer stack for the direct hetero-epitaxy growth of Ge on Si. (b) High resolution TEM micrograph showing the effectiveness of a pseudo-graded SiGe buffer in reducing the threading dislocation density within the Ge epilayer.

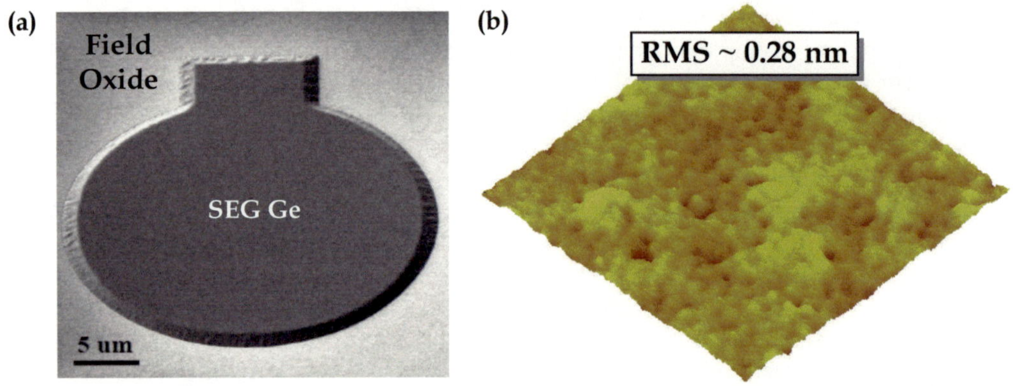

Fig. 2. (a) Scanning electron microscopy (SEM) image showing the achievement of excellent Ge epitaxy rowth and selectivity on SOI substrate. (b) Excellent Ge surface roughness of ~0.28nm was achieved, as determined using atomic force microscopy (AFM).

In addition, selective area growth of Ge on Si has also been developed using a cyclical deposition and etch back approach. In each deposition cycle, the Ge growth time is carefully optimized to avoid exceeding the incubation time needed for Ge seeds to nucleate on the dielectric film. After every Ge deposition cycle, a short etch back process using chlorine (Cl_2) precursor gas will then be introduced to remove possible Ge nucleation sites on the dielectric. This allows a highly selective Ge epitaxy process to be developed, along with the achievement of excellent surface roughness of ~0.28nm (Fig. 2).

3. High Performance Germanium *p-i-n* Photodetector

Due to its poor absorption coefficient as inherited by the large bandgap energy, silicon (Si) has been known to be prohibitive for the realization of photodetector that is capable of performing efficient optical detections at wavelengths commonly used in optical fiber communication (1.31~1.55μm). This can be addressed by introducing a new material with a smaller bandgap energy such as germanium (Ge) to provide favorable optical absorption property at these wavelengths. Recent research progress made in the photodetector technology development has clearly shown that Ge is attracting growing interest as the preferred photo-absorbing material due to its much higher absorption coefficient as compared to that of Si (Hartmann et al., 2002). In addition, its compatibility with current CMOS fabrication technology makes it an attractive material to enable the demonstration of high performance near-infrared photodetector (Soref, 2006).
However, the long absorption length in Ge at 1.55μm wavelength renders it difficult to meet the high quantum efficiency requirement for a surface illuminated photodetector. Despite the use of a ~1μm Ge active layer, Colace et al. (2007) reported the achievement of a maximum responsivity of merely 0.2A/W at 1.55μm wavelength. One way to overcome this constraint requires the growth of a thick Ge epilayer to enable full absorption at this wavelength. Unfortunately, hetero-epitaxy of Ge with such thickness imposes much process integration challenge such as high threading dislocation densities that would lead to increased leakage current and thus degrade the receiver sensitivity. An alternative approach to relax this requirement makes use of a waveguide based photodetector (Yin et al., 2007).

Be leveraging on the detector length, one would be able to achieve enhanced photo-detection efficiency and thus responsivity improvement. In addition, the bandwidth performance of the photodetector can also be simultaneously optimized by tweaking the Ge thickness to reduce the carrier transit time delay.

In this section, the different designs of waveguide integrated Ge photodetector featuring *p-i-n* configuration are discussed. The performance metrics such as dark current, responsivity, and bandwidth in these devices are evaluated and compared.

3.1 Ge-on-SOI Photodetector Designs and Fabrication

Two types of evanescent butt-coupled Ge-on-SOI photodetector design are shown in Fig. 3. Photodetectors featuring vertical *p-i-n* (VPD) and lateral *p-i-n* (LPD) configurations were fabricated, with Ge active layer selectively grown and integrated on a SOI micro-waveguide. For the VPD, the p+ and n+ junctions were formed in the Si and Ge regions, respectively, with the intrinsic region thickness ($t_{i\text{-}Ge}$) co-defined by the Ge thickness (t_{Ge}) and the thickness of the n+ implant region [Fig. 3(a)]. The width *W* and length *L* of this VPD design is 8μm and 100μm, respectively. For the LPD design, both the p+ and n+ junctions were formed in the Ge region, with the width of the intrinsic region ($w_{i\text{-}Ge}$) determined by defining the spacing of these alternating contacts [Fig. 3(b)]. Note that the width *W* and length *L* of this LPD design is 20μm and 100μm, respectively.

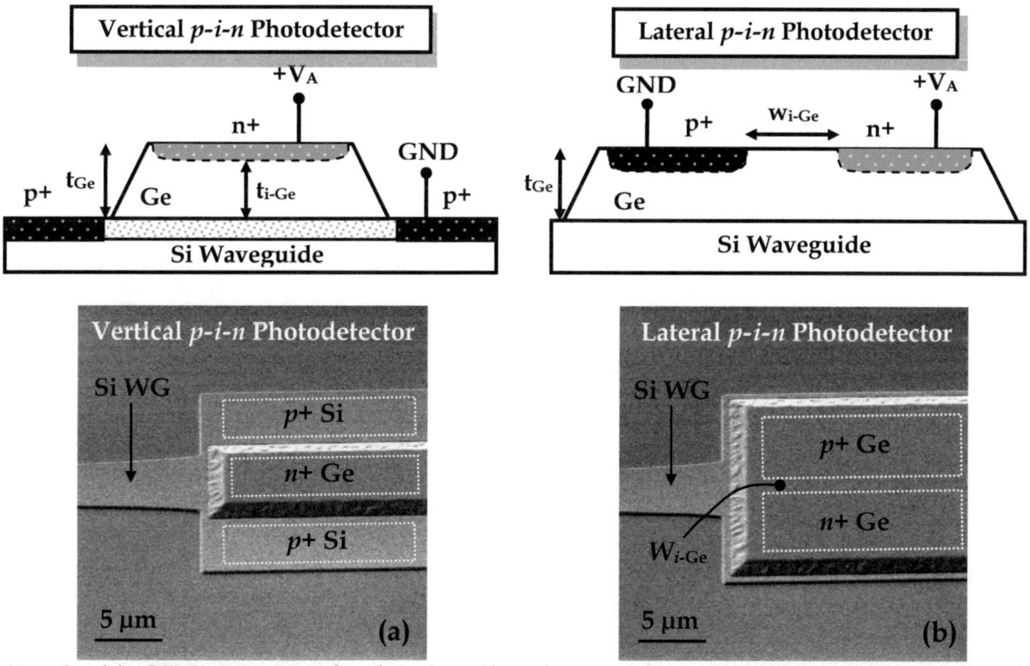

Fig. 3. (a) SEM micrograph showing the design of an evanescent butt-coupled Ge photodetector featuring vertical *p-i-n* configuration. (b) Ge photodetector design with a lateral *p-i-n* configuration.

By employing a waveguide-coupled design, the requirement for thick Ge epilayer to enhance the responsivity performance can be greatly relieved as one can leverage on the detector length to improve the optical absorption efficiency. As a result of the difference in the refractive index between Si and Ge, the incident photon traveling in the SOI micro-waveguide will be up-coupled into the Ge absorbing layer to allow optical signal to be encoded into its electrical equivalent efficiently. The insertion of a thick buried oxide (BOX) of ~2μm serves to confine the optical mode within the core of the channel waveguide so as to prevent leakage into the underneath Si substrate.

The fabrication process of the waveguide integrated Ge photodetector begins with the use of SOI substrate with a starting overlying Si thickness of ~220nm and a buried oxide (BOX) thickness of ~2μm. Channel waveguide with a nano-taper featuring a width of ~200nm was first formed by anisotropic dry etching to obtain a smooth sidewall profile for enabling low waveguide propagation loss. Ion implantation employing boron species was selectively done to form the Si anodes in a VPD detector. A moderately high p-type doping concentration was carefully chosen for the anode formation to ensure low series resistance while not impact the quality of the as-grown Ge epitaxy film. High dose p+ contact implant was subsequently performed and dopants are activated using rapid thermal anneal at 1030°C for 5s to obtain good Si ohmic contacts. After depositing a 600Å thick field oxide as passivation layer, a combination of anisotropic dry etch and followed by wet etch approach was adopted to preserve the top Si surface quality from possible damage by the reactive ion etching process. Hetero-epitaxy of Ge was then selectively grown in an ultra-high vacuum chemical vapor deposition (UHVCVD) epitaxy reactor. The selective Ge epitaxy process commenced with the deposition of a low temperature pseudo-graded silicon–germanium buffer (~20nm) and followed by a Ge seed layer with a thickness of ~30nm. A cyclical deposition and etch back approach was then used to raise the Ge thickness to ~500nm. Due to the achievement of low defects level within the Ge film, the high temperature post-epitaxy Ge anneal typically used for defects annihilation was skipped to reduce the overall thermal budget. High dose selective phosphorous and boron implants were then performed in a LPD and annealed at 500°C for 5 min to form good n-type and p-type Ge ohmic contacts, respectively. After the deposition of inter-layer dielectric (ILD), contact and metallization were subsequently done to complete the device fabrication. Fig. 3(a) and 3(b) show the top-view scanning electron microscopy (SEM) images of the VPD and LPD detectors, respectively.

3.2 Dark Current Characteristics

Dark current plays a vital role in affecting the shot noise (I_S) in a photodetector according to the following expression

$$I_S^2 = 2\, q\, B\, (I_D + I_B) \tag{1}$$

where q denotes the elemental charge, B the bandwidth, I_D the dark current of the detector, and I_B the photocurrent due to background radiation. Under a carefully controlled situation, I_B is usually small and can therefore be neglected. However, thermal generation and/or defects-assisted tunnelling current due to strong electric field give rise to considerable dark current which degrades the shot noise and thus affects the signal-to-noise (SNR) ratio.

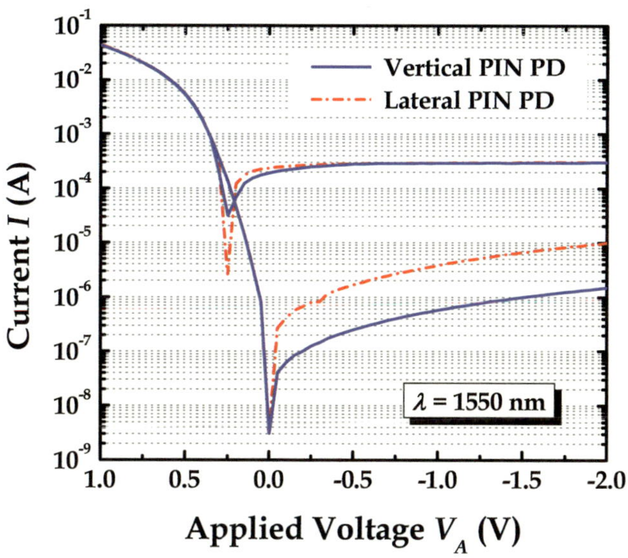

Fig. 4. The current-voltage characteristics of the VPD and LPD detectors measured under dark and illumination conditions.

Fig. 4 examines the current-voltage characteristics of the VPD and LPD detectors under dark and illumination conditions. Excellent rectifying characteristics were demonstrated in both the detectors, showing a forward-to-reverse current ratio of ~4 orders of magnitude. For a given applied bias of −1.0V, the dark current (I_{dark}) in a VPD was measured to be ~0.57μA (or ~0.7nA/μm^2), which is below the typical 1.0μA generally considered to be the upper limit for high speed receiver design. On the other hand, the dark current performance in a LPD showed a much higher I_{dark} value of ~3.8μA (or ~1.9nA/μm^2). In order to better understand the factors which affect the dark current density (J_{Dark}), let us review the expression that governs the leakage generation in a semiconductor diode

$$J_{Dark} = \frac{qn_i d}{\tau_{eff}} \tag{2}$$

where q denotes the elemental charge, n_i the intrinsic carrier density, d the depletion layer width, and τ_{eff} the effective carrier lifetime. Clearly, an increase in the depletion layer width leads to a detrimental impact on the dark current performance, which could possibly explain the higher I_{dark} experienced in a LPD detector. In addition, it is also important to note that I_{dark} exhibits a strong dependence on the effective carrier lifetime which is controlled by both the lifetime associated with the Shockley-Read-Hall recombination (τ_{SRH}) and the carrier drift time across the space charge region (τ_{drift}) as follow

$$\frac{1}{\tau_{eff}} = \frac{1}{\tau_{SRH}} + \frac{1}{\tau_{drift}} \tag{3}$$

$$\tau_{SRH} = \frac{1}{\sigma v_{th} N_{TD} N_D} \tag{4}$$

$$\tau_{drift} = \frac{d}{\mu E} \tag{5}$$

where σ denotes the capture cross section, v_{th} the carriers thermal velocity, N_{TD} the threading dislocation density, N_D the density of recombination centres, μ the low-field carrier mobility and E the electric field strength. It is obvious that the reverse dark current density should be proportional to the defects density within the Ge epilayer and a careful control of the epitaxy quality would be important to reduce the leakage current. Furthermore, an increase in the applied reverse bias has also resulted in an aggravated dark current degradation, which elucidates that I_{dark} has a strong dependence on the electric field strength. A further analysis on this phenomenon will be covered in a later discussion (see section 3.5).

3.3 Responsivity Characteristics

The responsivity ($\Re$) of a photodetector can be described using the following expression

$$\Re = I_{Photo} / P_{Opt} = \eta q / h\upsilon \tag{6}$$

where I_{Photo} denotes the photocurrent, P_{Opt} the incident optical power, η the quantum efficiency, q the elemental charge, h the Planck constant, and υ the frequency. In general, the incident photons which are absorbed in germanium generate electron-hole pairs which will be collected as photocurrent under applied electric field. This photocurrent is linearly dependent on the incident optical power before saturation is reached. Moreover, alike the quantum efficiency, the responsivity of the detector should be wavelength dependent. Hence, the responsivity of a detector will be significantly higher at wavelength where the photon energy enables electron-hole pair generation through direct transition.

In order to compare the responsivity performance between the VPD and LPD, optical measurements were performed by injecting an incident photon with a wavelength of 1550nm into the SOI micro-waveguide. The typical optical propagation loss in our SOI micro-waveguide under TE polarization mode is ~2dB/cm. No coupler was integrated with the Si waveguide and the incidence light was coupled through a single mode lensed fiber directly into the Si nano-taper. For an incident light power of ~300μW, optical measurements showed that both the VPD and LPD detectors achieved a comparable photocurrent level at high applied biases beyond -1.0V. Fig. 5 compares the responsivity of the detectors as a function of the applied voltages. It is interesting to note that the vertical PIN detector demonstrated a lower responsivity as compared to the lateral PIN detector for biases below −0.5 V. This could possibly be due to an enhanced carrier recombination process at the high density of defect centres near the Ge-Si heterojunction. This is set to compromise the absolute photocurrent value of a vertical PIN detector under low field influence. However with an increased electrostatic potential across the depletion layer, the photo-generated carriers can be assisted across the space charge region with enhanced mobility before they can recombine at these recombination centres.

For an applied bias larger than −1.0 V, a comparable responsivity was measured for both the vertical and lateral PIN detectors. Despite that the metallurgical junction is separated by merely 0.8μm, a lateral PIN detector showed a high absolute responsivity of ~0.9 A/W. The possible mechanisms accountable for such high responsivity could be attributed to the following reasons. Firstly, under high reversed bias, the intrinsic Ge region (i.e. between and beneath the metallurgical junction) was simulated to be totally depleted, as confirmed using MEDICI device simulator. When photon is absorbed to produce electron and hole pairs, the fringe field beneath the metallurgical junction enables the generated carriers to be collected by the electrode as photocurrent. Secondly, optical simulation shows that more than 80% of the incidence light travelling in the SOI waveguide is absorbed within the first 25 μm of the detector. Hence by leveraging on the long absorption length design, nearly all incidence photons will be expected to contribute to the achievement of high responsivity.

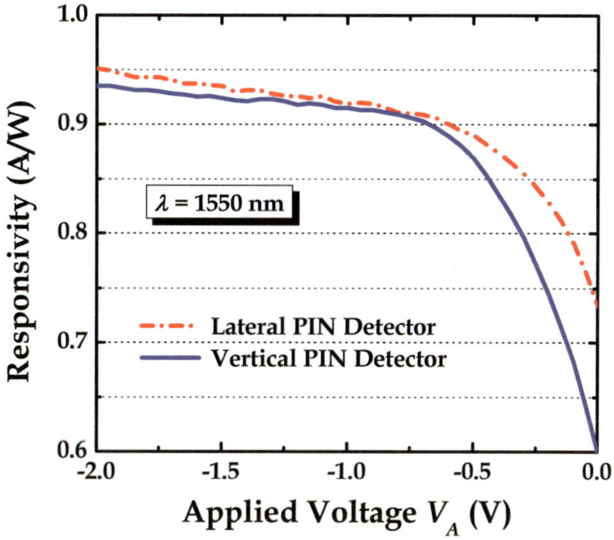

Fig. 5. Responsivity as a function of applied voltages for both the VPD and LPD detectors measured at a wavelength of 1550nm.

3.4 Impulse Response Characteristics

The impulse response of a photodetector is limited by both the carrier transit time ($f_{Transit}$) and the *RC* time constant (f_{RC}) which can be modelled using the following expressions

$$f_{Transit} = \frac{0.45\upsilon_{sat}}{t_{i-Ge}} \quad (7)$$

$$f_{RC} = \frac{1}{2\pi RC} \quad (8)$$

where υ_{Sat} denotes the carrier saturation velocity, *d* the depletion layer width, and *RC* the resistances and capacitances associated with the detector and its peripheral circuitry.

As described in these equations, the factors governing the fundamental response time limit of a detector lie with (1) the carrier drift time across the space charge region, and (2) the device junction capacitance. Drift of carriers is influenced by the electric field applied across the space charge region and can be expressed using

$$\upsilon_{drift} = \mu.E \tag{9}$$

where μ denotes the carrier mobility, and E the electric field. Clearly, increasing the electric field would enhance the drift velocity across the space charge region until a saturation velocity is reached. Sze (1981) showed that the carrier saturation velocity in germanium is on the order of ~10^7 cm/s. In addition, the higher carrier mobility in Ge as compared to that of Si makes it a material of choice to enable the realization of high speed photodetector. On the other hand, decreasing the junction capacitance would allow one to achieve a reduced RC time constant. The junction capacitance (C_j) which arises from the ionized donors (N_D) and acceptors (N_A) is expressed as follow

$$C_j = \varepsilon A\left[\frac{q}{2\varepsilon(V_o - V)}\frac{N_D N_A}{N_D + N_A}\right]^{1/2} = \frac{\varepsilon.A}{W} \tag{10}$$

where ε denotes the material permittivity, A the cross-sectional area of the detector, and W the depletion layer width. Intuitively, reducing the device area and increasing the depletion layer width are both beneficial to reduce the junction capacitance. However, adopting the former approach could lead to a compromise in the responsivity performance as the effective area for optical absorption is decreased. The latter approach in enlarging the depletion width could serve to enhance the response time as a result of lower junction capacitance. However, further increase in the depletion width would eventually lead to a degraded transit time across the space charge region. Therefore, an optimization of the RC time constant and the carrier transit time will be crucial in determining the overall bandwidth performance of the detector, as dictated using the following expression

$$f_{3dB} = \sqrt{\frac{1}{1/f_{Transit}^2 + 1/f_{RC}^2}} \tag{11}$$

In order to investigate the factors affecting the speed performance of the VPD and LPD detectors used in this study, impulse response measurements were performed at a photon wavelength of 1550nm. A pulsed laser source having a 80fs pulse width was used in the measurements. Both the detectors were characterized using microwave probes and the impulse responses were captured with a high speed sampling oscilloscope. Fig. 6 shows that a VPD detector achieved a smaller full-width-at-half-maximum (FWHM) pulse width of ~24.4ps as compared to that of LPD detector with a slightly larger FWHM of ~28.9ps. This could be attributed to the smaller depletion layer width design in a VPD detector which reduces the carrier transit time. The FWHM pulse width is related to the bandwidth and can be used as a metric to gauge the speed performance of the detectors. By performing a fast Fourier transform of the impulse responses, a −3dB bandwidth of ~11.3 and ~10.1 GHz were achieved in the VPD and LPD detectors, respectively.

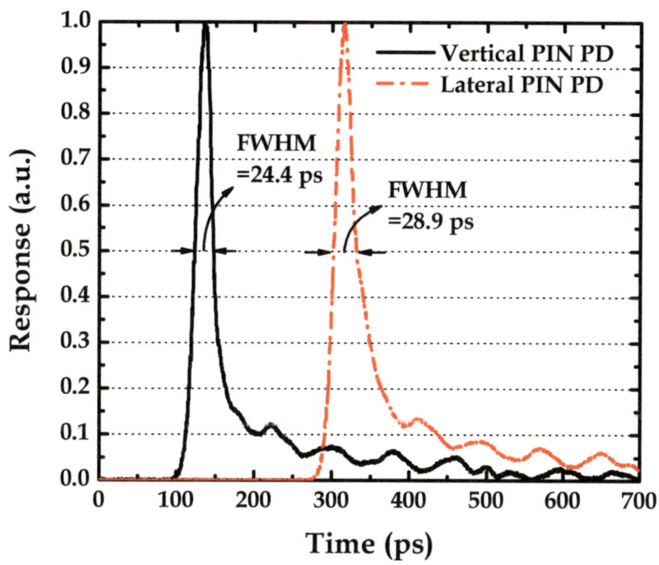

Fig. 6. Impulse responses of the VPD and LPD detectors measured at a wavelength of 1550nm. A smaller FWHM pulse width of ~24.4 ps was achieved in a VPD as compared to a LPD, which corresponds to a -3dB bandwidth of ~11.3 GHz.

The detector's bandwidth can be further evidenced by the eye patterns measurements (PRBS 2^7-1) done by directly connecting the output of the detector to the 50 Ω electrical input of the DCA. Fig. 7 shows that high sensitivity and low-noise photo-detection up to a bit-rate of 8.5Gb/s can be achieved by the VPD detector. The clean eye patterns clearly illustrate the low noise property of the detector. Higher speed measurements are possible through further scaling of the detector geometry to reduce the device capacitance.

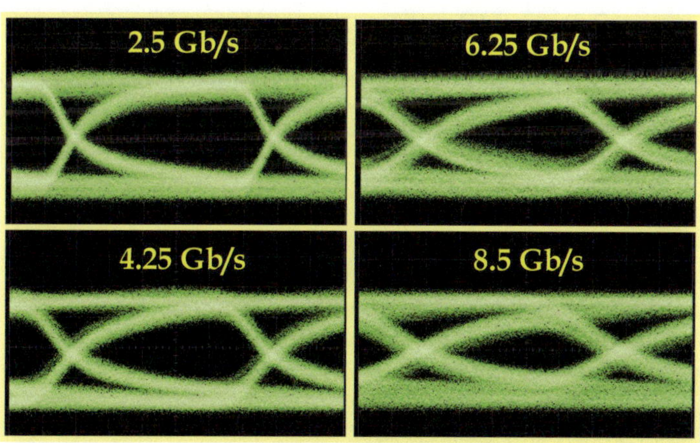

Fig. 7. Eye patterns (PRBS 2^7-1) measurements of the VPD at a bias of −1.0 V. The detector demonstrated high sensitivity and low-noise photodetection up to a bit-rate of 8.5Gb/s. The low noise property of the detector can be clearly illustrated by the clean eye patterns.

Fig. 8(a) shows the total device capacitance measured as a function of the applied reverse voltages for different detector geometry. Obviously, elongating the detector length increases the capacitance due to a larger effective detector area, as predicted in equation (9). Moreover, in the presence of increased reverse bias, the capacitance drops drastically and nearly plateaus off at high voltage regime. This is attributed to the widening of the depletion layer width as the applied bias is raised. Further increase in the bias across the alternating p+ and n+ junctions would lead to a total depletion of the intrinsic-Ge region, which causes the device capacitance to reach a saturation level.

The theoretical modelling results of the RC-time constant and the transit-time bandwidth are plotted in Fig. 8(b). Reducing the depletion spacing enhances the transit-time bandwidth performance significantly, but it leads to a degraded RC-time bandwidth. To overcome this limitation, one could scale the detector length to achieve lower capacitance for bandwidth improvement.

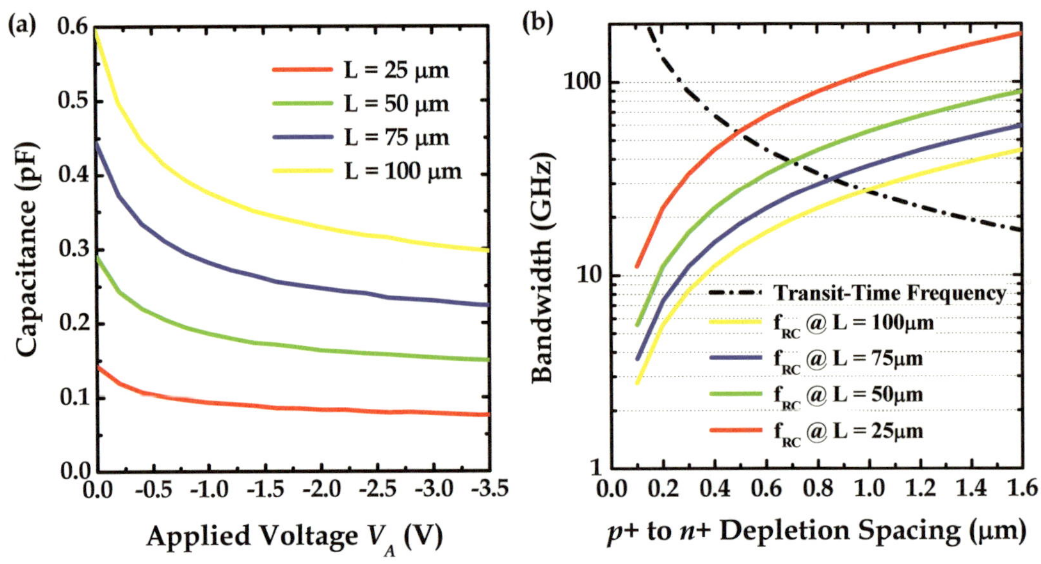

Fig. 8. (a) Measured capacitance as a function of applied voltage for Ge photodetector with various detector lengths. (b) Downscaling of detector length results in bandwidth enhancement due to a reduced device capacitance.

3.5 Impact of Band-Traps-Band Tunneling on Dark Current Generation

In order to gain insight into the leakage mechanism for the Ge detectors, an activation energy analysis of the dark current I_{Dark} was performed (Ang et al., 2009). In this analysis, the I_{Dark} can be modelled using the following functional form

$$I_{Dark} = BT^{3/2} e^{-E_a/kT} (e^{eV_a/2kT} - 1) \qquad (12)$$

where T denotes the temperature, Va the applied bias, and Ea the activation energy responsible for the leakage generation. Fig. 9(a) plots the bias dependence of dark current in a Ge p-i-n photodetector measured at increasing temperature range from 303K to 373K. As can be observed from this figure, temperature has a significant impact on I_{Dark}. Increasing the operating temperature and the applied bias are found to result in a higher I_{Dark}. The applied bias V_a used in this measurement ranges from -1.5V to +0.2V.

A semi-log plot of I_{Dark} as a function of 1/kT at various reverse bias voltages is shown in Fig. 9(b). A straight line fitting to this plot yields a gradient which corresponds to the activation energy E_a. At a fixed reversed bias of -0.5V, the extracted E_a is observed to be nearly half of the Ge bandgap energy E_g, which elucidates that the dark current mechanism is dominated by the Shockley-Read-Hall (SRH) process via deep levels in the Ge forbidden gap (Shockley & Read, 1952). This is not all unexpected as the large lattice mismatch between the two heterostructure materials could result in a Ge epitaxial film with high threading dislocation density. The existence of such defects has been shown to affect the effective carrier lifetime which causes an increase in the leakage current, as discussed earlier in Section 3.2.

Interestingly, the analysis also reveals a strong dependence of E_a on the electric field strength, as illustrated in Fig. 10(a). Increasing the field intensity across the depletion region is shown to result in a decreasing E_a responsible for the leakage generation. This in turn, contributes to an exponential increase in the dark current trend. The mechanism responsible for this is that electric field enlarges the band-bending which leads to an enhanced electrons and holes tunnelling from the resulting deep levels into the respective conduction and valence bands, thereby contributing to the dark current degradation. For instance, an increase in the electric field strength from 17kV/cm to 25kV/cm enhances the dark current from 0.27μA to 0.44μA, showing more than 60% I_{Dark} degradation.

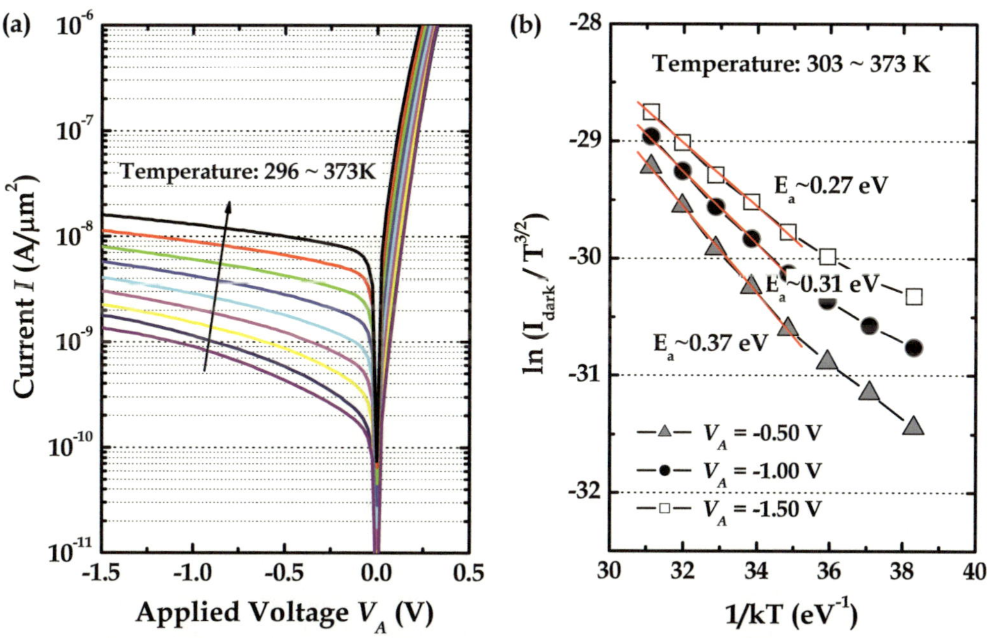

Fig. 9. (a) Plot of dark current characteristics as a function of applied bias for a Ge p-i-n photodetector with increasing temperature range from 303 K to 373 K. (b) An extraction of the activation energy for leakage generation as a function of applied bias.

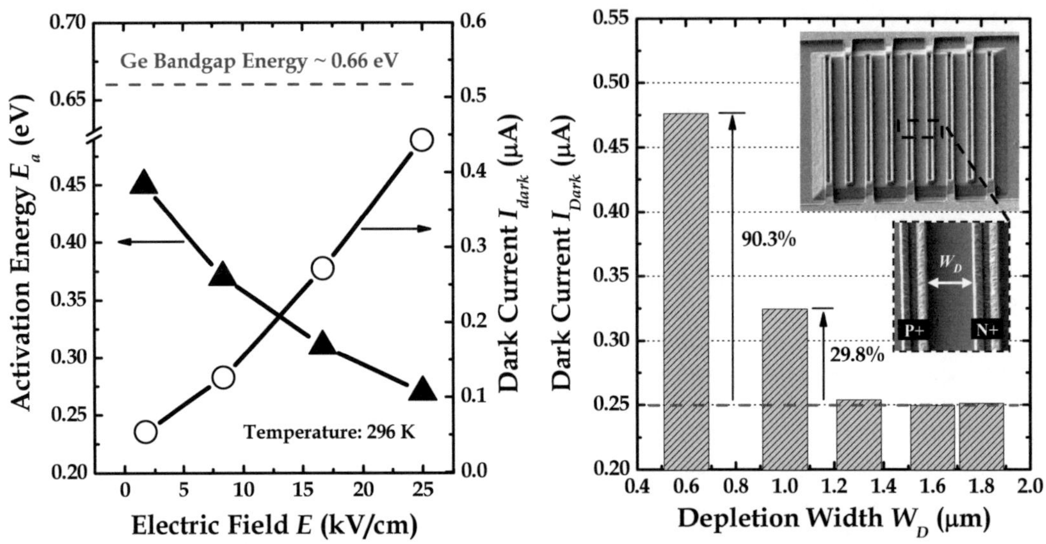

Fig. 10. (a) The activation energy E_a for leakage generation shows a strong dependence on the applied electric field, giving rise to a decreasing E_a trend with increasing field intensity. A reduced E_a at high field regime leads to an increased dark current generation. (b) Plot of dark current dependence on depletion width W_D of a Ge p-i-n photodetector. Scaling W_D leads to significantly higher dark current generation.

Such band-traps-band tunneling effect is observed to demonstrate a strong dependence on the depletion width W_D which separates the p+ and n+ metallurgical junctions. In this analysis, the area of the Ge detector is kept constant at 23 x 23 μm^2 while the W_D is varied from 0.6~1.8μm. To avoid a difference in the contact area due to a variation of intrinsic Ge width, the metal geometry is also altered such that the total metal contact region is comparable for all designs. Note that a reduced W_D is often desirable from the perspective of enhancing the detector's bandwidth performance. Fig. 10(b) shows that an aggressive downsizing of W_D results in a significant dark current degradation. Specifically, a reduction of W_D from 1.3μm to 1.0μm increases the dark current density by ~29%, which is further aggravated to ~90% when W_D reaches 0.6μm.

The underlying mechanism responsible for such phenomenon can be explained using the band diagrams as shown in Fig. 11. When operated in the high field regime, enlarged band-bending results in a more prominent Ge bandgap narrowing for a detector with an aggressively scaled W_D [Fig. 11(a)]. As a consequence, the occurrence of electrons and holes tunnelling from the resulting mid-gap trap levels could be further enhanced in the presence of strong electric field, which accounts for the achievement of higher dark current over a detector with wide W_D [Fig. 11(b)]. It is also noteworthy to highlight that the dark current density begins to plateau for W_D > 1.3μm, which implies that the influence of band-traps-band tunnelling on the leakage generation becomes relatively less prominent for wider W_D. This finding suggests that a design trade-off needs to be considered in the course of scaling W_D for enabling bandwidth enhancement as it would lead to a more pronounced dark current degradation.

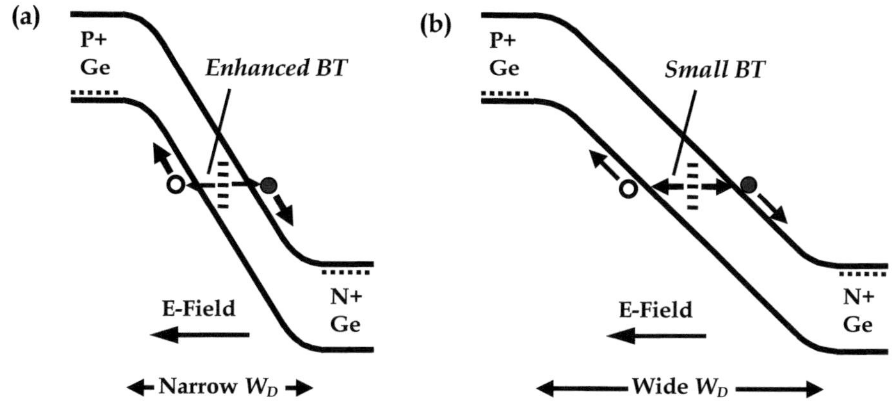

Fig. 11. Band diagrams illustrating the impact of scaling depletion width W_D on field-enhanced dark current generation. When operated at high field regime, enlarged band-bending results in a narrowing of Ge bandgap which enables electrons and holes tunnelling to occur via these defect centres. Such phenomenon is observed to become increasingly prominent for devices with (a) narrow W_D as compared to that with (b) wide W_D.

4. Schottky Barrier Engineered Germanium MSM Photodetector

In another photodetector scheme, a metal-semiconductor-metal (MSM) structure was utilized to leverage on the advantage of low capacitance and ease of process integration. However, high dark current issue experienced in these detectors imposes much concern for the achievement of poor signal-to-noise (SNR) ratio. This drawback would be further aggravated when a narrow bandgap material such as Ge is employed, where high dark current is predominantly attributed to the low hole Schottky barrier height as a result of Fermi level pinning near the valence band edge. Recent experimental demonstration showed that Ge MSM photodetector with an integrated SOI rib waveguide exhibited high dark current on the order of 150μA despite achieving impressive speed performance (Vivien et al., 2007). Such dark current level is way too high to be acceptable for high speed receiver design which typically tolerates a leakage current below 1.0μA.

This chapter aims to deal with this problem through the application of novel approaches to suppress the leakage current in Ge MSM photodetector. The concepts are based upon Schottky barrier modulation through bandgap engineering as well as Fermi level de-pinning by segregating valence mending adsorbate at the metal/germanium interface.

4.1 Schottky Barrier Modulation using Large Bandgap Material

The application of larger bandgap material for Schottky barrier modulation has been widely pursued to enable dark current suppression in Ge MSM photodetector. Oh et al. (2004) reported the fabrication of metal-Ge-metal photodetector featuring thin amorphous-Ge layer sandwiched between the metal and germanium interface to increase the Schottky barrier height. Using this approach, a substantial reduction of dark current by more than two orders of magnitude was achieved. Laih et al. (1998), on the other hand, adopted an amorphous-Si layer in a U-grooved metal–semiconductor–metal photodetector to enable dark current suppression by more than three orders of magnitude. In this work, a novel crystalline silicon-carbon (Si:C) epilayer was proposed for modulating the Schottky barrier height in a Ge MSM photodetector with an integrated SOI micro-waveguide.

The fabrication process begins with an 8-inch silicon-on-insulator (SOI) substrate with (100) surface orientation. The SOI substrate features a silicon body thickness of ~250 nm and a buried oxide thickness of ~1 μm. Si micro-waveguide was first formed by using anisotropic dry etching to achieve straight sidewall profile for enabling low propagation loss. After depositing a 120nm plasma enhanced chemical vapor deposition (PECVD) oxide as passivation layer, the Ge active regions were then patterned by reactive ion etching and cleaned with standard piranha solution (i.e. a mixture of sulfuric acid (H_2SO_4) with hydrogen peroxide (H_2O_2) for polymer removal. The wafers were subsequently cleaned with standard SC1 (NH_4OH : H_2O_2 : H_2O) and then subjected to a HF-last wet cleaning for oxide removal prior to the selective epitaxial growth of Ge in an ultra high vacuum chemical vapor deposition (UHVCVD) system. The epitaxy growth started with an in-situ baking in N_2 ambient at 800°C for native oxide removal and followed by the deposition of a ~5 nm thin Si buffer at 530°C. A thin SiGe buffer layer was then deposited to have a gradual transition from pure Si to pure Ge at the hetero-interface. A Ge seed layer with a thickness of ~30 nm was then grown using low temperature at 370°C before the growth of a ~300nm Ge epilayer at increased temperature. Precursor gases comprise of pure disilane Si_2H_6 and diluted germane GeH_4 (10% GeH_4 : 90% Ar) were employed for the hetero-epitaxy growth of SiGe and Ge layers. The defects density within the Ge epilayer was measured to be on the order of 10^7 cm^{-2} using etch-pit density (EPD) approach. Micro-Raman spectroscopy revealed a uniform distribution of residual tensile strain in the as-grown Ge film on Si substrate, which was attributed to the difference in the thermal expansion coefficient between Ge and Si during cooling.

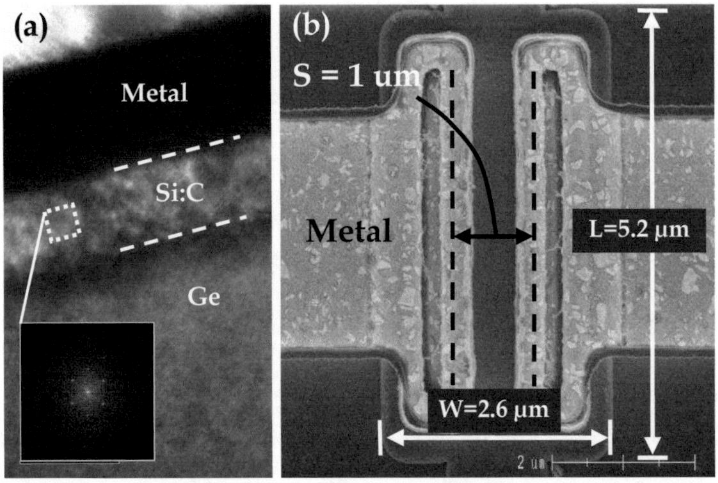

Fig. 12. (a) Silicon-carbon (Si:C) film is bi-dimensional and appears to be of good crystalline quality despite a substantial lattice mismatch between the Si:C barrier and the Ge epitaxial film. (b) SEM image of an evanescent coupled Ge-on-SOI MSM photodetector with an integrated Si micro-waveguide. The Ge detector features an effective device width *W* and length *L* of 2.6μm and 5.2μm, respectively. The metal contacts spacing *S* is ~1μm.

After contact hole patterning, a thin crystalline silicon-carbon (Si:C) epilayer of ~18nm was selectively deposited in the contact regions using disilane (Si_2H_6) and diluted mono-methylsilane (SiH_3CH_3) precursor gases. Such an optimum Si:C thickness was chosen based on the considerations for acting as a good barrier layer to suppress leakage current while achieving low defects density at the heterojunction. Chlorine (Cl_2) precursor gas was intermittently introduced to achieve selective epitaxial growth. The mole fraction of substitutional carbon incorporated in the Si:C film was measured to be ~1% based on the reciprocal lattice vector parameters obtained in X-Ray diffraction (Ang et al., 2007). Meanwhile, the total carbon concentration as obtained from SIMS analysis was found to be equal to ~1.3%, which means that around 0.3% of carbon was incorporated in the interstitial sites. Despite a substantial lattice mismatch, the Si:C layer is bi-dimensional and appears to be of good crystalline quality, as confirmed by the fast Fourier transform (FFT) diffractogram in Fig. 12(a). For comparison, Si:C epilayer was not deposited in a control sample. Metallization consisting of TaN/Al (250Å/6000Å) were subsequently deposited and patterned to complete the device fabrication. Fig. 12(b) shows the scanning electron microscopy (SEM) image of the evanescent coupled Ge-on-SOI MSM photodetector with an integrated Si micro-waveguide. The detector features an effective device length *L* and width *W* of 2.6 μm and 5.2 μm, respectively. The spacing *S* between the metal electrodes of the photodetector was lithographically defined to be ~1 μm.

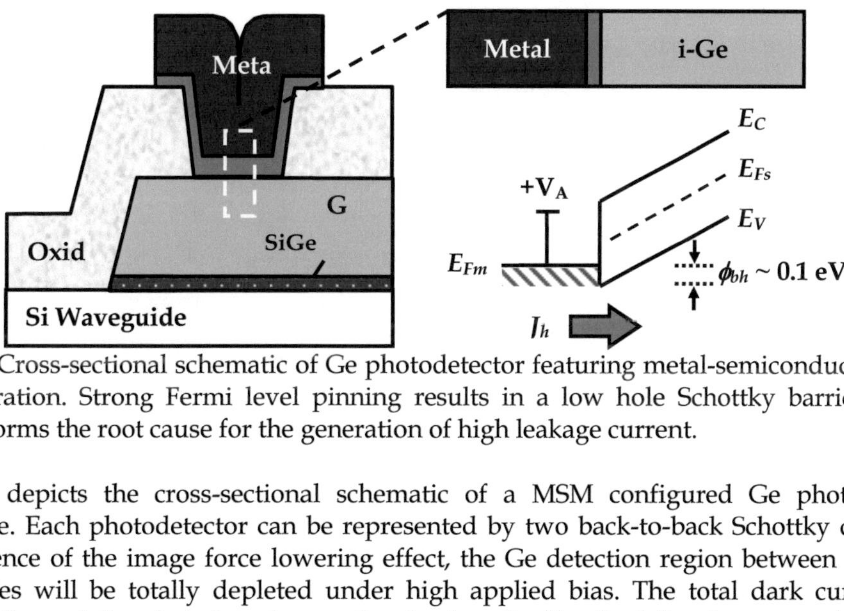

Fig. 13. Cross-sectional schematic of Ge photodetector featuring metal-semiconductor-metal configuration. Strong Fermi level pinning results in a low hole Schottky barrier height, which forms the root cause for the generation of high leakage current.

Fig. 13 depicts the cross-sectional schematic of a MSM configured Ge photodetector structure. Each photodetector can be represented by two back-to-back Schottky diodes. In the absence of the image force lowering effect, the Ge detection region between the metal electrodes will be totally depleted under high applied bias. The total dark current J_{Total} flowing through the photodetector can then be described by the following expression

$$J_{Total} = J_p + J_n = A_p^* T^2 e^{-q\phi_{bh}/kT} + A_n^* T^2 e^{-q\phi_{be}/kT} \tag{13}$$

where J_p (J_n) is the hole (electron) current injected from the anode (cathode), and A_p^* (A_n^*) is the Richardson's constant for hole (electron). Both the hole current and electron current are

observed to contribute to the dark current in the photodetector, which exhibit a strong dependence on the Schottky barrier for hole (ϕ_{bh}) and electron (ϕ_{be}), respectively. According to the Schottky-Mott theory on ideal metal-semiconductor system, the Schotkky barrier height (ϕ_B) can be determined from the difference of the metal work function (ϕ_m) and the electron affinity of the semiconductor (χ_S), i.e. $\phi_B=\phi_m-\chi_S$ (Tung, 2001). However, in practice, the presence of interface states has been shown to result in the Schottky barrier height less dependent on the metal work function. Strong Fermi level pinning feature of the metal/Ge or germanide/Ge junctions has been reported to contribute to high electron Schottky barrier ϕ_{be}, and thus leading to low ϕ_{bh} ~ 0.1eV (Dimoulas et al., 2006). As a result, hole injection is expected to dominate over electron injection in affecting the dark current of Ge MSM photodetector, as schematically illustrated in the band diagram of Fig. 13.
To circumvent this problem, a thin silicon-carbon (Si:C) Schottky barrier enhancement layer with bandgap energy E_g ~ 1.17eV (Soref, 1991) larger than that of Ge was inserted between the metal and Ge interface (Fig. 14). Due to Fermi level pinning away the valence band edge, an enhanced hole Schottky barrier height can be attained, as elucidated in the band diagram.

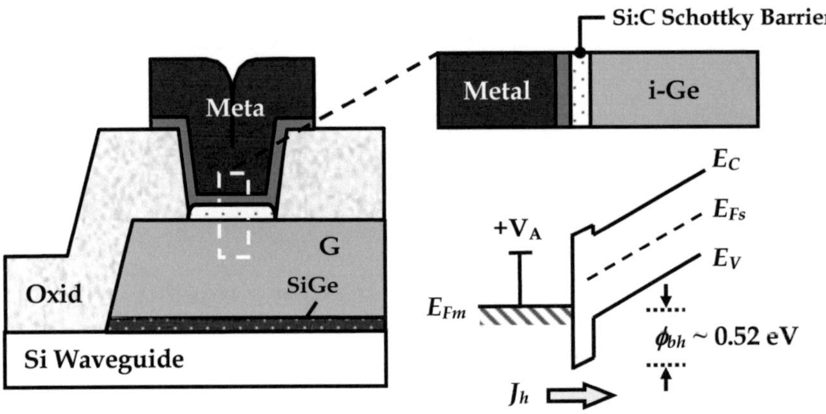

Fig. 14. Cross-sectional schematic of Ge MSM photodetector featuring silicon-carbon (Si:C) Schottky barrier enhancement layer. By inserting a large bandgap material between the metal and germanium interface, an enhanced hole Schottly barrier height can be achieved.

In order to accurately determine the enhancement of hole Schottky barrier due to the insertion of Si:C barrier layer, an activation energy measurement using low reverse bias voltage was employed. The temperature range used in the measurement varies from 303 K to 403 K, with an incremental increase of 10 K. Fig. 15 shows the Arrhenius plot for the extraction of activation energy under constant reverse bias V_R of 0.1V, according to the thermionic-emission model

$$\ln(\frac{I_R}{T^2}) = \ln(SA^*) - \frac{q\phi_{beff}}{kT} \tag{14}$$

where I_R denotes the reverse current, T the temperature, S the diode area, A^* the effective Richardson constant, q the elemental charge, k the Boltzmann constant, and $\phi_{b,eff}$ the effective Schottky barrier that includes a contribution from the image force barrier lowering. The temperature dependence of dark current was plotted in the inset of Fig. 15. An excellent fitting to the Arrhenius plot yields a gradient that is related to the Schottky barrier, showing a ϕ_{bh} of ~0.52eV above the valence band edge.

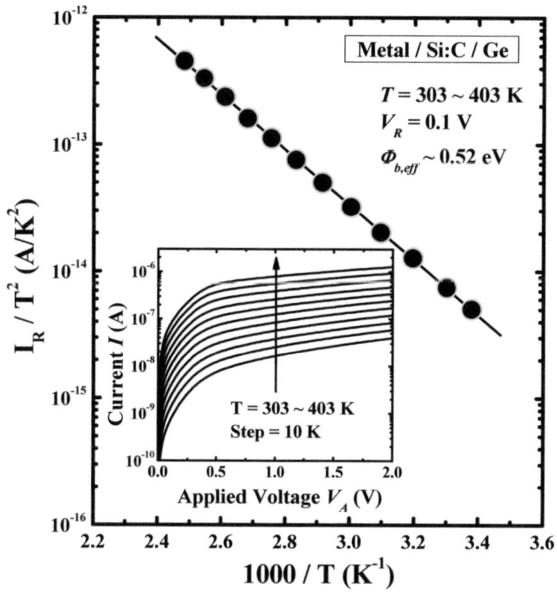

Fig. 15. Arrhenius plot for the extraction of activation energy under constant reverse bias of 0.1V according to the thermionic-emission model. The temperature dependence of dark current measurement was plotted in the inset.

Fig. 16(a) compares the dark current performance of both photodetectors with increasing reverse bias voltages. At a bias of 1.0V, high I_{Dark} of ~505.6µA (~984.2µA) was observed in a metal/Ge photodetector with a device width of 5.2µm (10.2µm). This I_{Dark} was substantially higher than the acceptable level of 1µA for typical receiver application. However, through an insertion of Si:C barrier layer between the metal/Ge junction, a significant dark current reduction by more than 4 orders of magnitude was achieved due to an enhanced hole Schottky barrier. At a bias of 1.0V, the I_{Dark} was measured to be ~11.5nA (~17.7nA) for a device width of 5.2µm (10.2µm), showing the effectiveness of Si:C barrier layer in suppressing the dark current of MSM photodetector. Fig. 16(b) plots the photo-response characteristics of Ge MSM photodetector with Si:C Schottky barrier layer. The optical measurement was performed by injecting an incident photon with a wavelength of 1550nm into the Si waveguide using a multimode lensed fiber. The incident light power is ~150µW. High photo-current on the order of ~10^{-4}A was demonstrated, leading to good internal responsivity and quantum efficiency of ~530mA/W and ~42.4%, respectively. These values were calculated based on an incident light power which excludes the fiber-to-waveguide coupling loss and the propagation loss components. Good signal-to-noise ratio of ~10^4 was also demonstrated in the MSM photodetector with Si:C barrier layer.

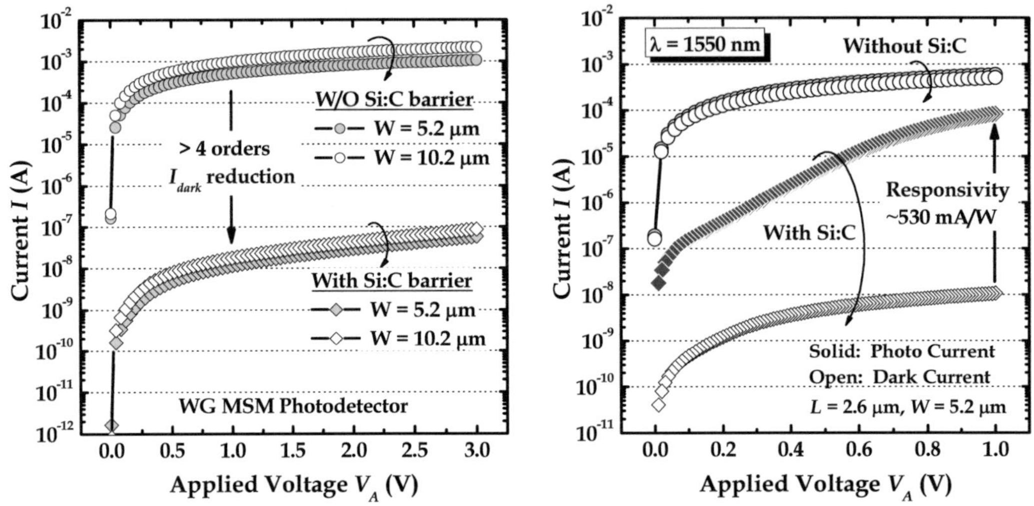

Fig. 16. (a) Dark current characteristics of Ge MSM photodetector with and without Si:C barrier layer. (b) Optical measurement results for both the detectors obtained at a photon wavelength of 1550nm.

A uniform spectral response of ~760mA/W was observed for wavelengths spanning from 1520-1560 nm, demonstrating an effective photo-detection for the entire C-band spectrum range [Fig. 17(a)]. In addition, the responsivities at wavelengths of 1570, 1590, and 1610 nm were measured to be 767, 602, and 290mA/W, respectively, suggesting an extended potential for L-band optical photo-detection applications. Fig. 17(b) plots the frequency response measured at a wavelength of 1550nm, as obtained from the Fourier transform of an impulse response. A -3dB bandwidth of 12GHz was achieved for a low applied bias of 1.0V.

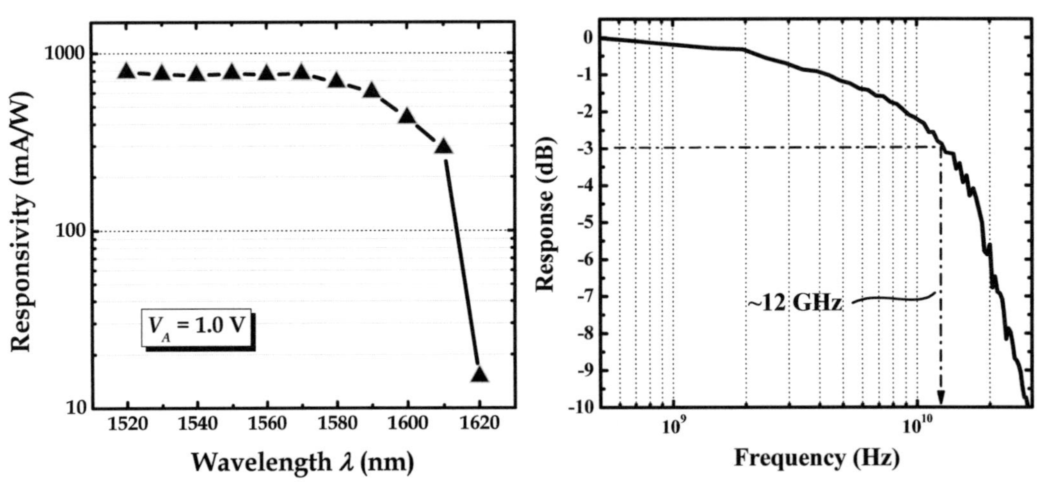

Fig. 17. (a) Spectral response as a function of wavelength for the novel MSM photodetector. (b) A -3dB bandwidth of ~12 GHz was achieved at a photon wavelength of 1550nm.

4.2 Schottky Barrier Modulation via Valence Mending Adsorbate Segregation

In the preceding section, the adoption of bandgap engineering approach for Schottky barrier modulation has been shown to enable effective dark current suppression. In yet another approach, the application of dopant-segregation technique in metal-germanium-metal photodetectors has also been reported to enable substantial dark current reduction. By implanting and segregating arsenic (As) and boron (B) dopants at the NiGe/Ge junction in an alternating electrode pattern, Zang et al. (2008) showed that an impressive dark current suppression by ~3-4 orders of magnitude can be achieved. In this work, a novel concept in employing asymmetrical Schottky barriers through sulfur co-implantation and segregation at the NiGe/Ge interface was proposed and demonstrated. Introducing sulfur (S) at the germanide-semiconductor interface has been shown to allow Fermi level pinning close to the conduction band edge, which results in an increased hole Schottky barrier height. This in turn, enables the achievement of substantial dark current reduction by more than three orders of magnitude in a Ge MSM photodetector.

Starting silicon-on-insulator (SOI) substrate was employed for the fabrication of Ge MSM photodetector with sulfur-segregation. Field oxide isolation was first formed by the deposition of a 120 nm plasma enhanced chemical vapor deposition (PECVD) oxide and followed by the Ge active window patterning. Reactive ion etching was employed to open up the active window and then cleaned with standard piranha solution (i.e. a mixture of sulfuric acid (H_2SO_4) with hydrogen peroxide (H_2O_2)) for polymer removal. Prior to the selective epitaxial growth of Ge in an ultra high vacuum chemical vapor deposition (UHVCVD) reactor, the wafers were subjected to standard SC1 (NH_4OH : H_2O_2 : H_2O) clean and a HF-last wet process. Selective Ge epitaxy growth first started with an in-situ baking in N_2 ambient at 800°C for native oxide removal and followed by the deposition of a ~5nm thin Si buffer at 530°C. A SiGe buffer layer of ~20nm was then deposited to have a gradual transition from pure Si to pure Ge at the interface. A low temperature growth at 370°C was subsequently employed to form a Ge seed layer of ~30nm before the high temperature Ge deposition at 550°C. Precursor gases comprise of pure disilane Si_2H_6 and diluted germane GeH_4 (10% GeH_4 : 90% Ar) were employed for the epitaxy growth of SiGe and Ge layers.

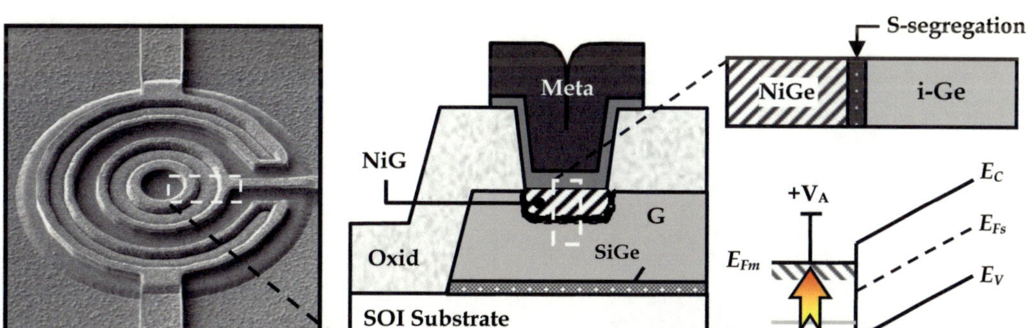

Fig. 18. Scanning electron microscopy (SEM) image of a Ge MSM photodetector with NiGe Schottky barrier. Co-implantating and segregating valence mending adsorbate such as sulfur at the NiGe/Ge interface leads to Fermi level de-pinning, which enables hole Schottky barrier enhancement.

On the device wafer, ion implantation of valence mending adsorbate such as sulfur (S) with a dose of 1x10^{15} cm^{-2} and an implant energy of 10KeV was selectively performed on one side of the contact regions (Fig. 18). On the control wafer, sulfur implantation was not performed. After diluted HF clean, a thin nickel (Ni) film with a thickness of 30 nm was deposited on both wafers. Germanidation process was then performed using rapid thermal annealing (RTA) at a temperature of 500°C for 30 sec in N_2 ambient to form nickel-monogermanide (NiGe) contacts. Diluted nitric acid (HNO_3:H_2O) with a concentration of (1:10) was employed to remove the unreacted Ni after germanidation. Metallization consisting of tantalum-nitride/aluminium (250Å/6000Å) were subsequently deposited and patterned to complete the device fabrication. Fig. 18 shows the scanning electron microscopy (SEM) image of the completed Ge MSM photodetector featuring surface illumination scheme with an effective diameter of 32μm.

Fig. 19(a) shows the high resolution transmission electron microscopy (HRTEM) image of the NiGe/Ge junction, showing an excellent interface quality with an approximately uniform NiGe thickness of ~70nm. An X-ray diffraction (XRD) analysis confirmed the formation of a nickel-monogermanide (NiGe) phase after a RTA of 500°C for 30 sec. The achievement of low sheet resistance and contact resistance of metal-germanide are beneficial for enabling high-speed performance. Fig. 19(b) shows the secondary-ion-mass spectroscopy (SIMS) depth profile of a sulfur-segregated NiGe Schottky contact. While germanidation front proceeds, the implanted sulfur atoms are observed to pile up at the interface between the NiGe and the Ge junction. Due to this segregation, the dangling bonds at the NiGe/Ge interface can be effectively passivated, which results in the pinning of germanide Fermi level close to the conduction band edge. Hence, an increase in the hole Shottky barrier can be achieved for suppressing hole injection which forms the dominant component responsible for the high leakage current in Ge MSM photdetector.

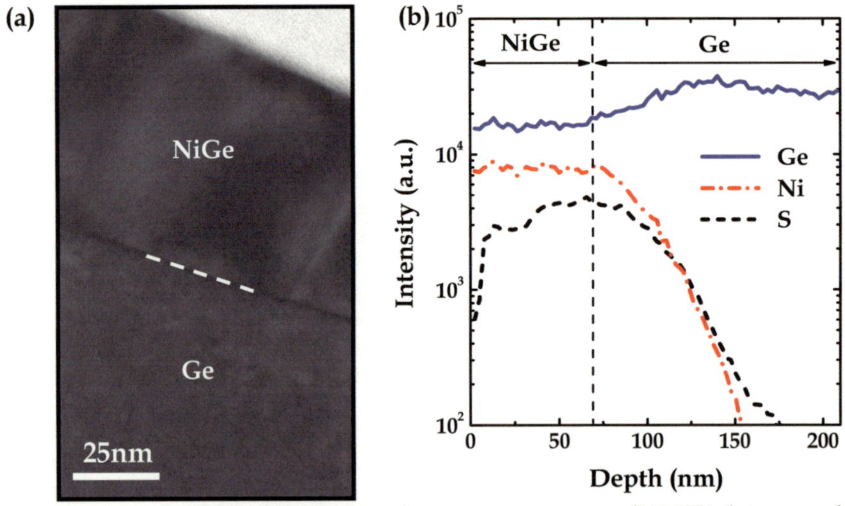

Fig. 19. (a) High resolution transmission electron microscopy (HRTEM) image showing the interface quality of the NiGe/Ge junction. (b) Secondary-ion-mass spectroscopy (SIMS) depth profile of a NiGe Schottky contact with sulfur-segregation.

Fig. 20 plots the room-temperature current–voltage (I–V) characteristics of the fabricated NiGe Schottky barrier Ge MSM photodetectors with and without sulfur-segregation. At an applied bias of 1.0V, high dark current (I_{Dark}) on the order ~1.69mA was measured in a conventional photodetector without sulfur-segregation. Such high dark current performance is predominantly attributed to the large hole injection as a result of small hole Schottky barrier height ϕ_{bh} of ~0.1eV. However, by introducing sulfur-segregation to passivate the dangling bonds at the NiGe/Ge interface, de-pinning of germanide Fermi level away the valence band edge was observed. This results in an increased hole Schottky barrier height (ϕ_{bh}) of ~0.49 eV, leading to a significant dark current suppression by more than three orders of magnitude. For an applied bias of 1.0V, the I_{Dark} of Ge MSM photodetector with sulfur-segregation was measured to be merely ~0.42μA.

An extraction of Schottky barrier height based on the thermionic-emission model (Sze, 1981) further confirms the achievement of hole Schottky barrier modulation from ~0.1 eV to ~0.49 eV due to sulfur-segregation at the NiGe/Ge interface, consistent with that reported in (Ikeda et al., 2006). This leads to the formation of asymmetrical Schottky barriers in the Ge MSM photodetector, where high and low hole barrier height are achieved for contacts with and without sulfur-segregation, respectively (see band diagrams in Fig. 20).

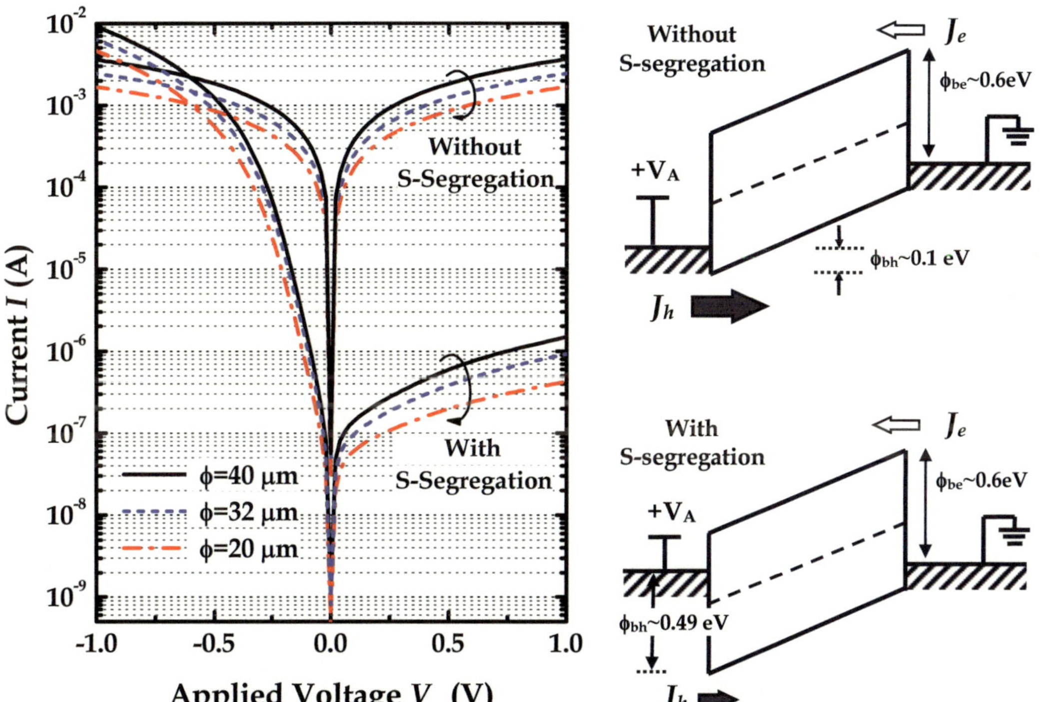

Fig. 20. Current-voltage (I-V) characteristics of the NiGe Schottky barrier photodetectors with and without sulfur-segregation. Significant reduction of dark current was achieved, predominantly attributed to the effective passivation of dangling bonds by the introduction of sulfur at the NiGe/Ge interface.

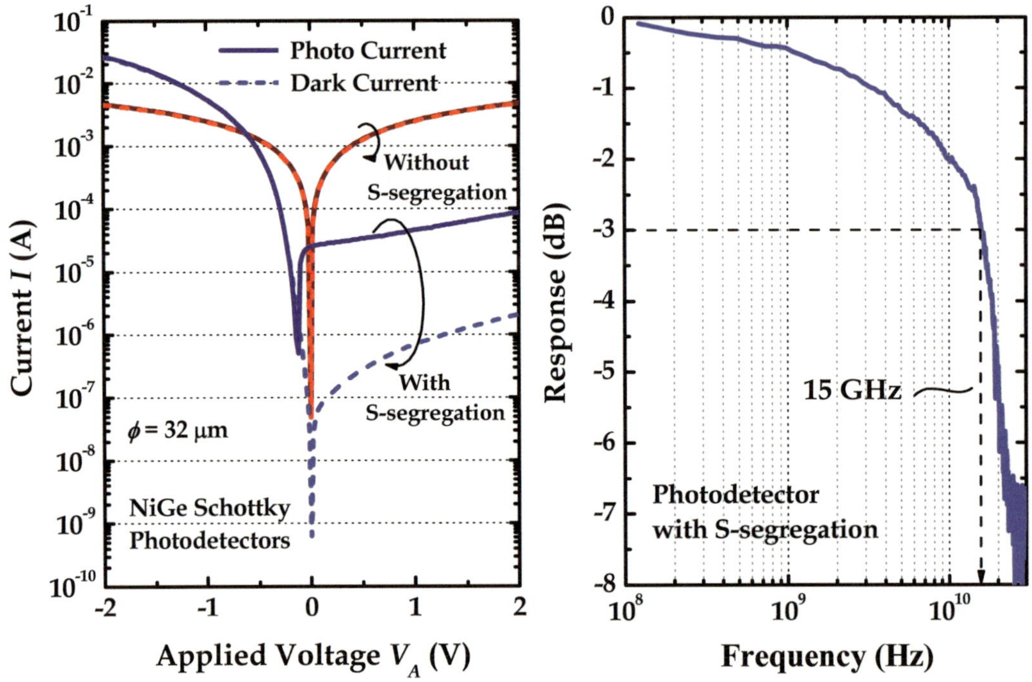

Fig. 21. (a) Photo-response characteristic of the Ge MSM photodetector with asymmetrical Schottky barriers. A responsivity of ~0.36A/W was achieved. (b) Normalized frequency response measured at a photon wavelength of 1550nm. A -3dB bandwidth of ~15GHz was achieved for a low applied bias of 1.0V.

Fig. 21(a) plots the photo-response characteristics of the NiGe Schottky photodetectors with and without sulfur-segregation. For a given applied bias of 1.0V, good optical response of ~0.36A/W was achieved in a Schottky barrier engineered detector, which corresponds to a quantum efficiency of ~34%. An appreciable signal-to-noise ratio of ~10^2 was also observed in these devices. On the contrary, a conventional NiGe Schottky MSM photodetector without S-segregation showed a poor signal-to-noise ratio. This is attributed to the high dark current on the order of few mA, which renders it impractical for high sensitivity receiver application.

Fig. 22(b) shows the frequency response of the photodetectors measured under normal incidence illumination at a wavelength of 1550nm, as obtained from the Fourier transform of an impulse response. A -3dB bandwidth of ~15 GHz was achieved for a low applied bias of 1.0V, showing comparable speed performance with that reported in (Vivien et al., 2007). On the contrary, the frequency response of a conventional photodetector could not be quantified for comparison as a result of its poor signal-to-noise ratio. Further enhancement in the responsivity and bandwidth performance could be achieved through a focused effort to optimize the Ge epilayer quality and the parasitic *RC* components. The adoption of smaller depletion layer width between the two metal electrodes could also contribute to a further bandwidth enhancement.

5. Ge/Si Avalanche Photodetector with High Gain-Bandwidth Product

Avalanche photodetectors (APD) are widely used in applications where high sensitivity is needed due to their internal carrier multiplication mechanism. Conventional APD receivers that are typically made of III-V compound semiconductors were shown to offer ~10dB sensitivity improvement over standard PIN receivers at a bit-rate of 10Gbs^{-1}. However, the limited gain-bandwidth product in III-V based APD (typically in the range of 100~150GHz) makes it a less attractive option for high bit-rate applications. Recent advancements have revealed that Group-IV based semiconductors such as germanium (Ge) and silicon (Si) are attracting growing interests for the realization of high performance APDs due to their favorable absorption coefficient, low excess noise and high impact ionization rates (Webb et al., 1974) properties. Using CMOS-compatible process technology, Kang et al., (2008) reported the fabrication of monolithic Ge/Si APDs with impressive gain-bandwidth product and sensitivity of 340GHz and -28dBm, respectively. In this demonstration, blanket Ge epitaxy growth and etch-back approach were employed to form the optical absorption region. Silicon, on the other hand, is used to form the carrier multiplication region.
In this section, the design and fabrication of avalanche photodetector using Group-IV based heterostructure materials are discussed. The performance metrics such as gain-bandwidth product, carrier multiplication gain, dark current and breakdown voltage thermal coefficient are examined.

5.1 Ge/Si Avalanche Photodetector Design and Fabrication

The Ge/Si APD design used in this study features a separate-absorption-charge-multiplication (SACM) configuration in which germanium and silicon are employed for light absorption and carrier multiplication, respectively. Unlike the approach reported in (Kang et al., 2008), this work employs a low thermal budget selective epitaxy growth of Ge (SEG Ge) to form high quality optical absorption region. The adoption of SEG Ge growth not only eliminates process complexity associated with Ge etch-back, but also prevents early edge breakdown due to the formation of beveled sidewall.

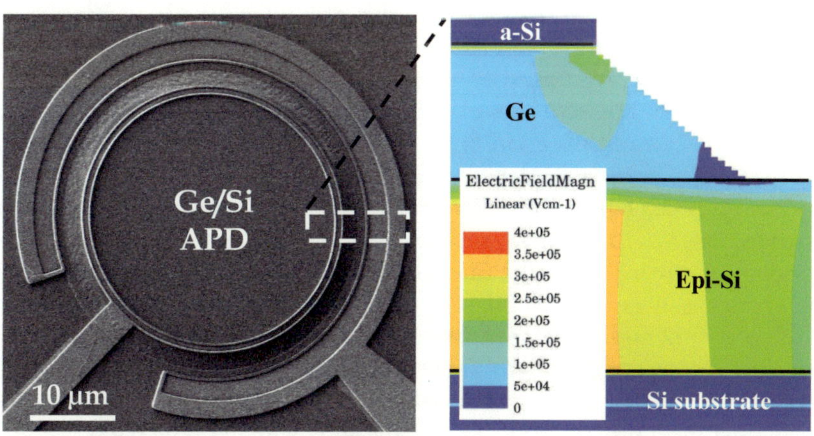

Fig. 22. Scanning electron microscopy (SEM) image of a surface illuminated Ge/Si avalanche photodetector. The simulated 2D electric field profile is plotted in the inset.

Fig. 22 shows the scanning electron microscopy (SEM) image of the Ge/Si APD demonstrated in this work. The device fabrication started with 8-inch p-type bulk Si wafers. Ion implantation employing arsenic dopants with a relatively high dose ($1\times10^{15}cm^{-2}$) was performed to form n-type contact layer with low series resistance. Blanket Si epitaxial layer with a thickness of 0.6μm was grown at a temperature of 640°C using an ultra high vacuum chemical vapor deposition (UHVCVD) reactor. This was followed by the p-type charge layer formation by implanting boron at $1.5\times10^{12}cm^{-2}$ onto the epi-Si surface. Circular mesas of diameter 28μm are patterned by anisotropic dry etching into the epi-Si layer. Oxide window was formed using a combination of dry etch and followed by wet etch approach to preserve the top Si surface quality from possible damage by the reactive ion etching process. Hetero-epitaxy of Ge was then selectively grown in an UHVCVD epitaxy reactor. The selective Ge epitaxy process commenced with the deposition of a low temperature pseudo-graded silicon–germanium buffer (~20nm) and followed by a Ge seed layer with a thickness of ~30nm. The temperature was then increased to ~550°C to complete the Ge growth using cyclical deposition and etch back approach. The top and sidewalls of the epi-Ge are then covered with 100nm of amorphous Si (a-Si) for passivation. The a-Si layer was implanted with high dose boron and dopants are activated at 750°C to form p-type ohmic contact. Nickel silicidation of the *p* and *n* contacts was done to further reduce the contact resistance. The fabrication process was completed after the formation of aluminium interconnects and an anti-reflective coating (ARC) layer.

The 2D contour map of the electric field profile as obtained using commercial MEDICI simulator is plotted in the inset of Fig. 22. The formation of bevelled Ge sidewall (~20°) is observed to contribute to a reduced electric field strength at the mesa edge which is advantageous in preventing early edge breakdown. This is true for both the base of the Ge absorption layer and the edge of the epi-Si layer. This eliminates the need for a floating guard ring to reduce the surface electric field strength at the silicon/insulator interface to prevent premature breakdown along the device perimeter (Kang et al., 2008).

5.2 Performance Metric of Ge/Si APD

Fig. 23 plots the current-voltage (I-V) characteristics of a Ge/Si APD with a cross-sectional diameter of 28μm measured under dark and illumination conditions. Using SACM configuration, the detector was observed to exhibit a typical rectifying characteristic at room temperature. Under low applied bias of less than -10V, the space charge region extends primarily within the Si multiplication layer. This results in a relatively insignificant dark current on the order of few nano-ampere (nA). Further increase in the applied reverse bias causes a widening of the depletion region into the Ge absorption layer. When this occurs, the dark current begins to increase rapidly due to trap-assisted generation within the depletion region as governed by the Shockley-Read-Hall (SRH) process. At -15V, the Ge absorption layer was observed to be fully depleted, which indicates the occurrence of punchthrough phenomenon. The dark current density at punchthrough was measured to be ~8.5mA/cm^2 but increases to ~320mA/cm^2 at 90% of the breakdown voltage. Such dark current level is competitive to that achieved in a state-of-the-art Ge p-i-n photodetector reported in the literature (Morse et al., 2006). This shows that enlarged electric field strength does not lead to a severe dark current degradation despite the possibility to experience defects-assisted tunneling current (Moll, 1964).

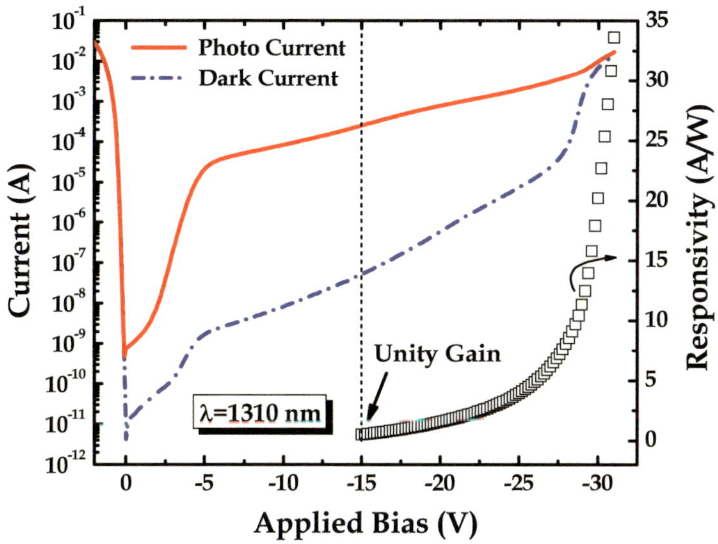

Fig. 23. Current-voltage characteristics of a Ge/Si APD with a cross-sectional diamter of 28μm. Due to an internal carrier multiplication gain, very impressive responsivity performance was demonstrated at a wavelength of 1310nm.

Normal incidence illumination with a wavelength of 1310nm was performed to extract the responsivity performance of the Ge/Si APD. Using conventional Ge p-i-n detector as a reference, the primary responsivity was measured to be ~0.51A/W, which corresponds to a quantum efficiency of ~48.2%. This is slightly smaller than the expected responsivity of 0.55A/W by assuming that the Ge absorption coefficient is 0.706μm^{-1} at 1310nm. A possible reason to explain this discrepancy is the occurrence of Ge-Si inter-diffusion along the hetero-interfaces which reduces the absorption coefficient. This is in good agreement with the Ge/Si APD responsivity at unity gain. Further increase in the applied bias causes the responsivity to rise rapidly. When operated at 90% of the breakdown voltage, a responsivity as high as ~4.2A/W was achieved in the APD (Fig. 23). Such high photoresponse is possible due to avalanche effects which enable carrier multiplication to take place within the Si multiplication layer.

By normalizing the APD's responsivity to the primary responsivity of a reference p-i-n detector, the multiplication gain factor can be calculated. Fig. 24 plots the dependence of multiplication gain (M) on the applied reverse bias of the Ge/Si APD at room temperature. Unity gain (M=1) was observed to occur at punchthrough voltage of -15V. With a further increase in the applied bias beyond punchthrough, the multiplication gain increases rapidly and could reach a value of ~8.2 at 90% of breakdown voltage.

However, the achievement of large multiplication gain could potentially impose a major concern for a degraded dark current which affects the APD receiver sensitivity. For a Ge/Si APD, the total measured dark current (I_{Total}) can be expressed in terms of the multiplied dark current ($I_{multiplied}$) and the unmultiplied dark current ($I_{unmultiplied}$) using

$$I_{Total} = I_{unmultiplied} + I_{multiplied} \cdot M \tag{15}$$

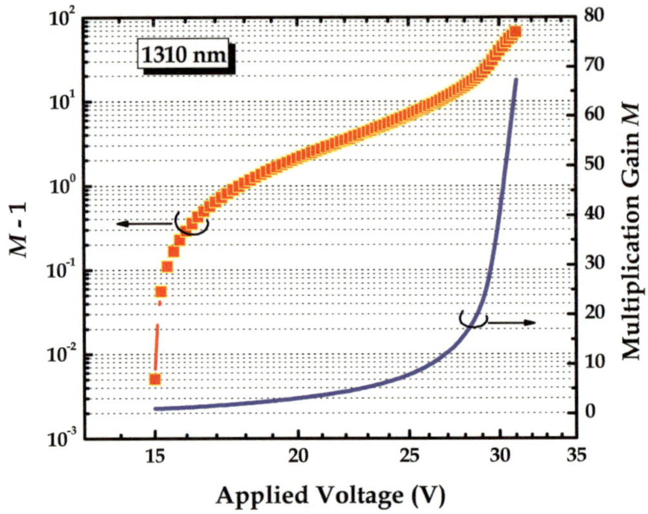

Fig. 24. The depedence of carrier multiplication gain on applied voltages. Higher gain is achieved with increasing biases.

As expected, increasing the carrier multiplication gain leads to higher dark current density, as shown in Fig. 25. At room temperature, the dark current is increased by one order of magnitude when the gain is increased from M=5 to M=25. Moreover, the dark current also demonstrates a strong dependence on temperature. Increasing the operating temperature from 23°C to 90°C has been shown to result in more than one order of magnitude increase in the dark current at a multiplication gain of 10.

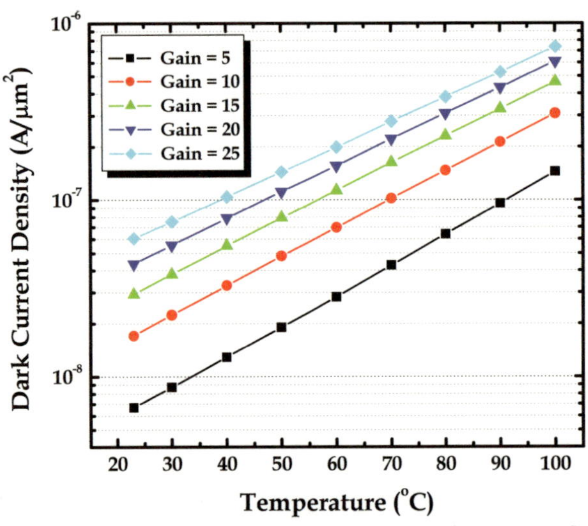

Fig. 25. The temperature dependence on dark current as a function of carrier multiplication gain. Both an increase in the multiplication gain and temperature results in a higher dark current density in a Ge/Si APD.

Fig. 26 plots the dependence of breakdown voltage (V_{bd}) on the operating temperature of a Ge/Si APD. The measurements are performed over a temperature range from 273K to 333K. A linear fitting to this plot yields a gradient which corresponds to the breakdown voltage thermal coefficient (δ) as defined by

$$\delta = \frac{\Delta V_{bd}}{\Delta T} \tag{16}$$

The extraction shows a δ value of 15mV/K, which is merely less than half of that obtained in a typical III-V based APD made of InP/InGaAs compound semiconductors (Ma et al., 1995). This makes Si a preferred material to meet the stringent requirement for thermal stability in APD receiver as it exhibits the least temperature sensitivity for breakdown voltage.

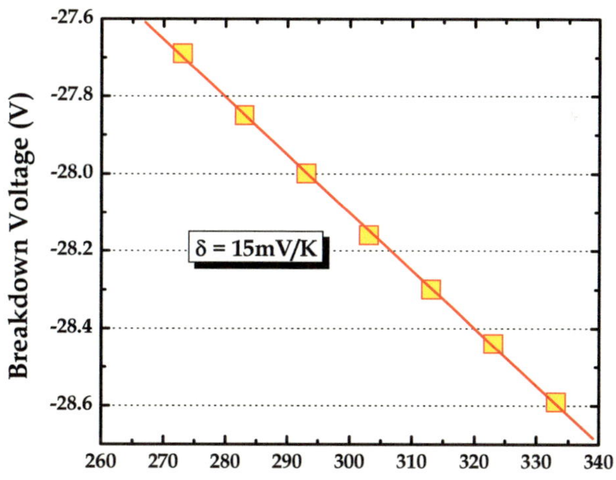

Fig. 26. The temperature dependence of breakdown voltage in Ge/Si APD. A small breakdown voltage thermal coefficient of 15mV/K was achieved.

The electrical -3dB bandwidth of the Ge/Si APD was measured using network analyzer at a wavelength of 1310nm. The frequency response was obtained from an APD with a diameter of 28μm. When the device is operated at low multiplication gain (M=10), a maximum -3dB bandwidth of ~7.5GHz was achieved, where RC time constant is found to be the dominant factor limiting the speed performance. This is lower than the expected theoretical bandwidth of 10GHz, which could possibly be due to parasitic effects such as high series resistance and capacitance between the metal pads. This is true as the total capacitance was measured to be ~230fF, in which 180fF was contributed by the metal pads while 50fF originates from the detector's junction capacitance. When operated in the high gain region, a slight decrease in the APD bandwidth is observed, which is predominantly attributed to the avalanche build-up time effect. For a multiplication gain of 23, a lower -3dB bandwidth of ~7.4GHz is achieved. This gives rise to a resulting gain-bandwidth product of 180GHz for the Ge/Si APD demonstrated in this work. Moreover, a CMOS TIA packaged Ge/Si APD achieved a very impressive sensitivity of -27.5dBm at a bit-rate of 10Gb/s, showing comparable performance to that of III-V based APD. Fig. 27 depicts a summary that benchmarks the gain-bandwidth product of III-V based and Si-based APDs reported in the literature. Clearly, Si-based APDs have out-performed III-V based APDs in terms of the gain-bandwidth performance, which makes it an extremely attractive option for 1.31μm fiber-link applications.

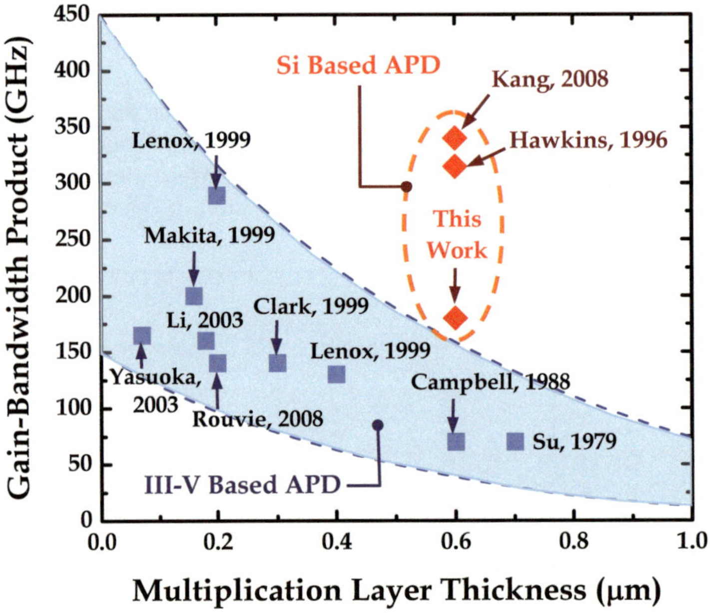

Fig. 27. A summary showing the benchmark of gain-bandwidth products as a function of multiplication layer thickness for both III-V based and silicon-based APDs.

6. Conclusion

In this chapter, the progress and development of Ge-based photodetector technologies for optical communication applications are discussed. To enable the realization of silicon photonics interconnect solutions, the development of Ge-on-Si hetero-epitaxy process that is compatible with existing CMOS fabrication technology is extremely important. Due to a large lattice mismatch between the two hetero-structure materials, high threading dislocations density within the Ge epilayer is not all unexpected. Various approaches proposed to overcome this technological challenge are discussed. Using low temperature pseudo-graded SiGe buffer engineering, low dislocations density on the order of ~10^7cm^{-2} has been achieved without the need to undergo additional high temperature annealing step. By leveraging on such high quality hetero-epitaxy platform, various state-of-the-art Ge-based photodetector schemes are proposed and demonstrated in this work. This includes the design and fabrication of Ge *p-i-n* photodetector with an integrated SOI micro-waveguide. The performance metrics of this detector are systematically discussed. Very impressive responsivity of ~0.9A/W (at 1550nm) and dark current density of ~0.7nA/μm^2 are demonstrated, along with the achievement of high bandwidth performance of >11GHz.

Parameter	Ge PIN PD	Ge MSM PD	Ge/Si APD
Responsivity (λ = 1550 nm)	Waveguided • 0.90 A/W Normal Incidence • 0.45 A/W	Waveguided • 0.76 A/W Normal Incidence • 0.36 A/W	Normal Incidence • > 5.0 A/W (Gain ~ 10)
-3dB Bandwidth	Waveguided • > 11 GHz Normal Incidence • > 9 GHz	Waveguided • > 15 GHz Normal Incidence • > 12 GHz	Normal Incidence • 7.5 GHz (G=10) • 7.4 GHz (G=23)
Gain-Bandwidth Product	• N.A.	• N.A.	• 180 GHz
Dark Current	Waveguided • 0.7 nA/μm^2 Normal Incidence • 1 nA/μm^2	Waveguided • 0.9 nA/μm^2 Normal Incidence • 1.1 nA/μm^2	Normal Incidence • 3.2 nA/μm^2 (@ 90% V_b)

Table 1. A summary showing the performance metrics of the various state-of-the-art Ge-based photodetectors developed in this work.

In addition, this chapter also deals with the high leakage current issue commonly seen in Ge photodetectors featuring metal-semiconductor-metal (MSM) configuration. Approaches reported in the literature to circumvent this problem are reviewed. By employing bandgap engineering approach through the use of new silicon-carbon Schottky barrier enhancement layer, significant dark current suppression by more than four orders of magnitude has been achieved. Novel concept in employing valence mending adsorbate segregation to de-pin the germanide Fermi level away the valence band edge is also proposed and demonstrated. The mechanism responsible for the hole Schottky barrier enhancement is elucidated.

Recent technological breakthrough in employing all Group-IV based materials for the realization of high performance Ge/Si avalanche photodetector is discussed. Using separate-absorption-charge-multiplication (SACM) configuration design, the Ge/Si APD demonstrated excellent gain-bandwidth product, large carrier multiplication gain, and low breakdown voltage thermal coefficient. These performances have been shown to be comparable, if not better, to the commercially available III-V based APDs.

Table 1 outlines a summary which compares the performance metrics of the various photodetectors demonstrated in this work. Each detector scheme has its own merits which can be leveraged to meet the stringent requirements of future integrated photonics circuits. A successful implementation of these state-of-the-art Ge-based photdetector technologies is poised to become an enabling technology to realize on-chip optical interconnects for next-generation high bandwidth computing applications.

7. References

Ang K.-W., Chui K.-J., Tung C.-H., Balasubramanian N., Samudra G. S., & Yeo Y.-C. (2007). Performance enhancement in uniaxial strained silicon-on-insulator N-MOSFETs featuring silicon-carbon source/drain regions, *IEEE Transactions on Electron Devices*, Vol. 54, No. 11, pp. 2910–2917.

Ang K.-W., Zhu S., Yu M., Lo G.-Q., & Kwong D.-L. (2008). High Performance Waveguided Ge-on-SOI metal-semiconductor-metal photodetectors with novel silicon-carbon (Si:C) Schottky barrier enhancement layer, *IEEE Photonics Technology Letters*, Vol. 20, No. 9, pp. 754-756.

Ang K.-W., Yu M.-B., Zhu S.-Y., Chua K.-T., Lo G.-Q., & Kwong D.-L. (2008). Novel NiGe MSM photodetector featuring asymmetrical Schottky barriers using sulfur co-implantation and segregation, *IEEE Electron Device Letters*, Vol. 29, No. 7, pp. 704–707.

Ang K.-W., Zhu S. Y., Wang J., Chua K. T., Yu M. B., Lo G. Q., & Kwong D. L. (2008). Novel silicon-carbon (Si:C) Schottky barrier enhancement layer for dark current suppression in Ge-on-SOI MSM photodetectors, *IEEE Electron Device Letters*, Vol. 29, No. 7, pp. 708–710.

Ang K.-W., Ng. J. W., Lo G.-Q., & Kwong D.-L. (2009). Impact of field-enhanced band-traps-band tunneling on the dark current generation in germanium p-i-n photodetector, *Applied Physics Letters*, Vol. 94, No. 22, pp. 223515-223517.

Ang K.-W., Liow T.-Y., Yu M.-B., Fang Q., Song J., Lo G.-Q., & Kwong D.-L. (2010). Low thermal budget monolithic integration of evanescent coupled Ge-on-SOI photodetector on Si-CMOS platform, to appear in *IEEE Journal of Selected Topics in Quantum Electronics*, Vol. 14, In Press.

Beals M., Michel J., Liu J. F., Ahn D. H., Sparacin D., Sun R., Hong C. Y., Kimerling L. C., Pomerene A., Carothers D., Beattie J., Kopa A., Apsel A., Rasras M. S., Gill D. M., Patel S. S., Tu K. Y., Chen Y. K., & White A. E. (2008). Process flow innovations for photonic device integration in CMOS, *Proceedings of SPIE*, Vol. 6898, p. 689804.

Campbell J. C., Tsang W. T., Qua G. J., & Johnson B. C. (1988). High-speed InP/InGaAsP/InGaAs avalanche photodiodes grown by chemical beam epitaxy, *IEEE Journal of Quantum Electronics*, Vol. 24, No. XX, pp. 496-500.

Clark W. R. (1999). Reliable, high gain-bandwidth product InGaAs/InP avalanche photodiodes for 10 Gbs^{-1} receivers, *Proc. Opt. Fiber Comm.*, Vol. 1, pp. 96-98.

Colace L., Masini G., & Assanto G. (1999). Ge-on-Si approaches to the detection of near-infrared light, *IEEE Journal of Quantum Electronics*, Vol. 35, No. 12, pp. 1843–1852.

Dimoulas A., Tsipas P., Sotiropoulos A., & Evangelou E. K. (2006). Fermi-level pinning and charge neutrality level in germanium, *Applied Physics Letters*, Vol. 89, No. 25, pp. 252110-252112.

Gunn C. (2006). CMOS photonics for high-speed interconnects, *IEEE Micro*, Vol. 26, No. 2, pp. 58–66.

Hartmann J. M., Abbadie A., Papon A. M., Holliger P., Rolland G., Billon T., Fedeli J. M., Rouviere M., Vivien L., & Laval S. (2004). Reduced pressure–chemical vapor deposition of Ge thick layers on Si(001) for 1.3–1.55-μm photodetection, *Journal of Applied Physics*, Vol. 95, No. 10, pp. 5905-5913.

Hawkins A. R., Wu W., Abraham P., Streubel K., & Bowers J. E. (1996). High gain-bandwidth-product silicon heterointerface photodetector, *Applied Physics Letters*, Vol. 70, No. 3, pp. 303-305.

Huang Z., Oh J., & Campbell J. C. (2004). Back-side-illuminated high-speed Ge photodetector fabricated on Si substrate using thin SiGe buffer layers, *Applied Physics Letters*, Vol. 85, No. 15, pp. 3286–3288.

Ikeda K., Yamashita Y., Sugiyama N., Taoka N., & Takagi S.-I. (2006). Modulation of NiGe/Ge Schottky barrier height by sulfur segregation during Ni germanidation, *Applied Physics Letters*, Vol. 88, No. 15, pp. 152115-152117.

Kang Y., Liu H.-D., Morse M., Paniccia M. J., Zadka M., Litski S., Sarid G., Pauchard A., Kuo Y.-H., Chen H.-W., Zaoui W. S., Bowers J. E., Beling A., McIntosh D. C., Zheng X., & Campbell J. C. (2008). Monolithic germanium/silicon avalanche photodiodes with 340 GHz gain–bandwidth product, *Nature Photonics*, Vol. 3, No. 1, pp. 59-63.

Kimerling L. C., Ahn D., Apsel A. B., Beals M., Carothers D., Chen Y. K., Conway T., Gill D. M., Grove M., Hong C.-Y., Lipson M., Liu J., Michel J., Pan D., Patel S. S., Pomerene A. T., Rasras M., Sparacin D. K., Tu K.-Y., White A. E., & Wong C. W. (2006). Electronic-photonic integrated circuits on the CMOS platform, *Proceedings of SPIE*, Vol. 6125, p. 12502.

Laih L.-H., Chang T.-C., Chen Y.-A., Tsay W.-C., & Hong J.-W. (1998). Characteristics of MSM photodetectors with trench electrodes on P-type Si wafer, *IEEE Transactions on Electron Devices*, Vol. 45, No. 9, pp. 2018-2023.

Lenox C., Nie H., Yuan P., Kinsey G., Homles A. L., Streetman B. G., & Campbell J. C. (1999). Resonant-cavity InGaAs–InAlAs avalanche photodiodes with gain–bandwidth product of 290 GHz, *IEEE Photonics Technology Letters,* Vol. 11, No. 9, pp. 1162-1164.

Li N., Sidhu R., Li X. W., Ma F., Zheng X. G., Wang S. L., Karve G., Demiguel S., Holmes A. L., & Campbell J. C. (2003). InGaAs/InAlAs avalanche photodiode with undepleted absorber, *Applied Physics Letters*, Vol. 82, No. 13, pp. 2175-2177.

Luan H.-C., Lim D. R., Lee K. K., Chen K. M., Sandland J. G., Wada K., & Kimerling L. C. (1999). High-quality Ge epilayers on Si with low threading-dislocation densities, *Applied Physics Letters*, Vol. 75, No. 19, pp. 2909–2911.

Ma C. L. F., Dean M. J., Tarof L. E., & Yu J. C. H. (1995). Temperature dependence of breakdown voltages in separate absorption, grading, charge, and multiplication InP/InGaAs avalanche photodiodes, *IEEE Transactions on Electron Devices*, Vol. 42, No. 5, pp. 810-818.

Makita K., Nakata T., Watanabe I. & Taguchi K. (1999). High-frequency response limitation of high performance InAlGaAs/InAlAs superlattice avalanche photodiodes, *Electron. Lett.*, Vol. 35, No. 25, pp. 2228-2229.

Mamor M., Perrossier J.-L., Aubry-Fortuna V., Meyer F., Bouchier D., Bodnar S., & Regolini J. L. (1997). Fermi-level pinning in Schottky diodes on IV-IV semiconductors: effect of Ge and C incorporation, *Thin Solid Films*, Vol. 294, No. 1/2, p. 141-144.

Moll J. L. (1964). *Physics of semiconductors,* McGraw-Hill, New York.

Morse M., Dosunmu O., Sarid G., & Chetrit Y. (2006). Performance of Ge-on-Si p-i-n Photodetectors for Standard Receiver Modules, *Proceeding of SiGe and Ge: Materials, Processing, and Devices*, Vol. 3, No. 7, pp. 75-84.

Oh J., Banerjee S. K., & Campbell J. C. (2004). Metal–germanium–metal photodetectors on heteroepitaxial Ge-on-Si with amorphous Ge Schottky barrier enhancement layers, *IEEE Photonics Technology Letters*, Vol. 16, No. 2, pp. 581-583.

Rouvie A., Carpentier D., Lagay N., Decobert J., Pommereau F., & Achouche M. (2008). High gain bandwidth product over 140 GHz planar junction AlInAs avalanche photodiodes, *IEEE Photonics Technology Letters*, Vol. 20, No. 5-8, pp. 455-457.

Shockley W. & Read T. W. (1952). Statistics of the recombinations of holes and electrons. *Physical Review*, Vol. 87, No. 5, pp. 835-842.

Soref R. A. (1991). Optical band gap of the ternary semiconductor $Si_{1-x-y}Ge_xC_y$, *Journal of Applied Physics*, Vol. 70, No. 4, pp. 2470-2472.

Soref R. (2006). The past, present, and future of silicon photonics, *IEEE Journal of Selected Topics in Quantum Electronics*, Vol. 12, No. 6, pp. 1678–1687.

Su Y. K., Chang C. Y., & Wu T. S. (1979). Temperature dependent characteristics of a PIN avalanche photodiode (APD) in Ge, Si and GeAs, *Opt. Quant. Electron.*, Vol. 11, No. 2, pp. 109-117.

Sze S. M. (1981). *Physics of semiconductor devices*, Wiley, New York.

Tung R. T. (2001). Recent advances in Schottky barrier concepts, *Mater. Sci. Eng.*, Vol. 35, No. 1, pp. 1-138.

Vivien L., Rouvière M., Fédéli J.-M., Morini, D. M., Damlencourt J.-F., Mangeney J., Crozat P., Melhaoui L. E., Cassan E., Roux X. L., Pascal D., & Laval S. (2007). High speed and high responsivity germanium photodetector integrated in a Silicon-On-Insulator microwaveguide, *Optics Express*, Vol. 15, No. 15, pp. 9843-9848.

Wada K., Luan H. C., Lim D. R. C., & Kimerling L. C. (2002), On-chip interconnection beyond semiconductor roadmap-Silicon microphotonics, *Proceedings of SPIE*, Vol. 4870, pp. 437-443.

Webb P. P., McIntyre R. J., & Conradi J. (1974). Properties of avalanche photodiodes, *RCA Review*, Vol. 35, pp. 234-278.

Yasuoka N., Kuwatsuka H., & Makiuchi M. (2003). Large multiplication-bandwidth products in APDs with a thin InP multiplication layer, *Proc. 16th IEEE Annual Meeting of LEOS*, pp. 999-1000.

Yin T., Cohen R., Morse M. M., Sarid G., Chetrit Y., Rubin D., & Paniccia M. J. (2007). 31GHz Ge n-i-p waveguide photodetectors on Silicon-on-Insulator substrate, *Optics Express*, Vol. 15, No. 21, pp. 13965–13971.

Zang H., Lee S. J., Loh W. Y., Wang J., Chua K. T., Yu M. B., Cho B. J., Lo G. Q., & Kwong D. L. (2008). Dark-current suppression in metal–germanium–metal photodetectors through dopant-segregation in NiGe–Schottky barrier, *IEEE Electron Device Letters*, Vol. 29, No. 2, pp. 161-164.

Lossless Operation in InP Mach-Zehnder Modulator Monolithically Integrated with Semiconductor Optical Amplifier

Yasuo Shibata and Takako Yasui
NTT Photonics Laboratories
Nippon Telegraph and Telephone Corporation
Japan

1. Introduction

There are a number of requirements for optical modulators used in large capacity photonic networks. They must be small, consume little power, operate at high speed, and be robust as regards fiber dispersion for long-haul transmission. Conventionally, lithium-niobate (LN) based modulators have been used in transmission systems. An InP Mach-Zehnder modulator (MZM) is attractive because it has a smaller chip size and lower driving voltage than an LN modulator [Rolland et al. (1993); Yoshimoto et al. (1999)]. Moreover, it has the potential for monolithic integration with a laser diode [J. S. Barton et al. (2003)]. We have been developing InP based MZMs, and have reported a 40 Gbit/s [Tsuzuki et al. (2006)] module and an 80 Gbit/s DQPSK modulator [Kikuchi et al. (2007)]. The drawback of the InP MZM is that the spot size of the device is only one-tenth that of an optical fiber, and the coupling loss between them through lenses cannot be reduced below 1.5 dB. Therefore, the insertion loss becomes larger than that when using LN modulators. Therefore we must find a way to compensate for both the coupling loss and the modulator insertion loss.

To achieve this, we monolithically integrated a semiconductor optical amplifier (SOA) with an InP n-p-i-n structure MZM. In this article, we describe the design, fabrication technology, and device performance of our SOA-integrated MZM (SOA-MZM). Fiber-to-fiber lossless operation is demonstrated across the entire C-band region. We also discuss the transmission characteristics of a 10-Gbit/s NRZ signal over a 100-km SMF, and the 200-km transmission of 10 Gbit/s optical duobinary signals.

This paper is organized as follows. Section 2 describes the device structure and fabrication processes. Section 3 presents the static characteristics of the fabricated device. Section 4 reports both dynamic and transmission characteristics. Section 5 contains our concluding remarks.

2. Device structure

The structure of the fabricated SOA-MZM device is shown schematically in Fig. 1. An SOA with a ridge waveguide structure and an MZM region with a high-mesa structure are monolithically integrated using the ridge-high-mesa butt joint technique [Kikuchi et al. (2002)]. The SOA has 8 periods of InGaAsP strained multiple quantum wells (MQWs) as an active layer, which is sandwiched between InGaAsP separated confinement heterostructure (SCH) layers. The MZM core is composed of 20 periods of InGaAsP/InP MQWs. The bandgap wavelengths of the MQW layers are set at 1550 nm for the SOA and 1430 nm for the MZM. In outline, the fabrication process is as follows: First, an SOA active layer and SCH layers are grown on an n-InP substrate. Next, all the layers are etched off except for the SOA region to form an SOA island, and then the MZM core layer, undoped InP cladding layer, thin p-doped layer, and n-doped cladding layer are butt-jointed. Next, a p-doped SOA cladding layer is grown. The p-cladding layer is removed from the MZM region, and an n-cladding layer is selectively grown. All crystal growth is performed using metalorganic vapor phase epitaxy (MOVPE). Then, an SOA stripe is dry etched using CH_4 reactive ion etching to form a ridge waveguide structure, and high-mesa passive waveguides forming a Mach-Zehnder interferometer are fabricated by using Br_2-N_2 reactive beam etching. The ridge waveguide of the SOA is 2.0 μm wide and 600 μm long. The width and height of the mesa in the MZM region are 2.0 and 3.5 μm, respectively. The SOA region and MZM region are joined using the ridge-high-mesa coupling technique. Both sides of the facet are anti-reflection (AR) coated. The 1.5-mm long dual phase modulator arms are equipped with electrodes with a lumped structure, which are terminated with a 50-Ωresistance. The total device size is 3.9 x 0.5 mm.

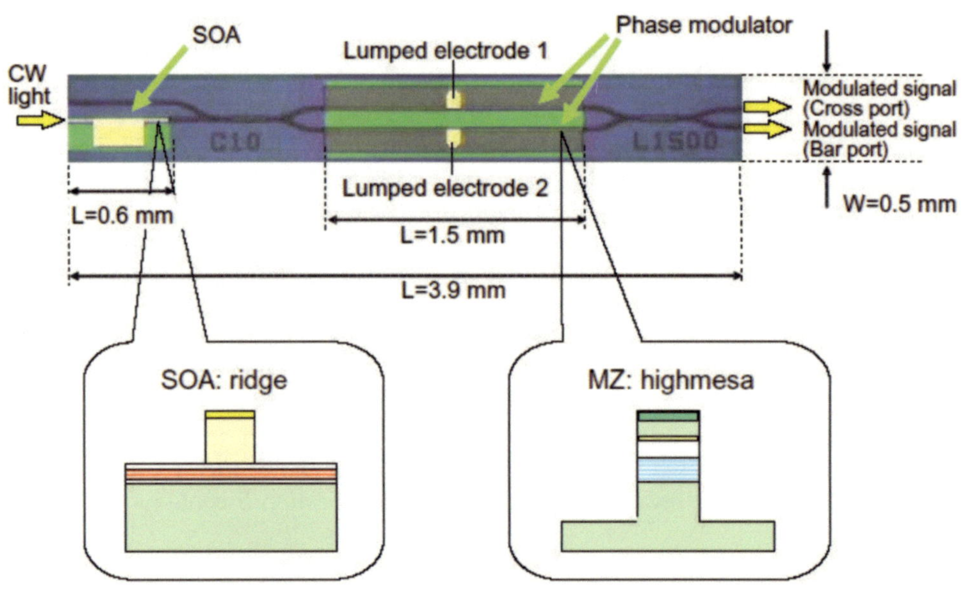

Fig. 1. Structure of the MZ-SOA

3. Static characteristics

The static optical characteristics of the monolithic device were investigated at 25 °C using TE polarized light focused with a lensed fiber on an AR-coated facet. The output light from the device was also collected by another lensed fiber. First, we examined the extinction characteristics under a single arm driving condition. Figure 2 shows the measured static extinction characteristics. Normalized optical transmissions from the MZM were plotted for input wavelengths of 1530, 1540, 1550 and 1560 nm as a function of the applied voltage (V_{b1}) to electrode 1 of phase modulator 1 in the upper arm while the bias voltage (V_{b2}) to electrode 2 of phase modulator 2 in the lower arm was held at 0 V. As shown in the figure, a half-wavelength voltage (V_{π}) of less than 2.5 V and a high extinction ratio (ER) of over 16 dB were obtained for the entire C-band region when the applied dc voltage was around 7 V. It is true that the extinction curves are asymmetric and the V_{π} value depends on both bias voltage and wavelength. This is because the refractive index change in a semiconductor material depends on the wavelength detuning between the bandgap wavelength and the input signal wavelength. It is also true that the detuning can be adjusted by adjusting the dc bias voltage, and therefore, the V_{π} value can be set at a constant value regardless of wavelength. Figure 3 shows an example of this condition. The normalized optical output of the SOA-MZM device for push-pull operation is shown in the figure. The horizontal axis indicates the difference between the applied voltage of the upper electrode and the bias voltage under a null condition in which the output optical power exhibits its minimum value. Under the null condition, the bias voltages applied to the upper and lower electrodes were 8.0 and 7.5 V, respectively, when the input wavelength was 1550 nm, and these values were adjusted according to the wavelength to keep the V_{π} value constant. From Fig. 3, V_{π} was 2.6 V and was constant for the entire C-band region.

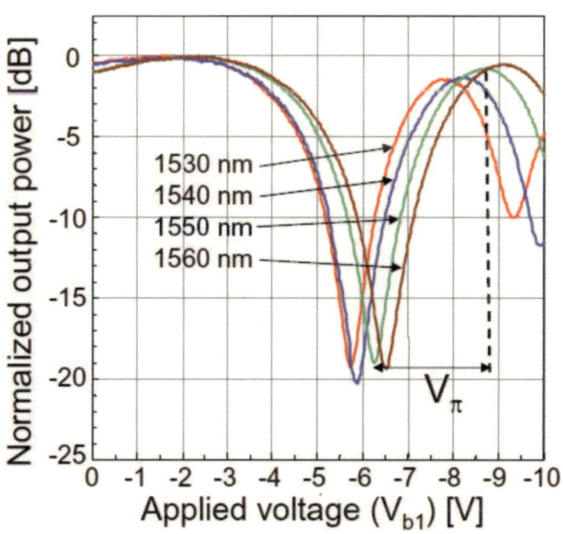

Fig. 2. Static extinction characteristics

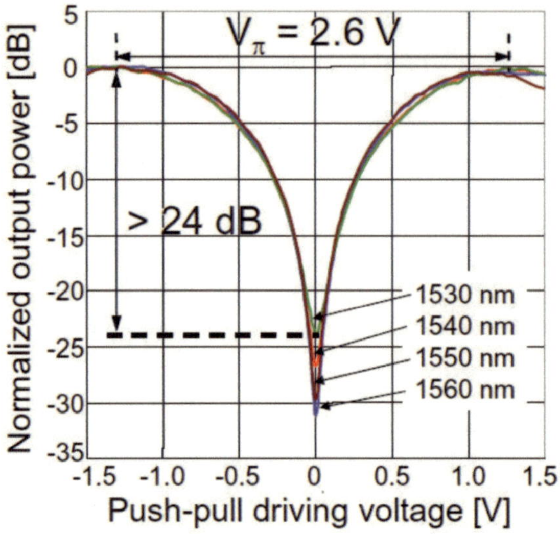

Fig. 3. Static extinction characteristics.

The extinction curves were completely symmetrical and identical, and the extinction ratio exceeded 24 dB for all wavelengths. These symmetrical characteristics indicate that the excess loss caused by the voltage-induced absorption (or electroabsorption) was negligible. These results reveal that this device has satisfactory characteristics for use in advanced modulation formats such as optical duobinary and differential phase shift keying signals.
We also investigated the gain characteristics of the integrated SOA. Figure 4(a) shows the fiber-to-fiber gain characteristics. The fiber output power was plotted against the fiber input power for a 1550-nm input signal while the upper and the lower arms of the MZM were held at 0 V. The coupling losses between the monolithic device and the lensed fiber were estimated to be 4 dB. The dashed line in the figure shows the fiber-to-fiber lossless condition, which means that the device gain exceeds 8 dB. The gain is sufficiently high to compensate for the 4 dB/facet fiber-coupling loss and the MZM insertion loss when the injection current exceeds 100 mA even when the input power is as high as -1.5 dBm. Furthermore, a large output power of over 6 dBm was obtained from the device with a 200 mA current injection when the input optical power into the device was -5.3 dBm. Figure 4(b) shows the device output power-SOA current characteristics for various input wavelengths across the C-band when the device input power was -8.6 dBm (the fiber input power was -4.6 dBm). We obtained both fiber-to-fiber lossless operation and operation with device gain for the entire C-band region at a current injection of 100 mA. And we obtained an output power of more than 4 dBm for all wavelengths with a 200 mA current injection.

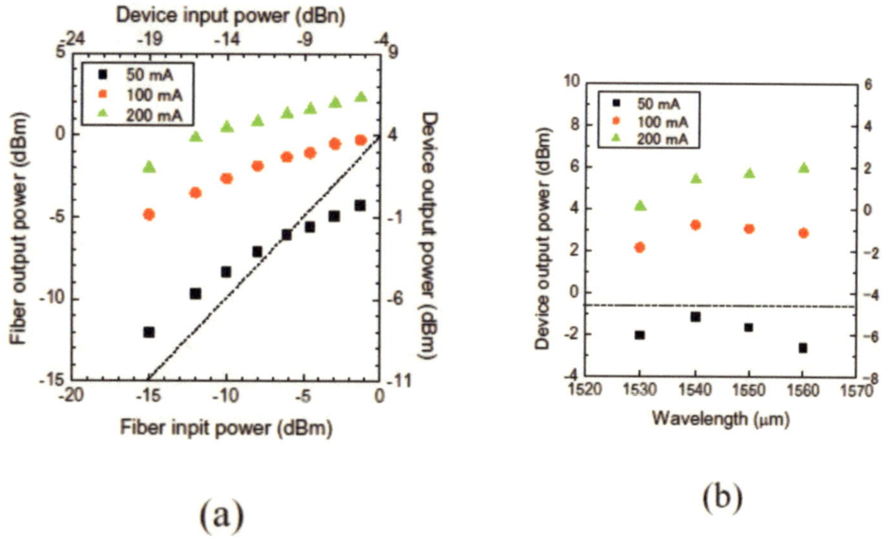

Fig. 4. Gain characteristics of the MZ-SOA. The input power dependence(a), and wavelength dependence(b).

4. Dynamic and transmission characteristics

The first topic in relation to the dynamic characteristics is the frequency response. Figure 5 shows the E/O response of the fabricated device. As seen in the figure, the 3-dB bandwidth was 7 GHz, which was large enough for 10-Gbit/s operation. Therefore, we then examined the transmission characteristics over an SMF at a bit rate of 10 Gbit/s.

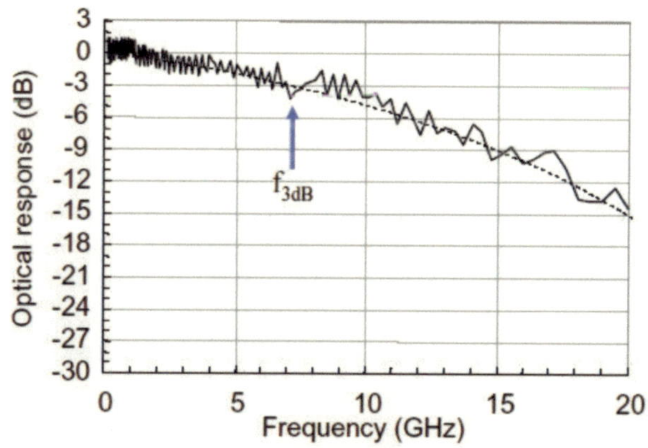

Fig. 5. Frequency response of the fabricated MZ-SOA.

The distance that optical signals can be transmitted through an SMF is limited by the group velocity (chromatic) dispersion of the SMF and the amount of spectral broadening (spectral width) of the optical signal. In principle, the optical spectrum is broadened when the optical intensity is coded by a high-speed digital signal. The spectrum experiences additional broadening as a result of the instantaneous frequency change of the signal light, a phenomenon normally designated as 'chirp'. Therefore, frequency chirping is an important issue in any discussion of the dynamic modulation characteristics of a modulator. The amount of chirp is expressed by using a chirp parameter α_{cp}, and the parameter is defined by [Kawanishi et al. (2001)]

$$\alpha_{\mathrm{cp}} = \Delta n / \Delta n' \tag{1}$$

where Δn and $\Delta n'$ are the relative changes in the real and imaginary parts of the complex refractive index, respectively.

It is well known that a negative chirp parameter of around -0.7 provides the best transmission distance results when using an NRZ intensity modulation format through an SMF fiber, while zero chirp is required for an advanced modulation format that utilizes phase information as well as intensity information such as the optical duobinary format. Therefore, we examined the dynamic modulation characteristics under two different driving conditions, namely a negative chirp condition and a zero chirp condition.

A. Negative chirp condition

We investigated the dynamic modulation characteristics when we applied a 2^{31}-1 pseudorandom binary sequence (PRBS) NRZ pattern at a bit rate of 9.953 Gbit/s to the MZM across the C-band. For the negative chirp driving condition, phase modulator 1 in the upper arm was dc biased and modulated with an NRZ signal of 3.0 V_{pp}, while phase modulator 2 in the lower arm was only dc biased to adjust the operation condition. The chirp parameter of the Mach-Zehnder modulator can be represented by

$$\alpha_{\mathrm{cp}} = \frac{V_2 - V_1}{V_2 + V_1} \tag{2}$$

where V_1 and V_2 are the modulation (RF) voltages applied to electrode1 of phase modulator 1 in the upper arm and to electrode2 of phase modulator 2 in the lower arm, respectively. Under this driving condition, since phase modulator 2 in the lower arm was only dc biased, $V_2 = 0$. Therefore, according to equation (2), the calculated chirp parameter α_{cp} was -1. The fiber input optical power was -4.6 dBm, and the bias current to the SOA was 100 mA. Under this condition, the monolithic device had a signal gain of over 10 dB. Figure 6(a) shows electrically filtered back-to-back eye diagrams for the measured wavelength region of 1530 to 1560 nm. The waveforms were almost the same for the entire C-band region and the dynamic ER exceeded 9.6 dB for all wavelengths.

The bit error rate (BER) performance is shown in Fig. 7. Error free operation was confirmed for every wavelength. The wavelength dependence of the variation in received power sensitivity at a BER of 10^{-12} was less than 0.5 dB for the measured wavelength region. The transmission characteristics were then investigated using a 100-km long SMF. Electrically filtered eye diagrams after a 100-km transmission are shown in Fig. 6(b). Clear eye opening was obtained even after 100-km transmission. The BER characteristics of the transmitted signal are also shown in Fig. 7. As shown in the figure, no floor was observed in the BER curve for any of the wavelengths, and error free operation after transmission was confirmed for every wavelength. The power penalty defined by the sensitivity degradation at a BER of 10^{-12} after a 100-km transmission was less than 1.5 dB across the C-band.

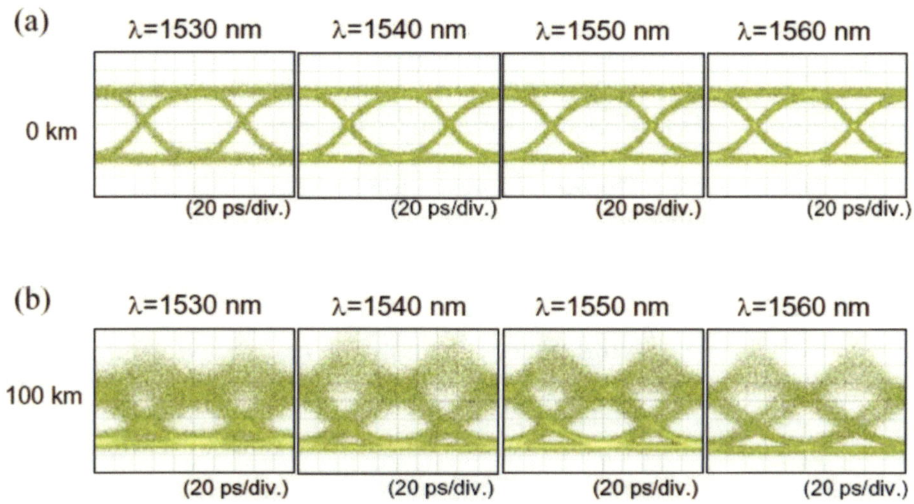

Fig. 6. Eye diagrams of back-to-back(a), and after 100-km transmission.

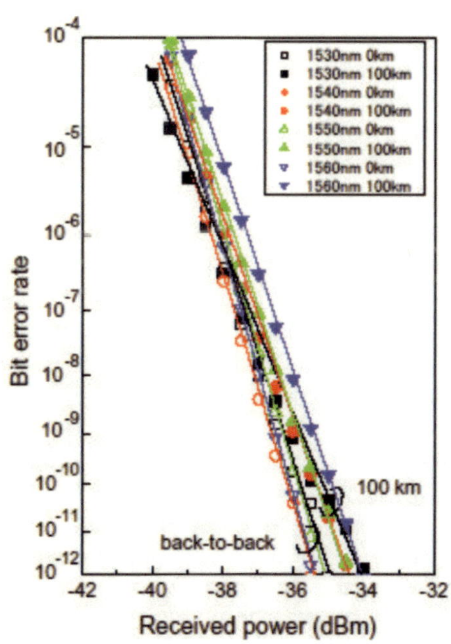

Fig. 7. Bit-error-rate characteristics for NRZ modulation.

B. Zero chirp condition

We examined the optical duobinary format for a zero-chirp driving condition [Yonenaga et al. (1997)]. The feature of an optical duobinary signal is its narrow spectral broadening compared with that of conventional non-return-to-zero (NRZ) modulation. An optical duobinary signal enables us to realize high-speed transmission with a large chromatic dispersion tolerance and a dense wavelength-division-multiplexing (DWDM) transmission with low adjacent channel crosstalk. Figure 8(a) is a diagram showing a method for generating an optical duobinary signal from a three-level electrical duobinary signal using an MZ modulator. A three-level electrical duobinary encoded signal ("-1", "0" and "1") is converted into a two-level optical duobinary signal ("1" and "0"), which is identical to the original binary signal inverted. The experimental setup is shown in Fig. 8(b). A TE polarized continuous-wave (CW) light is input into the dual-drive SOA-MZM using a lensed polarization maintaining fiber (PMF) with a coupling loss of 4 dB/facet. The optical duobinary signal is generated by differentially driving the MZM using three-level electrical signals with an amplitude of V_π while the MZM was dc biased at the null point where the optical output exhibits its minimum value without the RF modulation. As the modulation voltages were applied differentially, namely in a push-pull manner, $V_1 = V_2$, and therefore, according to equation (2), the calculated chirp parameter α_{cp} was 0. The three-level electrical signals were generated by low-pass filtering the 2^{31}-1 pseudorandom binary sequence (PRBS) NRZ signals at a bit rate of 9.953 Gbit/s with a cut-off frequency of 3 GHz. The output optical duobinary signals from the SOA-MZM were also coupled with the lensed SMF with a coupling loss of 4 dB. Following the 100-km SMF transmission, an erbium-doped fiber amplifier (EDFA) was employed to compensate for the fiber loss. The receiver consisted of an EDFA, an optical band-pass filter, a photodiode, and a clock and data recovery (CDR) circuit. The BER was measured using an error detector with the change of the input power into the receiver by a variable optical attenuator. The wavelengths of the CW light were 1550 nm. The temperature of the device was controlled at 25 °C. The transmission characteristics were investigated for SMF lengths of 0 (back-to-back), 60, 100, 160, and 200 km.

Figure 9(a) shows a back-to-back eye diagram of 10 Gbit/s optical duobinary signals using the SOA-MZM under the null condition shown in Fig. 3. Considering the electrical high-frequency loss at bias-T, and the connecting cables, the driving voltage was set at 2.8 peak-to-peak voltage (V_{pp}), which is slightly larger than the V_π value of 2.6 V shown in Fig. 3. The optical spectrum of the duobinary signal is shown in Fig. 9(b). We observed a narrow bandwidth with a 20-dB bandwidth of 13.6 GHz and no carrier frequency component, which is evidence that optical duobinary modulation was successfully realized with the fabricated MZ-SOA.

Figure 10(a) shows the measured eye diagrams for 10 Gbit/s optical duobinary signals after 60-, 100-, 160-, and 200-km SMF transmissions. Although chromatic dispersion degraded the eye opening as the transmission distance increased, clear eye opening was still obtained after a 200-km transmission. Figure 10(b) shows simulated results assuming a fiber dispersion of 16 ps/nm/km.

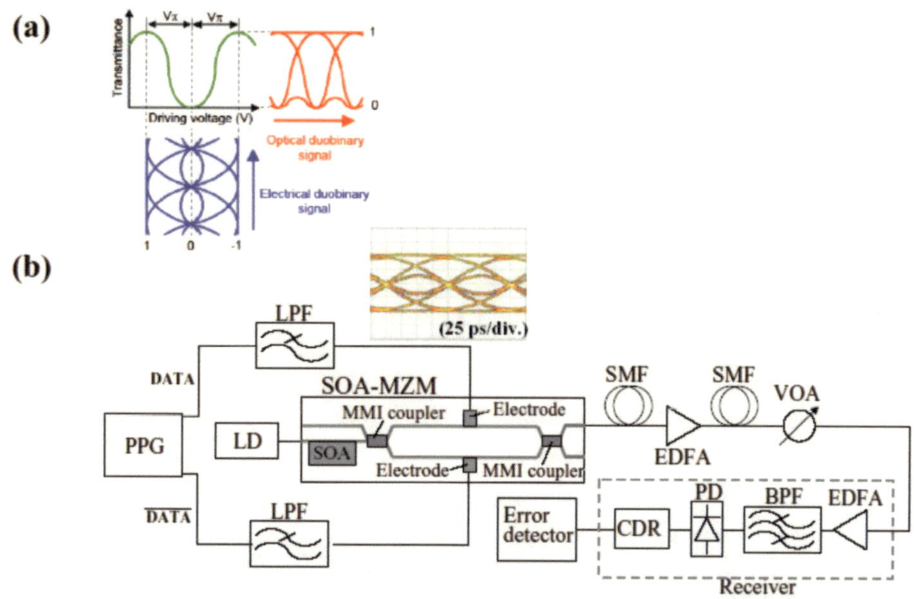

Fig. 8. Principle of generating optical duobinary signal (a), and experimental set-up (b).

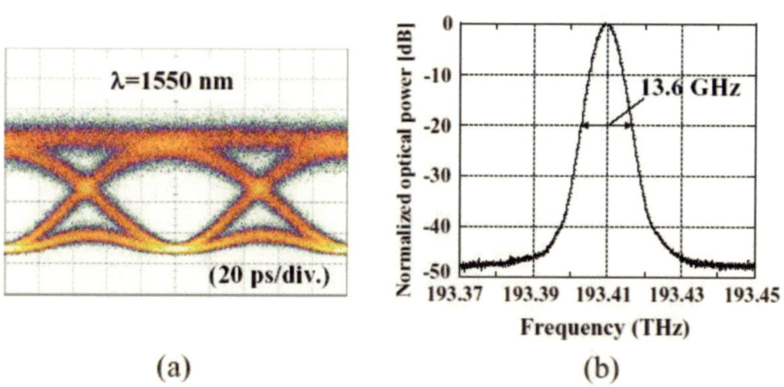

Fig. 9. Eye diagram (a) and optical spectrum (b) of the optical duobinary signal.

As can be seen, the experimental and simulation results are very similar for all distances, indicating the good quality of the modulator. Figure 11 shows the BER characteristics for all transmissions at a wavelength of 1550 nm. No floors were observed in the BER curves and error free operation was confirmed for all transmission distances. The power penalties defined by the sensitivity degradation at a BER of 10^{-12} were -0.7, -1.5, 0.4, 1.4 dB for 60-, 100-, 160-, and 200-km transmissions, respectively. These results are very similar to those obtained with a conventional (or non-integrated) MZ module [Kurosaki et al. (2007)], and therefore we can conclude that the monolithic integration process used in this work does not degrade the modulator quality. Figure 12 shows the wavelength dependence of the power penalty. The power penalty is less than -0.4, -1.4, 0.7, 2.3 dB for 60-, 100-, 160-, and 200-km transmissions, respectively, for the entire C-band region. These results prove that this compact lossless MZM performs sufficiently well for application to optical duobinary transmission systems.

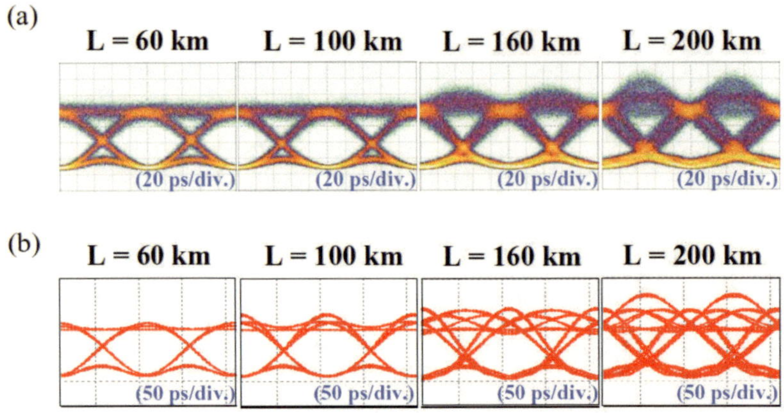

Fig. 10 Experimentally measured (a) and simulated (b) eye diagrams of optical duobinary signal after SMF transmission.

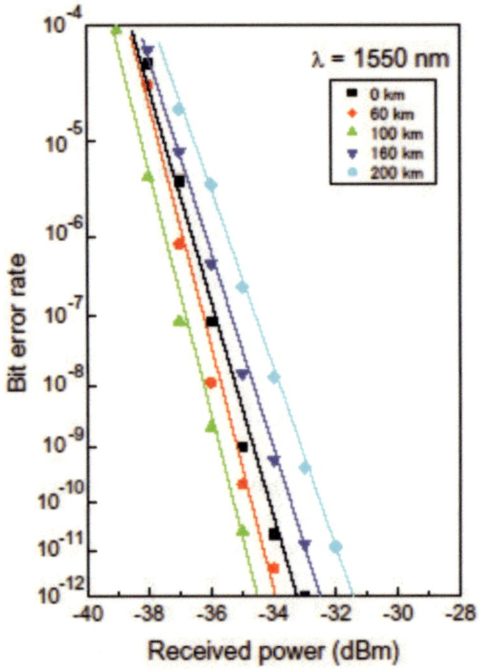

Fig. 11. Bit-error-rate characteristics for optical duobinary modulation.

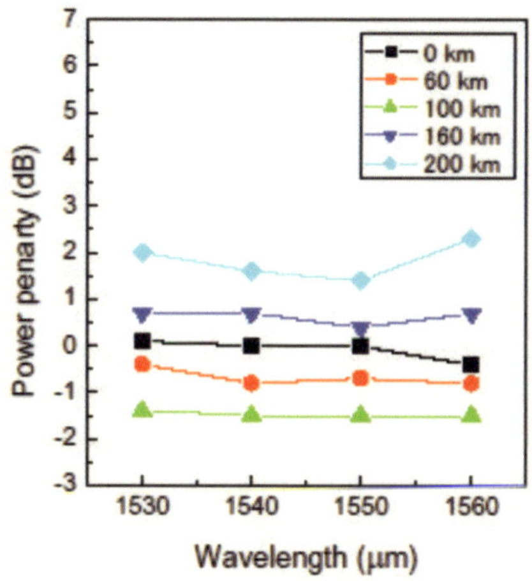

Fig. 12. Wavelength dependence of the power penalty.

5. Conclusion

An InP n-p-i-n MZM and an SOA were monolithically integrated to compensate for insertion loss. The device gain exceeded 10 dB, and fiber-to-fiber lossless operation was demonstrated for the entire C-band region. By using a lossless MZM, error free 100-km SMF transmissions were also demonstrated using an NRZ format at a bit rate of 10 Gbit/s for the entire C-band wavelength region. The measured power penalty after a 100-km transmission was only 1.5 dB.

10-Gbit/s optical duobinary transmissions were also demonstrated using the fabricated device. Lossless and error free operation was achieved with power penalties of less than -0.4, -1.4, 0.7, 2.3 dB for 0- (back-to-back), 60-, 100-, 160-, and 200-km SMF transmissions, respectively, at a low driving voltage of 2.8 V_{pp} for push-pull operation.

By comparing the experimental and simulation results, we confirmed that the modulation characteristics of this SOA-integrated lossless modulator are comparable to those of a discrete modulator. These results prove that this compact lossless MZM performs sufficiently well for application to optical duobinary transmission systems, and the integration process does not degrade modulator performance. These results constitute an important step towards achieving compact tunable light sources, realized by integrating a tunable laser and an MZM.

6. References

Barton, J. S.; Skogen, E. J.; Masanovic, M. L.; Denbaars, S. P. &Coldren, L. A. (2003). A widely tunable high-speed transmitter using an integrated SGDBR laser-semiconductor optical amplifier and Mach-Zehnder modulator. IEEE J. Sel. Topics Quantum Electron., 9, 1113-1117

Kawanishi, T.; Kogo, K.; Okikawa, S. &Izutsu, M. (2001). Direct measurement of chirp parameters of high-speed Mach-Zehnder-type optical modulators. Optics Communications, 195, 399-404

Kikuchi, N.; Shibata, Y.; Okamoto, H.; Kawaguchi, Y.; Oku, S.; Ishii, H.; Yoshikuni, Y. & Tohmori, Y. (2002). Error-free signal selection and high-speed channel switching by monolithically integrated 64-channel WDM channel selector. Electron. Lett., 38, 823-824

Kikuchi, N.; Sanjho, H.; Shibata, Y.; Tsuzuki, K.; Sato, T.; Yamada, E.; Ishibashi, T. & Yasaka, H. (2007). 80-Gbit/s InP DQPSK modulator with an n-p-i-n structure. 32nd European Conference on Optical Communication, Th10.3.1

Kurosaki, T.; Shibata, Y.; Kikuchi, N.; Tsuzuki, K.; Kobayashi, W.; Yasaka, H. & Kato, K. (2007). 200-km 10-Gbit/s optical duobinary transmission using an n-i-n InP Mach-Zehnder modulator. Proc. 19th IPRM, Matsue, Japan, May 14-18, WeB1-3

Rolland, C; Moore, R. S.; Shepherd, F. & Hillier, G. (1993). 10 Gbit/s, 1.56 μm multiquantum well InP/InGaAsP Mach-Zehnder optical modulator. Electronics Lett., 29, 471-472

Tsuzuki, K.; Kikuchi, N.; Sanjho, H.; Shibata, Y.; Kasaya, K.; Oohashi, H.; Ishii, H.; Kato, K.; Tohmori, Y. & Yasaka, H. (2006). Compact wavelength tunable laser module integrated with n-i-n structure Mach-Zehnder modulator. 31st European Conference on Optical Communication, Tu3.4.3

Yonenaga, K. & Kuwano, S. (1997). Dispersion-tolerant optical transmission system using duobinary transmitter and binary receiver. IEEE J. Lightwave Technol., 15, 1530-1537

Yoshimoto, N.; Shibata, Y.; Oku, S.; Kondo, S. & Noguchi, Y. (1999). Design and demonstration of polarization-insensitive Mach–Zehnder switch using a lattice-matched InGaAlAs/InAlAs MQW and deep-etched high-mesa waveguide structure. J.L.T, 17, 1662-1669

18

New Approach to Ultra-Fast All-Optical Signal Processing Based on Quantum Dot Devices

Y. Ben Ezra, B.I. Lembrikov
Holon Institute of Technology (HIT),P.O.Box 305, 58102, 52 Golomb Str., Holon Israel

1. Introduction

Fiber-optic technology is characterized by enormous potential capabilities: huge bandwidth up to nearly $50Tb/s$ due to a high frequency of an optical carrier, low signal attenuation of about $0.2dB/km$, low signal distortion, low power requirement, low material usage, small space requirement, and low cost Agrawal (2002), Mukherjee (2001).
However, the realization of these capabilities requires very high-bandwidth transport network facilities which cannot be provided by existing networks consisting of electronic components of the transmitters and receivers, electronic switches and routers Agrawal (2002). Most current networks employ electronic signal processing and use optical fiber as a transmission medium. Switching and signal processing are realized by an optical signal down-conversion to an electronic signal, and the speed of electronics cannot match the optical fiber bandwidth Ramamurthy (2001). For instance, a single-mode fiber (SMF) bandwidth is nearly $50Tb/s$, which is nearly four orders of magnitude higher than electronic data rates of a few Gb/s Mukherjee (2001). Typically, the maximum rate at which a gateway that interfaces with lower-speed subnetworks can access the network is limited by an electronic component speed up to a few tens of Gb/s. These limitations may be overcome by the replacement of electronic components with ultra-fast all-optical signal processing components such as fiber gratings, fiber couplers, fiber interferometers Agrawal (2001), semiconductor optical amplifiers (SOAs) Dong (2008), Hamié (2002), SOA and quantum dot SOA (QD-SOA) based monolithic Mach-Zehnder interferometers (MZIs) Joergensen (1996), Wang (2004), Sun (2005), Kanellos (2007), Wada (2007), Ben-Ezra (2008), Ben-Ezra (2009), all-optical switches based on multilayer system with enhanced nonlinearity and carbon nanotubes Wada (2007).
SOAs are among the most promising candidates for all-optical processing devices due to their high-speed capability up to $160Gb/s$, low switching energy, compactness, and optical integration compatibility Dong (2008). Their performance may be substantially improved by using QD-SOAs characterized by a low threshold current density, high saturation power, broad gain bandwidth, and a weak temperature dependence as compared to bulk and multi-quantum well (MQW) devices Bimberg (1999), Sugawara (2004), Ustinov (2003).
High-speed wavelength conversion, logic gate operations, and signal regeneration are important operations of the all-optical signal processing where SOAs are widely used Agrawal (2002), Ramamurthy (2001), Dong (2008).
A wavelength converter (WC) changes the input wavelength to a new wavelength without modifying the data content of a signal Agrawal (2002). Wavelength conversion is essential for optical wavelength division multiplexing (WDM) network operation Ramamurthy (2001).

There exist several all-optical techniques for wavelength conversion based on SOAs using the cross gain modulation (XGM) and cross phase modulation (XPM) effects between the pulsed signal and the continuous wave (CW) beam at the wavelength at which the converted signal is desired Agrawal (2002). In particular, MZI with a SOA inserted in each arm is characterized by a high on-off contrast and the output converted signal consisting of the exact replica of the incident signal Agrawal (2002).

All-optical logic operations are important for all-optical signal processing Sun (2005). All-optical logic gates operation is based on nonlinearities of optical fibers and SOAs. However, the disadvantages of optical fibers are weak nonlinearity, long interaction length, and/or high control energy required in order to achieve a reasonable switching efficiency Sun (2005). On the contrary, SOAs, and especially QD-SOAs, possess high nonlinearity, small dimensions, low energy consumption, high operation speed, and can be easily integrated into photonic and electronic systems Sun (2005), Hamié (2002), Kanellos (2007), Dong (2008).

The major problems of the improving transmission optical systems emerge from the signal-to-noise ratio (SNR) degradation, chromatic dispersion, and other impairment mechanisms Zhu (2007). For this reason, the optical signal reamplification, reshaping, and retiming (3R), or the so-called 3R regeneration, is necessary in order to avoid the accumulation of noise, crosstalk and nonlinear distortions and to provide a good signal quality for transmission over any path in all-optical networks Sartorius (2001), Zhu (2007), Leem (2006), Kanellos (2007). Optical regeneration technology can work with lower power, much more compact size, and can provide transparency in the needed region of spectrum Zhu (2007). All-optical 3R regeneration should be also less complex, and use fewer optoelectronics/electronics components than electrical regeneration providing better performance Leem (2006). All-optical 3R regenerator for different length packets at $40Gb/s$ based on SOA-MZI has been recently demonstrated Kanellos (2007).

We developed for the first time a theoretical model of an ultra-fast all-optical signal processor based on the QD SOA-MZI where XOR operation, WC, and 3R signal regeneration can be simultaneously carried out by ΛO-XOR logic gates for bit rate up to $(100-200)\ Gb/s$ depending on the value of the bias current $I \sim (30-50)\ mA$ Ben-Ezra (2009). We investigated theoretically different regimes of RZ optical signal operation for such a processor and carried out numerical simulations. We developed a realistic model of QD-SOA taking into account two energy levels in the conduction band of each QD and a Gaussian distribution for the description of the different QD size Ben-Ezra (2007), Ben-Ezra (2009), unlike the one-level model of the identical QDs recently used Berg (2004a), Sun (2005). We have shown that the accurate description of the QD-SOA dynamics predicts the high quality output signals of the QD SOA-MZI based logic gate without significant amplitude distortions up to a bit rate of about $100Gb/s$ for the bias current $I = 30mA$ and $200Gb/s$ for the bias current $I = 50mA$ being limited by the relaxation time of the electron transitions between the wetting layer (WL), the excited state (ES) and the ground state (GS) in a QD conduction band Ben-Ezra (2009).

The chapter is constructed as follows. The QD structure, electronic and optical properties are discussed in Section 2. The dynamics of QD SOA, XGM and XPM phenomena in QD SOA are described in Section 3. The theory of ultra-fast all-optical processor based on MZI with QD SOA is developed in Section 4. The simulation results are discussed in Section 5. Conclusions are presented in Section 6.

2. Structure, Electronic and Optical Properties of Quantum Dots (QDs)

Quantization of electron states in all three dimensions results in a creation of a novel physical object - a macroatom, or quantum dot (QD) containing a zero dimensional electron gas. Size

quantization is effective when the quantum dot three dimensions are of the order of magnitude of the electron de Broglie wavelength which is about several nanometers Ustinov (2003). An electron-hole pair created by light in a QD has discrete energy eigenvalues caused by the electron-hole confinement in the material. As a result, QD has unique electronic and optical properties that do not exist in bulk semiconductor material Ohtsu (2008).
QDs based on different technologies and operating in different parts of spectrum are known such as In(Ga)As QDs grown on GaAs substrates, InAs QDs grown on InP substrates, and colloidal free-standing InAs QDs. QD structures are commonly realized by a self-organized epitaxial growth where QDs are statistically distributed in size and area. A widely used QDs fabrication method is a direct synthesis of semiconductor nanostructures based on the island formation during strained-layer heteroepitaxy called the Stranski-Krastanow (SK) growth mode Ustinov (2003). The spontaneously growing QDs are said to be self-assembling. The energy shift of the emitted light is determined by size of QDs that can be adjusted within a certain range by changing the amount of deposited QD material. Smaller QDs emit photons of shorter wavelengths Ustinov (2003). The main advantages of the SK growth are following Ustinov (2003).

1. SK growth permits the preparation of extremely small QDs in a maskless process without lithography and etching which makes it a promising technique to realize QD lasers.
2. A great number of QDs is formed in one simple deposition step.
3. The synthesized QDs have a high uniformity in size and composition.
4. QDs can be covered epitaxially by host material without any crystal or interface defects.

The simplest QD models are a spherical QD with a radius R, and a parallelepiped QD with a side length $L_{x,y,z}$. The spherical QD is described by the spherical boundary conditions for an electron or a hole confinement which results in the electron and hole energy spectra $E_{e,nlm}$ and $E_{h,nlm}$ given by, respectively Ohtsu (2008)

$$E_{e,nlm} = E_g + \frac{\hbar^2}{2m_e}\left(\frac{\alpha_{nl}}{R}\right)^2 ; E_{h,nlm} = \frac{\hbar^2}{2m_h}\left(\frac{\alpha_{nl}}{R}\right)^2 \tag{1}$$

where

$$n = 1,2,3,...; l = 0,1,2,...n-1; m = 0,\pm1,\pm2,..\pm l \tag{2}$$

E_g is the QD semiconductor material band gap, $m_{e,h}$ are the electron and hole effective mass, respectively, $\hbar = h/2\pi$, h is the Planck constant, and α_{nl} is the n-th root of the spherical Bessel function. The parallelepiped QD is described by the boundary conditions at its corresponding surfaces, which yield the energy eigenvalues $E_{e,nlm}$ and $E_{h,nlm}$ given by, respectively Ohtsu (2008)

$$E_{e,nlm} = E_g + \frac{\hbar^2\pi^2}{2m_e}\left[\left(\frac{n}{L_x}\right)^2 + \left(\frac{l}{L_y}\right)^2 + \left(\frac{m}{L_z}\right)^2\right] \tag{3}$$

$$E_{h,nlm} = \frac{\hbar^2\pi^2}{2m_h}\left[\left(\frac{n}{L_x}\right)^2 + \left(\frac{l}{L_y}\right)^2 + \left(\frac{m}{L_z}\right)^2\right], n,l,m = 1,2,3,... \tag{4}$$

The density of states $\rho_{QD}(E)$ for an array of QDs has the form Ustinov (2003)

$$\rho_{QD}(E) = \sum_n\sum_m\sum_l 2n_{QD}\delta\left(E - E_{e,nlm}\right) \tag{5}$$

where $\delta\left(E - E_{e,nlm}\right)$ is the δ-function, and n_{QD} is the surface density of QDs.
The optical spectrum of QDs consists of a series of transitions between the zero-dimensional electron gas energy states where the selections rules are determined by the form and symmetry of QDs Ustinov (2003). The finite carrier lifetime results in Lorentzian broadening of a finite width Ustinov (2003).
Detailed theoretical and experimental investigations of InAs/GaAs and InAs QDs electronic structure taking into account their more realistic lens or pyramidal shape, size, composition profile, and production technique have been carried out Bimberg (1999), Bányai (2005), Ustinov (2003). A system of QDs can be approximated with a three energy level model in the conduction band containing a spin degenerate ground state GS, fourfold degenerate excited state (ES) with comparatively large energy separations of about $50 - 70meV$, and a narrow continuum wetting layer (WL). The electron WL is situated $150meV$ above the lowest electron energy level in the conduction band, i.e. GS and has a width of approximately $120meV$. In real cases, the QDs vary in size, shape, and local strain which leads to the fluctuations in the quantized energy levels and the inhomogeneous broadening in the optical transition energy. A Gaussian distribution may be used for the description of the QD sizes, and it shows that the discrete resonances merge into a continuous structure with widths around 10% Bányai (2005). The QDs and WL are surrounded by a barrier material which prevents direct coupling between QD layers. The absolute number of states in the WL is much larger than in the QDs. GS and ES in QDs are characterized by homogeneous and inhomogeneous broadening Bányai (2005). The homogeneous broadening caused by the scattering of the optically generated electrons and holes with imperfections, impurities, phonons, or through the radiative electron-hole pair recombination Bányai (2005) is about $15meV$ at room temperature Sugawara (2002). The inhomogeneous broadening in the optical transition energy is due to the QDs variations in size, shape, and local strain Bányai (2005), Sugawara (2004), Ustinov (2003).
In(Ga)As/GaAs QDs are characterized by emission at wavelengths no longer than $\lambda = 1.35\mu m$, while the InAs/InP QDs have been proposed for emission at the usual telecommunication wavelength $\lambda = 1.55\mu m$ Ustinov (2003).

3. Structure and Operation Mode of QD SOA

In this section, we will discuss the theory of QD SOA operation based on the electron rate equations and photon propagation equation Qasaimeh (2003), Qasaimeh (2004), Ben-Ezra (2005a), Ben-Ezra (2005b), Ben-Ezra (2007).

3.1 Basic Equations of QD SOA Dynamics

The active region of a QD SOA is a layer including self-assembled InGaAs QDs on a GaAs substrate Sugawara (2004). Typically, the QD density per unit area is about $\left(10^{10} - 10^{11}\right) cm^{-2}$. The bias current is injected into the active layer including QDs, and the input optical signals are amplified via the stimulated emission or processed via the optical nonlinearity by QDs Sugawara (2004). The stimulated radiative transitions occur between GS and the valence band of QDs. A detailed theory of QD SOAs based on the density matrix approach has been developed in the pioneering work Sugawara (2004) where the linear and nonlinear optical responses of QD SOAs with arbitrary spectral and spatial distribution of quantum dots in active region under the multimode light propagation have been considered. It has been shown theoretically that XGM takes place due to the coherent terms under the condition that the mode separation is comparable to or less than the polarization relaxation rate $|\omega_m - \omega_n| \leq \Gamma_g$ where $\omega_{m,n}$ are the mode frequencies and the relaxation time $\tau = \Gamma_g^{-1} = 130fs$ Sugawara

(2004). XGM is also possible in the case of the incoherent nonlinear polarization, or the so-called incoherent spectral hole burning Sugawara (2004). XGM occurs only for signals with a detuning limited by the comparatively small homogeneous broadening, and for this reason the ensemble of QDs should be divided into groups by their resonant frequency of the GS transition between the conduction and valence bands Sugawara (2004).
The phenomenological approach to the QD SOA dynamics is based on the rate equations for the electron densities of GS, ES and for combined WL and barrier serving as a reservoir. It is determined by electrons, because of the much larger effective mass of holes and their smaller state spacing Berg (2004a). Recently, an attempt has been carried out to take into account the hole dynamics for small-signal XGM case Kim (2009).
In the QD SOA-MZI, optical signals propagate in an active medium with the gain determined by the rate equations for the electron transitions in QD-SOA between WL, GS and ES Qasaimeh (2003), Qasaimeh (2004), Ben-Ezra (2005a), Ben-Ezra (2008). Unlike the model with the one energy level in the conduction band Berg (2004a), Sun (2005), we have taken into account the two energy levels in the conduction band: GS and ES Ben-Ezra (2007), Ben-Ezra (2009). The diagram of the energy levels and electron transitions in the QD conduction band is shown in Fig. 1.

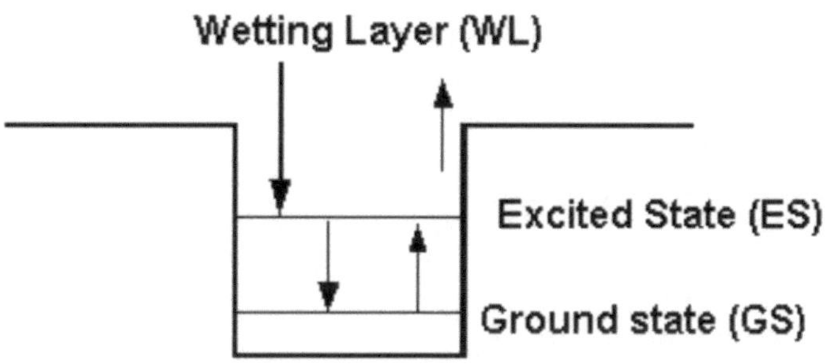

Fig. 1. Energy levels and electron transitions in a QD conduction band

The stimulated and spontaneous radiative transitions occur from GS to the QD valence band level. The system of the rate equations accounts for the following transitions:

1. the fast electron transitions from WL to ES with the relaxation time $\tau_{w2} \sim 3ps$;
2. the fast electron transitions between ES and GS with the relaxation time from ES to GS $\tau_{21} = 0.16ps$ and the relaxation time from GS to ES $\tau_{12} \sim 1.2ps$;
3. the slow electron escape transitions from ES back to WL with the electron escape time $\tau_{2w} \sim 1ns$.

The balance between the WL and ES is determined by the shorter time τ_{w2} of QDs filling. Carriers relax quickly from the ES level to the GS level, while the former serves as a carrier reservoir for the latter Berg (2001). In general case, the radiative relaxation times depend on the bias current. However, it can be shown that for moderate values of the WL carrier density $N_w \sim \left(10^{14} - 10^{15}\right) cm^{-3}$ this dependence can be neglected Berg (2001), Berg (2004b). The spontaneous radiative time in QDs $\tau_{1R} \gtrsim (0.4 - 0.5)\, ns$ remains large enough Sakamoto (2000), Qasaimeh (2003), Qasaimeh (2004), Sugawara (2004), Matthews (2005).
The carrier dynamics is characterized by slow relaxation processes between WL and ES. The rapidly varying coherent nonlinear population terms vanish after the averaging over the comparatively large relaxation time τ_{w2} ~several ps from the two-dimensional WL to the ES. We

have taken into account only incoherent population terms because for XGM between modes with the maximum detuning $\Delta\lambda_{\max} = 30nm$ within the especially important in optical communications conventional band of $\lambda = (1530 \div 1565)\ nm$ the condition $\omega_1 - \omega_2 > \Gamma_g^{-1}$ is valid even for the lowest relaxation time from the ES to GS $\tau_{21} = 0.16ps$, and the rapidly varying coherent beating terms are insignificant Sugawara (2004). The direct carrier capture into the GS is neglected due to the fast intradot carrier relaxation and the large energy separation between the GS and the WL and it is assumed that the charge neutrality condition in the GS is valid. The rate equations have the form Qasaimeh (2003), Qasaimeh (2004), Ben-Ezra (2007).

$$\frac{\partial N_w}{\partial t} = \frac{J}{eL_w} - \frac{N_w(1-h)}{\tau_{w2}} + \frac{N_w h}{\tau_{2w}} - \frac{N_w}{\tau_{wR}}, \tag{6}$$

$$\frac{\partial h}{\partial t} = \frac{N_w L_w (1-h)}{N_Q \tau_{w2}} - \frac{N_w L_w h}{N_Q \tau_{2w}} - \frac{(1-f)\,h}{\tau_{21}} + \frac{f(1-h)}{\tau_{12}}, \tag{7}$$

$$\frac{\partial f}{\partial t} = \frac{(1-f)\,h}{\tau_{21}} - \frac{f(1-h)}{\tau_{12}} - \frac{f^2}{\tau_{1R}} - \frac{g_p L}{N_Q}(2f-1)\, S_p \frac{c}{\sqrt{\varepsilon_r}} - \frac{g_s L}{N_Q}(2f-1)\, S_s \frac{c}{\sqrt{\varepsilon_r}}. \tag{8}$$

Here, S_p, S_s are the CW pump and on-off-keying (OOK) modulated signal wave photon densities, respectively, L is the length of SOA, g_p, g_s are the pump and signal wave modal gains, respectively, f is the electron occupation probability of GS, h is the electron occupation probability of ES, e is the electron charge, t is the time, τ_{wR} is the spontaneous radiative lifetime in WL, N_Q is the surface density of QDs, L_w is the effective thickness of the active layer, ε_r is the SOA material permittivity, c is the velocity of light in free space.The modal gain $g_{p,s}(\omega)$ is given by Uskov (2004)

$$g_{p,s}(\omega) = \frac{2\Gamma N_Q}{a} \int d\omega F(\omega)\, \sigma(\omega_0)\,(2f-1) \tag{9}$$

where the number l of QD layers is assumed to be $l = 1$, the confinement factor Γ is assumed to be the same for both the signal and the pump waves, a is the mean size of QDs, $\sigma(\omega_0)$ is the cross section of interaction of photons of frequency ω_0 with carriers in QD at the transition frequency ω including the homogeneous broadening factor, $F(\omega)$ is the distribution of the transition frequency in the QD ensemble which is assumed to be Gaussian Qasaimeh (2004), Uskov (2004). It is related to the inhomogeneous broadening and it is described by the expression Uskov (2004)

$$F(\omega) = \frac{1}{\Delta\omega\sqrt{\pi}} \exp\left[-\frac{(\omega-\overline{\omega})^2}{(\Delta\omega)^2}\right]. \tag{10}$$

where the parameter $\Delta\omega$ is related to the inhomogeneous linewidth $\gamma_{in\,\mathrm{hom}} = 2\sqrt{\ln 2}\Delta\omega$, and $\overline{\omega}$ is the average transition frequency.

3.2 XGM and XPM in QD SOA

XGM and XPM in QD SOA are determined by the interaction of QDs with optical signals. The optical signal propagation in a QD SOA is described by the following truncated equations for the slowly varying CW and pulse signals photon densities $S_{CW,P} =$

$P_{CW,P}/\left(\hbar\omega_{CW,P}\left(v_g\right)_{CW,P}A_{eff}\right)$ and phases $\theta_{CW,P}$ Agrawal (1989).

$$\frac{\partial S_{CW,P}\left(z,\tau\right)}{\partial z}=\left(g_{CW,P}-\alpha_{int}\right)S_{CW,P}\left(z,\tau\right) \tag{11}$$

$$\frac{\partial\theta_{CW,P}}{\partial z}=-\frac{\alpha}{2}g_{CW,P} \tag{12}$$

Here $P_{CW,P}$ are the CW and pulse signal optical powers, respectively, A_{eff} is the QD SOA effective cross-section, $\omega_{CW,P}$, $\left(v_g\right)_{CW,P}$ are the CW and pulse signal group angular frequencies and velocities, respectively, $g_{CW,P}$ are the active medium (SOA) gains at the corresponding optical frequencies, α_{int} is the absorption coefficient of the SOA material, α is a linewidth enhancement factor (LEF) which describes the coupling between gain and refractive index changes in the material and determines the frequency chirping Agrawal (2002). For the pulse propagation analysis, we replace the variables (z,t) with the retarded frame variables $\left(z,\tau=t\mp z/v_g\right)$. For optical pulses with a duration $T\gtrsim 10ps$ the optical radiation of the pulse is filling the entire active region of a QD SOA of the length $L\lesssim 1mm$ and the propagation effects can be neglected Gehrig (2002). Hence, in our case the photon densities

$$S_{CW,P}\left(z,\tau\right)=\left(S_{CW,P}\left(\tau\right)\right)_{in}\exp\left[\int_0^z\left(g_{CW,P}-\alpha_{int}\right)dz'\right] \tag{13}$$

can be averaged over the QD SOA length L which yields

$$S_{CW,P}\left(\tau\right)=\frac{1}{L}\left(S_{CW,P}\left(\tau\right)\right)_{in}\int_0^L dz\exp\left[\int_0^z\left(g_{CW,P}-\alpha_{int}\right)dz'\right] \tag{14}$$

Solution of equation (12) yields for the phases

$$\theta_{CW,P}\left(\tau\right)=-\left(\alpha/2\right)\int_0^L dz g_{CW,P}. \tag{15}$$

The time-dependent variations of the carrier distributions in the QDs and WL result in the strong phase changes (12) during the light propagation in the QD SOA Gehrig (2002). System of equations (6)-(8) with the average pump and signal photon densities (14) and phases (15) constitutes a complete set of equations describing XGM and XPM in QD SOA related by the LEF α as it is seen from equations (11), (12) and (15).

In order to investigate the possibility of XGM in QD SOAs due to the connections between different QDs through WL at detunings between a signal and a pumping larger than the homogeneous broadening we modified equations (6)-(8) dividing QDs into groups similarly to Sugawara (2002), Sugawara (2004), Sakamoto (2000). We consider a limiting case of the groups 1 and 2 with a detuning substantially larger than the homogeneous broadening, in order to investigate the possibility that they are related only due to the carrier relaxation from WL to ES Ben-Ezra (2007). The rate equations for such QDs take the form

$$\frac{\partial N_w}{\partial t}=\frac{J}{eL_w}-\frac{N_w\left(1-h_1\right)}{\tau_{w2}}+\frac{N_w h_1}{\tau_{2w}}-\frac{N_w}{\tau_{wR}}$$

$$-\frac{N_w(1-h_2)}{\tau_{w2}}+\frac{N_w h_2}{\tau_{2w}}, \tag{16}$$

$$\frac{\partial h_{1,2}}{\partial t}=\frac{N_w L_w(1-h_{1,2})}{N_Q\tau_{w2}}-\frac{N_w L_w h_{1,2}}{N_Q\tau_{2w}}$$

$$-\frac{(1-f_{1,2})h_{1,2}}{\tau_{21}}+\frac{f_{1,2}(1-h_{1,2})}{\tau_{12}}, \tag{17}$$

$$\frac{\partial f_{1,2}}{\partial t}=\frac{(1-f_{1,2})h_{1,2}}{\tau_{21}}-\frac{f_{1,2}(1-h_{1,2})}{\tau_{12}}-\frac{f_{1,2}^2}{\tau_{1R}}$$

$$-\frac{g_{p,s}L}{N_Q}(2f_{1,2}-1)S_{p,s}\frac{c}{\sqrt{\varepsilon_r}} \tag{18}$$

where the indices 1,2 correspond to the groups 1 and 2 of QDs. Equations (17)-(18) contain the electron occupation probabilities belonging to the same group and the photon density corresponding to the optical beam resonant with respect to this group, while the WL rate equation (16) includes the contributions of the both groups.

4. Theory of an Ultra-Fast All-Optical Processor

4.1 Theoretical Approach

The theoretical analysis of the proposed ultra-fast QD SOA-MZI processor is based on the combination of the MZI model with the nonlinear characteristics and the QD-SOA dynamics. The block diagram of the processor is shown in Fig. 2.

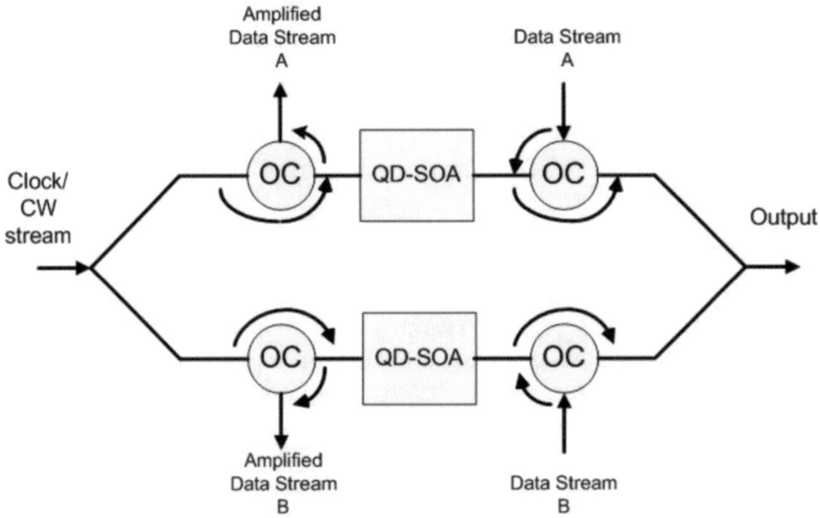

Fig. 2. A block diagram of the ultra-fast MZI processor containing in each arm a QD SOA, 50-50 3dB optical couplers, and optical circulators (OC)

At the output of MZI, the CW optical signals from the two QD SOAs interfere giving the output intensity Sun (2005), Wang (2004).

$$P_{XOR}=\frac{P_{in}}{4}\{G_1(t)+G_2(t)$$

$$-2\sqrt{G_1(t)\,G_2(t)}\cos\left[\phi_1(t)-\phi_2(t)\right]\} \tag{19}$$

where P_{in} is the CW or the clock stream optical signal divided and introduced via the symmetric coupler into the two QD-SOAs, $G_{1,2}(t) = \exp(g_{1,2}L_{1,2})$, $g_{1,2}$, $L_{1,2}$, and $\phi_{1,2}(t)$ are the time-dependent gain, the SOA gain, the active medium length, and phase shift, respectively, in the two arms of QD SOA-MZI. The phases $\phi_{1,2}(t)$ should be inserted into equation (19) from equation (15). When the control signals A and/or B are fed into the two SOAs they modulate the gain of the SOAs and give rise to the phase modulation of the co-propagating CW signal due to LEF α Agrawal (2001), Agrawal (2002), Newell (1999). LEF values may vary in a large interval from the experimentally measured value of LEF $\alpha = 0.1$ in InAs QD lasers near the gain saturation regime Newell (1999) up to the giant values of LEF $\alpha = 60$ recently measured in InAs/InGaAs QD lasers Dagens (2005). However, such limiting cases can be achieved for specific electronic band structure Newell (1999), Dagens (2005), Sun (2004). The typical values of LEF in QD lasers are $\alpha \approx (2-7)$ Sun (2005). Detailed measurements of the LEF dependence on injection current, photon energy, and temperature in QD SOAs have also been carried out Schneider (2004). For low-injection currents, the LEF of the dot GS transition is between 0.4 and 1 increasing up to about 10 with the increase of the carrier density at room temperature Schneider (2004). The phase shift at the QD SOA-MZI output is given by Wang (2004)

$$\phi_1(t) - \phi_2(t) = -\frac{\alpha}{2}\ln\left(\frac{G_1(t)}{G_2(t)}\right) \tag{20}$$

It is seen from equation (20) that the phase shift $\phi_1(t) - \phi_2(t)$ is determined by both LEF and the gain. For the typical values of LEF $\alpha \approx (2-7)$, gain $g_{1,2} = 11.5cm^{-1}$, $L_{1,2} = 1500\mu m$ the phase shift of about π is feasible.

4.2 Logic Gate Operation

Consider an AO-XOR gate based on integrated QD SOA-MZI which consists of a symmetrical MZI where one QD SOA is located in each arm of the interferometer as shown in Fig. 2. Two optical control beams A and B at the same wavelength λ are inserted into ports A and B of MZI separately. A third signal, which represents a clock stream of continuous series of unit pulses is split into two equal parts and injected into the two SOAs. The detuning $\Delta\omega$ between the signals A, B and the third signal should be less than the homogeneous broadening of QDs spectrum. In such a case the ultrafast operation occurs. In the opposite case of a sufficiently large detuning comparable with the inhomogeneous broadening, XGM in a QD SOA is also possible due to the interaction of QDs groups with essentially different resonance frequencies through WL for optical pulse bit rates up to $10Gb/s$ Ben-Ezra (2007). When $A = B = 0$, the signal at the MZI input port traveling through the two arms of the SOA acquires a phase difference of π when it recombines at the output port, and the output is "0" due to the destructive interference. When $A = 1$, $B = 0$, the signal traveling through the arm with signal A acquires a phase change due to XPM between the pulse train A and the signal. The signal traveling through the lower arm does not have this additional phase change which results in an output "1" Sun (2005). The same result occurs when $A = 0$, $B = 1$ Sun (2005). When $A = 1$ and $B = 1$ the phase changes for the signal traveling through both arms are equal, and the output is "0".

4.3 Wavelength Conversion

An ideal wavelength convertor (WC) should have the following properties: transparency to bit rates and signal formats, fast setup time of output wavelength, conversion to both shorter and longer wavelengths, moderate input power levels, possibility for no conversion regime, insensitivity to input signal polarization, low-chirp output signal with high extinction ratio and large SNR, and simple implementation Ramamurthy (2001). Most of these requirements can met by using SOA in the process of wavelength conversion. XGM method using SOAs is especially attractive due to its simple realization scheme for WC Agrawal (2001). However, the main disadvantages of this method are substantial phase distortions due to the frequency chirping, degradation due to spontaneous emission, and a relatively low extinction ratio Agrawal (2001). These parameters may be improved by using QD-SOAs instead of bulk SOAs due to pattern-effect-free high-speed wavelength conversion of optical signals by XGM, a low threshold current density, a high material gain, high saturation power, broad gain bandwidth, and a weak temperature dependence as compared to bulk and MQW devices Ustinov (2003). We combine the advantages of QD-SOAs as a nonlinear component and MZI as a system whose output signal can be easily controlled. In the situation where one of the propagating signals A or B is absent, CW signal with the desired output wavelength is split asymmetrically to each arm of MZI and interferes at the output either constructively or destructively with the intensity modulated input signal at another wavelength. The state of interference depends on the relative phase difference between the two MZI arms which is determined by the SOAs. In such a case the QD SOA-MZI operates as an amplifier of the remaining propagating signal. Then, the operation with the output "1" may be characterized as a kind of WC due to XGM between the input signal A or B and the clock stream signal. The possibility of the pattern-effect-free wavelength conversion by XGM in QD SOAs has been demonstrated experimentally at the wavelength of $1.3\mu m$ Sugawara (2004).

4.4 3R Regeneration

Short optical pulses propagating in optical fibers are distorted due to the fiber losses caused by material absorption, Rayleigh scattering, fiber bending, and due to the broadening caused by the material, waveguide, polarization-mode, and intermodal dispersion Agrawal (2001), Agrawal (2002). 3R regeneration is essential for the successful logic operations because of the ultra-fast data signal distortions. 3R regeneration requires an optical clock and a suitable architecture of the regenerator in order to perform a clocked decision function Sartorius (2001). In such a case, the shape of the regenerated pulses is defined by the shape of the clock pulses Sartorius (2001).

The proposed QD SOA-MZI ultra-fast all-optical processor can successfully solve three problems of 3R regeneration. Indeed, the efficient pattern–effect free optical signal re-amplification may be carried out in each arm by QD-SOAs. WC based on the all-optical logic gate provides the re-shaping since noise cannot close the gate, and only the data signals have enough power to close the gate Sartorius (2001). The re-timing in QD-SOA-MZI based processor is provided by the optical clock which is also essential for the re-shaping Sartorius (2001). Hence, if the CW signal is replaced with the clock stream, the 3R regeneration can be carried out simultaneously with logic operations. The analysis shows that for the strongly distorted data signals a separate processor is needed providing 3R regeneration before the data signal input to the logic gate.

5. Simulation Results and Discussion

The study of the ultrafast all-optical signal processor in different regimes such as XOR, WC, 3R signal regeneration requires the simultaneous analysis of the QD-SOA dynamics and MZI behavior. The MZI performance efficiency is determined by the combination of XGM and

XPM processes, and for this reason it strongly depends on the value of LEF which varies in an interval of $\alpha \sim (0.1 - 7)$ Newell (1999), Sun (2005). For small $\alpha < 1$ the XPM process is too weak, and the MZI efficiency is very low. However, the analysis shows that for $\alpha \geq 1$ and for the bias current values of $I = 30mA$ the system efficiency is large enough. The efficiency may be significantly increased by using larger values of the bias current for the same value of LEF α.

System of equations (6)-(8) with the average pump and signal photon densities (14) constituting a complete set of equations describing XGM and XPM in QD SOA are essentially nonlinear and extremely complicated. Their analytical solution in a closed form is hardly possible, and for this reason, the system of equations (6)-(8) has been solved numerically for the following typical values of the QD-SOA based AO-XOR Berg (2001), Uskov (2004), Qasaimeh (2003), Qasaimeh (2004): $L = 1500\mu m$, $L_w = 0.1\mu m$, the bias current $I = 30mA$, LEF $\alpha = 1$, the width of QD-SOA $W = 10\mu m$, $\Gamma \sim 3 \times 10^{-2}$, $\tau_{w2} = 3ps$; $\tau_{21} = 0.16ps$; $\tau_{12} = 1.2ps$; $\tau_{1R} = 0.4ns$; $\tau_{2w} = \tau_{wR} = 1ns$, $N_Q = 5 \times 10^{10} cm^{-2}$, $\alpha_{int} = 3cm^{-1}$, $\sigma_E = 30meV$, $\tau_{12} = \tau_{21}\rho \exp(\Delta E_{21}/k_B T)$. Here $\rho = 1$, $\Delta E_{21} = 50meV$ is the separation between the ES and GS energy levels, k_B is the Boltzmann constant, $T = 300K$ is the temperature. The situation when only one data signal interacts with the clock stream signal is shown in Fig. 3.

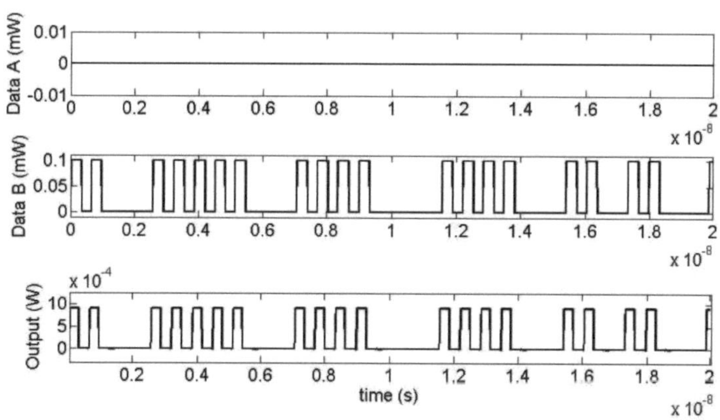

Fig. 3. Wavelength conversion realization by XGM between the data signal B with $\lambda_B = 1560nm$ and the clock stream signal with $\lambda_p = 1530nm$ for the signal bit rate $2.5Gb/s$

In such a case, wavelength conversion occurs between the optical signal B at the wavelength $\lambda_B = 1560nm$ propagating through the lower arm of QD SOA-MZI and the clock stream signal with $\lambda_p = 1530nm$. Pattern-effect free wavelength conversion can be realized for the bit rate up to $100Gb/s$ in the case of a detuning less than the homogeneous broadening, and up to $10Gb/s$ for a large detuning comparable to the inhomogeneous broadening Ben-Ezra (2007). Input optical signal distortions result in the output signal pattern-effect, significant pulse broadening and overlapping accompanied by information losses. In such a case the form of pulses can be essentially improved by using the 3R regeneration process shown in Fig. 4 where the RZ clock stream is fed into the input port instead of a CW signal.

It is seen in Fig. 4 that a distorted pulse duration is almost doubled while the regenerated signals shown with a solid line have a regular structure with the equal amplitudes and a shape defined by the clock.

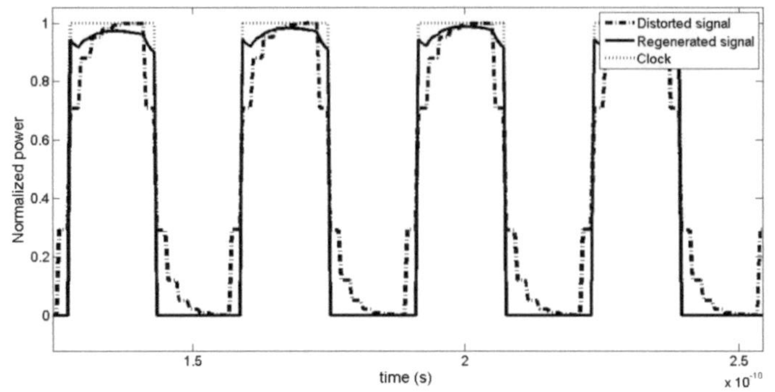

Fig. 4. Optical signal 3R regeneration process

The simultaneous XOR logic operation, wavelength conversion and 3R regeneration for the distorted at the input RZ signals A and B with the wavelengths $\lambda_A = 1550nm$ and $\lambda_B = 1560nm$ for the bit rate of 100Gb/s are shown in Fig. 5. Here the RZ clock stream is fed into the input port instead of a CW signal in order to carry out 3R regeneration.

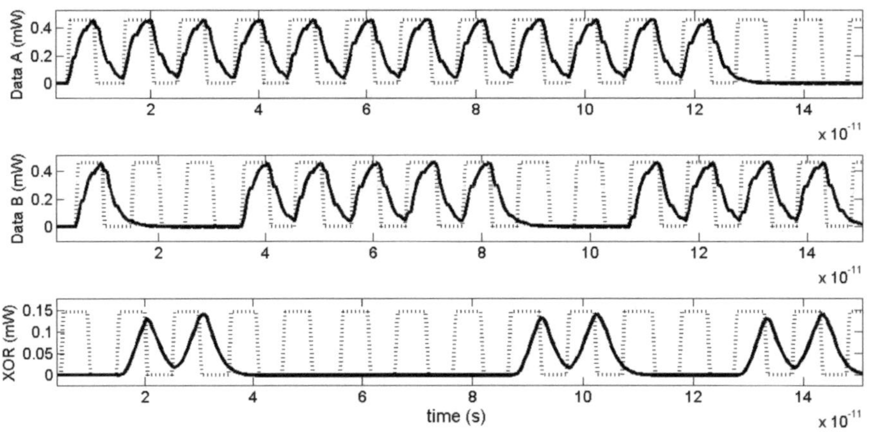

Fig. 5. Simultaneous logic XOR operation, wavelength conversion and 3R regeneration of the distorted RZ signals A, B with $\lambda_A = 1550nm$, $\lambda_B = 1560nm$, and the clock stream (dashed line) at the input port. The bit rate is $100Gb/s$.

The ultrafast all-optical signal processor operation performance is mainly determined by the electron dynamics in QD-SOA. In order to investigate the QD-SOA behavior in both arms of QD SOA-MZI we have solved numerically system of equations (6)-(8). The temporal dependence of the electron concentration in WL $N_w(t)$ and the electron occupation probability of GS and ES, $f(t)$ and $h(t)$, respectively, in the QD-SOA situated in the upper arm of the MZI for the signal repetition rates of $50Gb/s$, $100Gb/s$, and $250Gb/s$ are presented in Figs. 6, 7, 8, respectively.

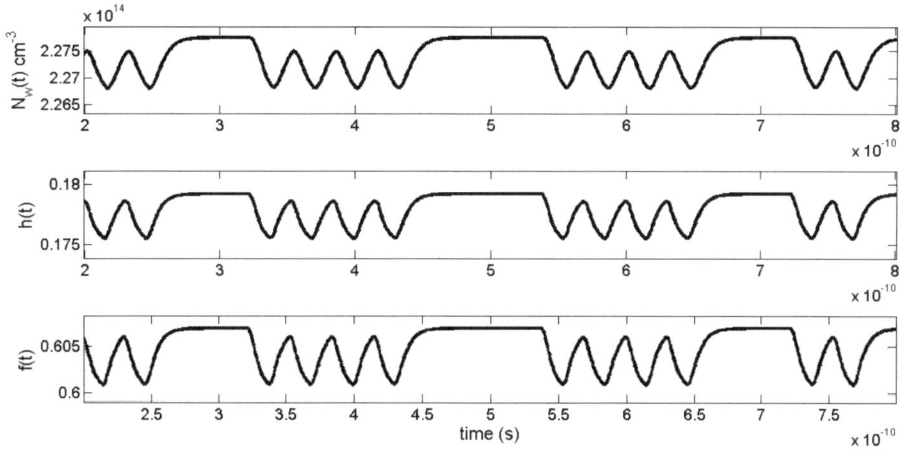

Fig. 6. emporal dependence of the electron concentration in WL $N_w(t)$, the electron occupation probability of GS $f(t)$, and the electron occupation probability of ES $h(t)$ for the QD SOA in the upper arm of the QD SOA-MZI with a clock stream at a data signal repetition rate of $50Gb/s$

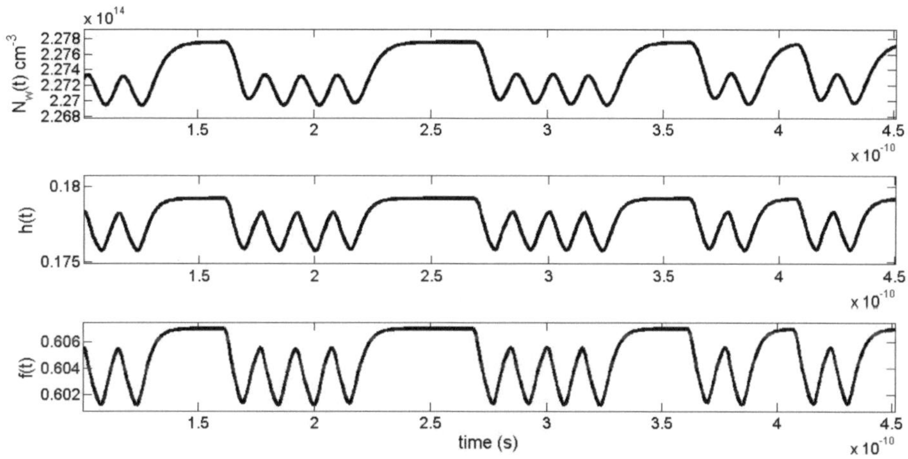

Fig. 7. emporal dependence of the electron concentration in WL $N_w(t)$, the electron occupation probability of GS $f(t)$, and the electron occupation probability of ES $h(t)$ for the QD SOA in the upper arm of the QD SOA-MZI with a clock stream at a data signal repetition rate of $100Gb/s$

The analysis of QD SOA dynamics for the signal detuning smaller than the homogeneous broadening clearly shows that the operation rate of QD-SOAs is limited by the relaxation time $\tau_{w2} \sim (3-5)\ ps$ for the electron transitions between WL and ES in the resonant QDs. At the bit rate of several dozens of Gb/s the oscillations of both WL and ES strictly follow the input optical signal variation.

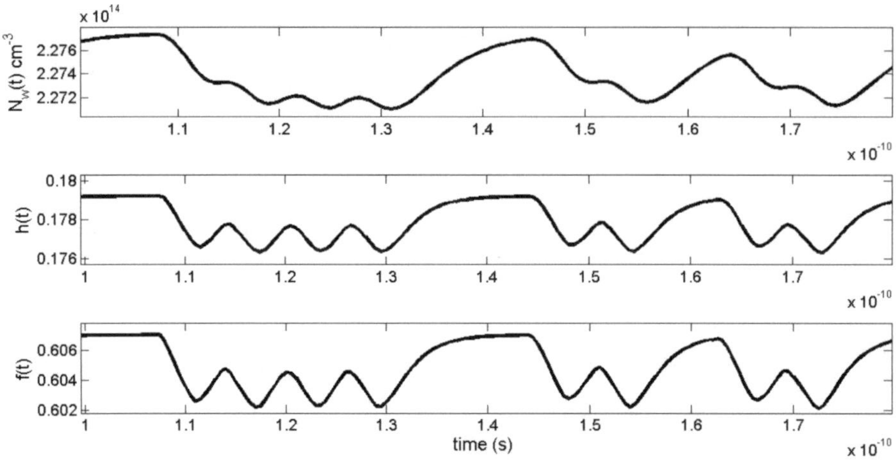

Fig. 8. emporal dependence of the electron concentration in WL $N_w(t)$, the electron occupation probability of GS $f(t)$, and the electron occupation probability of ES $h(t)$ for the QD SOA in the upper arm of the QD SOA-MZI with a clock stream at a data signal repetition rate of $250Gb/s$

The electron population of GS is supported at a comparatively high level $f \gtrsim 0.6$ due to the electron fast transitions between ES and GS. The ES population is low because ES level is emptied rapidly by the fast stimulated transitions to GS at the comparatively high input optical power $P_{in} \gtrsim 5mW$. The operation rate of QD-SOA can be increased due to the acceleration of the gain recovery at the high optical signal power and the large bias currents $I \sim 30mA$ Ben-Ezra (2008), Ben-Ezra (2005b). The filling of ES is determined by slower transitions from WL with the relaxation time $\tau_{w2} \sim (3-5)\,ps$. It is seen from the temporal dependences of $N_w(t)$, $h(t)$ and $f(t)$ in the lower arm QD-SOA for the pulse bit rate of $50Gb/s$, $100Gb/s$, and $250Gb/s$ shown in Figs. 6, 7, 8, respectively, that the WL electron population gradually fails to follow the rapid changes of the input optical signals. At $250Gb/s$ the oscillation form of the GS and ES occupation probabilities also deteriorates.
Consequently, the gain in the both arms of QD SOA-MZI as well as the input power P_{XOR} (19) and phase difference (20) cannot be controlled anymore. Indeed, as it is seen from Figs. 9, 10 for the operation rate of about $250Gb/s << \tau_{w2}^{-1}$ the performance of AO-XOR gate sharply deteriorates due to retardation of the QD-SOA dynamics. The increase of the bias current improves the system performance. However, typically the bias current values for QD-SOAs are of an order of magnitude of $50mA$, which corresponds to the current density of several hundred A/cm^2 Ustinov (2003).

6. Conclusions

We for the first time developed a theoretical model of a QD SOA-MZI based ultra-fast all-optical signal processor which under certain conditions can simultaneously carry out logic gates XOR operation, wavelength conversion, and 3R regeneration of the moderately distorted optical signals. The QD SOA-MZI operation has been analyzed theoretically by solving the rate equations of the QD-SOA dynamics, optical wave propagation equations in an active

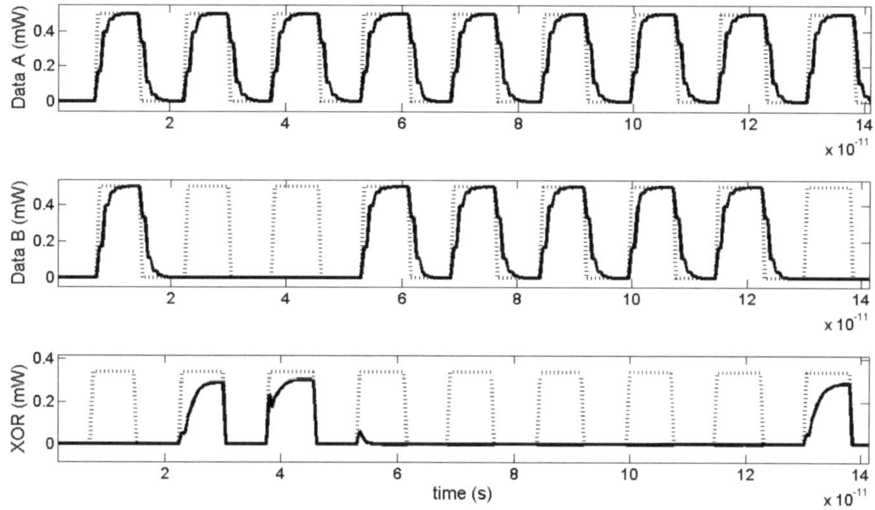

Fig. 9. XOR operation with a clock stream at a signal bit rate $100Gb/s$

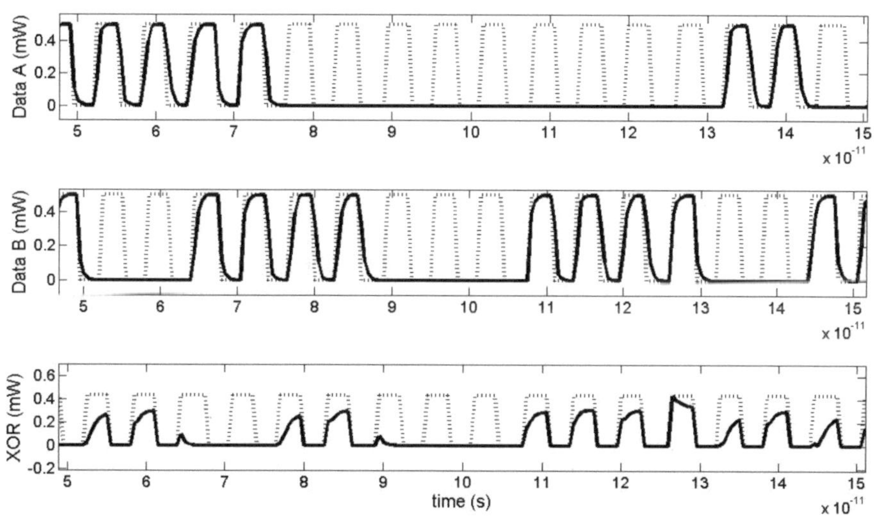

Fig. 10. XOR operation with a clock stream at a signal bit rate $250Gb/s$

medium, and the MZI equations. We have taken into account the two energy levels, namely, ES and GS in the QD conduction band and the inhomogeneity of the QDs size and the resulting inhomogeneous spectral broadening. For a high performance of the proposed processor, LEF should be $\alpha \gtrsim 1$. The QD SOA-MZI based all-optical processor operation rate is limited by the WL-ES electron transition relaxation time and sharply deteriorates with the increase of repetition rates. The limiting bit rate also depends on the bias current value. Analysis shows that for $I = 30mA$ and $I = 50mA$ the highest bit rates corresponding to the processor successful performance are $100Gb/s$ and $200Gb/s$, respectively.

7. References

Agrawal, G.P. & Olsson, N.A. (1989). Self-phase modulation and spectral broadening of optical pulses in semiconductor laser amplifiers. *IEEE Journal of Quantum Electronics*, Vol. 25, No.11, (November 1989) 2297-2306, ISSN 0018-9197

Agrawal, G.P. (2001). *Applications of Nonlinear Fiber Optics*. Academic Press, ISBN 0-12-045144-1, New York

Agrawal, G.P. (2002). *Fiber-Optic Communication Systems*. Wiley, ISBN 0-471-21571-6, New York

Bányai, L. & Koch, S. W. (2005). *Semiconductor Quantum Dots* (Second Edition). World Scientific, ISBN 9810213905, London,

Ben-Ezra, Y.; Haridim, M. & Lembrikov, B. I. (2005). Theoretical analysis of gain-recovery time and chirp in QD-SOA. *IEEE Photonics Technology Letters*, Vol. 17, No. 9, (September 2005) 1803-1805, ISSN 1041-1135

Ben-Ezra, Y.; Lembrikov, B. I. & Haridim, M. (2005). Acceleration of gain recovery and dynamics of electrons in QD-SOA. *IEEE Journal of Quantum Electronics*, Vol. 41, No. 10, (October 2005) 1268-1273, ISSN 0018-9197

Ben-Ezra, Y.; Lembrikov, B. I. & Haridim, M. (2007). Specific features of XGM in QD-SOA. *IEEE Journal of Quantum Electronics*, Vol.43, No. 8, (August 2007) 730-737, ISSN 0018-9197

Ben-Ezra, Y.; Haridim, M.; Lembrikov, B.I. & Ran, M. (2008). Proposal for All-optical Generation of Ultra Wideband Impulse Radio Signals in Mach-Zehnder Interferometer with Quantum Dot Optical Amplifier. *IEEE Photonics Technology Letters*, Vol. 20, No. 7 (April 2008) 484-486, ISSN 1041-1135

Ben-Ezra, Y.; Lembrikov, B.I. & Haridim, M. (2009). Ultra-Fast All-Optical Processor Based on Quantum Dot Semiconductor Optical Amplifiers (QD-SOA). *IEEE Journal of Quantum Electronics*, Vol. 45, No.1 (January 2009) 34-41, ISSN 0018-9197

Berg, T.W.; Bischoff, S.; Magnusdottir, I. & Mørk, J. (2001). Ultrafast gain recovery and modulation limitations in self-assembled quantum-dot devices. *IEEE Photonics Technology Letters*, Vol. 13, No. 6 (June 2001) 541-543, ISSN 1041-1135

Berg, T.W.; Mørk, J. & Hvam, J.M. (2004). Gain dynamics and saturation in semiconductor quantum dot amplifiers. *New Journal of Physics*, Vol. 6, No. 178, (2004) 1-23, ISSN 1367-2630

Berg, T.W. & Mørk, J. (2004). Saturation and Noise Properties of Quantum-Dot Optical Amplifiers. *IEEE J. of Quantum Electronics*, Vol. 40, No. 11, (November 2004) 1527-1539, ISSN 0018-9197

Bimberg, D.; Grundmann, M. & Ledentsov, N. N. (1999). *Quantum Dot Heterostructures*. John Wiley, ISBN 047 1973882, New York

Chen, H.; Zhu, G.; Wang, Q.; Jaques, J.; Leuthold, J.; Picirilli, A.B. & Dutta, N.K. (2002). All-optical logic XOR using differential scheme and Mach-Zender interferometer. *Electronic Letters*, Vol. 38, No. 21, (October 2002) 1271-1273, ISSN 0013-5194

Dagens, B.; Markus, A.; Chen, J.X.; Provost, J.-G.; Make, D.; de Gouezigou, O.; Landreau, J.; Fiore, A. & Thedrez, B. (2005). Giant linewidth enhancement factor and purely frequency modulated emission from quantum dot laser, *Electronics Letters*, Vol. 41, No. 6, (17th March 2005) 323-324, ISSN 0013-5194

Dong, J.; Zhang, X.; Fu, S.; Xu, J.; Shum, P. & Huang, D. (2008) Ultrafast all-optical signal processing based on single semiconductor optical amplifier. *IEEE Journal of Selected Topics in Quantum Electronics*, Vol. 14, No. 3 (May/June 2008) 770-778, ISSN 1077-260X

Gehrig, E. & O. Hess, O. (2002). Mesoscopic spatiotemporal theory for quantum-dot lasers. *Phys. Rev. A*, Vol. 65, No. 3, (March 2002) 033804-1-16, ISSN 1050-2947

Hamié, A.; Sharaiha, A.; Guégan, M. & Pucel, B. (2002). All-Optical logic NOR Gate Using Two-Cascaded Semiconductor Optical Amplifiers. *IEEE Photonics Technology Letters*, Vol. 14, No. 10, (October 2002) 1439-1441, ISSN 1041-1135

Joergensen, C. ; Danielsen, S.L.; Durhuus, T.; Mikkelsen, B. ; Stubkjaer, K.E.; Vodjdani, N.; Ratovelomanana, F.; Enard, A.; Glastre, G.; Rondi, D. & Blondeau, R. (1996). Wavelength conversion by optimized monolithic integrated Mach-Zender interferometer, *IEEE Photonics Technology Letters*, Vol. 8, No. 4, (April 1996) 521-523, ISSN 1041-1135

Kanellos, G.T.; Petrantonakis, D.; Tsiokos, D.; Bakopoulos, P.; Zakynthinos, P.; Pleros, N.; Apostolopoulos, D.; Maxwell, G.; Poustie, A. & Avramopoulos, H. (2007). All-optical 3R burst-mode reception at 40 Gb/s using four integrated MZI switches. *Journal of Lightwave Technology*, Vol. 25, No. 1, (January 2007) 184-192, ISSN 0733-8724

Kim, J.; Laemmlin, M.; Meuer, C.; Bimberg, D. & Eisenstein, G. (2009). Theoretical and experimental study of high-speed small-signal cross-gain modulation of quantum-dot semiconductor optical amplifiers. *IEEE J. of Quantum Electronics*, Vol. 45, No. 3, (March 2009) 240-248, ISSN 0018-9197

Leem, Y.A.; Kim, D. C.; Sim, E.; Kim, S.-B.; Ko, H.; Park, K.H.; Yee, D.-S.; Oh, J. O.; Lee, S. H. & Jeon, M. Y. (2006). The characterization of all-optical 3R regeneration based on InP-related semiconductor optical devices, *IEEE J. of selected topics in Quantum Electronics*, Vol. 12, No. 4, (July/August 2006) 726-735, ISSN 1077-260X

Matthews, D.K. ; Summers, H.D.; Smowton, P.M.; Blood, P.; Rees, P. & Hopkinson, M. (2005). Dynamics of the wetting-layer-quantum-dot interaction in GaAs self-assembled systems, *IEEE Journal of Quantum Electronics*, Vol. 41, No. 3, (March 2005), 344-350, ISSN 0018-9197

Mukherjee, B. & Zang, H. (2001). Introduction. Survey of State-of-the-Art, In: *Optical WDM Networks. Principles and Practice*, Sivalingam, K.M. & Subramaniam, S. (Ed.), 3-24, Kluwer, ISBN 0-7923-7825-3, Boston

Newell, T.C.; Bossert, D.J.; Stinz, A.; Fuchs, A. & Malloy, K.J. (1999). Gain and linewidth enhancement factor in InAs quantum-dot laser diodes, *IEEE Photonics Technology Letters*, Vol. 11, (November 1999), 1527-1529, ISSN 1041-1135

Ohtsu, N.; Kobayashi, K.; Kawazoe, T.; Yatsui, T. & Naruse, N. (2008). *Principles of Nanophotonics*, CRC Press, ISBN-13 978-1-58488-972-4, London

Qasaimeh, O. (2003). Optical gain and saturation characteristics quantum-dot semiconductor optical amplifiers. *IEEE J. of Quantum Electronics*, Vol. 39, No. 6, (June 2003) 793-798, ISSN 0018-9197

Qasaimeh, O. (2004). Characteristics of cross-gain (XG) wavelength conversion in quantum dot semiconductor optical amplifiers. *IEEE Photonics Technology Letters*, Vol. 16, No. 2, (February 2004) 542-544, ISSN 1041-1135

Ramamurthy, B. (2001). Swithches, wavelength routers, and wavelength converters. In: *Optical WDM Networks. Principles and Practice*, Sivalingam, K.M. & Subramaniam, S. (Ed.), 51-75, Kluwer, ISBN 0-7923-7825-3, Boston

Sakamoto, A. & Sugawara, M. (2000). Theoretical calculation of lasing spectra of quantum-dot lasers: effect of homogeneous broadening of optical gain, *IEEE Photonics Technology Letters*, Vol. 12, No. 2, (February 2000) 107-109, ISSN 1041-1135

Sartorius, B. (2001). 3R regeneration for all-optical networks, *Proceedings of 3rd International Conference on Transparent Optical Networks* (ICTON 2001), Th. A.4, pp. 333-337, ISBN 10 0780370961, Krakow, Poland, June 18-21, IEEE, Krakow

Schneider, S.; Borri, P.; Langbein, W.; Woggon, U.; Sellin, R.L.; Ouyang, D. & Bimberg, D. (2004). Linewidth enhancement factor in InGaAs quantum-dot amplifiers, *IEEE of Quantum Electronics*, Vol. 40, No. 10, (October 2004) 1423-1429, ISSN 0018-9197

Sugawara, M.; T. Akiyama, T.; N. Hatori, N. ; Y. Nakata, Y.; Ebe, H. & H. Ishikava, H. (2002). Quantum-dot semiconductor optical amplifiers for high-bit-rate signal processing up to 160 Gbs^{-1} and a new scheme of 3R regenerators, *Measurement Science and Technology*, Vol. 13, (2002), 1683-1691, ISSN 0957-0233

Sugawara, M.; Ebe, H.; Hatori, N.; Ishida, M.; Arakawa, Y.; Akiyama, T.; Otsubo, K. & Nakata, Y. (2004) Theory of optical signal amplification and processing by quantum-dot semiconductor optical amplifiers. *Phys. Rev.B*, Vol. 69, No. 23 (June 2004) 235332-1-39, ISSN 1098-0121

Sun, G.; Khurgin, J.B. & Soref, R.A. (2004). Design of quantum-dot lasers with an indirect bandgap short-period superlattice for reducing the linewidth enhancement factor, *IEEE Photonics Technology Letters,* Vol.16, No. 10 (October 2004), 2203-2205, ISSN 1041-1135

Sun, H.; Wang, Q.; Dong, H. & Dutta, N.K. (2005). XOR performance of a quantum dot semiconductor optical amplifier based Mach-Zender interferometer, *Optics Express*, Vol. 13, No. 6, (March 2005) pp.1892-1899, ISSN 10944087

Uskov, A.V. ; Berg, T.W. & Mørk, J. (2004). Theory of pulse-train amplification without patterning effects in quantum-dot semiconductor optical amplifiers. *IEEE J. of Quantum Electronics*, Vol. 40, No. 3, (March 2004) 306-320, ISSN 0018-9197

Ustinov, V.M.; Zhukov, A.E.; Egorov, A. Yu. & Maleev, N. A. (2003). *Quantum Dot Lasers,* Oxford University Press, ISBN 0 19 852679 2, Oxford

Wada, O. (2007). Femtosecond all-optical devices for ultrafast communication and signal processing, In: *Microwave Photonics,* Lee, C. H. (Ed), 31-75, CRC Press, ISBN-10: 0-8493-3924-3

Wang, Q.; Zhu, G.; Chen, H. ; Jaques, J. ; Leuthold, J.; Picirilli, A.B. & Dutta, N.K. (2004). Study of all-optical XOR using Mach-Zehnder interferometer and differential scheme. *IEEE J. of Quant. Electr.*, Vol. 40, No. 6, (June 2004) 703-710, ISSN 0018-9197

Zhu, Z.; Funabashi, M.; Zhong Pan; Paraschis, L.; Harris, D. & Ben Yoo, S.J. (2007). High-performance optical 3R regeneration for scalable fiber transmission system applications", *Journal of Lightwave Technology*, Vol. 25, No. 2, (January 2007) 504-511, ISSN 0733-8724

19

All-optical digital processing through semiconductor optical amplifiers: state of the art and perspectives

Gianluca Berrettini[2], Antonella Bogoni[1], Emma Lazzeri[2], Gianluca Meloni[2], Claudio Porzi[2], Luca Potì[1] and Mirco Scaffardi[1]

[1]*Consorzio Nazionale Interuniversitario per le Telecomunicazioni (CNIT), Pisa*
[2]*Scuola Superiore Sant'Anna, Pisa*
[1]*Italy*
[2]*Italy*

1. Introduction

Photonic digital processing is a capability which interest has been growing in for a long time. The main advantage of the photonic processing with respect to the electronic processing is the high speed (Tangdiongga et al., 2007; Liu et al., 2007). Nevertheless photonics is in its early stage and the realisation of complete all-optical computing system is still far, mainly due to the lack of efficient all optical memories. Despite the present limitations, photonic digital processing shows promise in the applications where fast computation speed is required. One of these scenarios is represented by the all-optical short-range photonic interconnection networks. The improvement of the present high performance computing systems is hampered by the bottleneck of the chip-to-chip and chip-to-memory communication. The limit is represented by the wiring density, the high power consumption and the throughput (Shacham et al., 2005). Photonic interconnection networks can overcome the limitations of the electronic interconnection networks by guaranteeing high bit rate communications, data format transparency and electromagnetic field immunity. Moreover they can reduce the wiring density and the power consumption. In such networks photonic digital processing can be the most suitable paradigm for simple and ultra-fast control and switching operations, since it reduces the packet latency to the optical time-of-flight.

Elementary logic gates as AND, NOR, NAND, NOT, XOR are the basic functions necessary to realizing digital processing. Nevertheless for the controlling of all-optical interconnection networks several complex functions are required. Among them, two important functions are the managing of the contentions and the controlling of the switch. Moreover for a more flexible and effective managing of the network the priority information has to be carried by the packet label. In case of packets directed to the same node output port, the priority field of the contending packet is compared. The packet with the highest priority is directed to the designated output port. The other packet is delayed or discharged. Therefore a complex photonic digital circuit, able to compare two boolean numbers, is mandatory (Andriolli et al., 2007). All-optical subsystems able to discriminate if an N-bit (with $N \geq 1$) pattern representing a boolean number is greater or lower than another one are not reported yet. Calculating the addition of boolean numbers is another important functionality to perform packet header processing. If some packets are routed to the wrong link or are mislabelled, they can be routed in circles without reaching the destination. These loops are a cause of network congestion and must be avoided. A Time-To-Live (TTL) field in the packet prevents the formation of loops. This field represents the maximum number of hops of a packet and it is decremented after each node. When the field value is zero, the packet is discharged (McGeehan et al., 2003). The implementation of this functionality requires an all-optical processing circuit able to perform the decrementing of the boolean number in the TTL field. This operation can be performed by means of an all optical full-adder applying the so called method of complements (Hayes 1998). Moreover the all-optical full-adder can find application in resolving the Viterbi algorithm in the Maximum-Likelihood Sequence Estimation (MLSE) (Forney 1973; Proakis 1996). This method requires performing fast additions. An all-optical implementation of this algorithm can improve its efficiency.

Other two important functions are analog-to-digital conversion (ADC) and digital-to-analog conversion (DAC). ADC is a key functionality which, converting continuous-time signals to digital binary signals, enables them to be transmitted through the modern digital communication networks. Applications regard e.g. radar signals, high-definition video, real-time signal monitoring, ultra-fast dispersion compensation. Although it didn't receive the same attention as all-optical analog-to-digital conversion, DAC and/or multilevel codification in the optical domain has also been extensively investigated in order to implement some ultra-fast signal processing functions. These functions include, for instance, pattern recognition for header extraction (Saida et al., 2001), amplitude multiplexing for increasing spectral efficiency (Abbade et al., 2005) or label/payload encoding techniques (Abbade et al., 2006), and waveform generation for radar and display applications (Yacoubian & Das, 2003).

The photonic digital processing is effective and attractive if it can be realised with integrated solutions. SOAs have shown to be attractive because of their compactness, stability, low switching energy and low latency. SOAs are reliable, relatively low cost devices which can be integrated within complex optical circuits with hybrid techniques (Maxwell, 2008; Lal et al., 2007; Kehayas et al., 2006 b). In this perspective, the possibility of using a single basic building gate for implementing all the complex logic functions is practical. In this chapter new schemes for the implementation of SOA-based reconfigurable logic gates, a photonic combinatorial network, a comparator, a full-adder, a digital-to–analog converter and an analog-to-digital converter will be presented.

2. Reconfigurable logic gates with a single SOA

Scheme of all-optical logic gates are reported in literature, using nonlinear effects in optical fibers (Bogoni et al., 2005; Chbat et al., 1992), in semiconductor devices (Ibrahim et al., 2003; Dorren et al., 2004) or in waveguides (Collecutt & Drummond, 2000). Moreover several efforts have been done to demonstrate the suitability of new structures for the realization of optical logic gates (Wu, 2005; Brzozowski & Sargent, 2001). In particular SOAs are very attractive nonlinear elements for the realization of different logic functions, due to their strong change of the refractive index together with high gain. Moreover, differently from fiber devices, SOAs allow photonic integration. In some schemes, SOA-based logic gates require interferometric structures including two or more devices with identical characteristics (Dorren et al., 2004) that need accurate control and stabilization. A practical and efficient solution makes use of a single SOA to realize the XNOR logic function, by exploiting simultaneously Four Wave Mixing (FWM) and Cross Gain Modulation (XGM) (Kumar & Willner, 2004; Berrettini et al., 2006 a). This architecture avoids using tricky interferometric configurations. Moreover it is easily reconfigurable: the same architecture can also implement NOT, AND, and NOR logic gates, merely turning on or off a probe signal. The use of a counter-propagating Continuous Wave (CW) light reduces the SOA response times, by increasing the maximum bit rate of the signals to be processed. In more details, the presence of the CW light allows to increase the mean saturation level of the SOA, thus reducing the dynamics of the gain recovery after the arrival of the input signals. In this way the signal distortions due to the SOA pattern effects can be minimized, thus allowing to process ultra-fast signals. All the new aspects of this scheme contribute to strongly reduce the complexity and the instability of the implementation, realizing an ultra-fast and reconfigurable logic gate.

2.1 Working principle

The working principle of the proposed XNOR gate is described in Fig. 1 (a). A and B are the signals that have to be processed, whose wavelengths are λ_A and λ_B respectively. The logic function is obtained exploiting simultaneously FWM between the two polarization-aligned signals A and B, and XGM on a co-propagating probe signal, whose wavelength is the same as one of the generated FWM term ($\lambda_{probe}=\lambda_{FWM}$).

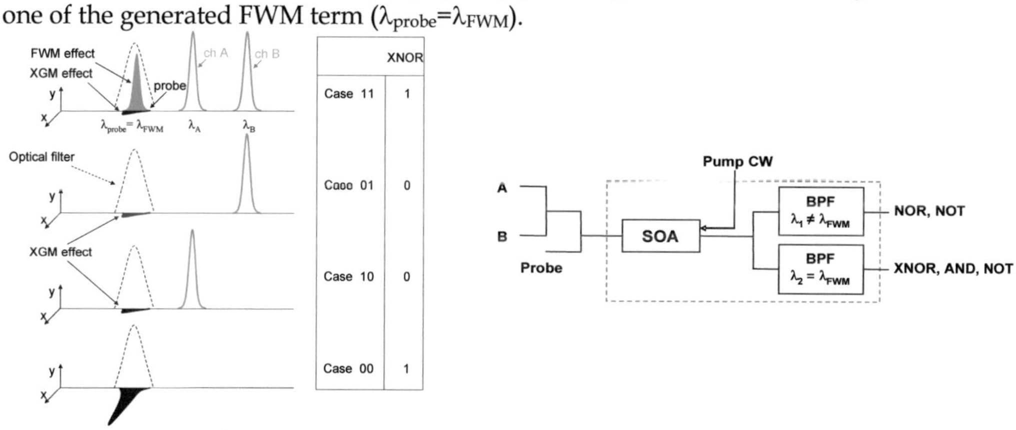

(a) (b)

Fig. 1. (a) Working principle of the XNOR gate. (b) Scheme of the reconfigurable logic gate.

When both signals are present into the SOA (case 11), the FWM component is generated and simultaneously the probe channel experiences a very low gain into the saturated device. At the SOA output the FWM term is generated and it can be optically filtered. Consequently the logic gate output is at the high level. In case both signals A and B are absent (case 00) the FWM effect does not take place and the SOA is not saturated. Therefore, the probe channel experiences a strong amplification. By opportunely setting the probe channel input power it is possible to equalize the high power level at the gate output in the case 00, which is equal to FWM component in the case 11. On the other hand, if only one of the signals A and B is present (cases 10 and 01) the FWM does not occur but the SOA is saturated, strongly reducing the gain of the probe. In these cases at the output the power level is low. In Fig. 1 (b) it is shown the scheme of the reconfigurable SOA-based logic gate. The proposed scheme can be exploited not only to realize the XNOR logic gate, but also AND, NOR and NOT functionalities, maintaining the same input conditions for the signals A and B. In particular, if the probe channel is turned off, the output of the optical filter centered at λ_{FWM} represents a FWM-based logic function AND. Moreover, by changing the wavelength of the probe channel so that $\lambda_{probe} \neq \lambda_{FWM}$ allows to extract the NOR signal, using an optical band pass filter centered at λ_{probe}. In this case, the NOR gate is based on XGM in the SOA. Finally, the NOT function can be obtained considering only one input signal in both gate output, by exploiting XGM on the probe channel. Therefore the simple and integrable scheme reported in Fig. 1 (b), which includes two different filters at the output of the SOA, can be easily reconfigured to obtain different logic gates, just controlling the wavelength of the probe channel or turning it off. The proposed scheme is suitable for both Non Return-to-Zero (NRZ) or Return-to-Zero (RZ) signals. In order to avoid the phase interference between probe and FWM component, the probe channel is launched into the SOA with orthogonal polarization with respect to the signals and consequently to the FWM term. The pulsed signals A and B have the same input peak power, and it is high enough to saturate the device and to induce a high-efficiency FWM effect. The probe peak power is sufficiently low to avoid SOA saturation. To eliminate pattern dependent signal distortions, a counter-propagating high-power CW light has been launched into the device, decreasing the mean life time of the carriers and keeping an optimum saturation level in the SOA.

2.2 Performance of the logic gate

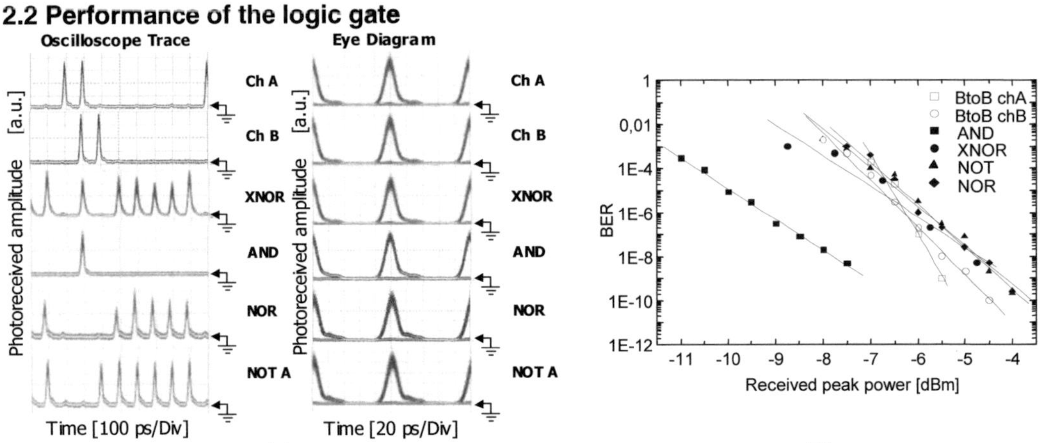

(a) (b)

Fig. 2. (a) Sequences of the input signals and of the corresponding logic gate output (left). Eye-diagram of the input signals and of the corresponding logic gate output (right). (b) BER of the signals A and B back-to-back (BtoB), and of the logic gate output.

The performance of the reconfigurable logic gate is summarized in Fig. 2 (a) and (b). The logic gate is tested with 10 Gb/s signals at λ_A=1550.9 nm, λ_B=1552.5 nm, and

$\lambda_{probe}=\lambda_{FWM}$=1549.3 nm or $\lambda_{probe}\neq\lambda_{FWM}$=1546.1 nm. The wavelength of the CW signal is 1544 nm. The SOA is a commercial polarization independent bulk SOA. The input average power is 3 dBm and 10 dBm for signals and CW respectively.
The corresponding signal pulse energy was 40 fJ. The probe mean power was -15dBm, corresponding to a pulse energy of 6.3 fJ, in order to obtain an output signal power equalized with the generated FWM component whose power was -13 dBm.
The input sequences for channel A and B respectively are shown in Fig. 2 (a)-left. All the cases 11, 01, 10, and 11 are considered. Fig. 2 (a)-left reports also the corresponding output functions, using the scheme as XNOR, AND, NOR and NOT respectively. The correct output sequences 1010111110, 0010000000, 1000111110, and 1001111110 for XNOR, AND, NOR and NOT gates respectively are generated. Fig. 2 (a) –right reports the eye diagrams at the input and at the output of each logic port considering 2^{31}-1 Pseudo-Random Bit Sequences (PRBSs) at the input of the logic gate. The high quality of the eye diagram for each logic gate demonstrates the effectiveness of the proposed reconfigurable scheme. Fig. 2 (b) shows the BER curves at the output of each logic port using for all cases the same 2^{7}-1 input sequence. The introduced penalty at 10^{-9} is lower than 0.5 dB with respect to the worst input signal, making the proposed scheme suitable for cascaded configurations.

3. Photonic combinatorial circuit for contention management in optical packet switched networks

A possible application of the optical digital processing is within the optical packet switching (OPS) networks. In fact high speed communication and processing are key features for future ultra-fast transmission and computing systems, within scenarios ranging from wide-area backbone networks to inter-chip interconnection networks. Nevertheless, few examples of the simple cascade of two logic gates have been demonstrated (Kehayas et al., 2006 a; Chan et al., 2003), while the design and the implementation of more complex digital circuits involving cascades of several logic functions have not been addressed yet.
Here it is presented an implementation of the combinatorial circuit for a 2×2 photonic node architecture, where all the forwarding functions (label recognition and processing, contention detection and resolution, switch control, switching, and regeneration) are optically performed by exploiting integrable solutions (Scaffardi et al., 2007).
Due to the full all-optical processing, the packet latency time approaches the passing-through time of light in the photonic switch (few ns, even less than one ns if optical integration can be achieved). The low latency, together with the high scalability, makes the 2×2 all-optical node suitable as a switching element in short-range multistage interconnection networks, aimed at connecting processors and memories of a high-performance computing cluster. Indeed in such a short-range scenario the network synchronization can be easily maintained and the absence of effective optical buffers is less problematic.

3.1 Architecture of the photonic node

Fig. 3 shows the architecture of the photonic node. The 2×2 switching element operates synchronously on fixed length packets. The switching operation is achieved by means of ultra-fast digital processing in the optical domain realized exploiting nonlinear effects in nonlinear devices. Since all the node operations and processing occur in the photonic domain, a simple label structure has been adopted. In this way the complexity of the all-optical packet processing is reduced and the packet self-routing in multistage node combinations is simplified. As shown in Fig. 3, the first bit PI (Packet Identifier) of the packet allows the packet recognition, while the path is defined associating the i-th switch with the i-th bit L_i of the label. Each bit of the label is read in the optical domain by exploiting nonlinear effects in semiconductor devices. The processing time in this case is less than 10 ns and it is limited by the propagation delay in the fiber pigtails. The contention detection is performed through the combinatorial network. Then, the contention is resolved by the cancellation (i.e. dropping) of contention-losing packets. The bar or cross state of the 2×2 fabric is set by means of a control signal represented by an optical gate (Malacarne et al., 2006). This gate lasts as long as the packet duration and, depending on its high or low power level, nonlinear effects respectively occur or do not occur in the 2×2 fabric. The packets are consequently switched to the proper output (Berrettini et al., 2006 b). The switching time, defined as the time needed to pass from 10% to 90% of the total swing, is lower than one bit time and it is limited by the transients of the switching control signal.

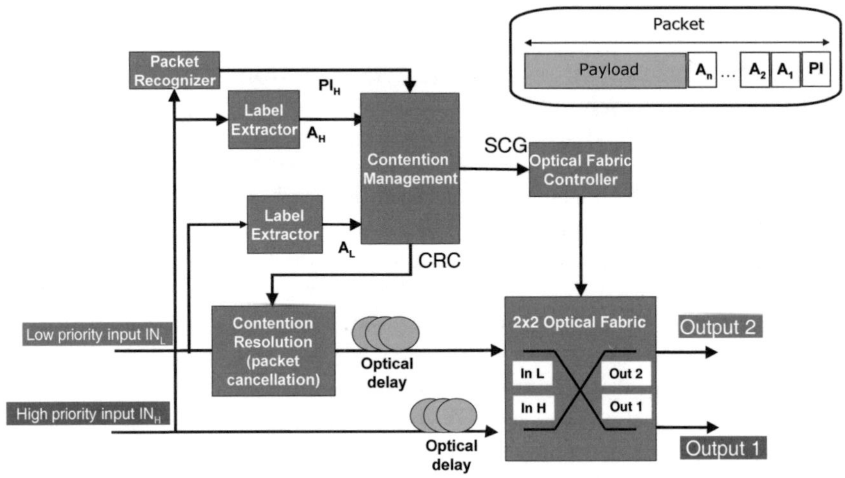

Fig. 3. Photonic node architecture and packet format (inset 1). A_H: high priority packet address bit; PI_H: high priority packet identifier bit; A_L: low priority packet address bit; OUT 1 identified by address '0'; OUT 2 identified by address '1'; CRC: Contention Resolution Control; SCG: Switching Control Generation.

3.2 Combinatorial network for contention management

All-optical packet contention management is addressed by means of a combinatorial network designed to process label information in order to properly configure the 2×2 all-optical switching node and to drive the contention resolution block. With reference to Fig. 3, the following hypotheses are considered: the two switching input ports have different priority (H: high, and L: low), due to the synchronous architecture, the packets reach synchronously the input ports and have the same time duration, the packet label is composed of one Packet Identifier (PI) bit and an N-bit address where each bit refers to one

of N network stages and its value univocally determines the packet route ('0' identifies output port 1 and '1' identifies output port 2 of the incoming switch).
Once the packet reaches the high priority input port, Label Extractor and Packet Recognizer block isolate the address bit A_H and the PI_H bit respectively, where subscript H stands for High priority. If the packet enters low priority input port, it is processed only by the Label Extractor in order to extract the address A_L, where L stands for Low priority. The PI is not necessary for low priority input port, being this port conditioned by the high priority input. A_H, A_L, and PI_H form the input signals for the combinatorial network whose outputs are the Switching Control Generation (SCG) and Contention Resolution Control (CRC) signals. The former is responsible for switching bar/cross configuration, the latter drives the contention resolution block.
For proper operation, the combinational network must preserve the packet incoming from the high priority input ($PI_H = 1$) and send it to the correct output port indicated into the address (A_H). At the same time, if contention is detected (CRC = 1 when $A_H = A_L$), it must be resolved by the devoted block. On the other hand when $PI_H = 0$ (high priority input packet not present), the low priority packet must be redirected to the proper output port (A_L). If we associate the values '0' and '1' of the SCG to the switch cross and bar states respectively, the truth table for the combinatorial network results as in Table 1. As first example, we consider the case $PI_H = 0$: it means that the packet is not present at the high priority input. Thus physically the value for the corresponding address bit A_H is '0'. For what concerns the low priority input, $A_L = 0$ states that the packet does not exist or that it must be routed to the output port '0'. In this case, no contention occurs (CRC = 0) and the switch must be set in the cross status (SCG = 0). If $PI_H = 1$ and both the addresses $A_H = A_L = 0$, a contention is detected (CRC = 1) and the circuit must guarantee the priority to the high priority input. This means bar configuration for the switch (SCG = 1). All the other cases can be easily determined following the previous examples. The truth table contains also two cases without physical sense: in fact, when $PI_H = 0$, no input packet flows through high priority port and therefore A_H can not assume the value '1'.

A_H	A_L	PI_H	CRC	SCG	
0	0	0	0	0	
0	0	1	1	1	
0	1	0	0	1	
0	1	1	0	1	
1	0	1	0	0	
1	1	1	1	0	
1	0	0	0	0	Impossible cases
1	1	0	1	0	

Table 1. Combinatorial network truth table: CRC=0 no contention; CRC=1 contention occurs; SCG=1 switch in bar configuration; SCG=0 switch in cross configuration.

The combinatorial network can be obtained by implementing the logic circuit in Fig. 4 (a) where the following logic gates are used: three NOR, two AND, and one OR. For this logic circuit we exploit a SOA-based implementation, which gives benefits in terms of compactness, stability and power consumption. One AND function is based on FWM in a

SOA. Two NOR functions are realized by means XGM induced in SOAs by the input signals on an auxiliary counter-propagating channel (aux). The cascade of one NOR and one AND function is obtained by means of XGM induced in SOA by two input signals on a third counter-propagating input signal. Finally the OR function is realized simply using a 3 dB coupler: this is possible just because its input ports can not be '1' at the same time. Since the signals fed into the 3 dB coupler must be at the same wavelength, a wavelength converted copy (PI_C) of PI_H is obtained by FWM in SOA using the auxiliary channel. The physical schematic setup is shown in Fig. 4 (b).

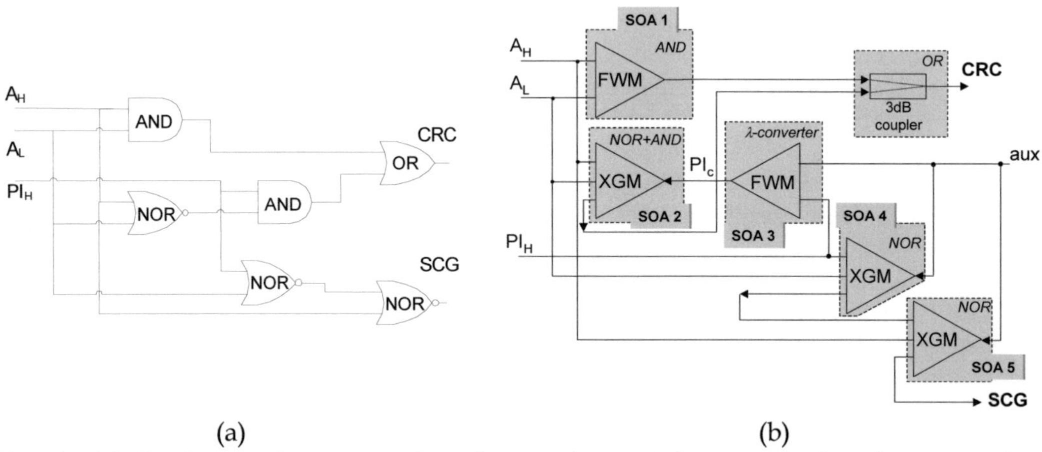

Fig. 4. (a) Logic circuit representing the combinatorial network for the contention management. (b) Physical schematic setup: triangles represent SOAs (exploited effect indicated inside); aux: probe signal; PI_C: wavelength converted PI_H.

The combinatorial circuit is implemented with commercial SOAs. The performance is measured with input signals at λ_H= 1550.9 nm (FWHM: 10 ps) for A_H and PI_H. A PRBS 2^7-1 is mapped onto the pulses. A_L is a converted (λ_L = 1552.5 nm) replica of A_H. A_H and A_L are fed into SOA1, both with a power of 9.6 dBm.

SOAs saturation level is biased through a Continuous Wave (CW) signal at a wavelength of λ_{CW} = 1540 nm. Tunable Optical Delay Lines (ODL) and 0.3 nm optical band-pass filters are used in order to properly synchronize and select the involved signals.

The main performances for the combinatorial network are summarized in Fig. 5 and Fig. 6 (a), where the sequences and the eye diagrams for the three input signals and for the two output signals are respectively shown. SCG and CRC sequences reveal that the logic circuit works properly, i.e. according with the truth table. The output sequences are not perfectly equalized due to residual patterning effect, but the Contrast Ratio (CR) between high and low level is between 8 and 9.3 dB for the SCG and between 8.4 and 10 dB for the CRC. These values could be further improved by using well-known pedestal suppressor schemes. The eye diagrams of the output signals look sufficiently clear, thus confirming the good performance of the implemented combinatorial network. BER measurements are after a pre-amplified receiver and shown in Fig. 6 (b). At BER=10^{-9} a negligible penalty is present for the SCG and a 5 dB power penalty for the CRC that can be mainly ascribed to the noise arising during the FWM process in SOA1 and SOA3.

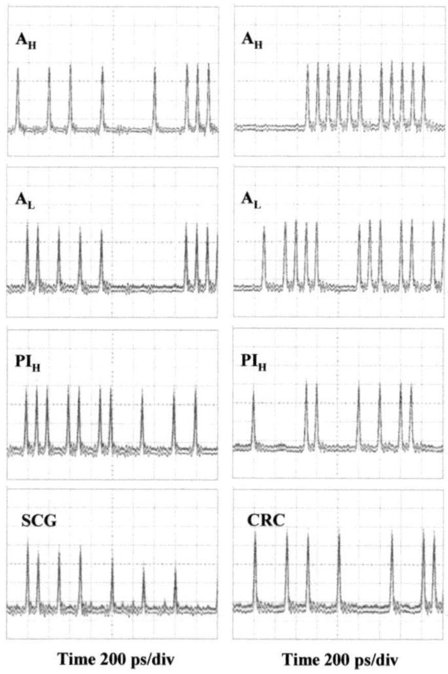

Fig. 5. Input sequences and corresponding SCG and CRC output signals.

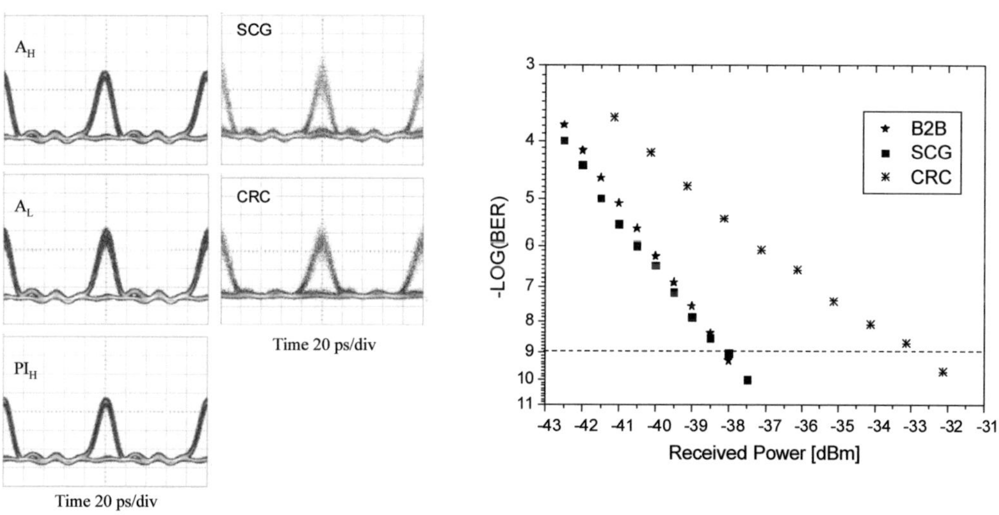

(a) (b)

Fig. 6. (a) Input eye diagrams for A_H, A_L, and PI_H (left) and output eye diagrams for SCG and CRC (right) in the case of BER = 10^{-9}. The input sequences are 2^7-1 PRBS. (b) BER measurements for Back-to-Back (B2B), SCG, and CRC. The input sequences are 2^7-1 PRBS.

4. N-bit comparator

In the previous section it is explained as for the controlling of all-optical interconnection networks several complex functions are required. Among them, two important functions are the managing of the contentions and the controlling of the switch. For a more flexible and effective managing of the network the priority information has to be carried by the packet label. In case of packets directed to the same node output port, the priority field of the contending packet is compared. The packet with the highest priority is directed to the designated output port. The other packet is delayed or discharged. Therefore a complex photonic digital circuit, able to compare two boolean numbers, is mandatory (Andriolli et al., 2007). Up to now some works report on the implementation of all-optical circuits for the pattern matching, i.e. able to determine if two boolean numbers are equal or not. Pattern matching by a XOR gate implemented with a nonlinear optical loop mirror is demonstrated in (Hall & Rauschenbach, 1996). In (Nielsen et al., 2002) pattern matching is obtained by combining AND and XOR gates in a single Semiconductor Optical Amplifier-Mach Zhender Interferometer (SOA-MZI). The cascade of SOA-MZI structures is used in (Martinez et al., 2006) in order to have a single output pulse in case of matching. With this last approach N-bit patterns require N SOA-MZIs. Multiple correlation of PSK-coded labels is demonstrated in (Wada et al., 2006) with an arrayed waveguide grating. In (Wang et al., 2007) an SOA-based all-optical circuit for the comparison of 1-bit boolean numbers is demonstrated. But all-optical subsystems able to discriminate if an N-bit (with N≥1) pattern representing a boolean number is greater or lower than another one are not reported.
In the following it is presented an all-optical N-bit comparator based on a basic building block, i.e. an SOA exploiting XGM between two counter-propagating signals (Scaffardi et al., 2008). The XGM-induced polarisation rotation is used for improving the output pulse extinction ratio. The N-bit all-optical comparator is able to compare two patterns A and B by computing the functions A>B, A<B and $\overline{A=B}$.

4.1 Working principle

Fig. 7 shows the logical representation of the comparator.

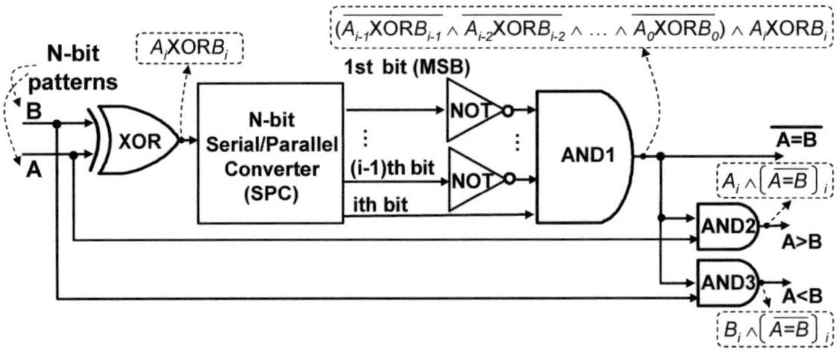

Fig. 7. Logical representation of the comparator.

The two N-bit patterns A and B are compared sequentially, starting from the Most Significant Bit (MSB). At the XOR output a Serial-to-Parallel Conversion (SPC) is performed. The AND (AND1) between the ith output of the XOR and the preceding (i-1) logically inverted bits is carried out. The AND1 output is a N-bit sequence which represents the function $\overline{A=B}$. If the patterns A and B are equal, the N bits are 0 at AND1 output. This is because the XOR output is 0 for each compared bit. If two patterns which differ at least for one bit are compared, the AND1 output becomes 1 when the first mismatch (at i[th] bit) occurs. Indeed the XOR output becomes 1, while the preceding (i-1) bits at the XOR output are 0. Consequently AND1 output results 1. For the remaining (N-i) bits AND1 output is 0, because at least one of its inputs is 0. The function A>B is obtained by exploiting the AND (AND2) between the AND1 output and A. While the pattern A and B match, AND1 output is 0, thus A>B is 0. When the first mismatch occurs, AND1 output becomes 1 as previously described. If the corresponding bit of A is 1, both the inputs of AND2 are 1, thus A>B is 1. Otherwise A>B is 0. For the following (N-i) comparisons, AND1 output is 0, i.e. A>B is 0. Similarly, A<B is obtained as AND (AND3) between the AND1 output and B. The outputs $\overline{A=B}$, A>B and A<B are sequences of N bits with no more than a single 1 if the logical function is true, and with all zeroes if the function is false. The position of the 1 in the output signals depends on the patterns to be compared. In order to align the 1s an SPC can be exploited. A guard time of N-1 bits is required between two consecutive comparisons in order to allow the depletion of the AND1 inputs. The boolean algebra table of the comparator for 3-bit input patterns is shown in Table 2 to summarising some cases.

A	B	$\overline{A=B}$	A>B	A<B
000	000	000	000	000
111	111	000	000	000
110	110	000	000	000
111	011	100	100	000
111	101	010	010	000
111	110	001	001	000
001	010	010	000	010
101	010	100	100	000

Table 2. Comparator boolean algebra table for 3-bit input patterns.

4.2 Implementation and performance

The implementation of the comparator is based on logic gates exploiting XGM and cross-polarisation rotation in SOAs. When two RZ signals enter the SOA in a counter propagating configuration, the gain saturation induced by the high power one (*Pump*) when its logic state is equal to 1 forces the output signal to 0. On the contrary, when the pump signal is not present, the low power signal (*Probe*) is amplified and its logic state is transferred to the SOA output. The basic gate realises, by means of XGM, the logic function $OUT = Probe\text{AND}\overline{Pump}$, where *Probe* and *Pump* are the logic input signals. It is noteworthy that if the *Probe* is a pulse train, the gate simply exploits the logic inversion of the *Pump*, thus working as a NOT. The counter propagating configuration allows the use of the same or different wavelength for the input signals; this feature is remarkable because it makes the scheme wavelength independent. A band-pass filter (BPF) at the gate output selects the *Probe* wavelength and cancels the excess ASE noise from the output signal.

Each logic operation shown in Fig. 7 can be implemented by exploiting XGM in SOAs. The comparator can be realised by means of six replicas of the SOAs-based logic gate, being the number of SOAs independent of the length of the input patterns. The basic scheme is shown in Fig. 8. The input patterns B and A are fed in SOA-1 and SOA-2. The output signals are coupled with orthogonal polarisation by means of a polarisation beam combiner, thus generating the function B XOR A. The approach used for implementing the XOR follows the one demonstrated in (Kim et al., 2002). Since the signals at the output of SOA-1 and SOA-2 have very clean one and zero level, and they are coupled with orthogonal polarisation, the XOR output is clean both on the zero and one level as well. The signal at the XOR output is fed into the SPC which splits it in N replicas fed in SOA-3. One of the replicas acts as probe. The other N-1 replicas, acting as pumps, are delayed progressively of a multiple of the bit time (T_B) with respect to the probe and coupled together. In this way the serial-to-parallel conversion is realised. The output of SOA-3, which corresponds to the output of AND1 in Fig. 7, is logically inverted in SOA-4. The logical inversion in SOA-4 is necessary to make SOA-5 and SOA-6 working as the AND2 and AND3 in Fig. 7. SOA-5 (SOA-6) must perform the AND between the ith bit of A (B) and the ith bit of $\overline{A=B}$. A and B act as probe, while the pump must be the inverted $\overline{A=B}$ in order to perform the correct logic operation. This is because, due to the XGM, the basic gate calculates the AND between the probe and the inverted pump. A>B and A<B are obtained at the output of SOA-5 and SOA-6 respectively.

The scheme is tested with 2-bit patterns at 10 Gb/s. The sequences of the two pattern A and B to be compared are generated starting from the same sequence. The sequence is obtained by modulating a 10 GHz RZ pulse train (FWHM ~30 ps) at 1556.55 nm. To consider all the possible cases, a proper sequence of 66 bits is mapped onto the RZ pulses. A logical 0 is inserted between two consecutive 2-bit patterns as guard bit. A total number of 22 2-bit patterns and 22 logical zero working as guard bits are generated. The sequence is split in two replicas A and B, with A delayed of 6 bit intervals with respect to B. Since the comparator is implemented for 2-bit patterns, the signal is split through two paths in the SPC before SOA-3. The pump is delayed of one bit time with respect to the probe. Because of the counter-propagating configuration in the SOAs, A and B can have the same wavelength. At the input of each SOA the power is about -9 dBm for the probes and 5 dBm for the pumps. A CW at 1540 nm is fed into the SOAs in order to minimise the pattern effect. A 0.6 nm band pass filter placed at the SOA output filters out the CW.

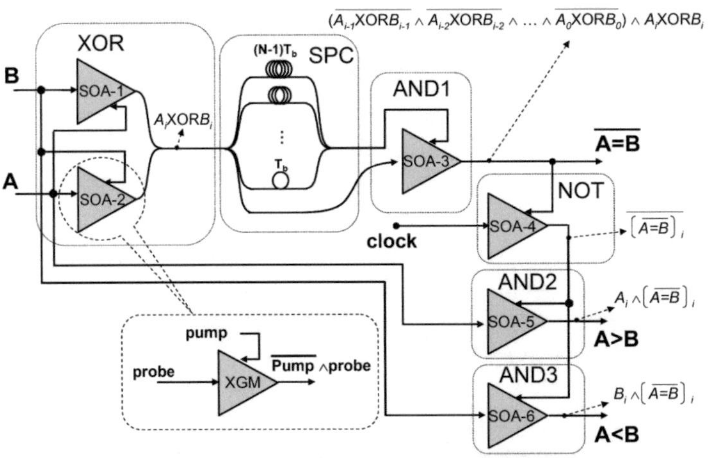

Fig. 8. Basic scheme of the comparator in the SOA-based implementation.

Fig. 9 (a) shows the output patterns $\overline{A=B}$, A>B and A<B together with the corresponding input patterns B and A. The guard bit between two patterns is labelled as g. When A and B are matched the output is 00 for all the three outputs. If A is higher than B, the outputs A>B and $\overline{A=B}$ become 1 as the first mismatching occurs. The other bit is 0. The same correct behaviour is observed for A<B, demonstrating the scheme works properly. The output eye diagrams, showed in Fig. 9 (b) bottom, look open. The measured eye opening is higher than 8 dB for $\overline{A=B}$, 8.6 dB for A>B and 8.4 dB for A<B. Since eye opening for the input patterns A and B is 9.9 dB, the maximum penalty introduced by the 2-bit comparator is 1.9 dB for $\overline{A=B}$, 1.3 dB for A>B and 1.5 dB for A<B. Fig. 10 (b)-top shows the BER as a function of the received peak power for the input and output signals. The measurements are performed with a pre-amplified receiver. Error-free operations are obtained for $\overline{A=B}$, A>B and A<B. The BER curves for the input sequences and the output sequences are shown on the same graph for convenience, but the probability of ones and zeroes in the output signals is different with respect to their probability in the input patterns. The input patterns undergo a sequence of logic operations generating output signals which are different from the input ones. The comparator outputs no more than a single one for each output signal. The lower sensitivity of A<B with respect to A>B is due to the worse performance of SOA-6, which output is noisier on the one level with respect to the signal at the output of SOA-5, resulting in a closer eye diagram. The back-to-back curve has a slightly lower sensitivity with respect to $\overline{A=B}$. This mostly comes from the spectrum of the SOA-3 output, which is better matched to the bandwidth of the optical filter at the receiver with respect to the spectrum of the back-to-back.

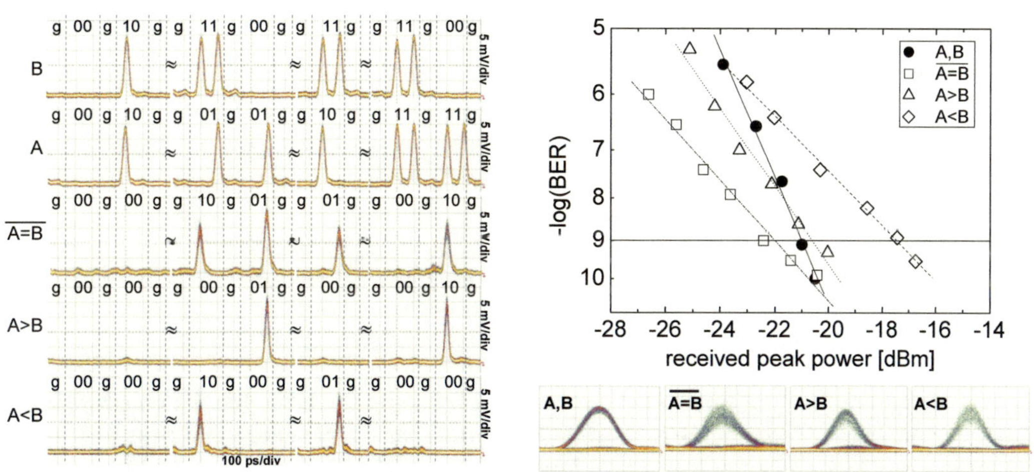

(a) (b)

Fig. 9. (a) Experimental results: input and output sequences. (b) Top: BER v.s. received peak power. Bottom: eye diagrams of the input and output signals.

5. Full-adder

Few works report on the implementation of all-optical full-adders. In (Poustie et al., 1999) an SOA is employed in a terabit optical asymmetric demultiplexer configuration. The reported operation speed is below 1 Gb/s. A faster full-adder is reported in (Kim et al., 2003) based on SOAs, but in that scheme the output sum depends directly on the carry in; moreover performances in terms of bit error-rate and eye opening are not reported. In the scheme presented in the following, both the sum and the output carry do not depend directly on the input carry. This helps improving the quality of the output signals in case of cascade of multiple full-adders.

5.1. Implementation and performance

In a full adder, the two input bits A and B are added to the third input bit ($Carry_{IN}$) that represents the carry of the previous addition. The outputs are the current sum and carry values, also expressed in binary digits. Table 3 shows the full adder truth table and Fig. 10the corresponding logical circuit.

A	B	$Carry_{IN}$	Sum	$Carry_{OUT}$
1	0	0	1	0
0	1	0	1	0
1	1	0	0	1
0	0	0	0	0
1	0	1	0	1
0	1	1	0	1
1	1	1	1	1
0	0	1	1	0

Table 3. Full-adder truth table.

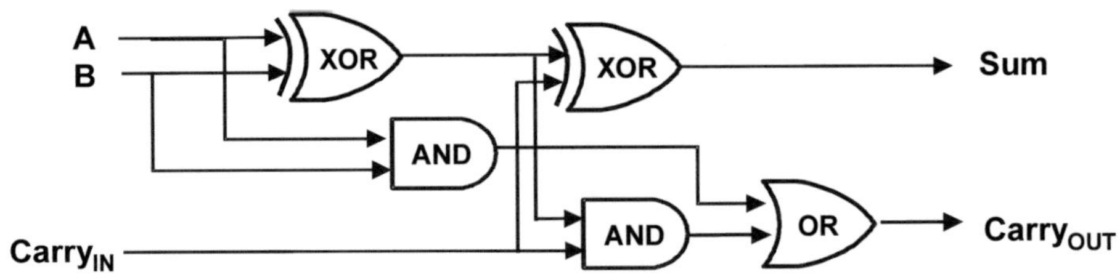

Fig. 10. Full-adder logical circuit.

As for the comparator, the full-adder can be build by using the basic block described in the previous paragraph. The scheme comprises 8 identical basic logic gates, and thus 8 SOAs, as shown in Fig. 11 The scheme uses three input bit sequences (A, B and $Carry_{IN}$) and a pulse train (probe). Where signals are to be coupled, a Polarisation Beam Combiner (PBC) is utilised to avoid beatings between parallel fields, while the resultant double polarised output is used in subsequent gates as pump signal only, in order to allow the full exploitation of the cross polarisation rotation phenomenon by means of the polarizer at the output of the basic gate output. By feeding the basic gate with a double-polarised probe would mean selecting only one of the coupled fields and thus losing information. Moreover, in case of a cascade configuration, the $Carry_{IN}$ coming from a previous full adder is treated as a double polarised signal and always used as pump as well. A particular sequence is mapped onto the RZ pulse train with a Mach Zehnder Modulator (MZM). A, B and Carry_{IN} are generating by splitting and opportunely delaying the pattern. In order to consider all the possible cases, a 24 bit sequence is produced and split in three replicas that are then delayed of 8 bit times with respect to each others. The signals have a repetition rate of 10 Gb/s at the wavelength of 1556.55 nm. The Band Pass Filters (BPF) in the setup have a bandwidth of 0.7 nm.

To assess the scheme performances, the BER and the Eye Opening (EO) have been evaluated. The receiver used was a pre-amplified one comprising an EDFA, a tunable 0.25 nm bandwidth BPF used to optimise the received signal, and a 12.3 Gb/s photo-receiver. The traces of input and output sequences reported in Fig. 12(a) demonstrate the correct behaviour of the circuit. A limited pulse broadening can be noticed, mostly due to cascading filters with slightly different centre wavelength. The BER measurements, shown in Fig. 12(b) - top show error-free operations both on the Sum and $\text{Carry}_{\text{OUT}}$ signal. Fig. 12(b) - bottom shows the input and output eye diagrams. Open eyes are obtained for both the output sequences even if the Sum is less equalised than the $\text{Carry}_{\text{OUT}}$ output signal, and suffer from slow polarisation fluctuations.

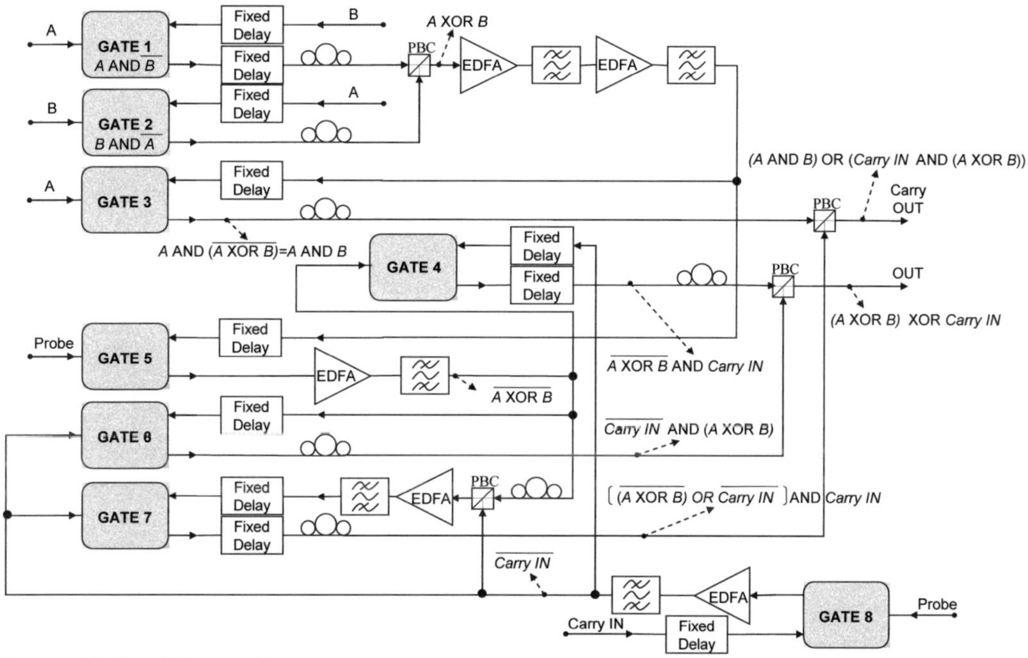

Fig. 11. Full-adder implementation.

These fluctuations are likely caused by mechanical stress on the fibres. The BER of the $\text{Carry}_{\text{OUT}}$ is measured in the best conditions, i.e. for the maximum eye opening. The eye closure penalty, measured in a long temporal scale, is 5.1 dB for the $\text{Carry}_{\text{OUT}}$ and 6.5 dB for the Sum. The polarisation stability of the system can be improved with an integrated implementation of the scheme.

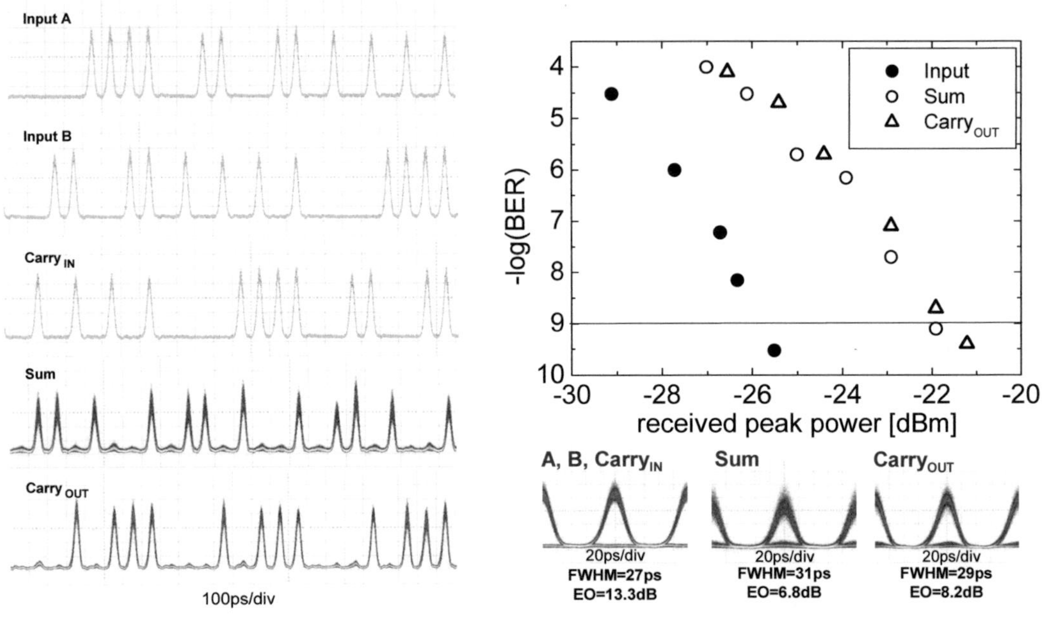

(a) (b)

Fig. 12. (a) Input and output sequences. (b) Top: BER measurements v.s. received peak power. Bottom: eye diagrams of the input and output signals.

6. Analog-to-digital converter

Electronic ADC is demonstrated up to 40 Gsamples/s with a 3-bit coding (Cheng et al., 2004). Nevertheless electronic ADC is mainly limited by the ambiguity of the comparators and jitter of the sampling window (Walden, 1999). The use of hybrid techniques employing an optical signal as sampling signal improves the performances. In (Li et al., 2005) polarization-differential interference and phase modulation is used. Optical sampling with amplitude modulators and time and wavelength-interleaved pulses is demonstrated in (Fok et al., 2004). In (Li et al., 2005 ; Fok et al., 2004) the quantizing and coding are exploited in the electronic domain. Besides the aforementioned hybrid techniques, all-optical ADC, i.e. optical sampling exploited together with optical quantising and coding, is being investigated. Optical quantising and coding allows higher processing speed as well as in principle low-cost implementations, avoiding parallel electronic ADC. In (Ikeda et al., 2006; Miyoshi et al., 2007) the periodical characteristic of the nonlinear optical loop mirror is employed for a 3-bit ADC at 10 Gsamples/s. Slicing of the spectrum broadened by SPM is exploited in (Nishitani et al., 2008; Oda & Maruta, 2005). In (Konishi et al., 2002) the soliton self-frequency shift followed by optical filtering is used. Nevertheless all these techniques, which exploit optical fiber, require high input power and are not suitable for integration.
The new approach proposed in the following realizes quantising and coding with modular blocks exploiting XGM in SOAs. In this way it is enabled analog-to-digital conversion with low optical power requirements with respect to the fiber-based implementations and allows integrated solutions (Scaffardi et al., 2009).

6.1 Working principle

Fig. 13 shows the proposed approach for generating the nonlinear characteristics of the encoders in the case of 3-bit (8-level) quantisation and coding. The multilevel pulsed input signal is split in 3 replicas fed into 3 encoders (a). The characteristic of the encoders (b) is obtained by combining step-like characteristics of nonlinear basic blocks with different thresholds (c), (d). Bit #1 is encoded by a nonlinear block with a step-like characteristic with threshold T4. If the power of the input pulse is less than T4, the output is a logical '1', otherwise it is '0'. Bit #2 is encoded by combining two characteristics with thresholds T2 and T6 (T2<T6). This is obtained by performing the logical AND between the inverted signal at the output of the nonlinear block with threshold T2 and the signal at the output of the nonlinear block with threshold T6. Bit #2 is '1' only if the input pulse power is between T2 and T6, because a '0' is present at the output of block with threshold T2 and a '1' is at the output of block with threshold T6. Otherwise bit #2 is '0'. Bit #3 is encoded by combination of four characteristics with thresholds T1, T3, T5 and T7 (T1<T3<T5<T7). The AND between the output of the nonlinear block with thresholds T3 and the inverted output of the nonlinear block with threshold T1 generates a characteristic which gives '1' if the input power is in the range [T1,T3] and '0' otherwise. In the same way the AND between the output of the nonlinear block with thresholds T7 and the inverted output of the nonlinear block with threshold T5 generates a characteristic which gives '1' if the input power is in the range [T5,T7] and '0' otherwise. The OR between the obtained signals produces the whole characteristic of encoder #3.

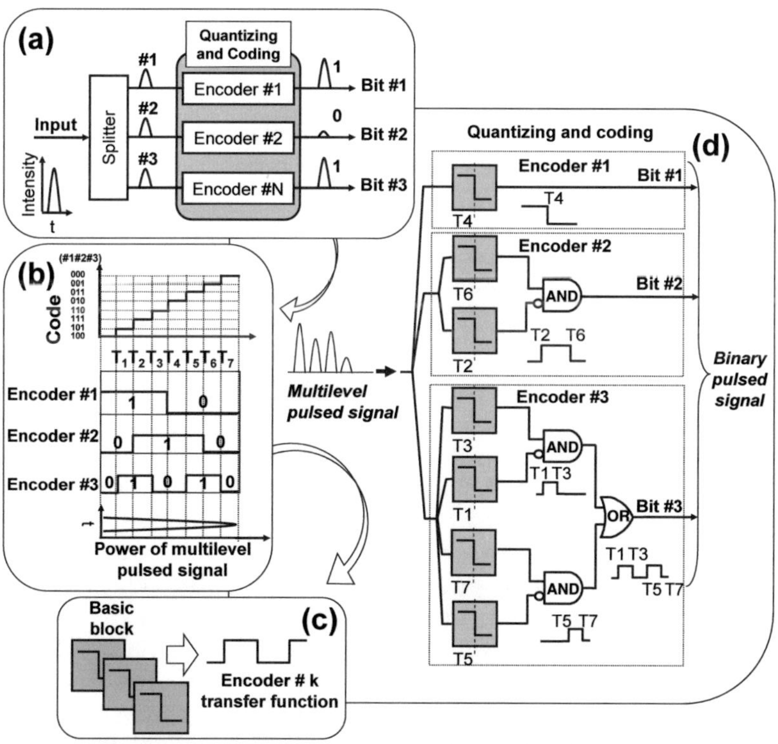

Fig. 13. Proposed scheme for 3 bit (8-levels) quantising and coding.

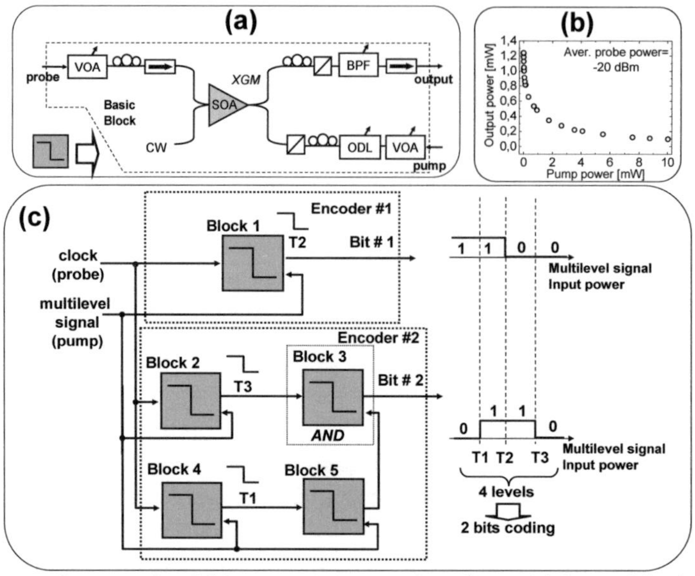

Fig. 14. Experimental setup for 2-bit quantising and coding; (a) basic block; (b) basic block characteristic; (c) overall scheme.

6.2. Implementation and performance

2-bit quantising and coding of a 20 Gsamples/s multilevel signal is implemented following the approach of Fig. 13. The experimental setup is shown in Fig. 14. The encoders characteristics are implemented by the same basic block, which exploits XGM in SOAs with the configuration shown in Fig. 14 (a). The SOA has two input signals: a probe and a counter-propagating pump which modulates the SOA gain saturation. At the SOA output a 1.3 nm-bandwidth filter cuts the out-of-band noise, while a polarisation controller followed by a polarizer are used to improve the output extinction ratio by taking advantage of the XGM-induced polarization rotation. Fig. 14 (b) shows the characteristic of the basic block for an average probe power of -20 dBm and a driving current of 240 mA. The threshold of the nonlinear block can be determined by setting the pump power and the SOA current which influence the working point on the characteristic.

In the implemented scheme, shown in Fig. 14 (c), the clock and the multilevel input signals are a 20 GHz pulse train and a 20 Gsamples/s signal respectively. They are generated starting from the same 10 GHz mode-locked fiber ring laser at 1550.5 nm. The pulsewidth is about 4 ps. Block 1 (I_{SOA1}=239 mA) encodes bit #1. Block 2 (I_{SOA2}=135 mA) generates a characteristic with threshold higher than the one generated by the cascade of block 4 (I_{SOA4}=351 mA) and block 5 (I_{SOA5}=357 mA). These last two blocks are cascaded in order to increase the steepness of the nonlinear function. Bit #2 is encoded by performing in block 3 (I_{SOA3}=377 mA) the AND **Error! Reference source not found.**between the bits at the output of block 2 and the logically inverted bits at the output of block 5. The pump power at SOAs input is in the range [3;9] dBm, while the power of the probe is in the range [-20;-11] dBm. A 4 dBm continuous wave (CW) is fed into the SOA of block 3 in order to reduce the noise on the output signal.

The multilevel input signal is a 4-level signal as shown in Fig. 15 (a). Each pulse is encoded with 2 bits. The signals at the output of encoder #1 (bit #1) and encoder #2 (bit #2) are shown in Fig. 15 (b) and (c) respectively. The peak power of pulse L3 is above threshold T2 and below T3, therefore the couple of output bits (bit #1, bit #2) is (0,1). The peak power of pulse L2 is below T2 and above T1, thus the output bits are (1,1). Pulse L4 has a peak power above T3, i.e. the output is (0,0). The peak power of pulse L1 is below T1 and the output results (1,0). Fig. 15 (d) shows the normalised pulse peak power for bit #1 and bit #2 as a function of the normalised input peak power. The extinction ratio is 6.8 dB and 3.6 dB for the outputs of encoder #1 and encoder #2 respectively. The thresholds Ti can be shifted by acting on the SOA driving current. Nevertheless by cascading nonlinear blocks, as for Encoder #2, the nonlinear characteristics becomes smooth, i.e. the thresholds shift towards higher values. By means of semiconductor devices working as zero-level suppressors, e.g semiconductor saturable absorbers, a sharp transitions of the characteristics can be obtained. Saturable absorbers can also help to improve the extinction ratio of the output pulses. The advantage of semiconductor-based schemes is that they enable optical analog-to digital conversion with integrated implementations. Since both nonlinear blocks and AND gates can be implemented with SOAs, the total SOAs number in the general case of N-bit quantisation is $\sum_{n=0}^{N-1}\left(2^{n}\right)+\sum_{n=1}^{N-1}\left(2^{n-1}\right)$ where the first term corresponds to the number of nonlinear blocks with step-like characteristic and the second term corresponds to the number of AND logic gates. N-bit A/D conversion with N>2, requires OR logic gates, which can be implemented by fiber or waveguide couplers.

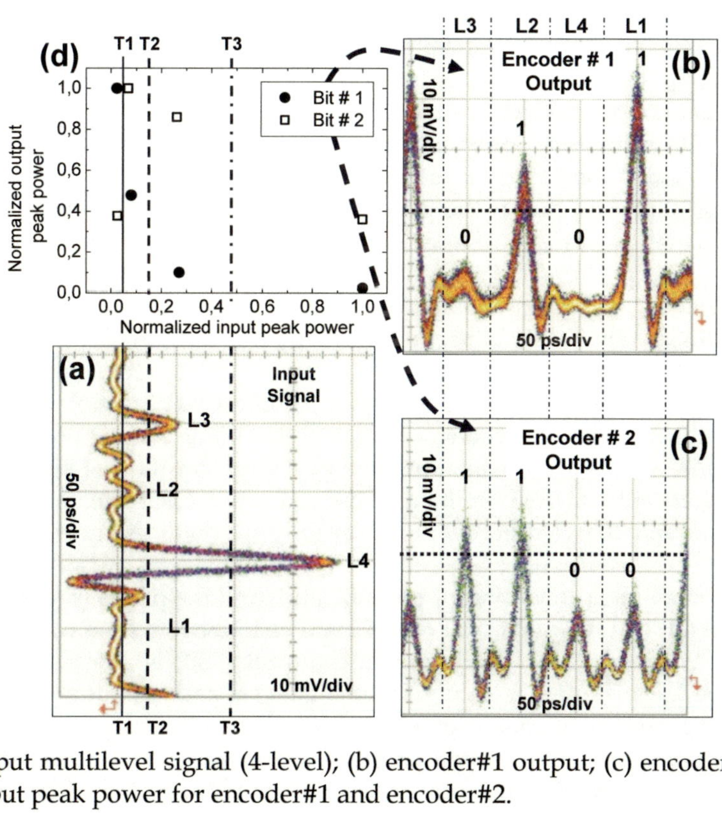

Fig. 15. (a) Input multilevel signal (4-level); (b) encoder#1 output; (c) encoder#2 output; (d) output vs. input peak power for encoder#1 and encoder#2.

7. Digital-to-analog converter

An all-optical DAC scheme that doesn't rely on coherent optical summation has been proposed (Saida et al., 2001). The advantage consisted of eliminating any needs for accurate phase control. In that work, for the 2-bit operation, three nonlinear optical loop mirror gates and a probe pulses train were employed in order to produce a quaternary ASK optical signal from an input OOK signal. Dynamic operation of the gate was not demonstrated, though.

In the following it is proposed a slightly different approach to realize phase-control-free all-optical 2-bit DAC by using two nonlinear gates, and no assist probe signal. Furthermore, the output quaternary signal is retrieved at the same wavelength of the input binary signal. Dynamic operation of the device is demonstrated. This implementation relies on cross-gain compression (XGC) in SOAs (Porzi et al., 2009).

7.1 Working principle

The operation principle of the proposed all-optical DAC scheme relies on the nonlinear gate shown in Fig. 16. Two OOK modulated signals, bit#1 and bit#2, are launched through an SOA in counter-propagating directions from the gate inputs IN_1 and IN_2, respectively. Variable attenuators (VAs) are used to control the power levels of the signals at the amplifier inputs and gate output. Optical paths are adjusted in order to synchronously deliver the two signals to the SOA. The pump bit#2, enters the SOA from port 2 of an optical circulator (OC), whereas the probe bit#1 exits the gate through port 3 of the OC. Thus, the gate output peak power is proportional to $a_{pb} \cdot P_{pb} \cdot G_{SOA}$, being a_{pb} the probe bit#1 logical value ($a_{pb} \in [0,1]$), P_{pb} the peak power associated to the "1" logical values of bit#1, and G_{SOA} the single-pass gain of the optical amplifier. The VA on bit#1 path is used to set P_{pb} much lower than the SOA saturation power, P_{sat}, whereas the VA on bit#2 path is used to adjust the peak power P_{pmp} of the "1" level of bit#2, in order to opportunely compress the SOA gain. Thus, if $a_{pb}=0$, the output of the gate is in the low state, whatever the pump bit#2 logical value a_{pmp} is ($a_{pmp} \in [0,1]$). If $a_{pb}=1$, and $a_{pmp}=0$ the gate output is proportional to $P_{pb} \cdot G_0$, being G_0 the unsaturated (small-signal) single-pass gain of the SOA. On the other hand, if $a_{pb}=1$, and $a_{pmp}=1$, the gate output is proportional to $P_{pb} \cdot G_s$, being G_s the saturated gain of the amplifier, with $G_s \leq G_0$. Thus, by controlling the pump peak power, exploiting the SOA gain saturation characteristic, it is possible to map any input binary signals combination into an analog output. The 2-bit photonic DAC is implemented having as binary input a sequence of two bit-long words, corresponding to four possible levels at the output of the DAC. By means of serial-to-parallel conversion the last significant bit (LSB) and the most significant bit (MSB) in the data are separated. Each stream is then split again into two paths and sent to two different replica of the basic nonlinear gate, named Gate1 and Gate2.

The LSB (MSB) enters the probe (pump) port of Gate1, and the pump (probe) port of Gate2. Thus, the output OUT_1 (OUT_2) of Gate1 (Gate2) will be in the low state when the LSB (MSB) is a logical "0", whatever the logical value of the MSB (LSB) is. For Gate1, if LSB=1 and MSB=0 the probe LSB experiences an unsaturated gain G_0, and the output peak power is $(1-\alpha_1) \cdot P_{pb} \cdot G_0$, being α_1 the attenuation coefficient of VA_1 ($0<\alpha_1<1$). The power level P_{pmp} of the pump MSB "1" pulse in Gate1 is high enough to strongly compress the amplifier. As a result, when LSB=1, and MSB=1, the gain experienced in SOA_1 by the probe LSB (with peak power P_{pb}) is negligible, since now $G(P_{pmp}) \approx 1$, and the output peak power of the gate is $(1-\alpha_1) \cdot P_{pb}$. Since $G_0 >> 1$, the optical power OUT_1 can now be assumed to be in the low state, when compared with the previous case.

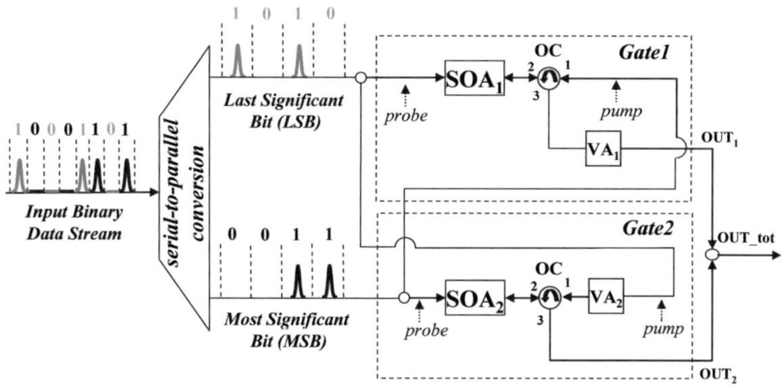

Fig. 16. Operation of the 2-bit all-optical DAC.

For Gate2, if MSB=1 and LSB=0, the output pulse peak power is $G_0 \cdot P_{pb}$, as discussed before. The attenuation coefficient α_2 of VA_2 on the pump LSB path is now adjusted in such a way that when MSB=1 and LSB=1, the probe MSB experiences a partially saturated gain G_s in SOA_2. The output pulse peak power is thus $G_s \cdot P_{pb}$. Table 4 summarizes the various gates' outputs for any input bits combination.

	MSB	LSB	OUT_1	OUT_2	OUT_tot
Level0	0	0	0	0	0
Level1	0	1	$(1-\alpha_1)\cdot P_{pb}\cdot G_0$	0	$(1-\alpha_1)\cdot P_{pb}\cdot G_0$
Level3	1	0	0	$P_{pb}\cdot G_0$	$P_{pb}\cdot G_0$
Level2	1	1	~0	$P_{pb}\cdot G_s$	$P_{pb}\cdot G_s$

Table 4. Input –output relations for the 2-bit DAC.

Since OUT_1 and OUT_2 are never simultaneously different from zero, the 2-bit DAC total output OUT_tot can be taken by combining OUT_1 and OUT_2 with a standard fiber coupler. From Table I it can be seen that 4 equally spaced Gary-coded levels representing the bit combinations 00 (level 0), 01 (level 1), 11 (level 2), and 10 (level 3), are generated at OUT_tot if the VAs in the gates are set in such a way that $\alpha_1 = (2/3)$, and $G_s = (2/3) \cdot G_0$. Alternatively, normal code could be implemented.

7.2 Implementation and performance

An all-fiber actively mode-locked laser at 10 GHz provided 3.5 ps-long optical pulses at ~1532 nm. The original pulses' bit rate was then lowered to avoid patterning effects originated by gain recovery time in the SOAs, by means of a pattern generator and an electro-optic modulator producing the bit sequence 1100 at 2.5 Gb/s. After the modulator, the signal was amplified by means of an EDFA followed by an optical filter and split into two different paths, to simulate the MSB and LSB data stream. Here, like in other reported works on DAC, we assumed that the MSB and the LSB have been previously parallelized

(Oda & Maruta, 2006). Both the LSB and MSB paths were subsequently split again in order to create the pump and probe signals for the two gates. VAs are used to adjust the different power levels at the gates' inputs and outputs, as described before. Optical delay lines were inserted to synchronize the signals and generate all the possible input bit combinations. A filter was employed after the output standard coupler. In the implementation we inserted a polarizer in the output path. In this way, by properly adjusting the two polarization controllers (PCs) on the probe input/output path, we further increased the suppression capability of the gate for the case *Level2* in Table I, by exploiting the nonlinear polarization rotation (NPR) effect in the saturated amplifier (Soto et al., 2001). By doing so we avoided any residual coherent interference at the standard output coupler, where OUT_1 and OUT_2 are combined, when both the LSB and MSB assume the "1" logical value. It should be noted that this alteration doesn't affect the operation principle of the proposed device, which could be implemented with any nonlinear switch for Gate1 which is able to provide a low enough output level for the case LSB=1, and MSB=1. In our experiment, the average power levels of the probe signals (for both gates) were ~ -18 dBm, whereas the pump average power was ~ -3 dBm and ~ -13 dBm at the SOAs inputs in Gate1 and Gate2, respectively. In the gates we used similar bulk amplifiers with a small signal gain of about 25 dB at the signal wavelength, a 3-dB output saturation power of ~11 dBm, and a gain recovery time constant of ~200 ps. Time traces of the signals at different stages of the 2-bit DAC are shown in Fig. 17 (a). From top to bottom, the data flows of LSB, MSB, OUT_1, OUT_2, and OUT_tot are shown, respectively. Input and level-converted output eye diagrams are reported in Fig. 17 (b) top and bottom respectively, showing a clearly open 4-level eye at the output of the device.

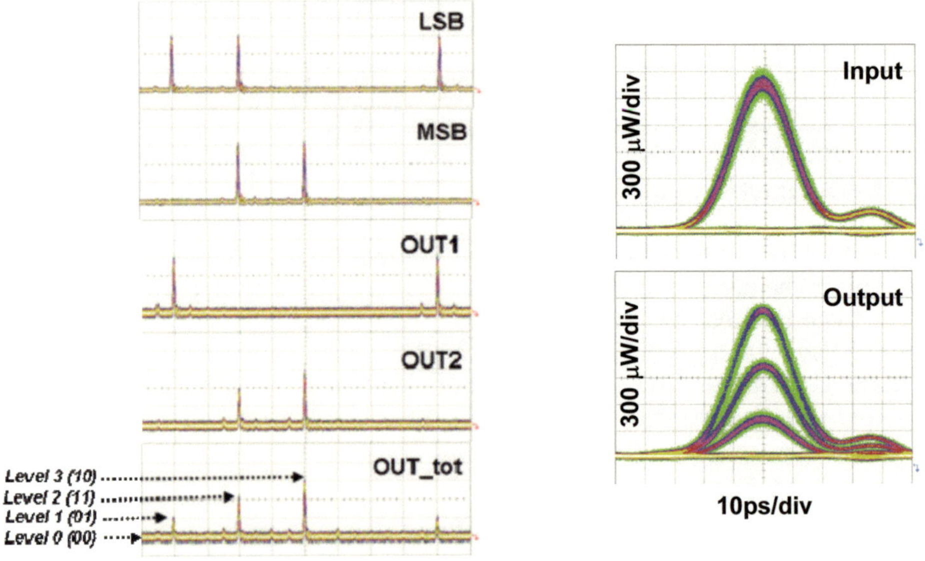

Fig. 17. (a) Time traces of the signals at different stages of the 2-bit DAC. From top to bottom: LSB, MSB, OUT1, OUT2, and OUT_tot data flows. (b) Input (top) and output (bottom) signals eye diagrams.

To the best of our knowledge, this is the first time that dynamic operation for a phase-control-free 2-bit all-optical DAC is experimentally demonstrated for single wavelength operation. We measured the Q factors for the input and level-converted output signals by taking the histograms of noise distributions with a standard sampling oscilloscope with 10-GHz of bandwidth. The input binary signal Q factor was measured to be 22, whereas the output Q factors were measured to be 10.2, 9.8, and 10.9 for $level_1$-to-$level_0$, $level_2$-to-$level_1$, and $level_3$-to-$level_2$ eyes, respectively.

8. References

Abbade M. L. F., Fagotto E. A. M., Braga R. S., Fonseca I. E., Moschim E. & Barbosa F. R., (2005). Optical Amplitude Multiplexing Through Four-Wave Mixing in Optical Fibers. *IEEE Photonics Technology Letters*, Vol. 17 No. 1 (Jan. 2005) pp. 151-153, 1041-1135.

Abbade M. L. F., Fagotto E. A. M., Braga R. S., Barbosa F. R., Moschim E. & Fonseca I. E., (2006). Quaternary Optical Packets Generated by Fiber Four-Wave Mixing. *IEEE Photonics Technology Letters*, Vol. 18 No. 2 (Jan. 2006) pp. 331-333, 1041-1135.

Andriolli N.,Scaffardi M., Berrettini G., Potì L. & Bogoni A.,(2007). Multistage interconnection network photonic controller exploiting a cascaded SOA-based ultrafast module. *Proceedings of ECOC'07*, P117, 978-3-8007-3059-9, Berlin, Germany, 16-20 September 2007.

Berrettini G., Simi A., Malacarne A., Bogoni A., & Poti L., (2006). Ultrafast Integrable and Reconfigurable XNOR, AND, NOR, and NOT Photonic Logic Gate. *IEEE Photonics Technology Letters*, Vol. 18 No. 8 (Apr. 2006) pp. 917-919, 1041-1135.

Berrettini G.,. Lauri E, Ghelfi P., Bogoni A. & Potì L., (2006). Ultra-Fast Integrable 2x2 All-Optical Switch for Optical Packet Routing. *Proceedings of ECOC'06*, We3.P.102, 978-2-912328-39-7, Cannes, France, 24-28 Sept. 2006.

Bogoni, A.; Poti, L.; Proietti, R.; Meloni, G.; Ponzini, E. & Ghelfi, P., (2005). Regenerative and Reconfigurable All-Optical Logic Gates for Ultra-Fast Applications. *IEE Electronics Letters*, Vol. 41, No, 7, (Mar. 2005) pp.435-436, 0013-5194.

L. Brzozowski & E. H. Sargent, (2001). All-optical analog-to-digital converters, hardlimiters, and logic gates. *IEEE Journal of Lightwave Technology*, Vol. 19, No. 1, (Jan. 2001) pp. 114–119, 0733-8724.

Chbat M. W.; Hong B. J.; Islam M. N., Coccoloch C. E. & Prucnal A. L., (1992). Ultrafast soliton-trapping AND gate. *IEEE Journal of Lightwave Technology*, Vol. 10, No. 12, (Dec. 1992) pp. 2011-2016, 0733-8724.

Chan L. Y., Qureshi K. K., Wai P. K. A., Moses B., Lui L. F. K., Tam H. Y. & Demokan M. S., (2003). All-Optical Bit-Error Monitoring System Using Cascaded Inverted Wavelength Converter and Optical NOR Gate. *IEEE Photonics Technology Letters*, Vol. 15 No. 4 (Apr. 2003) pp. 593-595, 1041-1135.

Cheng W., Ali W., Choi M.-J., Liu K., Tat T., Devendorf D., Linder L. & Stevens R., (2004). A 3b 40 Gsamples/s ADC-DAC in 0.12 μm SiGe. *Proceedings of IEEE Int. Solid-State Circuits Conference*, Vol. 1, pp. 262–263, 0-7803-8267-6, San Francisco, California, USA, 15-19 Feb. 2004.

Collecutt, G.R. & Drummond, P.D., (2000). Digital response in an all optical AND gate using parametric (χ (2)) solitons. *Proceedings of CLEO'00*, pp. 194, 1-55752-634-6, San Francisco, California, USA, 7-12 May 2000.

Dorren H. J. S., Lenstra D., Liu Y., Hill M. T. & Khoe G.-D., (2003). Nonlinear polarization rotation in semiconductor optical amplifiers: theory and application to all-optical flip-flop memories. *IEEE Journal of Quantum Electronics*, Vol. 39, No. 1, (Jan 2003) pp. 141-148, 0018–9197.

Dorren, H.J.S.; Mishra, A.K.; Li, Z.G.; Ju, H.K.; de Waardt, H.; Khoe, G.D.; Simoyama, T.; Ishikawa, H.; Kawashima, H. & Hasama, T., (2004). All-optical logic based on ultrafast gain and index dynamics in a semiconductor optical amplifier. *IEEE Selected Topics in Quantum Electronics*, Vol. 10, No. 5, (Sept./Oct. 2004) pp. 1079-1092, 1077-260X.

Fok M. P., Lee K. L. & Shu C., (2004). 4 2.5 GHz Repetitive Photonic Sampler for High-Speed Analog-to-Digital Signal Conversion. *IEEE Photonics Technology Letters*, Vol. 16 No. 3 (Mar. 2004) pp. 876-878, 1041-1135.

Forney G. D., (1973). The Viterbi algorithm. *Proceedings of IEEE*, Vol. 61, No. 3, (Mar. 1973) pp. 268–278, 0018-9219.

Hall K. L. & Rauschenbach K. A., (1996). All-optical bit pattern generation and matching. *IEE Electronics Letters*, Vol. 32, No, 13, (Jun. 1996) pp. 1214-1215, 0013-5194.

Hayes J. P., (1998). *Computer Architecture and Organization*. McGraw-Hill, 0-07-100479-3, New York.

Ibrahim, T.A.; Grover, R.; Kuo, L.C.; Kanakaraju, S.; Calhoun, L.C. & Ho, P.T., (2003). All-optical AND/NAND logic gates using semiconductor microresonators. *IEEE Photonics Technology Letters*, Vol. 15, No. 10, (Oct. 2003) pp. 1422-1424, 1041-1135.

Ikeda K., Abdul J. M., Tobioka H., Inoue T., Namiki S. & Kitayama K., (2006). Design Considerations of All-Optical A/D Conversion: Nonlinear Fiber-Optic Sagnac-Loop Interferometer-Based Optical Quantizing and Coding. *IEEE Journal of Lightwave Technology*, Vol. 24, No. 7, (Jul. 2006) pp. 2618-2628, 0733-8724.

Kehayas E., Seoane J., Liu Y., Martinez J. M., Herrera J., Holm-Nielsen P. V., Zhang S., McDougall R., Maxwell G., Ramos F., Marti J., Dorren H. S. J., Jeppesen P. & Avramopoulos H., (2006). All-Optical Network Subsystems Using Integrated SOA-Based Optical Gates and Flip-Flops for Label-Swapped Networks. *IEEE Photonics Technology Letters*, Vol. 18 No. 16 (Aug. 2006) pp. 1750-1752, 1041-1135.

Kehayas E. Tsiokos, D.I., Bakopoulos, P., Apostolopoulos, D., Petrantonakis, D., Stampoulidis, L., Poustie, A., McDougall, R., Maxwell, G., Liu, Y., Zhang, S. X., Dorren, H.J.S., Seoane, J., Van Holm-Nielsen, P., Jeppesen, P. & Avramopoulos, H., (2006). 40-Gb/s all-optical processing systems using hybrid photonic integration technology. *IEEE Journal of Lightwave Technology*, Vol. 24, No. 12, (Dec. 2006) pp. 4903-4911, 0733-8724.

Kim J. H., Jhon Y. M., Byun Y. T., Lee S., Woo D. H. & Kim S. H., (2002). All-optical XOR gate using semiconductor optical amplifiers without additional input beam. *IEEE Photonics Technology Letters*, Vol. 14 No. 10 (Oct. 2002) pp. 1436-1438, 1041-1135.

Kim J. H., Kim B. C., Byun Y. T., Jhon Y. M., Lee S., Woo D. H. & Kim S. H., (2003). All-optical full adder using semiconductor optical amplifiers. *Proceedings of ECOC'03*, We4.P75, Rimini, Italy, 21-25 Sept. 2003.

Konishi T., Tanimura K., Asano K., Oshita Y. & Ichioka Y., (2002). All-optical analog-to-digital converter by use of self-frequency shifting in fiber and a pulse-shaping technique. *Jornal of the Optical Society of American B*, Vol. 19, No. 11, (Nov. 2002), pp. 2817-2823, 0740-3224.

Kumar S. & Willner A. E., (2004). All optical gate using simultaneous four-wave mixing and cross-gain modulation in an SOA. *Proceedings of LEOS'04*, pp. 913, 0-7803-8557-8, Westin Rio Mar Beach, Rio Grande, Puerto Rico, 7-11 Nov 2004.

Lal V., Masanovic, M.L., Summers, J.A., Fish, G. & Blumenthal, D.J., (2007). Monolithic wavelength converters for high-speed packet-switched optical networks. *IEEE Selected Topics in Quantum Electronics*, Vol. 13, No. 1, (Jan./Feb. 2007) pp. 49-57, 1077-260X.

Li W., Zhang H., Wu Q., Zhang Z. & Yao M., (2005). All-Optical Analog-to-Digital Conversion Based on Polarization-Differential Interference and Phase Modulation. *IEEE Photonics Technology Letters*, Vol. 19 No. 8 (Apr. 2005) pp. . 625-627, 1041-1135.

Liu, Y., Tangdiongga, E., Li, Z., de Waardt, H., Koonen, A. M. J., Khoe, G. D., Shu, X., Bennion, & Dorren, H. J. S., (2007). Error-free 320-Gb/s all-optical wavelength conversion using a single semiconductor optical amplifier. *IEEE Journal of Lightwave Technology*, Vol. 25, No. 1, (Jan. 2007) pp. 103-108, 0733-8724.

Malacarne A., Bogoni A. & Potì L., (2006). An Optical Flip-Flop Based on Erbium-Ytterbium Doped Fibre. *Proceedings of ECOC'06*, We3.P.31, 978-2-912328-39-7, Cannes, France, 24-28 Sept. 2006.

Martinez M., Herrera J., Ramos F. & Marti J., (2006). All-optical address recognition scheme for label-swapping networks. *IEEE Photonics Technology Letters*, Vol. 18 No. 1 (Jan. 2006) pp. 151-153, 1041-1135.

Maxwell G., (2008). Hybrid Integration Technology for High Functionality Devices in Optical Communications. *Proceedings of OFC'08*, OWI3, 978-1-55752-856-8, San Diego, California, USA, 24-28 February 2008.

McGeehan, J.E.; Kumar, S.; Gurkan, D.; Nezam, S.M.R.M.; Willner, A.E.; Parameswaran, K.R.; Fejer, M.M.; Bannister, J. & Touch, J.D. (2003). All-optical decrementing of a packet's Time-to-Live (TTL) field and subsequent dropping of a zero-TTL packet. *IEEE Journal of Lightwave Technology*, Vol. 21, No. 11, (Nov. 2003) pp. 27466-2752, 0733-8724.

Miyoshi Y., Ikeda K., Tobioka H., Inoue T., Namiki S. & Kitayama K., (2007). All-Optical Analog-to-Digital Conversion Using Split-and-Delay Technique. *IEEE Journal of Lightwave Technology*, Vol. 25, No. 6, (Jun. 2007) pp. 1339-1347, 0733-8724.

Nielsen M. L., Fjelde T., Buron J. D. & Dagens B., (2002). All-optical bit-pattern recognition in data segments using logic AND and XOR in a single all-active MZI wavelength converter. *Proceedings of ECOC'02*, 1.4.6, 87-90974-63-8, Copenhagen, Denmark, 8-12 Sept. 2002.

Nishitani T., Konishi T. & Itoh K., (2008). Resolution Improvement of All-Optical Analog-to-Digital Conversion Employing Self-frequency Shift and Self-Phase-Modulation-Induced Spectral Compression. *IEEE Selected Topics in Quantum Electronics*, Vol. 14, No. 3, (May/Jun. 2008) pp. 724-732, 1077-260X.

Oda S. & Maruta A., (2005). A Novel Quantization Scheme by Slicing Supercontinuum Spectrum for All-Optical Analog-to-Digital Conversion. *IEEE Photonics Technology Letters*, Vol. 17 No. 2 (Feb. 2005) pp. . 465-467, 1041-1135.

Oda S. & Maruta A., (2006). All-Optical Digital-to-Analog Conversion Using Nonlinear Optical Loop Mirrors. *IEEE Photonics Technology Letters*, Vol. 18 No. 5 (Mar. 2006) pp. 703-705, 1041-1135.

Porzi, C., Nguyen, A.T., Poti, L. & Bogoni, A., (2009). Binary-to-Quaternary ASK Encoding in the Optical Domain With Semiconductor Optical Amplifiers. *IEEE Photonics Technology Letters*, Vol. 21 No. 10 (May. 2009) pp. 654 - 656, 1041-1135.

Poustie A.J., Blow K.J., Kelly A.E. & Manning R.J., (1999). All-optical full adder with bit-differential delay. *Optics Communications*, Vol. 168, (Sept. 1999) pp. 89-93, 0030-4018.

Proakis J. G., (1996). *Digital Communications*, McGraw-Hill, 978-0072321111, New York.

Saida T., Okamoto K., Uchiyama K.,. Takiguchi K, Shibata T. & Sugita A., (2001). Integrated optical digital-to-analogue converter and its application to pulse pattern recognition. *IEE Electronics Letters*, Vol. 37, No, 20, (Sept. 2001) pp. 1237-1238, 0013-5194.

Scaffardi, M; Andriolli, N; Meloni, G; Berrettini, G; Fresi, F; Castoldi, P (; Poti, L & Bogoni, A., (2007). Photonic combinatorial network for contention management in 160 Gb/s interconnection networks based on all-optical 2×2 switching elements. *IEEE Selected Topics in Quantum Electronics*, Vol. 13, No. 5, (Sept./Oct. 2007) pp. 1531- 1539, 1077-260X.

Scaffardi, M.; Ghelfi, P.; Lazzeri, E.; Poti, L. & Bogoni, A., (2008). Photonic processing for digital comparison and full addition based on semiconductor optical amplifiers. *IEEE Selected Topics in Quantum Electronics*, Vol. 14, No. 3, (May./Jun. 2008) pp. 826-833, 1077-260X.

Scaffardi, M; Lazzeri, E.; Fresi, F. ; Poti, L. & Bogoni, A, (2009). Analog-to-Digital Conversion Based on Modular Blocks Exploiting Cross-Gain Modulation in Semiconductor Optical Amplifiers. *IEEE Photonics Technology Letters*, Vol. 21 No. 8 (Apr. 2009) pp. 540-542, 1041-1135.

Shacham, A.; Small, B.A.; Liboiron-Ladouceur & O.; Bergman, K., (2005). A fully implemented 12x12 data vortex optical packet switching interconnection network. *IEEE Journal of Lightwave Technology*, Vol. 23, No. 10, (Oct. 2005) pp. 3066-3075, 0733-8724.

Soto H., Erasme D. & Guekos G., (1999). Cross-polarization modulation in semiconductor optical amplifiers. *IEEE Photonics Technology Letters*, Vol. 11 No. 8 (Aug. 1999) pp970-972, 1041-1135.

Soto H., Erasme D. & Guekos G., (2001). 5-Gb/s XOR optical gate based on cross-polarization modulation in semiconductor optical amplifiers. *IEEE Photonics Technology Letters*, Vol. 13 No. 4 (Apr. 2001) pp. 335-337, 1041-1135.

Tangdiongga E., Hansen Mulvad H.C., de Waardt H., Khoe G.D., Koonen A.M.J. & Dorren H.J.S., (2007). SOA-based Clock Recovery and Demultiplexing in a Lab Trial of 640-Gb/s OTDM and Demultiplexing in a Lab Trial of 640-Gb/s OTDM Transmission over 50-km Fibre Link. *Proceedings of ECOC'07*, PD 1.2, 978-3-8007-3059-9, Berlin, Germany, 16-20 September 2007.

Wada N., Cincotti G., Yoshima S., Kataoka N. & Kitayama K., (2006). Characterization of a full encoder/decoder in the AWG configuration for code-based photonic routers—part II: experiments and Applications. *IEEE Journal of Lightwave Technology*, Vol. 24, No. 1, (Jan. 2006) pp. 113-121, 0733-8724.

Walden R. H., (1999). Analog-to-Digital Converter Survey and Analysis. *IEEE Journal on Selected Areas in Communications,* Vol. 17, No. 4, (Apr. 1999) pp. 539-550, 0733-8716.

Wang Y., Zhang X., Dong J. & Huang D., (2007). Simultaneous demonstration on all-optical digital encoder and comparator at 40 Gb/s with semiconductor optical amplifiers. *Optics Express,* Vol. 15, No. 23, (Nov. 2007) pp. 15080-15085, 1094-4087.

Wu Y. D., (2005). All-optical logic gates by using multibranch waveguide structure with localized optical nonlinearity. *IEEE Selected Topics in Quantum Electronics,* Vol. 11, No. 2, (Mar. 2005) pp. 307-312, 1077-260X.

Yacoubian A. & Das P. K., (2003). Digital-to-Analog Conversion Using Electrooptic Modulators. *IEEE Photonics Technology Letters,* Vol. 15 No. 1 (Jan. 2003) pp. 117-119, 1041-1135.

Electro-Optical Monitoring and Analysis of Human Cognitive Processes

Erik Vavrinsky[1], Viera Stopjakova[1], Igor Brezina[2], Libor Majer[1], Petra Solarikova[2] and Vladimir Tvarozek[1]
[1]*Department of Microelectronics, Slovak University of Technology in Bratislava,*
[2]*Department of Psychology, Comenius University,*
Slovakia

1. Introduction

Miniaturization of biomedical sensors has increased the importance of microsystem technology in medical applications, particularly microelectronics and micromachining. Microelectronics enables a mass-production of planar microsensors with high reproducibility, low cost, advanced performance, and built-in smartness. Thin-film interdigital arrays (IDA) of microelectrodes are among the most commonly used periodic structures in a wide variety of sensors and transducers (Mamishev et al., 2004).

This chapter presents a new approach to biomedical monitoring and analysis of selected human cognitive processes. The proposed method measures human skin conductivity and heart rate using a dedicated monitoring system with the developed IDA microelectrodes (Sheppard, 1993) as well as skin temperature and reflectance using an optical spectrometer, which offers continuous monitoring and analysis of different electrophysiological aspects of human physiology in a completely safe and non-invasive manner. This technique also has no undesired influence on natural physiological processes. The main goal is monitoring of the psycho-galvanic reflex (PGR) of the human skin that might be very useful for the identification of psychical stress effect that is performed in different medical as well as psychological experiments.

A portable monitoring system, applicable also in wireless measurement environment, was designed and developed. Comparison to a standard laboratory-like bridge-based measurement system was done in terms of accuracy, sensitivity and other main features. The proposed mobile monitoring equipment utilizes microprocessors with an RF wireless communication modules used for data transfer between the measurement modules and a personal computer. A graphical user interface (GUI), developed in C++ under Windows XP platform, has been developed too, and is used to provide necessary calibration of the measurement as well as storage, displaying and postprocessing of the measured data (real and imaginary components of the skin impedance as well as phase of the measured impedance). The measurement method itself and the achieved results are discussed in the bellow sections.

The explosive growth in recent wireless communications has caused increased demands for wireless products that would be low-cost, low-power, and compact in size. Moreover, extensive research and development in semiconductor industry in latest years has led towards the novel technologies, and consequently, new products such as smart systems, system-on-chip (SoC), etc. have been widely developed and used. There are modern high-performance microcontrollers with many useful features (A/D converters, several types of interface, RF wireless transceivers, etc.) available and enabling the realization of modern compact and low-power portable measurement systems. Similarly, massive expansion in electronics trade has enabled to affiliate electronics to diverse domains like the health care that affords new opportunities. Different electronic-medical measurement systems and equipments are indispensable parts of various surgery procedures and help diagnostic a lot of affections. Therefore, biomedical monitoring can be very useful in the health care and psychology, and moreover, it can enhance the quality of life in areas of relaxing physiotherapy or professional and leisure sport activities. Even though measurements of the selected physiological parameters e.g. electrical conductivity (impedance) of the human skin surface and skin temperature have a long history, the way how physiological changes in the human tissue are reflected in electrical impedance has not been very cleared up yet (Olmar, 1998), (Wakelam, 2000).

The proposed biomedical monitoring approach employs advanced wireless technologies in controlling the measurement setup and transferring the measured data to a personal computer. There are many benefits of the proposed portable solution, such as wireless monitoring of the tested person, accurate and sensitive measurement, free movement of the person being tested, real life monitoring, no side effects (e.g. additional stress due to the fact that the person is in the laboratory conditions and become aware of being tested), etc. Furthermore, it can help to identify abnormal changes in human physiological and psycho-physiological reactions under certain psychical stress stimuli. This allows reliable diagnostics of the undesired stress that might be a very negative factor, influencing not only human performance but causing also serious health problems.

2. Stress and Physiological Processes

2.1 Stress Phenomena

What the psychical stress really is and may cause? Stress is a very troublesome and undesired factor dramatically affecting the central neural system and it might invoke significant psychical as well as health problems and inconveniences (Gerasimov, 2003). According to Canadian psychologist Hans Selye, stress is a physiological response of the body to certain physical demands. Such demands are known as stressors. There are three categories of psychological stressors:

- First category comprises stressors that cause frustration. We experience frustration when being blocked from reaching a goal. Our degree of psycho-physiological reactivity to frustrating situations may be affected by heredity.
- Second category of stressors comprises those that cause tension. We experience tension when we must fulfill responsibilities that tax our abilities.
- Third category of stressors comprises those that cause conflicts when we are torn by two or more potentional courses of actions.

Though major life changes are important stress loads, including events, they are not sole once. We can mention other social stress rating scales: from death of spouse (100), followed by marriage (50), change in family members (44), change of responsibility in work (29), change sleeping habits (19), vacations (15), Christmas (13), etc.
Generally, stress is defined by so called General Adaptation Syndrome. Author (Selye, 1976) believes that the syndrome represents body´s defence against the stress. Selye argues that the initial symptoms of almost any disease or trauma are virtually identical, that is, the body responds in the same way to any source of stress, whether it is external and environmental or whether it arises from within the body itself. He has defined stress as „the individual´s psycho-physiological response mediated largely by the autonomic nervous system and the endocrine system, to any demands made on the individual".
Stress has been related to a number of deleterious and costly individual problems (e.g., headaches, gastrointestinal disorders, anxiety, hypertension, coronary heart disease, depression) and organizational outcomes (e.g. job dissatisfaction, burnout, accidents, loss of productivity, absenteeism, turnover) (Ganster & Schaubroeck, 1991), (Gupta & Beehr, 1979), (Ironson, 1992), (Murphy et. al, 1995), (Quick et. al, 1987), (Sauter&Murphy, 1995). These consequences have sustained a continuing interest in assessing stress at work in order to understand the etiology of workplace stress, pinpoint sources of stress, and guide the use of stress reduction interventions. Current techniques for measuring stress fall into one of three categories: self - reports, behavioural and cognitive functions measures, and medical/biological measures (where our study is oriented). To facilitate non-invasive field and laboratory research, the present study has focused on the development of a self-reporting measure of work stress that can be administered rapidly and uniformly, using a structured, closed-ended response format. In developing this Stress in General (SIG) measure, we took a broad approach that avoided links to specific stressors or strains, in contrast to the many previous efforts to measure the work stress that counted on cataloguing the presence of various classes of stressors or short-term strains. The need for the present validation study is underscored by the existence of published studies that have used early prepublication versions of the scale. The instrument is clearly popular and practical even lacking published validity data.
Thus, monitoring of some high-risk groups of patients enables the effective prevention and remarkable reduction of possible health risks associated with the chronic consequences of the stress factors' influence. Stress monitoring might be very helpful also in a wide range of psychological applications, such as clinic psychology, treatment of drug-dependent people, monitoring of important and high-reliability jobs (dispatchers, pilots, drivers, etc.) and others. Thus, measurement of selected physiological variables offers experts a possibility to observe and analyze complex psycho-physiological processes that might considerably contribute to the optimalization of diagnostic and therapeutic procedures.

2.2 Psychogalvanic Reflex

Recently, an increased accuracy of biomedical experimental techniques brings great interest in the field of psycho-physiological correlatives. In general, it has been observed that psycho-galvanic reflex represents psycho-physiological activation, starting from the lowest amplitude in sleep up to the top amplitude under a strong activation. The amplitude fluctuation depends on the level of psychological activation, where the skin conductivity represents volume of sympathetic activity. It has been proven that so called psycho-galvanic reflex (a change in the human skin conductivity under stress influence), sensed continuously within a given time period, offers satisfactory information for stress, overwhelmed excitement, or a shock identification. From the accuracy and sensitivity points of view, the skin conductivity parameters are best sensed using microelectrodes (as explained bellow). The thermal sense represents one of the best characteristics of physical relaxation, where temperature changes are inflicted due to increased amount of blood flowing in the bloodstream and due to vasodilatation (Brezina, 2007).

Although, technical realization of these measurements might be very simple, in practice, there is a problem with measurement reproducibility and comparison. First, it was assumed that increase in the skin conductivity during a stress stimulus is only caused by the skin perspiration. Later, a very important factor of the *potential barrier* near the *stratum lucidum* layer, which thickness changes due to the nervous system, was discovered and proven (Olmar, 1998), (Weis et. al, 1995), (Shepherd, 2007), (Qubit systems, 2004).

If different electrodes are applied on human skin, various space distribution of electrical field into the skin can occurs.

In case of using macroelectrodes, when the distance between the coupled electrodes is greater than the thickness of electric active layers of skin h (*stratum corneum* (the outermost layer of the skin) with potential barrier) $d >> h$, the vector intensity lines of the electric field are enclosed perpendicular to the skin surface across the planar skin structures through *dermis* with high conductance (Fig. 1 a, b).

If microelectrode pairs are utilized, when the distance between the electrodes is less than thickness of electric active layers of skin: $d < h > s$ (s – thickness of *stratum corneum*), then the lines of electric field are enclosed in parallel direction relative to laminar skin structures of epidermis (of lower conductance) (in *stratum corneum*). From inner layers of skin, the electric field intensity lines are embossed to the surface (to the area with a lower conductivity) by the influence of the potential barrier which is generated by electrical double-layer around *stratum lucidum* (Fig.1c). Dynamic electrical properties of potential barrier reflect fact that it is responsible for trans-epidermal transports – substance exchange, water transport and thermo regulation (Fig. 1c). Under a stress stimulus the potential barrier narrows down and the electric field can reach inner layers of human skin with higher conductivity, and therefore, the total conductivity is increases (Fig. 1d). Such configuration is therefore ideal for the analysis of electrophysiological processes in human skin under stress (Vavrinsky et. al, 2008).

In case "too small" microelectrodes the vector intensity lines of the electric field are enclosed in top layers of *stratum corneum* and the flow of electric lines is independent on thickness of potential barrier (Fig. 1e, f). Such electrodes are more ideal for surface analysis in cosmetic.

The results of analytical analysis also showed that in case of non-symmetric coplanar electrodes (Fig. 2), the electric field is more enclosed in the outer layers of the skin laminar structures (stratus corneum). This system consists of the periodical electrode structure with different sizes. In a non-symmetric structure, the density of the electric field intensity lines along the planar structures of the skin is 30 % higher in outer layers (Ivanic et. al, 2003).

The greatest degree of conductivity change occurs in the skin of palms and bottom parts of fingers, but because of practical reasons, we put our microelectrodes on the second best sensitive part of human body – wrist bottom of non-dominant hand (left for right-handers) (Weis et. al, 1995), (Wakelam, 2000).

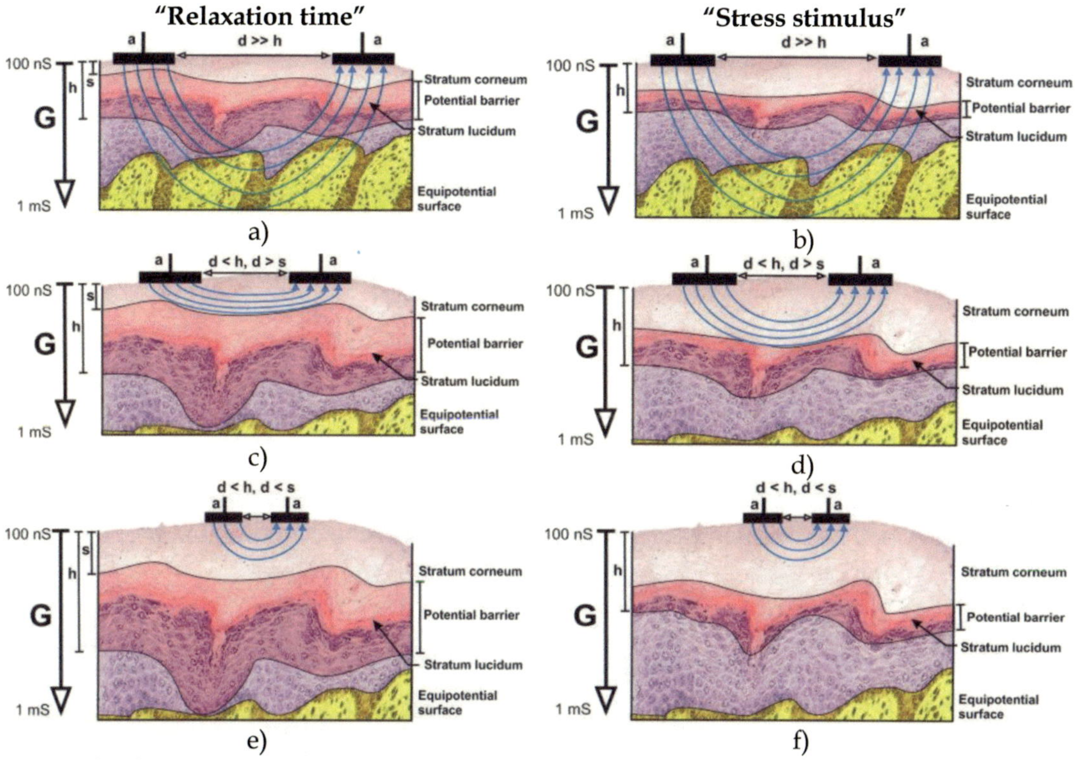

Fig. 1. The dominant vector intensity lines of the electric field in human skin:
- macroelectrodes: a) in relaxation time, b) under stress stimulus
- microelectrodes: c) in relaxation time, d) under stress stimulus
- "too small" microelectrodes: e) in relaxation time, f) under stress stimulus

2.3 Developed Microelectrodes

For non-invasive biomedical monitoring of psycho-physiological processes based on skin conductivity measurements, four types of IDA microelectrodes with the following configuration and sizes have been developed, produced and used:

- Non-symmetric configuration
  - 15 μm/25 μm /50 μm (finger/gap/finger) (Fig. 2)
- Symmetric configuration
  - 100 μm/100 μm (finger/gap dimensions)
  - 200 μm/200 μm
  - 400 μm/400 μm

The total size of the microelectrode chips is 10 mm x 15 mm. The microelectrodes were made from Pt or Au thin film to minimize the polarization effect (Fig. 2). The microelectrodes were fabricated by a standard thin film technology: Pt (Au) films (150 nm in thickness) underlaid by Ti film (50 nm) were deposited by rf sputtering on Al_2O_3 substrates and microelectrodes were lithographically patterned by lift-off technique.

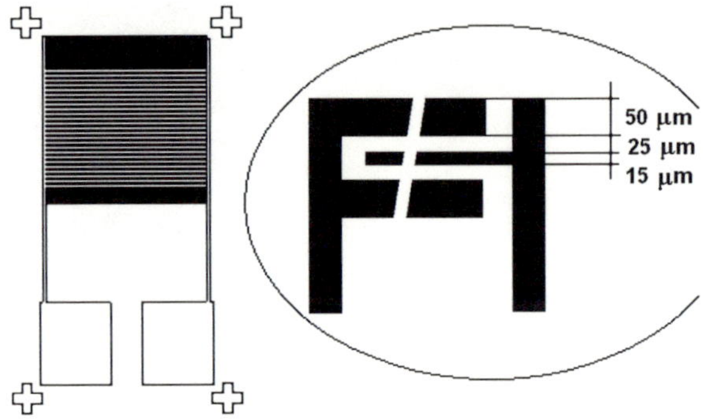

Fig. 2 Layout of the non-symmetric electrode chip

2.4 Experimental Skin Conductivity Measurements

The electrodermal response (EDR) to stress stimulus was in first experiments detected by variations in the skin (ΔG) using classical laboratory measurement equipment (Fig. 3).

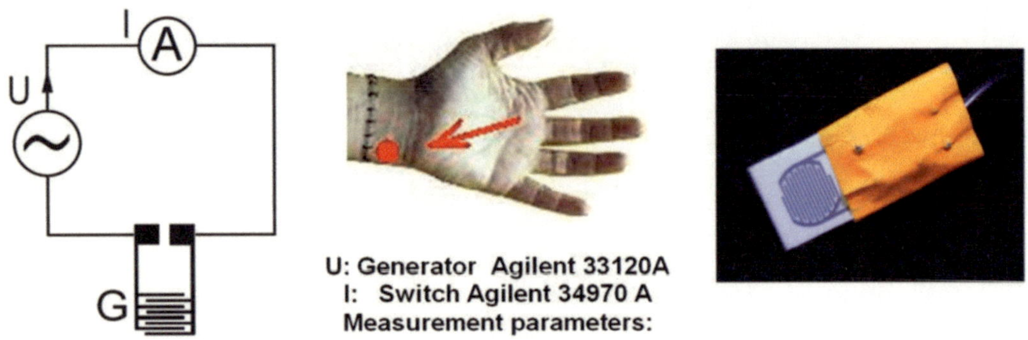

Fig. 3. Electrical laboratory equipment

During conductivity measurements, a drift of the output signals occurs due to polarization effects in the human skin - electrodermal phenomenon (EDF). We investigated the drift of output signals due to EDF and sweat hydration of the skin outer layers (stratus corneum and lucidum) in the time domain. These experiments were done using 200 µm/200 µm microelectrodes for dry skin (Fig. 4a) and for wet (sweaty) skin (Fig. 4b). Measured curves were interpolated by the exponential function $G(t) = A + B(1\text{-}e^{-t/C})$, where G is the skin conductivity, t is time and A, B, C are constants. This function is considered to be very suitable for description of polarization and hydration effects in the human skin. The performed experiments also showed that the signal stabilization (steady state) occurs sooner for hydrated skin (10 - 40 minutes) than for dry skin (40 minutes and more). For typical macroelectrodes, the increase in the conductivity due to the EDF effect is ended in 30 - 40 minutes (Weis et. al, 1995). During the several initial seconds, minutes of measurements it is possible to minimize and compensate the undesired output signal drift (caused by EDF, Fig. 5a)) by means of a proper software (Fig. 5b)). The software analysis program for the skin conductance activity was developed in the development environment Agilent VEE.

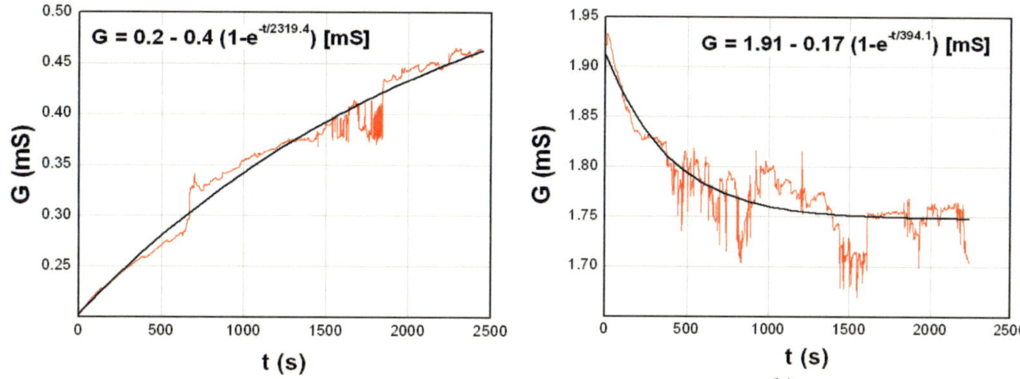

a) b)

Fig. 4. Electrodermal phenomenon and effect of sweat hydration: a) dry skin, b) sweat (hydrated) skin

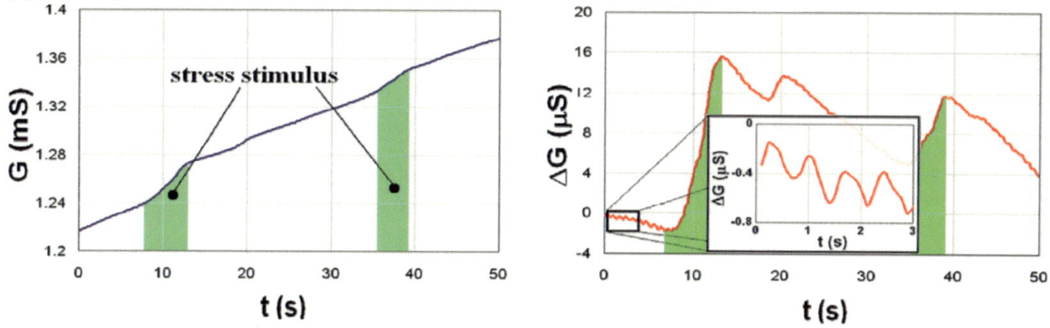

a) b)

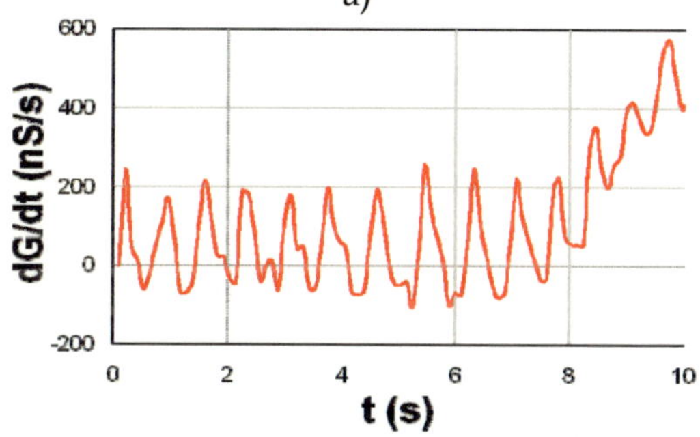

$$y = \sum_{t=0}^{N} \left\{ \left(BPF[\![FFT[x(t)]]\!]_{f_{Low_1}}^{f_{High_1}} \right) * * \left(BPF \left[\!\!\left[FFT \left[\frac{dx(t)}{dt} \right] \right]\!\!\right]_{f_{Low_2}}^{f_{High_2}} - A \right) \right\}$$

c) d)

Fig. 5. Typical time dependences of EDR a) uncorrected, b) corrected EDR and its zoom comprising the heart pulse, c) derived uncorrected signal comprising the heart pulse, d) heart pulse analysing formula

These experiments led to a very important result: the developed microelectrode probes are able to monitor electrodermal response as well as heart pulse simultaneously (Fig. 5b). The heart pulse was observable by derivation of the measured signal (Fig. 5c). For this purpose, dedicated software, based on the designed formula (Fig. 5d), has been used. In this formula, *BPF* stands for a band-pass filter in the frequency range from f_{LOW} to f_{HIGH}, *FFT* represents Fast Fourier Transform, and *A* is a constant. For our experiments, the following ideal set of

parameters has been found: $f_{Low1} = f_{Low2}$ = 0,5 Hz, $f_{High1} = f_{High2}$ = 3 Hz (heart-beat of 30-180 min^{-1}), N = 5, and A = 0,25. This software is under further development.
In the next experiments, the influence of microelectrodes size (symmetric: 200 μm/200 μm, 100 μm/ 100 μm, non-symmetric 15 μm/25 μm/50 μm) on the output signals were investigated. As mentioned above, the different types of microelectrodes generate the electrical field enclosed in various layers of the skin. Results were obtained using input signal amplitude and frequency of 3 V and 1 kHz, respectively, while employing the EDF correction. It is shown (Fig. 6), that the output signal measured by the non-symmetric IDA electrodes is very low in comparison to the results obtained by the symmetric IDA 100 μm/ 100 μm and 200 μm/200 μm microelectrode arrangement. It is inflicted by the fact that the penetration depth of the electric field generated by the non-symmetric microelectrodes is insufficient to reach the potential barrier of the skin, which is very sensitive to the detection of psycho-physiological processes like the psychical stress (Weis et. al, 1995), (Vavrinsky et. al, 2008). In case of the non-symmetric microelectrodes, most of the electrical field intensity lines (about 80 %) are enclosed in the depth of 0 - 25 μm , while the most important and sensitive layer of the potential barrier is placed more than 30 μm beneath the surface. The low response for the sensor is due to sweat in upper layer - stratum corneum.

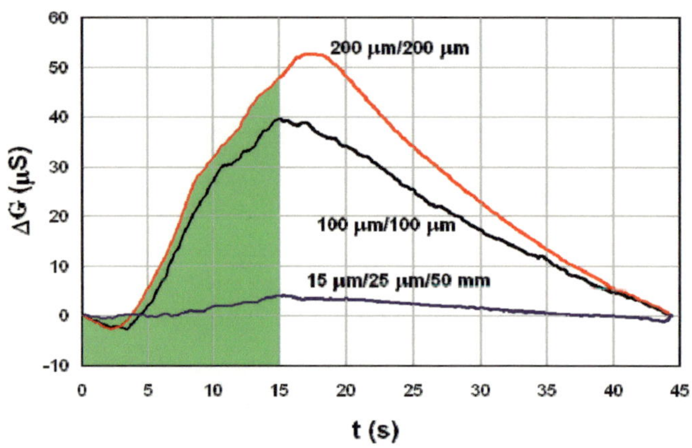

Fig. 6. Influence of size and configuration of microelectrodes on the output signals

We have also analyzed influence of sweat hydration on the output signals. For this purpose, we moistened the skin surface with NaCl solution of concentration, which was several times higher than the normal concentration of human sweat (0.3 – 0.8 %). The measurements were done using 100 μm/100 μm, 200 μm/ 200 μm and 400 μm/400 μm microelectrodes. Measured stress response was very small in case of the non-symmetric microelectrodes. Measurements showed that sweat hydration causes a noise-like influence on the signal, which gets more observable as the size of microelectrodes is smaller. Therefore, reading out the measured signal becomes more difficult (Fig. 7). Finally, one can say that sweat hydration increases the skin conductivity and brings noise to the output. Based on the experiment results, we can conclude that the skin hydration (sweat) has less influence on the output signal that might by expected, which is in correlation with the theory saying that sweat is not dominant for the current flow across the skin (Weis et. al, 1995).

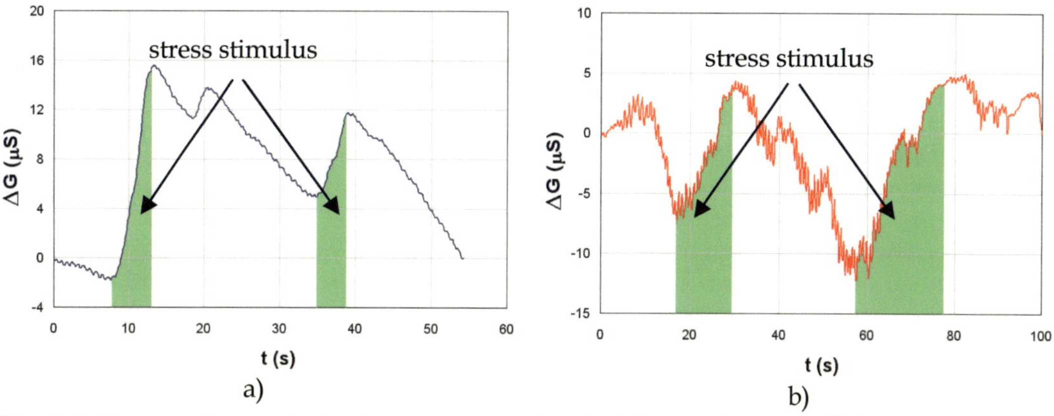

Fig. 7. Influence of sweat hydration on output signal of 200 μm/ 200 μm IDA microelectrodes: a) dry skin (G_0 = 1.22 mS), b) sweat skin (G_0 = 2.14 mS)

Finally, a comparison of our microelectrodes-based galvanic skin response (GSR) method to the commercial macroelectrode approach (Shepher, 2007), usually used in the laboratory medical or psychological experiments, was carried out. The comparison, performed using standard psychotests, shows that the responses given by both approaches were similar. However, the microelectrode signals are observed to be more stable with a shorter response time (Fig. 8).

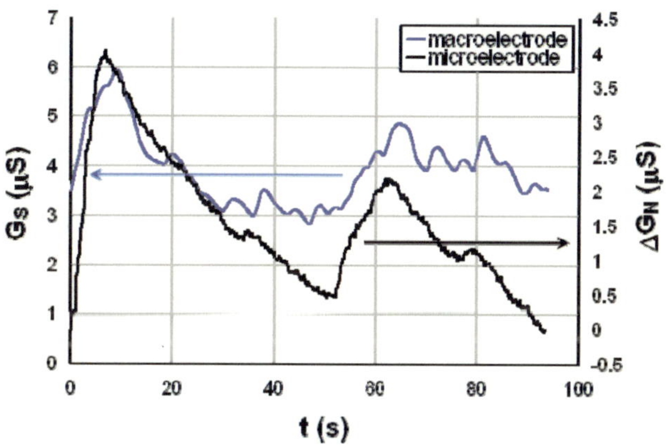

Fig. 8 Comparison of our microelectrode system and classical macroelectrode GSR method

After these experiments, the next step in our research was to develop a portable monitoring system offering sensitive and continuous measurement of PGR, with the measured data transfer to a personal computer.

3. Proposed Complex Measurement System

3.1 Measurement method

In the preliminary work, a thin film IDA microelectrodes system of different sizes and topologies has been designed, developed and realized. Then, experimental measurements of the electrodermal response (EDR) upon selected stress stimuli (invoked by different mental

tasks - standard psychotests) were performed using the developed microelectrode array. The obtained experimental results show that the optimal input signal amplitude should be selected from 1.5 V to 3V. The input signal frequency is not so critical, however, an optimal value in order of ones kHz has been proved. The most proper IDA microelectrode size is: 200μm/200μm (finger/gap ratio).
Several methods applicable to continuous measurement of the human skin impedance were analyzed first. As a result of this analysis, the auto-balancing bridge method was chosen because of few reasons (high accuracy, short time, high repeating rate of measurements, frequency and amplitude signal definition, possibility to measure both real and imaginary impedance components, controllability by a microprocessor, digital processing, etc.). Consequently, new measurement methods for both required versions of the proposed measurement equipment: a simple handy "stress- alarm" with limited features, functionality and accuracy as well as a precise laboratory-like monitoring system, have been developed. These methods are properly adjusted and modified in order to match for employing advanced available hardware accessories and peripherals. The stress-alarm method utilizes a microcontroller interface comparators for low-cost and small size. The complex laboratory measurement system is designed using precise A/D converters and with possibility to connect a lot of different integrated or external microsensors.

3.2 Developed Measurement Equipment

The complete measurement environment for continuous and non-invasive monitoring of the skin impedance has been developed. The first proposed measurement system, shown in Fig. 9, consists of the IDA microsensor described above, integrated circuit AD5933, microprocessor ADuC832, and a personal computer.

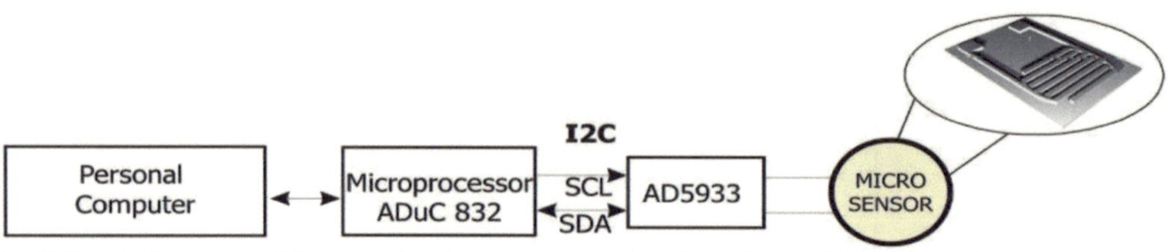

Fig. 9 Block diagram of the monitoring system using the planar IDA microelectrodes

The core of the proposed monitoring system is the integrated circuit AD5933 (Fig. 10) by Analog Devices (Analog Devices, 2007) that provides measurement of the skin impedance sensed by the developed microsensor system. The measurement process is controlled by the microprocessor ADuC832 (Analog Devices, 2007) via I2C interface. Using a serial interface RS232, the microprocessor then sends the measured data to a personal computer providing data storage. Additionally, the microcontroller also provides an initial configuration of the integrated circuit AD5933 that is needed at the measurement beginning. The configuration includes mainly setting the frequency and amplitude of the input signal used for measurement of unknown impedance. The microprocessor also controls time slots during which the measurements are performed. After the measurement in the respective time slot is done, the microprocessor reads and sends the measured data from AD5933 circuit to a PC, where data is stored and further processed.

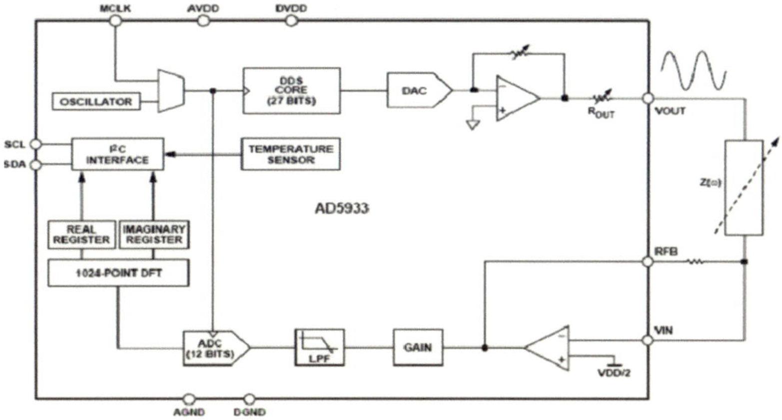

a)

Impedance Range	100 Ω ÷ 10 MΩ
Output Frequency Range	1k Hz ÷ 100 kHz
Internal Oscillator Frequency	16,777 MHz
Power Requirements	2,7 V ÷ 5,5 V
Total System Accuracy	0,5 %
Temperature range	-40 °C ÷ 125 °C

b)

Fig. 10 Integrated circuit AD5933 and its main specifications (Analog Devices, 2007)

The AD5933 circuit is composed of the following parts: an input signal generator, a 12-bit A/D converter, a DFT (Discrete Fourier Transform) circuit, a thermal sensor, and I2C interface. The generator provides a sine wave input signal of certain frequency and amplitude at the output VOUT. Unknown impedance is connected between VOUT and VIN terminals. Thus, the magnitude and phase of the current flowing through a load depend on its impedance. This current is then transformed to voltage that is converted into a digital signal by the D/A converter. Finally, the DFT circuit provides discrete Fourier transform of the converted signal. As a result, values of real and imaginary parts of a loaded admittance are measured. Photograph of the whole printed circuit board (PCB) of the monitoring system is shown in Fig. 11.

Fig. 11 PCB of the realized monitoring system.

The PCB has been realized on FR4 board by SMT technology with minimum strip width of 0.3 mm and minimum clearance width of 0.35 mm. The PCB is functionally divided into four parts: a voltage source, interface RS232, microprocessor ADuC832 and measurement module AD5933, and the power supply part providing two separate supply voltages for analog and digital parts. This is needed for the elimination of unwanted effects (e.g.

interference, noise, leakage, disturbance, etc.). Two SMD buttons ensure the software upload and microprocessor reset. The total size of the designed PCB is 50 × 60 mm.

3.3 Developed graphical user environment.

A graphical user interface (GUI), developed in C++ under Windows XP platform, provides both the necessary calibration of the measurement as well as storage, displaying, and post-processing of the measured data (real and imaginary components of the measured skin impedance). The developed software allows an easy and user-friendly control of the measurement process and data displaying and storage. From the measured data, absolute values of impedance and admittance as well as its phase are computed, and all these parameters can be displayed in several graphical and numeric modes.

3.4 Portable version of the Monitoring System

Long term biomedical monitoring plays an important role, especially, in improving diagnosis and therapeutic processes in contemporary medicine. The crucial step towards more exact and precise characterization of psychical stress influence on a monitored respondent in real life conditions is the respondent's free movement (out of a laboratory). This would enable continuous monitoring of the respondent, even during regular daily activities being carried out. From above considerations the demands on the monitoring equipment are as follows (Majer et. al, 2008):

- compact in size, low weight
- minimalization of connecting cables
- suitable sensor placement
- continual measurement ability and data storage
- low-power and battery supply

For these reasons, the developed measurement system should be small, portable and compact in size with possibilities of continuous measurement and wireless data transfer to a personal computer, eventually, storing data in memory. Moreover, this system should be able to measure other parameters (body temperature, heart pulse, etc.) for complex monitoring different psycho-physiological processes in humans. In many cases this measurement is needed to performed in plenty places of human body. The proposed complex system offers these opportunities (Fig. 12). It composed of sensing parts and the core (microcontroller with RF module, A/D, etc.) ensures measurement control, processing and transmit date. The several sensors are connected using matching circuits or wireless. One of the most important demands on the portable measurement system is the minimalization of connecting cables. Therefore, a wireless communication has been considered between the measurement equipment and the personal computer to provide necessary data transfer. The considered RF wireless communication module consists of a transmitter at the side of the measurement unit and a receiver at the PC side.

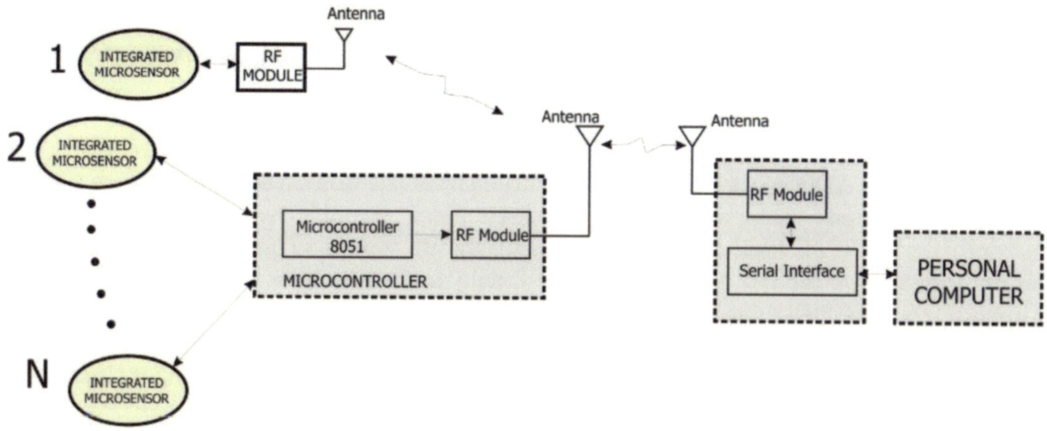

Fig. 12 A complex wireless biomedical monitoring system

Receiver module

The block diagram of the proposed receiver module, considered at the PC side, is shown in Fig. 13. There is microprocessor nRF24E1 (Nordic Semiconductors, 2008) with an integrated RF transceiver used to provide a simple, small, low-power, versatile solution at the receiver part. The RF module retrieves a signal received by the antenna and sends it to the PC via RS232 or USB interface. Microprocessor ensures control, converting, data transfer and power management.

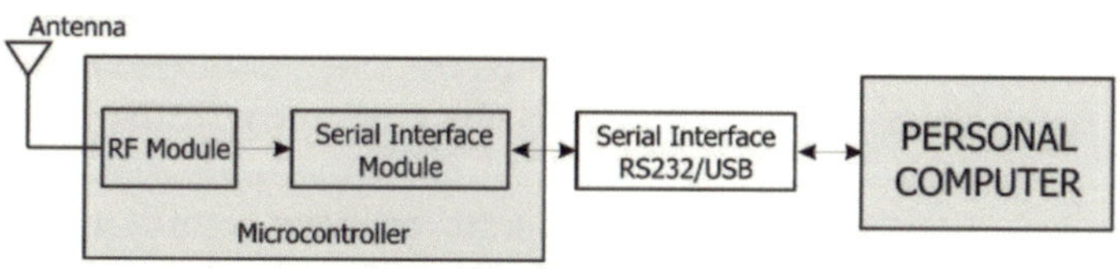

Fig. 13 Block diagram of the RF receiver communication module

Photograph of the whole printed circuit board (PCB) of the receiver module is shown in Fig. 14. The PCB has been realized on double layer FR4 board by SMT technology with minimum strip width of 0.2 mm and minimum clearance width of 0.2 mm. The total size of the developed receiver module is 11 mm × 17 mm (excluding the USB connector).

Fig. 14 PCB of a realized wireless/USB receiver module

Transmitter module

Several sensing systems with various conception and application way have been proposed. The first complete wireless measurement equipment consists of the planar microsensor, AD5933 circuit, and controlling microprocessor nRF24E1 with the RF communication module (Fig. 15). The core of the proposed portable monitoring system is again the integrated circuit AD5933 that provides measurement of the human skin impedance sensed by the developed microsensor. The measurement process is controlled by the microprocessor nRF24E1 via I2C interface. Using the RF wireless communication interface, the microprocessor then sends the measured data to the receiver part on the PC side. Consequently, the personal computer executes data storage and data post-processing. Additionally, the microcontroller also provides an initial configuration of integrated circuit AD5933 (setting the frequency and amplitude of the input signal, measurement time slots, power management, etc).

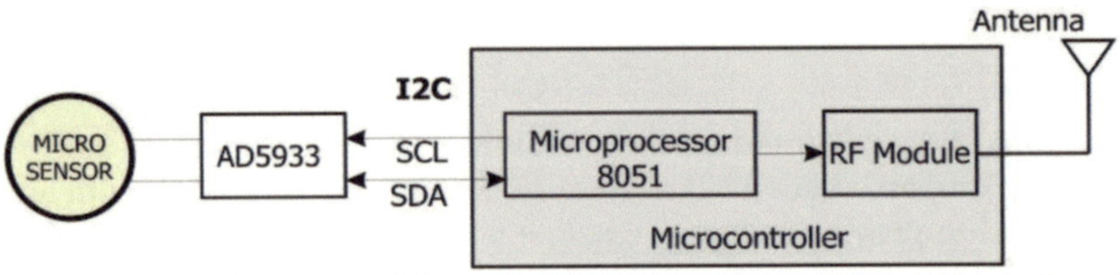

Fig. 15 Block diagram of the portable measurement system with a RF wireless transmitter communication module

The next work included mainly design and realization of the miniaturized portable version of the system with RF wireless data transfer. Thus, the integrated circuit AD5933 had to be removed, and its functions have been substituted by a microcontroller with A/D converter and RF transceiver nRF24E1 integrated within. At the same time, research and analysis of novel measurement methods, possibly applicable towards the significant miniaturization of the system and power consumption reduction (battery life time), have been performed. Two small-size monitoring equipments have been designed, applicable in both the stress alarm as well as in a complex precise laboratory measurement system.

The next step in the measurement system developing process will be led towards putting the whole system into a proper chase and making the microelectrode fixing more robust. By completing that, the laboratory or a stress alarm version of the monitoring system will be fully available.

Microcontroller nRF24E1 with RF transceiver

To reduce necessary measurement circuitry, and achieve selected features of the monitoring system (low-power, compactness, simplicity, and versatility), a modified solution of the measurement hardware has been proposed. Two microprocessors nRF24E1 (Nordic Semiconductors, 2008) with an integrated RF transceiver for the world wide 2.4 - 2.5 GHz ISM band have been employed. The RF24E1 microcontroller instruction set is compatible with the industrial standard 8051. The RF transceiver consists of a fully integrated frequency synthesizer, a power amplifier, a modulator, and two receiver units. Output power, frequency channels and other RF parameters are easily programmable by use of the RADIO register. RF current consumption is only 10.5 mA in the transmitting mode (output power -5dBm) and 18 mA in the receiving mode. The microcontroller clock is derived directly from the crystal oscillator. The nRF24E1 allows be set into a low-power down mode under program control, and also the ADC and RF subsystems can be turned on or off under. The current consumption in this mode is typically 2 µA. The device can exit the power down mode by an external signal, by the wakeup timer if enabled or by a watchdog reset.

With respect to all the features described above, this microcontroller has been chosen for a miniaturized portable version of the measurement system with RF wireless data transfer. In this case, the whole measurement equipment consists only of several sensors with necessary electronic circuits and microcontroller with the RF transceiver module.

A well-designed PCB is necessary to achieve good RF performance. We had to keep in mind that a poor layout may lead to loss of the performance, or even functionality, if due care is not taken. A fully qualified RF-layout for the RF communication module and its surrounding components, including matching networks, has been proposed as the key issue of the proper RF design. A PCB with two layers including a ground plane is needed for the optimum performance. It is designed by surface mount technology (SMT) to achieve the best possible performance. The device sizes 0603 and 0402 have been selected. The nRF24E1 supply voltage is filtered and routed separately from the supply voltages of any digital circuitry. Long power supply lines on the PCB are avoided. All device grounds, VDD connections and VDD bypass capacitors are connected as close as possible to the nRF24E1 circuit. The PCB antenna has been chosen for better miniaturization and compactness. In this configuration, the measurement system has the access range up to several tenths of meters.

Power management

The power management is another crucial issue to be taken into account, since it is the best important parameter determining the operating time, and secondary, also the size of the portable equipment. There are many factors influencing on the power consumption:

1) The input signal - the measured impedance exhibits its own power consumption that depends on the input signal parameters. In Fig. 16, curves of the measured human skin impedance versus the frequency of the input signal for different electrode arrangement are shown.

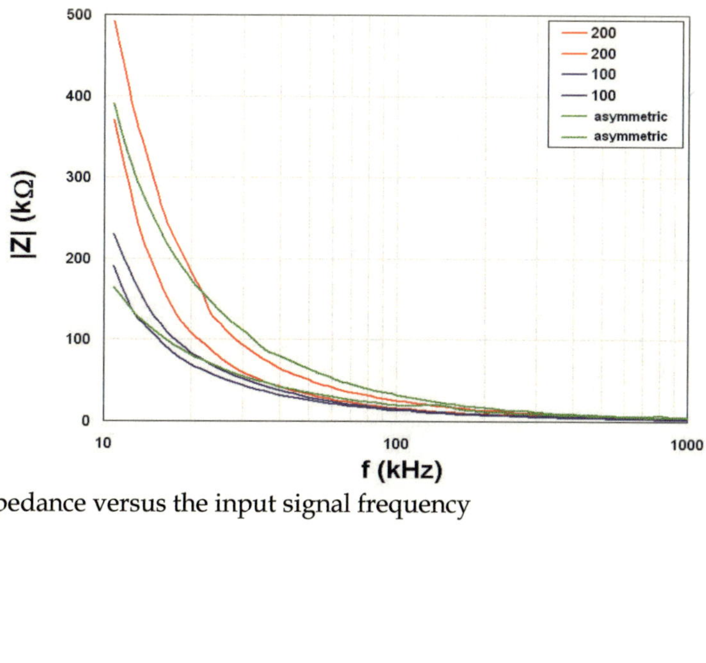

Fig. 16 Skin impedance versus the input signal frequency

The impedance value of skin is decreased as the input signal frequency arises. Other parameters influencing the power consumption are: actual value of the measured impedance, the input signal amplitude (possible range 1 - 3V), settling time after input signal is applied, sensing time of the measurement, and the frequency of sensing/sampling.

2) Power consumption of the measurement equipment – this portion is reduced by a simple design of the measurement system, and a proper power management of microcontrollers (standby mode, power down mode, etc.).

The total average power consumption is in order of ones up to tenths mW and it depends on the factors mentioned above.

3.5 Evaluation of monitoring system

Finally, experimental measurements and evaluation of the developed monitoring system have been carried out. Comparison of the results obtained by a standard HP 4284A laboratory bridge instrument measurement (G_1 - admittance) to the data measured by the developed monitoring system (G_2 and Φ - admittance amplitude and phase, respectively) is shown in Fig. 17. This measurements have been preformed in order verify realized system. The achieved results show that the developed monitoring system is sensitive to same applied stress stimuli. Moreover, the phase of the skin admittance may offer more sensitive monitoring of psycho-galvanic response than only a simple impedance amplitude sensing, since the phase reflects the admittance changes in much more significant way (high peaks in the lowest waveform).

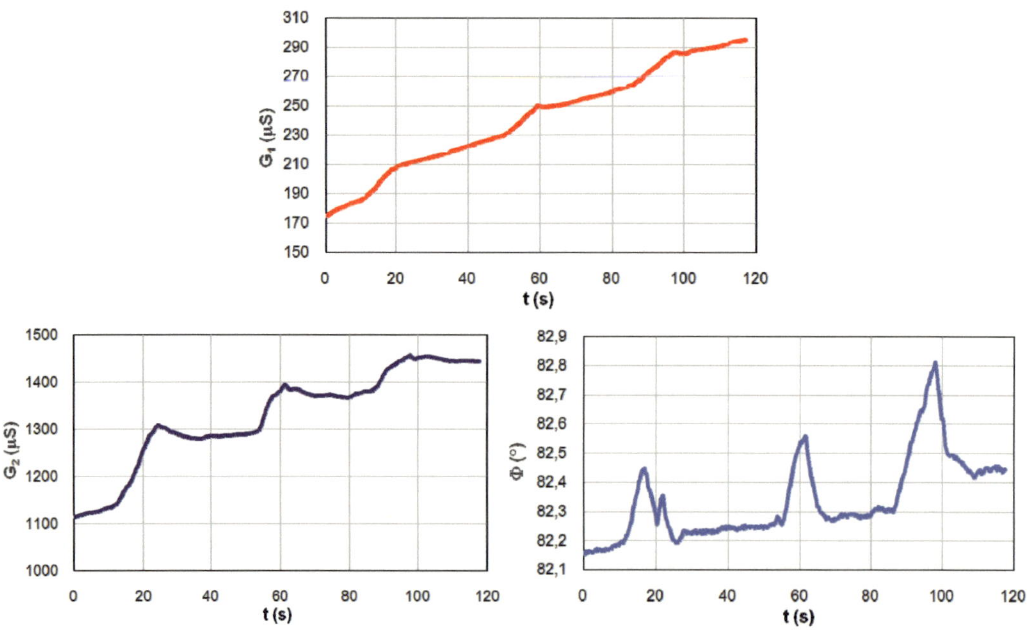

Fig. 17 Response of standard bridge instrument (G_1) versus developed system (G_2, □) in time

As the last experiment (Figure 18), our microelectrode approach was compared to a classical macroelectrodes GSR method (used on Faculty of Philosophy, Comenius University) (Shepher, 2007). The waveforms are as follows: *G3* - skin admittance obtained by macroelectrodes, *G2* and *Φ*- admittance and phase measured by the microelectrode approach. Certainly, for each method the input signal parameters were set in a proper way. In both cases, the physiological response has been evoked by the same stress stimuli. The standard psycho-tests performed have showed that the response signals obtained from both methods match and the microelectrode signals are more stable, accurate and with shorter time respond.

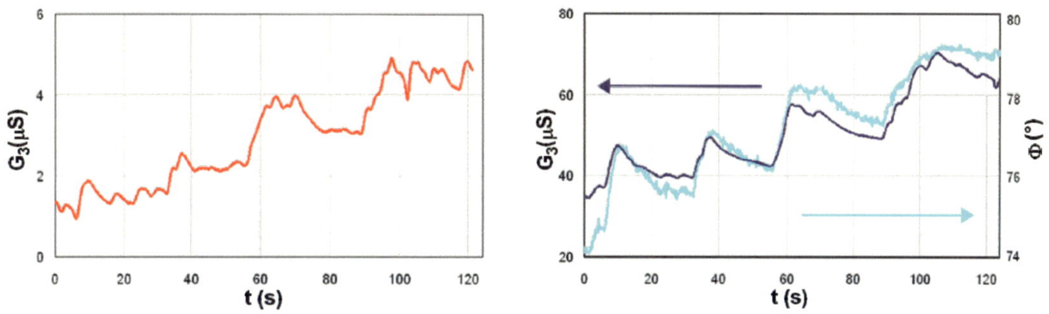

Fig. 18. Response of standard macroelectrode (G_3) versus microelectrode approach (G_2, □) in time

Fig. 19 shows curves of the measured skin admittance phase as an important factor for definition of the input signal with respect to the power consumption (stress influence periods marked by highlighted areas). It can be observed that the absolute value of the impedance decreases as the input signal frequency gets higher. Therefore, a certain trade-off between those two parameters is necessary. Moreover, the different character of the measured phase at different frequencies of the input signal has been observed, (for higher frequency - the phase decreases, for lower frequency - the phase has the increasing character).

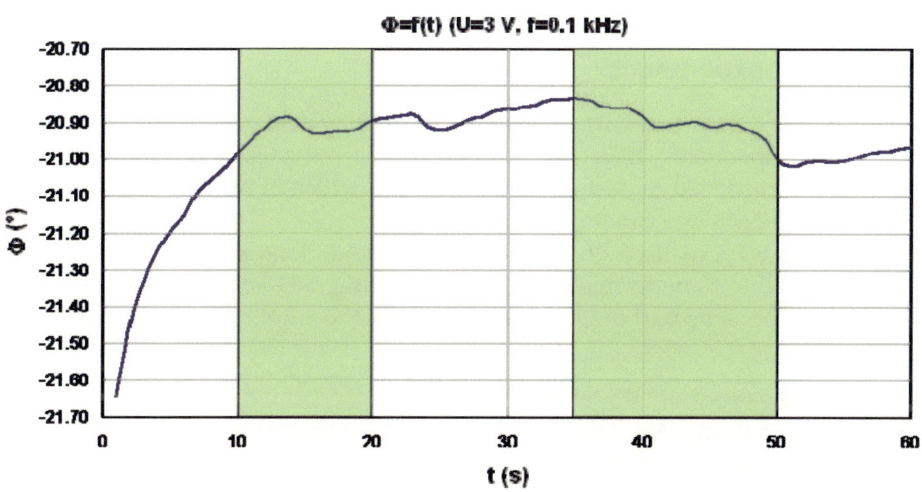

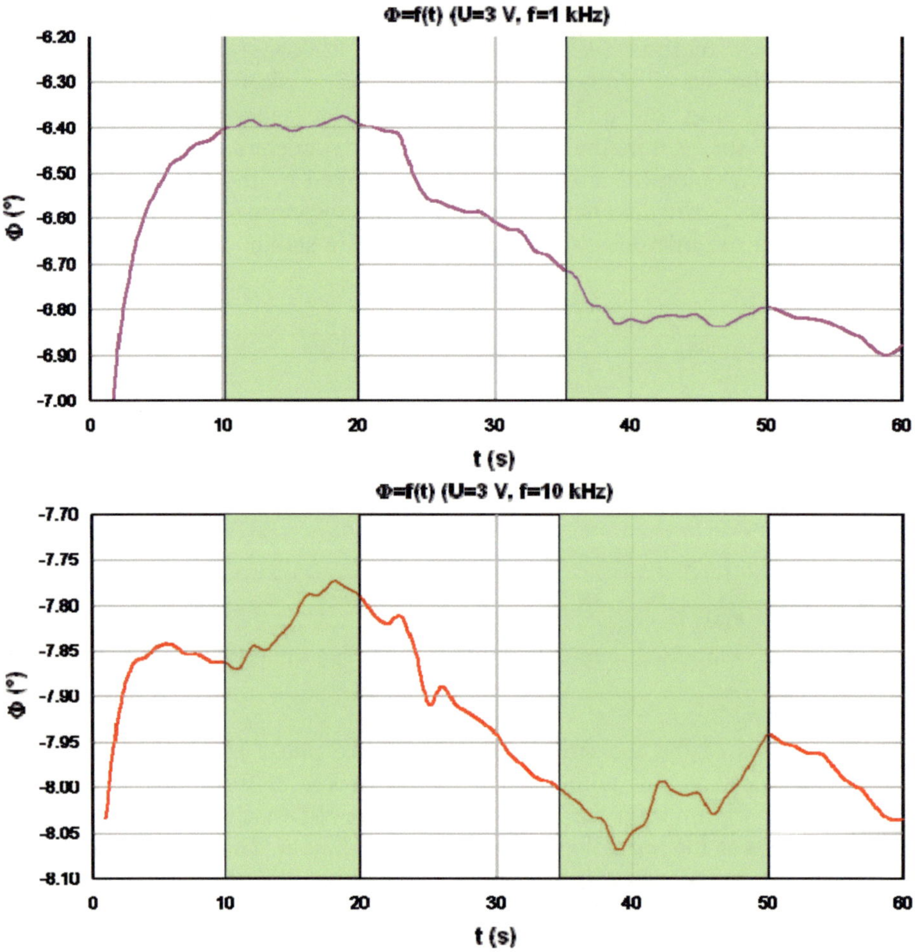

Fig. 19 Changes of admittance phase (its values and time development) at different frequencies of input signal

4. Psychological experiment

Additionally, further measurements of several selected physiological parameters, performed on a number of respondents, have been carried out as the next step of our research. The respondents were burdened by tests with various difficulties in order to invoke the proper stress conditions and achieve valuable results.

Psychological tests were focused on short-time stress situations with a large scale of tasks, which had the proband to perform simultaneously during the test. The tasks were chosen in the way to charge the same part of human brain in order to amplify the stress situation.

4.1 Psychological aspect of "Psychotest"

The presented experiment "Psychotest" was done in the laboratory of cognitive processes at Department of Psychology, Comenius University in cooperation with Department of

Microelectronics, Slovak University of Technology. This experiment has been done on a group of 35 probands in age between 19 and 30 years. The "Psychotest" was oriented to evoke short-timed stress stimulus with wide range of tasks, which a proband had to solve simultaneously. The software program Neorop II., test of distraction stress "SPEED" was used. All tasks are chosen in the way to charge the same part of human brain - in order to increase the overall stress activation (Brezina, 2007). Two types of "Psychotest" were used: "Test A" with lower stress activation and "Test B" with higher stress activation. Both tests are separated in two parts with different periodicities of tasks.

4.2 Measurement set-up

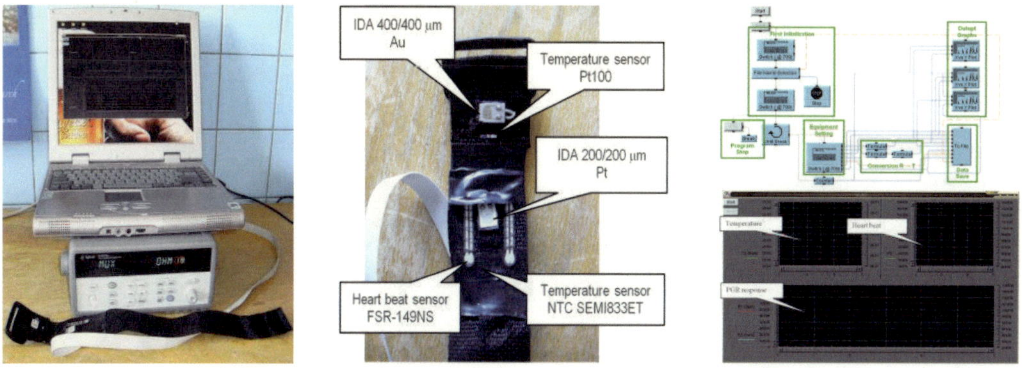

Fig. 20 Microelectrodes configuration detail

During the "Psychotest", psychogalvanic response (skin conductivity) and body temperature, sensed on the top as well as the bottom part of left hand wrist, have been monitored and analyzed. Pressure sensors FSR-149NS for heart pulse monitoring have been used too, but with unsatisfactory results. The measurements were done using 20-channel Switch Agilent 34970A millimetre controlled over data bus line GPIB by programmed software "PGR - switch" in the measurement environment Agilent VEE (Fig. 20).
Two different IDA microelectrodes and two types of the temperature microsensors have been employed. The skin conductance was monitored using a 400 / 400 μm (finger / gap dimensions) gold IDA microelectrodes (placed on the top part of wrist) and a 200 / 200 μm platinum IDA (bottom part of wrist). The body temperature was measured using NTC SEMI833ET (top) and Pt100 (bottom) microsensor.
We have minimized the drift effect like in chap. 2.4 and the final normalized signal ΔG corresponds to the difference between actual skin conductance and the approximation function: $\Delta G(t)=G(t)-G_N(t)$. The goal is to make psychogalvanic response easy readable. As the initial step of the experiment, we have tested a reference proband out of psychological test. In the first third of the experiment, the proband was without any activation. In the second third, he was very slightly activated through talking, and finally, in the last third, he was activated only via continual computer test of distraction stress "SPEED". As shown in Fig. 21, a consistent decreasing slope of temperature response and the respective consistent steep increase of the skin conductance response can be observed. Fig. 21 shows total variations of the measured parameters in different activations levels of the experiment.

An example of typical "Psychotest" responses is shown in Fig. 22, where one can observe variations in activation (stress stimulus) and relaxation (adaptation) phases.

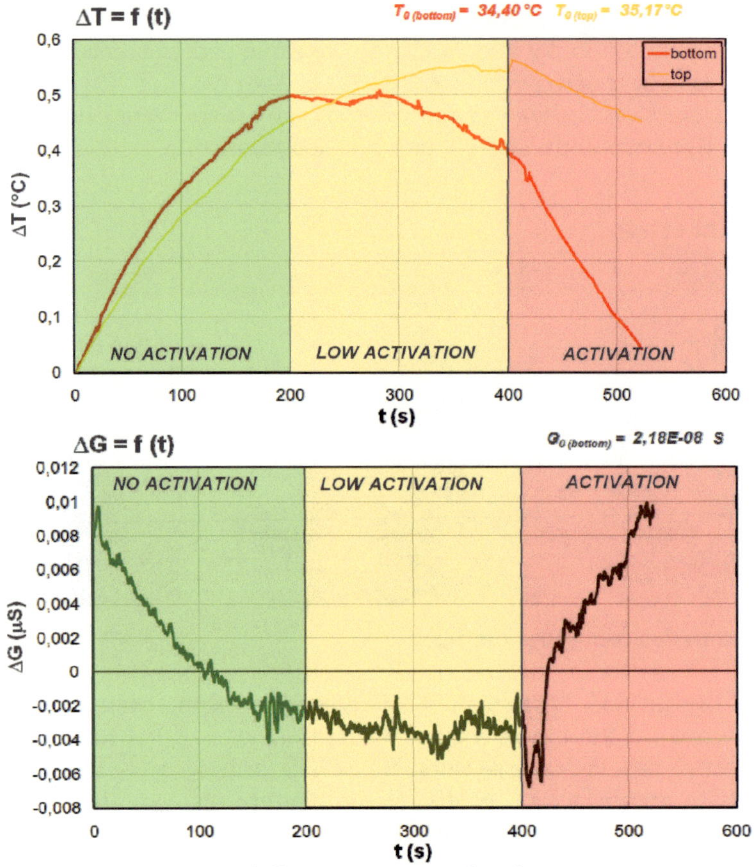

Fig. 21 Physiological responses on different activation levels:
Top - temperature deviation, bottom - normalized conductance EDR

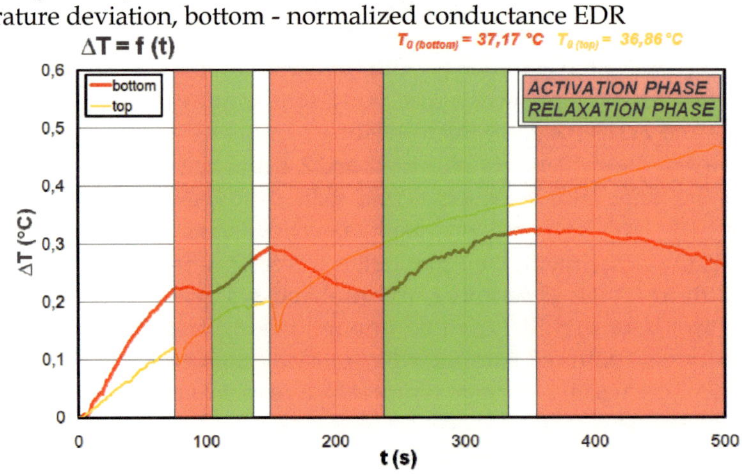

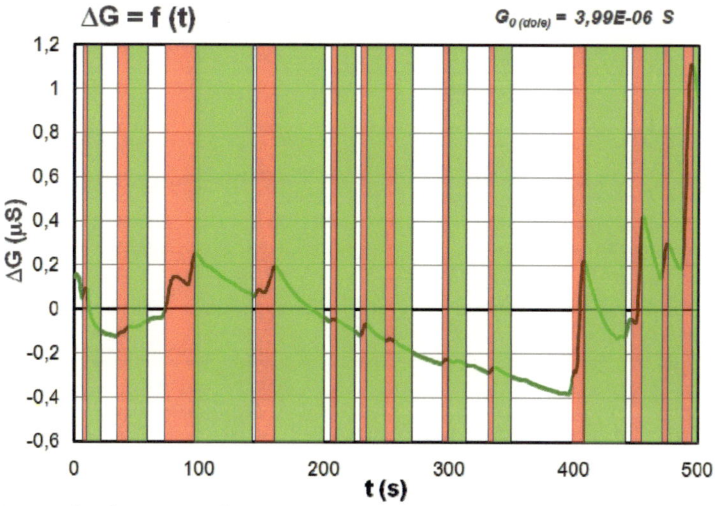

Fig. 22 Activation and relaxation phases:
Top - temperature deviation, bottom - normalized conductance EDR

Finally, the measured parameters of both tests: Test A - „lower stress activation" versus Test B - „higher stress activation" have been compared (Fig. 23). The presented results represent the average from all the entered probands' responses. One can observe a difference between the measured physiological variables depending on the stress activation level. Decreasing average temperature obtained within "Test B" probands is more significant as that of "Test A". Average increasing conductance of "Test B" probands is substantial as that of "Test A". This is due to the fact that "Test B" probands were psychologically more activated, and the resulting stress is rather substantial.

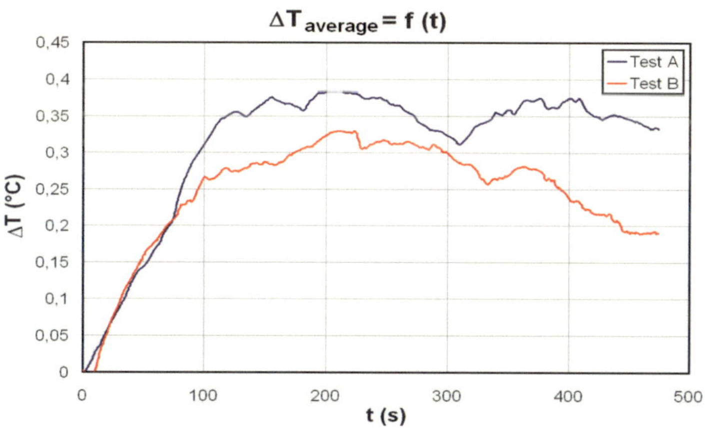

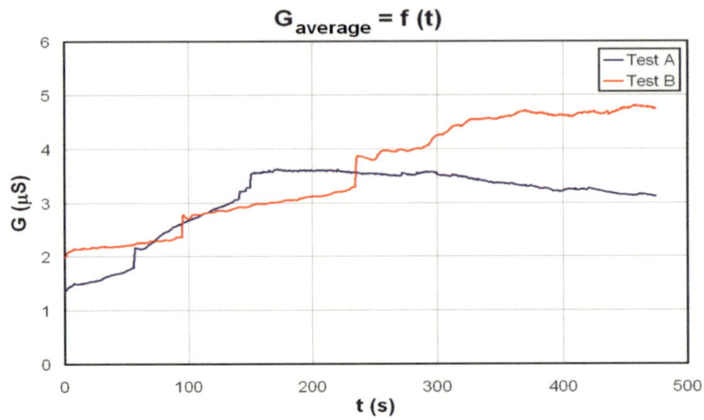

Fig. 23 Comparison of average EDR (Test A versus Test B)
Top - temperature deviation, bottom - conductance EDR (wrist bottom)

4.3 Achieved results

Psychological test experiments have been performed on a group of probands and two basic human skin parameters were sensed and analyzed. Psychological activation results to the temperature decreasing of human body and at the same time to the increasing of skin conductivity (Fig. 21). The particular parameter amplitude depends on the stress activation level and it is individual for each proband (Fig 22). Conductance EDR to the stress activation (stimulus) is shorter (2 - 5 seconds) in comparison to temperature response (20 - 120 seconds). The temperature response shows more integral character (Fig. 22). Activation phase of the parameter responses is followed by relaxation phase. The amplitudes values are getting back to their initial values. Skin conductance relaxation phases are 2 – 4 times longer than activation phases. Temperature relaxation phases are equal to the activation phases (Fig 22). During the performed "Psychotest", majority of probands exhibit increase of immunity after repeatable stress activations, and therefore, corresponding EDR amplitudes were decreasing. Both temperature and conductance sensors were more sensitive to psycho-physiological activations if placed on the bottom part of wrist compared to top part.

5. Optical reflectance of human skin

The dermis (deep inner layer of the skin) is heavily permeated with blood vessels containing haemoglobin. Haemoglobin is a protein contained in the red blood cells (95% of the dry mass of red cells). Haemoglobin binds very easily to oxygen, making it the ideal "vehicle" for the transportation of oxygen from the lungs to the tissue. Haemoglobin has an unique light absorption spectrum with characteristic absorption bands at 420 nm, and in the 545 - 575 nm wave length range, where the "W" pattern can be observed (Fig. 24b) (Angelopoulou, 1999), (Gerasimov, 2003).

These absorption bands occur only when haemoglobin is bound to oxygen. De-oxygenated haemoglobin exhibits rather shifted absorption bands, and the distinct pattern is no longer present (Angelopoulou, 1999).

Therefore, using optical (light reflectance) measurements, the quantity and oxygenation of haemoglobin in top layers of the human skin can be easily monitored that might offer another very important input factor in monitoring some psychosomatic processes. The advantage of the optical method is also in the contactless manner of monitoring, which is independent on the contact quality variations due to the possible physical activity of the respondent during testing.

5.1 Optical measurements

In the experiment for the optical method evaluation (Fig. 24, Fig. 25), the optical spectrum (reflectance), measured under certain stress stimulus in time domain again on the wrist bottom, has been analyzed. Experimental measurements were done using optical spectrometer AvaSpec 2048 and relevant optical sources (halogen + deuton light) (Fig 24a). The light was emitted and measured using an optical fibre at distance about 10 mm over the skin surface. Maximum changes of the skin reflectance occur in around 540 nm and 576 nm wavelengths, which perfect corresponds to the haemoglobin absorbance ranges (Angelopoulou, 1999), (Gerasimov, 2003) (Fig. 24b).

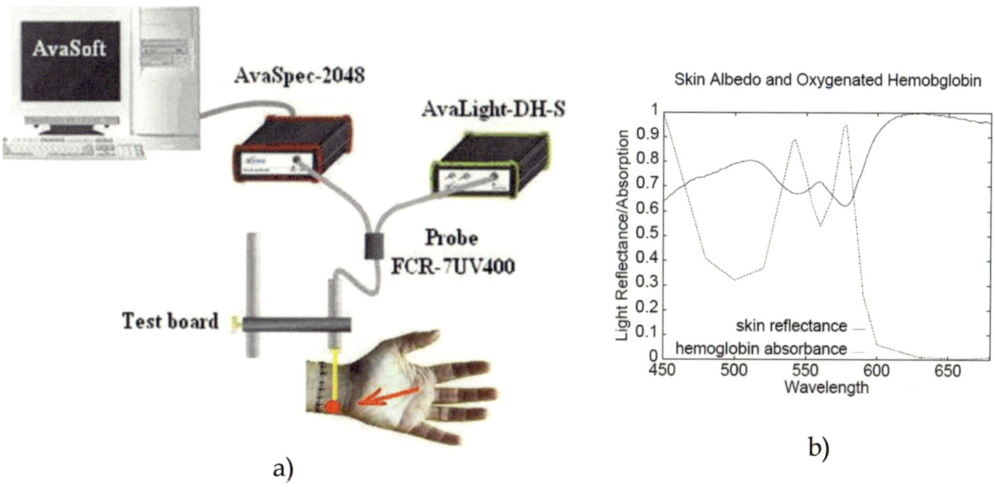

Fig. 24 Optical setup: a) equipment, b) haemoglobin absorbance and skin reflectance

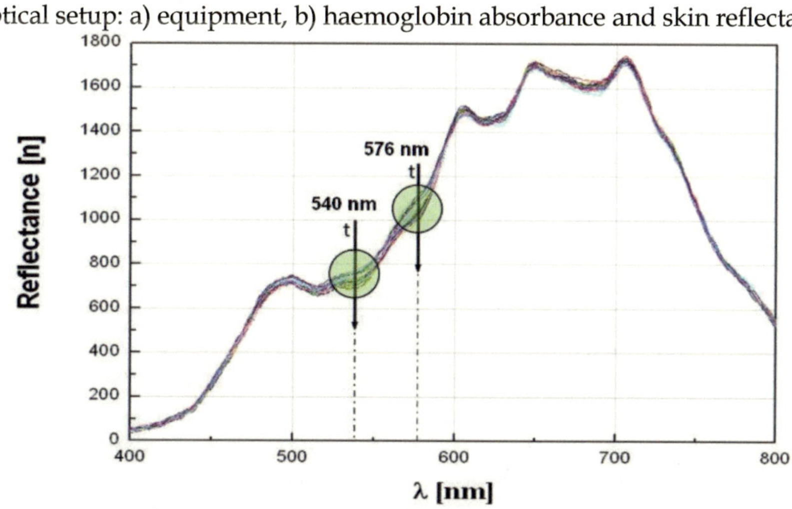

Fig. 25 Time response of skin optical spectrum to the stress stimulus

5.2 Electrical versus optical measurement

In the last experiment, both the electrical and optical measurements of psychosomatic processes are compared (Fig. 26). All the measurement were performed simultaneously, and evoked by the same stress stimuli (highlighted areas). The comparison shows rather perfect correlation of both methods in terms of the stress response monitoring ability and sensitivity.

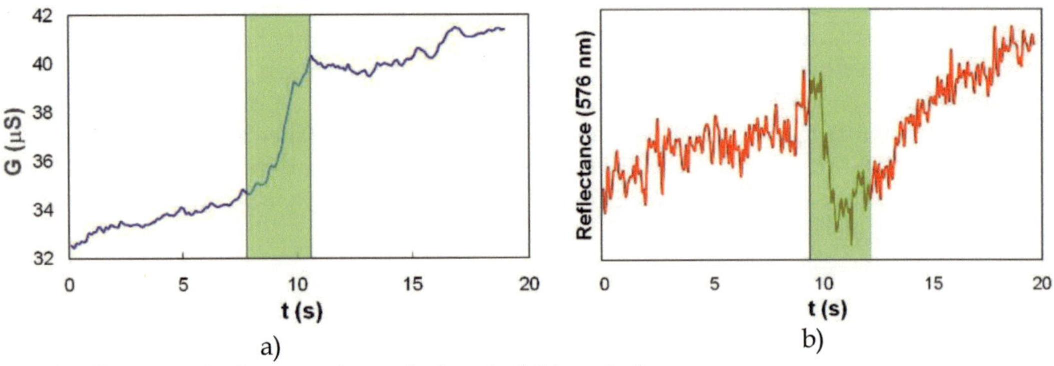

Fig. 26 Electro-optical comparison: a) electrical, b) optical

6. Conclusions

The achieved results have proven that the dedicated developed microelectrodes are able to sense the EDR in a very sensitive and fast way. Moreover, experiments, prior to the work presented here, prove that the achieved accuracy, input signal voltage and frequency ranges are suitable not only from biomedical monitoring point of view but also from the measurement system integration and miniaturization requirements.

A non-invasive portable measurement system for the reliable and precise stress detection, based on the psycho-galvanic reflex monitoring, has been designed and developed. The system is based on an auto-balancing bridge measurement method offering digital processing and displaying of the measured data in the developed GUI operating under Windows XP platform. The measurement equipment uses microelectrodes developed for this purpose, and utilizes microcontrollers containing RF wireless communication modules to transfer data between the measurement unit and a personal computer. The complete measurement system has been designed with respect to the system accuracy, sensitivity and power management. Interesting outcome has been observed - the psychogalvanic reflex might be much more accurately sensed by the skin admittance phase, since this parameter reflects the human skin conductivity changes more significantly. The achieved measurement accuracy, input signal voltage and frequency ranges are suitable not only from human biomedical monitoring point of view but also from the measurement system integration and miniaturization requirements.

Additionally, the proposed system is versatile and flexible, easy to be extended by other sensors' types (integrated or external with wireless communication feature) or different measurement approaches, which enables measurement of other physiologic parameters (e.g. body temperature, blood pressure, heart beat, etc.) There is no doubt that the developed measurement equipment offers new opportunities towards non-invasive wireless system for

continuous biomedical monitoring, applicable in diverse domains, such as clinic psychology, medicine or other everyday life areas. All these features and properties make the developed biomedical monitoring system very helpful also from the life quality enhancement point of view.
We have also analyzed a new approach to measurement of the psycho-galvanic reflex by skin optical reflectance method. This method seems to be very suitable for monitoring of the quantity and oxygenation of haemoglobin in top layers of the human skin, and implementation to the final integrated setup.
Motivated by the promising results achieved so far, the research will go on by the next step that is integration of the whole monitoring system into a single chip, working under low-voltage and low-power conditions that would meet basic requirements for modern portable wireless biomonitoring equipment.

7. Acknowledgement

This work has been done in Center of Excellence CENAMOST (Slovak Research and Development Agency Contract No. VVCE-0049-07) with support of projects AV4/0018/07 and APVV-20-055405, and grants VEGA 1/0285/09 and VEGA 1/0220/09.

8. References

Agilent Technologies. (2008). Agilent VEE Pro 9.0 and Agilent VEE Express 9.0 Data Sheet, 5989-9641EN, USA

Analog Devices. (2007). AD5933 Datasheet, http://www.analog.com/

Analog Devices. (2007). ADuC832 Datasheet, http://www.analog.com/

Angelopoulou, E. (1999). The Reflectance Spectrum of Human Skin, *Technical Report MS-CIS-99-29*, pp. 1-14, University of Pennsylvania, USA

Brewin, C. R., Andrews, B., & Valentine, J. D. (2000). Meta-analysis of risk factors for posttraumatic stress disorder in trauma-exposed adults, *Journal of Consulting and Clinical Psychology, 68(5)*, pp. 748-766

Brezina, I. (2007). Simultaneous performance of double controlled tasks and their impact of psychophysiological processes, *X. European congress of psychology, Book of papers*, Prague, Czech Republic

Cooper, C. L., Sloan, S. J., & Williams, S. (1988). *Occupational stress indicator management guide*, Oxford, UK: NFER-Nelson

Ganster, D. C. & Schaubroeck, J. (1991). Work and stress and employee health, *Journal of Management, 17*, pp. 235-271

Gerasimov, V. (2003). *Every sign of life*, Massachusetts Institute of Technology, USA

Gupta, N. & Beehr, T. A. (1979). Job stress and employee behaviour, *Organizational Behaviour and Human Performance, 23*, pp. 373-387

Ivanic, R.; Novotny, I.; Rehacek, V.; Tvarozek, V.; Weis, M. (2003). Thin film non-symmetric microelectrode array for impedance monitoring of human skin, *Thin Solid Films 433*, pp. 332-336

Nordic Semiconductor. (2008). nRF24E1 Datasheet, http://www.nordicsemi.com/

Majer, L. & Stopjaková, V. (2008). Portable Measurement Equipment for Continuous Biomedical Monitoring using Microelectrodes, *In: IEEE Workshop on Design and Diagnostics of Electronic Circuits and Systems - DDECS 2008*, pp. 26-29, April 2008 Bratislava, Slovakia

Manishev, A., Sundara-Rajan, K., Zang, F., Du, Yanqing, Zahn, M. (2004). Interdigital Sensors and Transducers, *Proceedings of the IEEE*, Vol. 92, No. 5, pp. 809-845

Murphy, L. R., Hurrell, J. J., Jr., Sauter, S. L., & Keita, G. P. (Eds.) (1995). Job stress interventions, American Psychological Association, Washington, DC

Olmar, S. (1998). *Bioelectrochemistry and Bioenergetics, 45. 2*, pp. 157-160

Qubit systems. (2004). *Qubit systems – Human electrophysiology*

http://www.qubitsystems.com/electro.html/

Selye, H. (1976). *The stress of life*, New York

Sheppard Jr., N.F.; Tucker, R.C. & Wu, Ch. (1993). Electrical conductivity measurements using microfabricated interdigitated electrodes, *Anal. Chem. 65*, 1199–1202

Shepherd, P. (2007). *The Galvanic Skin Response Meter*, http://www.trans4mind.com/psychotechnics/gsr.html/

Vavrinský, E.; Stopjaková, V.; Tvarožek, V.; Majer, L. & Brezina, I. (2008). Electro-Optical System for Continual Monitoring of Human Body Physiological Characteristics, *19th Biennial International Eurasip Conference Biosignal 2008*, ISBN 978-80-214-3612-1, Brno, University of Technology, Czech Republic

Wakelam, K. (2000). *Biofeedback and the human energy field*, pp. 7-15

Weis, M.; Danilla, T.; Matay, L.; Hrkút, P. & Kákoš, J. (1995). Noninvasive Biomedical Sensor s on the Biology - Interface of Human Skin, *7th International Conference on Measurement in Clinical Medicine*, pp. 89-91, Stará Lesná, Slovakia